Molecular and Cellular Neurobiology

Molecular and Cellular Neurobiology

Editors
M K Thakur
S Prasad

Narosa Publishing House
New Delhi Chennai Mumbai Kolkata

Editors
M K Thakur
S Prasad
Department of Zoology
Banaras Hindu University
Varanasi, India

NAROSA PUBLISHING HOUSE PVT. LTD.

22 Daryaganj, Delhi Medical Association Road, New Delhi 110 002
35-36 Greams Road, Thousand Lights, Chennai 600 006
306 Shiv Centre, D.B.C. Sector 17, K.U. Bazar P.O., Navi Mumbai 400 703
2F-2G Shivam Chambers, 53 Syed Amir Ali Avenue, Kolkata 700 019

www.narosa.com

Printed from the camera-ready copy provided by the Editors.

ISBN 81-7319-582-X

Published by N. K. Mehra for Narosa Publishing House, 22 Daryaganj, Delhi Medical Association Road, New Delhi 110 002 and printed at Chaman Enterprises New Delhi 110 002, India.

Preface

Neurobiology is one of the most exciting and rapidly advancing areas of science. It provides an account of the most complex organ of the body, the brain. Considering its importance, neurobiology is now taught as a full-fledged or an elective course in many colleges and universities throughout the world. There has been increasing demands from students and teachers for books dealing with recent information in neuroscience. The present edition is an attempt in this direction. This book is an account of the remarkable progress made so far. It provides critical and comprehensive overview of the wide range of topics covering different aspects of neurobiology. It will be a valuable source of recent information and reference for students, teachers, clinicians and scientists engaged in teaching and research in Neurobiology.

Moreover, the subject of neurobiology is multidisciplinary. For these reasons, the present book comprises of contributions from biochemistry, molecular biology, biophysics, anatomy and physiology. The text is designed to provide the subject material in five sections containing 25 chapters. The first section introduces the structure and development of the brain, showing how they are specialized for the functions they perform. The second section is concerned with hormones and neurotransmitters. The third section addresses various changes in the brain related to aging. The last two sections deal with different types of brain diseases including neurodegenerative diseases. All chapters have been written by experts in the field. They deserve special appreciation for their valuable contributions. We are very grateful to all contributors for their time and expertise. The publication work has been partly supported by grants from the Indian Council of Medical Research, New Delhi. Finally, we thank Mr. NK Mehra and other staffs of the Narosa Publishing House Pvt. Ltd., New Delhi, for their cooperation, encouragement and patience.

Editors

Contents

Section D: Brain Diseases

Section E: Neurodegenerative Diseases

Section A

Brain Structure and Development

Molecular and Cellular Neurobiology
Editors: M.K. Thakur and S. Prasad

Neuronal Plasticity in Developing and Adult Nervous System

B.S. Shankaranarayana Rao and T.R. Raju

Department of Neurophysiology, National Institute of Mental Health and Neuro Sciences
Bangalore-560 029, India

Abstract

Neuronal plasticity is an extraordinary property of the brain and refers to the potential for morphological, biochemical, behavioral and electrophysiological alterations in both the adult and developing nervous system. Although neuronal circuitry is capable of undergoing different types of alterations, synaptic changes are the most critical ones because they alter communication between neurons. Synaptic plasticity pertains to changes in the complex synaptic networks of individual brain circuits in response to various interventions. The brain is able to remodel its connections in order to adjust the organism's response to changing conditions. The brain shows greatest plasticity during early development when connections between neurons are being made for the first time. The connections are often formed under the guidance of sensory experience.

The role of activity-dependent synaptic plasticity in learning and memory is a central issue in neuroscience. Much of the relevant experimental work concerns the possible role of long-term potentiation (LTP) and long-term depression (LTD) in learning. The formation of long-term memory occurs gradually, over time, following learning. It has been shown that LTP can cause structural changes by strengthening the existing synapses or resulting in the formation of new synapses. It is evident that central nervous system retains a remarkable ability to modify the number, morphology and activity of synapses. The capability of synapses to alter their structure and number in response to physiological usage may play a critical role in plastic neurobehavioral processes. Gene knockout / transgenic technology has further enhanced the understanding of the role of various signaling molecules and proteins in neuronal plasticity. The capacity of the nervous system to reorganize the patterns of synaptic connectivity needs to be better understood in both health and disease states.

Key words: activity-dependent synaptic plasticity, behavioral plasticity, dendritic remodeling, learning and memory, transgenic and knock out mouse models

Introduction

Recent progress in our understanding of neuronal plasticity confirms that patterns of neuronal connectivity are not rigid. Merzenich and Kaas (1982) demonstrated that the sensory maps can be redrawn in adult brains and suggested that the brain is a network that is continually remodeling itself. Their findings drew to the idea that plasticity occurs in the adult nervous system and suggested that the adult brain can reorganize itself in areas that were long thought to be completely hard-wired. The concept of neuronal plasticity has been extensively used to deal with the enduring changes associated with brain's capacity to rewire, remodel the connectivity by experience (Shankaranarayana Rao et al 1998d, 2001c, Shankaranarayana Rao and Raju, 2001a).

Numerous studies have proven the biological basis of memory formation and have begun to identify the biochemical traces and cellular circuits that are formed by experience, and which participate in the storage of information in the brain, its retention for long durations, and its retrieval upon demand. Cells in the nervous system have the capability of undergoing extremely long-lasting alterations in response to hormonal, pharmacological, and environmental stimulations. The mechanisms underlying this neuronal plasticity are activated by experiential inputs and operate in the process of learning and in the formation of memories in the brain (Kutty et al 2003, Shankaranarayana Rao and Raju, 2001a).

I. Plasticity and Development of the Nervous System

Development and plasticity of the nervous system involves the coordinated actions of a huge number of individual neurons. Neurons of the appropriate types must be generated in appropriate places in optimal numbers. They must migrate to their final positions. They must differentiate into their final forms, growing out dendrites and axons. The axons must follow the right paths on the way to their targets, and when they reach there they must recognize the correct parts of the target structures and the appropriate kinds of cells with which to make contact. Finally, with all the connections approximately right, neurons must refine their synaptic connections to a high degree of precision, keeping or finding almost all the right ones and getting rid of all the wrong ones. After the brain is built, it must be able to change, dramatically during early life and to a lesser extent thereafter, to accommodate growth of the organism and new behavioral or intellectual capacities.

Molecular signals guide the early stages of brain development

The final stages of refinement of connections are orchestrated by an elaborate series of molecular signals. There is good evidence that some of the neurotrophins, brain-derived neurotrophic factor (BDNF) and neurotrophin 4/5 (NT4/5), play important roles in maintaining neural connections. The neurotrophin story is likely to flourish during the coming years and drugs directed at neurotrophin responses may come to play an important part in treating some debilitating diseases of nervous system.

Electrical activity and membrane depolarization have been shown to determine the survival and development of retinal ganglion cells (RGCs) (Govindaiah, 2001). Voltage activated ion channels of sodium, potassium and calcium underlie the genesis of spontaneous electrical activity of developing neurons. We have investigated the role of BDNF on the expression of voltage gated sodium, potassium and calcium channels in developing retina at various intervals, both qualitatively and quantitatively by measuring the immunofluorescence using confocal microscope. The expression of these channels followed a distinct developmental pattern suggesting that enhanced channel expression underlies the increased electrical activity observed during various developmental periods (Marita et al 2004). Our results demonstrated that trophic factors derived from the target nuclei would refine the neuronal circuitry. As trophic factors are considered as potential drugs, trophic therapy can be used to ameliorate certain neurodegenerative disorders. Our studies have shown that BDNF and CNTF attenuate neurofilament phosphorylation in the soma of cultured rat spinal cord neurons exposed to CSF from patients with amyotrophic lateral sclerosis (Shahani et al 2004).

Role of neural activity in refining connections in later stages of brain development

In the normal adult brain, many neural connections are refined to perfection, down to the level of individual cells. Much of this refinement takes place during critica periods in early development, in which many excess transient synaptic connections are made with the right kinds of cells in the appropriate areas, and these connections are maintained only if they are followed by appropriate patterns of neuronal activity. Our studies have shown that intrinsic / spontaneous electrical activity is crucial for the development of RGCs (Govindaiah, 2001, Govindaiah et al 2002). The RGCs displayed synchronous bursts of activity during initial stages of development and this increased firing frequency coincided with enhanced metabolic activity in these cells (Govindaiah et al 2000). Synchronized spontaneous rhythmic activity is a feature common to many parts of the developing nervous system and is necessary for the formation of precise connections. These findings indicate that RGCs *in vivo* displayed increased excitability during the period of synaptogenesis, which may lead to alterations in the expression of synaptic proteins (Dhingra et al 1997, 2001). Understanding the process of activity-dependent development could revolutionize our ability to identify, prevent, and treat developmental disorders resulting from disruptions of neural activity that interfere with the formation of precise neural circuits.

II. Morphological Plasticity

Dendritic plasticity

Dendritic outgrowth in hippocampal and motor cortical neurons in adult mammals is influenced by various factors such as nutritional status, hormones, aging, brain lesions, social isolation, sensory deprivation, exposure to complex or enriched environment, training for various tasks and electrical stimulation (Shankaranarayana Rao 2003b, Shankaranarayana Rao et al 2001c). Jones and Schallert (1994) have demonstrated use-dependent dendritic growth in layer V pyramidal neurons of the motor cortex. In addition to the effect of various environmental factors, treatment with (-) deprenyl, a selective monoamine oxidase-B inhibitor resulted in a significant increase in the dendritic arborization of CA3 hippocampal and layer III prefrontal cortical neurons in the monkey brain (Lakshmana et al. 1998b, Shankaranarayana Rao et al 1999c).

Dendritic outgrowth in adults can also occur by electrical stimulation. Rutledge et al (1974) observed additional outgrowth of apical dendrites in the cortex of adult cats after long-term electrical stimulation applied to the left suprasylvian gyrus. Following such stimulation, it was observed that apical dendrites of layer II and III pyramidal neurons have significantly more dendritic branches and spines in the cortex contralateral to the stimulated side. They interpreted this as evidence for induction of postsynaptic growth (remodeling) by increased use of specific pathways to and within the cerebral cortex. Furthermore, self-stimulation rewarding experience for 10 days resulted in an increase in the dendritic branching and dendritic length in CA3 hippocampal (Figure1) and layer V motor cortical pyramidal neurons (Shankaranarayana Rao et al 1993, 1994a) and such changes were found to be long-lasting (Shankaranarayana Rao et al 1998e).

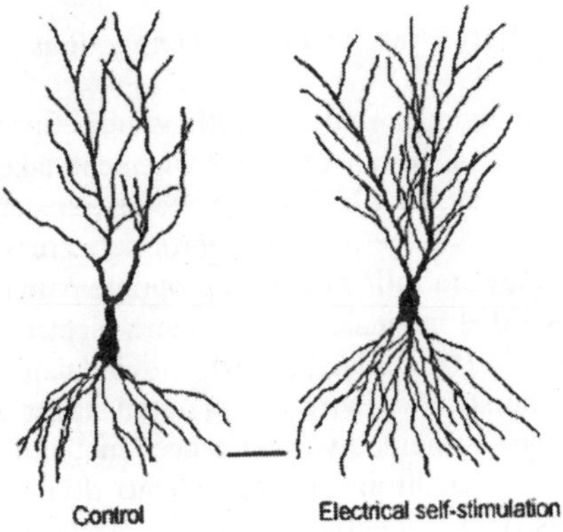

Fig. 1. Intracranial self-stimulation enhances dendritic arborisation of CA3 pyramidal neurons of the hippocampus. Scale bar = 50 μm. (From Shankaranarayana Rao, 1996).

In addition to dendritic growth, the size of brain regions is altered significantly by different experiences. For example, Walsh et al (1972) observed the medial area of the hippocampus was significantly thicker in animals reared in complex environments than those reared in isolation. Rosenzweig (1966) also reported that there was an increase in the size of hippocampus as a consequence of exposure to complex environment. Our studies have revealed a significant increase in the thickness of lacunosum, radiatum and lucidum laminae in the CA3 region of hippocampus in self-stimulation experienced rats (Shankaranarayana Rao et al 1993).

Severe traumatic stress or repeated exposure to stress can result in long-term deleterious effects leading to depression. It has been shown that restraint stress for 21 days causes a selective atrophy of apical dendrites of CA3 neurons of hippocampus (Shankaranarayana Rao et al 2001e, Vyas et al 2002) (Figure 2). As a compensatory mechanism, the density of dendritic spines was increased in these neurons to overcome decreased dendritic arborization (Sunanda et al 1995). Interestingly, the stress-induced atrophy of dendrites was completely reversed by 45 days of rehabilitation (Shankaranarayana Rao et al 2001e). Ramkumar et al. (2004b) have shown that self-stimulation rewarding experience reverses stress-induced dendritic atrophy in medial prefrontal cortical neurons in the rat. Chronic restraint stress is reported to induce sexually dimorphic neuronal remodeling in the CA3 region of the hippocampus (Mitra et al 2002, 2003). In contrast to the effects in the hippocampus, chronic immobilization stress results in dendritic hypertrophy in the basolateral nucleus of the amygdala (BLA) in male rats (Vyas et al 2002).

Subiculum, a critical structure of hippocampal formation is involved in hippocampal-dependent cognitive functions. Lesions of subiculum cause decreased cell density and dendritic atrophy of hippocampal and entorhinal cortical neurons (Govindaiah et al 1997, Latha Devi et al 2003, Shankaranarayana Rao et al 2001d, 2004f). Such morphological changes following subicular lesions are correlated with altered hippocampal theta activity and learning deficits (Laxmi et al 1997, 1999, 2000). Chronic glutamate ingestion also induces neuronal plasticity in hippocampal

CA1 and CA3 pyramidal neurons (Shankaranarayana Rao et al 1999d). It has been shown that long-term exposure to hypobaric hypoxia causes dendritic atrophy in CA1 and CA3 pyramidal neurons of the rat hippocampus (Titus et al 2004).

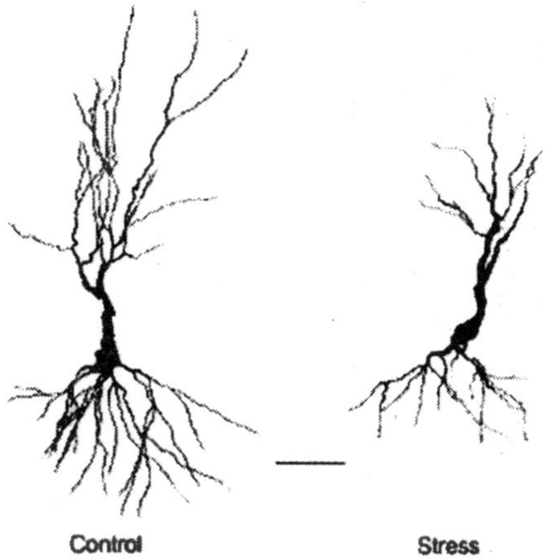

Fig. 2. Prolonged severe immobilization stress causes dendritic atrophy of CA3 pyramidal neurons of the hippocampus. Scale bar = 50 μm. (From Vyas et al. 2002).

Dendritic spine plasticity

Dendritic spines are tiny, specialized postsynaptic receptive sites that cover the surface of many neurons and they serve as major targets for excitatory synaptic inputs onto principal neurons in the hippocampus, the neocortex and other brain regions. The density of dendritic spines over any given length is subject to wide fluctuation depending on variety of environmental factors, electrical stimulation of afferent pathways, sensory deprivation, stress, social isolation, learning, hibernation, aging, hormones and radiation. Alterations in spine density were also observed in various pathological conditions such as brain tumour, alcohol consumption, undernutrition and sleep deprivation (Ramesh et al 1999). In addition, changes in both spine density and the appearance of abnormal spines have been reported in Down's syndrome and mental retardation.

Recent studies have thrown more light on the training induced changes in dendritic spine density in different regions of the brain. Changes in synaptic efficacy may be accomplished by alterations in synaptic density, which could change the strength of input to a particular neuron. Chang and Greenough (1984) have shown increases in the number of shaft and spine synapses, 10-15 min after induction of long-term potentiation (LTP), which persist up to 8hours. Desmond and Levy (1986) have also found a significant increase in the density of spines following LTP in the rat hippocampal dentate gyrus. We have also demonstrated a significant increase in the number of spines in different categories of dendrites in hippocampal and motor cortical neurons following brain stimulation reward (Shankaranarayana Rao et al 1998a, 1999b) (Figure 3).

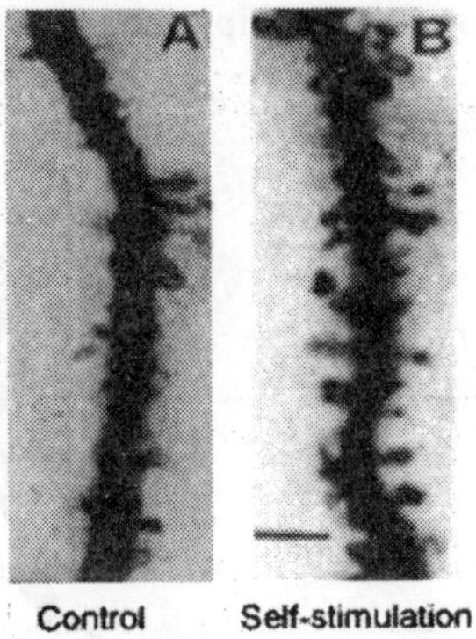

Fig. 3. Self-stimulation rewarding behavioral experience increases the number of dendritic spines in CA3 pyramidal neurons of the hippocampus. Note a significant increase in the spines in self-stimulation experienced rats (B) compared to control (A). Scale bar = 2.5 μm. (From Shankaranarayana Rao et al 1999b).

Synaptic plasticity

Synaptic plasticity denotes long-term changes in synaptic potency resulting from transient changes in synaptic activity. It includes the growth of new synapses or the activation of silent synapses, as well as changes in the efficacy of existing synapses. Synaptic density in various regions of the brain, including neocortex and hippocampus are altered in different conditions such as imbalance in hormonal levels, undernourishment, aging and Alzheimer's disease, exposure to drugs such as nimodipine, 5-hydroxytryptamine (5-HT) receptor agonist and antagonists, visual training and cerebral ischemia or lesions (Claverly and Jones, 1990). Our studies have shown an increase in the number of synapses in lucidum, radiatum and moleculare layers of CA3 hippocampal region and molecular layer of the motor cortex following self-stimulation rewarding experience (Shankaranarayana Rao et al 1999a) (Figure 4).

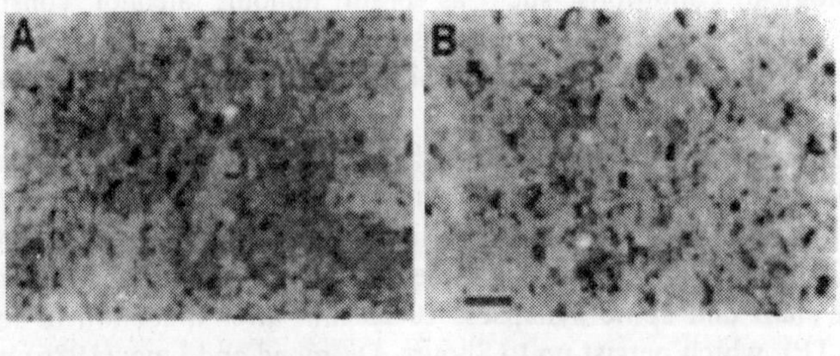

Fig. 4. Self- stimulation rewarding behavioral experience increases the number of synapses in CA3 region of the hippocampus. Note a significant increase in the synapses in self-stimulation experienced rats (B) compared to control (A). Scale bar = 2.5 μm. (From Shankaranarayana Rao et al 1999a).

There are various forms of synaptic plasticity, differing with respect to their persistence over time and their underlying induction and expression mechanisms. The best known of these is LTP (Bliss & Lømo, 1973) and is operationally defined as a long-lasting increase in synaptic efficacy in response to high-frequency stimulation of afferent fibers. An increase in synaptic efficacy persists from minutes to days and is thus a robust example of a long-term increase in synaptic strength. Most forms of LTP are glutamatergic and the most prominent form is induced following activation of the *N*-methyl-D-aspartate (NMDA) receptor. A train of 100 stimuli at 100 Hz can evoke a long-lasting potentiation of synaptic transmission in neurons of selected structures in the brain. This potentiation is called long-term potentiation.

Long-term depression (LTD) is a long-lasting activity-dependent decrease in synaptic efficacy that was first discovered in CA1 in vitro by Lynch et al. (1977). Depotentiation, the reversal of LTP, is also observed in vivo and in vitro. LTD requires a low rate of about 1 Hz stimulation for long periods (10-15 minutes). This pattern of activity depresses the activity of neurons for several hours. Having the capacity to bidirectionally modify synaptic efficacy improves the potential fidelity of memory recall in associative memory. Dendritic spines are morphological specializations that receive synaptic inputs. The morphology of spines are known to change after LTP, such as enlargements of the spine head and shortenings of the spine neck. It has often been pointed out that synaptic plasticity displays physiological properties that are highly suggestive of an information storage device (Bliss & Collingridge 1993). An increase in the population spikes in dentate gyrus induced by perforant path stimulation has also been observed on the completion of learning. Activity dependent synaptic plasticity has been implicated in a variety of physiologically and behaviorally induced changes in neuronal organization both during development and adulthood. Although extensive information is available about the mechanisms of synaptic plasticity in the adult mammalian brain, a coherent understanding is yet to emerge.

Cytoskeletal Plasticity

Experience-dependent structural plasticity is now a well documented phenomenon, however the central question about how functional demand operates at the cellular level to influence the expression of cytoskeletal proteins, namely, neurofilaments, still remains to be answered. Neuronal growth, whether it is dendritic growth or the formation of new synapses are found to be accompanied by changes in the expression of cytoskeletal proteins, which are responsible for the structural maintenance of neurons. Neurofilaments are neuron-specific intermediate filament proteins, which constitute a major component of the neuronal cytoskeleton and play a critical role in determining shape and volume of neuronal processes including complex dendritic arborization and axonal caliber. It has been suggested that the concentration of neurofilaments can serve as indicators for axonal development, neuronal plasticity, neuritic sprouting and nerve fiber regeneration. A few studies have also correlated the expression of neurofilaments and neuronal maturation during development.

Several studies have shown that under normal conditions the neurofilaments in the neuronal perikarya and dendrites are non - phosphorylated while in axons they are phosphorylated. Aberrant neurofilament phosphorylation in the soma of the ventral horn neurons of neonatal rat spinal cord was observed following exposure to cerebrospinal fluid (CSF) of patients suffering from amyotrophic lateral sclerosis (ALS). Hyperphosphorylation of neuronal soma suggests an initial stage of degenerative changes occurring in the motor (ventral horn) neurons following exposure to circulating toxic factor(s) in the CSF of patients with ALS (Rao et al 1995, Shahani et al 1998, 2001). Our recent studies have revealed an increased immunoreactivity for phosphorylated and non-phosphorylated forms of neurofilaments in self-stimulation experienced rats in hippocampal and motor cortical pyramidal neurons (Shankaranarayana Rao et al. 1997, 2004e). The increased dendritic arborization after self-stimulation experience (Shankaranarayana Rao et al 1993, 1994a) was associated with an enhanced expression of neurofilament proteins. These findings suggest a close relationship between the expression of cytoskeletal proteins and experience-dependent structural changes.

II. Neurochemical Plasticity

Neurotransmitters are known to play a critical role in the regulation of neuronal cytoarchitecture. Recent studies have begun to realize that neurotransmitters are not only involved in information coding but may also play an important roles in defining the structure of circuits in which they participate. Various neurotransmitters such as acetylcholine, noradrenaline, dopamine, gamma-amino butyric acid (GABA), glutamate, serotonin, somatostatin and neuropeptides are found to regulate the neuronal structure in the developing and adult nervous system (Mattson, 1998). Monoamines and acetylcholine are involved in the formation of specific dendritic morphology, lamination of cortex or the formation of topographical afferent and efferent projections. Our recent studies have correlated the relationship between increased dopamine metabolism (Lakshmana et al 1998a, 2004) and altered dendritic morphology in hippocampal and pre-frontal cortical pyramidal neurons in adult monkeys following chronic (-) deprenyl administration (Lakshmana et al 1998b, Shankaranarayana Rao et al 1999c). Glutamate plays a critical role in the morphological changes following LTP. Glutamate was also found to specifically affect the cytoarchitecture of dendrites of hippocampal pyramidal neurons in a graded manner, which suggests that glutamate may be involved in establishing hippocampal circuitry during the brain development; maintaining and modifying circuitry in the adult and inducing neurodegeneration in several disorders including epilepsy, stroke and Alzheimer's disease.

Neurotransmitters are prominent candidates for trans-cellular signals that influence the development of the CNS. Govindaiah et al (2002) have examined the effect of glutamate on survival, differentiation and metabolic activity of cultured rat retinal ganglion cells at 3 days in vitro. Retinal cultures from neonatal Wistar rats were treated with glutamate for 48 h. The metabolic activity was markedly increased in the retinal ganglion cells exposed to 20 μM glutamate. This was accompanied by an enhanced survival of these neurons. The number of differentiated retinal ganglion cells as determined by microtubule-associated protein-2 labeling was significantly increased following exposure to low but not higher doses of glutamate. Dopamine and GABA also modulated RGCs survival, diiferentiation and metabolic activity (Govindaiah et al 2004). These results show that neurotransmitters have a significant effect on survival, differentiation and metabolic activity. An increase in the metabolic activity indicates an enhancement in the electrical activity.

Acetylcholine is known to enhance neuritic outgrowth and in tuning the nerve growth cones. Acetylcholinesterase (AChE) can also regulate the neuritic outgrowth and survival of neurons. It has morphogenic and axogenic roles in the developing nervous system. Our studies have demonstrated an increased AChE activity following (-) deprenyl administration in hippocampus and pre-frontal cortex which was well correlated with an increase in the dendritic arborization in CA3 hippocampal and layer III cortical pyramidal neurons in adult monkeys (Lakshmana et al 1998b, Shankaranarayana Rao et al 1999c). It has also been shown that chronic caloric undernutrion alters the AChE activity and biogenic amines metabolism in different regions of the primate brain (Govindaiah et al 1994, Shankaranarayana Rao et al 1994b, 2004c-e). Light deprivation also alters the levels of dopamine and noradrenaline in rats (Shelke et al 1997).

Our studies have demonstrated enhanced levels of biogenic amines, amino acids and AChE activity following self-stimulation (Shankaranarayana Rao et al. 1998b,c, Ramkumar et al 2003a,b, 2004). On the other hand, different forms of chronic stress cause differential alterations of neurotransmitters in discrete regions of the rat brain (Nair, 2003, Nair et al 2004, Ramkumar et al 2003a,b, Shankaranarayana Rao et al 2003a, Sunanda et al 2000b). We have also demonstrated that the stress-induced dendritic atrophy is mediated by corticosterone (Sunanda et al 1998) and enhanced release of glutamate to hippocampus through entorhinal cortex (Sunanda et al 1997, 2000b). We have also demonstrated the role of dopaminergic transmitter system in stress-induced cognitive deficits (Srikumar et al 2004). Furthermore the biochemical changes are tightly correlated with the structural changes following either rewarding or non-rewarding experiences. These studies demonstrate the role of neurotransmitters in altering the morphology of neurons in adult nervous system. The correlation between neurochemical and structural changes indicate that neurotransmitters are likely to play an important role in bringing about structural changes in the adult brain for functional adaptation.

III. Behavioral Plasticity

The importance of experience for the development of behavior and for the maturation of the central nervous system is widely recognized. The recognition that environmental stimulation may be a potent factor governing an organism's capacity to compensate for brain injury raises certain hopes for the treatment of brain injury and also poses some basic questions about the mechanisms by which functions are restored in a damaged nervous system. Pharmacological manipulation of neurotransmitter systems may regulate the neuronal morphology and behavioral plasticity (Shankaranarayana Rao and Raju, 2001a).

Enriched environment (EE) has been shown to facilitate the locomotor activity, exploration and spatial behavior as well as in solving the spatial problems. It was demonstrated that in adult mice living in an enriched environment there was an increase in neurogenesis in the dentate gyrus of the hippocampus. Long-term environmental enrichment leads to increased neurotrophin levels in the rat brain and it also increased mRNA expression in the hippocampus for NGF, BDNF, and NT-3. Previous studies suggest that enriched environment induces neuronal plasticity by producing morphological changes in the hippocampus; increase in the hippocampal thickness, dendritic arborization and accompanied by behavioral improvement in spatial learning tasks. At the same time enriched living conditions in combination with daily rehabilitation improved forelimb motor function and enhanced dendritic growth after focal ischemic injury. Furthermore, EE attenuates cognitive deficits and preserves tissue integrity in traumatic brain injury. The mechanisms by which EE attenuates these cognitive deficits are unknown and this phenomenon is likely to be multifactorial. Our studies have shown that EE ameliorates the stress and subicular lesions induced cognitive deficits (Shankaranarayana Rao et al 2004g). Our earlier studies have demonstrated the regeneration of olfactory tract induced behavioral plasticity (Munirathinam et al 1997). It was also shown that spinal cord ischemia induced spinal motor neuron degeneration and locomotor deficits can be attenuated by (-) deprenyl (Kiran et al 1998, Ravikumar et al 1998). In addition, transplantation of embryonic tissues in to lesioned adult hippocampus and substantia nigra resulted in the recovery of functions such as learning and self-stimulation behavior (Rao, 1996). Furthermore, we have demonstrated that prior self-stimulation experience facilitated the operant and spatial learning tasks in adult rats (Yoganarasimha et al 1998) and self-stimulation also ameliorates fornix lesions induced behavioral deficits (figure 5) (Yoganarasimha and Meti, 1999, Yoganarasimha et al 2002).

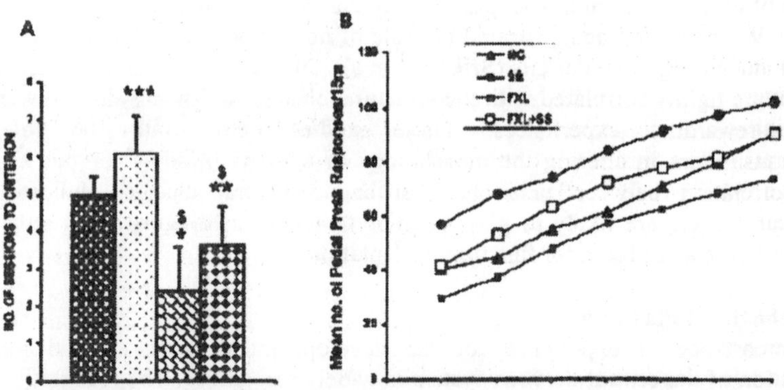

Fig. 5. Number of sessions to reach the criterion for the acquisition (A) and mean number of lever press responses over a period of 7 sessions (B) in the operant learning task in normal control (NC), fornix-lesioned (FXL), self-stimulation (SS) and FXL+SS groups of rats. Note a significant deficit in the acquisition and low performance in FXL rats and amelioration of lesion-induced deficits by exposure to ICSS experience in FXL+SS rats. NC vs. all other groups, $FXL vs. SS, FXL+ SS. * p<0.001, ** p<0.005, *** p<0.025, $ p<0.001. (From Yoganarasimha and Meti, 1999).

Subicular lesions are known to impair spatial learning abilities in rats (Laxmi et al 1999). These studies suggest that neuronal plasticity constitutes a potential mechanism for the recovery of functions after brain injury. In addition, chronic restraint stress is known to alter the acquisition and retention of spatial memory task in rats (Sunanda et al. 2000a). Interestingly, stress-induced cognitive deficits and cholinergic dysfunctions are reversible following 10 days of self-stimulation rewarding behavioral experience (Ramkumar et al 2004a, Shankaranarayana Rad 2003b, Shankaranarayana Rao et al 2004g-h). Our recent studies have demonstrated that herbal extract containing *Euphorbia hirta* reverses chronic stress-induced anxiety through $GABA_A$ receptor benzodiazepine receptor-Cl^- channel complex (Anuradha et al 2004). Further research should aim at developing strategies for amelioration of behavioral deficits in various neurodegenerative disorders.

IV. Experience-dependent plasticity

The ability of the nervous system to wire and rewire itself in response to lasting changes in experience is known as experience-dependent plasticity. The brain is able to remodel its connections in order to adjust the organism's response to changing conditions. The brain shows greatest plasticity during early development when connections between neurons are being made for the first time. The connections are often formed under the guidance of sensory experience. Work into experience-dependent plasticity provides a link between the synaptic mechanisms studied in vitro and the behavioral studies on learning and memory. One of the most spectacular properties of the human brain is its ability to store information. Underlying all memories are plastic changes in the ways in which neurons communicate with one another. Our investigations are designed to gain insight into how neurons change as a result of experience and whether those changes could be the mechanisms by which we learn and remember. By attaining this goal, we can begin to see how learning and memory could be improved, while at the same time identifying approaches that could prevent or reverse learning impairments.

Examining the neuronal basis of learning and memory is a daunting task, as there are several forms of learning and memory, and potentially as many ways for neurons to change their function. We are testing the hypothesis that long-term potentiation (LTP), an enduring change in the strength of communication signals between neurons, is the basis of learning in the mammalian central nervous system. Because this phenomenon is induced by brief, high frequency activation of synaptic afferents and produce changes in synaptic function for relatively long periods of time, it makes an attractive model for both learning-induced plasticity and experience-dependent plasticity.

Understanding the basis of neural plasticity has been the goal of considerable research. Recent studies demonstrated experience dependent structural changes in the presynaptic and post synaptic elements. The stage has been now set to accept the concept of neurogenesis in the adult neural system (Thomas and Raju, 2004). Some of the most detailed studies of experience dependent plasticity have been performed in the rodent barrel cortex. Manipulations of the sensory inputs such as, clipping the whiskers can change the receptive fields of cortical neurons. Robust experience dependent plasticity has been observed within 24 hrs of whisker clipping. A research problem of this complexity requires a variety of approaches and techniques. We have chosen to use morphological, biochemical, behavioral and electrophysiological techniques to assess learning and synaptic plasticity.

Molecular basis of plasticity

The molecular mechanisms involved in neuronal plasticity have been a topic or intensive research in recent years. Gene knock out technology has further enhanced the understanding of the role of genes and proteins in synaptic plasticity. Gene targeting offers an alternative way of investigating the relationship between synaptic plasticity and memory. Our collaborative research has significantly contributed the role of different signaling molecules in neuronal functions. We have studied the role of NMDA receptors in structural plasticity (Shankaranarayana Rao et al 2000a, 2001b, 2004a) and in stress (Govindarajan et al 2003a-b, 2004, Nair et al 2003). BDNF is critically involved in growth and maintenance of hippocampal and cortical neurons (Shankaranarayana Rao et al 2000). Hayashi et al. (2003a-b, 2004) have shown that p21-activated kinase (PAK) is associated with the postsynaptic density and also regulates the spine density of cortical neurons. In addition, we have also studied the role of calcineurin on dendritic arborization and spine morphology in knockout mice (Shankaranarayana Rao et al 2000). Furthermore, absence of presinilins causes synaptic dysfunction and neurodegeneration (Saura et al 2003, 2004). The above studies have provided the cellular and molecular basis of neuronal plasticity in hippocampal and cortical neurons in development, synaptic plasticity and in neurodegenerative disorders, namely aging and stress.

Activity dependent synaptogenesis is thought to be mediated by gene expression and protein translation. Activity dependent up regulation of BDNF leads to spine growth and recruitment of new synapses and proposed to enhance dendritic morphogenesis. It has been shown that neurotrophins are involved in processes of neuronal plasticity. There is increased evidence of neurotrophins and their receptor signaling in the production of LTP and activity-dependent plasticity associated with learning. BDNF expression and trkB signaling have been associated with dendritic and synaptic restructuring by means of regulating spine dynamics, functional maturation of presynaptic terminals, dendritic growth, triggering AMPA receptor proteins and in activity dependent conversion of silent synapses into functional ones. We have shown that BDNF is critical for the normal maintenance of dendritic arbors and synapses in hippocampal and cortical neurons (Govindarajan et al 2003a,b, 2004, Shankaranarayana Rao et al 2000). We have also shown that stress induces alterations in BDNF and glutamate receptor mRNA expression in the hippocampus (Mitra et al 1999).

Summary

It is becoming increasingly clear that a regenerative and continuing mechanism for adaptive reorganisation of the brain occurs because of the property of neuronal plasticity. This unique property of the nervous system may be responsible for the recovery of functions in injury as well as in various neurological disorders. However, the means to manipulate the promotors and inhibitors of neuronal plasticity are still in the process of development. However, neuroscience research will provide promising plasticity enhancers for clinical trials. As with the potential mechanisms for restoration of neuromorphology, the functional effectiveness of any biological manipulation will depend in part on the strategies employed to optimally induce activity in neuronal circuits. Our current research focuses on understanding the properties of neuronal plasticity in developing and adult nervous system by various behavioral manipulations and recovery of functions.

"Brain is an 'enchanting loom', the threads of the loom can be broken either by internal perturbations like lesion or by external perturbations like learning, new threads then form, branch out, and give a different pattern- always shifting, but a meaningful pattern."

Sir Charles S Sherrington (1906)

Acknowledgements

Supported by National Institute of Mental Health and Neuro Sciences (NIMHANS), Bangalore.

References

Anuradha H, Srikumar BN, Deepti N, Shankaranarayana Rao BS and Lakshmana M (2004) Proc Int Neurosci Conf of Indian Acad of Neurosciences and Society for Neurochemistry, Hyderabad. pp 79.

Bliss TVP and Collingridge GL (1993) Nature **36**:31–39.

Bliss TVP and Lomo T (1973) J Physiol **232**:331-356.

Calverley RKS and Jones DG (1990) Brain Res Rev **15**:215-249.

Chang F-LF and Greenough WT (1984) Brain Res **309**:35-46.

Desmond NL and Levy WB (1986) J Comp Neurol **253**:476-482.

Dhingra NK, Ramamohan Y and Raju TR (1997) Dev Brain Res **102**:267-273.

Dhingra NK, Reddy R, Govindaiah, Hemavathy U, Raju TR and Ramamohan Y (2001) Int J Dev Neurosci **19**:533-540.

Govindaiah, Lakshmana MK, Shankaranarayana Rao BS, Dhingra NK, Sudha S, Ravikumar R, Meti BL and Raju TR (1994) J Neurochem **63** (Suppl. 2):S19D.

Govindaiah, Shankaranarayana Rao BS, Raju TR and Meti BL (1997) Brain Res **745**:121-126.

Govindaiah, Shankaranarayana Rao BS, Ramamohan Y, Singh YK, Dhingra NK and Raju TR (2000) Dev Brain Res **124**:117-120.

Govindaiah (2001) Role of neurotransmitters in developing retinal neurons: An in vitro study on neuronal survival differentiation and metabolic activity. PhD Thesis, NIMHANS Deemed University, Bangalore.

Govindaiah, Shankaranarayana Rao BS and Raju TR (2002) Neuroscience **113**:547-553.

Govindaiah, Shankaranarayana Rao BS and Raju TR (2004) Neuroscience Submitted.

Govindarajan A, Shankaranarayana Rao BS, Nair D, Vyas A, Mawjee N, Lee C, Tonegawa S and Chattarji S (2004) Science Submitted.

Govindarajan A, Nair D, Shankaranarayana Rao BS, Mawjee N, Tonegawa S and Chattarji S (2003a) Proc Neurobiol Symp Networks Behav B3.

Govindarajan A, Nair D, Shankaranarayana Rao BS, Mawjee N, Lovett C, Lee C, Tonegawa S and Chattarji S (2003b). Soc Neurosci Abs **33**:926.13

Hayashi ML, Choi S, Shankaranarayana Rao BS, Jung H, Chattarji S, Kirkwood A and Tonegawa S (2003a). Soc Neurosci Abs **33**:123.3

Hayashi ML, Choi S-Y, Shankaranarayana Rao BS, Chattarji S, Kirkwood A, Tonegawa S (2003b) Proc. The Cell Biology of the Neuron- In celebration of the 15th Anniversary of Neuron, November 7, 2003, JW Marriott, New Orleans, LA, USA

Hayashi ML, Choi S-Y, Shankaranarayana Rao BS, Jung H-Y, Lee H-K, Zhang D, Chattarji S, Kirkwood A and Tonegawa S (2004) Neuron **42**: 773-787.

Jones TA and Schallert T (1994) J Neurosci **14**:2140-2152.

Kiran B, Shankaranarayana Rao BS, Raju TR and Bindu PN (1998) Med Sci Res (London) **26**:89-92.

Kutty BM, Shankaranarayana Rao BS and Raju TR (2003) Neurophysiological aspects of plasticity of the brain. Proc Second National Workshop in Clinical Neuropsychology National Institute of Mental Health and Neuro Sciences, Bangalore, India. pp 1-7.

Lakshmana MK, Shankaranarayana Rao BS, Dhingra NK, Ravikumar R, Govindaiah, Sudha S, Meti BL and Raju TR (1998a) Neurochem Res **23**: 1031-1037.

Lakshmana MK, Shankaranarayana Rao BS, Dhingra NK, Ravikumar R, Govindaiah, Ramachandra, Meti BL and Raju TR (1998b) Brain Res **796**:38-44.
Lakshmana MK, Shankaranarayana Rao BS, Dhingra NK, Ravikumar R, Govindaiah, Sudha S and Raju TR (2004) Neuroscience Submitted.
Latha Devi, Diwakar L, Raju TR, Kutty BM (2003) Brain Res **960**:9-15.
Laxmi TR, Shankaranarayana Rao BS, Raju TR and Meti BL (1997) Indian J Physiol Pharmacol **41**(5): 452A.
Laxmi TR, Bindu PN, Raju TR and Meti BL (1999) Brain Res **816**:245-248.
Laxmi TR, Meti BL and Bindu PN (2000) Brain Res **869**:236-240.
Lynch GS, Dunwiddie T, Gribkoff V (1977) Nature **266**:737–739.
Marita KG, Chatterji S and Raju TR (2004) Vis Res (Submitted).
Mattson MP (1988) Brain Res Rev **13**:179-212.
Merzenich MM and Kaas JH (1982) Trends Neurosci **5**: 434-436.
Mitra R, Shankaranarayana Rao BS, Vyas A, Sharath BS and Chattarji S (1999) Proc Int Colloq Brain Res B24.
Mitra R, Wood GE, Shankaranarayana Rao BS, Chattarji S and McEwen BS (2002) Soc Neurosci Abs **32**: 866.16
Mitra R, Wood GE, Shankaranarayana Rao BS, Chattarji S and McEwen BS (2003) Proc Neurobiol Symp Networks Behav B8.
Munirathinam S, Rao MS, Ramamohan Y and Raju TR (1997) Exp Neurol **144**: 174-182.
Nair D, Shankaranarayana Rao BS, Nakazawa K, Tonegawa S and Chattarji S (2003) Proc Neurobiol Symp Networks Behav B11.
Nair D (2003) Stress-induced alterations in the levels of amino acid neurotransmitters and acetylcholinesterase activity in discrete regions of the rat brain. Mphil Dissertation, NIMHANS Deemed University, Bangalore. India.
Nair D, Raju TR and Shankaranarayana Rao BS (2004) Proc Int Neurosci Conf of Indian Academy of Neurosciences and Society for Neurochemistry, Hyderabad. pp 80.
Ramesh V, Lakshamana MK, Shankaranarayana Rao BS, Raju TR and Mohan Kumar V (1999) Sleep Res Online **2**: 49-55.
Ramkumar K, Shankaranarayana Rao BS and Raju TR (2004a) Neuroscience Submitted.
Ramkumar K, Venkatasubramanian D, Siva R, Shankaranarayana Rao BS and Raju TR (2004b) Proc. 4th IBRO School of Neuroscience, Hong Kong.
Ramkumar K, Shankaranarayana Rao BS and Raju TR (2003a) Proc International IBRO/FENS Summer School "Development & Plasticity of the Human Cerebral Cortex", Dubrovnik/Zagreb, Croatia. 37.
Ramkumar K, Shankaranarayana Rao BS and Raju TR (2003b) Proc Neurobiol Symp Networks Behav B15.
Rao MS, Gourie-Devi M, Nalini A, Shahani N and Raju TR (1995) Neurodegeneration **4**: 397-401.
Rao MS (1996) Experimental study of neural transplants in behavioural recovery and patterns of survival and degeneration. PhD. thesis, NIMHANS Deemed University.
Ravikumar R, Lakshmana MK, Shankaranarayana Rao BS, Meti BL, Bindu PN and Raju TR (1998) Exp Neurol **149**: 123-129.
Rosenzweig MR (1966) Am Psychol **21**: 321-332.
Rutledge LT, Wright C and Duncan J (1974) Exp Neurol **44**:209-228.
Saura CA, Choi S, Malkani S, Shankaranarayana Rao BS, Chattarji S, Duff K, Kirkwood A and Shen J (2003) Soc Neurosci Abs **33** : 239.2
Saura CA, Choi S-Y, Beglopoulos V, Malkani S, Zhang D, Shankaranarayana Rao BS, Chattarji S, Kelleher III RJ, Kandel ER, Duff K, Kirkwood A Shen J (2004) Neuron **42** :23-36.
Shahani N, Nalini A, Gourie-Devi M and Raju TR (1998) Exp Neurol **149**:295-298.
Shahani N, Gourie-Devi M, Nalini A, Raju TR (2001) J Neurol Sci **185**:109-118.
Shahani N, Gourie-Devi M, Nalini A, Priti YR Shobha K, Harsha HN and Raju TR (2004 ALS and other motor neuron disorders. In press.
Shankaranarayana Rao BS, Desiraju T and Raju TR (1993) Brain Res **627**:216-224.
Shankaranarayana Rao BS, Desiraju T, Meti BL & Raju TR (1994a) Indian J Physiol Pharmacol **38**:23-28.
Shankaranarayana Rao BS, Lakshmana MK, Dhingra NK, Sudha S, Govindiah, Ravikumar R, Meti BL and Raju TR (1994b) J Neurochem **63** (Suppl.2): S23A.
Shankaranarayana Rao BS (1996) Neuronal plasticity induced by self-stimulation rewarding behavioural experience: Morphological, biochemical and immunocytochemical evaluation. Ph.D. Thesis, NIMHANS Deemed University, Bangalore. India.

Shankaranarayana Rao BS, Shahani N, Raju TR and Meti BL (1997) Proc 15th Annual Conf Indian Acad Neurosci p21.
Shankaranarayana Rao BS, Raju TR and Meti BL (1998a) Brain Res **804**:320-324.
Shankaranarayana Rao BS, Raju TR and Meti BL (1998b) Neurochem Res **23**:1053-1059.
Shankaranarayana Rao BS, Raju TR and Meti BL (1998c) J Neurochem **70** (Suppl.2):S32A.
Shankaranarayana Rao BS, Meti BL and Raju TR (1998d) Neuronal plasticity - A unique property for Neurorehabilitation. In: Neurorehabilitation: Principles and Practice, (Taly AB, Nair KPS and Murali T eds) National Institute of Mental Health and Neurosciences, Bangalore, India, pp. 36-49.
Shankaranarayana Rao BS, Raju TR and Meti BL (1998e) Brain Res Bull **47**:95-101.
Shankaranarayana Rao BS, Raju TR and Meti BL (1999a) Neuroscience **91**: 799-803.
Shankaranarayana Rao BS, Raju TR and Meti BL (1999b) Neuroscience **89**: 1067-1077.
Shankaranarayana Rao BS, Lakshmana MK, Meti BL and Raju TR (1999c) Brain Res **821**: 218-223.
Shankaranarayana Rao BS, Santhosh MG, Devendra kumar, Selvamurthy W and Raju TR (1999d) Indian J Physiol Pharmacol **43**:576A.
Shankaranarayana Rao BS, Deepti N and Chattarji S (2000) Annu Rev Res Tata Institute of Fundamental Research, Mumbai.
Shankaranarayana Rao BS and Raju TR (2001a) Proc Indian Natnl Sci Acad **B67**: 155-188.
Shankaranarayana Rao BS, Deepti N, Tonegawa S and Chattarji S (2001b) Soc Neurosci Abs **31**: 902.10
Shankaranarayana Rao BS, Meti BL and Raju TR (2001c) Neuronal plasticity - A unique property for Neurorehabilitation In : Neurorehabilitation : Principles and Practice, (Taly AB, Nair KPS and Murali T eds) 2nd ed, Ahuja Book Company, New Delhi, India, pp.27-37.
Shankaranarayana Rao BS, Govindaiah, Laxmi TR, Meti BL and Raju TR (2001d) Neuroscience **102**:319-327.
Shankaranarayana Rao BS, Madhavi R, Sunanda and Raju TR (2001e) Curr Sci **80**:653-659.
Shankaranarayana Rao BS, Deepti N, Prabhu BM and Raju TR (2003a) Soc Neurosci Abs 33:713.9
Shankaranarayana Rao BS (2003b) Role of intracranial self-stimulation and BDNF in reversal of stress induced deficits. Proc Brain Storming Session on Neurobiology in the Post-Genomic Era, (Thakur MK, ed), Banaras Hindu University, Varanasi, India. pp. 39-43.
Shankaranarayana Rao BS, Deepti N, Tonegawa S and Chattarji S (2004a) Nature Neurosci submitted.
Shankaranarayana Rao BS, Lakshmana MK, Dhingra NK, Govindaiah, Ravikumar R and Raju TR (2004b) Neurochem Res Submitted.
Shankaranarayana Rao BS, Lakshmana MK, Dhingra NK, Govindaiah, Ravikumar R, Sudha S and Raju TR (2004c) Neurochem Res Submitted.
Shankaranarayana Rao BS, Lakshmana, MK, Dhingra NK, Govindaiah, Ravikumar R, Sudha S and Raju TR (2004d) Neurochem Res Submitted.
Shankaranarayana Rao BS, Meti BL and Raju TR (2004e) Brain Res Submitted.
Shankaranarayana Rao BS, Govindaiah and Raju TR (2004f) Neuroscience Submitted.
Shankaranarayana Rao BS, Ramkumar K and Raju TR (2004g) J Neurochem **88** (Suppl. 1): 17.
Shankaranarayana Rao BS, Ramkumar K, Srikumar BN, Deepti N And Raju TR (2004h) Proc Int Neurosci Conf of Indian Acad of Neurosciences and Society for Neurochemistry, Hyderabad. pp 22.
Shelke RJ, Lakshmana MK, Ramamohan Y and Raju TR (1997) Int J Dev Neurosci **15**: 139-143.
Srikumar BN, Raju TR and Shankaranarayana Rao BS (2004) Proc Int Neurosci Conf of Indian Academy of Neurosciences and Society for Neurochemistry, Hyderabad. pp 50.
Sunanda Rao MS, Raju TR (1995) Brain Res **694**:312-317.
Sunanda, Meti BL and Raju TR (1997) Brain Res **770**:302-306.
Sunanda, Shankaranarayana Rao BS and Raju TR (1998) Brain Res **791**:295-298.
Sunanda, Shankaranarayana Rao BS and Raju TR (2000a) Curr Sci **79**:1581-1584.
Sunanda, Shankaranarayana Rao BS and Raju TR (2000b) Neurochem Res **25**:1547-1552.
Thomas SK and Raju TR (2004) Nat Neurosci Submitted.
Titus ADJ, Harsha HN, Ramkumar K, Singh SB, Chatterji S, Shankaranarayana Rao BS and Raju TR (2004) Proc Int Neurosci Conf of Indian Academy of Neurosciences and Society for Neurochemistry, Hyderabad. pp 82.
Vyas A, Mitra R, Shankaranarayana Rao BS and Chattarji S (2002) J Neurosci **22**:6810-6818.
Walsh RN, Bedtz-Olsen OE, Penny JE and Cummins RA (1972) J Comp Neurol **137**:361- 366.
Yoganarasimha D, Shankaranarayana Rao BS, Raju TR and Meti BL (1998) Behav Neurosci **112**:725-729.
Yoganarasimha D and Meti BL (1999) Brain Res **845**:246-251.
Yoganarasimha D, Shankaranarayana Rao BS and Raju TR (2002) Soc Neurosci Abs **32**:184.5

Molecular and Cellular Neurobiology
Editors: M.K. Thakur and S. Prasad

Neuronal Plasticity in Adult Brain: Role of Polysialylated Neural Cell Adhesion Molecule in Cell-Cell Interactions

Gurcharan Kaur

Department of Biotechnology, Guru Nanak Dev University, Amritsar, India

Abstract

Synaptic remodeling is thought to reflect functional responses of neural tissue to specific physiological and pathological stimuli. A central problem in neuroscience concerns the elucidation of molecular interactions that contribute to the activity-dependent synaptic remodeling. Understanding the molecular mechanisms implicated in these aspects of plasticity and identifying molecules that might represent potential targets for promoting repair is a potential area of research in neuroscience. Neural cell adhesion molecule (NCAM) and its polysialylated form, PSA-NCAM is one such molecule. NCAM mediates neuronal and glial interaction during development of the nervous system and thereby affects processes such as neurite fasciculation, axonal path finding and synaptogenesis. It has been suggested that in adult brain, regulation of adhesion alters the morphological and physiological properties of synapses through effects on membrane interactions. Changes in the levels of expression of NCAM and PSA-NCAM may contribute to the maintenance of synaptic contacts and may be relevant to adult forms of activity-dependent plasticity.

Key words: Neuronal plasticity, Neural Cell Adhesion Molecule, Polysialylated Neural Cell Adhesion Molecule, Synaptic Remodeling, Synaptogenesis.

Introduction

The notion of neuronal plasticity in the postnatal mammalian brain is now a well established phenomenon, having been demonstrated in the hypothalamus, hippocampus and olfactory bulb. Plastic morphological rearrangements of the brain cells are thought to reflect functional responses of neural tissue to specific physiological and pathological stimuli. A major aim in neuroscience research concerns the elucidation of molecular interactions that contribute to the activity-dependent remodeling and repair of nervous system. Although central nervous system is known for having limited capacity for self-repair, evidence indicates that brain plasticity is nevertheless quite substantial, and may contribute to recovery of function. There is, therefore considerable interest in understanding the molecular mechanisms implicated in these aspects of plasticity and in identifying molecules that might represent potential targets for promoting repair. In this context, molecules such as growth factors, proteases or cell-cell adhesion molecules that shape the formation of neuronal networks during development are particularly interesting candidates. Recent data suggest that these molecules play a role in several aspects of cellular and synaptic plasticity mechanism that are proposed to underlie the reorganization of cortical maps associated with functional recovery. Here the focus will be on one such molecule, the neural cell adhesion molecule (NCAM) and its polysialylated form, PSA-NCAM, and review the evidence that suggests that this molecule is implicated in neuronal network plasticity.

Molecular composition of NCAM and PSA-NCAM

NCAM, a member of the immunoglobulin superfamily of adhesion molecules, is characterized by several immunoglobulin (Ig)-like domains or Ig modules (Edelman, 1986). The extracellular part of NCAM consists of five of these Ig modules and two fibronectin type III homology regions (Cunningham, 1987). NCAM is coded by a single copy gene composed of 26 exons (Goridis and Brunet, 1992). However, at least 20-30 distinct isoforms can be generated by alternative splicing and by post-translational modifications (Goridis and Brunet, 1992). Three major isoforms are generated via alternative splicing of a primary transcript from a single gene (Owens et al 1987). Two isoforms are transmembrane forms of 140 kDa and 180 kDa and the third of 120 kDa is attached to the cell membrane via a glycophosphatidyl inositol linkage. Through its extracellular region, NCAM mediates homophilic and heterophilic, cis and trans interactions (Hoffman and Edelman, 1983).

Recent work has revealed the crystal structure of the N-terminal extracellular domain of NCAM (Kasper et al 2001). It appears that the packing of this structure can form a cross shaped antiparallel dimer that could mediate trans-cellular homophilic recognition by NCAM. The molecule, however, can also bind to a number of extracellular matrix components or other cell adhesion molecules through heterophilic interactions. Studies utilizing normal and transfected cells together with antibodies, peptides, and recombinant Ig domains suggest that all five Ig domains are involved in NCAM homophillic binding and that Ig III has a special role (Watanabe et al 1986; Zhou et al 1993). A model for NCAM homophillic binding has emerged from these studies which proposes an antiparallel interdigitation of the Ig domains so that Ig I binds Ig V, Ig II binds Ig IV, and Ig III binds to Ig III among molecules on apposing cells, and that one or more of the other domains (unspecified) interacts cis with NCAM molecules on the same. Figure 1 (left panel) illustrates the possible mechanism of homophillic interaction of NCAM at the cell surface.

The capacity for neuronal plasticity of NCAM has been linked to the expression of polysialic acid (PSA) residues on extracellular domains of NCAM. (Sunshine et al 1987). PSA is a polymer of highly charged α-2,8 sialic acid, which is post-translationally added to NCAM by a specific polysialyltransferase (PST) enzyme (Eckhardt et al 1995; Nelson et al 1995). PSA is attached to two asparagines in the Ig5 module of the extracellular part of NCAM. A single PSA chain may consists of more than 50 monomers, but the chain length may vary substantially in NCAM isolated from various sources (Rougon, 1993). The amount of PSA on the NCAM molecule modulates its adhesion properties and a convergent set of data suggests that PSA-NCAM supports structural plasticity both in the developing and in the adult nervous system. Figure 1 (right panel) illustrates the possible mechanism of how the presence of PSA on NCAM reduces the possibility of homophillic interactions of NCAM. By increasing the intermolecular space through hydration, PSA may affect not only the homophilic and heterophilic binding of NCAM but also a broad spectrum of cellular interactions. Addition of PSA to NCAM takes place, as for other protein glycolysation pathways, in the trans golgi compartment. Two sialyltransferases, which have been cloned, appear to be responsible for the synthesis of PSA-NCAM: they are referred to as ST8SiaIV/ PST and ST8SiaII/STX (Kitagawa and Paulson, 1994; Nakayama et al 1995; Eckhardt et al 2001). These two enzymes seem to be the regulating factors for PSA- synthesis and their activity is controlled at the mRNA level (Eckhardt et al 1995). The concentration of PSA at the cell surface probably depends upon two major regulatory mechanisms: (I) the biosynthesis of PSA-NCAM, that is itself dependent upon the level of transcription and / or activity of the sialyltransferases and (ii) the intracellular trafficking that affects the turnover of the molecule at the cell surface (Kiss and Rougon, 1997).

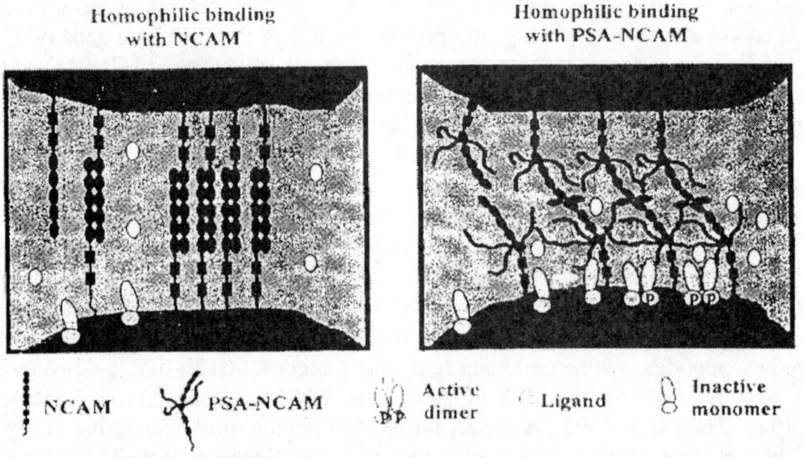

Fig. 1. Hypothetical model illustrating the possible mechanism of interaction between NCAMs and receptor tyrosine kinases. Left: homophillic binding of NCAM may stabilize cell-cell contacts through interactions involving the five Ig modules of NCAM. Right: the presence of PSA in the NCAM reduces possibilities of homophillic interactions that are limited to a cross shaped anti-parallel binding of the first two Ig modules and at the same time, favours cis interaction and the dimerisation and phosphorylation (P) of receptor tyrosine kinases such as the FGF receptor or the BDNF receptor Trk-B, leading in this way to an increased sensitivity to the released growth factors. In addition. PSA on NCAM could also act by trapping positively charged ligands such as BDNF in the proximity of their receptors. [Kiss et al Brain Res Rev 36 (2001) 175-184].

Mechanism of NCAM and PSA-NCAM action

The polysialylation of NCAM facilitates events such as cell migration, axon growth, remodeling of cell contacts and synaptic reorganization. The postnatal loss of PSA on NCAM follows a timing that is highly region specific and this decrease has been consistently shown to be associated with stabilization of cellular interactions and synapse formation (Szele et al 1994). In the adult brain, high levels of PSA-NCAM are retained only in certain brain areas undergoing structural reorganization (Rousselot et al 1995), in response to injury (Sandi and Loscertales, 1999; Sandi et al 2001; Singh et al 2003) and in areas exhibiting morphological remodeling in response to physiological changes. This latter case can be exemplified by the hypothalamo-hypophyseal-gonadal axis, which undergoes important morphological modifications under physiological stimuli (Theodosis and Poulain, 1993; Nothias et al 1997; Kaur et al 2002).

A traditional view of NCAM is that cell-cell and cell to matrix interactions mediated by these molecules establish a physical anchorage of cells to their environment that is required for proliferation, growth and migration of cells. Adhesion and signaling are thus mutually interdependent. All NCAM isoforms can mediate cis-, trans-, homo- and heterophillic interactions (Crossin et al., 2001). In these interactions, the large negatively charged PSA chain could function as a spacer that reduces adhesion forces between cells, allowing dynamic changes in the membrane contacts (Rutishauser et al 1988). Furthermore, aggregation of PSA-NCAM could be much reduced as compared to NCAM without PSA (Hoffman and Edelman, 1983). One possibility is that when homophillic binding is mediated by NCAM, all five Ig domains are involved in antiparallel topology (Fig.1). The presence of long PSA chains could change this interaction in the way that only the first or second Ig domains may participate in binding, forming a cross shaped antiparallel dimerisation (Kasper et al 2001). Thus, the PSA chain could affect trans interactions and reduces adhesion forces between opposing membranes, but it could also modify lateral, cis-interactions resulting in multimerization and cluster formation. For example, a homomeric multimerization of L1 through the third fibronectin domain has been reported (Silletti et al 2001).

A large number of experimental data suggest that NCAM mediated interactions at the cell surface activate intracellular signaling cascades that are fundamental to the biological functions of the molecule (Crossin and Krushel, 2001). Two major signaling cascades, the Ras/MAP kinase pathway and the phospholipase-C-γ (PLC-g)/ diacyl-glycerol/Ca^{2+}/ protein kinase C pathway, are believed to play a critical role in NCAM-induced neurite outgrowth. Ras/MAPK cascade has also been implicated in the mechanisms of gene expression contributing to long-term potentiation form of synaptic plasticity. PSA-NCAM, in addition to its putative role of relieving adhesion between opposing membranes, also acts through the modification of intracellular signaling via brain derived neurotrophic factor (BDNF) receptor Trk B. Just as PSA-NCAM, BDNF and its high affinity Trk B receptor are required for activity-dependent plasticity. This cross talk between PSA-NCAM and BDNF signaling could mediate the activity-dependent aspects of neuronal network remodeling and, thus, confer specificity to these mechanisms. This leads to the conclusion that NCAM and PSA-NCAM are critically important for and contribute to the mechanisms of activity-dependent forms of synaptic plasticity that are believed to underlie information processing by the brain. How these molecules contribute to synaptic plasticity still remains essentially unknown.

The modulatory role of PSA-NCAM on BDNF/TRK B signaling could play a crucial role in the developmental processes and adult plasticity, during which cells are competing for limited amount of this factor (Katz and Shatz, 1996). Neurons or individual synapses with more PSA at their surface would be more responsive to BDNF and may have a clear advantage over neurons that express a lower level of or no PSA. One way to increase the concentration of PSA-NCAM at the cell surface as well as to increase the secretion of BDNF is neuronal activation. Thus, PSA-NCAM might be a critical partner of BDNF in regulating activity-dependent competition and plasticity of the neurons (Katz and Shatz, 1996; Boulanger, 1999).

Functions of NCAM in synaptic plasticity and learning

Activity-dependent modulation of cell adhesion molecule expression and function blocking studies using antibodies, peptides or gene knockout approaches have demonstrated the involvement of cell adhesion molecules in both behavioral and synaptic plasticity (Murase and Schuman, 1999). The observation that many adhesion molecules are expressed both pre-synaptically and post-synaptically raises the possibility that information about synaptic activity might simultaneously be communicated to both sides of the synapse, circumventing the need for distinct anterograde and retrograde messengers. Some studies have suggested that the neuronal Ig superfamily is involved in both synaptogenesis and synaptic plasticity. Mice with a targeted deletion of the NCAM gene showed an impairment of axonal growth and synaptogenesis in the hippocampal mossy fiber axon system (Cremer et al 1997; Cremer et al 1998). Mossy fibers are usually restricted to the proximal dendrites of CA3 cells. Axons in mutant CA3 hippocampus are no longer fasiculated and mossy fiber presynaptic terminals are present ectopically in the CA3 cell body layer and throughout the CA3 region. Despite these morphological abnormalities, basal synaptic transmission and short-term synaptic plasticity were normal in the mutant mice. LTP however, was abolished in these mice in both area CA3 and CA1 (Cremer et al 1998; Muller et al 1996). These observations are supported by an earlier study showing that application of function-blocking antibodies to either NCAM or L1, an Ig family member, were found to inhibit LTP induction in CA1 (Luthi et al 1994).

In addition, NCAM knockout mice exhibited deficits in performance on the Morris water maze, a spatial memory task (Cremer et al 1994). In this task, animals are required to swim and find a platform hidden below the surface of an opaque pool. Control animals show improved performance, (e.g. decreased latency to find the platform) with repeated trials. In addition, on probe trials after training, following removal from the pool, control animals spend more time in the vicinity of the previous platform location. The NCAM knockout mice exhibited longer latencies to find the platform during training, as well as less time in the correct platform location following training (Cremer et al 1994).

It is still not known, however, whether these molecules mediate a rearrangement of synaptic structures during memory formation. If the expression of cell adhesion molecules can be regulated by activity, then newly synthesized molecules might participate in the formation or modification of synapses. Indeed, the expression levels of some cell adhesion molecules have been reported to change in an activity-dependent manner. For example, the expression pattern of N-cadherin, NCAM and L1 can be regulated by distinct patterns of action potentials (Itoh et al 1995). In addition, the promoter of the NCAM gene is sensitive to AMPA (a-amino-3-hydroxy-5-methyl-isoxazole-4-proprionic acid) receptor activity (Holst et al 1998). Consistent with these observations, immunostaining following LTP has shown that proportion of NCAM containing synapses is increased (Schuster et al 1998). Antibodies that specifically block the function of NCAM have been shown to alter retinal layer formation, development of the retinotectal projection, neurite outgrowth, neuroblast migration, and hippocampal LTP (Doherty et al 1995; Chiba and Keshishian, 1996). During peripheral nerve regeneration, dramatic changes in N-CAM and L1 expression occur even in sites distant from the site of perturbation, including the target muscle and spinal cord (Covault and Sanes, 1985).

The PSA-NCAM expression in neuroendocrine plasticity

Growing evidence indicates that neuroendocrine plasticity is not only limited to prenatal development, but continues post-natally and even into adulthood. A well-documented example is the neurohypophyseal-magnocellular system, which undergoes dramatic morphological changes in response to physiological stimuli such as suckling and dehydration (Theodosis and Poulain, 1992). Another instance occurs within hypothalamic centers that govern reproduction. In the rat and monkey, estradiol induces a synaptic reorganization within the basal hypothalamus (Witkin et al 1991; Parducz et al 1993). With regard to GnRH neurons, the preovulatory GnRH surge of the rat is associated with decreased glial ensheathment of GnRH terminals that contact portal vascular loops in the median eminence (King and Rubin, 1995). Gonadal steroids influence innervation and glial ensheathment of GnRH perikarya of monkeys (Witkin et al 1991) and recently we have reported that GFAP expression declines significantly in preoptic area (POA) of hypothalamus having GnRH perikarya (Kaur et al 2002; Heera and Kaur, 2003) in both proestrous phase of cycling rats and EBP-primed OVX rats. Synaptic input to GnRH neuron changes in association with puberty in monkeys and with reproductive aging in rats (Perera and Plant, 1997).

Despite extensive evidence for neuroplasticity within the adult GnRH system, the underlying molecular basis remains unknown. In the neurohypophyseal-magnocellular system, proteins that influence neuronal remodeling, in particular NCAM is implicated in plasticity during periods of altered neurosecretion (Theodosis et al 1998a,b). In a recent study from our lab, the immunohistochemical staining for GnRH using anti GnRH LR1 antibody showed a marked increase in the length and branching of GnRH axon terminals in the ME-ARC region in the proestrous phase. On qualitative analysis, many GnRH terminals were seen to emit very thin processes in the direction of the perivascular space in the proestrous rats indicating that these are the sites of active GnRH secretion (fig. 2). The results of this study (Kaur et al 2002) provide confirmatory Immunoblot data on the presence of PSA-NCAM in the ME-ARC region and its changing expression in different phases of the rat estrous cycle (fig. 3). Cerebral hemispheres and dorsal hippocampus region were immunonegative for PSA-NCAM upon western blot analysis.

While the stage and spatial specificity of expression of the multiple NCAM isoforms has been documented in variety of tissues, the functional significance of the presence of these molecules is still speculative. It has been proposed that their structure allows them to differ in subcellular localization, turnover rate, ability to cluster at the cell surface, cell contact stabilization and ability to respond to cell signaling processes. The pronounced increase in the hypothalamic contents of the NCAM-180 and NCAM-140 isoforms (which are associated with PSA expression) observed in the proestrous as compared to the diestrous phase (Kaur et al 2002), could promote a significant degree of neuronal remodeling in neuroendocrine pathways concerned with pulsatile GnRH release.

In the Siberian hamster, photoperiodic manipulation that cause seasonal shifts in GnRH secretion have been seen to alter the relative amounts of PSA and other NCAM isoforms in the basal hypothalamus. In vivo enzymatic removal from the mouse suprachiasmatic nucleus seem to significantly impair the photic induction

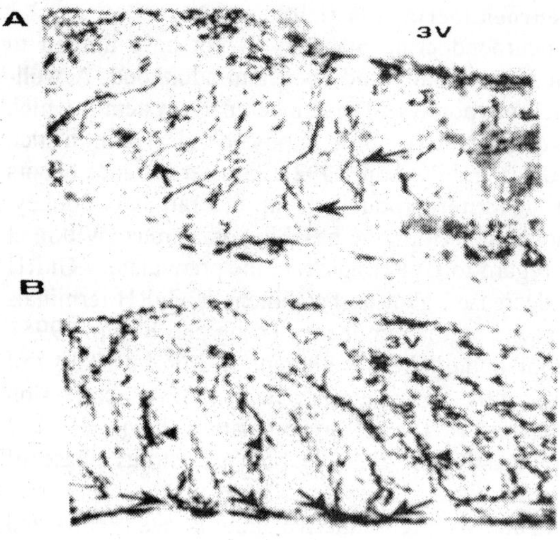

Fig.. 2. **GnRH containing fibers in the ME-ARC region in representative section of the ME-ARC (A) in the diestous phase and (B) in the proestrous phase of the estrous cycle. Note the presence of darkly stained GnRH axons processes extending towards the pericappillary space in B (arrows) and enhanced density of GnRH –ir fibers (arrow heads) in B. The GnRH axonal immunoreactivity was much less in the diestrous phase as shown in A (arrows). Magnification, X400 [Kaur et al, Brain Res 954 (2002) 21-31]**

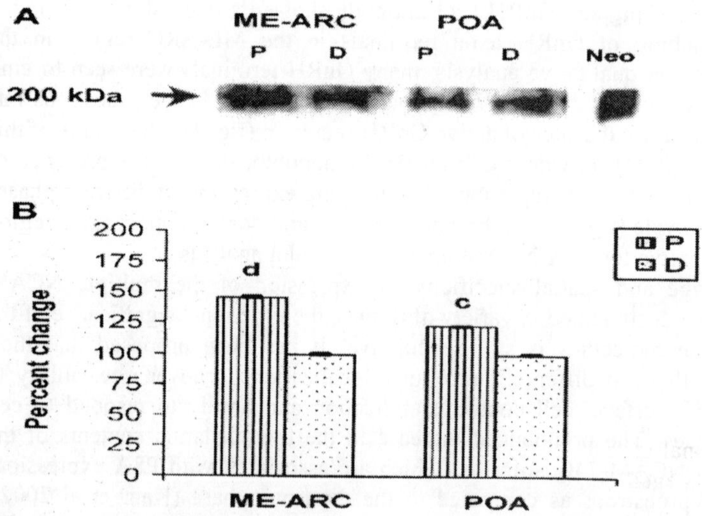

Fig.3. **Representative western blot hybridization signals using antibody specific for PSA-NCAM from median eminence (ME-ARC) preoptic area, (POA), from proestrous (P) and diestrous (D) phase of estrous cycle and neonate pups (Neo). (B)Mean values of diestrous and proestrous phase rats expressed as percentage change. Stastical analysis performed using student's 't' test, c= p<0.01, d-=p<0.001. [Kaur et al, Brain Res 954 (2002) 21-31].**

of Fos protein, whereas, genetic deletion of PSA and its NCAM carrier attenuated light induced circadian phase shifting. Another study in sheep further demonstrates that the expression of PSA-NCAM is intimately associated with GnRH neurons and their terminals and this expression varied with the seasonal shift in reproductive neuroendocrine activity. The present and the previous findings along with the well-established role of PSA-NCAM in promoting neuroplasticity strongly support the notion that PSA-NCAM may provide a molecular substrate for structural remodeling of the GnRH system of adults.

Of interest, NCAM and PSA-NCAM, which promotes plasticity by modifying the stability of cell-cell contacts, are expressed along the prenatal migratory path of GnRH neurons (Schwanzel-Fukuda et al 1992). The importance of these molecules to the migratory process is suggested by findings that GnRH neuronal migration is perturbed by immunoneutralization of NCAM or by enzymatic alteration of its PSA isoform (Schwanzel-Fukuda et al 1994). Both NCAM and PSA-NCAM continue to be expressed in the adult brain and are found in regions that contain GnRH perikarya, fibres, and terminals (Alonso et al 1997). All of these findings are consistent with a role for NCAM and PSA-NCAM in remodeling of neuroendocrine system.

NCAM and functional recovery in animals

Although regenerative potential of adult mammalian brain is highly restricted, previous studies have revealed that adult CNS is inherently capable of regrowth following injury (Yi Li et al 1998). The expression of NCAM molecule has been shown to decrease following different brain injuries like chronic exposure to elevated corticosterone levels (Sandi and Loscertales, 1999) and contextual fear (Sandi et al 2001) in adult rat brain as well as in long-term cultured astrocytes (Yazaki et al 1995).

In a recent study from our lab, we have shown that after insulin induced hypoglycemia, the expression of 140 kDa and 180 kDa isoforms of NCAM decreased in different brain regions such as cerebral hemispheres, brain stem and diencephalon (Singh et al 2003). Such a decrease may be attributed to loss of synaptic connections, which ultimately leads to neuronal cell death in acute hypoglycemic condition. NCAM expression was increased in cerebellum region, which may suggest that the cerebellum region has higher capability for regeneration. This finding is also supported by previous studies (Agardh et al 1981; Mohseni, 2001) demonstrating cerebellum region is more resistant to energy failure mediated neuronal damage mechanisms.

Although axonal growth in the adult mammalian brain is primarily hampered by adverse environmental conditions, the success of regenerative process is dependent on the capability of injured neurons to express the intrinsic molecular machinery required for neurite elongation. Another study undertaken in our lab investigated how monosodium glutamate (MSG) induced excitotoxic injury affects the expression of NCAM and PSA-NCAM in neonate rat brain from four brain regions namely cerebral hemispheres, cerebellum, diencephalon and brain stem (Kaur et al 2003). The results of this study demonstrate for the first time, acute increases in expression of proteins associated with neuronal plasticity and synaptic functions following MSG induced brain injury in neonate rat brain. The present data may suggest that the enhanced expression of PSA-NCAM in the neonate brain after MSG treatment is related to the capacity of immature neurons to actively regenerate their lesioned axons, since the whole brain expresses high levels of PSA-NCAM during development (Butler et al 1999).

Future prospects

The mechanisms that regulate the development and plasticity of the nervous system are traditionally viewed as separate from those that drive the evolution of brain pathology and disease. It is well known, for example, that neurotrophic factors are critical for the development, maintenance and plasticity of the nervous system. It is becoming increasingly clear that cell adhesion molecules (CAMs) also regulate neuronal development, maintenance and plasticity. Together these are primary extracellular signaling systems controlling such functions.

While it is clear that neurotrophic factors are one of the molecular classes that continue to regulate brain plasticity in the adult CNS, less appreciated but equally profound is the role of NCAM in plasticity mechanism such as long term potentiation, preservation of neurons and regeneration. Ironically, however, NCAM can also reorganize the extracellular space and cause disturbances that drive the development of brain pathology in conditions such as Alzheimer's disease and multiple sclerosis.

Once all the putative contributors to cell-cell adhesion are known following the sequencing of the relevant genomes, the challenges in the post genome era will remain to determine how NCAM is regulated and interact to affect morphology during neural development, regeneration and disease. Clearly, future research will be necessary to identify the precise mechanisms whereby NCAM may be involved in synaptic remodeling and whether NCAM is involved in modifying synaptic activity pre or postsynaptically? Does this involve dentritic and /or axonal sprouting changes in the shape of synaptic contacts, branching pattern, elaboration of spines and finally cell specific connectivity? In addition, the contribution of glial cells in interaction with synaptic membranes needs to be scrutinized. Because cell adhesion molecules are known to influence intracellular signaling cascades of second messenger systems, an intriguing question is how NCAM interacts with signal transduction pathways known to be crucial in synaptic strength. Finally, the key question is what are the principle mechanisms by which cell adhesion molecules interact with other events to stabilize or destabilize neural networks and are there unitary concepts, for example, transition to various long term changes?

References

Agardh CD, Kalimo H, Olsson Y and Siesjo BK (1981) J Cereb Blood Flow Metab **1**:71-84.

Alonso G, Prieto M. Legrand, A and Chauvet N (1997) J Comp Neurol **384**: 181-199.

Boulanger L and Poo MM (1999) Nat Neurosci **2**:346-351.

Butler AK, Uryu K, Rougon G and Chesselet MF (1999) Neuroscience **4**: 1169–1181.

Chiba A and Keshishian H (1996) Dev Biol **180**: 424-432.

Covault J and Sanes JR (1985) Proc Natl Acad Sci (USA) **82**: 4544-4548.

Cotman CW, Hailer NP, Pfister KK, Soltesz I and Schacner M (1998) Prog Neurobiol **55**: 659-669.

Cramer H, Lange R, Christoph A, Ploman M, Vopper G, Roes J, Braun R, Baldwin S, Kraemer P, Scheff S, Barthels D and Rajewsky K (1994). Nature **367**: 455-459.

Cremer H, Chazal G, Goridis C and Represa A (1997) Mol Cell Neurosci **8**: 323-335.

Cremer H, Chazal G, Carleton A, Goridis C, Vincent JD and Liedo PM (1998) Proc Natl Acad Sci. (USA) **95**: 13242-13247.

Crossin KL and Krushel J (2001) Dev Dyn **218**: 260-279.

Cunnigham BA Hemperly J, Murray BA, Prediger EA, Brackenbury R and Edelman GM (1987) Science **236**: 799-806.

Doherty P, Fazeli MS and Walsh FS (1995) J Neurobiol **26**: 437-444.

Eckhardt M, Bukalo O, Chazal G, Wang L, Goridis C, Schachner M, Gerardy-Schahn R, Cremer H and Dotyatev A (2001) J Neurosci **20**: 5234-5244.

Eckhardt M, Muhlenhoff M, Bethe E, Koopman J Frosch M and Gerardy-Schahn R (1995) Nature **373**: 715-781.

Edelman GM (1986) Annu Rev Cell Biol **2**:189-197.

Goridis C and Brunet JF (1995) NCAM Structural diversity, function and

Heera PK and Kaur G (2003) J Neurol Sci (Turkish*)* **20**: 3-13.

Hoffman S and Edelman GM (1983) Proc Natl Acad Sci (USA) 80: 5762-5766.

Holst BD, Vanderklish PW, Krushel LA, Zhou W, Langdon RB, Mc-Whirter JR, Edelman GM and Crossin KL (1998) Proc Natl Acad Sci (USA) **95**: 2597-2602.

Itoh K, Stevens H and Fields RD (1995) Science **270**: 1369-1372.

Kasper C, Rasmussen H, Kastrup JS, Ikemizu S, Jones EY, Berezin V, Bock E and Larsen IK (2001) Nat Struct Biol **1 7**: 389-393.

Katz LC and Shatz CJ (1996) Science **274**:133-1138.

Kaur G, Heera PK and Srivastava LK (2002) Brain Research **954**: 21-31.

Kaur G, Basu A and Kaur A (2003) J Neurol Sci (Turkish*)* **20**: 245-255

Kings, JC and Rubin, BS (1995) Cell Mol Neurobiol **15**: 89-106.

Kiss JZ and Rougon G (1997) Curr Opin Neurobiol 7: 640-647.

Kitgawa H and Paulson JC (1994) J Biol Chem **269**: 17872-17878.

Kobayashi NR, Fan DP, Giehl KM, Bedard AM, Wiegand SJ and Tetzlaff W (1999) J Neurosci **17**: 9583-9598.

Luthi A, Laurent JP, Figurov A, Muller D and Scachner M (1994) Nature **372**: 777-779.

Mohseni S (2001) .Acta Neuropathol (Berl) **102**:413-421.

Muller D, Wang C, Skibo G, Toni N, Cremer H, Calaora V, Rougon G and Kiss JZ (1996) Neuron **17**: 413-422.

Murase S and Schuman EM (1999) Curr Opinion Cell Boil **11**:549-553.

Nakayama J, Fukuda MN, Fredette V, Ranscht V and Fukuda M (1995) Proc Natl Acad Sci (USA) **92**: 7031-7035.

Nelson RW, Bates PA and Rutishauser U (1995) J Biol Chem **270**: 17171–17179.

Nothias B, Vernier P, Boxberg YV, Mirman S and Vincent JD (1997) Euro J Neurosci 89: 1553–1565.

Owens GC, Edelman GM and Cunningham BA (1987) Proc Natl Acad Sci (USA) **84**: 294-8.

Párducz A, Perez J and Garcia-Segura LM (1993) Neurosci **53**: 395-401.

Perera AD and Plant TM (1997) J Comp Neurol **385**: 71-82.

Rougon, G. (1993) Eur. J Cell Biol **61**: 197-207.

Rousselot P, Lois B and Alvarez BA (1995J Comp Neurol **35**: 52 – 61.

Rutishauser U, Aacheson A, Hall AK, Mann DM and Sunshine J (1988) Science **240**: 53-57.

Sandi C and Loscertales M (1999) Brain Res **828**:127-134.

Sandi C, Merino JJ, Cordero MI, Touyarot and K Venero C (2001) Neurosci **102**: 329-339.

Schuster T, Krig M, Hassan H and Schachner M (1998) J Neurobiol **37**: 359-372.

Schwanzel-Fukuda MS, Abraham S, Crossin KL, Edelman GM and Pfaff, DW (1992 J Comp Neurol **321**: 1-18.

Schwanzel-Fukuda MS, Reinhard GR, Abraham S, Crossin, KL, Edelman GM and Pfaff, DW (1994) J Comp Neurol **342**: 174-185.

Silletti S, Mei F, Sheppard D and Montgomery AM (2001) J Cell Biol **149**: 1485-1502.

Singh P, Pawan Kumar Heera and Gurcharan Kaur (2003) Mol Cell Biochem **247**: 69-74.

Sunshine J, Balak K, Rutishauser U. and Jacobson, M. (1987) Proc Natl Acad Sci (*USA*). **84**: 5989–5990.

Szele FG, Dowling JJ, Gonzales C, Thevenian M, Rougon G and Chesselet, M F (1994) Neurosci **60**: 133– 144.

Theodosis DT, and Poulain DA (1993) Neurosci 57: 501– 535.

Theodosis DT, and Poulain DA (1992) (Marsh J ed) pp 209-232, Wiley and Sons, Chichechester

Theodosis DT, Bonhomme R, Vitiello S, Rougan G and Poulain DA (1998a) J Neurosci **19**: 10228-10236.

Theodosis DT, El Majdoubi M, Pierre K, and Poulain DA (1998b) Cell Mol Neurobiol **18**: 285-298.

Watanabe M, Frelinger AL and Rutishauser U (1986) J Cell Biol **103**: 17212-1727.

Witkin, JW, Ferin M, Popilskis SJ and Silverman J (1991) Endocrinology **129**: 1083-1092.
Yang P, Yin X and Rutishauser U (1992) J Cell Biol **116**: 1487–1496.
Yazaki T, Asou H, Arimoto K, Toya S and Uyemura K(1995) Neuroreport **6**:1085-1088.
Yi Li MD, Ning Jiang MD, Cecylia Powers BS, Michael and Chopp PhD(1998) Stroke **29**:1972-1981.
Zhou H, Fuks A, Alcaraz G, Bolling TJ and Stanners CP (1993) J Cell Biol **122**: 951-960.

Molecular and Cellular Neurobiology
Editors: M.K. Thakur and S. Prasad

Role of Thyroid Hormone on the Maturation of Astrocytes in the Developing Brain

Sumantra Das and Pranab Kumar Sarkar

Neurobiology Division, Indian Institute of Chemical Biology, 4, Raja S.C. Mullick Road
Kolkata-700 032, India

Abstract

Astrocytes represent bulk of the neural cells in the mammalian brain. These cells have multifarious functions and astrocyte development is an important criterion for the proper function of the adult brain. The effect of TH on the differentiation and maturation of astrocytes in the developing brain has been described. Various target genes, which are regulated by TH during development and their mode of regulation, has been reviewed. Studies on primary cultures of astrocytes have conclusively demonstrated that TH is essential for the final step of maturation of the cells from flat polygonal to process bearing form. Cytoskeletal proteins, especially actin and vimentin, have been shown to play important and pivotal role during this maturation process. Reports that the β-adrenergic system play an important role in the downstream of TH action offers scope for future studies in the signaling mechanisms in TH-induced maturation of astrocytes.

Key words: Thyroid hormone, astrocyte, maturation, gene expression, β-adrenergic

Introduction

Nervous system contains more than 500 billion glial cells, which make up 50% of the total cell volume in the cerebral cortex (Bass et al 1971, Pope, 1978). Astrocytes, one of the most prominent classes of glial cells, makeup half of the glial cell population (Hansson, 1986). There are two main types of astrocytes. Protoplasmic or type I astrocytes are located in the gray matter and have end feet that contact synapses and blood vessels. Fibrous astrocytes or type II are located in the white matter and have end feet that contact nodes of Ranvier and blood vessels (Peters et al 1991). Besides those there are also some specialised astrocytes like Bergmann cells in cerebellum, Muller cells in retina, and pituicytes in the pituitary glands (Shao and McCarthy, 1994). Studies suggest that astrocytes may originate from radial glia (Choi et al 1983, Misson et al 1988, Culican et al 1990) which in turn is generated from the ventricular zone along with neurons around the same time (Levitt et al 1981). Raff and his colleagues revealed that both type I and type II astrocytes contain glial filaments. The type II astrocytes were found to share some properties with neurons in terms of binding to tetanus toxin, gangliosides and monoclonal antibody A2B5 (Bevan and Raff, 1985) whereas type I was A2B5- (Miller et al 1986). It was also reported that type I astrocytes appeared early at embryonic day 16 (E16) and type II astrocytes between postnatal days 7 to 10 (Miller et al 1985).

Astrocytes were initially regarded as the stuctural matrix for neurons. It is now known that radial glial fibres provide the primary migratory pathway for neuronal migrations in the developing brain (Rakic, 1972, Rakic, 1978, Fishell and Hatten, 1991, Hatten, 1993). Cultured astroglia express a myriad of receptors e.g. adrenergic, glutamatergic, GABAergic, purinergic, cholinergic, serotonergic etc. and also receptors for peptide transmitters like opioid receptors (Porter and McCarthy, 1997) and they are linked to second messengers including cAMP (McCarthy and deVellis, 1978) and Ca2+ (McCarthy and Salm, 1991).

Neonatal thyroid hormone (TH) deficiency causes severe and frequently irreversible mental retardation (Grave, 1977, McFaul et al 1978, Grant and Smith, 1988). Most of the earlier investigations on the mechanism of these effects of altered TH-deficiency used neurons and oligodendrocytes as target cells, neglecting the possible direct effect of the hormone on astrocytes. Recent observations indicate that astrocytes contain TH receptor isotypes TRα1, TRα2 and TRβ1 (Lebel et al 1993; Leonard et al 1994). These evidences strongly suggest astrocytes as target cells of TH.

Role of Thyroid hormone on differentiation and maturation of astrocytes

The obligatory role of TH in normal development of vertebrate brain has been long considered to be due to its influence on neuronal differentiation and maturation (Legrand, 1982-1983) and for survival of developing oligodendrocytes (Jones et al., 2003). However, several other reports indicated a role of TH on astrocyte development. These include (a) the effect of pre- and postnatal hypothyroidism on glial cell proliferation (Behnam-Rassoli et al. 1991) and maturation of Bergmann glia in cerebellum (Clos et al. 1980), (b) a decline in GFAP-immunoreactivity in white matter tracts and in radial fibres of neocortex (Kalman et al 1991), (c) a significant increase in the density of mature astrocytes by T_3 in forebrain and hippocampus (Gould et al 1990; Cordero et al 1988) and (d) decrease in the proportion of mature glial fibres in iodine deficiency (Martinez-Galan et al 1997).

Genes regulated by thyroid hormone in astrocytes

Table I summarises the genes, related to astrocyte differentiation, which are regulated by TH. However, two effects of the hormone, both mediated by actin polymerization, do not involve new gene expression.

Table I: Target genes for thyroid hormones in astrocytes

Target genes	Experimental system	Mode of regulation	Reference(s)
(a)Tubulin and actin	Primary cultures of astrocytes and organ cultures of neonatal rat cerebra	Primarily post transcriptional for tubulin and transcriptional for actin	De et al 1991, Bhattacharya et al 1991, Chaudhury and Sarkar, 1983; Chaudhury et al 1985, Poddar et al 1996.
(b)TH receptor beta gene	Astrocyte cultures from rat cerebral hemispheres	Five fold increase in mRNA by T_3 within 4 hrs suggest transcriptional regulation	Lebel et al 1993.
(c)Type II iodothyronine 5'-Deiodinase	Cultured astrocytes	Inactivation of the enzyme by promoting endocytosis mediated by T_4 (not T_3)-induced actin polymerization	Farwell et al 1996.
(d)Type III Deiodinase	Cultured astrocytes	Not determined	Esfandiari et al 1994
(e)Glial fibrillary acidic protein	Astroglial primary cultures	Transcriptional regulation	Andres-Barquin et al 1994; Beslin et al 1995; Lima et al 1997.
(f)Secretory phospholipase A(SPLA)GrIIA & GrV	Cultured rat astrocytes	Tumor necrosis factor alpha enhances and TH declines both isoforms	Thomas et al. 2000
g)Laminin	Cultured astrocytes	Mediated by T_4-induced actin polymerization	Farwell et al 1995; Farwell et al 1999.
(h)Metallothionein-3	Primary cultures of rat astrocytes	Not reported	Yelser, 1999.
(i)Vimentin	Primary cultures of rat astrocytes and immunostaining of cerebellar sections	Primarily post-transcriptional effects leading to appearance of new variants of vimentin and their phosphorylation by T_3	Paul et al 1999.
(j)Tumor necrosis factor beta and EGF	Cultures of astrocytes in conditioned medium	Not investigated	Gomes et al 1999.
(k)Na-K-ATPase	Primary cultures of astrocytes	Transcriptional regulation	Banerji and Chaudhury 2001, Bajpai and Chaudhury, 1999.

(a) ***Tubulin and actin*** : As early as 1991, De et al. reported the ability of exogenous T_3 to stimulate tubulin and actin in primary cultures of both neuronal and glial cells derived from neonatal rat brain. The sensitivity to T_3 disappeared by postnatal day 3, which coincided with the appearance of endogenous T_3 in the system indicating that the hormonal effect is a natural ontogenic phenomenon (Chaudhury and Sarkar, 1983). Analysis of the short term kinetics on the stimulation of tubulin by T_3 showed that the induction is a dual effect - an increase in the rate of synthesis as well as decrease in the rate of turnover of tubulin (Chaudhury et al 1985). Detailed mechanism concerning the mode of regulation of tubulin and actin by TH was studied by Poddar et al. (1996). Fig 1 shows that in the normal brain, the mRNA levels for both proteins increased progressively reaching a peak during the mid phase of synaptogenesis (5-10 days postnatal) and then declined rapidly whereas in the hypothyroid brain, the relative mRNA content for both tubulin and actin was suppressed significantly during the first week.

Finally studies on the relative rates of transcription in vitro by nuclear run off assay using nuclei from normal and hypothyroid rats of postnatal days 1, 5 and 15 revealed that at all ages, hypothyroidism resulted in a marked depression (30-60%) of actin gene transcription but had no significant effect on the transcription of tubulin genes.

Fig. 1

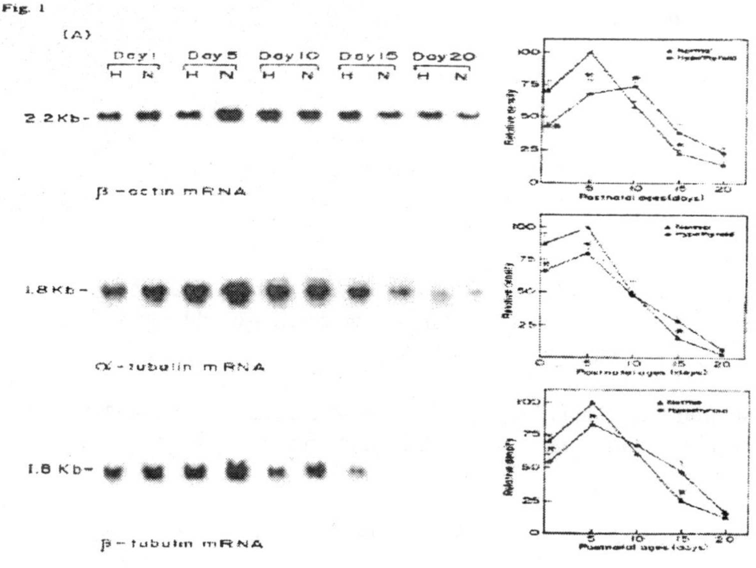

Fig.1. Northern blot analysis of actin and tubulin mRNAs from cerebra of 1, 5, 10, 15 and 20 day old normal (N) and hypothyroid (H) rats (left). Densitometric analysis of the corresponding blots (right). Equal amounts of total RNA (20 µg) were loaded in each lane and blots were probed with β-actin, α-tubulin CDNAs as indicated.

(b) ***Thyroid hormone receptors***: It is currently believed that four different forms of thyroid hormone receptors (TR) viz. TRα1, TRα2, TRβ1 and TRβ2 arise from two distinct TR genes - alpha and beta. Of these, TRα1, TRβ1 and TRβ2 display T_3

binding activity whereas TRα2 represents the non T_3-binding receptor variant. Lebel et al (1993) reported that in cultures of astrcytes, T_3 treatment led to a 5-fold increase in TRβ1 mRNA without any significant effect on other isoforms.

(c) ***Type II iodothyronine 5'-deiodinase (DII)***: Farwell et al (1993) and Leonard et al. (2000) demonstrated that T_4, but not T_3, modulates DII which is known to convert T_4 to T_3, by actin based endocytosis.

(d) ***Typé III Deiodinase***: Type III deiodinase, catalyses the degradation of thyroid hormone in brain. In astrocytes, multiple factors viz. cAMP, TPA, fibroblast growth factor (Fgf), retinoic acid (RA) and thyroid hormone induce the enzyme. The effect of RA and thyroid hormone are synergistic.

(e) ***Glial fibrillary acidic protein (GFAP)***: Treatment of astrocytes with T_3 or conditioned medium from T_3-treated cells induce perinuclear distribution of cytoskeletal GFAP as well as fibronectin (Lima et al 1997; Gomes et al 1999).

(f) ***Group IIA and Group V low molecular weight phospholipases***: Two isoforms of secretory phospholipase A(2), mainly Gr IIA and Gr V, are enhanced by tumor necrosis factor alpha (TNFα) and the TNFα effects are partially reversed by thyroid hormone (Thomas et al 2000).

(g) ***Laminin***: Farwell et al (1995, 1999) demonstrated that the matrix protein laminin, cell surface receptor integrin and the microfilament network play a fundamental role in the process of neuronal migration and that T_4 treated astrocytes readily attach laminin which gets deposited on the surface of the astrocytes. T_3 also has a profound effect on extracellular matrix (ECM) expression and organization in cerebellar astrocytes (Trentin et al 2003).

(h) ***Metallothionein-3(MT-3)***: Yelser et al (1999) demonstrated that in PTU-induced hypothyroid pregnant dams, MT-3 mRNA increased 4-6 fold compared to that in the normal pups between postnatal days 1-3.

(i) ***Vimentin***: Paul et al (1999) reported that maturation of astrocytes is associated with appearance of several acidic and basic and new phosphorylated variants of vimentin.

(j) ***Tumor necrosis factor beta and epidermal growth factor***: These two factors were identified in T_3-treated conditioned medium (T_3CM) from cultures of astrocytes as active agents involved in the T_3-induced proliferation of astrocytes. They were also shown to mediate the inductive effect of the T_3CM on neuronal cultures (Gomes et al 1999).

(k) ***Sodium-potassium ATPase***: Bajpai and Chaudhury (1999) demonstrated that TH regulated the expression of the enzyme in the developing rat brain at the transcriptional level. Banerjee and Chaudhury (2001) showed a reduced expression and altered distribution of this enzyme in TH-deficient astrocytes.

Effect of TH on the morphological differentiation of astrocytes *in vitro* – correlation with biochemical events

Earlier studies demonstrated a role of TH in astroglial differentiation and maturation using serum free chemically defined medium supplemented with the hormone (Aizenman and de Vellis, 1987, Gavaret et al 1991). Aizenman and de Vellis (1987) reported that insulin, TH and hydrocortisone act synergestically in promoting morphological changes in astrocytes. Moura Neto and his group (Trentin et al., 1995; Moura Ńeto et al 1996) also reported a morphological effect of TH on cultured cortical astrocytes. They recently reported (Lima et al 1997) that astrocytes cultured from different regions of the brain are not homogeneous in their responsiveness to T_3.

Our laboratory has also been involved in undertaking a comprehensive study of astrocyte maturation, Paul et al (1996) employed primary cultures of astrocytes derived from neonatal normal and hypothyroid rat brain and maintained them in normal and TH-deficient serum containing medium respectively for a period of 30 days, sufficient to cover the critical period of rat brain development. Comparison of the staining pattern (Fig.2) with anti-GFAP antibody showed that astrocytes obtained from euthyroid cerebra, maintained in normal serum (normal astrocytes), developed a star like appearance, comparable to radial glia in vivo, within 3-5 days of plating (Fig.2, left panel). Two distinct stages of morphological differentiation followed during maturation. By day 7 the radial glia were transformed into flat polygonal cells. The second and final step, initiated by day 10, involved the transformation of the flat polygonal cells into mature process bearing astrocytes (day 20). In the hypothyroid astrocytes the first transformation was delayed occurring only at day 10 instead of day 7 (Fig.2, right panel) and the second morphological transformation was totally arrested .

Our studies also showed that alterations in the expression and intracellular distribution of actin represents the two key steps associated with the maturation of astrocytes (Paul et al 1996). Vimentin and glial fibrillary acidic protein (GFAP) represents the two important cytoskeletal constituents of in astrocytes (Galou et al 1996). Recent studies by Paul et al (1999) revealed TH-induced expression of new variants of vimentin and their phosphorylation during maturation of astrocytes whose expression was significantly delayed or reduced in the TH-deficient astrocytes.

Role of β-adrenergic receptor and other agents on astrocyte maturation

In addition to TH, various other extrinsic factors including growth factors (Miller et al 1995), neurotransmitters (Abe and Saito, 1998), transcriptional regulators (Wohl and Weiss, 1998), c-AMP and its derivatives (Won and Oh, 2000), phorbol esters (Narumi et al., 1978) have been suggested to have profound effect in the regulation of astrocyte differentiation *in vitro*. Of the various neurotransmitter receptors present in astrocyte, the β-adrenergic receptors (β-AR) are abundantly present in astroglial cells both in culture (Salm and McCarthy, 1990) and in adult brain (Shao and

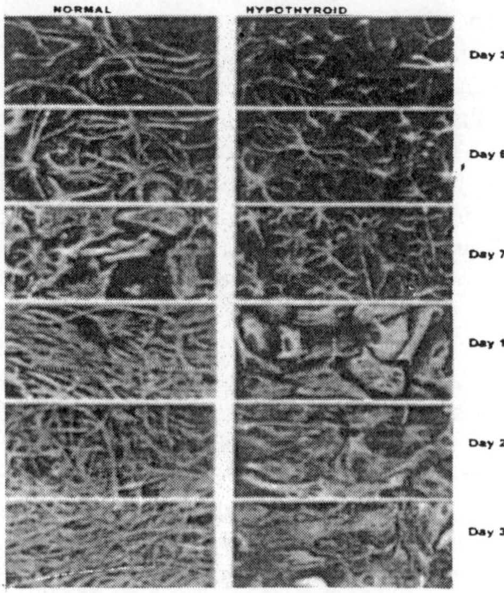

Fig.2. Maturation of astrocytes in culture and effect of thyroid hormone deficiency. Astrocytes from cerebra of neonatal normal and hypothyroid rats were cultured for 30 days in normal (left column) or TH-deficient serum containing medium (right column). Indirect immunofluorescent staining of astrocytes were carried out at days 3, 5, 7, 10, 20 and 30 using monoclonal antibody to GFAP (GA-5 clone) and FITC-goat antimouse IgG. Note the two distinct stages of morphological transformation : from radial glia (day 5) to flat epithelioid form (day 7) and subsequently to mature process bearing form (days 10 to 30) in the case of normal astrocytes and the arrested transformation of the flat epithelioid form in hypothyroid astrocytes (days 10 to 30). Magnification, x 400.

McCarthy, 1994). β-AR agonists can also transform astroglial cells as observed in primary culture system (Narumi et al 1978) and in C6 glioma cells (Oey, 1975). We investigated the role of the β-AR agonists in TH-induced maturation of astrocytes since several line of evidences, including ours, suggested a neuromodulatory link between TH and β-AR system in the brain (Gross et al 1980; Atterwill et al 1984; Das and Paul 1994; Tejani-Butt and Yang, 1994;.Wagner et al 1994; Rozanov and Dratman, 1996). We observed (Gharami and Das, 2000) that the β.AR antagonist, propranolol as well as other β-AR antagonists, selectively block the differentiation induced by TH as well as the β-AR agonist, isoproterenol (ISP). Addition of TH or ISP to hypothyroid cultures is also associated with a decrease in the level of filamentous cytoskeletal (F*i*) actin and an increase in the level of actin mRNA. While addition of propranolol inhibited the decline in the level of F*i* actin in the TH or ISP supplemented cells as well as the induction of actin mRNA by TH, it partially inhibited the ISP-induced actin mRNA in these cultures. The hormone-induced maturation appears to be selectively β_2-AR. The overall results indicate that β-adrenergic system plays an obligatory role in promoting TH-induced maturation of astrocytes by regulating the hormone-induced expression of actin and its intracellular organization in a way conducive to the morphological differentiation of the cells.

Conclusion

It is now established that TH plays an important role in driving the maturation of astrocytes in the developing brain. The evidence reviewed here led to the identification of the critical step of maturation affected in TH-deficiency, viz. the morphological transformation of the flat polygonal astrocytes into matured process bearing stellate form. Although alterations in the TH induced biogenesis, assembly and phosphorylation of two major cytoskeletal proteins, viz. actin and vimentin, have been shown to play a key role in the maturation process, evidence presented here on the role of β-adrenergic receptor system in this process offers newer insight on the mechanism of TH action on the morphological maturation of astroglial cells. The β-AR is known to act through classical G-protein coupled receptor system linked to signal transduction pathways. Recent observations on the role of protein kinase A (PKA) in TH induced actin gene expression (Sarkar et al 1999) are a point in this direction. Very recently we have shown that a ISP causes a PKA dependent delayed but long lasting activation of p-ERK activity which appears to coincide with initiation of morphological differentiation of the astrocytes and may be critical for the transformation of the cells (Gharami and Das, 2004). The molecular mechanisms underlying the selective T_3 induced gene expression and the cascade of signal transduction events involved in these processes is currently under active investigations.

References

Abe K and Saito H (1998) Brain Res **804**: 63-71.

Aizenman J and de Vellis J (1987) Brain Res **414:** 301-308.

Andres–Barquin PJ (1994) Neurochem Res **19:** 65-69.

Atterwill CK, Bunn SJ, Atkinson DJ, Smith SL and Heal DJ (1984) J Neural Transm **59:** 43-55.

Bajpai M and Chaudhury S (1999) Neuro Report **10:** 2325-2328.

Banerjee B and Chaudhury S (2001) Life Sci **69:** 2409-2417.

Bass NH, Hess HH, Pope A and Thalheimer C J (1971) Comp Neurol **143:** 481-90.

Behnam-Rassoli M, Herbert LC, HowardV, Pharoah PO and Stanisstreet M (1991) Neuroendocrinol **53:** 321-327.

Beslin A, Vie MP, Blondeauu JP and Francon (1995) J Biochem J **305:** 729-737.

Bevan S and Rafff MC (1985) Nature **315:** 229-232.

Bhattacharya S, Saha S and Sarkar PK (1991) Mol Brain Res **10:** 347-352.

Chaudhury S and Sarkar PK (1983) Biochem Biophys Acta **763:** 93-98.

Chaudhury S, Chatterjee D and Sarkar PK (1985) Brain Res **339:** 191-194.

Choi BH, Kim RC and Lapham LW(1983) Dev Brain Res **8** 119-130.

Clos J, Legrand C and Legrand J (1980) Dev Neurosci **3:** 199-208.

Cordero ME, Zvaighaft A, Gonzales F, Prado R and Muzzo S (1988) Early Human Dev **17:** 7-18.

Culican SM, Baumrind NL, Yamamoto M and Pearlman AL (1990) J Neurosci **10:** 684-92.

Das S and Paul S (1994) Life Sci **54:** 621-629.

De A, Chaudhury S and Sarkar PK (1987) Int J Dev Neurosci **9:** 381-390.

Esfandiari A, Gagelin C, Gaveret JM, Pavelka S, Lennon AM, Pierra M and Courtin F (1994) Glia **3:** 256-261.

Farwell AP and Dubord Tomasetti SA (1999) Endocrinology **140:** 5014-5022.

Farwell AP, DiBenedetto DJ and Leonard JL (1993) J Biol Chem **268:** 5055-5062.

Farwell AP, Tranter MP and Leonard JL (1995) Endocrinology **136:** 3909-3915.

Farwell AP, Safran M, Dubord S, Leonard JL (1996) J Biol Chem **271:** 16369-71634.

Fishell G and Hatten ME (1991) Development **113:** 755-765.

Galou M, Coucci-Guyon E, Ensergueix D, Ridet JL, Ribotta MG, Privat A, Babinet C and Dupouey PJ (1996) Cell Biol **133:** 853-863.

Gavaret JM, Delbauffe DT, Chalaye DB, Pomerance M and Pierce M (1991) Dev Brain Res **58:** 43-49.

Gharami K and Das S (2000) J Neurochem **75:** 1962-1969.

Gharami K and S Das (2004) J Neurochem **88:** 12-22.

Gomes FC, Maia CG, de Menezes JR and Moura Neto V (1999) Glia **25** 247-255.
Gould E, Frankfurt M, Westlind-Danielsson A and McEwen BS (1990) Glia **3**: 283-292.
Grant DB and Smith I (1988) Br Med J **296**: 1335-1358.
Grave GD (1977) Thyroid hormone and brain development, Raven Press, New York.
Gross G, Brodde OE and Schumann HJ (1980) Eur J Pharmacol **61**: 191-194.
Hansson E (1986) In: S. Federoff and A. Vernadakis (Eds.), Astrocytes: development, morphology and regional specialisation of astrocytes, Vol. 1, Academic, FL, pp.225-238.
Hatten ME (1993) Curr Opinion Neurobiol. **3**: 34-44.
Jones SA, Jolson DM, Cuta KK, Mariash CN and Anderson GW (2003) Molec Cell Endocrinol **199**: 49-60.
Kalman M, Moskovkin GN and Martinez K (1991) Mol Chem Neuropathol **15**: 103-116.
Lebel JM, Lehrault S, Dussault JH and Puymirat J (1993) Glia **9**: 105-112.
Legrand J (1982-1983) J Physiol (Paris) **78**: 603-652
Leonard JL, Farwell AP, Yen PM, Chin WW and Stula M (1994) Endocrinology **135**: 548-555
Leonard DM, Stachelek DM, Safran SJ, Farwell M, Kowalik AP, and Leonard JL (2000) J Biol Chem **275**: 25194-25201.
Levitt P, Cooper ML and Rakic P (1981) J Neurosci **1**: 27-39.
Lima FR, Trentin HG, Rosenthal D, Chagas C Moura and Neto V (1997) J Endocrinol **154**: 167-175.
Martinez-Galan JR, Pedraza P, Santacana M, Escobar del Ray F, Morreale de Escobar G and Ruiz-Marcos A (1997) J Clin Invest **99**: 2701-279.
Maura Neto V, Trentin A, Lima F, Gomes F, Goncalves N, Chamas LV, Lins C, Garcia-Abreu J, Rosenthal D and Chagas C (1996) Rev Bras Biol **56** 123-134.
McCarthy KD and de Vellis J (1978) J Cyclic Nucleotide Res **4**: 15-26.
McCarthy KD and Salm AK (1991) Neuroscience **41**: 325-33.
McFaul R, Dorner S, Brett EM and Grant DB (1978) Archs Dis Childh **53**: 611-619.
Miller RH, Davis S, Patel R, Abney ER and Raff MC (1985) Dev. Biol. **111** 35-41.
Miller RH, Abney ER, Davis S, French F, Constant C, Lindsay R, Patel R, Stone J and Raff MC (1986) J Neurosci **6**: 22-29.
Miller S, Romano C and Cotman CW (1995) J Neurosci **15**: 6103-6109.
Misson JP, Edwards MA, Yamamoto M and Caviness VS Jr. (1988) Dev Brain Res **44**: 95-108.
Narumi S, Kimelberg HK and Bourke RS (1978) J Neurochem **31**: 1479-1490.
Oey J (1975) Nature **257**: 317-319.
Paul S, Das S, Poddar R and Sarkar PK (1996) Eur J Neurosci **8**: 2361-2370.
Paul S, Gharami K, Das S and Sarkar PK (1999) J Neurochem **73**: 1964-1972.
Peters A, Palay SL and Webster HF (1991) The fine structure of the nervous system : The neurons and supporting cells, Oxford Univ. Press, Oxford.
Poddar R, Paul S, Chaudhury S and Sarkar PK (1996) Mol Brain Research **35**: 111-118.
Pope A (1978) In: E. Schoffeniels, G Frank, DB Towers and L Hertz (Eds.), Dynamic properties of glial cells, Pergamon, New York, , pp.13-20.
Porter JT and McCarthy KD (1997) Prog Neurobiol **51**: 439-455.
Rakic P (1972) J Comp Neurol **145**: 61-84.
Rakic P (1978) Postgrad Med J **54**: 25-40.
Rozanov CB and Dratman MB (1996) Neurosci **74**: 897-915.
Salm AK and McCarthy KD (1990) Glia **3**: 529-38.
Sarkar S, Biswas S, Chatterjee O and Sarkar PK (1999) Molec Brain Res **67**: 158-164.
Shao Y and McCarthy KD (1994) Glia **11** 147-155.
Smith R.M, Patel AJ, Kingsbury AE Hunt A and Balazs R (1980) Brain Res **198**: 375-387.
Tejani-Butt SM and Yang J (1994) Neuroendocrinol **59**: 235-44.
Thomas G, Bertrand F and Saunier B (2000) J Biol Chem **275**: 10876-10886.
Trentin AG and Moura Neto V (1995) NeuroReport **6**: 293-296.
Trentin AG, De Aguiar CB, Garcez RC, Alvarez-Silva M (2003) Glia **42**: 359-369.
Wagner JP, Seidler FJ, Lappi SE, McCook EC and Slotkin TA (1994) J Pharmacol Exp Ther **271**: 472-483.
Wohl CA and Weiss S (1998) J Neurobiol **37**: 281-290.
Won C-K and Oh YS (2000) Brain Res **887**: 250-258.
Yelser EC, Fitch CA, Horning MS, Rutkoski N and Levenson CW (1999) Dev Brain Res **115**: 195-200.

Section B

Hormones and Neurotransmitters

Molecular and Cellular Neurobiology
Editors: M.K. Thakur and S. Prasad

Role of Norepinephrine in Thermoregulation during Rapid Eye Movement Sleep and its Deprivation

Birendra N. Mallick, Sushil K. Jha and Vibha Madan

School of Life Sciences, Jawaharlal Nehru University, New Delhi-110 067, India

Abstract

Sleep is a natural reversible behavioural phenomenon primarily present in higher living species. Based on characteristic electrophysiological signals it has been classified into rapid eye movement (REM) sleep and non-REM sleep. Mammalian sleep is influenced by diverse factors including ambient- and body-temperatures that affect the quality and quantity of sleep in distinctive ways. The body temperature, on the other hand, changes during sleep, waking and REM sleep period. The brainstem reticular sleep and waking areas modulate each other and also the REM sleep related as well as thermosensitive neurons in the brain. The medial preoptico-anterior hypothalamic (mPOAH) area in the basal forebrain modulates sleep-wakefulness as well as body temperature. It has been reported that level of norepinephrine (NE) in the brain plays a significant role in the regulation of body temperature as well as REM sleep. Isolated studies have reported that on one hand, NE is increased in the brain after REM sleep loss and on the other hand, NE activates thermosensitive neurons and causes hypothermia. Normally the body temperature rises during REM sleep and it reduces during REM sleep loss. Based on existing information it has been proposed in this chapter that altered level of NE is likely to play a significant role in modulating the body temperature during REM sleep and its loss/deprivation. The possible neural mechanism(s) of action has been discussed.

Key Words: locus coeruleus; norepinephrine; preoptic area; REM sleep; thermoregulation

Introduction

Sleep is a reversible behavioural state of perceptual disengagement from and unresponsiveness to the environment when the consciousness remains in a subdued state. Since it is a behavioural phenomenon, electrophysiological criteria are used for its unbiased objective identification, analysis and quantification. Based on such electrophysiological signals viz. electroencephalogram (EEG), electrooculogram (EOG) and electromyogram (EMG), sleep-wakefulness has been classified into five stages (Fig. 1). Subsequently, in early 1950s sleep has been objectively classified into Rapid Eye Movement (REM) sleep and non-REM (NREM) sleep (Aserinsky and Kleitman 1953, Aserinsky 1999, Jouvet, 1999). REM sleep is defined by EEG activation (low-voltage and high frequency desynchronized EEG), muscle atonia and episodic bursts of rapid eye movements. Human (most other higher animals as well) REM sleep is associated with dreaming based on vivid dream recall from approximately 80% of arousal from this state of sleep (Dement and Kleitman, 1957). The concept that the brain remains in an inactive state during sleep has turned out to be a myth. In fact, several areas in the brain are more active during sleep. It has also been reported that some neurons cease (or almost cease) activity exclusively during REM sleep, the REM-OFF neurons, while some others become active (almost exclusively) during REM sleep, the REM-ON neurons (Hobson et al 1974; Sakai et al 1983, El Mansari et al 1989). Based on the characteristic whether one has the ability to maintain body temperature within physiological limit, animals have been

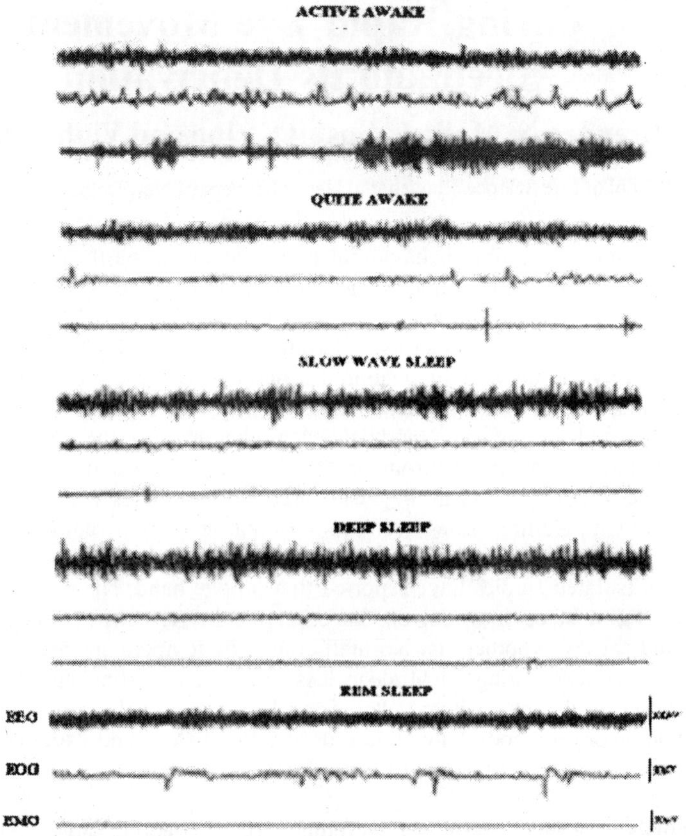

Fig. 1

Fig 1. Polygraphic traces of electrophysiological signals viz. electroencephalogram (EEG), electrooculogram (EOG) and electromyogram (EMG) from a freely moving normally behaving rat are shown. Based on characteristic electrophysiological signs, the S-W stages of the rat have been classified into five stages as shown in this figure.

classified into homeotherms and poikilotherms. For homeotherms, increase and decrease of body temperature follows a circadian rhythm. Similarly, occurrence of sleep also follows a circadian rhythm. Thus, both sleep as well as body temperature follow circadian pattern. It has been reported that sleep as well as body temperature may influence each other. They may also be influenced simultaneously or independently by various other factors. The brain regulates the body temperature as well as both NREM and REM sleep. It is also known that some common sites in the brain and some common neurotransmitter(s) are involved in the regulation of both, the sleep as well as the body temperature. Hence, it is quite likely that changes in either of those functions per se or any one or more of the neurotransmitters involved in either of their regulations would modulate the other function, possibly as a compensatory mechanism. Possible cellular mechanism of action for changes in body temperature during REM sleep and its deprivation with emphasis on changes in a neurotransmitter, norepinephrine (NE), will be discussed in this chapter.

Neural regulation of REM sleep

Pons and REM sleep generation: The REM sleep is classically identified by characteristic electrophysiogical signals in the EEG, EOG and EMG. Each of these signals is modulated by the neurons in the brain. Similar to that of other studies for identification of brain area(s) controlling a particular physiological phenomenon, electrical as well as chemical stimulation and lesion and single neuronal activity studies have been used to localize the areas in the brain regulating REM sleep. Studies showed that REM sleep associated electrophysiological sign(s) was expressed in animals in which transection was made in front of the pons, suggesting that the midbrain was not essential for REM sleep regulation. Similarly, it was also shown that the spinal cord was not necessary for REM sleep control. Further, by separating the brain stem at the junction of the pons and the medulla it was possible to localize the brain regions responsible for REM sleep. The results showed that in a situation when the pons was connected to the medulla, REM sleep signs were expressed in the pontine and medullary regions, while when the pons was connected to the forebrain, REM sleep signs were observed in the pons and forebrain. These transection studies, therefore, led to the conclusion that the pontine region in the brain stem is most critical for REM sleep regulation (Siegel 1983, 1985, 1989, Siegel et al 1984).

Locus Coeruleus and REM sleep generation: Search continued to locate a critical site within the pons, if any, which may be responsible for the regulation of REM sleep. Attention was drawn towards locus coeruleus (LC), the primary site of noradrenergic neurons in the brain. Destruction of LC although did not have lasting effect on REM sleep loss (Jones et al 1977), other studies found that there was irreversible disappearance of REM sleep atonia (Jouvet and Delorme 1965, Henley and Morrison 1974, Sakai 1980). Sastre et al (1979) also demonstrated that REM sleep was completely suppressed during two post lesion months after the destruction of both the LC-proper plus the LC-α along with LC-peri-α. Braun and Pivik (1981) from their lesion study in the rabbits suggested that the LC regions are essential for the integrity of sleep and are especially important for the control of motor mechanisms during sleep. Further, local cooling of the LC led to increase in non-REM sleep as well as REM sleep (Cespuglio et al 1982). These studies supported the role of LC within the pons in regulation of REM sleep. Injections of cholinergic agonist mimicking the actions of the neurotransmitter, acetylcholine, into the nucleus reticularis pontis oralis, including LC, triggered REM sleep (Baghdoyan et al 1984, Hobson et al 1993, Mallick et al 2001). This suggested that acetylcholine is normally released in this area for the initiation of REM sleep.

The role of LC in REM sleep regulation was further substantiated by the fact that there are neurons in this area, which are active continuously through waking and NREM sleep but cease firing only during REM sleep (Chu and Bloom 1973, 1974, Aston-Jones and Bloom 1981). Since these neurons ceased firing exclusively during REM sleep, they have been termed as REM-OFF neurons. Subsequently, the role of these REM-OFF neurons in REM sleep was confirmed by the fact that their firing rates were reduced during REM sleep deprivation possibly as a compensatory mechanism (Mallick et al 1989). Subsequently it was proposed that cessation of the REM-OFF neurons in the LC may be a prerequisite for REM sleep generation. As a corollary, it was hypothesized that if the REM-OFF neurons in the LC could be kept active, REM sleep generation could be prevented or at least reduced, simulating a condition similar to that of REM sleep loss or REM sleep deprivation. Hence, experiments were designed to electrically stimulate the LC bilaterally in freely moving normally behaving rats and study the effects on sleep-waking cycle. Under surgical anesthesia rats were prepared with bilateral bipolar stimulating electrodes in the LC and other electrodes for recording EEG, EOG and EMG. After recovery of the rats from surgery, the neurons in the LC were stimulated bilaterally with very low intensity current and at an average frequency at which the LC neurons normally fire. This was done to keep the LC neurons constantly active and not to allow them to cease firing through the period of experiment. It was found that the REM sleep was reduced significantly, which gradually returned to normal levels after cessation of stimulation (Singh and Mallick 1996).

A second major REM sleep related cell type is the REM-ON neurons (Hobson et al 1974), which are almost exclusively active during REM sleep. These neurons have been identified mainly in the pedunculo-pontine tegmentum and laterodorsal tegmental region in the brain. Initially these neurons were thought to be cholinergic, however, later studies suggest that some of the REM-ON neurons could be non-cholinergic as well.

How do the LC neurons get inhibited during REM sleep?: It has been mentioned earlier that functionally there are at least two types of REM sleep related neurons in the brain, the aminergic REM-OFF type, that cease firing during REM sleep and the presumably cholinergic REM-ON type, that are active or increase firing during REM sleep. It was proposed that there are likely to be reciprocal connections between these two groups of REM sleep related REM-ON and REM-OFF neurons (Hobson et al 1975, Sakai 1988). Isolated and independent studies showed that activation of REM-OFF neurons, either by electrical stimulation (Singh and Mallick 1996) or by blocking the inhibitory neurotransmitter action on them by applying a chemical blocker (Kaur et al 1997), reduced REM sleep. Similarly, activation of the area possessing the REM-ON neurons by electrical (Thakkar et al 1996) or chemical (Baghdoyan et al 1984, Thankachan et al 1997) stimulation increased REM sleep. However, it was shown that the LC neurons were excited and not inhibited by the cholinergic agonist, carbachol (Egan and North 1985). Hence, it was proposed that there is likely to be GABA-ergic inhibitory interneuron that transduces the cholinergic excitatory input into an inhibition on the REM-OFF neurons (Alam et al 1993, Mallick et al 2001). This view was supported by the fact that GABA-ergic receptors, interneurons and terminals were reported in the LC region (Iijima and

Ohtomo, 1988, Jones 1990, Luque et al 1994), GABA levels increased in LC during REM sleep (Nitz and Siegel 1997) and micro-iontophoretic application of GABA in LC inhibited noradrenergic neurons (Gervasoni et al 1998). Experimental design involving either anatomical, including electron microscopic, electrophysiological or microinjection of single chemical (agonist or antagonist) study was not sufficient to confirm the proposed model involving GABA interneuron. This is because; each of those studies would have provided either micro-anatomic connections or chemical nature of the neurons for modulation of behaviour. Hence, it needed micro-chemo-anatomo-behavioural study. Subsequently, our hypothesis that GABA in LC plays a key role in REM sleep induction, possibly by inhibiting the REM-OFF neurons, was confirmed as a result of extensive studies and based on those results a model showing connections between the neurons has been put forward (Mallick et al 2001).

The model was further supported by the finding that blocking of GABA receptors in LC reduced REM sleep as well as increased Na-K ATPase activity in the rat brain (Kaur et al 2003) as has been reported in case of REM sleep deprived rats (Gulyani and Mallick 1993, 1995). Furthermore, it was also observed that area prepositus hypoglossi provided GABA-ergic inputs to the LC and that also modulated REM sleep (Kaur et al 2001). Thus, our studies suggest that REM-OFF neurons in the LC receive GABA-ergic inputs from local interneurons as well as from prepositus hypoglossi and their activation increases REM sleep most likely by inhibition of the REM-OFF neurons.

Do the sleep and waking areas in the brain modulate REM sleep?: Under normal condition the REM sleep signs do not appear during wakefulness. Another fact is that although REM sleep appears after a period of NREM sleep, every NREM sleep is not followed by an episode of REM sleep. It is also a fact that an episode of REM sleep may terminate in wakefulness or NREM sleep. Hence, it was proposed that neurons in the REM sleep controlling area possibly are inhibited by the neurons located in the wakefulness inducing area and when certain conditions are satisfied, the neurons in the NREM sleep inducing area(s) excite them for the generation of REM sleep. In order to understand the mechanism, experiments were conducted in freely moving normally behaving cats. Under surgical anaesthesia, the experimental cats were prepared with electrodes for recording EEG, EMG, EOG and pontogeniculo-occipital (PGO) waves for identification of sleep-wakefulness. Three stimulating electrodes were implanted, one, in the wakefulness inducing area in the midbrain reticular formation, second, in the caudal brain stem reticular formation hypnogenic area and third, in the preoptico-anterior hypothalamic basal forebrain hypnogenic area. A pair of microdrive was also implanted for introducing insulated flexible micro-wires (25-32 μm) for recording single neuronal activities from freely moving normally behaving cats (Mallick et al 1998). Single neuronal activities were recorded and correlated with simultaneous changes in the sleep-waking pattern. The neurons were classified based on their firing rate during different sleep-waking states. Such studies revealed that midbrain reticular formation (MRF), the waking area in the brainstem, excited the REM-OFF neurons but inhibited the REM-ON neurons (Thankachan et al 2001). On the other hand, although the caudal brainstem reticular formation (CRF) sleep inducing area excited the REM-ON neurons and inhibited the REM-OFF neurons, the basal forebrain hypnogenic area did not significantly affect those neurons (Mallick et al 2004b). Thus, our studies showed for the first time that the wakefulness and sleep inducing areas in the brainstem have opposite influences on the REM sleep related neurons viz. the REM-ON and the REM-OFF neurons.

How is REM sleep generated?: Based on the available knowledge including those from our laboratory, the mechanism of generation of REM sleep may be as follows. Normally, during wakefulness the wake-active neurons in the brainstem are more active and the sleep-active neurons are less active. Such behaviour of the awake- and sleep-active neurons keep the REM-OFF neurons active but the REM-ON neurons suppressed during wakefulness. During sleep, the wake-active neurons gradually slow down causing withdrawal of inhibitory and excitatory effects from the REM-ON and REM-OFF neurons, respectively. Subsequently, during NREM sleep the sleep-active neurons become active. Thereafter, at certain stage of deep sleep, when some yet unknown conditions are fulfilled, the sleep active neurons actively excite the REM-ON neurons and thus, the REM sleep is initiated (Fig 2) Although the neurochemical nature of all the neurons are not fully known, based on the recent studies it may be said that at least the cholinergic input from the REM-ON neurons to LC is mediated through GABA and the action is mediated by GABA-A receptor for the regulation of REM sleep (Mallick et al 2002b).

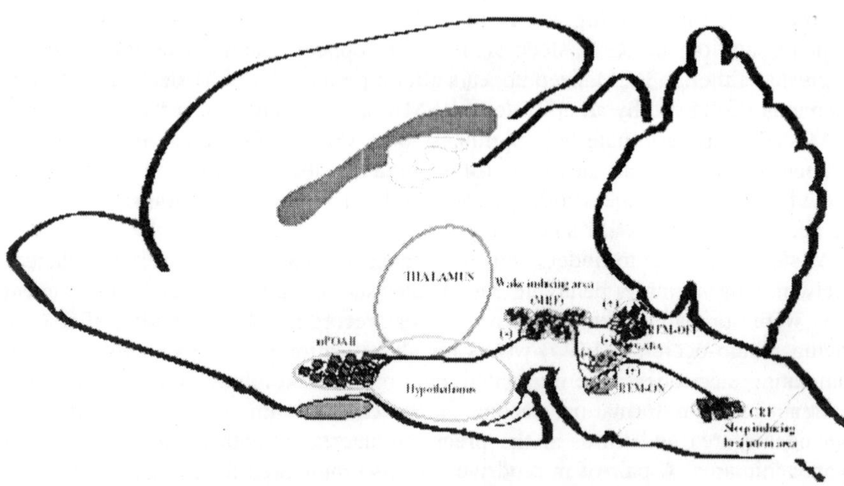

Fig 2

Fig 2. **Diagrammatic representation of connections between NE-ergic REM-OFF (blue), ACh-ergic REM-ON neurons (green) and GABA-ergic neurons (red) in the brainstem are shown in this figure. Inputs from the brainstem wakefulness (MRF) and sleep (CRF) inducing areas and also from the mPOAH sleep inducing area are shown. The abbreviations are same as in the text. (+) indicates excitation, while (-) shows inhibition. The detail mechanism of modulation of activities of these neurons for the regulation of REM sleep has been explained in the text.**

REM sleep deprivation and Brain Neurotransmitters

Norepinephrine (NE): As mentioned earlier, the NE-ergic neurons cease firing during REM sleep. It is also known that there is a decrease in NE in the brain during REM sleep (Shouse et al 2000). Pujol et al (1968) reported an increased turnover of NE in the forebrain after REM sleep deprivation, which was later confirmed by Stern et al (1971). There are other evidences also that support the findings that there is increased turnover of NE after REM sleep deprivation. There is increase in NE synthesizing enzyme, tyrosine hydroxylase (TH) (Sinha et al 1973, Basheer et al 1998, Majumdar and Mallick 2003), its mRNA (Porkka-Heiskanen et al 1995) and NE-transporter mRNA (Basheer et al 1998) after REM sleep deprivation. Following REM sleep deprivation there is decrease in monoamine oxidase-A (MAO-A), the enzyme responsible for the breakdown of NE (Thakkar and Mallick 1993a), which was later confirmed by Perez and Benedito (1997). Normally, the NE-ergic neurons in the LC cease firing during REM sleep (Jacobs 1986), however, during REM sleep deprivation they continue firing (Mallick et al 1989). Further, by using immunohistochemical detection of Fos and NGFI-A, increased activity of the LC neurons after spontaneous and short period of sleep deprivation has been observed (Tononi et al 1994). These findings offer explanation for possible mechanism of increased levels of NE in the brain after REM sleep deprivation (Mallick et al 2002b). Neurotransmitters are known to act through respective receptors. It is likely that alterations in the levels of neurotransmitters might alter the receptor concentrations or their sensitivity after sleep or REM sleep deprivation. It has been proposed that REM sleep and its deprivation might alter the sensitivity of adrenergic receptors (Siegel and Rogawski 1988). However, isolated studies on sensitivity and receptor densities of aminergic receptors after REM sleep deprivation have reported different results (Mogilnicka 1981, Serra et al 1981, Hipolide et al 1998, Tsai et al 1993).

Acetylcholine (ACh): At least some of the REM-ON neurons are reported to be presumably ACh-ergic (El Mansari et al 1989). Normally, these REM-ON neurons are inactive during wakefulness, however, they continue firing during wakefulness after REM sleep deprivation, possibly a compensatory phenomenon (Mallick et al 1989). ACh release during REM sleep in the dorsolateral pons, a REM sleep generating site, has been reported to be about two times higher than during other sleep-waking states (Lydic et al 1991). The detail role of cholinergic mechanism in the brainstem for the regulation of REM sleep has been extensively reviewed earlier (Rye 1997, Capece et al 1999). The levels of acetylcholinesterase (AChE), an ACh hydrolyzing enzyme that maintains the levels of ACh, increased after REM sleep deprivation (Thakkar and Mallick 1991, Mallick and Thakkar, 1991). Subsequently, the AChE results have been confirmed (Benedito and Camarini 2001). These suggested that effective ACh levels would decrease after REM sleep deprivation, possibly as a compensatory phenomenon.

GABA: In addition to ACh and NE, GABA has been implicated in the regulation of REM sleep (Mallick et al 1999). The GABA level was found to increase in locus coeruleus during REM sleep (Nitz and Siegel 1997). An interaction between the NE, ACh and GABA, at least in locus coeruleus, for the regulation of REM sleep has been shown (Mallick et al 2001). Hence, it is expected that GABA levels would alter during REM sleep deprivation or related disorders. Although the GABA levels increase in the LC during spontaneous REM sleep, as mentioned above, its level after REM sleep deprivation is yet to be estimated. However, the concentration of glutamic acid decarboxylase (GAD), the enzyme responsible for GABA synthesis, increased within the LC only after REM sleep deprivation (Majumdar and Mallick, 2003).

REM Sleep deprivation and its effects on neuronal properties

It has been hypothesized that one of the functions of REM sleep is to maintain brain excitability status (Mallick et al 1994). A series of REM sleep deprivation studies has been carried out in this laboratory to explore the possible mechanism of action (Gulyani et al 2000, Mallick et al 2002b). Although recording of transmembrane potential before and after REM sleep deprivation in freely behaving conditions would have been the ideal choice, such kind of study has not been possible due to technical limitations.

It was proposed that if one of the functions of REM sleep was to maintain neuronal excitability, its deprivation would alter factor(s) affecting the excitability. One of the primary components involved in maintenance of neuronal excitability is the Na-ion concentration within the neuron and Na-K ATPase largely maintains its concentration gradient. After REM sleep deprivation, Na-K ATPase activity was found to increase in the rat brain (Gulyani and Mallick 1993). As mentioned earlier, spontaneous firing of NE-ergic REM-OFF neurons ceases during REM sleep and their activity continues during REM sleep deprivation. It was also shown that presumably non-cessation of those neurons, by electrical stimulation (Singh and Mallick 1996) or by neuro-receptor antagonist, also induced conditions similar to that of REM sleep deprivation (Kaur et al 1997). Since non-cessation of the NE-ergic neurons during REM sleep deprivation is likely to increase NE in the brain, the increased NE was thought to be the possible cause for increased Na-K ATPase activity in the brain after REM sleep deprivation. Our *in vivo* as well as *in vitro* studies confirmed that REM sleep deprivation induced increase in Na-K ATPase activity in rat brain was actually due to NE and the effect was mediated through α-1 adrenoceptor (Gulyani and Mallick 1995). This was further confirmed by the fact that reduction in REM sleep by neuro-receptor antagonist in LC also increased the Na-K ATPase activity in the rat brain (Kaur et al 2003). Ca-ion, that is inhibitory to Na-K ATPase activity (Davis and Vincenzi 1971), was also affected after REM sleep deprivation (Mallick and Gulyani 1996). In an attempt to find the relationship between them it was observed that increased NE releases membrane bound calcium and also increased Na-K ATPase activity (Mallick and Adya 1999). Further studies to understand the intracellular mechanism of action it was found that NE acted on α-1A adrenoceptor that stimulated phospholipase-C pathway, activated calcium-calmodulin and dephosphorylated Na-K ATPase to increase its activity (Mallick et al 2000). Subsequently, prior to investigate the molecular basis of increase in Na-K ATPase activity, its kinetics were studied. It was observed that after REM sleep deprivation there was an increase in both the K_m as well as V_{max} of the enzyme (Adya and Mallick 2000). This suggested that REM sleep deprivation increases synthesis of the enzyme as well as causes post-translational modulation in the enzyme molecules resulting in increased enzyme activity and both are likely to be increased by NE. To confirm, an increased synthesis of the Na-K ATPase molecules after REM sleep deprivation has already been observed by us (Majumdar et al 2003). It has also been noted earlier that some of the neuronal firing rate changes during REM sleep and its deprivation. These changes in firing rate are likely to alter neuronal membrane permeability, which in turn might affect neuronal excitability. Also, since REM sleep deprivation decreases fluidity of the neuronal membrane (Mallick et al 1995), it may affect neuronal permeability and excitability as well.

Preoptico-Anterior Hypothalamic area (POAH) and thermoregulation

The early clinico-pathological studies of the brain of humans suffered from lethargic encephalitis showed that there were lesions around the anterior hypothalamus (von Economo 1926). Cats with bilateral lesions around the same area were found to be hyperthermic and lacked panting as well as sweating when brought into heated room (Clar 1979). Changing the local temperature of the POAH elicited various thermoregulatory responses. Warming and cooling of the POAH evoked heat loss and heat production responses, respectively (Hellstrom and Hamel 1967, Boulant, 1980, 1991). Mechanical or electrolytic lesions of POAH and adjoining areas elicited hyperthermic responses across species (Gamble and Patton 1953, Lipton, 1968, McGinty and Sterman 1968, Squires and Jacobson 1968, Roberts and Robinson 1969, Sakaguchi et al 1979, Nagel and Satinoff 1980, Szymusiak et al 1985, Szymusiak and Satinoff 1984, Boulant 1991). Bilateral POAH cell loss led to an impaired heat defense response with an associated elevation in the hypothalamic temperature threshold for panting (Feng et al 1995). Recently the effect of cell specific damage to the ventrolateral and ventromedial preoptic area on thermoregulatory behaviour was examined. It was observed that lesions in the ventromedial preoptic nucleus, caused loss of fine tuning of body temperature whereas the loss of ventrolateral preoptic area neurons does not alter mean body temperature (Lu et al 2000). Neural activation has been studied using fos-immunohistochemical technique after 3 h, 24 h and 14 days of clod exposure. Such studies showed that there are a significant number of Fos positive cells in the medial, lateral and ventral preoptic area of anterior hypothalamus, paraventricular thalamic nucleus, lateral lemniscus, tegmental nucleus, dorsal parabrachial nucleus and cerebellum (Miyata et al 1995).

The medial POAH (mPOAH) mediated influence on thermoregulation

Anatomically the POAH is divided into medial and lateral portions. Careful and systematic studies from our laboratory showed that the medial POAH (mPOAH) is more effective than the lateral part for sleep and body temperature regulations (Alam and Mallick 1990, 1991, Mallick and Alam, 1991). Unlike earlier, our studies showed that reversible inactivation (by marcain) of the neurons in the medial and the lateral portions of the mPOAH altered sleep-wakefulness and thermoregulation. The advantage of the reversible inactivation study was that it enabled us to study the functions during recovery also as compared to that of lesion studies where the effects during and after recovery could not be studied. In addition to sleep sensitive neurons as reported earlier, the mPOAH also contains two populations of thermosensitive neurons, one strongly responding to increased temperature while the other to decreased temperature (Boulant and Dean 1986).

The thermoregulatory and sleep-regulatory functions of the mPOAH are closely associated (Glotzbach and Heller 1984, Parmeggiani et al 1987, McGinty and Szymusiak 1990, Szymusiak 1995, Nakao et al 1995). Alterations in one of the functions e.g. a change in the body temperature or change in the sleep-wakefulness have been found to influence the other function as well. Some of the thermosensitive neurons in the mPOAH are also sleep related. These observations highlighted a close link between the mPOAH-mediated influences on the two functions. It has been proposed that although the two functions viz. sleep-wakefulness and body temperature are simultaneously modulated, they may have independent regulation by the mPOAH (Mallick and Alam 1991).

Neurotransmitters and thermoregulation

The homeothermic animals maintain a constant internal body temperature, independent of external temperature, by a variety of regulatory mechanisms. Several neurotransmitters in the POAH have been reported to modulate sleep-wakefulness and body temperature (Fox 1974, Szymusiak 1995). Application of NE and its agonist into POAH induced wakefulness and alerting responses in cats and rats (Yamaguchi et al 1963, Poole and Stephenson 1979, Day et al 1979, Tsoucaris-Kupfer and Scmitt 1972). In an attempt to investigate the role of different neurotransmitters in mPOAH in sleep-waking and thermoregulation, agonist or antagonist of either NE, acetylcholine and GABA was microinjected into the mPOAH and the effects were recorded. It was observed that NE modulated body temperature by acting on α_1, but affected sleep and waking by acting on α_2 and β-adrenoceptors, respectively (Mallick and Alam 1992). Subsequently iontophoretic studies have confirmed that the sleep related neurons in the mPOAH possess α_2 adrenoceptors (Osaka and Matsumura 1994, 1995), while the temperature sensitive neurons possess α_1 adrenoceptors (Mallick et al 2002a).

The cholinergic modulation of sleep-wakefulness by the hypothalamic basal forebrain region including the POAH was known since long (Hernandez-Peon 1962, Hernandez-Peon et al 1963). We studied the role of acetylcholine in the mPOAH on sleep-wakefulness and thermoregulation. It was observed that ACh by acting on muscarinic receptors (Mallick and Joseph 1997, 1998) modulated sleep-wakefulness and body temperature in rats. A single oral dose of Gaboxadol (THIP), a partial GABA-A receptor agonist, was found to increase sleep significantly (Faulhaber et al 1997) in humans while intraperitoneal administration of muscimol, a GABA-A receptor agonist, in the rats induced an increase in sleep (Lancel 1999). It was reported by us that GABA in the mPOAH also modulates sleep-wakefulness and body temperature (Ali et al 1999, Jha et al 2001b). Subsequently, at the cellular level we investigated the role of GABA-A receptors on the thermosensitive neurons. It has been reported that microiontophoretic application of picrotoxin, a GABA-A antagonist, exerted opposite effects on warm and cold sensitive neurons in the mPOAH (Jha et al 2001a).

Thermoregulation during sleep-wakefulness-REM sleep and its deprivation

A rise in body temperature is associated with wakefulness while a fall with sleep (McGinty and Szymusiak 1990). On the other hand, an increased body temperature (or warming) favours sleep (Roberts and Robinson 1969, Nakao et al 1995), while a decreased body temperature (or cooling) induces wakefulness (Szymusiak and Satinoff, 1984). Recent findings that wakefulness inducing area in the brainstem excites the warm sensitive neurons but inhibits the cold sensitive neurons (Mallick et al 2004a) may offer explanation at the neuronal level. Lesion of the mPOAH produced insomnia and increased body temperature (McGinty and Sterman 1968) while electrical as well as thermal stimulation of this area induced sleep (Sterman and Clemente 1962, Roberts and Robinson 1969). Isolated studies have reported that a majority of the neurons in this area are sleep related (Mallick et al 1983, Kaitin, 1984) and thermosensitive (Boulant 1980, 1991). Although there is brain and body cooling during non-REM sleep (McGinty and Szymusiak 1990), brain and body temperatures increase during REM sleep (Wehr 1992). It has been reported that temperature sensitive neurons loose their thermosensitivity during REM sleep (Parmeggiani et al 1987). A short-term poikilothermic like state has been observed just after paradoxical sleep onset in humans (Dewasmes et al 1997), which could possibly be due to loss of thermosensitivity of the temperature sensitive neurons during REM sleep.

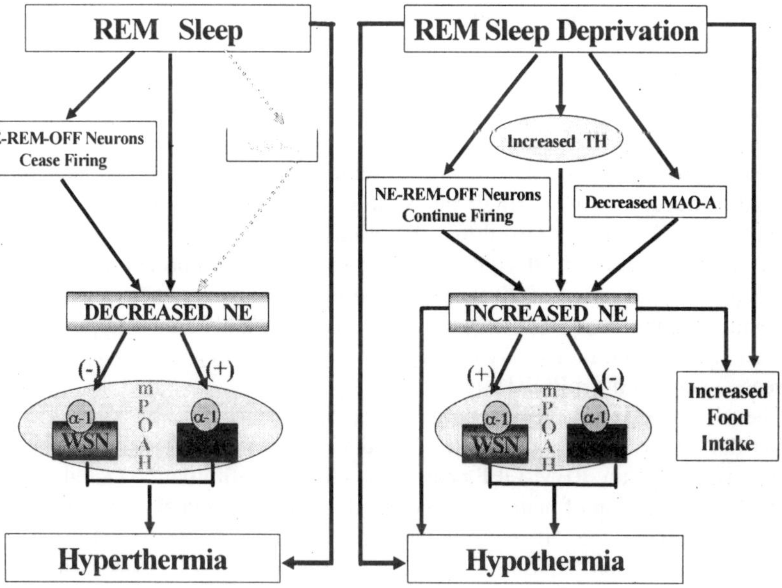

Fig 3

Fig 3. **Schematic representation of possible mechanism of changes in the level of norepinephrine during normal REM sleep and after its deprivation and the mechanism of NE modulating the body temperature during those stages have been shown.**

During REM sleep deprivation there is progressive increase in heat production and increased food intake (Rechtschaffen et al 1989), however, it is associated with progressive decrease in brain and body temperatures (Landis et al 1992). The activities of some of the glucose metabolic enzymes in rat brain also change after REM sleep deprivation (Thakkar and Mallick 1993b). REM sleep deprived rats did not appear to have a raised thermoregulatory set point (T_{set}). Although REM sleep deprived rats prefer and select increased ambient temperature, quantitative analysis indicated that these were targeted to compensate for falling intraperitoneal temperature rather than an increased T_{set}. After total sleep deprivation of rats, which were also REM sleep deprived, the body temperature decreased although the energy expenditure increased. Sleep deprived rats chose higher ambient temperature (>50^0C) to preclude the associated heat loss. However, given self-choice, sleep deprived rats preferred mean cage temperature at 37^0C compared to the baseline mean 26^0C (Rechtschaffen et al 2002). Also, there was increased food intake in those rats. However, the increased food intake did not increase their bodyweight, it was not dissipated in locomotion or any other waste. Loss of heat to the water under either disk or flowerpot has been ruled out, as the control rats under similar conditions also showed a similar heat loss. It has been suggested that REM sleep loss was the major factor for excessive heat loss in case of total sleep deprivation in rats (Rechtschaffen et al 2002).

Role of NE in thermoregulation during REM sleep and its deprivation - A Synthesis

It has been reported that the mPOAH receives NE-ergic inputs from the brainstem (Tanaka et al 1992) and the NE excites the thermosensitive neurons (Sun et al 1997) by its action on the α-1 adrenoreceptors (Mallick et al 2002a). On the other hand, it has been shown that the NE in the mPOAH induces hypothermia by its action on the α-1 adrenoreceptors, while their blockade caused hyperthermia (Mallick and Alam, 1992). Isolated studies showed that the brainstem stimulation increased NE level in the POAH (Herbison et al 1990) and also excited and inhibited WSNs and CSNs, respectively (Mallick et al 2004a). Also, since adrenoceptor numbers are altered in relation to changes in NE concentration (Dausse et al 1982), it is possible that the altered NE concentrations may modulate thermosensitivity by altering the number of adrenoceptors on the mPOAH neurons.

The NE-ergic REM-OFF neurons cease firing during REM sleep but continue firing during REM sleep deprivation (Mallick et al 1989). Also, it has been shown that activation of the REM-OFF neurons by electrical stimulation (Singh and Mallick 1996) or by injection of neurotransmitter antagonist (Kaur et al 1997, 2003), which is likely to activate the NE-ergic REM-OFF neurons, decreased REM sleep. The implications of such behaviour of the NE-ergic REM-OFF neurons are that there would be reduced NE in the projected area(s), including the mPOAH, during REM sleep but an increased NE in those areas during REM sleep deprivation. Also, an increased NE level in the brain during REM sleep deprivation may be supported by the fact that there is decreased monoamine oxidase levels in the brain (Thakkar and Mallick 1993a) and an increased synthesis of NE (Majumdar and Mallick 2003) after REM sleep deprivation.

A reduced concentration of NE during REM sleep would stimulate the CSNs (Mallick et al 2004a) and that is possibly the reason for increased body temperature during normal REM sleep (McGinty and Szymusiak 1990). However, during REM sleep deprivation there would be increased NE in the mPOAH, which would activate the WSNs (Mallick et al 2004a) and that is possibly the reason and mechanism for induction of heat-dissipation causing a decrease in body temperature during REM sleep deprivation (Fig 3). Additionally, the increased concentrations of NE in the hypothalamus (that includes the mPOAH) is likely to be involved in the REM sleep deprivation induced compensatory responses to glucose deprivation. Circulating glucose and glucose-mobilizing hormones are increased in response to microinjection of NE into the medial hypothalamus (Chafetz et al 1986, Steffens et al 1988). Injection of NE receptor agonist into ventromedial nucleus of hypothalamus stimulated food intake (Grandison and Guidotti 1977, Kelly et al 1979, Leibowitz, 1986). Hence, we propose that the increased concentration of NE is possibly the cause of increased food intake and hypothermia during REM sleep deprivation.

Summary

The NE-ergic neurons in the LC cease firing during REM sleep but continue firing during REM sleep deprivation. There is increased synthesis of NE in the LC neurons during REM sleep deprivation. Such behaviours of these neurons result in reduced NE during spontaneous REM sleep but an increased NE during REM sleep deprivation in the projected areas of those neurons in the brain including the medial preoptic area, the site for thermosensitive neurons in the brain. NE in the preoptic region is reported to cause hypothermia by acting on alpha-1-adrenoceptors present on the thermosensitive neurons and also increases food intake. Thus, the findings reviewed here suggest that an increased body temperature during spontaneous REM sleep compared to preceding NREM sleep period is due to decreased NE in the mPOAH, while an increased NE during REM sleep deprivation is the likely cause for hypothermia and increased food intake associated with REM sleep deprivation.

References

Adya HV and Mallick BN (2000) Neurochem Int **36**: 249-253.

Alam MN, Kumari S and Mallick BN (1993) Sleep Res **22**: 541.

Alam MN and Mallick BN (1990) Brain Res **525**: 242-248.

Alam MN and Mallick BN (1991) Brain Res **566**: 303-307.

Ali M, Jha SK, Kaur S and Mallick BN (1999) Neurosci Res **33**: 245-250.

Aserinsky E and Kleitman N (1953) Science **118**: 273–274

Aserinsky E (1999) In: Rapid Eye Movement Sleep (Mallick BN and Inoue S, eds) pp 1–16, Marcel Dekker: New York.

Aston-Jones G and Bloom FE (1981) J Neurosci **13**: 876–886.

Baghdoyan HA, Rodrigo-Angulo ML, McCarley RW and Hobson JA (1984) Brain Res **306**: 39-52.

Basheer R, Magner M, McCarley RW and Shiromani PJ (1998) Brain Res Mol Brain Res **57**: 235-240.

Benedito MA and Camarini R (2001) Braz J Med Biol Res **34**: 103-109.

Boulant JA (1980) In: Hand Book of Hypothalamus (Morgane PJ and Panksepp J, eds), pp. 1-82. Marcel Dekker: New York.

Boulant JA and Dean JB (1986) Ann Rev Physiol **48**: 639-654.

Boulant JA (1991) In: Basic mechanisms and management (Mackowiak P ed), pp. 1-22, Raven Press: New York.

Braun CMJ and Pivik RT (1981) Brain Res **230**: 133–151.

Capece ML, Baghdoyan HA and Lydic R (1999) In: Rapid Eye Movement Sleep, (Mallick BN and Inoue S, eds), pp. 123-141, Marcel Dekker, New York.
Cespuglio R, Gomez ME, Faradji H and Jouvet M (1982) EEG Clin Neurophysiol **54**: 570–578.
Chafetz MD, Parko K, Diaz S and Leibowitz SF (1986) Brain Res **384**: 404-408.
Chu NS and Bloom FE (1973) Science **179**: 908-910
Chu NS and Bloom FE (1974) J Neurobiol **5**: 527–544.
Clar HE (1979) Acta Neurochir (Wien) **50**: 153-199.
Dausse JP, Le Quan-Bui KH and Meyer P (1982) Eur J Pharmacol **78**: 15-20.
Davis PW and Vincenzi FF (1971) Life Sci **10**: 401-406.
Day TA, Willoughby JO and Geffen LB (1979) Brain Res **174**: 175-179.
Dement W and Kleitman N (1957) J Exp Psychol **53**: 339-346.
Dewasmes G, Bothorel B, Candas V and Libert JP (1997) J Sleep Res **6**: 252-258.
Egan TM and North RA (1985) Br J Pharmacol **85**: 733-735.
El Mansari M, Sakai K and Jouvet M (1989) Exp Brain Res **76**: 519-529.
Faulhaber J, Steiger A and Lancel M (1997) Psychopharmacology (Berl) **130**: 285-291.
Feng PF, Bergmann BM, Rechtschaffen A (1995) Brain Res **703**: 93-99.
Fox RH (1974) In: Recent Advances in Physiology, vol **9** (Linden RJ, ed), pp. 340-405, Churchill Livingstone, London.
Gamble JE and Patton HD (1953) Am J Physiol **172**: 623-631.
Gervasoni D, Darracq L, Fort P, Souliere F, Chouvet G and Luppi PH (1998) Eur J Neurosci **10**: 964-970.
Glotzbach SF and Heller HC (1984) Brain Res **309**: 17-26.
Grandison L and Guidotti A (1977) Neuropharmacology **16**: 533-536.
Gulyani S and Mallick BN (1993) J Sleep Res **2**: 45-50.
Gulyani S and Mallick BN (1995) Neuroscience **64**: 255-260.
Gulyani S, Majumdar S and Mallick BN (2000) Sleep and Hypnosis **2**: 49-68.
Hellstrom, B and Hammel HT (1967) Am J Physiol **213**: 547-556.
Henley K and Morrison AR (1974) Acta Neurobiol Exp (Warsz) **34**: 215–232.
Herbison AE, Heavens RP and Dyer RG (1990) Neuroendocrinology **52**: 161-168.
Hernandez-Peon R (1962) EEG Clin Neurophysiol **14**: 423-424.
Hernandez-Peon R, Chavez-Ibarra G, Morgane PJ and Timo-Iaria C (1963) Exp Neurol **8:** 93-111.
Hipolide DC, Tufik S, Raymond R and Nobrega JN (1998) Neuroscience **86**: 977-987.
Hobson JA, Datta S, Calvo JM and Quattrochi J (1993) Prog Brain Res **98**: 389-404.
Hobson JA, McCarley RW, Pivik RT and Freedman R (1974) J Neurophysiol **37**: 497–511.
Hobson JA, McCarley RW and Wyzinski PW (1975) Science **189**: 55–58.
Iijima K and Ohtomo K (1988) Am J Anat **181**: 43-52.
Jacobs BL (1986) Prog Neurobiol **27**: 183-194.
Jha SK, Islam F and Mallick BN (2001b) J Neurobiol **48**: 291-300.
Jha SK, Yadav V and Mallick BN (2001a) Pharmacol Biochem Behav **70**: 115-121.
Jones BE, Harper ST and Halaris AE (1977) Brain Res **124**: 473–496.
Jones BE (1991) Neuroscience **40**: 637-56.
Jouvet M (1999) In: Rapid Eye Movement Sleep (Mallick BN and Inoue S, eds) pp v-ix, Marcel Dekker: New York.
Jouvet M and Delorme F (1965) Comp Ren Soc Biol **159**: 895–899.
Kaitin KI (1984) Exp Neurol **83**: 347-357.
Kaur S, Saxena RN and Mallick BN (1997) Neurosci Lett **223**: 105-108.
Kaur S, Saxena RN and Mallick BN (2001) Synapse **42**: 141–150.
Kaur S, Panchal M, Faisal M, Madan V, Nangia P and Mallick BN (2004) Behav Brain Res **151**: 185-190.
Kelly J, Rothstein J and Grossman SP (1979) Physiol Behav **23**: 1123-1134.
Lancel M (1999) Sleep **22**: 33-42.
Landis CA, Bergmann BM, Ismail MM and Rechtschaffen A (1992) Sleep **15**: 13-20.
Leibowitz SF (1986) Fed Proc **45**: 1396-1403.
Lipton JM (1968) Physiol Behav **3**: 165-169.
Lu J, Greco MA, Shiromani P and Saper CB (2000) J Neurosci **20**: 3830-3842.
Luque JM, Malherbe P and Richards JG (1994) Mol Brain Res **24**: 219-226.
Lydic R, Baghdoyan HA and Lorinc Z (1991) Am J Physiol **261**: R766-R770.
Majumdar S and Mallick BN (2003) Neurosci Lett **338**: 193-196.
Majumdar S, Faisal M, Madan V and Mallick BN (2003) J Neurosci Res **73**: 870 - 875.
Mallick BN and Adya HV (1999) Neurochem Int **34**: 499-507.
Mallick BN and Alam MN (1991) Brain Res Bull **26**: 215-218.

Mallick BN and Alam MN (1992) Brain Res **591**: 8-19.
Mallick BN and Gulyani S (1996) Neuroscience **75**: 729-736.
Mallick BN and Joseph MM (1997) Brain Res **750**: 311-317.
Mallick BN and Joseph MM (1998) Pharmacol Biochem Behav **61**: 193-199.
Mallick BN and Thakkar M (1991) Neurosci Lett **130**: 221-224.
Mallick BN, Adya H and Faisal M (2000) J Neurochem **74**: 1574-1578.
Mallick BN, Chhina GS, Sundaram KR, Singh B and Kumar VM (1983) Exp Neurol **81**: 586-597.
Mallick BN, Jha SK and Islam F (2002a) Neuropharmacology **42**: 697-705.
Mallick BN, Jha SK and Islam F (2004a) Synapse **51**: 59-70.
Mallick BN, Kaur S and Saxena RN (2001) Neuroscience **104**: 467-485.
Mallick BN, Kaur S, Jha SK and Siegel JM (1999) In: Rapid Eye Movement Sleep (Mallick BN and Inoue S, eds) pp 153-166, Marcel Dekker: New York.
Mallick BN, Majumdar S, Faisal M, Yadav V, Madan V and Pal D (2002b) J Biosci **27**: 539-551.
Mallick BN, Siegel JM and Fahringer H (1989) Brain Res **515**: 94–98.
Mallick BN, Thakkar M and Gulyani S (1994) In Environment and Physiology (Mallick BN and Singh R, eds) pp 196-203, Narosa Publishing House, New Delhi.
Mallick BN, Thakkar M and Gangabhagirathi R (1995) Neurosci Res **22**: 117-122.
Mallick BN, Thankachan S and Islam F (1998) Sleep Res Online **1**: 132-146.
Mallick BN, Thankachan S and Islam F (2004b) J Neurosci Res **75**:133-42.
McGinty DJ and Sterman MB (1968) Science **160**: 1253-1255.
McGinty D and Szymusiak R (1990) Trends Neurosci **13**: 480-487.
Miyata S, Ishiyama M, Shido O, Nakashima T, Shibata M and Kiyohara T (1995) Neurosci Res **22**: 209-218.
Mogilnicka E (1981) Pharmacol Biochem Behav **15**:149-151.
Nagel JA and Satinoff E (1980) Science **208**: 301-303.
Nakao M, McGinty D, Szymusiak R and Yamamoto M (1995) Jpn J Physiol **45**: 291-309.
Nitz D and Siegel JM (1997) Neuroscience **78**: 795-801.
Osaka T and Matsumura H (1994) J Neurosci Res **19**: 39-50.
Osaka T and Matsumura H (1995) Neurosci Res **21**: 323-330.
Parmeggiani PL, Cevolani D, Azzaroni A and Ferrari G (1987) Brain Res **415**: 79-89.
Perez NM and Benedito MA (1997) Pharmacol Biochem Behav **58**: 605-608.
Poole S and Stephenson JD (1979) Br J Pharmacol **65**: 43-51.
Porkka - Heiskanen T, Smith SE, Taira T, Urban JH, Levine JE, Turek FW and Stenberg D (1995) Am J Physiol **268**: R1456-R1463.
Pujol JF, Mouret J, Jouvet M and Glowinski J (1968) Science **159**: 112-114.
Rechtschaffen A, Bergmann BM, Everson CA, Kushida CA and Gilliland MA (1989) Sleep **12**: 68-87.
Rechtschaffen A, Bergmann BM, Everson CA, Kushida CA and Gilliland MA (2002) Sleep **25**: 68-87.
Roberts WW and Robinson TC (1969) Exp Neurol **25**: 282-294.
Rye DB (1997) Sleep **20**: 757-788.
Sakaguchi S, Glotzbach SF and Heller HC (1979) Am J Physiol **237**: R80-R88.
Sakai K (1980) In: The reticular formation revisited (Hobson JA and Brazier MAB, eds) pp. 427–447, Raven Press, New York.
Sakai K, Vanni-Mercier G and Jouvet M (1983) Exp Brain Res **49**: 311-314.
Sakai K (1988) Arch Ital Biol **126**: 239–257.
Sastre JP, Sakai K and Jouvet M (1979) C R Seances Acad Sci D.**289**: 959–964.
Serra G, Melis MR, Argiolas A, Fadda F and Gessa GL (1981) Eur J Pharmacol **72**:131-135.
Shouse MN, Staba RJ, Saquib SF and Farber PR (2000) Brain Res **860**: 181-189.
Siegel JM (1983) In: Sleep Mechanism and Functions in Humans and Animals: an Evolutionary Perspective (Mayes A and Reinhold VN, eds), pp. 217-231, University Press, Cambridge.
Siegel JM (1985). In : Brain Mechanism of Sleep (McGinty DJ, Drucker Çollin R, Morrison, A and Parmeggiani PL, eds), pp. 157-174, Raven Press, New York.
Siegel JM (1989) In : Principles and Practice of Sleep Medicine (Kryger MH, Roth T and Dement WC, eds) pp. 104-120, Saunders WB, Philadelphia.
Siegel JM and Rogawski MA (1988) Brain Res **472**: 213-233.
Siegel JM, Nienhuis R and Tomaszewski KS (1984) Neurosci Lett **45**: 241-246.
Singh S and Mallick BN (1996) Neurosci Res **24**: 227–235.
Sinha AK, Ciaranello RD, Dement WC and Barchas JD (1973) J Neurochem **20**: 1289-1290.
Squires RD and Jacobson FH (1968) Am J Physiol **214**: 549-60.
Steffens AB, Scheurink AJ, Luiten PG and Bohus B (1988) Physiol Behav **44**: 581-589.

Sterman MB and Clemente CD (1962) Exp Neurol **6**: 103-117.
Stern WC. Miller FP, Cox RH and Maickel RP (1971) Psychopharmacologia **22**: 50-55.
Sun JR, MaYC, Xu ZH, Zhao WJ and CaiYP (1997) Sheng Li Xue Bao **49**: 666-670.
Szymusiak R (1995) Sleep **18**: 478-500.
Szymusiak R and Satinoff E (1984) Brain Res Bull **12**: 295-305.
Szymusiak R, DeMory A, Kittrell EM and Satinoff E (1985) Am J Physiol **249**: R219-R227.
Tanaka J, Nishimura J, Kimura F and Nomura M (1992) Neuroreport **3**: 946-948.
Thakkar M and Mallick BN (1991) Pharmacol Biochem Behav **39**: 211-214.
Thakkar M and Mallick BN (1993a) Neuroscience **55**: 677-683.
Thakkar M and Mallick BN (1993b) Sleep **16**: 691-694.
Thakkar M, Portas C and McCarley RW (1996) Brain Res **723**: 223-227.
Thankachan S, Islam F and Mallick BN (1997) Sleep Res **26**: 54.
Thankachan S, Islam F and Mallick BN (2001) Brain Res Bull **55**: 43-49.
Tononi G, Pompeiano M and Cirelli C (1994) Brain Res Bull **35**: 589-596.
Tsai LL, Bergmann BM, Perry BD and Rechtschaffen A (1993) Brain Res **602** : 221-227.
Tsoucaris-Kupfer D and Scmitt H (1972) Neuropharmacology **11**: 625 - 635.
von Economo (1926) In: Handbuch der normalen und pathologischen Physiologie. (Bethe VA, Bergmann GV, Embden G and Ellinger A, eds) vol 17, pp. 591-610, Springer, Berlin.
Wehr TA (1992) Neurosci Biobehav Rev **16**: 379-397.
Yamaguchi N, Ling GM and Marczynski TI (1963) Recent Adv Biol Psychiat **84**: 9-20.

Acknowledgement

Financial supports from CSIR, DBT, DST, ICMR and UGC are acknowledged.

Molecular and Cellular Neurobiology
Editors: M.K. Thakur and S. Prasad

Estrogen Effects in the Brain: Implications for Aging and Alzheimer's Disease

M.K. Thakur, P.K. Sharma and Soumi Ghosh

Biochemistry & Molecular Biology Laboratory, Department of Zoology
Banaras Hindu University, Varanasi-221 005, India

Abstract

Sex hormone estrogen exerts a variety of effects in the brain. The interaction between estrogen and nerve cells is highly complicated, each type of neuron in each brain region may react in a different way to estrogen, depending on species, sex, age, cell type and the presence of neuropathology. Therefore, understanding the ways in which estrogen acts in the brain is complex and has been challenging. Particularly, the presence of two types of estrogen receptor and a number of co-regulators add to the complexity of estrogen action. In this article, we present evidences for the wide-ranging actions of estrogen in the brain at cellular as well as molecular levels and discuss their implications during aging and Alzheimer's disease.

Key words: estrogen action, gene expression, brain function, aging, AD

Introduction

Aging of the brain is associated with an overall reduction in brain volume and weight, enlargement of the brain ventricles, and subtle morphological and functional changes in specific neurons and neuronal circuits. There also occurs atrophy of pyramidal neurons and synapse, decrease of striatal dopamine receptors, accumulation of fluorescent pigments, cytoskeletal abnormalities, and reactive astrocytes and microglia. The selected brain areas show loss of synapses and dendritic pruning (Anderton 2002). The profile of gene expression also varies remarkably. Using oligonucleotide arrays, Lee et al (2000) have detected differences in the expression of genes between the neocortex and cerebellum of aging mice. Their results indicate an inflammatory response, oxidative stress and reduced neurotrophic support in both brain regions. These alterations affect the brain function and lead to impairments in cognitive and motor skills, and account for several neurological disorders including Alzheimer's disease (AD).

AD is the most common cause of progressive decline of cognitive function in aged people. It is characterized by the presence of numerous senile plaques composed mainly of 40-42 amino acid heterogeneous peptides collectively known as β-amyloid. These peptides are produced by proteolytic cleavage of a large transmembrane protein, amyloid precursor protein (APP), into various smaller fragments that are either soluble or aggregating. Such processing of APP is inherently important for the initiation and progression of AD. Further studies suggest that it is not the formation of β-amyloid plaques per se, but the altered processing of APP and subsequent loss of soluble APP, that sets the stage for the massive neuronal loss during AD.

This chapter is dedicated to Professors MS Kanungo and Jim Edwardson who introduced us to the area of Aging and Alzheimer's disease, respectively.

The age of onset and incidence of AD vary with sex, indicating the role of sex steroids in the pathophysiology of AD. A growing body of evidence suggests that postmenopausal estrogen deficiency accelerates brain aging and increases the risk of various neurodegenerative processes including AD (Compton et al 2001). Clinical studies also show that estrogen has positive effects on brain homeostasis by preserving neural plasticity and neurotransmitter pathways involved in learning, memory and balance (Birge 2003). The involvement of estrogen in the brain development and function has thus long been known, but only recently its role in the age-dependent deterioration of mental function has attracted more careful attention (Thakur 2003 a, b).

This article reviews the remarkable progress that has been made during the past decades in the area of estrogen action in the brain, particularly during aging and AD. Mainly the following major issues have been addressed: How does estrogen act? What are functions of estrogen in the brain? What changes occur in the expression of estrogen receptor (ER) and other factors involved in signal transduction during the brain development and aging? How do these changes influence different physiological processes and lead to age-associated diseases like AD? Can estrogen replacement therapy (ERT) prevent or delay these diseases? The work done in the author's laboratory and reported from elsewhere in this context is discussed below.

Molecular mechanism of estrogen action

Specific ER was observed in the central nervous system as early as 1970s, but until recently little was known about the molecular mechanism of estrogen action. Now it is established that estrogen acts through three mechanisms – (i) receptor-mediated genomic mechanism, (ii) cross-talk between ER and members of second messenger pathways such as cAMP, mitogen-activated protein kinase (MAPK), IP3-protein kinase C (PKC), and intracellular calcium signal, and (iii) non-genomic mechanism involving membrane receptors or channels. Genomic action of estrogen leads to altered synthesis of proteins, which after axonal or dendritic transport may participate in pre- or postsynaptic events. However, in non-genomic mechanism, estrogen acts at the pre- or postsynaptic membrane and alters permeability and function of neurotransmitter receptors.

The genomic action of estrogen is mediated by two receptors, ERα and ERβ, which are members of the nuclear receptor superfamily. Both these receptors are coded by separate gene, which is localized on chromosome 6 for ERα and on chromosome 14 for ERβ (Warner et al 1999). Initially, ERα was cloned from a human breast cancer cell line and ERβ from a rat prostate cDNA library. Later they were identified in other target organs including the brain. Both receptors are composed of three independent but interacting functional domains: amino-terminal transactivation domain (ATD), central DNA-binding domain (DBD) and carboxyl-terminal ligand-binding domain (LBD). Both ERα and ERβ show a highly conserved DBD but bear no homology in ATD (Fig. 1). They mediate transcriptional activation by two activation function (AF) domains; AF-1 in ATD which is regulated by growth factors through MAPK pathway, and AF-2 in LBD which is necessary for mediating ligand-dependent transactivation. Both receptors also interact with each other for determining the effect of estrogen (Nilsson and Gustafsson 2000). In many contexts, ERβ appears to antagonize the actions of ERα (yin/yang relationship), although some genes are specifically regulated by one of the two receptors (Weihua et al 2003).

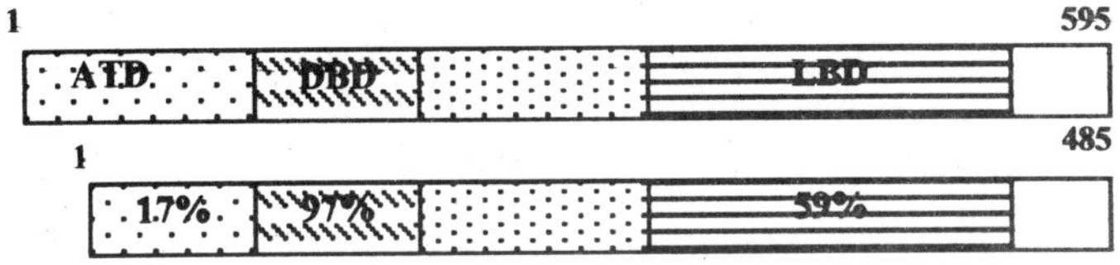

Fig 1. Schematic representation of ER functional domains. Number inside the box represents percentage amino acid identity between ERα (upper) and ERβ (lower).

Upon binding of estrogen to ER, many proteins associated with the inactive receptor (such as heat shock protein HSP) are dissociated. Then the receptor becomes activated and forms a dimer. Now the hormone-receptor complex binds to specific consensus nucleotide sequences known as estrogen response elements (ERE) located in the promoter regions of target genes. This is followed by recruitment of a number of co-regulators and interaction with basal transcription factors (like TATA-binding protein TBP and TBP-associated factors TAFs) to modulate the transcriptional activity of specific gene (Fig. 2). Estrogen is reported to regulate its own receptor as well as steroid receptor co-activator (SRC)-1 transcript level in the rat brain cortex (Mitev et al 2003). SRC-1 amplifies ER-dependent transcription in a temporally and spatially coordinated manner and provides functional specialization of brain areas. Other co-activators include ER associated proteins (ERAP) 160 and 140, receptor intermediary protein (RIP) 160, 140 and 80 and transcriptional intermediary factor (TIF) 2 (McKenna et al 1999). Additionally, phosphorylated CREB-binding protein (CBP) and related p300 also act as ERAP (Katzenellenbogen et al 2000). As a consequence of interaction of hormone receptor complex with different co-regulators, the level of different gene products alters and this leads to changes in cellular functions. Such mechanism of estrogen action, however, neither explains the rapid effects of estrogen nor accounts for its wide range of actions in the brain. Therefore, another mechanism is proposed, in which ER interacts with various growth factors including neurotrophins and leads to the stimulation of oncogenes src and ras, resulting in neuroprotection. Thus estrogen activates many members of MAPK cascade and influences genes that don't contain a canonical ERE (Singh et al 2000, Ivanova et al 2001). In addition to this, estrogen also acts at the cell membrane through the non-genomic pathway (Ábrahám et al 2003). Though estrogen binding sites in the plasma membrane raise the possibility of novel ER subtypes, experimental evidence in this favor is currently lacking.

Actions of estrogen in the brain

Estrogens are involved in embryonic and fetal development, female secondary sexual characteristics, behavior and reproductive functions, and cardiovascular and neuronal physiology. Their effects have been linked to several human diseases including breast and endometrial cancer, cardiovascular disease and osteoporosis. In addition, estrogen exerts a variety of important effects throughout the brain. It improves cerebral blood flow, increases secretion of neuromediators, stimulates development of gliacyte and formation of new synapses, and activates genes responsible for production of anti-apoptotic proteins and growth factors. It regulates a large number of genes in different areas of the brain (Table 1), which are associated with different brain functions such as learning, cognition, memory, awareness, fine motor skills, depression, mood and temperature regulation (Thakur 1999, McEwen 2001).

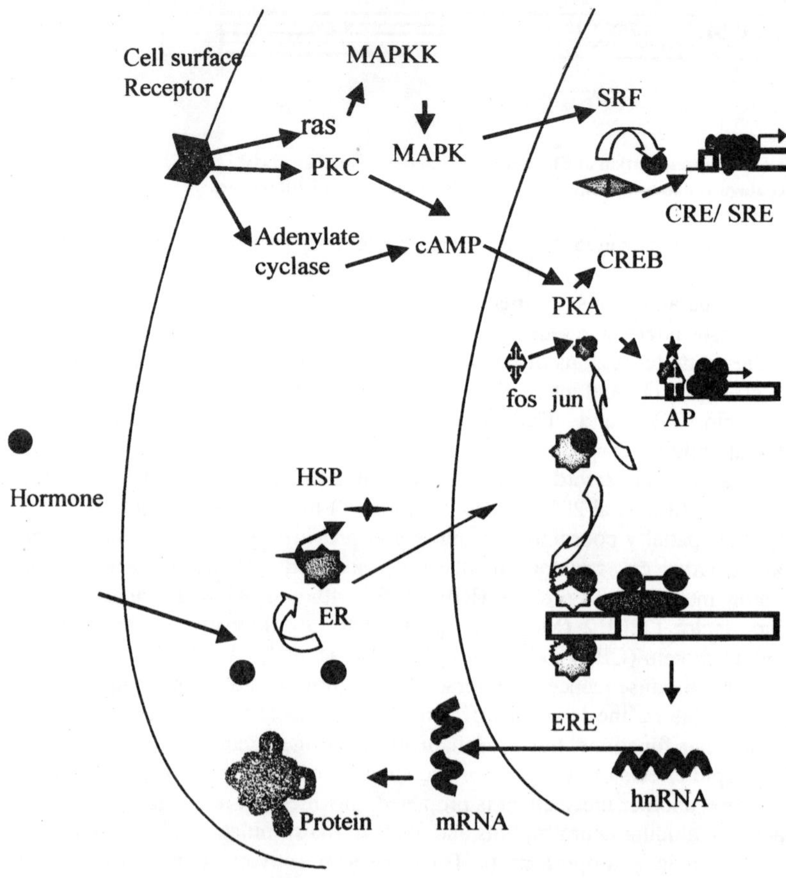

Fig. 2. Molecular mechanism of estrogen action.

The diverse biological effects of estrogen are mediated through ERα and ERβ. The human brain shows a distinct distribution pattern of these receptors. ERβ mRNA is most abundant in the hippocampus, claustrum and cerebral cortex. In some brain regions like hypothalamic preoptic area, bed nucleus of the stria terminalis and medial amygdaloid nucleus, ERα and ERβ are co-expressed (Hrabovszky et al 1998) and form heterodimers, thus giving rise to additional possible variants for gene regulation (Pettersson et al 1997). The mRNA and protein of both receptors are distributed throughout the brain in neurons of the rat cerebral cortex, hippocampus, cerebellum, hypothalamus and olfactory bulb. Non-neuronal cells like glial cells of the rat brain also manifest the expression of both subtypes of the receptor (Platania et al 2003).

ERα has been localized within the pre-synaptic terminals and spines of pyramidal neurons in the hippocampus where it may mediate the effects of estrogen on synaptic plasticity. ERα expression in these areas is regulated in response to circulating estrogen concentration, and the responsiveness to estrogen decreases with age in parallel with a reduction in the number of spines containing ERα. In contrast to ERα,

Table 1. Estrogen regulated genes in different areas of the brain

(a) Genes which are upregulated

Genes	*Brain regions*	*Animal*	*Ref*
ERα and β	hypothalamus	rat	Chakraborty et al (2003)
Norbin	brain cortex	mouse	Mani et al (2001)
Tryptophan hydroxylase	medial basal hypothalamus	guinea pig	Lu et al (1999)
BDNF	pyriform cortex	rat	Gibbs (1999)
ApoE	cortex and hypothalamus	mouse	Levin-Allerhand et al (2001)
AchE	striata terminalis	rat	Luine and Rhodes (1983)
NMDA receptor	hypothalamus	rat	Watanabe et al (1999)
CREB	amygdala	rat	Carlstrom et al (2001)
IGF-1	cortex	monkey	Cheng et al (2001)
NO synthase	cerebral blood vessel	rat	McNeil et al (2002)
AR	brain cortex	mouse	Kumar and Thakur (2004)
Glut3	brain cortex	monkey	Cheng et al (2001)
Glut4	brain cortex	monkey	Cheng et al (2001)
CIF	neuron	human	Zhang et al (2001a)

(b) Genes which are downregulated

Genes	*Brain regions*	*Animal*	*Ref*
ER	brain cortex	mouse	Asaithambi et al (1997)
PS1	brain cortex	mouse	Thakur et al (2004)
PS2	brain cortex	mouse	Thakur et al (2004)
nip-2	neuroblastoma	cell culture	Meda et al (2000)
$p75^{NTR}$	cholinergic and basal forebrain	rat	Ping et al (2002)
APP	brain cortex	mouse	Asaithambi et al (1999)
MAO-A	hypothalamus,	rat amygdala	Holschneider et al (1998)

ERβ level is low in the hypothalamus and amygdala. The expression of both ER subtypes varies with sex (Zhang et al 2002) and is regulated by estrogen in brain region-specific manner (Nomura et al 2003). ERβ knockout (BERKO) mice show several morphological abnormalities (Wang et al 2001), which become evident as early as two months. When BERKO mice grow, the neuronal deficit becomes more pronounced and by two years of age, there is degeneration of neuronal cell bodies throughout the brain. This shows the importance of ERβ for neuronal survival.

Further findings also support the role of ERβ in late embryonic development of the brain and in neuronal migration and apoptosis (Wang et al 2003).

In the brain, estrogen acts in several ways and involves different areas and neurotransmitter systems including serotonin pathways, catecholaminergic neurons and the basal forebrain cholinergic system (McEwen 2002). It increases blood flow and hence the supply of oxygen and glucose that the brain needs to function. It also increases the number of synapses or connections between nerve cells and thus helps neurons to grow and regenerate. It raises the level of serotonin, which is helpful in maintaining a good mood. The serotonin (5-HT) system in hypothalamus, hippocampus and cortex is involved in the modulation of reproduction, mood, sleep and cognition. The level and activity of serotonin in the brain is regulated by the concentration of serum estrogen. The substitution of estrogen in ovariectomized rats positively affects the serotonergic system. It also increases the expression of tryptophan hydroxylase, the rate-limiting enzyme for serotonin synthesis. It reduces the level of $5\text{-}HT_{1A}$ autoreceptor subtype in the dorsal raphe nucleus of spayed monkeys, suggesting that estrogen may facilitate 5-HT neurotransmission. Estrogens also influence other transmitter systems such as vasopressin, oxytocin, somatostatin, galanin, GABA and glutamate system. In addition, estrogen is involved in following processes:

Brain development: The mammalian brain begins life as a bipotential organ, equally capable of assuming a male or a female phenotype. The development of such sex-specific structure in the brain is under profound influence of steroid hormones. Particularly, estrogen has key role in developing a "male" brain. The ultimate determinant of brain phenotype is dimorphism in the steroid hormone milieu during a perinatal sensitive period. Differential activation of the embryonic testis or ovary results in dramatic sex differences in circulating gonadal steroids in a variety of mammalian species including humans. In rats, aromatase level is high in neurons and this enzyme converts testosterone into estradiol that induces male-specific brain differentiation (Hutchinson 1997). Two sex-specific cDNAs have been isolated from the fetal mouse brain. They are F29 and M17 corresponding to the female-specific Xist gene and the male-specific Smcy gene, respectively (Eriksson et al 1999). In the developing nervous system of rodents and primates, sex steroids exert a trophic influence by stimulating axonal and dendritic growth, modulating synapse formation and neurotransmitter activity.

Neuronal growth and plasticity: During development, estrogen plays a central role in maturation and sexual differentiation of the mammalian central nervous system. In rodents, estrogen directly affects the differentiation and maturation of cortical and hypothalamic neurons during the perinatal period. It is involved in the modulation of morphologic changes in neurons, regulating both axonal sprouting and dendritic plasticity (Kadish and Groen 2002). In rats, dendritic spines, which are necessary for optimal communication between neurons, shrivel as the level of estrogen goes down but regrow when the hormone is supplemented. It selectively enhances the growth and differentiation of axons and dendrites. Thus estrogen distinctly enhances neurite growth and may therefore contribute to the sexual differentiation of cognitive functions. The expression of mice brain norbin (neurite-outgrowth-related brain protein) increases in male but decreases in female whereas its transcript increases in both sexes during aging (Mani et al 2001).

Estrogen shows synergistic effect with insulin growth factor (IGF)-1 in regulating neuronal development, neural plasticity, neuroendocrine events and the response of neural tissue to injury (Cardona-Gomez et al 2000, 2002). Both ER and IGF-IR are co-expressed in the same nerve cells. ERα, in particular, is involved in differentiation of hippocampal neurons *in vitro* (Audesirk et al 2003). Furthermore, in embryonic mouse midbrain neurons, estrogen stimulates the expression of brain-derived neurotrophic factor (BDNF) through a membrane-mediated and calcium-dependent mechanism (Kuppers et al 2001). It also regulates BDNF mRNA and protein level in the rat hippocampus during development (Solum and Handa 2002). BDNF plays a role in plasticity and neuron function. Estrogen induces synaptic plasticity in areas with high relevance to human cognitive processes. It regulates hippocampal synaptic spine density by increasing the level of N-methyl-D-aspartate (NMDA) receptors and thereby providing a continuous excitatory neuronal stimulus. Several key elements of the mechanisms involved in estrogen-mediated regulation of synaptic plasticity are altered in aged animals (Bi et al 2003).

Learning and memory: The relationship between estrogen and memory function was noted when it was demonstrated that verbal memory declined with the loss of estrogen and was restored after estrogen treatment following menopause (Philips and Sherwin 1992). Estrogen can prevent the deterioration of both short and long-term memory that occurs with advancing age. In rats, removal of ovary results in impaired learning due to decline in choline uptake and choline acetyl transferase (ChAT) activity. Estrogen substitution following ovariectomy induces ChAT activity in the rat basal forebrain and hippocampus. Thus it boosts the amount of acetylcholine that plays a critical role in memory. Estrogen has been reported to reverse several hippocampal markers of age-related memory impairment (Foster et al 2003).

Both ER subtypes are involved in learning and memory. The co-localization of ERα with nerve growth factor (NGF) receptors in cholinergic neurons of the rat basal forebrain and the stimulation of NGF receptor and ChAT mRNA by estrogen in the rat basal forebrain suggests a possible role for ERα in learning and memory. On the other hand, the predominant expression and localization of ERβ in rat neocortex, hippocampus and basal forebrain nuclei suggests an important role for ERβ in learning and memory. The predominant presence of ERβ message in the human hippocampus, entorhinal cortex and thalamus also supports a putative role of ERβ in cognition, memory and motor functions.

Cognition: Estrogen modulates neuronal activity in those brain regions, which are highly implicated in cognitive functions. Its deprivation is likely to initiate or enhance degenerative changes caused by oxidative stress, and reduces the brain's ability to maintain synaptic connectivity and cholinergic integrity leading to the cognitive decline seen in aged and disease-afflicted individuals (Gandy 2003). The importance of ERα in the establishment of cognitive functions is evident from the observation that ERα knock out mice display impaired performance on cognitive tasks. However, many epidemiological and clinical studies show conflicting results of estrogen effects on cognition (Asthana 2003). Some reports favor beneficial effects of ERT in women with menopausal symptoms (LeBlanc et al 2001) and in aged monkeys (Rapp et al 2003). In contrast, another randomized double-blind placebo-controlled study shows that ERT given for one year does not slow cognitive decline in women with AD (Mulnard et al 2000). The level of estrogen in the serum also does not correlate with the extent of cognitive function (Thal et al 2003). However, Gibbs and Gabor (2003) suggest that ERT may be beneficial with respect to age-related cognitive decline, provided the therapy is administered in an appropriate regimen and is initiated within a window of time following the loss of ovarian function.

Stroke: The incidence of stroke is lower in premenopausal women than in men. However, after menopause, the incidence of stroke rises in women, suggesting that estrogen plays a protective role. Estrogen can prevent neuronal damage from stroke. It reduces consequences of the injury by enhancing synaptogenesis and dendritic arborization, stimulating antioxidant mechanism, and reducing excitotoxicity. Estrogen can dilate cerebral blood vessels through increased synthesis of nitric oxide and stimulation of such compounds as prostacycline and a potent vasodilator, epoxyeikosotrienoic acid (McNeil et al 2002). Dubal et al (2001) have further demonstrated that the deletion of ERα completely abolishes the protective actions of estradiol in all regions of the brain, whereas the absence of ERβ totally preserves the ability of estradiol to protect against brain injury. This finding clearly suggests that ERα mediates the protective effects of estradiol in brain injury. It binds estrogen and accelerates genes that help the cells survive or suppresses genes that can harm the cells.

Oxidative stress and cell death: Estrogen protects neuronal damage through many ways (Brinton 2001). It reverses oxidative stress in neuronal cell culture by increasing intracellular concentration of glutathione, a natural free radical scavenger. It modulates the activity of antioxidant enzymes such as superoxide dismutase, catalase and glutathione peroxidase, all of which are important cellular redox regulators. The antioxidant capacity of estrogen has been related primarily to its basic chemical property such as the presence of an intact hydroxyl group in steroid ring A of the estrogen molecule. The modification or absence of this group leads to a loss of the neuroprotective effect (Behl and Lezoualc'h 1998). Another mechanism by which estrogen prevents neuronal death is through inhibition of apoptosis. It activates PKC which plays a key role in many cellular events including regulation of apoptosis (Cordey et al 2003). In the culture of neuroblastoma and rat embryo neurons, estradiol prevents cell death by blocking the increase in nip-2 mRNA, which is induced by apoptotic stimuli (Meda et al 2000). Studies of Zhang et al (2001a) demonstrate that 17β-estradiol induces a caspase inhibitory factor (CIF) that prevents caspase-mediated neuronal death. The induction of CIF occurs within 10 min of estradiol exposure to neurons, does not require *de novo* protein synthesis, and involves MAPK activation, thus supporting a receptor-mediated non-genomic pathway. Further studies of Zhang et al (2001b) indicate that estrogen may protect neurons partly via ER-mediated protein kinase B activation.

Estrogen protects against ischemia-induced neuronal damage by inhibiting the release of free calcium from intracellular stores and the influx of calcium from the extracellular space to maintain non-toxic intracellular calcium concentration and to prevent activation of apoptotic signaling cascades (McCullough and Hurn 2003). The pro-apoptotic neurotrophin receptor $p75^{NTR}$ and caspase-3 are upregulated by ischemia, and their increase is attenuated by exogenous estrogen administration. The anti-apoptotic activity of estrogen is also confirmed by increasing bcl-2 protein activity, which prevents cell apoptosis in ischemic region. Estrogen also protects rat cerebral cortical neurons from ischemic injury (Shughrue and Merchenthaler 2003). However, a clinical trial with ERT in elderly women having a history of cerebrovascular disease shows that this hormone is unable to protect against reoccurrence of ischemia and it also does not reduce the incidence of mortality compared to a placebo (Stein 2001).

Estrogen effects in brain aging

Estrogen exerts a multitude of effects during brain aging in region and sex-dependent manner. It is reflected from a significant sex and age differences in ER density in several brain areas. The reduced level of ER in the brain has been found to correlate with the low serum levels of estrogen in old female rats (Kanungo et al 1975). Particularly, in the mouse brain cortex, the expression of ER decreases drastically in old age and it is downregulated by estrogen (Asaithambi et al 1997). The presence of low level of estrogen and ER in old age is responsible for reproductive aging in females, characterized by irregular estrus cycles and a decline in fertility, pseudopregnancies and sexual behavior (Thakur 1988).

The proportion of two forms of ER shows variation with respect to sex, age and brain region. Whereas ERα is expressed in the fetal brain of both sexes, ERβ shows a different expression pattern. The mRNA level of ERβ is high in the male hypothalamus and preoptic area from 17 days post-coitum until day 15 after birth, indicating a possible function in sex-specific brain development. Interestingly, a selective reduction in ER expression has been suggested to be the cause of sexual dimorphism in the developing brain (Karolczak and Beyer 1998). The level of ERβ is relatively lower than that of ERα in the mouse brain cortex and the amount of both receptors decreases in old age (Thakur et al 2004). Age-related decrease in ERα level causes reduced responsiveness of hippocampal synapses to estrogen in aged rats (Adams et al 2002). Expression of ERα and ERβ mRNA and protein is differentially modulated in the aging rat and mice brain and changes are region-specific (Wilson et al 2002, Sharma et al 2003). A significant age-related decrease is observed in ERβ cell number in the anteroventral periventricular nucleus, but not in the bed nucleus of the stria terminalis. Estrogen also has no significant effect on ERβ cell number in either brain region at any age (Chakraborty et al 2003).

In addition to age-related changes in ER expression, the response to estradiol also varies with age. For instance, in the rat brain, estradiol stimulates covalent modifications of histones and non-histone chromosomal proteins by phosphorylation, acetylation and methylation in adult but not in old age (Kanungo and Thakur 1979a, Thakur and Kanungo 1981). Furthermore, the rate of transcription increases after estradiol treatment in immature and adult rats (Thakur et al 1978, Kanungo and Thakur 1979b). Both testosterone and estradiol modulate the synthesis and phosphorylation of brain cortex proteins in age- and sex-specific manner (Mukherjee et al 1999). Such age-dependent changes in the covalent modifications of chromosomal proteins and DNA brought by estrogen in the rat brain attribute to alterations in cellular functions during aging. Sex-specific alteration in chromatin conformation of the aging mice brain also suggests the role of sex steroids in chromatin organization and gene expression in the brain (Thakur et al 1999). Estradiol decreases the level of androgen receptor (AR) and increases its phosphorylation with age in the mice brain cortex (Thakur et al 2000a). Its supplementation significantly increases AR mRNA level and decreases AR promoter methylation in the brain cortex of adult but not old mice (Kumar and Thakur 2004). These findings further establish the reduced responsiveness of brain to estrogen in old age.

Estrogen has been shown to improve many deteriorating brain functions that result due to hormonal deficiency in old age. Alkayed et al (2001) have observed that young adult female rats sustain smaller infarcts after experimental stroke than age-matched males. This sex-difference disappears after surgical removal of ovary or after reproductive senescence in middle-aged females. However, it can be restored by estrogen treatment, suggesting a protective role for hormonal therapy in postmenopausal women. Studies using rats have shown that estrogen exerts similar effects and can enhance cell survival and induce synaptic plasticity (Wise and Dubal 2000). In a population of older women, lifetime ERT has been found to be associated with improvement and protection against cognitive decline during aging (Carlson et al 2001). Estradiol is reported to protect verbal memory and possibly frontal lobe mediated functions in older women. In contrast to the positive findings in women, endogenous sex steroids do not appear to be closely linked to better cognition in old men (Wolf and Kirschbaum 2002).

Role of estrogen in AD

There are ample evidences from epidemiological and clinical studies suggesting that estrogen has beneficial and protective role in reducing the risk of AD (Yaffe et al 1998, Thakur 2000, 2002, Birge 2003). Postmenopausal women show permanent hypoestrogenicity and suffer from increased risk of brain injury associated with AD. However, ERT decreases the risk and severity of neurodegenerative conditions. It is observed that women receiving estrogen replacement are 40% less likely to have AD and related dementia than women who do not receive estrogen. Thus estrogen may serve as an important therapeutic agent in the maintenance of normal neural function during aging and after injury. The altered expression of ER in brain regions highly implicated in AD also indicates a role for this hormone in AD. The presence of ERβ in senile plaques points to a direct effect of estrogen on amyloid pathology. Also ERβ immunoreactivity is higher in the hippocampus of AD patients than in control subjects, indicating a role for ERβ-mediated estrogen effects in the disease. Hippocampus, which is intimately involved in memory, is smaller in AD. However, it is larger in volume in women taking estrogen than in women who don't take estrogen. A number of other reports suggest that estrogen reduces the risk of AD and vascular dementia (Sherwin 2003). While estrogen's benefits for women with AD remain controversial, evidence is accumulating that bathing the brain in estrogen enhances cognitive function in healthy older women. Two randomized, double-blind, placebo-controlled, parallel-group trials found that both short-term (Henderson et al 2000) and long-term (one year) estrogen therapy (Mulnard et al 2000) did not improve the symptoms of most women with AD. Thus, it appears that estrogen maintains and sustains neuronal viability to prevent degenerative disease, but it does not reverse the degenerative disease.

The neuroprotective action of estrogen during the course of neurodegenerative disorders may be mediated through different ways such as antioxidative properties of estrogen, increased expression of anti-apoptotic proteins, enhanced cell viability, and direct regulatory effects on the processing of amyloid-β peptide (Savaskan 2002). Amyloid peptides account primarily for the neurotoxicity in AD and induced cell death which is blocked by estrogen via ERα-dependent pathways (Kim et al 2001, Shah et al 2003). Both ERα and β are upregulated in AD (Ishunina and Swaab 2001). Studies of Petanceska et al (2000) suggest that cessation of ovarian estrogen production in postmenopausal women facilitates amyloid-β deposition by increasing the local concentrations of amyloid-β 40 and 42 peptides in brain and such modulation of amyloid-β metabolism may be one of the ways by which ERT prevents and/or delays the onset of AD in postmenopausal women. Clinical studies further show that lower estrogen in cerebrospinal fluid is inversely correlated with amyloid-β 1-42 concentration. After estrogen treatment, microglia show an increased internalization of amyloid-β (Li et al 2000), providing an additional mechanism for reducing the amyloid-β load. In guinea pig brain, ovariectomy increases 1.5 fold amyloid-β level whereas estrogen treatment significantly reverses this increase (Gandy and Duff 2000). Also in cell culture, estrogen significantly reduces the generation of amyloid-β peptide, suggesting that estrogen may prevent the accumulation of cerebral plaques in AD. Hosoda et al (2001) found that amino acid residues of 25 to 35 (amyloid-β 25-35) are more cytotoxic than the full-length protein (amyloid-β 1-40) and these residues induce DNA fragmentation typical for apoptosis. In addition, estrogen inhibits calcium influx and cytochrome C release induced by amyloid-β 25-35. Since these sequential events cause apoptosis, the protective effect of estrogen may be mediated through blockade of the mitochondrial apoptotic pathway induced by amyloid-β.

Overall the literature suggests that estrogen can help to reduce the incidence or severity of cognitive decline occurring during aging and neurodegenerative diseases. ERT improves cognitive function and reduces the risk of AD-type senile dementia in women. It regulates the enzymatic actions of crucial neurotransmitter systems including the cholinergic system, which is involved in higher cognitive functions. The nucleus basalis of Meynert, a distinct population of basal forebrain neurons, is the main source of cholinergic projections in the cerebral cortex and is primarily devastated in AD. This cholinergic deficit is associated with the decline in learning and memory in AD patients. Estrogen is reported to delay the decline in basal forebrain cholinergic functions during aging and neurodegenerative disorders and thereby reduces the risk and severity of AD. In addition to cholinergic neurons, serotonergic, noradrenergic and dopaminergic neurons show responsiveness to estrogen.

Estrogen is also involved in the modulation of expression of different genes associated with AD, viz., APP, presenilin (PS) and apolipoprotein (apo) E. The level of APP mRNA is higher in the brain of old than adult mice and it is downregulated by estrogen (Asaithambi et al 1999). Estrogen stimulates the formation of vesicles containing APP from the trans-Golgi network and reduces the secretion of amyloid peptide from neuronal cells (Greenfield 2002). It induces α -secretase activity which cleaves APP to produce 100 kDa non-amyloidogenic soluble APP (sAPPα) and reduces amyloid-β secretion via MAPK and PL-3 kinase signal transduction pathways (Goodenough et al 2003). Thus it influences the processing of APP and this effect is dependent on sex and age (Mani 2004). In the mouse brain, apoE has been implicated in neuronal protection and repair. Its level is higher in old than adult and it is upregulated by estrogen (Levin-Allerhand et al 2001). The level of PS mRNA and protein is also modulated by estrogen (Thakur et al 2004). All these experimental evidences support the role of estrogen in AD.

Selective estrogen receptor modulators (SERM)

ERT has been used to improve the impairments of functions of brain and other organs that result due to deficiency of estrogen with increasing age. However, recent studies caution against this therapy because it can increase the risk of breast and endometrial cancer in postmenopausal women (Thakur and Ghosh 2003). Particularly, the report of Women's Health Initiative of the National Institutes of Health is of much concern, though this study used horse estrogen (prempro) instead of human estrogen, and combined it with synthetic progesterone that did not act like natural progesterone. Nevertheless, to avoid any risk, a new class of compounds called SERM is being considered as a possible alternative to ERT. SERM are drugs that bind to ER and act like estrogen (agonists) in some tissues while oppose the action of estrogen (antagonists) in other tissues. For example, tamoxifen acts as an estrogen antagonist in the breast and is used to prevent and treat breast cancer, but it acts as an estrogen agonist in the endometrium where it can induce cancer.

Different SERM acts differently in different brain regions. For example, tamoxifen increases ERβ mRNA in hypothalamus whereas raloxifene increases it in amygdala (Zhou et al 2002). With the recognition that SERM have differential tissue-dependent effects on ER function, there is much interest in studying the effects of raloxifene, tamoxifen and other modulators on cognition and other brain functions that deteriorate during aging and AD (Yaffe 2001). Researchers have sought to determine whether raloxifene can provide beneficial cognitive effects. It was observed that the woman who received raloxifene showed improvement in delayed verbal recall, which usually declines in AD patients (Asthana et al 2001). However, in case of aged rhesus monkeys, Lacreuse et al (2002) found that estradiol but not raloxifene enhanced spatial working memory. The raloxifene is particularly interesting because, like estrogen, it improves lipid metabolism and reduces bone loss, without adverse effects in the breast or uterus. In addition, isoflavone phytoestrogens which show antiestrogenic effects on ER-dependent gene expression in the brain and estrogen-dependent behavior, may reduce the risk of cancer, osteoporosis and heart disease (Patisaul et al 2001). Therefore, isoflavone dietary supplements are becoming increasingly popular and are frequently advertised as natural alternatives to ERT.

Concluding remarks and future perspectives

Estrogen effects in the brain are complicated and multifaceted. The complexity is reflected at the molecular level with more genes being expressed in the brain than in any other tissues and a large number of genes being regulated by estrogen. The brain shows a highly complex pattern of estrogen signaling in a region-specific manner, and the detailed mechanism is poorly understood. Studies evaluating estrogen as a treatment for aging and AD have resulted with mixed findings. While a few studies have found modest improvements, others have largely been disappointing. Evidence from both animals and humans supports the role of estrogen as neuroprotective and neuroregenerative agents in stroke, traumatic brain injuries and neurodegenerativo diseases. Although estrogen appears to be beneficial, evidence that hormone therapy improves cognition or prevents dementia is limited, controversial and inconclusive. However, hormone replacement therapy is found to improve the overall quality of life and increases life expectancy for most menopausal women.

Majority of the studies on estrogen action in the brain suffer from substantial methodological problems that leave the findings in doubt. Well-controlled studies using larger sample sizes and for longer treatment durations are warranted to provide definite answers to many questions in the near future before the therapeutic potential of estrogen replacement for elderly women can be firmly established. The development of new ER ligands without side effects is also very valuable. This area of research demands attention since estrogen may be an important therapeutic agent in the maintenance of normal neural function after injury and during aging.

Acknowledgements
The author's research work is supported by financial assistance from the University Grants Commission, Council of Scientific & Industrial Research, Department of Biotechnology and Department of Science & Technology, Government of India.

References

Ábrahám IM, Han SK, Todman MG, Korach KS and Herbison AE (2003) J Neurosci **23**:5771-5777.
Adams MM, Fink SE, Shah RA, Janssen WGM, Hayashi S, Milner TA, McEwen BS and Morrison JH (2002) J Neurosci **22**:3608-3614.
Alkayed NJ, Goto S, Sugo N, John HD, Klaus J, Crain BJ, Bernard O, Traystman RJ and Hurn PD (2001) J.Neurosci **21**: 7543-7550.
Anderton BH (2002) Mech Age Dev **123**:811-817.
Asaithambi A, Mukherjee S and Thakur MK (1997) Biochem Biophys Res Commun **231**: 683-685.
Asaithambi A, Mukherjee S and Thakur MK (1999) Mol Biol Rep **26**: 179-184.
Asthana S (2003) J Gerontol **58A**:322-323.
Asthana S, Baker LD, Craft S, Stanczyk FZ, Veith RC, Raskind MA and Plymate SR (2001) Neurology **57**: 605-612.
Audesirk T, Cabell L, Kern M and Audesirk G. (2003) Neuroscience **121**: 927-934.
Behl C and Lezoualc'h F (1998) Restor Neurol Neurosci **12**:127-134.
Bi R, Foy MR, Thompson RF and Baudry M (2003) Neurobiol Aging **24**:977-983.
Birge SJ (2003) Clin Geriatr Med **19**: 617-627.
Brinton RD (2001) Learning & Memory **8**:121-133.
Cardona-Gomez GP, Mendez P, DonCarlos LL, Azcoitia I and Garcia-Segura LM (2001) Brain Res Rev **37**: 320-334.
Cardona-Gomez GP, Mendez P, DonCarlos LL, Azcoitia I and Garcia-Segura LM (2002) J Steroid Biochem Mol Biol **83**: 211-217.
Carlson C, Zandi PP, Plassman BL, Tschanz JT, Welsh-Bohmer KA, Steffens DC, Bastian LA, Mehta KM and Breitner JCS (2001) Neurology **57**:2210-2216.
Carlstom L, Ke ZJ, Unnerstall JR, Cohen RS and Pandey SC (2001) Neuroendocrinology **74**: 227-243.
Chakraborty TR, Ng L and Gore AC (2003) Endocrinology **144**: 4164-4171.
Cheng CM, Cohen M, Wang J and Bondy CA (2001) FASEB J **15**: 907-915.
Compton J, van Amelsvoort T and Murphy D (2001) Br J Clin Pharmacol **52**:647-653.
Cordey M, Gundimeda U, Gopalakrishna R and Pike CJ (2003) J Neurochem **84**:1340-1348.
Dubal DB, Zhu H, Yu J, Rau SW, Shughrue PJ, Merchenthaler I, Kindy MS and Wise PM (2001) Proc Natn Acad Sci (USA) **98**:1952–1957.
Eriksson A, Wahlestedt C and Nordqvist K (1999) Molec Brain Res **74**:91-97.
Foster TC, Sharrow KM, Kumar A and Masse J (2003) Neurobiol Aging **24**:839-852.
Gandy S (2003) Neurochem Res **28**:1003-1008.
Gandy S and Duff K (2000) Exp Gerontol **35**:503-511.
Gibbs RB (1999) Brain Res **844**:20-27.
Gibbs RB and Gabor R (2003) J Neurosci Res **74**:637-43.

Goodenough S, Schafer M and Behal C (2003) J Steroid Biochem Molec Biol**84**:301-305.
Greenfield JP, leung LW, Cai D, Kassik K, Gross RS, Rodriguez-Boulan E Greengard P and Xu H (2002) J Biol Chem **277**:12128-12136.
Henderson VW, Paganini-Hill A, Miller BL Elble RJ, Reyes PF, Shoupe D, McCleary CA, Klein RA, Hake AM and Farlow MR (2000) Neurology**54**:295-301.
Holschneider DP, Kumazawa T, Chen K and Shih JC (1998) Life Sci **63**:155-160.
Hosoda T, Nakajima H and Honjo H (2001) Neuroreport **12:**1965-1970.
Hrabovszky E, Kalló I, Hajszán T, Shughrue PJ, Merchenthaler I and Liposits Z (1998) Endocrinology **139**:2600-2604.
Hutchinson JB (1997) Cell Molec Neurobiol **17**:603-626.
Ishunina TA and Swaab DF (2001) Neurobiol Aging **22**:417-426.
Ivanova T, Karolczak M and Beyer C (2001) Brain Res **889**:264-269.
Kadish I and Groen TV (2002) J Neurosci **22**:4095–4102.
Kanungo MS, Patnaik SK and Koul O (1975) Nature **253**:366-367.
Kanungo MS and Thakur MK (1979a) Biochem Biophys Res Commun **86**:14-19.
Kanungo MS and Thakur MK (1979b) J Steroid Biochem **11**:879-888.
Karolczak M and Beyer C (1998) Neuroendocrinology**68**: 229-234.
Katzenellenbogen BS, Motano MM, Ediger TR and Sun J (2000) Rec Prog Hor Res**55**:163-193.
Kim H, Bang OY, Jung MW, Ha SD, Hong HS, Huh K, Kim SU and MookJung I (2001) Neurosci Lett **302**:58-62.
Kumar RC and Thakur MK (2004) Neurobiol Aging **25**: 925-933 .
Kuppers E, Engele J and Beyer C (2001) J Neurosci Res**66**: 221-230.
Lacreuse A, Wilson ME and Herndon JG (2002) Neurobiol Aging**23**:589-600.
LeBlanc ES, Janowsky J, Chan BKS and Nelson HD (2001) JAMA **285**:1489–1499.
Lee CK, Weindruch R and Prolla TA (2000) Nat Genet**25**:294-297.
Levin-Allerhand J, McEwen BS, Lominska CE, Lubahn DB, Korach KS and Smith JD (2001) J Neurochem **79**:796-803.
Li R, Shen Y, Lang LB, Lue LF, Finch C and Rogers J (2000) J Neurochem **75**: 1447-1454.
Luine VN and Rhodes JC (1983) Neuroendocrinology**36**:235-241.
Lu NZ, Shales TA, Gundlah C, Dziennis SE, Lyle RE and Bethea CL (1999) Endocrine **11**:257-267.
Mani ST (2004) Ph.D thesis submitted to Banaras Hindu University, Varanasi
Mani ST, Kumar RC and Thakur MK (2001) Neurosci Lett**308**:57-59.
McCullough LD and Hurn PD (2003) Trends Endocrinol Metab **14**:228-235.
McEwen BS (2001) J Appl Physiol **91**:2785-2801.
McEwen BS (2002) Recent Prog Horm Res **57**:357-384.
McKenna NJ, Lanz RB and O'Malley BW (1999) Endocrine Rev **20**: 321-344.
McNeil Am, Zhang C, Stanczyk FZ, Duckles SP and Krause DN (2002) Stroke **33**:1685-1691.
Meda C, Vegeto E, Pollio G, Ciana P, Patrone C, Pellicciari C and Maggi A (2000) J Neuroendocrinol **12**:1051-1059.
Mitev YA, Wolf SS, Almeida OFX and Patchev VK (2003)FASEB J **17**:518-519.
Mukherjee S, Asaithambi A and Thakur MK (1999) Mech Age Dev **111**:13-22.
Mulnard RA, Cotman CW, Kawas C, van Dyck DC, Sano M, Doody R, Koss E, Pfeiffer E, Jin S and Gamst A(2000) J Amer Med Assoc **283**:1007-1015.
Nilsson S and Gustafsson JA (2000) Breast Cancer Res **2**:360-366.
Nomura M, Korach KS, Pfaff DW and Ogawa S (2003) Molecular Brain Res **110**:7-14.
Patisaul HB, Dindo M, Whitten PL and Young LJ (2001) Endocrinology**142**: 2946-2952.
Pentanceska SS, NagyV, Frail D and Gandy S (2000) Exp Gerontol**35**: 1317-1325.
Pettersson K, Grandien K, Kuiper GGJM and Gustafsson JA (1997) Mol Endocrinol**11**:1486-1496.
Philips SM and Sherwin BB (1992) Pschyconeuroendocrinology**17**:485-495.
Ping SE, Greferath U and Barrett GL (2002) J Neurosci Res **69**:51-60.
Platania P, Laureanti F, Bellomo M, Giuffrida-Stella AM, Catania MV and Sortino MA (2003) Neuroendocrinology **77**: 334-340.
Rapp PR, Morrison JH and Roberts JA (2003) J Neurosci **23**: 5708-5714.
Savskam E (2002) XX vs. XY **1**:32-38.
Shah RD, Anderson KL, Rapoport M and Ferreira A (2003) Mol Cell Neurosci**24**:503-516.
Sharma PK, Ghosh S, Mani ST and Thakur MK (2003) In: Int Symp on Building the Brain, National Brain Research Center, Haryana, 15-17 Dec, pp 48.

Sherwin BB (2003) Endocr Rev **24**:133-151.
Shughrue PJ and. Merchenthaler I (2003) Neuroscience **116**: 851-861.
Singh M, Sétáló G J, Guan X, Frail DE, and Toran-Allerand DC (2000) J Neurosci **20:** 1694-1700.
Solum DT and Handa RJ (2002) J Neurosci **22**:2650-2659.
Stein DG (2001) Trends Neurosci **24**: 386-91.
Thakur MK (1988) Mech Age Dev **45**:93-110.
Thakur MK (1999) J Anti-Aging Med **2**:127-132.
Thakur MK (2000) Curr Sci **79**: 101-108.
Thakur MK (2002) In: Alzheimer's Disease in India (ed. K Bagchi) pp 83-103, New Delhi.
Thakur MK (2003a) In: Modulating aging and longevity (ed SIS Rattan) pp 219-238, Kluwer Academic Publisher, Netherlands.
Thakur MK (2003b) Proc Ind Natn Sci Acad (special issue on Aging) **B69**: 179-190.
Thakur MK, Asaithambi A and Mukherjee S (1999) Mol Biol Rep **26**:239-347.
Thakur MK, Asaithambi A and Mukherjee S (2000) Mol Cell Biochem **23**:95-101.
Thakur MK, Das R and Kanungo MS (1978) Biochem Biophys Res Commun **81**:828-831.
Thakur MK and Ghosh S (2003) Aging and Society **13**:78-102.
Thakur MK and Kanungo MS (1981) Exp Gerontol **16**:331-337.
Thakur MK, Sharma PK, Ghosh S and Mani ST (2004) Geriat Gerontol Int **4**: S259-S261.
Thal LJ, Thomas RG, Mulnard R, Sano M, Grundman M, Schneider L (2003) Arch Neurol **60**:209-212.
Wang L, Andersson S, Warner M and Gustafsson JA (2001) Proc Natn Acad Sci (USA) **98**:2792-2796.
Wang L, Andersson S, Warner M and Gustafsson JA (2003) Proc Natn Acad Sci (USA) **100**:703-708.
Warner M, Nilsson S and Gustafsson JA (1999) Curr Opin Obs Gyn **11**:249-254.
Watanabe T, Inoue S, Hiroi H, Orimo A and Muramatsu M (1999) Brain Res Mol Brain Res **63**:375-379.
Weihua Z, Andersson S, Cheng G, Simpson ER, Warner M and Gustafsson JA (2003) FEBS Lett **546**:17-24.
Wilson ME, Rosewell KL, Kashon ML, Shughrue PJ, Merchenthaler I and Wise PM (2002) Mech Age Dev **123**:593-601.
Wise PM and Dubal DB (2000) Biol Reprod **63**: 982-985.
Wolf OT and Kirschbaum C (2002) Horm Behav **41**:259-266.
Yaffe K (2001) Ann N Y Acad Sci **949**:215-222.
Yaffe K, Sawaya G, Lieberburg I and Grady D (1998) JAMA **279**: 688-695.
Zhang JQ, Cai WQ, Zhou de S and Su BY (2002) Brain Res **935**: 73-80.
Zhang L, Rubinow DR, Xaing G, Li BS, Chang YH, Maric D, Barker JL and Ma W (2001b) Neuroreport **12**: 1919-1923.
Zhang Y, Tounekti O, Akerman B, Goodyer CG and LeBlanc A (2001a) J Neurosci **21**:RC176.
Zhou W, Koldzic-Zivanovic N, Clarke CH, Beun R de, Wassermann K, Bury PS, Cunningham KA and Thomas ML (2002) Neuroendocrinology **75**:24-33.

Molecular and Cellular Neurobiology
Editors: M.K. Thakur and S. Prasad

Neurotransmitter Receptor Gene Expression: Insulin Secretion and Cell Proliferation

C.S. Paulose, T.R. Renuka and P.N. Eswar Shankar

Molecular Neurobiology and Cell biology Unit, Centre for Neuroscience, Department of Biotechnology, Cochin University of Science and Technology, Cochin-682 022, India

Abstract

Brain neurotransmitters and their receptors play an important role in regulating various cellular activities of an organism. The consequences of the neurotransmitter-receptor function can influence the regulation of metabolic manifestations in hypothyroidism, hypertension, diabetes and cell proliferation directly by central nervous system function or through the hypothalamic-pituitary-end organ axis. The functional difference of neurotransmitters and hormones through their receptor subtypes can lead to differential gene expression. Hormones such as insulin, glucagon, thyroxine, tri-iodothyronine, glucocorticoids function as growth regulators. These hormonal functions can be dictated by the neurotransmitters and their receptors. Neurotransmitters- Norepinephrine, Dopamine, Serotonin, Gamma aminobutyric acid, Acetylcholine and their receptor subtypes studies at the molecular level have shown that the functional difference of these neurotransmitters through their receptor subtypes can control insulin synthesis and release, cell proliferation and aging. Thus during developmental periods, neurotransmitter - hormonal function is critical for the cell proliferation, cell differentiation and functional integration of tissues and organs.

Introduction

Neurotransmitters like norepinephrine (NE), serotonin (5-HT), dopamine (DA), gamma aminobutyric acid (GABA) and acetylcholine (ACh) are involved in the regulation of endocrine function through the mediation of hypothalamus and pituitary. The hypothalamus is the main endocrine centre in the brain where the hormones of the target glands interact with the neurotransmitters (Beilezikian, 1987; Paulose et al 1984, 1999). Neurotransmitter acts by binding to specific membrane bound receptors and neurotransmitter receptor studies revealed that the functional regulation of various neurotransmitters likes NE, GABA, 5-HT, DA and ACh have important role in cell proliferation and insulin secretion (Fig. 1).

Neurotransmitters and Insulin Secretion

The pancreatic islets are innervated by parasympathetic, sympathetic and sensory nerves. Several neurotransmitters are stored within the terminals of these nerves. The preganglionic fibres of the parasympathetic system originate from perikarya located in the dorsal motor nucleus of the vagus and possibly also in the nucleus ambigus (Ionescu et al 1983; Sharkey & Williams, 1983; Ahren, 1986; Louis-Sylvestre, 1987; Chen et al 1996). The sympathetic innervation of the pancreas originates from preganglionic perikarya located in the thoracic and upper lumbar segments of the spinal cord (Furuzawa et al 1996; Gilon & Henquin, 2001). Stimulation of autonomic nerves and treatment with neurotransmitters affect islet hormone secretion (Ahren, 2000). Glucose is an important regulator of various β-cell processes including insulin biosynthesis and release. Glucose, over short intervals stimulates insulin biosynthesis at the level of translation (Permut et al

1972). Studies of insulin gene expression in primary cultures of rat islets transfected Insulin I gene 5'-flanking sequence suggested that metabolic signal from glucose influx is transmitted through the insulin enhancer (German et al 1990). Glucose induced insulin secretion is modulated by neural, hormonal and paracrine factors (Porte et al 1975). Catecholamines (CA) exert a direct and dual effect on the β-cell to induce either inhibition or stimulation of insulin secretion through their interaction with their receptors. The pancreatic islet β-cells have more α_2 than α_1 adrenergic receptors (Ahren et al 1984, 1986, 2000). Although the effect of the sympathetic nervous system (SNS) and of circulating CA on islet physiology has been studied in both normal and pathological states, little is known about the possible participation of endogenously-generated CA in the control of islet function. Islet cells have been shown to contain enzymes involved both in the synthesis of CA – tyrosine hydroxylase (TH) and dihydroxyphenylalanine (DOPA) decarboxylase and in their inactivation – monoamine oxidase (Alpert et al 1987; Hanahan, et al 1993; Thibault et al 1993; Gagliardino et al 1997)

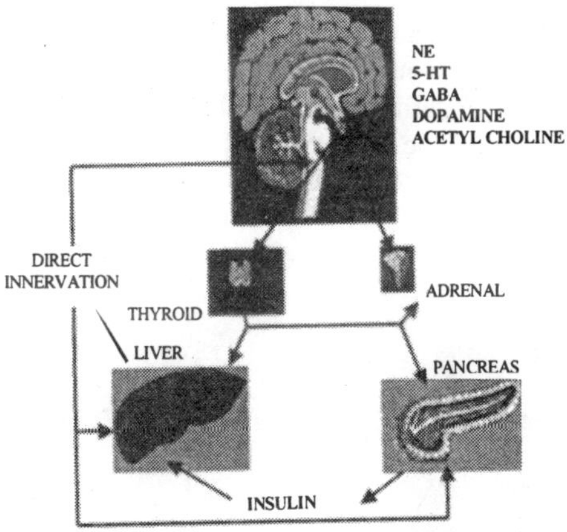

Fig. 1. Neurotransmitter receptor regulation of cell proliferation and insulin secretion.

Norepinephrine is the principal neurotransmitter of sympathetic nervous system. It inhibits insulin secretion, both *in vivo* and *in vitro* (Renstrom et al 1996; Porte, 1967) and exerts opposite effects on peripheral glucose disposal and glucose stimulated insulin secretion (Avogaro et al 1996). Norepinephrine and epinephrine (EPI), the flight and fright hormones, are released in all stress conditions and are the main regulators of glucose turnover in strenuous exercise (Simartirkis et al 1990). Neurotransmitter receptor studies (Dakshinamurti, et al 1985, 1988; Viswanathan et al 1988, 1990; Paulose & Dakshinamurti 1985) and its regulation of hypothyroidism (Dakshinamurti et al 1986; Tessy et al 1997) leading to sympathetic stimulation and hypertension in pyridoxine deficient rats (Paulose et al 1988 & Dakshinamurti et al 1990a, b) which in turn lead to diabetes have been reported. In severe insulin-induced hypoglycaemia, a 15 to 40 fold increase of epinephrine plays a pivotal role in increasing glucose production independently of glucagon (Gauthier et al 1980).

In addition, it has been reported that epinephrine enhances glycolysis through an increased activation of phospho-fructokinase. In humans, EPI stimulates lipolysis, ketogenesis, thermogenesis and glycolysis and raises plasma glucose concentrations by stimulating both glycogenolysis and gluconeogenesis. EPI is, however, known to play a secondary role in the physiology of glucose counter-regulation. Indeed, it has been shown to play a critical role in one pathophysiological state, the altered glucose counter-regulation in patients with established insulin-dependent diabetes mellitus (Cryer, 1993). The inhibitory effect of EPI upon insulin secretion induced by glucose was reported by Coore and Randl (1964), who incubated pancreatic tissue from the rabbit. As judged by Malaisse, et al (1967), the inhibitory effect of EPI on glucose-induced insulin secretion is mediated through the activation of α-adrenoreceptors. Studies reported (Ani, 2000) that central α_1 adrenergic receptor gene expression increased and α_2 adrenergic receptor gene expression decreased in partially pancreatectomised rats during pancreatic regeneration when the insulin secretion is maximum (Fig. 2).

Serotonin is an endogenous amine involved in diverse biological processes within the central and peripheral nervous system and the cardiovascular and gastrointestinal and respiratory systems (Fozard, 1989; Hindle, 1994). This diversity of actions is made possible because of the existence of specific 5-HT cell surface receptor subtypes and their coupling to distinct intracellular messenger systems or ion channels (Hoyer et al 1994). Serotonin content is decreased in the brain regions during diabetes (Jackson & Paulose, 1999; Sumiyoshi et al 1997; Sandrini et al

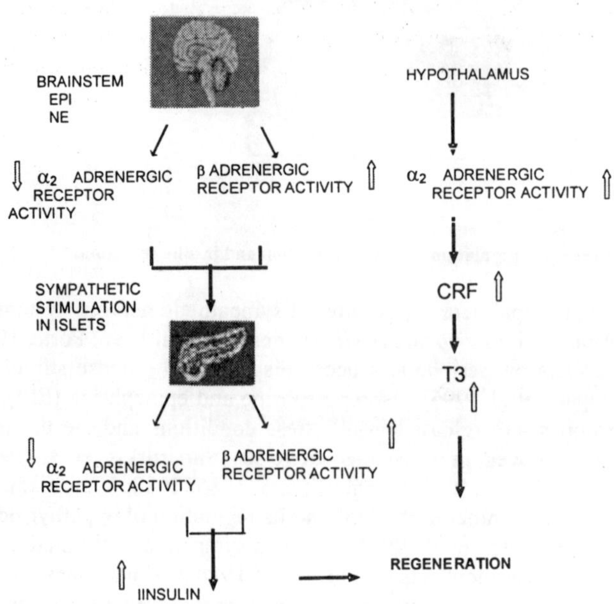

Fig 2. Adrenergic regulation of pancreatic regeneration and insulin secretion.

1997) but there are reports suggesting an increase in brain 5-HT content during diabetes (Chen & Yang, 1991; Lackovic et al 1990). Ohtani et al (1997) have reported a significant decrease in extracellular concentrations of NE, 5-HT and their metabolites in the ventro medial hypothalamus (VHM). The ratio of 5-HIAA/5-HT was increased. A similar observation was reported by Ding, et al (1992) with a decrease in 5-HT in cortex (19%) and 5-HT turnover (5-HIAA/5-HT) that increased by 48%. Chu et al (1986) has reported lower 5-HT levels in both hypothalamus and brain stem but not in corpus striatum. Insulin treatment brought about an increase in the cerebral concentration of 5-HIAA and accelerated the cerebral 5-HT turnover (Juszkiewicz, 1985). The 5-HIAA concentration was reported to be approximately twice as high as the controls regardless of duration of treatment. Brain tryptophan, the precursor of 5-HT, was also reduced in brain regions during diabetes (Jamnicky et al 1991).

Insulin treatment was reported to reverse this reduced tryptophan content to normal (Jamnicky, et al 1999). 5-HT_{2A} receptors are upregulated in the brain stem of streptozotocin (STZ) induced diabetic rats. In the cerebral cortex the affinity of these receptors increased (Jackson & Paulose, 1999, 2001). There was a significant increase in 5-HIAA observed at 2-6 hours after insulin administration (Kwok & Juorio, 1987). Affinity of the serotonergic receptors was reported as decreased in the STZ induced diabetic rats (Pius & Paulose, 1999).

Gamma aminobutyric acid is the main inhibitory neurotransmitter in the central nervous system. GABA is reported to be present in the endocrine pancreas at concentrations comparable with those found in central nervous system. The highest concentration of GABA within the pancreatic islet is confined to β-cells (Sorenson, 1991). Glutamate decarboxylase, the primary enzyme that is involved in the synthesis of GABA, has been identified as an early target antigen of the T-lymphocyte mediated destruction of pancreatic β-cells causing insulin-dependent diabetes mellitus (Baekkeskov, 1990). Glutamate dehydrogenase activity is reported to be increased in the brain of diabetic rats (Biju & Paulose, 1998) and the combined administration of pyridoxine and insulin found to be more effective in the control of diabetes (Aswathy, et al 1998). GABA, through its receptors, has been demonstrated to attenuate the glucagon and somatostatin secretion from pancreatic α-cells and δ-cells respectively (Gaskins, 1995). It is present in the cytoplasm and in synaptic-like microvesicles (Reetz, 1991) and is co-released with insulin from β-cells in response to glucose. The released GABA inhibits islet α and δ-cell hormonal secretion in a paracrine manner. During diabetes the destruction of β-cells will lead to decrease in GABA release resulting in the enhancement of glucagon secretion from α-cells leading to hyperglycaemia. The brain GABAergic mechanisms also play an important role in glucose homeostasis. Inhibition of central $GABA_A$ receptors increases plasma glucose concentration (Lang, 1995). Thus, any impairment in the GABAergic mechanism in the central nervous system and/or in the pancreatic islets is important in the pathogenesis of diabetes.

Cholinergic system plays an important role in physiological and behavioural functions. ACh acts by binding to specific membrane receptors and can be divided into muscarinic and nicotinic receptors. Cholinergic stimulation of pancreatic β-cells increases insulin secretion (Kaneto et al 1967). This effect is mediated by muscarinic receptors (Grill & Ostenson, 1983; Henquin & Nenquin, 1988) and is dependent on extracellular glucose concentration (Henquin et al 1988). Acetylcholine stimulated insulin secretion coupling is mediated by complex mechanisms of signal transduction. It has been proposed that ACh activates phospholipid turnover and thereby increases the intracellular calcium level. Normal β-cells voltage-dependent sodium channels are important for membrane depolarisation. ACh increases sodium influx in to the cells (Henquin et al 1988). ACh hyperpolarises the cell by increasing potassium permeability. Quist (1982) reported that cholinergic agonist, carbachol, causes Ca^{2+}-dependent stimulation of phosphate incorporation into phosphatidyl inositol phosphates in the canine heart. Muscarinic M_1 and M_3 receptors were found to be stimulatory to insulin secretion in rat pancreatic islets *in vitro* (Renuka, 2003).

The muscarinic acetylcholine receptors are widely distributed throughout the body and subserve numerous vital functions in both the brain and autonomic nervous system (Hassal et al 1993). Central muscarinic acetylcholine receptors regulate secretion of insulin from pancretatic islets and maintain normal glucose level. When carbachol, muscarine, bethanechol, methacholine or neostigmine was injected into the third cerebral ventricle, it caused a dose-dependent increase in the hepatic venous plasma glucose concentration. However, in the case of 1, 1-dimethylphenyl-4-piperazinium iodide (DMPP) or nicotine, the level of hepatic venous glucose did not differ from that of the saline-treated control rats. The increase in glucose level caused by neostigmine was dose-dependently suppressed by co-administration of atropine. These facts suggest that cholinergic activation of muscarinic receptors in the central nervous system plays a role in increasing hepatic glucose output. Neostigmine-induced increments in glucose did not occur in adrenalectomized rats. This suggests that the secreted epinephrine acts directly on the liver to increase hepatic glucose output (Iguchi et al 1986). The injection of adrenaline and carbachol into the third cerebral ventricle resulted in a marked hyperglycaemia associated with increased immunoreactive glucagon. Adrenaline-induced hyperglycaemia was not affected by bilateral adrenalectomy, while carbachol-induced hyperglycaemia was completely inhibited by adrenalectomy (Iguchi et al 1985). eostigmine-induced epinephrine and glucagon secretion and increased hepatic glucose output stimulated by direct neural innervation to liver is mediated by central muscarinic receptor in fed rats (Iguchi et al 1990).

The ventro medial hypothalamus, lateral hypothalamus, paraventricular hypothalamus and median site of the lateral-preoptic area were involved in increasing the plasma levels of glucose and epinephrine by cholinergic stimulation (Honmura et al 1992). Studies by Iguchi et al (1992) suggest that the glucoregulatory hippocampal activity evoked by the acetylcholine esterase inhibitor, neostigmine transmitted to peripheral organs *via* the ventromedial hypothalamus. Takahashi et al (1993) reported that neostigmine induced hyperglycaemia affects not only the cholinergic system but also the noradrenergic and dopaminergic systems in the hypothalamus (Takahashi et al 1993). Muscarinic cholinergic system is reported to participate in the $HgCl_2$-induced central hyperglycaemic effect through the function of the adrenal medulla. Norepinephrine and dopamine content were found to be decreased suggesting that their neurons may also be related to hypothalamic glycoregulation (Takahashi et al 1994).

Cholinergic synapses in the ventromedial hypoythalamus participate in a central glucoregulatory system that increases hepatic glucose production mainly through a stimulation of adrenal medulla epinephrine secretion (Brito et al 1993). Insulin-induced hepatic glucose uptake depends on the sensing by muscarinic, intra-hepatic nerves of a glucose concentration gradient between portal vein and hepatic artery. The function of these intrahepatic nerves is impaired in diabetic animals (Stumpel et al 1998). Insulin partly reversed the changes observed in the STZ-treated rats. There was a decrease in the muscarinic receptor number and axonal transport of receptor-bound opiate in STZ induced hyperglycaemia suggesting that impaired axonal transport of receptors partly involved in the neurological disturbance which is seen in diabetic patients (Laduron & Janssen, 1986). Muscarinic receptor number increased in the pancreatic islets of diabetic rats. Cholinergic-induced insulin release was also higher in STZ induced diabetes than in normal islets (Ostenson & Grill, 1987).

Dopamine is a major neurotransmitter in the central nervous system, and its receptors are associated with a number of neuropathological disorders such as Parkinson's disease and Schizophrenia. It also plays a major role in the regulation of appetite and growth hormone. Dopamine and diabetes mellitus are reported to have close link between them. Studies on the effect of dopamine has revealed that co-administration of dopamine D_1 and D_2 agonists additively inhibits the feeding effect and body weight as a result of their combined activation mediated by the action of hypothalamic neuropeptide (NPY). This reveals the efficiency in which D_1/D_2 agonist combination improves hyperphagia in diabetic animals (Kuo, 2002). The allelic variants of dopamine receptor D_2 locus is reported to be associated with weight and height with a linkage disequilibrium with allelic variants of the DRD_2 gene that play a major role in the regulation of weight (obesity) and height that serves as a risk factor in late-onset non-insulin-dependent diabetes mellitus (NIDDM) (Gysin et al 1993).

Diabetes is reported to damage dopaminergic function as a result of hyperglycemia. Insulin pathways in the brain may play an important role in regulating dopamine transporter (DAT) activity (France et al 2003). The DAT regulates extracellular DA levels and during diabetes there is a significant decrease in their number as a result of hypoinsulinemia, which damages the dopaminergic activity (Figlewicz et al 1996). It is reported that midbrain dopamine neurons, implicated to be critical in the mediation of motivational and reward aspects of stimuli, are affected by alterations in insulin levels. As an approach to evaluate this hypothesis, double-labeling fluorescence immunohistochemistry was used to determine whether the midbrain dopamine neurons express insulin receptors. Extensive co-expression of tyrosine hydroxylase (a marker for dopamine neurons) with insulin receptor was observed in the ventral tegmentum and substantia nigra. These findings suggest that midbrain dopamine neurons are direct targets of insulin and they participate in mediating the effects of these hormones on reward-seeking behaviour (Figlewicz et al 2003).

Experiments carried out in the striata of diabetic rats on the behavioural response, dopamine metabolism and characteristics of dopamine subtypes revealed that diabetes caused an increase in the on-set and duration of cataleptic behaviour. Dopamine metabolites, dihydroxyphenylacetic acid (DOPAC) and homovanillic acid (HVA), were significantly reduced in the striata of hyperglycaemic rats while level of DA was significantly increased. The ratio of DOPAC and HVA to DA were decreased, suggesting decreased turnover of DA. The affinity of striatal D_1 receptors was significantly increased without changes in the number of binding sites, while the maximum binding number of D_2 receptors was significantly increased without affecting its affinity in the diabetic rats (Ho et al 1995). Diabetes causes a significant increase in locomotary activities as a result of decreased dopamine D_1 receptors. This is believed to be the major cause of hyporesponsiveness leading to dopamine-related locomotor impairment (Kamei et al 1998). Dopamine receptor changes have been established in various neuroleptic disorders (Chiu et al 1981 a, b).

The CNS monoamine cell groups that project to the pancreatic parasympathetic preganglionic neurons have been identified with the use of that originate exclusively from the dorsal vagal motor nucleus and some of these are found to be dopamine neurons because as they were tyrosine hydroxylase immunopositive. Transneuronally labelled aminergic neurons are also found throughout the medulla oblongata (Loewy et al 1994). Electron microscopic autoradiography studies using tritiated compounds have revealed that $[^3H]$ DA formed from administered $[^3H]$ DOPA are present in the beta cells of the islet. In the β-cells the $[^3H]$ DA-grains were observed to be associated with the secretory granules.

The pancreatic β-cells have ability to store substantial amounts of calcium dopamine and serotonin apart from epinephrine and norepinephrine (Ahren et al 1985, 1989). Dopamine accumulation is also observed to inhibit the insulin secretion in the pancreatic islets, which involves interference with a calcium translocation. This is a consequence of a complex interaction between the accumulated dopamine and a pool of Ca^{2+} mainly confined to the secretory granules. Dopamine accumulation initially causes a transient increase in cytosolic Ca^{2+} accompanied by insulin release. The increasing cytosolic Ca^{2+} as a result of dopamine accumulation makes the cell more sensitive to a concomitant stimulation with glucose and the release of insulin is triggered. A long-term dopamine accumulation on the other hand decreases the granular Ca^{2+} pool inhibiting the insulin release. Thus, studies also suggest an extra-neuronal source of dopamine in addition to its occurrence in adrenergic nerves, which effects insulin secretion.

Dysfunction of pancreatic islets plays a crucial role in the etiology of diabetes. Chronic hyperglycaemia or hyperlipidaemia impairs islet function. Studies have demonstrated that dopaminergic agonists ameliorated hyperglycaemia and hyperlipidaemia in obese and diabetic rodents. The effect of dopamine D_2/D_1 receptor agonists (bromocriptine/SKF38393, BC/SKF) on islet dysfunction in diabetic mice markedly reduced hyperglycaemia and hyperlipidaemia, and significantly improved isle dysfunction demonstrated by an increase of secretagogue-stimulated insulin release from islets of diabetic mice (Cincotta et al 1998, 2001; Boiadzhieva. 1990). These actions are mediated via alterations in the hypothalamic-neuroendocrine axis, which drives metabolic changes in peripheral tissues leading to a marked reduction in hyperglycemia and hyperlipidemia and corrects autonomic control of islet function. Thus the systemic sympatholytic dopaminergic therapy that attenuates hyperglycemia and hyperlipidemia improves islet function in ob/ob mice by improving aberrations in the α-cell glucose-sensing apparatus, enhancing insulin storage and/or retention, and stabilizing hyperplasia, thus reducing basal insulin levels reducing the β-cell hyperplasia. These studies underline the importance of dopamine as a neuromediator in the regulation of pancreatic insulin secretion.

In the exocrine pancreas studies have sited dopamine receptors involved in the stimulation of cellular cyclic AMP. Dopamine is reported to elicit concentration-dependent stimulation of cellular cyclic AMP with a maximal increase occurring at a concentration of 0.1 mM (EC_{50}=1μM). Studies with various agents of dopamine depended on their affinities to stimulate cellular cyclic AMP formation or to inhibit dopamine-stimulated cellular cyclic AMP formation by inhibiting the binding of [^{3}H] dopamine to pancreatic acini, providing evidence that dopamine binding sites are receptors that mediate the action of dopamine on cAMP accumulation (Ribet et al 1982, 1986). Thus, he central dopamine inhibits pancreatic exocrine secretion via D1-like receptors and that the inhibitory effect is mediated via sympathetic nerves, especially α-adrenoceptors. The dopamine D_2 receptors increased in the corpus striatum and cerebral cortex but decreased in the hypothalamus and brain stem indicating their involvement in regulating insulin secretion (Eswar, 2003).

The dopamine D_2 receptors increased in the corpus striatum and cerebral cortex but decreased in the hypothalamus and brain stem indicating their involvement in regulating insulin secretion. The mechanism of stimulation of insulin at low concentrations and the inhibition has not been well established. Further studies in this aspect with a focus on the second messenger system involved would help in the elucidating the role of dopamine and its various agonists on insulin secretion. Role of dopamine through it D_2 receptors is reported to mediate mitogenesis. Dopamine D_2 receptors are members of the G protein-coupled receptor superfamily and are expressed on both neurons and astrocytes. It has been observed that in rat C6 glioma cells stably expressing the rat D2L receptor, dopamine (DA) can activate both the extracellular signal-regulated kinase (ERK) and c-Jun NH2-terminal kinase (JNK) pathways through a mechanism involving D_2 receptor-G protein complexes and the Ras GTP-binding protein. Agonist binding to D_2 receptors rapidly activated both kinases within a short time demonstrating the role of D_2 receptor-stimulated MAPK pathways (Roth et al 1998). Similarly a thorough study and research in the pancreatic islets on the involvement of dopamine and its receptors in cell proliferation will be of immense clinical significance in the treatment of diabetes.

Neurotransmitters and Cell Proliferation

Neurotransmitters stimulate or inhibit cell proliferation in non neuronal cells by activating receptors coupled to various second messenger pathways (Kluess et al 1991). Norepinephrine is reported to amplify the mitogenic signals of both EGF and hepatocyte growth factor (HGF) by acting through the α_1 adrenergic receptor. It induces the production of EGF and HGF at distal sites and also enhances the response to HGF at target tissues (Broten et al 1999). Norepinephrine rises rapidly in the plasma within one hour after partial hepatectomy (Knopp et al 1999). It also suppresses the mito-inhibitory effects of transforming growth factor β_1 (TGF-β1) on cultured hepatocytes isolated from the early stages of regeneration (Michalopoulose & DeFrancis, 1997). Prazosin, a specific antagonist of α_1-adrenergic receptor, as well as sympathetic denervation greatly decrease DNA synthesis at 24 hours after partial hepatectomy (Cruise et al 1989). Addition of NE to hepatocytes stimulates Ca^{2+} mobilisation or phosphoinositol turnover and either or both of these processes was proposed to be involved in the mitogenicity of NE (Exton, et al 1981 1988; Nagano et al 1999). Rat hepatomas lacked the α_{1A} and α_{1B} mRNA and receptor binding, while in the human hepato-cellular carcinoma cell line, HepG2, their expression is high but they lack receptor binding (Kost et al 1992). Hepatic neoplasm are characterised by an increase in α_2 and β adrenergic receptors and a concomitant decline in α_1 receptors (Sanae et al 1989). Studies have shown that proliferation and insulin secretion of foetal rat β-cells could be significantly suppressed by α-adrenergic stimulation. When α-adrenergic agonists were given together with Sp-cAMP[S] or to pertussis toxin-pretreated islets, the suppressed β-cell proliferation and insulin secretion were partially prevented, suggesting that α adrenergic stimulation represses β-cell growth and hormone release in part by interfering with GTP binding proteins that connect cell surface receptors to adenylate cyclase (Sjoholm, 1991).

Serotonin has been implicated as a potential mitogen (Seuwen & Pouyssegur, 1990) and was shown to have effects on morphogenesis and neuronal development (Lauder, 1990). 5-Hydroxytryptamine has been recognised to cause proliferation of a variety of cells in culture including vascular smooth muscle cells and hepatocytes (Fanburg & Lee, 1997).

5-Hydroxytryptamine mediates mitogenic effects in many cell types (Garnovskaya, et al 1996 & Cowen et al 1996). The mitogenic action of 5-HT, first identified in bovine aortic smooth muscle cells by Nemeek et al (1986), may bear a relationship to the stimulatory effect of 5-HT on neuroembryogenesis. In cultured rat pulmonary artery smooth muscle cells (SMC), 5-HT induces DNA synthesis and potentiates the mitogenic effect of platelet-derived growth factor (Eddahibi, et al 1999). 5-Hydroxytryptamine's effects on cell proliferation may involve the phosphorylation of GTPase-activating protein (GAP), an intermediate signal in 5-HT-induced mitogenesis of SMC (Lee et al 1997).

In pancreatic cell line, activation of pertussis toxin insensitive 5-$HT_{1A/1B}$ receptors stimulate proliferation through the activation of PLC and PKC that resulted in the down regulation of cAMP (Ishizuka et al 1992). 5-HT_{1A} receptor agonist 8-OHDPAT inhibited the DNA synthesis in rat hepatocytes *in vitro*. Studies using mesulergine, 5-HT_{2C} antagonist revealed that 5-HT_{2C} receptors are stimulatory to hepatocyte cell division. RT-PCR studies revealed that 5-HT_{1A} receptor mRNA decreased in the brain regions and liver during liver regeneration (Pyroja, 2002).

There is a synergistic effect of 5-HT with more traditional protein growth factors, such as platelet derived growth factor, fibroblast growth factor, and insulin like growth factor and with ADP, ATP, thromboxaneA2 (Crowley et al 1994, Stroebel & Groppelt-Struebe, 1994). 5-HT in concentrations as low as 0.1–1 mM stimulates both proliferation and hypertrophy of SMC in culture. Furthermore, the mitogenic action of 5-HT is synergistic with that of conventional peptide growth factors. All agents that block transport of 5-HT block the proliferative response (Lee et al 1991, 1994).

The biological mechanism used by 5-HT to change cell morphology and induce proliferation may directly target the cytoskeleton. The main component of the cytoskeleton is microtubules which gives cells their shape. These microtubules consist of long polymers of tubulin, which spontaneously depolymerise if they are not actively polymerising (Mitchison & Kirschner, 1984). Tan and Lagnado (1975) found effects of 5-HT and related indole alkaloids on brain microtubular proteins. Several years later, it was found that 5-HT is taken up by endothelial cells and binds to stress fibers (Alexander, et al 1987). Here 5-HT induces actin polymerisation and affects changes in the cytoskeleton. Thus, there is evidence that 5-HT has a direct role in regulating and maintaining microtubules and microfilaments. The changes reported in 5-HT induced cytoskeletal stability may be partially mediated by microtubule-associated proteins (MAPs). MAPs serve to stabilize the cytoskeleton by binding to tubulin polymers and inhibiting their depolymerisation. In undifferentiated human neuroblastoma cells (LAN-5), high levels of 5-HT (50 μM) induce a decrease while low levels of 5-HT (50 nM) induce an increase in the cytoplasmic tau protein, a MAP found in high concentrations in the axon.

The effects of 5-HT as a mitogen and/or growth factor have been documented as a part of neuromodulator substances acting via G protein coupled receptor signaling pathway (Seuwen et al 1988, 1990; Julius, 1991). Specifically, 5-HT has been shown to increase DNA synthesis in rat pulmonary vascular smooth-muscle cells in culture (Pitt, et al 1994). Introduction of functional 5-HT_{2A} receptor and 5-HT_{2C} receptor into NIH 3T3 cells results in generation of transformed foci at high frequency (Julius et al 1989). The long-term maintenance of the transformed state requires continued activation of these 5-HT receptors, indicating that they may represent conditional proto-oncogenes (Julius, 1991). 5-HT is mitogenic for bovine pulmonary artery SMC producing both hyperplasia and hypertrophy through its action on a 5-HT membrane transporter, with a rapid elevation in tyrosine phosphorylation (Tyr-P) of GTPase-activating protein (GAP) (Lee et al 1997) and early inductions of c-myc (Lee et al 1994). 5-HT, acting via the 5-HT_{2A} receptor, is a known activator of the ERK pathway in vascular smooth muscle cells (Watts, 1996; Banes et al 1999).

Mobilisation of 5-HT in intestine and its accumulation in liver and spleen tissues were observed at the initial periods after partial hepatectomy (Kulinskii, et al 1983). 5-HT caused a dose-dependent increase in DNA synthesis in primary cultures of rat hepatocytes in the EGF and insulin, as measured by $[^3H]$thymidine incorporation (Sudha & Paulose, 1998). 5-HT and monoamine oxidase inhibitor o-chlorpargyline injected alone or combined increase the endogenous 5-HT level in the regenerating liver and stimulates mitotic activity. The tryptophan hydroxylase inhibitor p-chlorophenylalanine and reserpine decrease both the endogenous 5-HT level and the mitotic index. There is a close correlation between the endogenous 5-HT level and the mitotic index (Kulinskii et al 1983).

Gamma aminobutyric acid is the principal inhibitory neurotransmitter of the mammalian brain. GABA inhibits the growth of murine squamous cell carcinoma and HeLa cell lines (Boggust & Al-Nakib, 1986). Gliomas with high proliferation rate lack the expression of functional GABA binding sites (Labrakakis et al 1998). GABA also plays an important role in terminating the growth of rapidly developing tissues *in utero* (Gilon et al 1987). Studies from lab have shown that hypothalamic GABergic system plays an important role in the neoplastic transformation of rat liver. $GABA_A$ receptor agonist muscimol, dose dependently inhibited EGF induced DNA synthesis and enhanced the TGF β_1 mediated suppressed DNA synthesis in rat primary hepatocyte culture (Biju et al 2001, 2002). Increased $GABA_A$ receptor activity inhibits proliferation of HepG2, human hepatocyte carcinoma cell line. The inhibition is prolonged in the cell line co-transfected with $GABA_A$ receptor β_2 and γ_2 subunit genes (Zhang et al 2000). $GABA_B$ receptors were increased in neoplastic rat liver (Biju et al 2002). $GABA_A$ receptor agonist, muscimol, dose dependently inhibited EGF induced DNA synthesis and enhanced the TGFβ1 mediated DNA synthesis suppression in primary hepatocyte cultures. $GABA_A$ receptor acts as an inhibitory signal for hepatic cell proliferation (Biju et al 2001). Serum GABA levels were increased in partial hepatectomy (PH), lead nitrate (LN) treated and N-nitrosodiethylamine (NDEA) treated rats. GABA content decreased in the hypothalamus of PH and NDEA treated rats, while it increased in LN treated rats. $GABA_A$ receptor number and affinity in hypothalamic membrane preparations of rats showed a significant decrease in PH and NDEA treated rats, while in LN treated rats the affinity increased without any change in the receptor number. The $GABA_B$ receptor number increased in PH and NDEA treated rats, while it decreased in LN treated rats. The affinity of the receptor also increased in NDEA treated rats. Liver cell proliferation is influencing the hypothalamic GABAergic neurotransmission (Biju et al 2001) and these changes in turn regulate the hepatic proliferation (Figure 3). The mitogenic effect of acetylcholine has been studied in different cell types. Acetylcholine analogue carbachol stimulated DNA synthesis in primary astrocytes derived from perinatal rat brain (Ashkenazi, 1989). Acetylcholine is reported to induce proliferation of rat astrocytes and human astrocytoma cells (Guzzetti et al 1996).

Muscarinic M_1 and M_3 receptors were differentially expressed in the brain regions during pancreatic regeneration in partially pancreatectomised young rats. RT-PCR analysis revealed that central muscarinic M_1 receptor mRNA was decreased at the time of regeneration while muscarinic M_3 receptor mRNA was increased. It is also

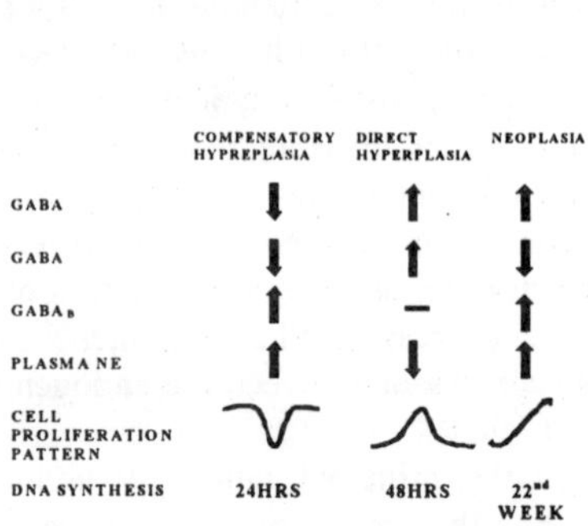

Fig 3. Liver cell proliferation GABAergic regulation in brain during hepatic proliferation

found that in the pancreatic islets, both M_1 and M_3 receptors were increased at the time of regeneration. Muscarinic M_1 receptor antagonist pirenzepine and M_3 receptor antagonist 4-DAMP mustard inhibited EGF induced DNA synthesis in primary culture of rat pancreatic islets (Renuka, 2003). Muscarinic acetylcholine receptors activate many downstream signalling pathways, some of which can lead to mitogen activated protein kinase (MAPK) phosphorylation and activation. MAPKs play a major role in regulating cell growth, differentiation and synaptic plasticity. Both Gi and Gq coupled muscarinic receptors have been shown to activate MAPK in various systems. Muscarinic M_3 receptors activate MAPK in the oligodendrocyteprogenitors (Ragheb et al 2001). Berkeley et al (2000) reported the involvement of M_1 receptors in activation of MAPK in PC12 cells. Acetylcholine analogue carbachol stimulated DNA synthesis *via* muscarinic receptors in primary astrocytes derived from perinatal rat brain. Carbachol is also mitogenic in certain brain derived astrocytoma and neuroblastoma, as well as in Chinese hamster ovary (CHO) cells expressing recombinant muscarinic receptors (Ashkenazi 1989). Proliferation experiments with subtype specific antagonists in astrocytes suggest that cell proliferation is induced by the activation of M_3 receptors (Guizette, 1996). Proliferative signalling has been generally associated with polypeptide growth factor receptors which possess an intrinsic protein tyrosine kinase activity (Yarden et al 1986). In NIH 3T3 cells transfected with human muscarinic m1 receptor gene carbachol stimulate DNA synthesis. The effect of carbachol was blocked by atropine further demonstrating the role of muscarinic receptors. The MAPK activity by muscarinic receptors is dependent on PKC and EGF receptor mediated signalling pathways. PKC inhibitors, or down regulation of PKC by long term exposure to phorbol esters, completely inhibited MAPK activation in response to carbachol in SH-SY5Y and SK-N-BE2(C) human neuroblastoma cells which express endogenous M_3 receptors (Offermanns et al 1993, Kim, et al 1999). The MAPK activation by M_3 receptor stimulation is inhibited by two pathways: one dependent on PKC and the other mediated *via* the EGF receptor and Src (Slack, 2000).

Conclusion

Neurotransmitters - norepinephrine, serotonin, dopamine, gamma aminobutyric acid and acetylcholine are involved in the regulation of endocrine function through the mediation of hypothalamus and pituitary. Neurotransmitter acts by binding to specific membrane bound receptors. Neurotransmitter receptor studies revealed that the functional balance of various neurotransmitters likes NE, GABA, 5-HT, DA and ACh through their receptor sub types have important role in cell proliferation and insulin secretion. These results have immense clinical significance in the diagnosis and management of diseases like hypothyroidism, hypertension, diabetes and cancer.

Acknowledgement

C. S. Paulose thanks DAE, DST, DBT, ICMR, Govt of India for research grants.

References

Ahren B and Lundquist I (1985) Pharmacology **30**:71-82

Ahren B, Taborsky GJ Jr and Porte D Jr (1986) Diabetologia **29**:827–836

Ahren B (2000) Diabetologia **43**: 393-410

Ahren B, Lundquist I and Jarhult J (1984) Acta Endocrinol (Copenh) **105**:78–82

Alexander J S, Hechtman HB and Shepro DP (1987) Biochem Biophys Res Commun **143**: 152-158

Ani Das V (2000) Ph.D. Thesis submitted to Cochin University of Science and Technology, Cochin, India.

Asha A and Paulose C S (1999) J Biochem Mol Biol & Biophys 3: 171-176

Ashkenazi A, Ramachandran, J and Capon DJ (1989) Nature **340**:146-150

Aswathy R Nair, Biju MP and Paulose CS (1998) Biochimica et Biophysica Acta **1381**: 351-354

Avogaro A, Toffolo G, Valerio A and Cobelli C (1996) Diabetes **45**:1373-1378

Baekkeskov S, Anastoot HJ, Christgua S, Reetz A, Solimena M, Cascalho M Folli F, Olesen H and Camilli PD (1990) Nature **347**: 151-156.

Banes A, Florian, JA and Watts SW (1999) J Pharmacol Exp Ther **291**: 1179-1187.

Beilezikian JP (1987) (PA Insel, Ed.), 37- 66.

Biju MP, Pyroja S, Rajeshkumar NV and Paulose CS (2001) Hepatology Research **21**:136-146.

Biju MP, Pyroja S, Rajeshkumar NV & Paulose, CS (2002) Journal of Biochemistry, Molecular Biology and Biophysics, 6: 209-214.
Biju MP, Pyroja S, Rajeshkumar, NV and Paulose, CS (2001) Mol Cell Biochem **216**: 65-70.
Biju, MP, and Paulose CS (1998) Biochemistry and Molecular Biology International **44**: 1-7.
Boggust WA and Al-Nakib T (1986) IRCS Medical Sci **14**: 174-175.
Boiadzhieva N (1990) Eksp Med Morfol **29**:20-6.
Borelli MI, Villar MJ, Orezzoli A and Gagliardino JJ (1997) Diabetes Metab **23**: 161-163.
Brito NA, Brito MN, Kettelhut IC and Migliorini RH (1993) Brain Res **626**: 339-342.
Broten J, Michalopoulos G, Petersen B and Cruise J (1999) Biochem Biophys Res Commun. **262**: 76-79.
Chen C and Yang J (1991) Brain Res **552**: 175-179.
Chen XH, Itoh M, Sun W Miki Tand Takeuchi Y (1996) J Auton Nerv Syst **59**: 12-16.
Chiu S, Paulose CS and Mishra RK (1981) Science **214**: 1261-1262.
Chu P, Lin M, Shian, L and Leu S (1986) Diabetes **35**: 481-485.
Comings DE, Flanagan SD, Dietz G, Muhleman D, Knell E and Gysin R (1993) Biochem Med Metab Biol. **50**:176-85.
Coore HG and Randle PJ (1964) Biochem J **93**: 66-72.
Cowen DS, Sowers RS and Manning DR (1996) J Biol Chem **271**: 22297-22300.
Crowle ST, Dempsey EC, Horwitz KB and Horwitz LD (1994) Circulation **90**: 1908-1918.
Cruise JL, Muga SJ, Lee Y and Michalopoulose GK (1989) J Cell Physiol **140**: 195-201.
Cryer PE (1993) Int J Obes Relat Metab Disord **17**: S43-6.
Dakshinamurti K, Paulose CS and Vriend J (1986) Journal of Endocrinology **109**: 345-345.
Dakshinamurti K, Paulose CS, Viswanathan M, Siow YL and Sharma K (1990) Ann New York Acad of Science **585**: 128-144.
Dakshinamurti K, Paulose CS, Thliveris JA and Vriend J (1985) J Endocr **104**: 339-344.
Dakshinamurti K, Paulose CS, Viswanathan M and Siow YL (1988) Neurosci & Biobehavioral Review **12**: 189-193.
Dakshinamurti K, Paulose CS and Viswanathan M (1990) Ann New York Acad of Science **585**:241-249.
Ding A, Nitsch R and Hoyer S (1992) J Cereb Blood Flow Metab **12**: 103-109.
Eddahibi S, Fabre V, Boni C, Martres MP, Raffestin B, Hamon M and Adnot S (1999) Circ Res **84**:329 - 336.
Eswar Shankar PN (2003) *Ph.D.* Thesis submitted to Cochin University of Science and Technology, Cochin, India.
Exton JH (1988) Hepatology **8**: 152-166.
Exton JH (1981) Molec Cell Endocrinol **2**:233-264.
Fanburg B L and Lee SL (1997) Am J Physiol 272:L795 -L806.
Figlewicz DP, Brot MD, McCall AL, Szot P (1996) Brain Res **736**:54-60.
Figlewicz DP, Evans SB, Murphy J, Hoen M and Baskin DG (2003) Brain Res **964**:107-15.
Fozard R, Mir AK and Middlemiss D (1987) J Cardiovasc Pharmacol **9**: 328-347.
Fozard YR (1989) Oxford: Oxford University Press.
Furuzawa Y, Ohimori Y and Watanabe T (1996) J Vet Med Science 58:243-248.
Galici R, Galli A, Jones DJ, Sanchez TA, Saunders C, Frazer A, Gould GG, Lin R Z and France CP (2003) Neuroendocrinology **77**:132-140.
Garnovskaya MN, Biesen TV, Hawe B, Ramos SC, Lefkowitz RJ and Raymond, JR (1996) Biochemistry 13716-13722.
Gaskins H, Baldeon M, Selassie L, and Beverly J (1995) J Biol Chem **270**: 30286-30289.
Gauthier C, Vranic M and GF Hetenyi J(1980) Am J Physiol **238**: E131-E140.
German MS, Moss LG and Rutter WJ (1990) Diabetes **265**: 22063-22066.
Gilon P, and Henquin JC (2001) Endocrine Reviews **22**: 565-604.
Gilon P, Reusens-Billen B, Remacle C, deV Ph J, Pauwels G, and Hoet JJ (1987) Cell Tissue Res. **249**: 593-600.
Grill V and Ostenson CG (1983) Biochim Biophys Acta **756**: 159-162.
Guizzetti M, CP, Peters J, and Costa LG (1996) Eur J Pharmacol **297**: 265-273.
Hassall C, Stanford C, Burnstock and Buckley NJ (1993) Neuroscience **56**: 1041-1048.
Henquin JC and Nenquin M (1988) FEBS Lett **15**:89-92.
Henquin M, Gracia MC, Bozem M, Hermans MP and Nenquin M (1988) Endocrinology **122**: 2134-2142.
Hindle AT (1994) Br J Anaesth **73**: 395-407.
Honmura A, Yanase M, Saito H and Iguchi A (1992) Endocrinology **130**: 2997-3002.
Hoyer D, Clarke DE, Fozard JR, Hartig PR, Martin GR, Mylecharane EJ, Saxen PR and Humphrey PPA (1994) Pharmacol Rev **46**:157-203.
Iguchi A and Yatomi AN (1990) Brain Res **507**:295-300.

Iguchi A, Gotoh M, Matsunaga H, Yatomi A, Honmura A, Yanase M and Sakamoto N (1986) Am J Physiol **251**:E431-437.
Iguchi A, Uemura K, Kunoh Y, Miura H, Ishiguro T, Nonogaki K, Tamagawa T, Gotoh M and Sakamoto N (1991) Neuropharmacology **30**:1129-1131.
Iguchi A, Uemura K, Miura H, Ishiguro T, Nonogaki, Tamagawa T, Goshima K and Sakamoto Nl (1992) Neuroendocrinology **55**: 44-50.
Iguchi A, Matsunaga H, Gotoh M, Nomura T, Yatomi A and Sakamoto N (1985) Acta Endocrinol (Copenh) **109**:440-445.
Ionescu E, Rohner-Jeanrenaud F, Berthoud HD and Jeanrenaud B (1983) Endocrinology **112**: 904-910.
Ishizuka J, Beauchamp RD, Townsend CM Jr, Greeley GH Jr and Thompson JC (1992) J Cell Physiol 150: 1-7.
Iturriza F C and Thibault J (1993) Neuroendocrinology **57**:476–480.
Jackson J and Paulose C S (2001) J Biochem Mol Biol & Biophys **5**: 1-7 .
Jackson J and Paulose CS (1999) Molecular Cell Biochem **199**: 81-85.
Jackson J, Pius SP, Thomas P and Paulose CS (1997) Current Sci **72**: 137-139.
Jamnicky B, Muck-Seler D and Slijepcevic M (1993) Comp Biochem Physiol Comp Physiol **105**: 267-273.
Jamnicky B, Slijepcevic M, Hadzija M, Juretic D and Borcic O (1991) Acta Diabetol Lat **28**: 11-18.
Jetton TL, Liang Y and Cincotta AH (2001) Metabolism **50**:1377-84
Julius D (1991) Annu Rev Neurosci **14**: 3350-360.
Julius D, Livelli TJ, Jessell TM and Axel R (1989) Science **244**: 1057-1062.
Juszkiewicz M (1985) Pol J Pharmacol Pharm **37**:591-600.
Kaneto A, kosaka K and Naka K (1967) Endocrinology **80**: 530 -536.
Kim JY, Yang MS, Oh CD, Kim KT, Ha MJ, Kang SS and Chun JS (1999) Biochem J **337**:275-280.
Kluess C, Hescheler J, Ewel C, Rosenthal W, Schultz G and Wittig B (1991) Nature **353**: 43-48.
Knopp J, Gesova D, Rusnak M, Jaroscakova I, Farkas R and Kvetnasky R (1999) Endocr. Regul. **33**:145-153.
Kost DP, DeFrances MC, Lee Chi-Ru and Michalopoulos GK (1992) Pathobiology **60**:303-308.
Kulinskii AS, Saratikov AS, Vstavskaia Iu A and Udovitsina TI (1983) Farmakol Toksikol **46**: 92-95.
Kulinskii VI, Udovitsina TI, Vstavskaia Iu A and Rykov SA (1983) Vopr Med Khim **29**:104-107.
Kuo Y (2002) J Biomed Sci **9:126**-32
Kwok R and Juorio A (1987) Neuroendocrinol **45**: 267-273.
Labrakakis C, Patt S, Hartmann J and Kettenmann H (1998) Eur J Neurosci **10**: 231-238.
Lackovic Z, Salkovic M, Kuci Z and Relja M (1990) J Neurochem **54**: 143-47.
Laduron PM and Janssen PF (1986) Brain Res **380**: 359-62.
Lang C (1995) Brain Res Bull **37**: 611-616.
Lauder JM (1990) In The Neuropharmacology of Serotonin. ed. Whitaker-Azmitia PM and Peroutka SJ, pp. 297-314. New York: The New York Academy of Sciences
Lee SL, Wang WW, Lanzillo JJ and Fanburg BL (1994) Am J Physiol **266**: L46-L52.
Lee SL, Wang WW and Fanburg BL (1997) Am J Physiol **272**: C223-C230.
Lee SL, Wang WW, Moore BJ and Fanburg BL (1991) Circ Res **68**: 1362-1368.
Liang Y, Jetton TL, Lubkin M, Meier AH and Cincotta AH (1998) Cell Mol. Life Sci **54**:703-11.
Lim DK, Lee KM and Ho IK (1994) Arch Pharm Res **17**: 398-404.
Loewy AD, Franklin MF, Haxhiu MA (1994) Brain Res **638**: 248-60.
Louis-Sylvestre J (1987) Diabetes Metab **13**: 63-73.
Lundquist I, Ahren B, Hansson C and Hakanson R (1989) Pancreas **4**: 662-7.
Luo Y, Kokkonen, GC, Wang X, Neve KA, Roth GS (1998) J Neurochem **71**: 980-90.
Malaisse W, Malaisse-Lagae F, Wright PH and Ashmore J (1967) Endocrinology **80**: 975-978.
Malaisse WJ (1972) Washington D C: Am Physiol Soc EB pp. 237-260
Marcel Dekker, Berkeley JL and Levey AI (2000) J Neurochem **75**: 487-493
Michalopoulose GK and DeFrancis MC (1997) Science **276**: 60-66.
Mitchison T and Kirschner M (1984) Nature **312**: 237-242.
Nagano Sato R, Matsuda H and Aramaki T (1999) Nippon Ika Daigaku Zasshi **66**: 127-33.
Nemeck GM, Coughlin SR, Handley DA and Moskowitz MA (1986) Proc Natl Acad Sci (USA) **83**: 674-678.
Offermanns S, Bombien E and Schultz G (1993) Biochem J **294**: 545-550.
Ohtani N, Ohta M and Sugano T (1997) J Neurochem **69**: 1622-1628.
Oestenson C and Grill V (1987) Endocrinology **121**: 1705-1710.
Paulose CS, Dakshinamurti K, Packer S and Stephens NL (1988) Hypertension **11**: 387-391.

Paulose CS, Padayatti PS and Sudha B (1999) Comparative Endocrinology and Reproduction KP Joy, A Krishna & C Haldar (Eds). Narosa Publishing House, New Delhi, India. 559-568.
Paulose CS and Dakshinamurti K (1985) J Neuro Sci Res **14**:263-270.
Paulose CS and Dakshinamurti K (1984) Neurosci letters **8**:311-316.
Permutt MA and Kipnis DM (1972) J Biol Chem **247**: 1200-1207.
Pipeleers DG, Schuit FC, In't Veld PA, Maes E, Hooghe-Peters EL, van de Winkel M and Gepts W (1985) Endocrinology **117**:824—833.
Pitt BR, Weng W, Steve AR, Blakely RD, Reynolds I and Davies P (1994) Am J Physiol **26**: L178-L186.
Pius S, Padayatti and Paulose CS (1999) Life Sciences **65**: 403-414.
Porte DJ, Graber AL, Kuzuya T and Williams RH (1966) J Clin Invest **45**: 228-236.
Porte J (1967) J Clin Invest **46**: 86-94.
Porte D, Jr Woods SC, Chen M, Smith PH and Ensinc JW (1975) Pharmacol Biochem Behav **3**:127-133.
Pyroja S (2002) Ph.D. Thesis submitted to Cochin University of Science and Technology, Cochin, India.
Quist E (1982) Biochem Pharmacol **31**: 3131-3133.
Ragheb F, Molina-Holgado E, Cui QL, Khorchid A, Liu HN Larocca JN and Almazan G (2001) J Neurochem **77**:1396-406.
Reetz A, Solimena M, Matteoli M, Folli F, Takei K and Camilli P (1991) EMBO J **10**: 1275-1284.
Renstrom E, Ding W, Bokvist and Rorsman P (1996) Neuron **17**: 513-522.
Renuka TR (2003) Ph.D. Thesis submitted to Cochin University of Science and Technology, Cochin, India.
Saitoh A, Morita K, Sodeyama M and Kamei J (1998) Pharmacol Biochem Behav **60**:161-6.
Sanae F, Miyamoto KI and Koshiyura R (1989) Cancer Res. **49**: 6242-6246.
Sandrini M, Vitale G, Vergoni A, Othani A and Bertolini A (1997) Life Sci **60**: 1393-1397.
Seuwen K and Poussegur J (1990) Biochem Pharmacol **39**: 985-990.
Seuwen K, Magnaldo and Poussegur J (1988) Nature **335**: 254-256.
Sharkey KA and Williams RG (1983) Neurosci Lett **42**: 131-135.
Simartirkis E, Miles PDG, Vranic M, Hunt R, Gougen-Rayburn R, Field CJ and Marliss EB (1990) Clin Invest Med **13**: 134.
Sjoholm A (1991) Biophys Biochem Res Commun **180**: 152-155.
Skoglund G, Lundquist I, and Ahren B (1986) Pancreas **1** : 415—420.
Sorenson R, Garry D and Brelje T (1991) Diabetes **40**: 1365-1374.
Stroebel M and Groppelt-Struebe M (1994) J Bio Chem **269**: 22952-22957.
Stumpel F, Scholtka B and Jungermann K (1998) FEBS Lett **436**: 185-8.
Sudha B and Paulose CS (1998) Hepatology **27**: 62-66.
Sumiyoshi T, Ichikawa J and Meltzer H (1997) Neuropsycopharmacol **16**:183-190.
Takahashi A, Ishimaru H, Ikarashi Y and Maruyama Y (1993) Neurosci Lett.**156**: 54-56.
Takahashi A, Ishimaru H, Ikarashi Y and Maruyama Y (1994) Brain Res Bull **34**: 47-52.
Tan LP and Lagnado JR (1975) Biochem Soc Trans **3**: 121-124.
Teitelman G, Alpert S, Polak JM, Martínez A and Hanahan D (1993) Development **118**: 1031—1039.
Teitelman G, Lee JK and Alpert S (1987) Cell Tissue Res **250**: 435—439.
Tessy TM, Sudha B and Paulose CS (1997) Life Sciences **21**: 1867-1874.
Vaysse N, Laval J, Senarens C, Esteve JP and Ribet A (1982) Biochem Biophys Acta **720**: 378-83.
Vayssette J, Vaysse N and Ribet A (1986) Eur J Pharmacol **122**: 321-8.
Viswanathan M, Paulose CS, Lal KJ, Sharma SK and Dakshinamurti K (1990) Neurosci Letters **111**: 201-205.
Viswanathan M, PauloseCS, Siow YL, and Dakshinamurti K(1988) Brain Research **473**: 37-42.
Watts SW (1996) J Pharmacol Exp Ther **279**: 1541-1550.
Woods SC and Porte D Jr (1974) Physiol Rev **54**:596—619.
Yarden Y, Escobedo JA, Kuang WJ, Yang-Feng TI, Harkins RN, Francke U, Fried VA, Ullrich A and William LT (1986) Nature **323**: 226-232.

Molecular and Cellular Neurobiology
Editors: M.K. Thakur and S. Prasad

$5\text{-}HT_{1A}$ Receptor Mediated Signal Transduction: Effect of Antidepressants

M.N. Subhash

Neurochemistry Department, National Institute of Mental Health and Neurosciences
Bangalore, India

Abstract

$5\text{-}HT_{1A}$ receptor density was measured in brain regions of rats treated with imipramine (5 mg/kg body wt), desipramine (10 mg/kg body wt), clomipramine (10 mg/kg body wt) and trazodone (10 mg/kg body wt) for 40 days, using $[^3H]$8-OH-DPAT. It was observed that chronic exposure to antidepressants (ADs) results in significant downregulation of $5\text{-}HT_{1A}$ receptors (30-80%) in cortex with all the drugs. Interestingly, in hippocampus, imipramine treatment increased the $5\text{-}HT_{1A}$ receptor density (48%). With trazodone, however, there was a significant decrease in the density of $5\text{-}HT_{1A}$ receptors in hippocampus (79%). The affinity of $[^3H]$8-OH-DPAT was increased with imipramine treatment both in cortex and hippocampus. With trazodone treatment the affinity of $[^3H]$8-OH-DPAT to $5\text{-}HT_{1A}$ receptors was significantly decreased only in cortex. There was no change in the basal adenylyl cyclase (AC) activity after exposure to any of the drugs. However, the 5-HT sensitive AC activity was significantly increased in cortex with imipramine (72%), clomipramine (20%) and trazodone (50%) treatment, whereas in hippocampus only imipramine (50%) and trazodone (60%) treatment significantly increased AC activity. In conclusion, chronic treatment with ADs results in downregulation of cortical $5\text{-}HT_{1A}$ receptors along with concomitant increase in 5-HT stimulated AC activity, suggesting the involvement of cortical $5\text{-}HT_{1A}$ receptor-mediated AC responses in the mechanism of action of ADs.

Key Words: $5\text{-}HT_{1A}$ receptors, $[^3H]$8-OH-DPAT, antidepressants, rat brain, adenylyl cyclase.

Introduction

Advances made in the field of Serotonin (5-HT) research during the past decade have been a direct result of more than 50 years of 5-HT research. The 5-HT neurons are located in the raphe nuclei of the brain stem but the 5-HT nerve terminals can be found throughout the brain, suggesting the involvement of 5-HT in many brain functions. The effects of 5-HT on a given neuron are dependent on the different types of receptors expressed and the location of that neuron. It is also well established that different receptor subtypes can exhibit different affinities for endogenous 5-HT and recently, neurons expressing different receptor mRNAs have been identified using various molecular biology techniques.

Considerable evidence is now available to support the pivotal role of serotonin system in exerting the antidepressant response in humans. Deficits in the function of 5-HT receptors have long been implicated in many psychiatric disorders, such as depression (Meltzer and Lowy, 1993), suicide (Cheetham et al., 1990), obsessive-compulsive disorders (OCD) (Murphy et al., 1989) and migraine (Murphy, 1990). Different types of antidepressant treatments have been shown to enhance 5-HT neurotransmission via different pre- or post-synaptic mechanisms. Chronic treatment with antidepressants (ADs), such as selective serotonin re-uptake inhibitors (SSRIs), has been shown to alter the functioning of 5-HT neurons in brain. The hypothesis that an important function of ADs is to restore a disturbed balance between different 5-HT receptor subtypes in depressed patients has been proposed (Berendsen, 1995). The alteration in $5\text{-}HT_{1A}$ receptor mediated cAMP pathway has also been shown to be involved in neuropsychiatric disorders such as affective disorders and schizophrenia (Lieberman et al.,1998).

Numerous observations support the notion that serotonin and its multiple receptor subtypes are linked not only to the biological basis of affective disorders, but also to the mechanism of action of ADs (Blier and de Montigny, 1994). At least seven types of 5-HT receptors have been identified in the central nervous system using pharmacological and molecular cloning techniques. Among 5-HT receptors, $5\text{-}HT_1$ and $5\text{-}HT_2$ receptor subtypes have received considerable attention owing to their involvement in the affective disorders (Meltzer and Lowy, 1993). $5\text{-}HT_1$ receptors, which constitute a major class of 5-HT receptors, are negatively coupled to adenylyl cyclase (AC) via Gi-protein. $5\text{-}HT_1$ receptors have been further subdivided into 5 subtypes, namely, $5\text{-}HT_{1A}$, $5\text{-}HT_{1B}$, $5\text{-}HT_{1D}$, $5\text{-}HT_{1E}$ and $5\text{-}HT_{1F}$ sites (Peroutka, 1986; Humphrey et al., 1993). It has also been shown that $5\text{-}HT_{1A}$ receptors have two affinity sites, designated as high affinity and low affinity sites (Nenonene et al., 1994). High affinity (classical $5\text{-}HT_{1A}$) sites, but not low affinity sites, are coupled to AC via Gi-protein.

Apart from $5\text{-}HT_2$ receptors, $5\text{-}HT_1$ receptor subtypes, especially $5\text{-}HT_{1A}$ receptors, have been shown to be involved in the mechanism of action of various ADs (Hamon et al., 1988). Physiological studies have implicated $5\text{-}HT_{1A}$ receptors in therapeutic mechanism of action of ADs (Chaput et al., 1991). Over the last decade, growing evidence has suggested that $5\text{-}HT_{1A}$ receptor ligands represent a new class of mixed anxiolytic antidepressant drugs. Buspirone, ipsapirone and gepirone, $5\text{-}HT_{1A}$ receptor partial agonists, have been reported to have antianxiety effects both in pre-clinical and clinical studies (Traber and Glaser, 1987; Eison and Eison, 1994). Evidence also indicates a significant role for $5\text{-}HT_{1A}$ receptors in the mechanism of action of antidepressant (Newman et al., 1993; Stahl, 1994; Le Poul et al., 2000) and antimanic drugs (Subhash et al., 1999).

Attempts to delineate the molecular mechanisms of action of ADs on the basis of acute studies have limited value and increasing attention is being turned on to resultant adaptive changes stemming from their chronic treatment. In this, post-synaptic changes are more pertinent to the mechanism of action of ADs. The time course for the occurrence of these adaptive changes in the brain of laboratory animals is consistent with the delayed onset of the antidepressant response in humans. The elucidation of these mechanisms of action has lead to the development of new pharmacological strategies to potentiate the therapeutic effect of the currently available drugs and the identification of novel targets to accelerate and further improve the treatment response in depression (Blier and de Montigny, 1998). The best evidence that 5-HT neurons may be involved in mediating antidepressant effects is the antidepressant activity of selective 5-HT uptake blockers (Cowen, 1990).

Although SSRIs have been recently found to be more effective with lesser side effects, tricyclic antidepressants (TCAs) are still considered for treatment in many cases (Delgado et al., 1992). The TCAs, imipramine (IMI), desipramine (DMI) and clomipramine (CMI) are chemically closely related but have distinct pharmacokinetic and pharmacodynamic properties. All three drugs are known to be non-specific monoamine inhibitors that inhibit both norepinephrine (NE) and 5-HT uptake with different potencies. Both IMI and DMI have been demonstrated to be effective in treating long-term depression (Wroblewski et al., 1996). CMI, which increased a long lasting cortical suppression of firing elicited by dorsal raphe nucleus, seems to act on 5-HT_{1A} receptor subtype (Contreras et al., 1993). Murphy et al (1989), in their clinical study, have shown that CMI is choice of drug in treating OCD.

Trazodone (TZD), a chlorophenylpiperazine derivative, is an atypical AD drug that is commonly referred to as SSRI (Asberg et al., 1987). The most discussed mechanism of action for antidepressant effects of TZD is selective blockade of 5-HT uptake. TZD inhibits the uptake of 5-HT with high efficacy when compared to its efficacy towards NE or DA re-uptake inhibition (Richelson, 1988). Among 5-HT receptors, TZD has been shown to bind to 5-HT_2 receptors with higher affinity than 5-HT_1 receptors (Peroutka and Snyder, 1980). Apart from its weak, but selective inhibition of 5-HT uptake, other most likely mechanism of action of TZD is the formation of metabolite, meta-chlorophenylpiperazine (mCPP), a 5-HT_1 receptor agonist (Caccia et al., 1982). Chronic administration of TZD has been shown to affect 5-HT_{1A} mediated inhibition of forskolin stimulated AC activity, suggesting the direct involvement of 5-HT_{1A} receptors in the mechanism of action of TZD (Tsuchiyama et al., 1991; Newman et al., 1993).

A general hypothesis of 5-HT receptor dysfunction in depression suggests that 5-HT_{1A} receptors may be downregulated, where as 5-HT_2 receptor may be upregulated and display inadequate ability to convert receptor occupancy by 5-HT into adequate physiological response. Pre-clinical studies have shown that numerous ADs downregulate both 5-HT_{1A} and 5-HT_2 receptors (Stahl, 1994). Decreased 5-HT_{1A} receptors have been shown autoradiographically in various brain regions of rats treated chronically with IMI and other drugs (Mizuta and Segawa, 1989). Similarly significantly decreased 5-HT_{1A} receptors with concomitant increase in 5-HT stimulated AC activity has been shown after fluoxetine treatment, suggesting that fluoxetine, which has high affinity for these receptors, acts by modulating 5-HT_{1A} receptor mediated response in brain (Subhash et al., 2000). However, there are contradicting reports suggesting no alteration in 5-HT_{1A} receptors and also in second messenger system following chronic treatment with ADs (Odagaki et al., 1991; File et al., 1999).

Although the signal transduction pathway of 5-HT receptor subtypes is well known, the agonistic or antagonistic activity of variety of chemical agents at these receptors is not clearly understood because of lack of selective ligands to study these receptors. The availability of 8-hydroxy-2 [di-n-propylamino] tetralin (8-OH-DPAT), 5-HT_{1A} selective agonist, has helped in characterizing the 5-HT_{1A} receptors in various regions of brain. Radioligand binding is a technique that measures the binding of a labeled agonist or antagonist to its receptor, either in tissues or in purified form. However, membrane preparations are widely used. This technique allows number of receptors to be quantified with the determination of the affinity of ligand/drug to the receptor, which was not possible earlier. It also helps in studying the dissociation and association kinetics and integration between ligand/drugs, effector systems and G-proteins (Challis, 1997). In our lab this radioligand binding technique is used to study these parameters of 5-HT_{1A} receptors in regions of rat brain using highly specific radioligand $[^3H]$8-OH-DPAT, which has very high affinity to this subtype. The detailed procedure employed and the results observed in cortex and hippocampus of rats chronically (40 days) treated with IMI, DMI, CMI and TZD on 5-HT_{1A} receptors and linked second messenger system (cAMP) is presented. The study of long-term AD treatment provides an insight into the mechanism of actions of these drugs in modulating the serotonin signal transduction and will also help in understanding the diverse actions of ADs in different mood disorders.

Materials and Methods

Materials

[^{3}H] 8-hydroxy-n-dipropylaminotetralin ([^{3}H]8-OH-DPAT, s.a. 221 Ci/mmol) was obtained from Amersham Int. (UK). Imipramine, desipramine, trazodone, 5-HT, theophylline, ATP and pargyline were obtained from Sigma Chemicals (USA). GF/B filters were obtained from Millipore (UK). Clomipramine was a gift from M/S Sun Pharmaceuticals (India).

Animals and administration of drugs

Adult male Sprague Dawley rats, weighing 200-240 Gms, were used for all experiments. Animals (20 for each drug), procured from Central Animal Research Facility (CARF), were housed in cages (4 rats per cage) and exposed to regular day/night period with food and water *ad libitum*.

Imipramine (5 mg/kg body wt), desipramine (10 mg/kg body wt), clomipramine (10 mg/kg body wt) and trazodone (10 mg/kg body wt), were injected intraperitoneally, once daily, in the morning for a period of 40 days. Animals were sacrificed by decapitation under ether anesthesia, 24 hours after the last injection. Control rats received 0.5 ml of normal saline for the same period and were sacrificed along with the experimental rats. Brains were removed and cerebral cortex and hippocampus were dissected out on ice-cold petri dish. Tissues obtained from three rats were pooled for receptor binding and AC assay.

Membrane preparation

Crude membrane preparation was obtained by homogenizing the brain tissue in 20 volumes of Tris-HCl buffer (50 mM, pH 7.4) containing 0.32 M sucrose, following the procedure described elsewhere (Subhash et al., 1998). For [^{3}H]8-OH-DPAT binding the pellet was resuspended in 50 mM Tris-HCl buffer (pH 7.4). Protein concentration was estimated by Lowry's method (1951) and adjusted to 1 mg/ml using same buffer.

Receptor binding assay

5-HT_{1A} receptors were labeled with [^{3}H]8-OH-DPAT (0.06-1.0 nM), following essentially the procedure described in detail earlier (Subhash et al., 1998). Non-specific binding was defined by using 10 μM 5-HT. After the incubation of membranes (200 μg protein) at 37^{0}C, with 6-8 concentrations of the ligand, the reaction was stopped by the addition of 2 ml ice-cold Tris-HCl buffer (pH 7.4). The reaction mixture was rapidly filtered through GF/B filters under vacuum. Filters were dried and transferred to scintillation vials containing 5 ml scintillation cocktail and allowed to equilibrate overnight. Radioactivity was counted using LKB beta counter with 53% efficiency.

Adenylyl cyclase assay

Tissues, obtained from cerebral cortex and hippocampus, were homogenized in 10 volumes of Tris-HCl buffer (50 mM, pH 7.4) and centrifuged at 3800 rpm to remove the cell debris. The supernatant obtained was used for AC assay. Both basal and 5-HT (10 μM) stimulated AC activity was estimated using [^{3}H]ATP, following essentially the procedure described by Malnoe et al (1990) and employed earlier (Subhash et al., 1998). AC activity was expressed as the amount of [^{3}H]cAMP (pmoles) formed per milligram of protein. Protein content was estimated by Lowry's method.

Data analysis

The data from the binding experiments was analyzed using computerized program "LIGAND" (McPherson, 1983) to obtain the equilibrium dissociation constant (Kd), the density of receptors (Bmax) and the Hill co-efficient. The Bmax values were expressed in fmol/mg protein and Kd in nM. Student's "t" test was used for statistical analysis.

Results

In vivo effect of ADs on 5-HT$_{1A}$ receptors

The density of cortical 5-HT$_{1A}$ receptors was decreased significantly with IMI

Table 1. Effect of ADs on 5-HT$_{1A}$ receptors in regions of rat brain

5-HT$_{1A}$ Receptors	Control	IMI	DMI	CMI	TZD
Cortex					
Bmax	24.2±2.1	9.5±1.0*	16.5±1.6*	15.3±2.3*	7.2±0.5*
Kd	0.47±0.02	0.26±0.02@	0.52±0.05	0.47±0.12	0.60±0.06**
Hippocampus					
Bmax	30.2±3.8	44.6±3.8**	30.5±1.6	28.5±3.1	6.3±1.2*
Kd	0.34±0.05	0.26±0.04@	0.32±0.07	0.31±0.05	0.40±0.05

Values, expressed as mean ± SD (n=6), were obtained from three experiments, each assayed in duplicate. The density of 5-HT$_{1A}$ receptors was estimated, as described, in brain regions of rats treated with IMI (5 mg/kg body wt), DMI (10 mg/kg body wt), CMI (10 mg/kg body wt) and TZD (10 mg/kg body wt.), i.p., for 40 days. Bmax = fmol / mg protein and Kd = nM. *p<0.0001, **p<0.001, @p<0.01, compared to control group.

(60%; p<0.0001), DMI (32%; p<0.0001), CMI (36%; p<0.0001) and TZD (70%; p<0.0001) treatment. The affinity of [^{3}H]8-OH-DPAT to cortical 5-HT$_{1A}$ receptors, as reflected by the Kd values, was significantly increased (p<0.01) in IMI treated rats (Kd=0.26±0.02 nM) and was significantly decreased (p<0.001) in TZD treated rats (Kd=0.60±0.06 nM) as compared to control (K_d = 0.47±0.02) rats (Table 1).

The density of 5-HT$_{1A}$ receptors was decreased significantly in hippocampus (79%; p<0.0001) after chronic treatment with trazodone. However, there was no significant change in 5-HT$_{1A}$ receptor density in hippocampus of DMI and CMI treated rats. The receptor density was, however, increased significantly (48%; p<0.001) in hippocampus of IMI treated rats with a significant decrease in Kd (p<0.01) values. The affinity of [^{3}H]8-OH-DPAT to 5-HT$_{1A}$ receptors in hippocampus was not altered in rats treated with other drugs (Table 1).

Hill co-efficient values were near to unity in both the regions of experimental rats, suggesting that at this concentration, $[^3H]$8-OH-DPAT binds to a single class of high affinity 5-HT_{1A} receptors, as in control rats.

Table 2. Adenylyl cyclase activity in brain regions of rats after chronic treatment with ADs

	Control	IMI	DMI	CMI	TZD
Cortex					
Basal	7.2±0.9	5.5±1.3	7.6±0.3	8.6±0.4	6.4±0.3
+ 5-HT	12.7±1.3	21.9±2.5*	12.1±0.3	15.2±1.8**	18.8±1.8*
Hippocampus					
Basal	7.1±0.5	7.2±1.5	8.0±0.5	6.2±0.6	6.8±0.6
+ 5-HT	12.1±1.0	18.1±1.2**	12.3±0.9	12.0±0.7	19.5±2.3*

Values, expressed as mean ± SD (n=6), were obtained from three experiments, each assayed in duplicate. Both basal and 5-HT (10 μM) stimulated AC activity was assayed, as described, in brain regions of rats treated with IMI (5 mg/kg body wt.), DMI (10 mg/kg body wt.), CMI (10 mg/kg body wt.) and TZD (10 mg/kg body wt.) , i.p., for 40 days. Activity was expressed as pmoles of cAMP formed per mg protein. * $p<0.0001$, ** $p<0.001$, compared to control group.

Adenylyl cyclase activity

Both basal and 5-HT (10 μM) stimulated AC activity was assayed in cerebral cortex and hippocampus of experimental and control rats. In control cortex the basal AC activity (7.2±0.9 pmol/mg protein) was stimulated by nearly 75% with 10 μM 5-HT. A similar increase in AC activity was observed in the hippocampus (Table 2).

After chronic AD treatment, the basal AC activity remained unaltered in both the regions. However, the 5-HT sensitive AC activity was significantly increased by 72% ($p<0.0001$) in cortex and by 50% ($p<0.001$) in hippocampus after IMI treatment. After CMI treatment, the 5-HT stimulated AC activity was significantly increased only in cortex by 20% ($p<0.001$). The 5-HT stimulated AC activity was significantly ($p<0.0001$) increased in cortex (50%) and hippocampus (60%) after TZD treatment. However, no significant alterations were observed in 5-HT sensitive AC after DMI treatment, in both regions (Table 2).

Discussion

Effects of chronic treatment of IMI, DMI, CMI and TZD (ADs) on cortical and hippocampal 5-HT_{1A} receptors and linked second messenger system, AC were studied using radioligand binding technique and enzymatic assay, respectively. The results demonstrate that chronic administration of these ADs has differential effect on 5-HT_{1A} receptors in cortex and hippocampus. The ADs used are chemically closely related and are non-specific monoamine uptake blockers that block uptake of 5-HT and NE. IMI and CMI are equipotent in blocking 5-HT and NE uptake and DMI has been shown to be approximately 40 fold more potent in blocking NE than 5-HT uptake (Auerbach et al., 1995). Trazodone, an atypical antidepressant, inhibits the reuptake of 5-HT and this relative specificity of TZD in blocking the 5-HT uptake was the first suggested mechanism of action for its antidepressant effects (Richelson and Pfenning, 1984).

The major finding of the present study is the significant downregulation of cortical 5-HT_{1A} receptors in rat brain after chronic treatment (40 days) with ADs. However, only TZD treatment showed significant decrease in hippocampal 5-HT_{1A} receptor density. Interestingly, IMI treatment produced a significant increase in the density of

$5\text{-}HT_{1A}$ receptors in hippocampus. The affinity of the ligand ($[^3H]$8-OH-DPAT) to both cortical and hippocampal $5\text{-}HT_{1A}$ receptors was altered only in IMI treated rats. The affinity however, was decreased only in cortex of TZD treated rats, as seen by increased K_d values. These results are in agreement with earlier reports (Pandey et al., 1991; Lund et al., 1992; Wieland et al., 1993). However, Odagaki et al. (1991) have observed no significant changes in $5\text{-}HT_{1A}$ receptors after chronic treatment with ADs and Papp et al. (1994) failed to show any correlation between $5\text{-}HT_{1A}$ receptor density and unpredictable stress after chronic treatment with IMI.

The AC activity in cortex and hippocampus was affected in differential manner after AD treatment. The Basal AC activity was unaltered after treatment with ADs in both the regions. The 5-HT sensitive AC activity was significantly increased in cortex and hippocampus of IMI and TZD treated rats and only in cortex of CMI treated rats. Variable effects of AD treatment on AC activity have been reported. Newman et al. (1990) reported reduced degree of inhibition of forskolin stimulated AC activity by 5-HT after chronic treatment with DMI. However, Odagaki et al. (1991) have observed no significant changes in AC activity in rat hippocampus after chronic treatment with ADs. Similarly, Sapena et al. (1994) reported that DMI treatment did not alter the AC activity in rat cortical slices. In this study also we observed no alterations in hippocampal 5-HT sensitive AC activity after DMI and CMI treatment, which might be due to no change in the $5\text{-}HT_{1A}$ receptor density in this region.

Basic and clinical research has indicated that the sensitivity of $5\text{-}HT_{1A}$ receptor may be reduced in depression and that agonists acting at this receptor subtype may have antidepressant properties (Cheetham et al., 1993; Thielen and Frazer, 1995 ; deVry, 1995). Reduced hypothermic response in depressed patients has been demonstrated following administration of buspirone and ipsapirone indicating a sub sensitivity of $5\text{-}HT_{1A}$ receptors (Lesch et al., 1990). These studies have implicated that $5\text{-}HT_{1A}$ receptors play an important role in the pathophysiology of depression and in the mechanism of action of ADs.

In this study, the density of $5\text{-}HT_{1A}$ receptors has decreased only in cortex after chronic treatment with ADs, where as in hippocampus the density remained unchanged, except in IMI and TZD treated rats. Functionally also, only in IMI and TZD treated rats, the 5-HT stimulated AC activity was significantly altered indicating the sensitivity of hippocampal $5\text{-}HT_{1A}$ receptors to IMI and TZD. This differential effect of ADs on $5\text{-}HT_{1A}$ receptors might be due to the direct interaction of the ADs with $5\text{-}HT_{1A}$ receptors or due to the complex mutual inhibitory effects of 5-HT receptor subtypes, which respond differentially to the individual drugs (Berendsen, 1995). The significant downregulation of nearly 75% of $5\text{-}HT_{1A}$ receptors observed in both cortex and hippocampus with TZD treatment could be due to the agonistic action of the metabolite mCPP formed in vivo from trazodone. This agonistic effect is also suggested from the study of linked AC response. The basal AC activity did not alter in TZD treated rats. However, there was a significant increase in the 5-HT stimulated AC activity in both cortex and hippocampus after TZD treatment. This alteration in AC response could be suggestive of downregulation of $5\text{-}HT_{1A}$ receptors.

Few studies have been undertaken to study the effect of TZD on $5\text{-}HT_{1A}$ receptor-mediated response (Tsuchiyama et al., 1991; Newman et al., 1993). However, in majority of these studies the effect of chronic administration (30-40 days) has not been studied. Selective blockade of 5-HT_2 receptors, relative to 5-HT_1 receptors may be another potential mechanism of AD drug effects. Lakoski and Aghajanian (1985) found that the 5-HT_2 selective antagonist, ketanserin, enhanced the inhibitory effects of 5-HT in forebrain areas. Subsequent work has suggested that stimulation of 5-HT_2 receptors may functionally oppose the effects of 5-HT_{1A} receptors in postsynaptic areas (Sheldon and Aghajanian, 1990). These electrophysiological findings raise the possibility that the clinical effects of ADs could be related to antagonist action at 5-$HT_{2/2C}$ receptors and agonist action at 5-HT_1 receptors.

These observations support the hypothesis that the important function of ADs is to restore a disturbed balance between 5-HT_1 and 5-HT_2 receptors (Stahl, 1994). Another explanation for the selective desensitization of 5-HT_{1A} receptors could be the differential coupling of 5-HT_{1A} receptors to different G-proteins, leading to a decreased efficiency in triggering the second messenger system and therefore cause a desensitization of 5-HT_{1A} receptors (Li et al., 1997). The same theory can also explain the increase in the density of 5-HT_{1A} receptors in hippocampus after treatment with IMI. Evidence for this hypothesis comes from the study of Lason and Przewocki (1993) where, they have shown that long-term treatment with IMI decreased the G_O mRNA without affecting G_S mRNA levels. Thus suggesting that the effect of AD treatment can also be seen at intracellular signal transduction level and their gene expression.

The results of this study suggest that chronic treatment with ADs such as IMI, DMI, CMI and TZD significantly downregulates cortical 5-HT_{1A} receptors and the effect on AC activity is region specific. This suggests the involvement of cortical 5-HT_{1A} receptor-mediated actions in the mechanism of action of ADs. The possibility of significant downregulation of 5-HT_{1A} receptors by TZD observed in this study might be due to the direct action of TZD on these receptors as an agonist, apart from its 5-HT uptake inhibition, which cannot be ignored. Thus TZD seems to produce its therapeutic effects via downregulation of 5-HT_{1A} receptors as produced by other classes of ADs. The effect of ADs on the second messenger system linked to 5-HT_1 receptor subtypes may be understood well by studying the forskolin stimulated AC system in presence of GTP and 5-HT_{1A} agonists, and attempts in this regard are underway.

References

Asberg C, Eriksson B, Marensson B, Traskman-Bendz L and Wagner A (1987) Jr Clini Psychiat **47**: 23-35.

Auerbach SB, Lundberg JF and Hjorth S (1995) Neuropharmacol **34**: 89-96.

Berendsen HH (1995) Pharmacol Ther **66**: 17-37.

Blier P and de Montigny C (1994) Trends Pharmacol Sci **15**: 220-227.

Blier P and de Montigny C (1998) Biol Psychiat **44** : 313-23

Caccia S, Fong MH, Garrattini S and Zanini MG (1982) Jr Pharmacol **34**: 605-606.

Challis RAJ (1997) In: Methods in molecular biology (ed.) **83**, pp 1-24. Humana Press, New Jersey,

Chaput Y, de Montigney C and Blier P (1991) Neuropsychopharmacol **5**: 219-229.

Cheetham SC, Crompton MR, Katona CLE and Burton RW (1990) Psychopharmacol **102**: 544-548.

Cheetham SC, Viggers JA, Slater NA, Heal DJ and Buckett WR (1993) Neuropharmacol **32**: 737-743.

Contreras CM, Marvan ML and Munoz-Mendez A (1993) Neuropsychobiol **27**: 199-203.

Cowen PJ (1990) Pharmacol Ther **46**:43-51

De Vry J (1995) Psychpharmacol **121**: 1-26.

Delgado PL, Price LH, Heninger GR and Charney DS (1992) In: Handbook of affective disorders(Paytkel ES, ed.) **83**, pp 1-24,Churchill Livingstone, New York.

Eison AS and Eison MS (1994) Prog Neruopsychopharmacol Biol Psych **18**: 47-62.

File SE, Ouagazzal AM, Gonzalez LE and Overstreet DH (1999) Pharmacol Biochem Behav **62** : 695-701.
Hamon M, Fattaccini CH, Adrien J, Gallisot MC, Martin P and Gozlan H (1988) J Pharmacol Expt Therp **24** : 745-752.
Humphrey PPA, Hartig P and Hoyer D (1993) Trend Pharmacol Scien **14**: 233-236.
Lakoski J and Aghajanian GK (1985) Neuropharmacology **24**: 265-273.
Lason W and Przewocki R (1993) Pol J Pharmacol **45**: 219-226.
Le Poul E, Boni C, Hanoun N, Laporte AM, Laaris N, Chauvean J, Hamon M and Lanfumey L (2000) Neuropharmacol **39**: 110-122.
Lesch KP, Mayer S, Disselkamp-Tietz J, Hoh A, Schoellnhammer G and Schulte HM (1990) Life Sci **46**: 1271-1277.
Li Q, Battaglia G, Louis D and Van deKar LD (1997) Brain Res **769**: 141-151.
Lieberman JA, Mailman RB, Duncan G, Sikich L, Chakoc M, Nichols DE and Kraus JE (1998) Biol Psychiat **44** : 1099-1117.
Lowry OH, Rosebrough NJ, Farr AL and Randall RJ (1951) Jr Biol Chem **193** : 265-273.
Lund A, Mjellem-Joly N and Hole K (1992) Neuropharmacol **31** : 25-32.
Malnoe A, Milon H and Reme C (1990) Jr Neurochem **55** : 1480-1484.
McPherson GA (1983) Comput Prog Biomedic 17: 107-114.
Meltzer HY and Lowy T (1993) In: Psychopharmacology: The third generation of progress. (Herbert HY ed) pp.513-525, Raven Press, New York.
Mizuta T and Segawa T (1989) Jpn Jr Pharmacol **50** : 315-326.
Murphy DL, Zohar J, Benkelfat C, Pato MT, Pigott TA and Insel TR (1989) Brit J Psychiat **155**: 15-24.
Murphy DL (1990). Neuropsychopharmacol **3**: 457-471.
Nenonene EK, Radja F, Carli M, Grondin L and Reader TA (1994) J Neurochem 62: 1822-1834.
Newman ME, Drummer D and Lerer B (1990) J Pharmacol Expt Ther **252**: 826-831.
Newman ME, Lerer B and Shapira B (1993) Biol Psychiat **17**: 1-19.
Odagaki Y, Koyama T and Yamashita I (1991) J Neur Transm **86**: 85-96.
Pandey SC, Isaac L, Davis JN and Pandey GN (1991) Eur J Pharmacol 202: 221-225.
Papp M, Klimek V and Willner P (1994) Psychopharmacol **114**: 309-314.
Peroutka SJ (1986) Jr Neurochem **57**: 529-540.
Peroutka SJ and Snyder SH (1980) Science **210**: 88-90.
Richelson E. (1988) McLean Hosp Jour **13** : 67-88.
Richelson E and Pfenning M (1984) Eur Jr Pharmacol **104**: 277-286.
Sapena R, Morin D, Zini R and Tillement JP (1994) Drug Expt Clin Res **20**: 93-102.
Sheldon PW and Aghajanian GK (1990) Brain Res 506: 62-69.
Subhash MN, Srinivas BN, Vinod KY and Jagadeesh S (1998) Neurochemic Resear **23** : 1321-1326.
Subhash MN, Vinod KY and Srinivas BN (1999) Neurochem Int **35** : 247-253.
Subhash MN, Srinivas BN, Vinod KY and Jagadeesh S (2000) J Neural Transm **107**: 377-87.
Stahl S (1994) Psychopharmacol Bull **30**: 39-43.
Thielen RJ and Frazer A (1995) Life Sci **56**:163-168.
Traber J and Glaser T (1987) Trend Pharmacol Sci **8**: 431-437.
Tsuchiyama K, Nagayama H, Akiyoshi J, Yamada K, Miyamoto T, Yamauchi C, Oba A, Ito Y, Okada K and Kojima K (1991) Jap Jr Psychiat Neurol **45**: 139-140.
Wieland S, Fischette CT and Lucki I (1993) Neruopharmacol **32**: 561-573.
Wroblewski BA, Joseph AB and Cornblatt RR (1996) J Clin Psychiat **57**: 582-58.

Section C

Brain Aging

Molecular and Cellular Neurobiology
Editors: M.K. Thakur and S. Prasad

Putative Biomarkers of Neuronal Ageing: Effects of Drugs on Lipofuscin and Lipid Peroxidation

Mahdi Hasan*, Achla Gupta+ and Abbas Ali Mahdi¶

*Department of Anatomy, ¶Department of Biochemistry, K.G. Medical University Lucknow-226 003, +Interdisciplinary Brain Research Centre, IN Medical Collage Aligarh Muslim University, Aligarh-202 002, India

Abstract

There is emerging consensus that the increasing accumulation of lipofuscin pigment within postmitotic cells is a recognized hallmark of ageing. This wear and tear pigment distorts and displaces the protein synthesizing machinery of the cell, thus, causing its dysfunction and atrophy. Furthermore, the non-functioning or "deafferented" neurons exhibit significant increment of lipofuscin. Biochemically, accumulation of lipofuscin has been correlated with corresponding increase in the rate of lipid peroxidation. Consequently, significant rise in lipid perioxidation products and lipofuscin have been used as reliable biomarkers of ageing, for testing the efficacy of drugs thereon. Lipofuscin inhibitors, lipofuscinolytic drugs and antioxidants have been successfully tried by a number of investigators. Numerous reports on the efficacy of dimethyl aminoethyl parachlorophenoxyacetate (hilfergin, centrophenoxine, lucidril) are available. Also, an Ayurvedic herbal mixture, Maharshi Amrit Kalash, has yielded encouraging results.

Key Words: lipofuscin, lipid peroxidation, centrophenoxine, antioxidants, MAK.

Introduction

Increasing intraneuronal accumulation of lipofuscin pigment has been linked with ageing process by a strong linear correlation between the degree of accumulation and the chronological age (Glees and Hasan, 1976). Recently, Brunk and Terman (2002) have further strengthened the earlier reports by their conclusion that the accumulation of lipofuscin within postmitotic cells is a recognized hallmark of ageing, occurring with a rate inversely related to longevity.

Hasan and Glees (1973a) classified the lipofuscin granules in the undermentioned six types: 1. Homogeneous,2. Finely granular;3. Vacuolated; 4. Lamellated or banded; 5. Coarsely granular and 6. Compound or heterogeneous. Apparently, there is an increase in structural complexity with increase in age. More recently, Boldrini et al (2001), in concordance with the observations of Hasan and Glees (1973a), concluded that at the ultrastructural level it is possible to distinguish 5. different patterns of osmiophilic lipopigments: the usual lipofuscin, finger-print deposits, granular profiles, curvilinear bodies and microtubular aggregates. In spite of recent advances in immunohistochemical identification of biochemical markers, the ultrastructural identification of lipofuscin pigments remains the gold standard to identify neuronal ceroid lipofuscinosis, together with clinical aspects and respective gene defect (Boldrini et al 2001). On the other hand, Ledda et al (1999) comprehensively studied quantitative aspects of the age-related accumulation of lipofuscin in the rabbit spinal ganglia by stereological methods using the electron microscope. In neurons, the shape and structure of lipofuscin bodies in aged rabbits (79-months old) differed significantly in comparison with young adults (12 month old) but substantial changes were not observed in satellite cells. Lipofuscin bodies were scattered singly in young adults but were often clustered in old animals. Whereas lipofuscin occupied an average of 0.36% of neuronal perikaryal volume in young adults, its volume increased to 2.55% in the aged (more than 7 time) (Ledda et al 1999).

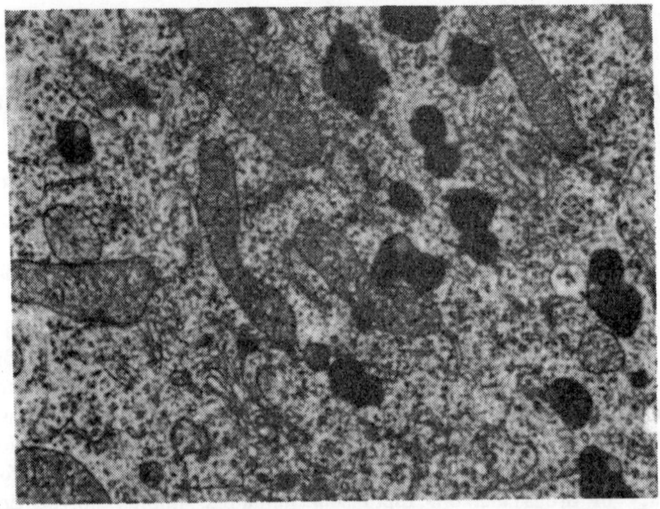

Fig. 1. Photomicrograph depicting pleomorphic lipofuscin profiles mingled with mitochondria (Scale = 1 μm)

Several aspects of genesis of lipofuscin, its chemical nature and its possible reversibility have been exhaustively reviewed by Hasan and Glees (1972b), Sohal (1981), ZsNagy (1987) and Patro (2002). The main objectives of the present communication are to briefly review the putative morpholgical and biochemical markers of neuronal ageing and to evaluate changes, if any, following therapy with both the synthetic and herbal drugs. The age-pigment, lipofuscin is characterized by its wide range of morphological and physico-chemical properties. Due to its heterogeneous chemical composition (Tappel; 1975, 1980) a single method of detection cannot represent a correct measure.

Functional significance of lipofuscin

That a functional diminution of cellular activity takes place due to lipofuscin deposition is obvious (Patro and Patro, 1992; Hasan, 1996). With increasing order of pigment accumulation as age advances, a decline in RNA staining and degeneration of Nissl substance are prominently discernible in the neurons. Pyroninophobic nature of cytoplasm in the immediate vicinity of lipofuscin and loss of Nissl substance substantiate the view that the pigment bodies displace or destroy the protein synthesizing machinery, resulting in cell dysfunction, atrophy and lysis of the remaining cell components (Patro and Sharma, 1984; Patro et al 1989). To study the functional significance of lipofuscin, Hasan and Glees (1973a) used the lateral geniculate nucleus of *Macaca nemestrina* as an experimental model. The lateral geniculate body of the monkey, as in man, has a six layered arrangement of neurons

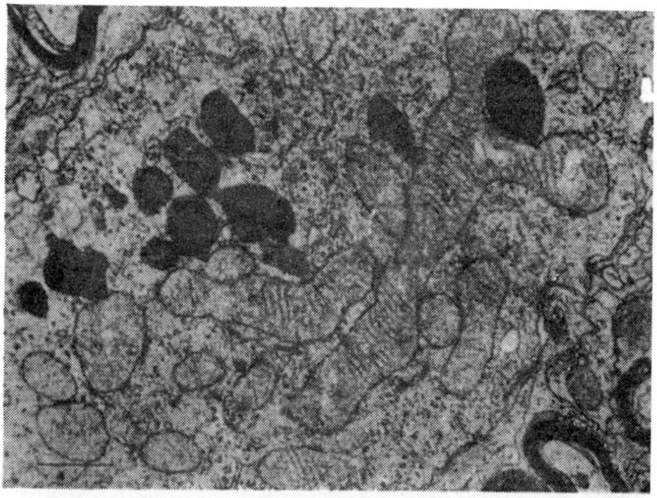

Fig. 2. Electron micrograph showing lipofuscin granules, intermingled with mitochon-dria, one being surrounded by a mitochondrial profile (Scale = 1 μm)

which contain a significant amount of lipofuscin with advancing age (Figs. 1). Layers 2, 3 and 5 receive afferents from the ipsilateral eye, while the nerve impulses from the contralateral retina are relayed in layers 1, 4 and 6. Following eye enucleation, the lateral geniculate body serves as "experimental model" for studying the effects of deafferentation on the neurons of the affected layers. The normal layers of the same nucleus serve as control. An apparent increase in lipofuscin granules was observed in a significant majority of deafferented neurons (Fig. 2). Lipofuscin presents the morphological picture of intracellular debris, waiting to be removed. These pigment profiles were commonly observed mingled with mitochondria and at times within mitochondria (Fig. 3).

Lipofuscin, mitochondria and lipid peroxidation

Gupta and Hasan (1996) have shown that mitochondria play an important role in ageing. Most of the existing hypothesis of ageing are limited to descriptive phenomenology of ageing organisms without offering molecular insights. Mitochondrial oxidation of malonaldehyde is substantially lost with age, a cellular explanation for the accumulation of malonaldehyde and lipofuscin like materials (Hasan, 1996). The close relationship of lipofuscin granules and mitocondria is striking (Fig. 2). An essential role played by mitochondria is the detoxification of O_2 via its reduction to H_2O by cytochrome oxidase systems. It has been demonstrated that mitochondria generate a steady state of superoxide radicals. Many of the compounds localized in the mitochondrial electron transport chain (NAD, FAD, coenzyme Q) when reduced, may react with O_2 to release O_2^- and this may be considered a univalent leak. It may be reasonable to suggest that under normal intracellular intramitochondrial O_2 tensions, this univalent leak can be controlled by the free radical scavenging mechanisms available within mitochondria. During incomplete ischemia with the passage of time, however, this univalent leak may increase and mitochondrial damage may thus be dependent on the integrity of scavenging mechanisms.

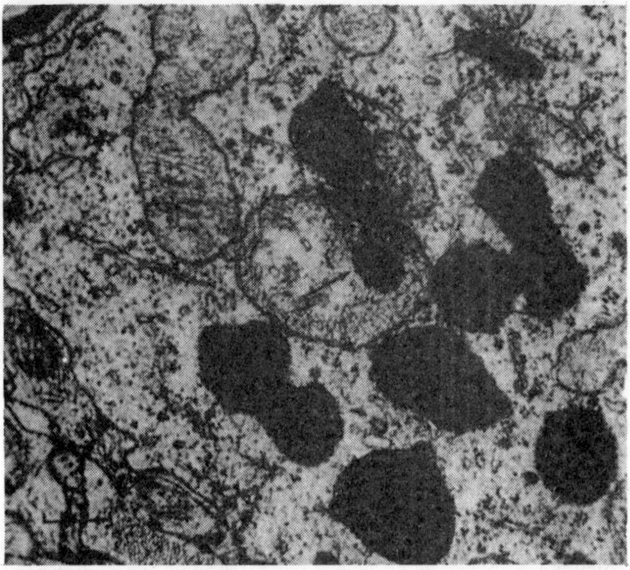

Fig. 3. Electron micrograph revealing one pigment granule (arrow) within mitochondrial profile (Scale = 1 μm).

Several studies suggest that age related fluorescent pigment, lipofuscin and cross-linked proteins increased with lipid peroxidation in tissue (Kikugawa, 1986, Kikugawa and Beppu, 1987; Patro, 2002). The most accepted mechanism of lipofuscin formation is that it is the fluorescent end-product of free radical induced cross-linking of proteins with oxidized lipids, including membrane proteins and lipids. The rate of lipid peroxidation can be estimated by various methods Kikugawa and Beppu, 1987; Kikugawa, 1988). The most common method is the measurement of thiobarbituric acid-reactive substances (TBA-RS). A red pigment with an absorption maximum at 532 nm is produced when 2-thiobarbituric acid (TBA) reacts with malondialdehyde (MDA) in an acidic medium (Ohkawa et al., 1979, Uchiyama and Mihara, 1978; Sharma and James, 1991). But the TBA reaction is not specific to MDA or its derivatives, as many other lipid peroxidation products and intermediate substances give positive reactions (Marcuse and Johnsson, 1973, Kikugawa, 1986; Kosugi and Kikugawa, 1986). However, the best-studied effect of a free radical attack is the one causing lipid peroxidation, i.e. oxidation of α-methylene bridges of unsaturated fatty acids, resulting in the formation of lipid peroxides and hydroperoxides, finally leading to the fragmentation of lipids (Rehncrona et al 1980). Gupta et al (1991) investigated age-related elevation of lipid peroxidation products in discrete regions of the rat brain. Their findings are summarized in histograms 1 to 4. It is thus clearly demonstrated that estimation of lipid peroxidation products, i.e., lipid peroxides, lipid hydroperoxides and conjugated dienes can easily be used as an index or marker of ageing.

Histogram 1. Lipofuscin levels in different regions of the brain of aged rats

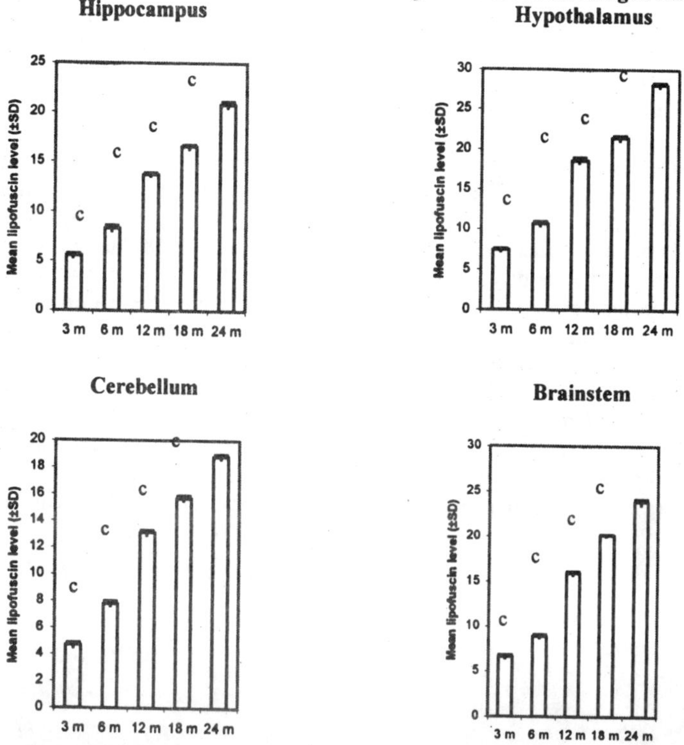

Results, expressed as units/g tissue weight, are mean ± SE of 6 rats. a-c statistical analysis of values obtained in different age groups a=<0.05; b=<0.01; C-0.001.

Effect of drugs on neurolipofuscin

Lipofuscin appears to be one of the morphologically observable endoproducts of lipid peroxidation. It has been used by numerous investigators as a biomarker for evaluating the efficacy of drugs, both for its retardation and removal. (Nandy and Bourne, 1966; Nandy, 1968; Varkonyi et al 1970, Hasan and Glees, 1972a; Hasan and Glees 1972b, Hasan et al 1973b, Hasan et al 1974a & Spoerri and Glees, 1974; Spoerri et al 1974, Low et al 1974; Aloj-Totaro et al 1981, Bernocchi et al 1983; Pisanti et al 1983; Ohtani and Kawashima, 1983; Roy et al 1983; Patro and Sharma, 1984; Aloj-Tataro et al 1985a & b; Patro et al 1988, Peretz, 1989, Maurya et al 1990, Patro, 1990; Bhardwaj, 1996, Singhal, 1996; Kawashima, 2003).

Antioxidants

The rate of free radical reactions in cells are reduced by anti-oxidants (2-mercapto-ethylamine, 2 MEA and butylated hydroxytoluene, BHT), capable of trapping free radical intermediates. Harman's (1969) observation that diets including these agents increase the mean life span of mice by 20-40%, without affecting the maximum life-span, is not surprising.

Lipofuscin inhibitors/eliminators

Nandy and Bourne (1966) observed that meclofenoxate (dimethyl amino-ethyl p-chlorophenoxy acetate, centrophenoxine), a drug known to be beneficial in conditions of presenile and senile confusion, decreased the amount of brain lipofuscin in aged guinea-pigs. A marked elimination of the pigment (Figs. 4,5) was noticeable in animals under treatment for 12 weeks. Nandy (1968) has suggested that the diversion of the oxidation of glucose from the Krebs cycle and glycolytic pathway to pentose cycle may play an important role in the elimination process of wear and tear pigment, lipofuscin, from the neurons of senile guinea pigs. The findings of Nandy and Bourne (1966) were later confirmed by the histological studies of Chemnitius et al (1970) and combined histochemical and electron microscopic study of Meier and Glees (1971), who demonstrated vacuolation of pigment granules following centrophenoxine treatment. The precise mode of dissolution and removal of lipofuscin by hilfergin (dimethyl-amino-ethyl p-chlorophenoxyacetate) was however, clearly demonstrated by Hasan and Glees, 1974b). They conducted

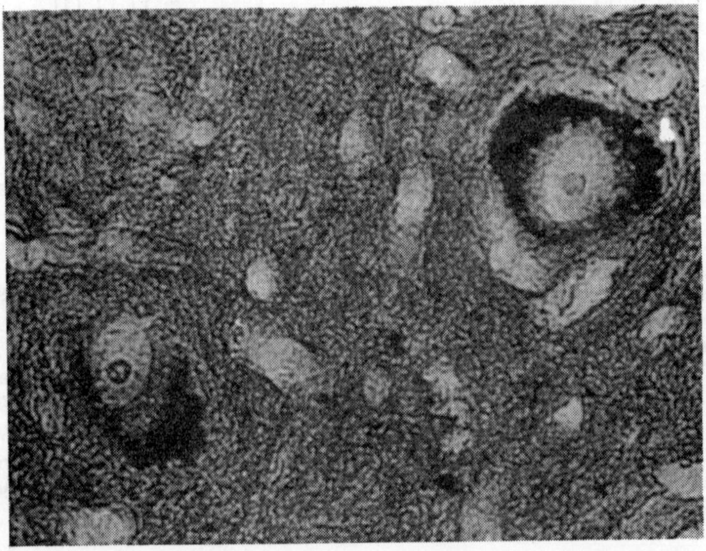

Fig. 4. Photomicrograph showing perinuclear lipofuscin granules in neurons of reticular formation of senile guinea pig (4 y) x 600 (Courtesy: Dr K. Nandy)

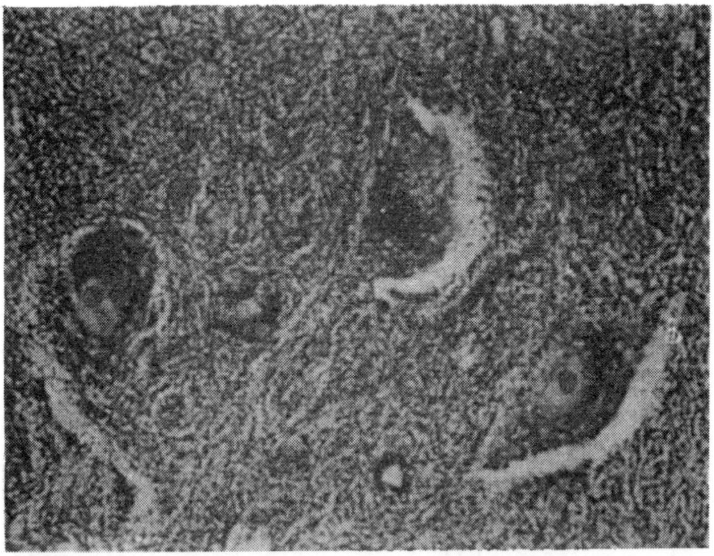

Fig. 5. Photomicrograph exhibiting dissolution of lipofuscin following 12 weeks' treatment with centrophenoxine 80 mg/kg body weight x 900 (Courtesy Dr K. Nandy)

extensive electron microscope examination on hippocampus, midbrain and area postrema of guinea pigs following the administration of centrophenoxine, 80 mg/kg body weight) for 30 to 56 days. After 4 weeks of administration of the drug, the pigment masses appeared to lose their electron density and large electron lucent areas (vacuoles) became visible. Incorporation of altered lipofuscin granules in the cytoplasm of "reactive" phagocytic cells (microglia) occurred. Such cells were found in close proximity to capillaries and altered lipofuscin granules were present in pericytes and endothelial cells of the capillaries in the treated animals; these findings revealed the ultimate participation of capillary endothelium in the removal of lipofuscin. Furthermore, Spoerri and Glees (1974) used electron microscopy to investigate the formation and distribution of lipofuscin in cultured neurons and satellite cells of the rat spinal ganglia. Following daily application of centrophenoxine to the culture for two to three weeks, a "fading out" and a couspicuous vacuolation of the pigment could be observed. This was followed by disintegration of the pigment granules. Further removal of pigment occurred via phagocytes and proliferating endothelial cells of residual capillaries (Spoerri and Glees,1974).

Also Ohtani and Kawashima (1983) cultured for 32 days cells from neonatal rat cerebral hemispheres, after their dispersal by trypsin. Their combined histochemical, fluorescence and electron microscopic analysis demonstrated that lipofuscin increased inneuronal and non-neuronal cells in primary culture accoding to the lapse of time. When centrophenoxine (10^{-4} or 5×10^{-4} M) or chlorpromazine (10^{-6} or 10^{-5} M) was added to the medium, the accumulation of lipofuscin in neurons was significantly reduced. However, the effect of these agents was not detected in non-neuronal cells. In this way, the *"in vitro"* studies confirmed the two earlier *"in vivo"*

Histogram 2. Lipid hydroperoxide levels in various regions of the rat brain at different ages

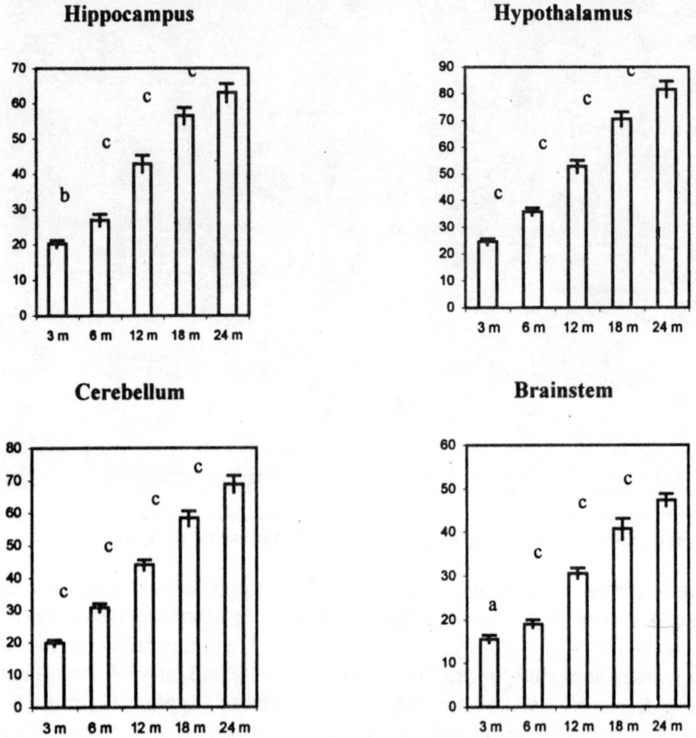

Results, are expressed as n mole of cumine hydroperoxide/g tissue, each value is a mean ± SE of 6 rat brain. a-c statistical analysis of values obtained in different age groups. a=<0.05; b=<0.01; c=<0.001.

reports of Hasan *et al.* (1974 a & 1974b). It ought to be noted, however, that a much higher drug concentration (0.1%) was used in tissue cultures than in the *in vivo* studies (0.008%), following intramuscular injection or intraperitoneal injection of 80 mg/kg b.w. Furthermore, hypertrophy of the endoplasmic reticulum could be detected in the treated neurons (Glees and Hasan 1976), which might be an indication of regenerative attempt on the part of the cultured cells.

Some other pharmacological products, such as potassium orotate (25 mg/day) removed lipofuscin from neurons of nucleus reticularis gigantocellularis (Low et al 1974). Also, nordihydroguaiaretic acid (NDGA), a natural antioxidant, observed to partially correct and/or prevent vitamin-E deficiency symptoms in chicks restricted pigment accumulation in natural death mutant of Neurospora Crassa (Munkres and Rana 1978). A free radical scavenger, citiolone (N-homocysteine thiolactone) has also been observed to retard pigment accumulation and favour its removal (Aloj-Totaro et al 1981). Patro (1990) made a comparative study of the under-mentioned four neurotropic agents for their lipofuscinolytic activity. Dimethylaminothanol, chlorpromazine, centrophenoxine and encephabol depleted the lipofuscin in Purkinje

Histogram 3. Changes in levels of lipid peroxide in rat brain of different age groups

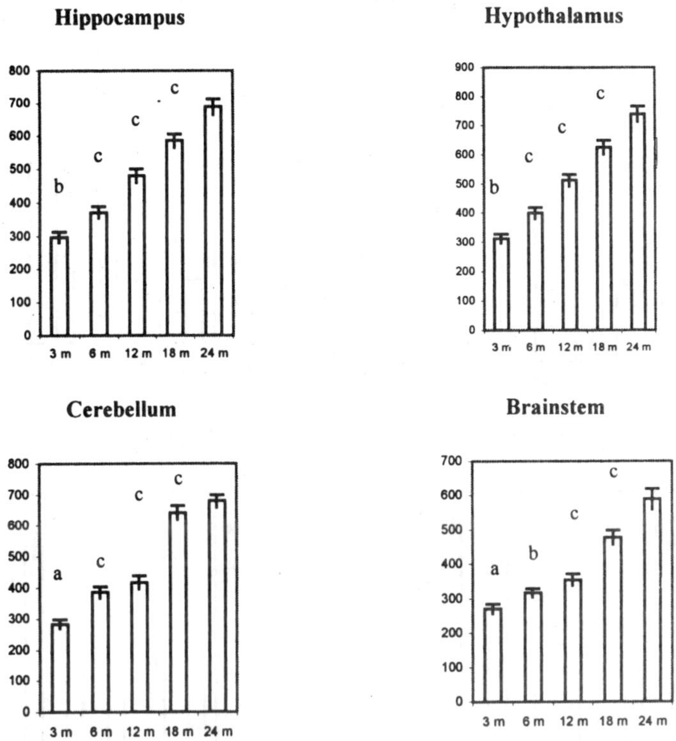

Results, are expressed as n mole of MDA/g tissue weight; each value is a mean ± SE of 6 rat brains. a-c statistical analysis of values obtained in different age groups. a=<0.05; b=<0.01; c=<0.001.

neuronal area by 21%, 27%, 42% and 61% respectively after 3 months of treatment.

On the other hand, Roy et al (1983) proposed yet another mechanism of action of centrophenoxine. They observed that this drug is capable of increasing the activity of antioxidative enzymes or also acts by regenerating protein synthesizing machinery. The latter finding supports the earlier electron microscopically obtained results of Hasan et al (1974a).

Unfortunately, not much work has been done on the effects of Ayurvedic herbal remedies in this thrust area of research. Recently, Vohra et al (2001) have reported their findings related to the effect of Maharshi Amrit Kalash (MAK), an Ayurvedic herbal mixture, on two biomarkers of ageing, lipid peroxidation and lipofuscin, on ageing guinea pig brain. Brain regions studied were cerebral cortex, hypothalamus, cerebellum and spinal cord. Treatment with MAK, at a dose of 500 mg/kg body weight, daily for two months, reduced the lipid peroxide and lipofuscin pigment significantly in brain regions and also helped in restoring normal oxygen consumption in the older animals. They concluded that their results indicated anti-oxidant properties of MAK.

Histogram 4. Levels of conjugated dienes in various regions of the rat brain of different age groups

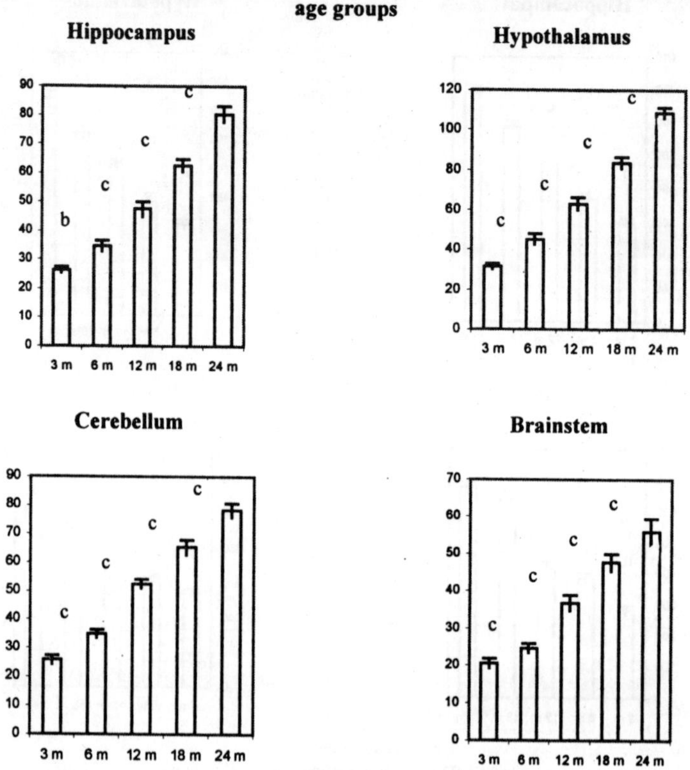

Results, are expressed as μ mole/100g total lipids each value is a mean ± SE of 6 brains. a-c statistical analysis of values obtained in different age groups. a=<0.05; b=<0.01; c=<0.001.

Acknowledgement

The authors are indebted to the Indian National Science Academy, New Delhi (India), for their financial support in the form of a Senior Scientist position to MH. Furthermore, the research facilities provided to the senior author by Prof. Ashok Sahai, HOD, Anatomy, K.G. Medical University, Lucknow and Dr. V.K. Bajpai, Electron Miscroscope Division, CDRI, Lucknow are gratefully acknowledged.

References

Aloj-Totaro EA, Pisanti FA, Russo E and Russo G (1981) Boll Soc H Biol Sper **LVIII**: 2047-2052.

Aloj-Totaro EA, Pisanti FA, Contillo A and Libevatri E (1985a) Res Commun Chem Pharmacol **47**: 415-426.

Aloj-Totaro EA, Pisanti FA and Libevatri E (1985b) Arch Geront Geriatr **4**: 67-72.

Bhardwaj SK (1996) Ph.D. Thesis, Jivaji University, Gwalior, India.

Boldrini R, Biselli R, Santorelli FM, Bosman C (2001) Ultrastruct Pathol **25**: 51-58.

Brunk UT and Terman A (2002) Free Radic Biol Med **33**: 611-619.

Chemnitius KH, Machnik G, Löw O, Amrich M and Urban J (1970). Exp Path **4**: 163-167.
Glees P and Hasan M (1976) Lipofuscin in neuronal ageing and diseases. Stuttgart, Georg Thieme Publishers, pp. 1-11.
Gupta A and Hasan M (1996) In: Vinod Kumar (ed.) "Aging: Indian perspective and global scenario". Proc. International Symposium on Gerontology and Seventh Conference of Association of Gerontology (India), pp. 419-422.
Harman D (1969) J Amer Geriatrics Soc **17**: 721-735.
Hasan M (1996) Association of Gerontology (India), pp. 423-427.
Hasan M and Glees P (1972a) Exp Geront **7**: 345-351.
Hasan M and Glees P (1972b) Gerontologia **18**: 217-236.
Hasan M and Glees P (1973a) Acta Anat **84**: 85-95.
Hasan M and Glees P (1973b) Exp Geront **8**: 75-83.
Hasan M and Glees P (1974a) Exp Geront **9**: 153-159.
Hasan M and Glees P (1974b) Cell Tissue Res **150**: 369-375.
Kawashima S (2003) Proc. of International Conference on Role of Free Radicals and Antioxidants in Health and Disease, CSM Medical University, Lucknow (India).
Kikugawa K (1986) Adv Free Radical Biol Med **2**: 389-417.
Kikugawa K (1988) In: Lipofuscin-1987, Zs.Nagy, I. Ed. Elsevier Science Publisher, Amsterdam, pp. 51-68.
Kikugawa K and Beppu M (1987) Chem Phys Lipids **44**: 277-296.
Kosugi H and Kikugawa K (1986) Lipids **21**: 537-542.
Ledda M, Barni L, Altieri L, Altieri L, Pannese F (1999) J Submicrosc Cytol Pathol **31**: 237-246.
Low VO, Machmik G, Amrich M, Urban J and Lemke H (1974) Zbl Phar **113**: 701-704.
Marcuse R and Johansson L (1973) J Amer Oil Chem Soc **50**: 387-391.
Maurya AK, Sharma D and Singh R (1990) In: Perspectives of Aging Research. Singh R and Singhal GS (eds.) "Todays and Tomorrow" Printers and Publishers, New Delhi, 163-165.
Munkres KD and Rana RS (1978) Mech Ageing Dev **7**: 399-406.
Nandy K (1968) J Geront **23**: 82-92.
Nandy K and Bourne G (1966) Nature (Lond) **210**: 313-314.
Ohkawa H, Ohishi N and Yagi K (1979) Anal Biochem **95**: 351-358.
Ohtani R and Kawashima S (1983) Exp Gerontol **18**: 105-112.
Patro IK (1990) In: Perspectives in Aging Research. Singh, R. and Singhal, G.S. (eds.). Today and Tomorrow's Printers and Publishers, New Delhi, pp. 133-136.
Patro IK (2002) Indian J Gerontol **16**: 45-68.
Patro IK and Patro N (1992) Ind Rev Life Sci **12**: 133-146.
Patro IK and Sharma SP (1984) Exp Gerontol **19**: 241-252.
Patro IK, Sharma SP and Patro N (1988) Ind Rev Life Sci **8**: 95-12.
Patro IK, Sharma SP and Patro N (1989) Proc Natl Acad Sci (India) **59**: 287-293.
Peretz B (1989) In: The cellular basis of neuronal plasticity: Morphology and biochemistry of molluscan neurons. Bullock, A. (ed.) University of Manchester Press, Manchester, pp. 200-220.
Piasanti EA, Fascotare S, Aloj-Totaro E and Crici A (1983) J Gerontol **31**: 785-795.
Rehncrona S, Smith DS, Akessm B, Westerberg E and Siesjo BK (1980) J Neurochem **34**: 1630-1634.
Roy D, Pathak DN and Singh R (1983) Exp Gerontol **18**: 185-197.
Sharma SP and James TJ (1991) Free Radical Biol & Med **10**: 443-444.
Singhal P (1996) Ph.D. Thesis, Jiwaji University, Gwalior (India).
Sohal RS (1981) Age pigments, Elsevier/North Holland, Biomedical Press, Amsterdam.
Spoerri PE and Glees P (1974) Mech Ageing Dev **3**: 131-155.
Spoerri PE, Glees P and El-Ghazzawi E (1974) Mech Ageing Dev **3**: 311-321.
Tappel AL (1975) In: Pathology of Cell Membranes 1. Trump,BF and Astila, A.D (eds.), Academic Press, New York, pp. 145-170.
Tappel AL (1980) In: Free Radicals, in Biology. W.A. Pryor (ed.), Vol. 4, Academic Press, New York, 1-47.
Uchiyama M and Mihara M (1978) Anal Biochem **86**: 271-278.
Várkonyi T, Domokos H, Maurer M, Zoltán ÖT, Csillik B and Földi M (1970) Z Gerontol **3**: 245-260.
Vohra BP, Sharma SP, Kansal VK, Gupta SK (2001) Indian J Exp Biol **39**: 355-359.
Zs-Nagy I (1987) In: Lipofuscin-1987, Zs.-Nagy, I. (ed.) Elsevier Science Publications, Amsterdam.

Molecular and Cellular Neurobiology
Editors: M.K. Thakur and S. Prasad

Oxidative Stress in Brain Aging

Sasanka Chakrabarti and Uttara Chatterjee[1]

Department of Biochemistry, [1]Department of Pathology, University College of Medicine
Calcutta University, 244B, Acharya J.C. Bose Road, Calcutta-700 020, India

Abstract

Reactive oxygen species produced from metabolic pathways can induce membrane lipid peroxidation and varied forms of protein and DNA damage. Such molecular damage manifests at the cellular level in the form of altered membrane properties, inactivation of enzyme or receptor functions, loss of Ca^{2+} homeostasis, induction of apoptosis and genetic instability of different kinds. The brain is highly vulnerable to oxidative damage because of readily available oxidizable substrates and an impaired anti-oxidant defence. The aged brain shows characteristic morphological and histological changes and an accumulation of oxidative damage products of lipid, protein and DNA. This is associated with a decline of several key metabolic and transport enzymes and some impairment of mitochondrial functionwhich are presumably related to oxidative stress. Further, in comparison to brains from young animlas, the aged brain produces more of toxic oxygen radicals while its antioxidatnt defence is more compromised. This implies that oxidative damage mecahanisms are likely to play a central role in the phenomenon of brain aging. However, this aspect needs further explorations with regard to identification of specific oxidative lesions in target proteins (channels, receptors, enzymes etc.) from aged brain and its possible consequence on neuronal functions.

Keywords: reactive oxygen species, 8-hydroxy deoxyguanosine, mitochondrial dysfunction, apoptosis, brain aging.

Introduction

The interplay of biochemical changes, cellular responses and actions of specific genes during the life-time of an organism ultimately leads to the generation of aging phenotypes in which the fitness of the organism is grossly compromised. The phenomenon of aging affects cells, tissues and organisms as a whole. The molecular and cellular events involved in the phenomenon of aging especially in complex multicellular organisms are not clearly understood and this has given rise to a multiplicity of theories which overlap with each other to a considerable extent (Holliday, 1995). A recent hypothesis based on Darwinian concept of evolution proposes that aging is the result of activities of certain genes selected during evolution because of their beneficial effects in young organisms but the same genes produce deleterious and damaging effects in old organisms. This concept has been termed antagonistic pleiotropy (Campisi, 2003). Attractive though this theory is, there are other theories such as spontaneous and induced somatic mutations and DNA damage, protein error theory and immunological theory, which are also of great importance. It is not necessary to assume that there is one single genetic mechanism involved in the aging process. Moreover, the importance of genetic mechanisms has offen been overstated ignoring the fact that a central role is played by many external or endogeneous modulatory factors in the generation of aging phenotypes. Among the external modulatory factors the damage by short-lived free radicals of oxygen is generally considered to be of utmost importance in the phenomenon of aging and age-related degenerative disorders (Ames et al 1993, Sohal and Weindruch, 1996).

The original hypothesis on the role of oxygen free radicals or reactive oxygen species (ROS) in aging has gained much support from recent work. Thus overexpression of anti-oxidant enzymes extends the life of transgenic *Drosophila melanogaster*. The susceptibility to oxidative damage and the levels of anti-oxidant enzymes correlate nicely with the organism's life-span and the caloric restriction lowers the extent of oxidative stress and increases the life-span of mammals (Harman, 1981; Sohal and Weindruch, 1996; Calabrese et al 2000). The rates of mitochondrial generation of reactive oxygen species and accumulation of oxidized products of lipid and protein increase exponentially with age (Sohal and Weindruch, 1996). In mutant houseflies with larger life-expectancy, a lower rate of mitochondrial production of ROS and a decreased level of oxidative protein and DNA damage and higher activities of anti-oxidant enzymes have been observed compared to age-matched short-lived flies (Sohal and Windruch, 1996). These results have led to a resurgence of research activity on the role of free-radicals in the phenomenon of aging. It is interesting to point out that oxyradical damage theory of aging overlaps to a great extent with other theories such as somatic mutation or protein error in as much as free radicals of oxygen can damage DNA or proteins extensively.

In the context of brain aging, the role of reactive oxygen species (ROS) can hardly be over-emphasized because of special vulnerability of brain to free-radical induced injury. Of all organs, the brain has the highest utilisation of oxygen per gram of tissue and the membrane phospholipids of brain are very rich in polyunsaturated fatty acids, which are particularly sensitive to attack by oxygen radicals (Halliwell, 1992; Joseph et al 1996; Halliwell and Gutteridge, 1999). The brain contains many oxidizable substrates like catecholamines and several regions in brain (globus pallidus substantia nigra etc) are enriched in transition metals which can trigger formation of ROS (Halliwell, 1992; Halliwell and Guteridge, 1999). The high content of ascorbate in brain may further aggravate the situation by keeping the transition metal in the reduced state or releasing them in reduced form from storage proteins e.g. ferritin. Moreover, the brain is somewhat underdefended against attack by ROS, since it is deficient in catalase activity and has only moderate amounts of superoxide dismutase and glutathione peroxidase (Halliwell and Gutteridge, 1999). These three anti-oxidant enzymes act in concert to protect against attack by oxygen free radicals (Fig. 1).

This review will briefly discuss the various forms of reactive oxygen species *in vivo* and their damaging actions on cellular components in general (Section I) and then highlight the general features of brain aging and its relation to oxidative damage (Section II). The results obtained from our study in this area will be briefly presented along with future research prospects in this field in post-genomic era. (Section III).

Section I

Nature and Sources of Reactive Oxygen Species (ROS)

The most important members of ROS are superoxide radical ($O°_2$), H_2O_2, °OH radical and singlet oxygen. The major sources of superoxide radicals *in vivo* are electron transport chains in mitochondrial and endoplasmic reticulum, respiratory burst of phagocytic cells, metal-catalysed substrate oxidation, oxidation of xanthine by xanthine oxidase and decomposition of lipid hydroperoxides by transition metals (Halliwell and Gutteridge, 1984; Halliwell and Gutteridge, 1999). In hydrophobic environment $O°_2^-$ radicals are highly reactive and act as a strong base and a nucleophile. In aqueous environment, however, it is poorly reactive and unlikely to cause any significant damage in biological systems *in vivo*. It is converted to H_2O_2 by spontaneous dismutation or enzymatically by superoxide dismutase. In presence of nitric oxide (NO) it produces highly toxic peroxynitrite radical which represents an important mechanism of toxicity of NO.

Fig. 1
Oxygen Radicals : Pathways of Formation and Removal

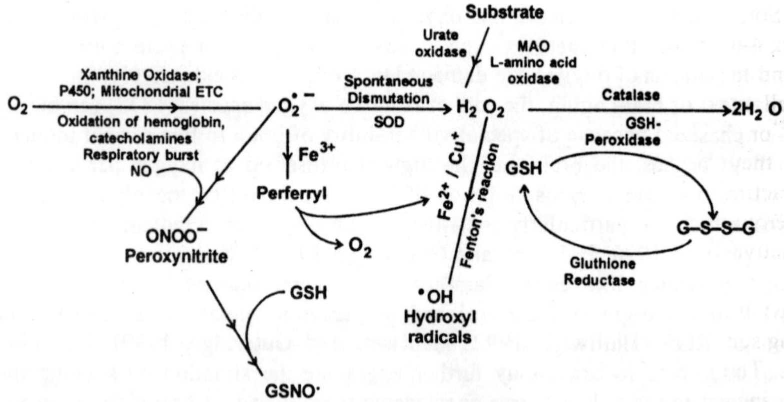

The major source of H_2O_2 in tissue is the spontaneous or enzymatic dismutation of superoxide radicals, but enzymes like monoamine oxidase (MAO), L-aminoacid oxidases urate oxidase etc. also produce H_2O_2. It is a weak oxidizing agent and has limited direct toxicity. It can, however, cross biological membranes and in the presence of transition metal ions, can give rise to highly reactive hydroxyl radicals (°OH). Hydrogen peroxide is removed by either catalase or selenium containing glutathione peroxidase (StClair and Chow, 1996; Halliwell and Gutteridge, 1999).

The most highly reactive member of ROS is °OH radical which in aqueous solution may be generated by the action of ionising radiation but the major source of hydroxyl radicals *in vivo* is Fenton's reaction in which H_2O_2 is broken down by transition metals like Fe^{2+} or Cu^+ (Halliwell and Gutteridge, 1984; Halliwell and

Gutteridge, 1999). The inter-related reactions involving O°_2, H_2O_2, $^{\circ}OH$ are shown in Fig. 1. Singlet oxygen, another member of ROS, is not a free-radical since it has nc unpaired electrons. Within living system, singlet oxygen can be generated by photodecomposition of chlorophylls, retinal flavins or porphyrins. Singlet oxygen can combine chemically with biomolecules or can transfer its excitation energy to them.

Molecular Targets of ROS

The various members of ROS family and especially the $^{\bullet}OH$ radicals can avidly react with critical biomoleules like phospholipids, proteins and DNA producing an array of damaged products, many of which are non-functional molecules but some of them are reactive and highly toxic.

***Phospholipids*:** The phospholipids in biological membranes contain considerable amounts of polyunsaturated fatty acids which are prone to attack by reactive oxyradicals. The free-radicals of oxygen set off a chain reaction by abstracting a H-atom from a polyunsaturated fatty acid moiety and toxic intermediates like peroxyl radicals, lipid hydroperoxides, lipid peroxides and finally smaller compounds like aldehydes, ketones, carboxylic acids and hydrocarbons like ethane, pentane etc. are formed (Fig.2). This process known as lipid peroxidation, proceeds in three stages of chain initiation, propagation and termination (Slater, 1984; Halliwell and Gutteridge, 1999). Some of the products of lipid peroxidation like malondialdehyde (MDA), acrolein, 4-hydroxynonenal (4-HNE) etc. can form adducts rapidly with proteins and DNA and have cytotoxic and mutagenic effects (Slater, 1984; Sullivan et al 1991; Halliwell and Gutteridge, 1999; Picklo and Montine, 2001). The transition metals like Fe^{2+}or Cu^{+} can promote lipid peroxidation in biological systems in several ways. Firstly, they can take part in the formation of reactive oxyradicals, which are H-abstracting species. Secondly, they can cause catalytic decomposition of lipid hydroperoxides to generate peroxy and alkoxy radicals which further propagate the peroxidative process.

***Proteins*:** Varied forms of oxidative protein modifications have been identified by exposing subcellular components or purified protein preparations to ROS generating systems. The loss of protein thiols, fragmentation or cross-linking of polypeptide chains, incorporation of carbonyl groups in proteins, formation of protein hydroperoxides, specific amino-acid modifications and glycation of proteins can occur as a result of oxidative damage (Hunt et al 1988; Stadtman, 1990; Dean et al 1997; Griffiths, 2000). Oxidation of amino-acids like phenylalanine, tyrosine, histidine etc. can give rise to hydroxylated products e.g. protein-bound DOPA, while methionine residues in proteins can be oxidatively modified to methionine sulphoxide. The formation of bityrosine residues in proteins from $^{\cdot}OH$ mediated damage or the generationof 3-nitrotyrosine from interaction with peroxynitrite radical are important oxidative damage markers of proteins (Dean et al 1997; Halliwell and Gutteridge, 1999). A heterogeneous group of fluorescent damage products is also derived from the interaction of proteins with lipid peroxidation enc products (Dillard and Tappel, 1984; Shimasaki, 1994). These products contribute to ceroid, lipofuscin and other pigments found in cells.

Fig. 2
Outline of Lipid peroxidation

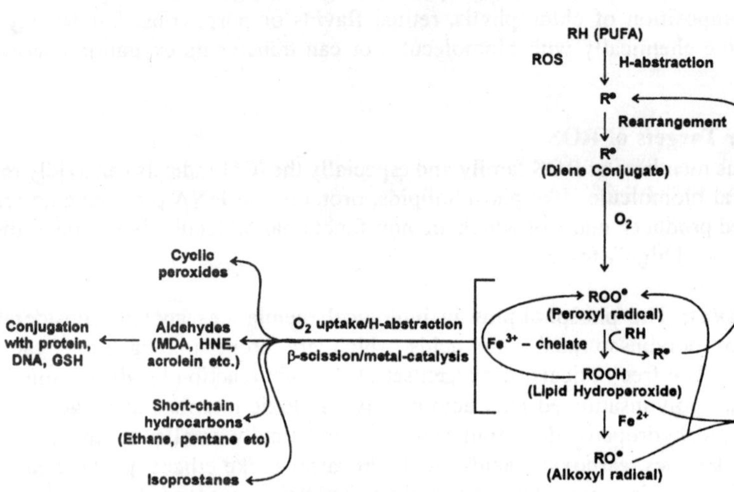

Some forms of oxidative protein damage are primary in nature and result from a direct attack by oxygen radicals while secondary damage is the consequence of reactions between radical intermediates or reactive end products of lipid peroxidation (MDA or 4-HNE) and functional groups of polypeptide chains (Shacter et al 1996). Oxidative protein damage such as from metal-catalyzed oxidation or several mixed-function oxidation system can lead to the inactivation of enzymes e.g. glutamine synthetase pyruvate kinase, phosphoglycerate, kinase etc. (Levine, 1983; Fucci et al 1983; Halliwell and Gutteridge, 1999). Multicatalytic proteasome (MCP) which is responsible for intracellular degradation of damaged or oxidized protein, is also highly suceptible to oxidative stress (Friguet and Szweda, 1997). Non-enzymes like, □-proteinase inhibitor, neutrophil serine-proteinase inhibitor etc are also inactivated by metal catalyzed oxidation (Halliwell and Gutteridge, 1999).

DNA

DNA is a primary target of attack by ROS, expecially the ˙OH radicals and singlet oxygen. The oxidation products include hydroxylated derivatives of guanine and adenine at positions 8, 5and 4 and imidazole ring-opened derivatives. An attack by ˙OH radicals at position C-8 of guanine produces derivatives like 8-hydroxydeoxyguanosine (*8-oxodG*) and 2, 6-diamino-4-hydroxy-5-formamidopyrimidine (*FAP$_{Y}$G*). 8-hydroxy-deoxyguanosine is widely used as a marker of oxidative DNA damage. The reaction of ˙OH radicals with adenine produces similar products. The pyrimidine bases can undergo oxidative damage tc produce a multitude of products like thymine glycol, cytosine glycol, 5,6 dihydroxycytosine etc (Simandan et al 1998; Halliwell and Gutteridge, 1999). The deoxyribose sugar in DNA is also damaged by radical attack to produce a variety of carbonyl derivatives. DNA fragmentation, single-strand breaks, depurination and

DNA-protein cross-links are also major consequences of ROS attack on nucleic acids (Dizdaroglu, 1992; Halliwell and Gutteridge, 1999). Several other types of oxidative DNA lesions such as type II I compounds and exocyclic DNA adducts have been identified by ^{32}P-post-labelling techniques (Nair et al 1995; Joseph et al 1996; Flora et al 1996). It has been shown that oxidative modifications of DNA also results in a block of PCR amplification of target sequences and ligation-mediated polymerase chain-reaction (LMPCR) has been used to monitor the frequency and distribution of oxidative lesions induced by ROS in DNA. The damage to mitochondrial DNA by oxidative stress is far more extensive than nuclear DNA partly because of proximity of mt DNA with oxidant generating electron-transport chain of mitochondria and partly owing to lack of mitochondrial DNA repair enzymes (Yakes and Van Hauten, 1997).

Cellular Effects of Oxidative Damage: Lipid peroxidation, protein oxidation and oxidative DNA lesions cause extensive structural and functional derangement of cells through inter-related pathways. The oxidative insult causes decreased membrane fluidity, altered membrane permeability and excitability as well as loss of functions of soluble and memebrane-bound enzymes (e.g. Na^+,K^+-ATPase, Ca^{2+}-ATPase), transport proteins and receptors (Vladimirov et al 1980; Muakkassah-Kelly et al 1982; Heikkila and Cabbat, 1983; Levine, 1983; Morel et al 1998; Halliwell and Gutteridge, 1999; Calabrese et al 2000; Miralles et al 2001). Some of the these

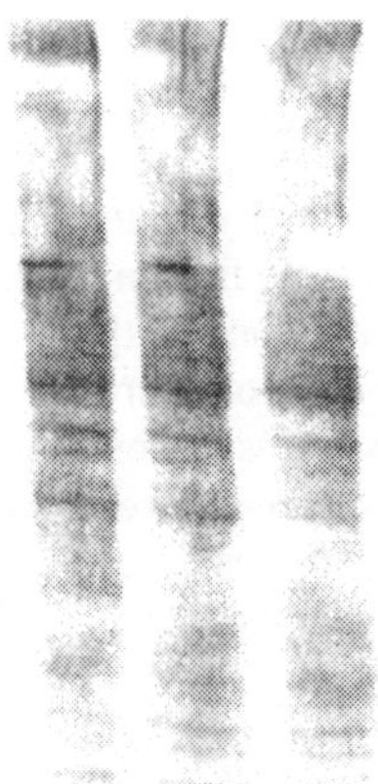

Fig 3. Immuno-detection of protein carbonyls in rat brain crude synaptosomes (P_2 pellet) after in vitro incubation with Fe^{2+} (100μM) and ascorbate 2 (mM). Lanes A or B : P_2 pellet unincubated or incubated with buffer only. Lanes C – E : P_2 pellet + Fe^{2+} + ascorbate (Lane C), P_2 pellet + Fe^{2+} + ascorbate + 50 μg/ml of catalase (Lane D), P_2 pellet + Fe^{2+} + ascorbate + 20 mM mannitol (Lane E).

effects are due partly to disruption of phospholipid bilayer architecture by peroxidative process but direct free-radical damage (oxyradicals or lipid peroxy radical) to proteins or modification of protein chains by reactive end products of lipid peroxidation (MDA, 4-HNE etc) may account for other effects in a large measure. Further, the oxidised proteins, may be recognized by the immune system triggering antibody formation and auto-immune response as happens in rheumatoid arthritis (Halliwell and Gutteridge, 1999). The multicatalytic proteasome (MCP) responsible for non-lysosomal degradation of proteins inside cells is also inactivated by oxidative insult and this in turn may lead to accumulation of oxidized and damaged proteins in tissue. Damage to DNA-repair enzymes e.g. endonucleases, ligase etc.could aggravate the accumulation of oxidative DNA lesions and increase mutation frequency. The damage to DNA by ROS in the form of single or double strand breaks, modified bases, abasic sites or adducts-of DNA with MDA or 4-HNE may lead to mutations directly or during replication or repair of damaged DNA. It is conceivable that some formsof DNA damage will block replication and transcription as well (Halliwell and Gutteridge, 1999).

The inter-related damage pathways triggered by oxidative stress in cells have several other manifestations such as an increased intra-cellular Ca^{2+} level, mitochondrial dysfunction and activation of apoptotic pathways. The raised level of Ca^{2+} inside cells can cause membrane 'blebbing', activation of Ca^{2+}– dependent protease (e.g. calpain) and nucleases (Halliwell and Gutteridge, 1999). Mitochondrial dysfunction manifests in the form of altered membrane potential, opening of permeability transition pores and damage to respiratory enzymes (Zhang et al 1990; Bernardi, 1996; Bailey, 2003). The activation of apoptosis in various cultured cell lines is often accompanied by excess generation of ROS or depletion of anti-oxidants and one mechanism by which oxyradicals activate apoptosis in cells is by releasing 'pro-apoptotic protein' in cytosol e.g. cytochrome c. from mitochondria (Halliwell and Gutteridge, 1999; Macho et al 1999; Kannan and Jain, 2000; Kwon et al 2001). The anti-apoptotic protein bcl-2 protects against apoptosis induced by H_2O_2, t-butyl hydroperoxide, ionising radiation etc, by several mechanisms including one that prevents release of pro-apoptotic protein from mitochondria.

The net effect of ROS in cells or tissues is, therefore, loss of membrane integrity, genetic instability and finally apoptosis or necrosis. The anti-oxidant defence inside cells comprise of several enzymes like superoxide dismutase, catalase, glutathione peroxidase and several smaller molecules like □-tocopherol, carotenoids, reduced glutathione, ascorbate etc. which can scavenge various toxic radicals involved in initiation and propagation of oxidative damage.

Section II

Brain Aging

General Features of the Aged Brain: In brain aging even in the absence of any distinctive neurodegenerative disease, typical morphological, biochemical and metabolic alterations take place along with cognitive deficits. The gross morphological changes include reduction in brain volume, gyral atrophy and enlargement of ventricles. This is associated with some neuronal loss, glial reactivity and loss of synaptic connectivity (Anderten, 1997; Mrak et al 1997; Schochet, 1998). However, recent data indicate that in aged brain neuronal loss is much less than what was suggested from earlier studies and further neuronal plasticity continue even in very aged brain (Mrak et al 1997). Histologically, some characteristic changes are manifested in aged brain which include increasing number of corpora amylacea within astrocytic processes, accumulation of 'age-pigments' (lipofuscin), granulovacuolar degeneration of neurons, variable but diffuse deposition of β-amyloid and appearance of neurofibrillary tangles especially in amygdala and hippocampus (Anderten, 1997; Yasha et al 1997; Schoehet, 1998). There are also some breakdown of myelin seahth, changes in oligodendrocytes and loss of white matter as seen in the brain of aged non-human primates (Peters, 2002).

The metabolic and biochemical changes in aged brain present as a spectrum in which exist a variety of DNA lesions, inactivation of enzymes, mitochondrial dysfunction, alteration of membrane properties, accumulation of oxidized products of lipids and protein including advanced glycation end-products and some alterations in the expression of specific genes including those coding for apoptosis-related proteins. Many but not all of these changes have been linked to oxidative stress.

***Oxidative Damage to lipid, protein and DNA in aged brain*:** The special vulnerability of brain to oxidative injury has already been discussed. In aged brain, the production of reactive oxygen species especially in mitochondria is further enhanced (Muscari et al 1990; Sohal and Sohal 1991; Driver et al 2000). The peroxidizability of brain membranes is also increased during aging and coupled with it has been a decline in the level of anti-oxidant enzymes (Viani et al 1991; Tian et al 1998). The stage thus is set precisely for the propagation of oxidative damage pathways in brain during aging.

A large body of evidence indicates an accumulation of oxidatively modified proteins and an enhanced membrane lipid peroxidation in brain during aging. An increased production of MDA, fluorescent lipid peroxidation products, protein carbonyls and nitrotyrosine and a loss of protein thiols have been reported in brain homogenate and subcellular fractions from aged animals compared to those from young ones (Sawade and Carlson, 1987; Smith et al 1991; Tian et al 1998; Calabresc et al 2000; Chakraborty et al 2003). The levels of oxidized proteins in brain increase progressively with age and may represent 30-50% of total protein in old aminals (Calabrese et al 2000). The oxidative modifications of proteins and lipids are accompanied by inactivation of several enzymes or transport proteins e.g. glutamine synthetase, cytosolic neutral protease, glucose-6-phosphate dehydrogenase, multicatalytic proteasome (MCP), Na^+,K^+ – ATPase, glutamate transporter etc in aged brain (Wheeler and Ondo, 1986; Viani et al 1991; Calabrese et al 2000; Keller et al 2000). It is also possible, however, that apart from oxidative inactivation, expression of some of these enzymes or proteins is also impaired in aged brain. The increased peroxidative damage to membrane lipids results in an altered lipid micro-environment, and an increased rigidity of membranes, along with alteration in membrane potential, ion-transport and inactivation of membrane-bound enzymes in nerve terminals of aged brain (Tanaka and Ando, 1990; Benzi et al 1994; Choi and Yu, 1995; Chakraborty et al 2003).

DNA lesions in aged brain take varied forms. In permeabilized neurons from aged rat brain, an accumulation of single and double-stranded breaks, which are mediated by a non-apoptotic mechanism, has been reported (Mandavilli and Subba Rao, 1996a; Mandavilli and Subba Rao, 1996b; Subba Rao et al 2000). The authors, however, have not directly implicated oxidative mechanism with such DNA damage in aged brain, However, other kinds of DNA lesions which are well-known markers of oxidative damage such as 8-oxo-2-deoxyguanosine (oxo-8dG), DNA-protein cross-links, MDA-DNA adducts or certain types of I-compounds also increase consistently in aged brain (Adelman et al 1988; Randerath et al 1993; Cai et al 1996; Izzotti et al 1999; Hamilton et al 2001). Despite a high rate of mitochondrial turnover, oxidative DNA lesions accumulate with age in mt DNA and this also explains the high frequency of mt DNA mutation in aged tissues (Calabrese et al 2000).

***Mitochondrial dysfunction, in brain aging: relation to oxidative stress*:** Several forms of mitochondrial dysfunction e.g. a decrease in cytochrome oxidase activity, a reduction in the level of cytochrome a /a3 without changes in ubiquinone and cytochrome b population, an enhanced activation of mitochondrial transition pore, loss of mitochondrial GSH, altered mitochondrial transmembrane potential, decrease in the activity of aconitase and carnitine acetyl-transferase and point and deletion mutations of mitochondrial DNA (mt DNA) have been reported in brain during aging by various groups (Schapira and Cooper, 1992; Benzi et al 1992; Garcia et al 1996; Mather and Rottenberg, 2000; Sastre et al 2002). These changes in part are related to oxidative stress in aged brain. The block in mitochondrial respiratory chain during aging as a result of oxidative stress can accelarate the production of further ROS by leakage of electrons from the autoxidizing chain components (e.g. ubiquinone, cytochrome b etc). Using long-extension polymerase chain reaction (LX-PCR) several varieties of mt DNA rearrangements have been identified in mouse and human brain during aging. Such mitochondrial genome rearrangements possibly are related to oxidative stress as has been implicated in ANT (adenine nucleotide translocator) mutant mice (Melov et al 1997; Melov et al 1999).

Table 1. Lipid peroxidation and protein oxidation in rat brain crude synaptomal-mitochondrial fraction induced by Fe^{2+} (100 μm) and ascorbate (2 mM), Values are mean ± S.E. of 4 observations.

Incubation (2h)	Lipid peroxides nmoles MDA/mg protein	Fluorescent peroxidation products (Relative fluorescence units)	Protein thiols(nmoles/mg /protein)	Protein carbonyls (nmoles/mg. protein)
P_2 alone (control)	0.66±0.03	0.15±0.021	98.8±0.50	4.45±0.78
$P_2 + Fe^{2+}$ + ascorbate	18.08±0.91	0.628±0.03	58.4±0.54	17.13±1.02
$P_2 + Fe^{2+}$+ ascbroate + BHT (0.2 mM)	0.49±0.02	0.228±0.05	80.2±0.72	3.79±0.37

Apoptosis, oxidative stress and brain aging

The relationship of oxidative stress with the activation of apoptotic cascade is well-established (Adams et al 1996; Halliwell and Gutteridge, 1999; Knon et al 2001; Single et al 2001; Pollock and Leeuwenburgh, 2001). In aged brain, however, the evidence in favour of apoptotic neuronal death is scanty and inconclusive, although in several age-related diseases like Alzheimer's disease or Parkinson's disease, the activation of apoptotic cascades in neuronal death has been established.

In general, some studies have indicated the activation of apoptotic pathways in aged brain, while others have reported that markers of apoptosis are present in a low level in aged rat brain compared to developing brain of early post-natal rats (Kiatipattanasakul et al 1996; White and Barone, 2001; Martin et al 2002). The expression of apoptosis linked gene ALG-2 has been shown to be similar in young and aged rat brain (Venn and Conway, 1998). More interestingly, mitochondrial bcl 2 protein which protects cells from apoptosis is over expressed in aged brain and the phenomenon can be reversed by treatment with free-radical spin-trap like N-tert-butyl-l-phenyl nitrone [Kaufmann et al 2001]. The anti-oxidant effect of bcl-2 has also been demonstrated in another study [Howard et al 2002]. It is plausible that bcl-2 plays an important protective role to prevent oxidative stress mediated neuronal death in aged brain and indirectly this also lends support to the recent-evidence that very little neuronal loss takes place in brain during aging. On the other hand, in the absence of apoptosis, many damaged neurons would persist in aged brain thereby deranging the neuronal circuitry and impairing the function of brain.

***Ca^{2+}* - homeostasis, brain aging and oxidative hypothesis:** In brain aging plasma membrane associated Na^+, K^+ – ATPase, Ca^{2+} – ATPase have been shown to be decreased in a complex way presumably involving oxidative inactivation [Tanaka and Ando, 1990; Benzi et al 1994; Calabrese et al 2000; Chakraborty et al 2003] Moreover, several methionine residues of calmodulin get oxidized to methionine sulphoxides during aging resulting in impaiment of its function with regard to activation of target protein such as plasma membrane Ca^{2+} – ATPase [Calabrese et al 2000]. Thus, in senescent neurons, an increased level of Ca^{2+} exists causing activation of Ca^{2+} , dependent proteases, endonucleases etc. with consequent damage to cells.

Section III

Results

In our study, rat brain crude synaptosomes or purified mitochondria were exposed to an oxidative insult *in vitro* in the form of a mixture of Fe^{2+} (100 □M) and ascorbate (2 mM) for varying periods of time. The oxidative damage to lipid and protein components of brain membranes including synaptosomal transport enzymes and mitochondrial respiratory chain components were assessed. Under such conditions of incubation production of MDA and fluorescent damage products of peroxidation increased dramatically in brain membranes (Table 1). This was associated with loss of protein thiols, increased incorporation of protein carbonyls and non-disulphide co-valent cross-linking of membrane proteins (Table 1, Fig. 3). The lipid peroxidation and oxidative protein modifications were not prevented by hydroxyl radical scavenger or anti-oxidant enzymes like SOD or catalase implying that active oxygen radicals per se are not involved in the process [Chakraborty *et al.*, 2001]. On the other hand, □-tocopherol and butylated hydroxytoluene (BHT), both lipid soluble anti-oxidants, could prevent lipid peroxidation as well as protein oxidation [Chakraborty et al 2001]. The synaptosomal Na^+, K^+– ATPase activity was also markedly inhibited by incubation *in vitro* in presence of iron/ascorbate [Chakraborty et al 2003]. The enzyme inactivation again was not-prevented by radical scavenger or antioxidant enzyme catalase but BHT had a significant protective effect (Table 2).

Table 2. Inactivation of rat brain (P_2-pellet) Na+, K+-ATPase during *in vitro* incubation with Fe^{2+} (100 □M) and ascorbate (2 mM), Values are mean ± S.E. of 4 observations.

Incubation (2h)	Na^+, K^+-ATPase activity (□moles Pi liberated/mg protein/h)
P_2 pellet alone (control)	11.22±0.39
$P_2 + Fe^{2+}$ + ascorbate	2.07±0.36
$P_2 + Fe^{2+}$ + ascorbate + catalase (50 □g/ml)	3.73±0.68
$P_2 + Fe^{2+}$ + ascorbate + mannitol (20 mM)	2.42±0.32
$P_2 + Fe^{2+}$ + ascorbate + BHT (0.2 mM)	11.67±0.13

The effect of iron/ascorbate induced oxidative stress on isolated rat brain mitochondria was also investigated. The electon transport chain *in vivo* is the major source of reactive oxygen species and lipid, protein and DNA of mitochondria could be early targets of oxidative injury. Interestingly, under our experimental conditions, Fe^{2+} (50 □M) and ascorbate (2 mM) did not cause any significant inhibition of complex I and IV activity of incubated mitochondria up to 2h, although both MDA and protein carbonyls accumulated in considerable amounts (results to be published elsewhere). However, the electron transport capacity of intact mitochondria in presence of substrate like succinate (5 mM) as measured by MTT reduction assay [Cohen et al 1997] was inhibited to a considerable extent by prior exposure to iron/ascorbate, Mitochondrial transmembrane potential measured by confocal microscopy using fluorescent cataionic dye was also diminished following incubation with Fe^{2+} (50 □M) and ascorbate (2 mM) for 1h. It is not clear whether such mitochondrial damage induced by Fe^{2+} and ascorbate is related to oxidative stress. Using different radical scavengers and other anti-oxidants we have attempted to elucidate the role of different members of oxygen free-radical family in causing injury to mitochondrial function.

It has been generally accepted that catecholamines like dopamine can undergo spontaneous or enzymatic oxidation to produce a variety of reactive oxygen species which contribute to vulnerability of brain to oxidative damage. We, therefore, attempted to find the effects of dopamine oxidation products on lipids, proteins and enzymes associated with crude synaptosomal-mitochondrial fraction from rat-brain. Although dopamine produced H_2O_+ and $^{\cdot}OH$ radicals *in vitro*, it did not promote either lipid peroxidation or formation of protein carbonyls in incubated brain membrane fractions. However autoxidizing dopamine caused cross-linking of membrane proteins, inhibition of Na^+, K^+ – ATPase and several forms of mitochondrial dysfunctions which are mediated primarily by quinone products of dopamine oxidation and not by reactive oxygen radicals [Khan et al 2001; Khan et al 2003].

The experiments *in vitro* provided data that demonstrated various kinds of oxidative derangements in brain synaptosomes and mitochondria. The same parameters were then examined in brain membrane fractions from young and aged rats. The levels of fluorescent peroxidation products and protein carbonyls were considerably higher in crude synaptosomal fraction from aged rat brain, compared to that from young rats, but a loss in protein thiols was not, however, apparent in the aged brain (Table 3). The high molecular weight cross-linked proteins so chorasteristically formed after brain membranes were subjected to *in vitro* oxidative stress was also not visible in the aged rat brain. The activity of Na^+, K^+ – ATPase was however, significantly impaired in brain crude synaptosomal fraction from aged rats [Chakraborty et al 2003]. These results show that some of the oxidative damages identified *in vitro* accumulate in aged brain but there are repair or rectification mechanisms for other kinds of lesions. Further, in conformity with results of our *in vitro* experiments, no significant impairment of brain mitochondrial complex I or complex IV activity was detected in aged rat compared to young ones (data to be published elsewhere). At present in our lab, several parameters of mitochondrial functions, activation of apoptotic cascade as well oxidative lesions in Na^+, K^+ - ATPase are being examined in aged rat brain.

Table 3. Oxidative damage markers in brain crude synaptosomal fraction from young and agedrats, Values are mean ± S.E. of number of animals indicated in parentheses.

	Protein thiols (nmoles/mg protein) (10)	Protein carbonyls (nmoles/mg protein) (10)	Fluorescent peroxidation products (Relative fluorescence units) (4)
Young (4-6 months)	100±5.07	4.57±0.29	0.16±0.01
Aged (18-22 months)	104±7.84	5.95±0.33	0.28±0.02

Conclusion and Future Prospects in Post-genomic Era

The large body of evidence presented in the review is convincing enough to indicate the central role of oxidative damage in brain aging. However, many lacunae exist in corelating the genomic damage, altered protein functions, neuronal dysfunction and finally cognitive decline.

One important task in this direction would be to purify several target proteins e.g. membrane associated channels, receptors, transport enzymes from aged brain and to identify the specific oxidative molecular lesions in them and to relate these to altered electrophysiology of neurons. Secondly, the different kinds of oxidative DNA lesions in aged brain may impair the rate of transcription of specific genes or result in the synthesis of non-functional or dysfunctional proteins. We have very little data on this and a major thrust of the future research would be to link the DNA damage with altered or impaired functions of proteins in aged brain. Such proteins may be neurotrophic factors or members of cell signaling pathway or enzymes involved in neurotransmitter synthesis and degradation or receptors on post-synaptic membranes It is important to remember that agonist stimulation of neuronal receptors especially in hippocampus and striatum is diminished during aging and this may have a direct bearing on the cognitive decline seen in this condition.

Endogenous or exogenous oxidant stress is known to increase the expression of genes for AP-1, NFK-, HSF etc. which modulate the response of the cells to inflammation and stress. The NFK- activation is particularly seen in brain following stroke, seizure or traumatic injury [Calabrese et al 2000]. To what extent such oxidant-modulated genes are involved in the process of brain aging remains largely unexplored at this moment but this could well be an area for intense future research.

References

Adams JD, Mukherjee SK, Klaidman, LK, Chang ML and Yasharel R (1996) Ann NY Acad Sci **786**: 135 - 151.

Adelman R, Saul RL and Ames BN (1988) Proc Natl Acad Sci USA **85**: 2706 - 2708.

Ames BN, Shingenaga MK and Hagen TM (1993) Proc Natl Acad Aci USA **90**: 7915 - 7922.

Anderton BH (1997) B Biol Sci **1363**: 1781 - 1792.

Bailey SM (2003) Free Radic Res **37**: 585 - 596

Benzi G, Gorini A, Arnaboldi R, Ghigini B and Villa RF (1994) Neurochem Int **25**: 61 - 67.

Benzi G, Pastoris O, Marzatico F, Villa R.F, Dagani F and Curti D (1992) Neurobiol Aging **13**: 361 - 368

Bernardi P (1996) Biochim Biophys Acta **1275**: 5 - 9.

Cai Q, Tian L and Wei H (1996) Gerontol **31**: 373 - 385

Calabrese V, Bates TE and Stella AMG (2000) Neurochem Res **25**: 1315 - 1341.

Campisi J (2003) Expt Gerontol **38**: 5 – 11.

Chakraborty H, Ray SN and Chakrabarti S (2001) **39**: 311 - 317.

Chakraborty H, Sen P, Sur A, Chatterjee U and Chakrabarti S (2003) Expt. Gerontol. **38**: 705 - 710

Choi, JH and Yu, BP (1995) Free Radic Biol Med **18**: 133 - 139.

Cohen G, Farooqui, R and Kesler, N (1997) Proc Natl Acad Sci USA **94** : 4890 - 4894.

Dean RT, Fu S, Stocker R and Davies MJ (1997) Biochem J **324**: 1 - 18

Dillard CJ and Tappel AL (1984) Methods Enzymol **105**: 337 - 341.

Dizdaroglu, M (1992) Mutat Res **275**: 331 - 342.

Driver AS, Kodavanti PR and Mundy WR (2000) Teratol **22**: 175 - 181

Flora SD, Izzotti A, Randerath K, Randerath E, Bartsch H, Nair J, Balansky R, Van Schooten F, Degan P, Fronza G, Walsh D and Lewtas J (1996) Mutat Res **366**: 197 - 238.

Friguet B and Szweda LI (1997) FEBS Lett **405**: 21 - 25

Fucci L, Oliver CN, Coon MJ and Stadtman ER (1983) Proc Natl Acad Sci USA **80**: 1521 - 1525

Garcia de La A, Millan A, Pla R, Bruseghini L, Esteras A, Pallardo FV, Sastre J and Vina J (1996) FASEB J **10**: 333 - 338

Gieseg SP, Simpson JA, Charlton TS, Duncan MW and Dean RT (1993) Biochemistry **32**: 4780 - 4786.

Griffiths HR (2000) Free Radic Res **33**: s47 - s 58.

Halliwell B (1992) J Neurochem **59**: 1609 - 1623

Halliwell B and Gutteridge JMC (1984) Biochem J **219** : 1 - 14.

Halliwell B and Gutteridge JMC (1999) Free radicals in biology and medicine, Oxford University Press, New York.

Hamilton ML, van Remmen H, Drake JA, Yand H, and Guo ZM, Kewitt K, Walter CA and Richardson A (2001) Proc Natl Acad Sci USA **98**: 10469-74.

Harman D (1981) Proc Natl Acad Sci USA **78**: 7124 - 7128

Heikkila RE and Cabbat FS (1983) J Neurochem **41**: 1384 - 1392

Holliday R (1995) Understanding ageing, Cambridge University Press.

Howard S, Bottino C, Brooke S, Cheng E, Giffard RG and Sapolsky R (2002) J Neurochem **83**: 914 - 923.

Hunt JV, Simpson JA and Dean RT (1988) Biochem J **250**: 87 - 93.

Izzotti A, Cartiglia C, Taningher M, De Flora S and Balansky R (1999) Mutat Res **446**: 215 - 223.

Joseph JA, Denisova N, Villalobos- Molina R, Erat S and Strain J (1996) Mol Chem Neuropathol **28**: 35 - 40.

Kannan K and Jain SK (2000) Oxidative stress and apoptosis. Pathophysiology **7**: 153 – 163.

Kaufmann JA, Bickford PC and Taglialatela G (2001) J Neurochem **76**: 1099 – 1108.
Keller JN, Hanni KB, Markesbery WR (2000) Mech Ageing Dev **113**: 61 - 70.
Khan FH, Saha M and Chakraborti S (2001) Brain Res **895**: 245 - 249
Khan FH, Sen, T and Chakrabarti S (2003) Free Radic Res **37**: 597 - 601.
Kiatipattanasakul W, Nakamura S, Hossain MM, Nakayama H, Uchino T, Shumiya S, Goto N and Doi K (1996) Acta Neuropathol **92**: 242 – 248.
Kwon D, Choi C, Lee J, Kim KO, Kim JD, Kim SJ and Choi IH (2001) J Neuroimmunol **113**: 1 - 9.
Levine, RL (1983) J Biol Chem **258**: 11823 – 11827.
Macho A, Calzado MA, Munoz - Blanco J, Gomez - Diaz C, Gajate C, Mollinedo F, Navas P and Munoz E (1999) Cell Death Differ **6**: 155 – 165.
Mandavilli BS and Subba Rao K (1996) J Neurochem **67**: 1559 – 1565.
Mandavilli BS and Subba Rao KS (1996) Biochem Mol Biol Int **40**: 507 - 514
Martin DS, Lonergan PF, Boland B, Fogarty MP, Brady M, Horrobin DF, Campbell VA and hynch MA (2002) J Biol Chem **277**: 34239 - 34246
Mather M and Rottenberg H (2000) Res Commun **273**: 603 – 608.
Melov S, Hinerfeld D, Esposito L, Wallace DC (1997) Nucleic Acid Res **25**: 974 - 982.
Melov S, Schneider JA, Coskun PE, Bennett DA and Wallace DC (1999) Neurobiol Aging **20**: 565 – 571.
Miralles VJ, Martinez-Lopez I, Zaragoza R, Borras E, Garcia C, Pallardo FV and Vina JR (2001) Brain Res **922**: 21 – 29.
Morel P, Tallineau C, Pontcharraud R, Piriou A, Huguet F (1998) Neurochem Int **33**: 531 - 540
Mrak RE, Griffin ST and Graham DI (1997) J Neuropathol Exp Neurol **56**: 1269 – 1275.
Muakkassah - Kelly SF, Anderson JW, Shih JC and Hochstein P (1982) Biochem Biophys Res Commun **104**: 1003 –1010.
Muscari C, Frascaro M, Guarnieri C, Caldera CM (1990) Biochem Biophys Acta **1015** : 200 – 204.
Nair J, Barbin A, Guichard Y and Bartsch H (1995) Carcinogenesis **16**: 613 - 617.
Peters A (2002) Brain Res **136**: 455 - 465.
Picklo MJ and Montine TJ (2001) Biophys Acta **1535**: 145 - 152.
Pollack M and Leeuwenburgh C (2001) Biol Sci Med Sci **56**: B475 - 482.
Randerath K, Yang PF, Danna TF, Reddy R, Watson WP and Randerath E (1991) Mutat Res **250**: 135 - 144.
Randerath K, Putman, KL, Osterburg HH, Johnson SA, Morgan DG and Finch CE (1993) Mutat Res **295**: 11 - 18.
Sastre J, Pallardo FV, Garcia de la A and Vina J (2002) Neurochem Int **40**: 69 - 78.
Sawada M and Carlson JC (1987) Mech Ageing Dev **41**: 125 - 137.
Schapira AHV and Cooper JM (1992) Mutat Res **275**: 133 – 143.
Schochet SS Jr (1998) Neurol Clin **16**: 569 - 580.
Shacter E, Williams JA, Stadtman ER and Levine RL (1996) Free radicals: a practical approach (Punchard NA and Kelly FJ eds.) pp 159-170, Oxford Unversity Press Inc, New York.
Shimasaki H (1994) Methods Enzymol **233**: 328 - 346.
Simandan T, Sun J and Dix TA (1998) Biochem J **335**: 233 - 240.
Single B, Leist M and Nicotera P (2001) Expt Cell Res **262**: 8 - 16.
Slater TF (1984) Biochem J **222**: 1 – 15.
Smith CD, Carney JM, Stark - Reed PE, Oliver CN, Stadtman ER, Floyd RA and Markesbery WR (1991) Proc Natl Acad Sci USA **88**: 10540 – 10543.
Sohal RS, and Sohal, BH (1991) Mech Aging Dev **57**: 187 - 202.
Sohal RS and Weindruch R (1996) Science **273**: 59 - 63.
Stadtman ER (1990) Free Radic Biol Med **9**: 315 - 325
St Clair, DK and Chow, CK (1996) Free radicals: a practical approach (Punchard, NA and Kelly, FJ eds.) pp 227-240, Oxford University Press Inc, New York.
Subba Rao K, Annapurna VV, Raji NS and Harikrishna T (2000) Mol Brain Res **85**: 251 - 259.
Sullivan SJ, Roberts RJ, Spitz DR (1991) J Cell Physiol **147**: 427 - 433.
Tanaka Y and Ando S (1990) Brain Res **506**: 46 - 52.
Tian L, Cai Q and Wei H (1998) Biol Med **24**: 1477 - 1484.
Venn MK and Conway EL (1998) Neuroreport **22**: 1981 – 1985.
Viani P, Cervato G, Fiorilli A, Cestaro E (1991) J Neurochem **56**: 253 - 258.
Vladimirov Yu A, Olenev VI, Suslova TB and Cheremisina ZP (1980) Adv Lip Res **17**: 173 - 249.

Wheeler DD and Ondo JG (1986) Expt Gerontol **21**: 159 – 168.
White LD and Barone S Jr(2001) Cell Death Differ **8**: 345 - 356.
Yakes FM and Van Houten B (1997) Proc Natl Acad Sci USA **94**: 514 - 519
Yasha TC, Shankar L, Santosh V, Das S and Shankar SK (1997) Ind J Med Res **105**: 141 - 150.
Zhang Y, Marcillat O, Giulivi C, Ernster L and Davies KJA (1990) J Biol Chem **265**: 16330 - 16336.

Molecular and Cellular Neurobiology
Editors: M.K. Thakur and S. Prasad

Synergistic Effects of Physical Exercise and Antioxidants in Aging Mammalian Brain

S. Asha Devi and S. Prathima

Gerontology Lab, Department of Zoology, Bangalore University, Bangalore, India

Abstract

The aging brain is subjected to several factors such as oxidative stress (OS) that cause damage to its cellular constituents and thereby impair its cognitive ability. Advances in neurobiological research have led to development of several genetic factors that can ameliorate age-related neurodegeneration. The present review focuses on two such important factors viz. physical exercise and antioxidants and their positive synergistic effects on the aging brain and their implications in decreasing its susceptibility to several age-related neurodegenerative disorders such as Alzheimer's disease.

Key Words: Age, Antioxidants, Brain, Exercise, Free Radicals, Neurodegeneration, Oxidative Stress.

Introduction

A progressive decline in the efficiency of various physiological processes is normally observed with aging. One of the several explanations put forth to explain the nature of mechanisms underlying senescence, point towards loss of functional capacity as attributable to molecular oxidative damage. This view is based on the fact that although oxygen is necessary for their immediate survival, it may be hazardous for the long-term existence of aerobes. Oxygen qualifies as a free radical and reactive oxygen species (ROS) or pro-oxidants such as superoxide anion, singlet oxygen, hydroxyl radicals and hydroperoxides can impair cell function and lead to subsequent cell damage. Given the potential of these free radicals in causing homeostatic disturbances, cells are equipped with several naturally occurring mechanisms for free radical injury prevention, interception and repair. The formation of pro-oxidants is balanced by a similar rate of their consumption by antioxidants (AO). Any disturbance in this balance in favor of pro-oxidants leads to 'Oxidative Stress' (OS).

Evidences suggest that exhaustive exercise can cause OS and enhance free radical production by various mechanisms (Davies et al 1982). This aspect is of particular concern and has largely been in focus recently, as exercise is no more a mere recreational activity but is established to have several diverse therapeutic values. Several exercise-related studies have been carried out in animal models to elucidate the effects of physical activity on various parameters in different organs (Table 1).

Although our current knowledge on the significance of brain aging is enormous, we are still groping to find sufficient evidences to conclude the deficits in cognitive abilities are age-related. An important direction of future neurobiology of aging lies in a thorough evaluation of the plausible mechanisms that can explain the anatomical and physiological paucities in this field. Studies comparing cerebral cortex (CC) and medulla oblongata (MO) in rats of different age groups revealed a linear decrease in glucose content with age (Anitha and Asha, 1997). Glucose utilization in rat brain regions have been reported to decrease between 3 and 12 mo of age and remain unchanged during second and third year of their life span. Although the glucose needs for the brain in old rats does not differ from young adults, glucose influx in to the brain is insignificantly less in the old. Some subtypes of glucose nutrients such as tyrosine, tryptophan, folic acid, choline and lecithin have significant effects on the aging brain. Tyr and Try are precursors of serotonin. Availability of Try to the brain neurons that synthesize serotonin directly influences the rate at which it is converted to a neurotransmitter. Administration of the amino acid itself or diets that enhance the access of Try to serotonin neurons result in stimulation of serotonin production. Try supplementation brings about an elevation in serotonin release in neurons that are actively firing (Sharp et al 1992). These treatments can cause changes in brain function. Tyr is the precursor to dopamine, norepinephrine and epinephrine. Elevation in tyrosine content in brain catecholamine neurons can stimulate neurotransmitter productions. Intake of folic acid has been associated with several psychiatric conditions as well as to deficits in learning and memory especially in the elderly. Choline and lecithin are studied as potential memory enhancers, based on the fact that acetylcholine neurons in the hippocampus (HC) play an important role in memory and enhancing the transmitter production can improve memory (Pernstorm, 2000).

Old age is characterized by a decline in cognitive function. In animal models, age-related deficits in learning have been demonstrated in tests of maze learning, discrimination reversal learning, passive avoidance conditioning and response extinction. The fact that old animals are highly susceptible to interferences from conflicting experiences than the young, adversely affects their learning ability.

Old age is also associated with a reduction in memory and ability to recall specific events. An impaired HC function could be a contributing factor here, as there exists a well-documented relationship between memory loss and HC damage (Wincour, 1986). Neuroanatomical and physiological evidence reveal that the HC system is particularly vulnerable to the aging process and learning deficiencies in the aged may in part be related to HC function deterioration.

With aging nerve cell mass is lost. This causes brain atrophy. The number of neurons decreases and demyelinization occurs which slows down the speed of impulse transmission. Aging also elevates the latency period of the neurons. Aging in rats is associated with reduced spatial memory. Although little is known about the nature of age-related alterations in the information encoded by the HC, the scope of information encoded by the HC was reduced in memory-impaired aged subjects, although no changes were observed in the number of neurons responsive to environmental stimuli. After repeated cell manipulations, memory-intact rats like their young counterparts altered their spatial representations, while memory-impaired aged rats showed reduced plasticity. Hence aging in rats results in spatial memory and learning impairment (Tanila et al 1997). Nutritional intervention may play an important role in protecting against and possibly reversing the cognitive declines seen due to aging. Free radical generation is thought to be a contributory factor to the decline in memory and motor performance seen in aging.

The current objective is to present a comprehensive report on the cellular damage pertaining to oxidative stress in the critical regions of the brain and the impact of physical exercise and antioxidants on ameliorating the distress of neuronal aging.

Table Relevance of Exercise Type in Rodent Model of Aging

Rodents	Gender	Supplements	Exercise	Reference
Wistar rats	M	Vitamin E	Swimming	Asha and Kiran 2004
Wistar rats	M	Vitamin E	Swimming	Asha et al 2003
Wistar rats	M		Swimming	Asha et al 2002
Wistar rats	M		Swimming	Sharon and Asha, 2001
Albino rats	M		Treadmill	Indira and Jhansi, 2001
C 57BL/Gjnia mice	M	Vitamin E and Coenzyme Q10		Lass and Sohal, 2000
Wistar rats	M		Swimming	Prathima and Asha, 1999
Sprague Dawley	M		Treadmill	Chillibeck et al 1998
Wistar rats	M		Treadmill	Somani and Rybak, 1996
Wistar rats	M		Swimming	Venditti and Meo, 1996
Wistar rats	F	Vitamin E	Treadmill	Tiidus and Houston, 1994
Wistar rats	F	Vitamin E deprived, estrogen-treated		Tiidus and Houston, 1993
Sprague Dawley	M		Treadmill	Starnes et al, 1983
Long Evan rats	M	Vitamin E	Treadmill	Davies et al 1982
Swiss Albino mice	F	Vitamin E		Kruk and Enesco, 1982

1. Physical Exercise and Brain

1.1 Exercise and Free Radical Generation

Exercise is known to exert numerous physiological changes in the vital organ systems of the body. Among these changes, the most important is the enhanced respiration and utilization of oxygen in the body. Increased oxygen influx during exhaustive exercise may be potentially harmful to the body (Somani, 1996). Thus an understanding of the possible mechanisms of exercise-induced oxidative stress and designing measures to minimize such stress is important. Several mechanisms may exist that can potentially lead to the generation of free radicals during exercise. Oxygen consumption in the exercising muscle can increase as much as ten to forty folds during exercise, over the resting state (Leaf et al 1997). Regular strenuous exercise has been found to increase in both the number and size of mitochondria. Thus, increased flow and metabolism of oxygen in the exercising muscle can enhance the production of $O_2^{\cdot-}$ in the mitochondria. This may lead to enhanced generation of H_2O_2 and highly reactive $OH^{\cdot}$ radicals.

During exercise, blood flow is directed towards the exercising muscle, resulting in a markedly reduced supply of blood and hence oxygen to other organs, particularly the liver, kidney and intestine. After the cessation of exercise, blood flow to these organs resumes. This mimics the ischemia-reperfusion (I-R) phenomenon. Evidence suggests that (I-R) causes excessive production of free radicals leading eventually to extensive tissue damage (Coghlan, 1991).

Severe physical stress and shearing forces can cause significant damage to the muscles during exercise. This tissue damage may attract neutrophils, leading to generation of free radicals (Wetzstein, 1998). The mitochondria may be a potential site for free radical generation during intensive muscle activity. The increased energy demand may lead to increased accumulation of hydrogen ions as well as an increased electron flow through the electron transport chain (ETC) in the mitochondria.

Endurance running and other sports with high mechanical impact cause destruction of RBC, which then may serve as a potential source for free iron. An already initiated free radical attack on the cell membrane, causing leakage from cell cytosol, increases the availability of free iron, which in turn leads to free radical formation (Halliwell and Gutteridge, 1989).

1.2 General Effects of Exercise on Brain Aging

Chronic stress is known to act detrimentally on the brain health and memory performance. The brain is subjected to multiple factors that result in damage to its cellular constituents, the neuron and supporting cells, and the neural networks that form the bases of cognitive ability. Like other systems, the brain has a remarkable capacity to repair that damage and to adapt or compensate for the loss of neurons and disruption of neural architecture. Brain aging and dementia can be conceptualized as a balance between neuronal injury and repair. This balance relies on several genetic, age-related and environmental factors. Advances in neurosciences have paved the way for the development of pharmacological agents that can ameliorate effects of genetic and age-related factors (Ball and Birge 2002).

Research has shown that exercise protects against an array of chronic diseases and slows the effects of aging. Exercise also improves certain psychological traits such as energy levels and mild-to-moderate depression. Van Praag et al (1999) demonstrated that adult mice exercised on a wheel were more effective learners than sedentary mice. Mice were trained to swim for 6 days to locate a slightly submerged platform in a murky mouse-size pool. After training, the exercising mice swam straight to the platform when released into the pool, while the sedentary mice took wandering paths and longer time to find the platform. The exercising mice also had 2.5 times the growth of new neurons in the dentate gyrus, an area believed to be important in memory. In rats, improved aerobic fitness promotes development of capillary networks (Jones et al 1998), enhances cortical high-affinity choline uptake, increases dopamine receptor density and increases brain-derived neurotrophic factor gene expression (Fordyce and Farrar, 1991).

Exercise has several positive effects on the nervous system. Significantly it benefits the neurotrophic factors of the brain. Neurotrophic factors protect the neurons from damage, promote their function and growth and maintain their long-term plasticity of the nervous system. Initial studies on swim-trained rats have indicated age-related biochemical adaptations in the CC compared to their sedentary counterparts (Anitha and Asha, 1997). Common neurotrophic factors include brain-derived neurotrophic factor (BDNF), nerve growth factor (NGF) and fibroblast growth factor (FGF). The CC and HC, which serve as areas of memory and high mental function, express higher levels of BDNF and NGF (Cotman and Neeper,

1996). A strong correlation exists between the distance the animal runs and brain levels of BDNF mRNA.

Similar stimulatory effects of exercise are also noted in other neural characters. Exercise-dependent increase in these neural factors could be the basic molecular mechanism for use-dependent changes in the brain's cellular structure. The resulting enhanced neurons could probably strengthen initiation of exercise behaviors (Cotman and Neeper, 1996). Chronic stress could be detrimental to brain health and memory performance. Animal studies involving prolonged exposure to stress hormones reveal adverse effects on the hippocampus-the brain region involved in memory and learning (Sapolsky, 1999). Studies on humans involving exposure to high levels of cortisol-the stress hormone have reported memory impairment (Newcomer et al 1999). Chronic stress is a pivotal factor in depression and anxiety disorders which interfere with normal memory processing, particularly with advancing age. Hence minimizing stress could have positive impact on brain health. Studies on people who have been physically active between 20 to 60 years have a significantly low risk of Alzheimer's disease (AD) in later life (Small, 2002). Physical exercise increases cerebral blood flow and promotes nerve cell growth. Laboratory animals that were exercised regularly developed new neurons in the HC compared to their sedentary counterparts (Gage, 2002).

2. Antioxidants and Brain Function

The past few decades have witnessed the importance of dietary factors for a healthy brain, in terms of either macro or micronutrients. OS as well as the expression of genes is most often held as causative factors for aging. It is known that endogenous ROS that are generated as a result of aerobic metabolism are potent candidates of much of the age-related neuronal damage. Brain is most vulnerable to oxidative damage due to its high oxygen consumption rate, high lipid content and low AO capacity compared to other tissues. The organ is also characterized by low linoleic acid and high amounts of fatty acids with 4, 5, and 6 double bonds, a fact that also accounts for the high lipid peroxidations. Given the susceptibility of this organ to OS, several AOs exist to prevent the damage caused by free radicals. These may be non-enzymic and/or enzymic in nature.

2.1 Non-enzymic antioxidants

Carotene, a precursor of Vitamin A is known to have excellent antioxidant properties. People with low levels of β-carotene are found to be more susceptible to various types of cancer. Protection by β-carotene and related compounds against O_2-mediated cytotoxicity and genotoxicity w.r.t. its implications on carcinogenesis and anticarcinogenesis have been demonstrated. Adipose tissue, liver and plasma serve as major pools of β-carotene in humans. Orally administered β-carotene is known to have photoprotective function. The AO mechanism of β-carotene is attributed to its free radical scavenging singlet oxygen quenching and breaking LP chain reactions. One-sixth of total β-carotene is absorbed by the intestine and converted to retinal, while in the case of other provitamin-A carotenoids, less than only one-twelfth is absorbed. When β-carotene supplemented animals were exposed to gamma radiation, no significant changes were noted in the levels of LP products such as MDA. This suggests the protective effects of the AO in preventing radiation-induced LP in the brain (Manda et al 2000).

Vitamin E is the generic term for a group of naturally occurring tocopherol and tocotrienol derivatives of which RRR-alpha-tocopherol has the highest bioactivity. Vitamin E is an essential nutrient in humans and functions as a natural antioxidant, scavenging free radicals in cell membranes and protecting unsaturated fatty acids from lipid peroxidation. This vitamin therefore plays an important role in the CNS with regard to metabolic stability and physiological functions.

An age-dependent accumulation of autofluorescent age pigments such as lipofuscin (LF) is considered to be a morphological marker of neuronal lipid peroxidation caused by free radical attack. Dietary Vitamin E is known to have significant effect on the LF accumulation with age in the rat brain. Mean total area of LF and ceroid pigments are shown to increase in CA_3 hippocampal pyramidal neurons with aging as well as in Vitamin E deficient animals. Hence absence of tocopherol can be one of the mechanisms initiating deposition of these age pigments (Fattoretti et al 2002). The same effect was limited and dietary Vitamin E had no decisive effect on LF accumulation in the latter half of their life. These results may be attributable to age-dependent alterations in free radical production and compensatory systems of antioxidant mechanisms (Monji et al 1994). Studies on the effect of Vitamin E on LF accumulation with age in HC, inferior olive and CB of young, middle-aged and old rats revealed remarkable effect of Vitamin E on LF accumulation in the rat brain till middle-age, an effect that became less distinct in the old (Monji et al 1994).

Levels of Vitamin E in brain tissue were low in animals fed with a low Vitamin E diet while animals on Vitamin E enriched diet showed a significant increase in vitamin E concentration between 500 to 900 % compared to low Vitamin E groups. Maximum incorporation of Vitamin E was seen in HC. However, Vitamin C concentration showed no significant changes between ages (Antonio et al 2001). The detoxification mechanism by Vitamin E in specific regions of the brain has been demonstrated *in vivo* (Fattoretti et al 2002) and *in vitro* (Kan et al 1991) studies.

Centrophenoxine, a dimethylaminoethanol and chlorophenoxy acetic acid ester finds wide application in psychodegenerative treatment of various disorders in the elderly. It is also known to delay hypoxic amnesia, improve learning and memory. Although this drug has no proven stimulant or sedative effects, it is known to alter cerebral energy metabolism and glucose utilization. Beneficial effects of this drug have also been documented on improving learning and memory. Centrophenoxine treatment has shown to significantly increase protein and RNA content and decrease LF concentration in the cerebrum and CB while the MO revealed no significant changes. Acetylcholine esterase shows a remarkable decline in the cerebrum. Centrophenoxine induces decrease in LF storage due to LP inhibition resulting from an over all increase in AOE activity. Centrophenoxine lipofuscinolytic activity is attributed to the presence of dimethyl aminoethanol moiety in its structure (Varma and Bhatia, 2000).

Aging in female individuals is highlighted by reduced synthesis of female sex hormones. Brain cells are reported to be affected by declining estrogen levels during aging. Injection of estrogen to ovarectomised mice significantly increased acid phosphates while lipid content and MDA decreased in the HC, cerebral hemisphere, corpora quadrigemina (CQ), CB and MO. Alterations in estrogen levels in ovarectomised mice cause lipid mobilization leading to increased LP. Decrease in acid phosphatases further complicates the scenario by causing MDA accumulation. Thus administration of estrogen before menopause can adversely affect brain cells (Ashokan and Pillai, 2000).

Several herbal products are known to be effective in breaking various nervous system disorders. Studies have indicated role of *Withania somnifere* in protecting CA_2, CA_3 and dg in hippocampus sub-regions from stress-induced degenerative changes (Jain et al 1998). Phytochemicals that are isolated from blueberry, spinach or strawberry and fed to old rats of 19 months of age have suggested the beneficial effects in reversing the onset and course of neuronal and behavioral aging (Joseph et al 1999). Observations have indicated that although docosahexaenoic acid (DHA) is deleterious due to its capacity to increase lipid peroxides (LPOs), it can under certain circumstances function as an AO (Hossain et al 1998). DHA induces antioxidative defense against ROS by increasing the activities of catalase, gluathione peroxidase and glutathione in the cerebral region (Hossain et al 1999).

2.2 Enzymic Antioxidants

Several antioxidant enzymes (AOEs) such as superoxide dismutase (SOD), catalase (CAT) and glutathione peroxidase (GSH-Px) protect cells from free radical injury. Aging is characterized by a decrease in the antioxidant defense capacity, the system that prevents active oxygen radical formation. The resulting imbalance increases free radical generation and decreases AO functional capacity, which is suggested to be one of the significant factors contributing to the aging process. Although an acute bout of exercise is known to augment the activities of AOE, the threshold and magnitude of exercise required for activation varies among tissues and enzymes. Experiments conducted on young and old treadmill exercised rats reveal a decrease in SOD and CAT activities in the old while these changes are reversed in the animals subjected to 3 month endurance exercise. Xanthine oxidase (XO) activity was greater in the brain of young rats than the old after three months. Treadmill exercise also increased LP in the young rats than the old. Aging and exercise significantly affect the AO capacity in rats. Exercise training protects the individual against OS by inducing antioxidant capacity (Indira and Lakshmi 2001).

Brain cells are at a higher risk of free radical damage. PUFA, the major components of cell membranes serve as substrates for LP. Brain regions such as Globus Padillus (GP) and Substantia Nigra (SN) are known to accumulate iron, which aids in the formation of cytotoxic radicals. Paucity of defense mechanisms in the brain is another cause for concern. The brain contains relatively low levels of CAT, GSH, GSH-Px and Vitamin E. Neurons of SN pars compacta are particularly susceptible to oxidative stress owing to oxidative metabolism of dopamine that can generate cytotoxic free radicals. Clinical trials of AO as possible neuroprotective therapy for neurodegenerative disorders are in progress (Olanow, 1993). The brain is an organ particularly susceptible to oxidative stress owing to its high oxygen consumption rate, abundant lipid content and relatively low concentration of AOEs. Metabolism-generated ROS contribute significantly to age-related neuronal damage. Basal MDA levels show significant increase with aging. Brains of adult mice are suggested to have the ability of sequestering exogenous free radical stimuli, a feature not found in the aged mice, resulting in enhanced MDA generation (Leutner et al 2001).

3. Neurodegeneration and aging

The varied structural and functional alterations that occur in the aging brain results from deleterious changes in the activity and/or abundance of molecules responsible for the integrity and functioning of nerve cells. Cellular controls governing genetic activity may be relaxed during aging and hence the enzymatic machinery responsible for repair and excision of faulty DNA becomes less efficient with aging.

3.1 Changes in the course of normal brain aging

The aging process is characterized by deterioration in a variety of CNS functions. Several behavioral and biochemical changes also accompany aging such as decrease in movement and impairment in learning and memory. Working memory, the temporary storage of information, information analysis, decision-making and information selection occurs in the HC and frontal cortex. Long-term memories are stored in different locations of the brain. Although there is a loss in brain cells with age, the process does not accelerate with age and the areas of cell loss are not necessarily associated to memory. With age, the way the brain processes information is slowed affecting the rate at which new information can be incorporated into permanent memory. Intracellular accumulation of damaged and abnormal proteins is a common feature of neurodegenerative and other age-related disorders. Increasing the activity of the intracellular proteolytic system, normally responsible for the removal of these abnormal proteins might lower the severity of such disorders (Martin et al 2002). Earlier studies have demonstrated an age-related deficit in antioxidant enzymes (AOEs) in the CC and HC, which can be overcome through Vitamin E plus exercise and thereby suggesting the rationale for looking at these markers of oxidative stress in several age-associated neuronal diseases (Asha and Kiran, 2003).

One of the several causes contributing to these alterations could be variation in brain biogenic amines with aging. Levels of various biogenic amines such as noradrenaline (NA), dopamine (DA), serotonin (5-HT) and the metabolite of 5-HT-5-hydroxyindolacetic acid (5-HIAA) have significant roles in locomotion (Baskar et al 1995).

Cellular aging is characterized by the slowing of protein synthesis. Age-related decline, organ efficiency in the old is related to loss of post-mitotic cells. Reduction in Nissl substances in areas of neuronal loss is the most important feature of the aging brain. Age pigment granules like lipofuscin (LF) begin to accumulate before the actual loss of Nissl substance occurs. Accumulation of such pigments leads to intracellular catabolism, intralysosomal peroxidation and polymerization of the undigested material, eventually leading to cellular impairment. A negative correlation has been evidenced between protein content and MDA in different regions of the brain in female mice. Decline in the rate of protein synthesis with advance in age can be largely attributed to the increased accumulation of the age pigment in the brain cells. LF accumulation in the cerebral cortex of mice has been reported to increase with age from 3 days to 35 days in primary cultures maintained for a period of 32-35 days (Kan et al 1991) (Fig. 1). This accumulation however does not interfere with the activity of certain enzymes like creatinine phosphokinase (CPK) (Asha et al, 1988).

Proteins undergo qualitative and quantitative changes with aging and the rate of protein synthesis decreases in aging mammalian brain. Dysfunction of protein catabolism in aging individuals could be one of the reasons contributing to the decreased cellular and tissue vitality. This is largely due to the fact that protein catabolism plays a pivotal role in maintaining cellular homeostasis by either regulating enzyme activity, mobilizing protein stores during nutrient deprivation or eliminating unwanted proteins. Caloric restriction is focused as one method of decelerating aging process by retarding age-related decrease in protein synthesis. Studies have indicated high protease activity in old animals. This is suggestive of high rate of protein degradation with advancing age. Changes in protein content in the old can also be due to a change in protein species or changes in the relative abundance of specific proteins at different ages (Venkatesh et al 1997).

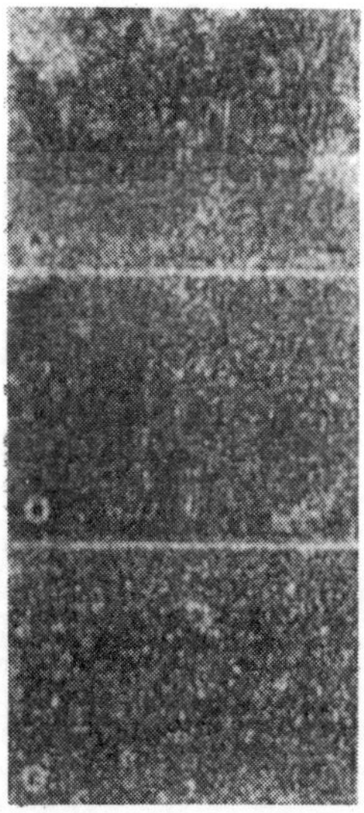

Fig. Phase contrast photomicrographs of mice cerebral cortical cells in primary culture (A) culture age =12 hours. Spherical dissociated cells along with clumps of undissociated cells are visible. Many cells have a 'neuronal' appearance. (B) By 20 days, neurons in 'pyramidal form' with extending processes are seen. (C) 23 days confluent culture consisting of neurons and non-neuronal cells such as glia and blood vessel elements. Bar: 100 um

3.2 Alzheimer's disease

Alzheimer's disease (AD) is a progressive, irreversible disorder damaging the brain cells and eventually killing its victims. Symptoms of AD include a gradual loss of memory, confusion, impaired judgement, personality and behavioral changes, inability to carry out routine tasks of living. The likelihood of contacting AD is high as the person ages. It strikes most frequently between the ages of 65 and 75. Women appear to be two to three times more likely than men to develop AD. The fact that women tend to live longer than men makes more likely to reach that age where risk of AD is high. Major site of oxidative damage in AD is the neuronal cytoplasm as evidenced by RNA and oxidative damage (Perry et al 2002). Damage to cell bodies is restricted to vulnerable neurons in AD. Conspicuous changes in iron are detected in the cytoplasm of vulnerable neurons (Smith et al, 1999; Mori et al 1987). Levels of choline and ethanolamine are significantly lower in AD patients. Both of these are used for synthesis of phospholipids - the major components of cell membrane of each nerve cell. Stress and depression may cause some forms of memory loss. Prolonged stress leads to elevated levels of cortisol, which appears to shrink the HC, the central unit of the brain related to AD.

In normal aging, production and deposition of amyloid-β successfully treats age-related redox imbalances, but in AD, which is characterized by severe redox imbalance, high levels of β-amyloid also prove insufficient (Smith et al 2002, Yankner, 1996). Protein oxidation, LP, free radical formation, advanced glycation end products and DNA oxidation are markers of oxidative stress for AD. The question of what causes AD has led to a number of studies that analyze the chemistry and molecular biology of the β-amyloid deposits. DNA mutations that occur within the gene is known to give rise to the beta-amyloid precursor protein i.e, beta-APP. Persons who inherit AD due to mutations in genes for presenilin-1, -2 and APP always have excessive amounts of β-amyloid (Selkoe, 2001).

Vitamin E prevents oxidative damage induced by β-amyloid in cell culture and delays memory deficits in animal models. Vitamin E and other AOs effectively improve cognitive performance in aged animals and prevent oxidative damage in animal models of AD (Behl and Moosmann, 2002; Socci, 1995). Results of clinical trials indicate that Vitamin E treatment delays disease progression in patients with moderately severe AD (Grundman, 2000). Studies involving subjects treated with Vitamin E indicate that this potent AO slows functional deterioration. β-amyloid can induce cytotoxicity through a mechanism involving oxidative stress and hydrogen peroxide (H_2O_2) production. Vitamin E can block H_2O_2 production and the resulting cytotoxicity (Behl et al 1994). The effects of glycation, oxidation and free radical formation promote irreversible cross-linking of tau proteins and β-amyloid. AO such as Vitamin E prevent this cross-linking thereby protecting the brain (Smith et al 1995).

Proteomics has become an invaluable resource in the investigation of neurodegenerative diseases. Expression of proteomics enables to compare and contrast various conditions and reproducibly map, screen, quantitate and identify a vast number of proteins at once. Such principles assist in detecting specific alterations in the protein expression levels of various regions of AD brain compared to control brain which in turn facilitates an explanation of mechanism by which the disease progresses. The accumulation of glycolytic enzymes such as α and γ-enolase and glyceraldehydes-3-phosphate dehydrogenase confirms that glucose metabolism is affected in AD. An increase in the expression of several AO proteins is also observed in AD brain (Butterfield et al 2003). An increase in carbonyl reductase (CBR), alcohol dehydrogenase (ADH), peroxiredoxin-I (Prx) and –II and Cu/Zn SOD was detected. Carbonyls are generally toxic metabolic intermediates that serve as markers of oxidative stress. An elevation in the expression of the enzymes that are involved in the removal of these cytotoxic intermediates suggest adaptive accumulation in these species (Balez et al 2001). Apoptosis plays a major role in the neuronal loss exhibited in AD. The pro-apoptotic proteins zipper-interacting protein kinase (ZIPK) and receptor interacting protein (RIP)-like kinase was increased in AD brain. Classes of specifically oxidatively modified proteins identified thus far in AD deal with energy metabolism, excitotoxicity, membrane structure, apoptosis and neural communication.

Proteomics is emerging as an important tool towards an understanding of the implications of defective synthesis of specific proteins involved in AD. Post-translational modifications of proteins that can be identified through proteomics can form the key to unlock the molecular mechanisms of this dementing disorder. Proteomic studies in animal models relevant to AD and cell cultures will provide further insight into the role of β-amyloid in the progression of the disease. Rodents with over expressed knocked-in or knocked-out proteins that are significant in AD have an important role in a better understanding of AD. The future of therapy lies on the information obtained by proteomic analysis of the mechanisms of neurodegeneration in AD.

Questions relating to the applicability of various therapies for overcoming age-related dementia and the ravages of neurodegenerative diseases such as AD, ranging from physical exercise and clinical implications of vitamins to enzyme inhibitors are being vigorously pursued. A logical conclusion based on parallel studies on human and animal models suggests that aerobic exercise training when initiated at an early age and continued for rest of the life can prevent dementia, improve mental faculties and cognitive ability in old age. However, lifestyle improvements and exercise alone may not suffice without a dietary intervention in the elderly. Hence a combination of physical exercise and AO such as vitamins can be an effective, rational and integrative approach towards overcoming mild-to moderate-forgetfulness in the aged.

References

Anitha V and Asha DS (1997) Ind J Exp Biol **35**: 280-284.

Asha DS, Jolitha AB and Prathima S (2002) Biogerontology **3**:1-6.

Asha DS, S Kan and Kawashima S (1988) Zoo Sci **5**: 1284.

Asha DS and Kiran TR (2004) Neurobiol Aging **25** : 501-508

Asha DS Prathima S and Subramanyam MVV (2003) Exp Geron **38**: 285-290.

Antonio MP, Ronald PL, Barbaro SH, Cao G and James JA (2001) Effect of fruits, vegetables or vitamin E rich diet on vitamin E and C distribution in peripheral and brain tissues: Implications to brain function. Tektran (Agricultural Res. Service) USA.

Ashokan KV and Pillai MM (2000) Ind J Geron **14**: 86-91

Balez B, Kirchner L, Cairns N, Fountoulakis M and Lubee G (2002) J Neural Transm **32**: 1050-1060.

Ball LJ and Birge SJ (2002) Clin Geriatr Med **18**: 485-503.

Baskar R, Uma Devi P, Nalini N and Shivananda B (1995) Ind J Geron **9**: 51-59.

Behl C, Davis JB, Lesley R and Schubert D (1994) Cell **77**: 817-82.

Behl C and Moosman B (2002) Free Rad Biol Med **33**:182-191.

Butterfield DA, Boyd-Kimball D and Castegna A (2003) J Neurochem **86**: 1313-1327.

Chilibeck PD, Bell GJ, Farrar RP and Martin TP (1998) Can J Physiol Pharmacol **76**: 891-894.

Coghlan JG, Flitter WD and Ilsley CD (1991) Lancet **338**: 1145-1146.

Cotman CW and Neeper S (1996) Activity-dependent plasticity and the aging brain. In: Handbook of the Biology of Aging. 4th edition. E.L. Schneider and JW Rowe(eds). pp281-299, Academic Press, San Diego.

Davies KJA, Quintanilha AT, Brooks GA and Packer L (1982) Biochem Biophy Res Commun **107**:1198 1205.

Fattoretti P, Bertoni-Freddari C, Casoli T, Di Stefano G, Solazzi M and Corri E (2002) Arch Geron Geriatrics **34**:263-268.

Fordyce DE and Farrar RP (1991) Behav Brain Res **43**:115-123.

Gage FH (2002) J Neurosci **22**:612-613.

Grundman M (2000) *Am J Clin Nutr* **71**: 6305-6365.

Halliwell B and Gutteridge JMC (1989) Free radicals in biology and medicine II ed., Oxford Claredon Press.

Hossain S, Hashimoto M and Masumura S (1998) Neurosci Lett **244**: 157-160.

Hossain S, Hashimoto M, Gamoh S and Masumura S (1999) J Neurochem **71**:1133-1138.

Indira SK and Jhansi L (2001) Curr Sci **80**:921-923.
Jain S, Shukla SD, Sharma K, Bhatnagar M (1998) Ind J Geron **12**:60-69.
Jones TA, Hawrylak N and Klinstova AY (1998) Ment Retard Dev Disabil Res Rev **4**: 231-235.
Joseph JA, Shukitt-Hale B, Denisova NA, Bielinski D, Martin A, McEwen JJ and Bickford PC (1999) J Neurosci **19**: 8144-8121.
Kan S, Asha DS and Kawashima S (1991) Exp Gerontol **26**: 365-374.
Kruk P and Enesco HE (1982) Experentia **37**:1301-1302.
Lass A and Sohal RS (2000) FASEB J **15**: 87-94.
Leaf DA, Kleinman MT, Hamilton M and Barstow TJ (1997) Med Sci Sports Exerc **29**:1036-1039.
Leutner S, Eckert A and Muller WE (2001) J Neural Transm **108**: 955-967.
Manda K, Sharma M, Sisodia R and Bhatia AL (2000) Ind J Geron **14**: 10-14.
Martin A, Joseph JA and Cuervo AM (2002) J Neurochem **82**:538-549.
Monji A, Morimoto N, Okuyama I, Yamashita N and Tashima N (1994) Brain Res **634**: 62-68.
Mori H, Kondo J, Ihara Y (1987) Science **235**:1641-1644.
Newcomer JW, Selke G, Melson AK, Hershey T, Craft S, Richards K (1999) Arch Gen Psychiatry **56**: 527-533.
Olanow CW (1993) TINS **16**:11-15.
Pernstorm JD (2000) Am J Clin Nutr **71**:1669S-1673S.
Perry G, Nunomura A, Hirai K, Zhu X, Perez M, Avila J, Castellani RJ, Atwood CS, Alier G, Sayre LM, Takeda A and Smith A (2002) Free Rad Biol Med **33**: 1475-1457.
Prathima S and Asha DS (1999) Mech Age Dev **108**: 61-95.
Sapolsky RM (1999) Exp Gerontol 721-732.
Selkoe DJ (2001) Physiol Rev **81**: 741-766.
Sharon C and Asha D S (2001) Ind J Exp Biol **39**: 29-34.
Sharp T, Bramwell SR, Grahame-Smith DG (1992) Life Sci **50**:1215-1223.
Small GW (2002) Br Med J **324**:1502-1505.
Smith MA, Casadescus G, Joseph JA and Perry G (2002) Free Rad Biol Med **33**: 1194-99.
Smith MA, Harris PLR, Sayre LM, Beckman JS and Perry G (1999) J Neuroscience **19**:1959-1964.
Smith MA, Sayre LM, Monnier VM and Perry G (1995) Trends Neurosci **18**: 172-176.
Starnes JW, Edington DW and Beyer RE (1983) J Gerontol **38**: 660-665.
Socci DJ, Crandall BM and Arendash GW (1995) Brain Res **693**: 88-94.
Somani SM (1996) Exercise, drugs and tissue specific antioxidant enzymes. In: Pharmacology in exercise and sports. (Somani, S.M. Ed.) pp 57-96, Boca Raton, FL CRC Press.
Somani SM and Rybak L (1996) Ind J Physiol Pharmacol **40**: 205-219.
Tanila H, Shapiro M, Gallagher M and Eichenbaum H (1997) J Neurosci **17**: 5155-5166.
Tiidus P M and Houston ME (1994) Med Sci Sports Exerc **26**: 354-359.
Tiidus P M and Houston ME (1993) J Nutr **123**: 834-840.
Van Praag H, Christie BR and Sejnowski TJ (1999) Proc Natl Acad Sci **96**: 13427-13431.
Varma A, Bhatia AL (2000) Ind J Geron **14**: 67-80.
Venditti P and Meo SD (1996) Ind J Geront **11**: 8-15.
Wetztein CR, Shern-Brewer RA, Santanam N and Green NR (1998) Free Rad Biol Med **24**:679-682.
Wincour G (1986) J Geront **41**: 758-763.
Yankner BA (1996) Neuron **16**: 921-932.

Molecular and Cellular Neurobiology
Editors: M.K. Thakur and S. Prasad

Electrophysiological Ageing of the Brain and Pharmacology of Ageing

Rameshwar Singh and Deepak Sharma

School of Life Sciences, Jawaharlal Nehru University, New Delhi-110 067, India

Abstract

During ageing there is a general decline in nervous system functions due to many structural, molecular (including genomic), biochemical and physiological changes. Biological changes implicated in ageing of the nervous system are not restricted to metabolic and biochemical measures of neuronal activity but include also changes in neural electrical activity. The electrical activity basically resulting from the movement of ions across membranes (thus producing electric current) is the functional basis of the nervous system. Ageing of the nervous system is associated with changes in electrophysiological activity. Neurons being electrically active cells use electrical signals generated by them to transmit and process the external and internal information, and any disturbance or disorganization of electrical activity will lead to alterations in physiological as well as behavioral functions. For example, inability to produce bursts of high frequency firing in aged tissue may influence neurotransmitter release thus adversely affecting synaptic plasticity and information processing. The main electrical signals produced by nerve cells are action and synaptic potentials. Age-related changes in these potentials may increase Ca^{2+} currents that may alter the intraneuronal Ca^{2+} homeostasis which may lead to ageing of the nerve cell. Age-related alterations in neuronal excitability and action potential conduction velocity are other parameters of interest. Changes in the spectral composition of electrical activity signals such as electroencephalographic activity are indicative of the altered degree of cerebral activity and of cerebral functions. Multiple-unit activity (action potentials simultaneously derived from many neurons) represents an electrophysiological marker of cellular firing activity of the concerned neuronal population. The decline in the spontaneous multiple-unit activity with age in brain areas can be considered as a measure of neuronal impairment and thus as an important parameter of electrophysiological ageing. How age-related alterations in electrical activity might arise from biochemical changes in nerve cells is another important aspect to consider. Age-related changes in membrane lipid peroxidation and sodium potassium adenosine triphosphatase (Na^+,K^+-ATPase) activity contribute to alterations in membrane electrophysiology.

The development of antiageing therapies is a matter of great public interest. Since the oxidative stress caused by reactive oxygen species has a role in the ageing process, strategies aimed at reducing or preventing it or enhancing the host defense mechanisms involved in coping with it should provide significant antiageing effects. Pharmacological interventions may possibly be of value in interfering with age-related decline in neural functions and thus provide a suitable antiageing strategy to reverse neurobiological deterioration and disease. In this article, we outline the effects of three drugs acetyl-L-carnitine, L-deprenyl and centrophenixine on multiple-unit action potentials and related biochemical parameters in several brain areas. These drugs prevent the ageing-related declines in brain electrical activity and biochemical parameters.

Introduction

For the purposes of this write-up we shall define electrophysiological ageing as changes occurring with ageing in neural electrophysiologic parameters such as action potentials, synaptic potentials, spontaneous field potentials (EEG) etc, and such changes being signifying ageing-related impairment and disorganization of electrical signals leading to decrements in neurological functions. Neurons are electrically active cells. They use electrical signals generated by them to transmit information (Lisman 1997,Ludwig and Pittman 2003). In the human brain, there are about 10,000 million neurons that function by generating and transmitting electrical signals, which are put together and from this integration higher neural and behavioural functions emerge. The main electrical signals produced by nerve cells are action potentials and synaptic potentials. Action potentials conduct information along axons while synaptic potentials are involved in the transmission of information at synaptic junctions.

Another kind of electrical activity that is often studied is the brain electroencephalographic activity (Buzsaki et al 1988, Duffy et al 1993,1996), which is a compounded waveform, derived from complex neuronal tissue network of the cerebral cortex. It represents integrated postsynaptic potentials and reflects the state of neurophysiological activity of the cerebral cortex. Electrophysiology using scalp-recordings has even been widely used to assess the heritability of brain physiology (Winteres and Goldman 2003). A higher frequency of electroencephalographic activity indicates increased cerebral activity and lower frequencies may be associated with cognitive decline.

Experimentally-produced electrophysiological signals such as the sensory evoked potentials (Skrandies 1998, Tandon and Mahajan 1999, Verleger et al 1996, Zumsteg and Wieser 2002, Yordanova et al 2004), and the excitatory post-synaptic potentials associated with long-term potentiation (LTP) represent neurophysiological electrical activity (Gubellini et al 2001, Lisman and Morris 2001, Frankland et al 2001, Almaguer et al 2002, Zhao et al 2004, Lees and Dougalis 2004) which provide information about mechanisms of higher neural functions. In general, neural electrophysiological activity signals reflect various facets and status of neuronal tissue physiological activity (Malmo and Malmo 1977; Musha et al 2002).

Ageing is characterized by a progressive decline in physiological function of cells and consequently of the whole organism (Franceschi et al 2000; Kanungo 2003, Kyriazis 2003). Evidence suggests that the nervous system may be a primary target of cellular ageing as nerve cells are particularly vulnerable to the effects of ageing (Sapolsky 2003, Sowell et al 2003). For example, in some cases of familial amyotrophic lateral sclerosis, gene products (misfolded superoxide dismutase) of the mutated superoxide dismutase gene affect the motor neurons in the motor cerebral cortex, brain stem and spinal cord more severely than other cells to cause the disease (Jonsson et al 2004). Furthermore, the nervous system may even be considered as a central regulator of the organismal lifespan (Boulianne 2001). This is indicated by the following. Expression of a human superoxide dismutase transgene in wild type adult fruit fly (*Drosophila*) motor neurons increased its life span, but the same enzyme's overexpression in fly muscles did not. A particular consequence of the ageing process in the nervous system is change in electrophysiological activity (Duffy et al 1993). Ageing-related alterations in neuronal electrophysiologic parameters across species have not been extensively studied. Such data will help in determining that a particular change in brain functions might also represent a meaningful change in several species including man (Potier et al 1993).

Age-related Changes in Action and Synaptic Potentials

The resting membrane potential of nerve cells is generally not affected by age (Frolkis et al 1984,Potier et al 1993), but the synaptic resting membrane potential decreases with ageing (Tanaka and Ando 1990). The excitability of neurons may decrease in aged neurons. The rheobase was found to significantly increase in the CA1 hippocampal neurons in aged Wistar and Fisher 344 rats (Potier et al 1993). Species and regional differences however, are often found. For example, age-related increases in the excitability of CA1 and CA3 hippocampal neurons (Barnes et al 1987) and granule cells in the rat hippocampus were also reported (Papatheodropoulos 1996). The excitability of the cortical motor neurons in animals and humans have been reported to decrease with ageing (see Yordanova et al 2004).Blockage of sodium channel but not the input resistance or membrane potential was found to be responsible for the decrease in excitability in hippocampal neurons (Leung and Yim 1991). The axonal excitability properties are also of interest, and they are sensitive to membrane potential, nodal and internodal ion channels, nodal width, electrical isolation between the internodal and nodal compartments, altered myelination etc (see Nodera et al 2004). The action potential spike amplitude is not altered with ageing, but some other characteristics of the action potentials show changes (Bunting and Scott 1989). For example, age-related increases in action potential overshoot, prolonged after hyperpolarization, and duration of the action potential spike (Pitler and Landfield 1990, Landfield and Pitler 1984) have been reported. Increase in the after hyperpolarization duration, however, has been thought not to be a reliable marker of neuronal ageing (Potier et al 1993). In CA1 cells of the rat hippocampus, the amplitude of sodium and calcium spikes was found not to change with age while the mean duration of after hyperpolarization significantly decreased in aged Wistar rats (Potier et al 1993). The nerve impulse conduction velocity in peripheral nerve fibres may show decrease during ageing (Birren and Wall 1956, Sato et al 1985, Johnson and Murray 1992). Factors such as axon diameter, myelin thickness and internodal distance may influence conduction velocity (Nodera et al 2004). Thus, axon shrinkage or demyelination occurring in the ageing nerve may account for the slowed conduction (Johnson and Murray 1992). The velocity of nerve impulse (action potential) in brain fibers may also be affected by ageing (Aston-Jones et al 1985).

Neurons integrate synaptic activity, and as they are also endowed with complex intrinsic firing properties (see Destexhe and Sejnowski 2003) they exhibit widely varying patterns of firing action potentials (Arechiga et al 1993). As the intensity of a stimulus is coded by frequency of firing, action potentials may be fired as: a train of individual action potentials that slow down in frequency or as a train of tonic action potentials; bursts (Lisman 1997) of action potentials (each burst consisting of 3 or more action potentials) even when the cell is depolarized for a short period of time; high frequency trains of action potentials; or trains of action potentials at a very low frequency in a pace-maker (Smith and Thomson 1987) fashion. In the medial septum and the nucleus of diagonal band of Broca of rats, for example, there are present both rhythmically and nonrhythmically bursting neurons (Apartis et al 2000). The former category includes those neurons, which require synchronization inputs from other brain regions as well as those in which bursting is intrinsically driven. The rhythmic bursting activity appears necessary for memory and attentional purposes, the rhythmic bursting activity is impaired by ageing while the nonrhythmic activity is not affected (Apartis et al 2000). Some neurons are endowed with the ability to change their pattern of firing action potentials during different states of arousal. Neurons, which fire nonrhythmically during slow wave sleep, adopt burst-firing patterns during waking and rapid eye movement sleep. This functional plasticity , however, may decrease with ageing (Apartis et al 2000).

The postsynaptic sensitivity to neurotransmitters may be affected in aged subjects. As a consequence, the inhibitory and excitatory postsynaptic potentials (IPSP and EPSP) may show age-related alterations. Changes (Shen and Barnes 1996, Taylor and Griffith 1993) will depend upon the number of binding sites (receptor density) available, and the presynaptic release mechanisms. Variable results have been reported concerning N-methyl-D-aspartate (NMDA), and or glutamate binding sites in the hippocampus of aged rats (Baudry et al 1981, Bonhaus et al 1990, Wenk et al 1991). For example, in aged rats the hippocampus shows a 30-40% decrease in NMDA receptor binding density (Candy et al 2001). During the ageing process a decline was found in the levels of NMDA receptors in CA1 and CA3 and subiculum areas, but not in the dentate gyrus (Wenk and Barnas 2000, Clark et al 1992, Riedel et al 2003). The decline was correlated with the ability to learn (Adams et al 2001). An age-related decrease in the amplitude and duration of γ-aminobutyric acid (GABA) -related IPSP was found in CA1 neurons in three strains of rats although the glutamate-related EPSP's were not significantly altered in aged rats (Potier et al 1993). Aged synapses may show deficient synaptic plasticity (Almaguer et al 2002,Zhao et al 2004). For example, even if the amount of long-term potentiation may be the same in aged and young rats, the decay of these responses back to baseline occurs more rapidly in aged rats (Davis et al 1993).

Molecular mechanisms of electrogenesis are complex and involve ion movements through membrane channels and pumps. The membrane pumps, channels, receptors are protein molecular entities and are a part of the biological structure and functional organization. Therefore, alteration in molecular functions resulting from changes in genetic expression/genes may be causal to changes in electrophysiological signals. Because channel currents contribute to the generation of potentials, age-related alterations in the genetic expression of channels (number of channels) and their subunit composition (Murchison and Griffith 1995) may affect electrical signaling. Changes in channel current density would have profound impact on cell functioning. For instance altered Ca^{2+} ion homeostasis resulting from altered Ca^{2+} channel currents, may contribute to cell ageing (Murchison and Griffith 1995).

Age-related Changes in the Electroencephalogram

Electrical changes of the brain cerebral cortex are often recorded as the surface or cortical electroencephalogram (EEG) which is made up of a compounded waveform consisting of multiple frequencies, and is derived mainly from summation of the synaptic activity in the pyramidal cells of the cerebral cortex (Salek-Haddadi et al 2003). Spectral content of the EEG do reflect changes occurring in brain subcortical structures, and several electrophysiologic measures appear genetically heritable (Winterer and Goldman 2003). Alterations in the damage to the forebrain nucleus basalis whose cholinergic neurons (Frick et al 2002) have a facilitatory effect on cortical neuronal activity, result in increased power of slow waves and a higher incidence of immobility-associated high voltage spindles (Buzsaki et al 1988a,1988b) in the EEG. The low frequency EEG may be indicative of a cognitive decline (Radeck et al 1994), while increases in alpha and beta activity may reflect vigilance-promoting influences (Siwak et al 2000).The human electroencephalogram is reported to show altered frequencies and synchronization with ageing (Duffy et al 1996;Mc Evoy et al 2001). Ageing reduces power in the human sleep EEG and also causes frequency specific changes in the brain topography (Landolt and Borbely 2001). EEG studies in animals have indicated that spectral composition of the base line EEG is likely to change with age. As a dog ages, there is an increase in low frequencies and a decrease in higher frequencies of EEG (Siwak et al 2000). In aged rats, there is a depression of the power of alpha and beta activity (Buzsaki et al 1988) and the slowing of the EEG (Fig.1) which may be related to decreased brain activity associated with ageing (Tani and Ishihara 1988). Although a number of studies on the human electroencephalogram reported a slowing of frequencies during old age, several recent studies seem to contradict these findings and show that frequency changes are different for different EEG spectral bands and there is an increased desynchronization (fast activity) with age (Duffy et al 1996). The EEG slowing seen in various earlier studies on old human subjects may be more reflective of age-related pathology (disease) than of the normal age-related process (Duffy 1996).

Age-related Changes in Multiple-Unit Activity

Electrophysiological changes in the ageing brain can also be assessed by measuring cellular-level electrophysiological activity i.e. unit and multiple-unit action potentials. Multiple-unit activity (MUA) alterations may reflect the biochemical, physiological and behavioural changes of neurons (Mizumori et al 1996, Barnes et al 1987, Malmo and Malmo 1982). Therefore, an age-related change in spontaneous (basal) neuronal firing is of interest as one of the neuroelectric indicator of electrophysiological functional alteration in the brain. Age-related declines in spontaneous neuronal firing activity have been found in several brain areas: inferior and superior colliculi of rat (Malmo and Malmo 1982), locus coeruleus neurons of rat (Olpe and Steinmann 1982), forebrain neurons of rat (Jones and Olpe 1983), the cerebral cortex of rat (Roy and Singh 1988, Abdulla et al 1995), the hippocampus , thalamus, and striatum of rat (Lippa et al 1986,Kaur et al 1998, Sharma et al 1993, Kaur et al 2001,2003). In some studies the absence of age-related decreases in action potential activity has been reported (Jones and Olpe 1984, Lamour et al 1987, Barnes et al 1983). Figures 1 and 2A show the age-related decline in MUA in four brain regions of rats.

Electrophysiological Ageing

The age-related alterations in electrophysiological activity briefly outlined above appear to be related to the process of ageing in the nervous system. The alterations indicate ageing-related loss, dysfunction and disorganization of electrophysiological activity and thus represent electrophysiological ageing of the nervous system. The disorganization of electrophysiological signals itself would result in derangement of neural functions since information is conducted and integrated as electrical signals.

For example, deficiency of, or absence of, synchronous action potential bursts can alter presynaptic neurotransmitter release consequently influencing synaptic plasticity, information processing, learning etc (Apartis et al 2000; Lisman 1997).

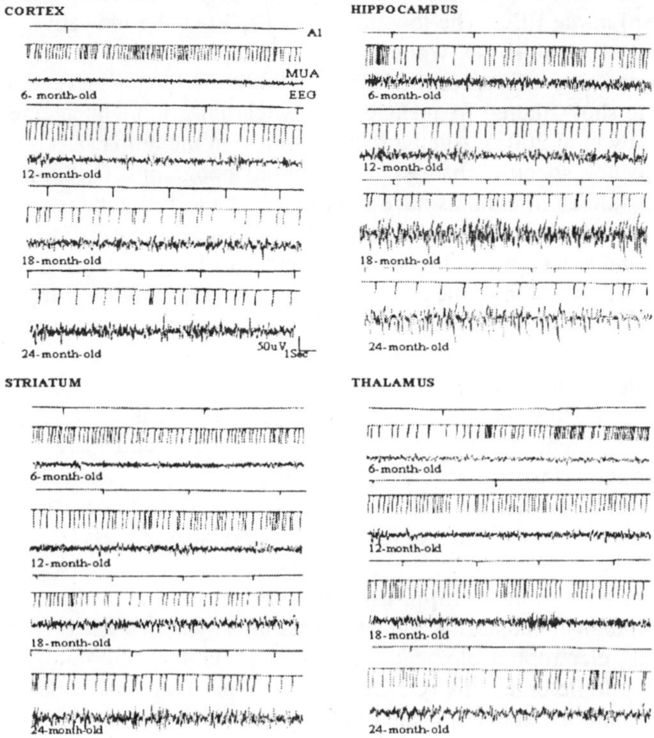

Fig.1 Representative polygraphic records of multiple-unit activity (MUA) and corresponding electroencephalograms (EEG) from four brain regions showing age-related decline in MUA and slowing of the EEG in rats. Recordings are from conscious unrestrained animals implanted with chronic electrodes. Location of recording electrodes in brain regions : parietal cortex 2.00 mm lateral to midline, 2.00 mm posterior to bregma; CA_3 hippocampal subfield (AP –4.3, L 4.3, V 4.5); striatum-caudate-putamen (AP –0.3, L 3.3, V 4.5); thalamus-ventroposterior thalamic nuclei- (AP –3.3, L –2.5, V 6.0). AI refers to cumulative amplitude integration marks of corresponding EEGs. Data in this and subsequent figures are adapted from Kaur et al (1998, 2001, 2003).

Functional cellular changes are likely even in the absence of macroscopic pathologic changes, and thus age-related alteration in electrophysiological response can occur even in the absence of cell deformation or loss etc. Single cell recordings may show alterations of electrophysiological responses even in the absence of cellular degenerative changes (Raz 2001). The decline in spontaneous MUA with age, like other neurophysiological signals such as EEG (Musha et al 2002) can be considered as a measure of neuronal impairment and thus as an important parameter of electrophysiological ageing. In this article, we have focused on this parameter together with a number of related biochemical parameters.

Electrical Activity and Biochemical Parameters

The age-related loss or disorganization of the ability to fire electrical signals may be related to several biochemical factors since biochemical and electrical signals are linked to each other (Katz and Clemens 2001). Membrane lipid peroxidation is known to increase with age (Fig.2B). It causes oxidative stress genomic damage and ageing (Blair 2001, Chatgilialogla and O'Neill 2001). It also significantly influences membrane electrophysiology by altering action potential production (Pellmar 1986), membrane excitability (Pellmar 1987,1995, Seutin et al 1995), and synaptic transmission (Avshalumov 2000, Gilman 1992, Zoccarto et al 1995), and can also lead to generation of epileptiform electrical activity (Singh and Pathak 1990). As lipid peroxidation causes the membrane's lipid environment to change, it also adversely affects the membrane's sodium pump (Na^+,K^+-ATPase) activity (Mattson 1998, Mate's et al 1999). Age-related alterations in Na^+,K^+ ATPase activity may also result from : the direct effects of free radicals (Lees 1991); and or a decrease in ATP availability consequent to the age-related mitochondrial bioenergetic impairment (Tretter and Adam-Vizi 1996, Martinez et al 2000, Shigenaga et al 1994, Sapolsky 2003).The enzyme Na^+,K^+-ATPase is involved in the maintenance of ionic gradients of Na^+ and K^+ across the plasma membrane and the resting membrane potential. Age-related changes in the sodium pump activity may, thus, alter membrane electrophysiology. Changes in Na^+,K^+-ATPase activity have indeed been found to be associated with alterations in neuronal action potential firing (Beal 1995,Deboer and Tobler 1995, Riddle et al 1993, Huttenlocher and Rawson 1968). It would thus also appear clear that age-related alterations in MUA would be related to these two biochemical parameters. Correlation of age-related changes in MUA (Fig.3) have actually been found with both lipid peroxidation and Na^+,K^+-ATPase activity (Kaur et al 1998). The action of antilipidperoxidative enzymes like glutathione peroxidase, glutathione-s-transferase (Fig.2) would also influence the lipid peroxidation status of a tissue and consequently the degree of oxidative damage (Ames et al 1993, Harman 1956, Davydov et al 2004), . Changes of these enzymatic parameters may also thus become related to age-related changes in electrical activity. Some oxidative stress parameters, however, may also appear to be beneficial rather than neurotoxic (Rice et al 2002, Zs-Nagy 2002) and regulate processes such as LTP (Knapp and Klann 2002, Klann et al 1998), neurotransmitter release (Chen et al 2001,2002, Auerbach et al 1997) and signaling between neurons and glia (Atkins et al 1999).

Pharmacology of Ageing

In this paper, we have focused on some aspects of the age-related disorganization of electrophysiologic and biochemical parameters that are likely to underlie electrophysiological dysfunction. It will be of interest to examine whether drugs that appear to retard ageing also counter the ageing-related electrophysiological and biochemical derangements. There is tremendous public interest in the development of antiageing therapies, and the idea that the ageing process can be reversed is emerging (Kyriazis 2003,Zs-Nagy 2002, de Grey 2003). A variety of substances and drugs have been investigated and continue to be studied for their antiageing effects. Even the possibility of using gene transfer to target the delivery of therapeutic proteins to specific regions of the nervous system to treat ageing-related diseases appears promising (Mata et al 2003). Strictly speaking pharmacological antiageing action may not involve modification of the ageing process or processes (Hayflick 2000) but

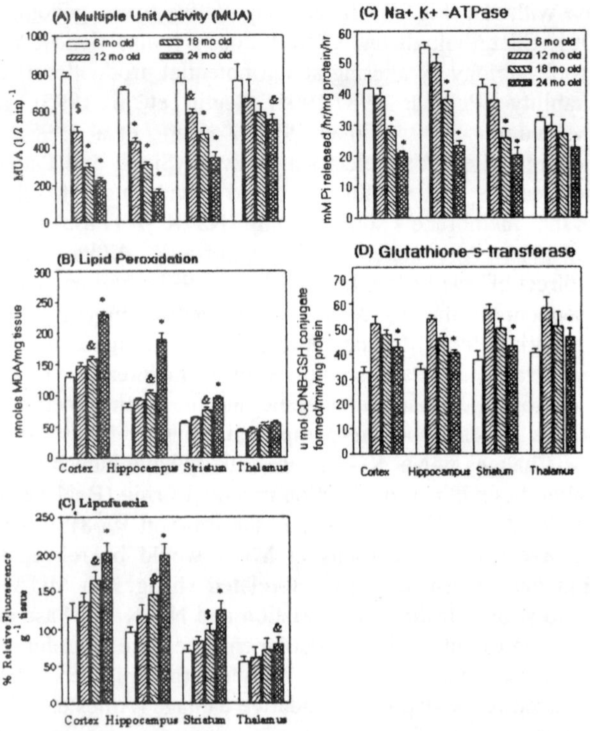

Fig 2 Age-related decrease in multiple-unit activity and the activity of the enzymes Na^+,K^+,-ATPase and glutathione-s-transferase; and age-related increase in lipid peroxidation and lipofuscin content in regions of rat brain during ageing from 6 to 24 months of age (Thalamus does not show changes in Na^+,K^+ -ATPase and lipid peroxidation). * $p< 0.01$, \$ $p<0.01$ and & $p<0.05$.

actually it may refer to the reversal of, or countering, the age-associated deficits. The human ageing process is an extremely complex process and as yet we do not know how it can be retarded or slowed although effects which caloric restriction has on the ageing phenomenon indicate the possibility by which ageing process might possibly be slowed (Tavernarakis and Driscoll 2002, Remmen et al 2001,Mattson et al 2001,Hayflick 2000, Masaro 1993,2000, Subba Rao et al 1996). Furthermore, the goal of biogerontological research is largely to increase active longevity or functional lifespan free from disability and functional dependence (Hayflick 2000, Bickford et al 1997). In this regard pharmacological and dietary manipulations may be of value for prolonging the life of healthy or even that of the aged individuals because these may help to alter the time course of biological ageing to enhance both the quality and quantity of life (Bickford et al 1997, 2000). It is of interest to note that certain diets can reverse the age-induced declines in β-adrenergic receptor function (Bickford et al 2000). In this paper, we outline the electrophysiological and biochemical effects of a few drugs namely acetyl-L-carnitine, L-deprenyl and centrophenoxine that are indicative of their antiageing potential.

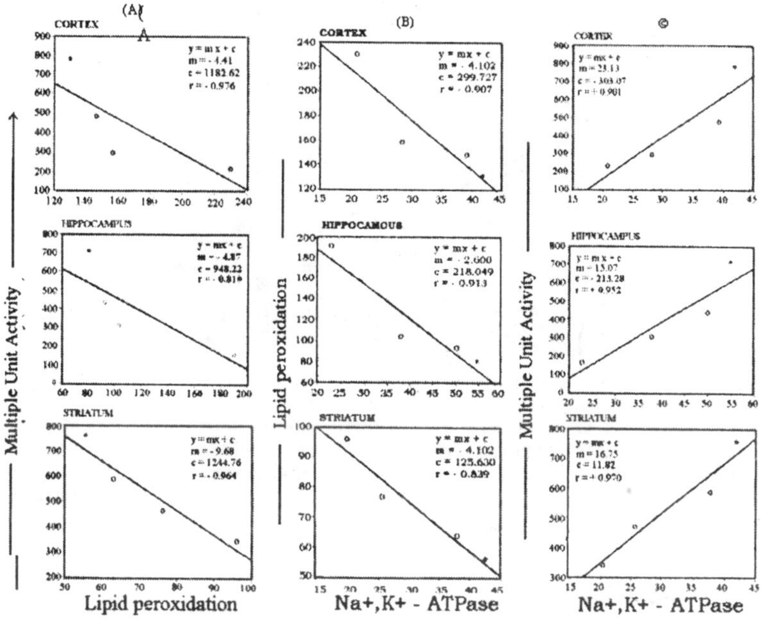

Fig 3 (A) Age-related decline in MUA is correlated (Pearson's correlation) with age-related increase in lipid peroxidation in three brain regions, (B) Age-related increase in lipid peroxidation is correlated with age-related decline in Na^+,K^+ -ATPase in three brain regions and (C) The drug-induced elevation in MUA is positively correlated with the drug induced increase in Na^+,K^+ -ATPase activity in three brain regions.

Acetyl-L-carnitine

Acetyl-L-carnitine (ALC) is a normal physiological component of the inner mitochondrial membrane and is endogenously formed in the brain from the reversible acetylation of carnitine (Shigenaga et al 1994, Kaur et al 2001). Carnitine-acetylcarnitine exchange across the inner mitochondrial membrane is involved in shuttling long-chain fatty acids into mitochondria for β-oxidation. ALC administration in experimental subjects significantly increases brain carnitines (Maccari et al 1990). Pharmacologically, carnitine esters are more effective than carnitine and thus ALC exerts beneficial effects in the nervous system (Calvani et al 1999, Taglialtela et al 1994,1996). It reverses many age-associated deficits in cellular functions (Shigenaga et al 1994,Athanassakis et al 2002), and increases ATP production, and this property probably contributes to its beneficial effects. The drug can be considered as a putative antiageing substance. Experimentally, when chronically fed to rats, it reduces accumulation of lipofuscin (Fig 4) in brain neurons in the brain (Dowson et al 1992, Kaur et al 2001), and attenuates age-related decrements in a variety of physiological, behavioral and biochemical parameters (see Shigenaga et al 1994). Carnitines increase membrane stability and have the ability to cause differentiation and growth of neural cells (Athanassakis et al 2002).

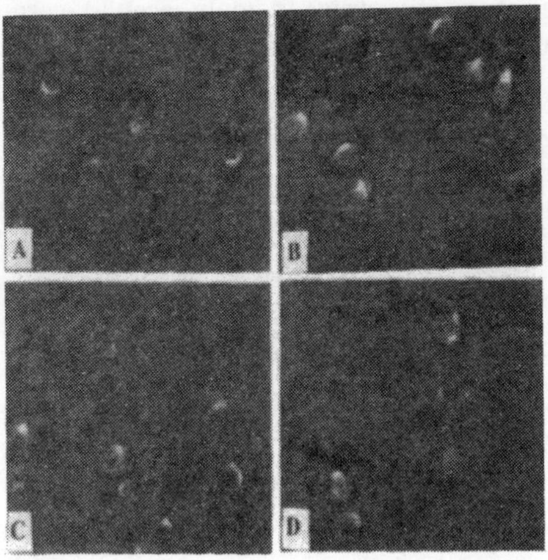

Fig.4 Fluorescence photomicrographs showing age-related increase in lipofuscin, and the effect of acetyl- L-carnitine and L-deprenyl on lipofuscin content: (A) lipofuscin deposits in cerebral cortex neurons at 6 months of age; (B) lipofuscin deposits in cerebral cortex neurons at 24 months of age; and (C) - lipofuscin deposits at 24 months of age after ALC treatment . (D) lipofuscin deposits at 24 months of age after L-deprenyl treatment. Drug treatment lasted 3 months beginning at the age of 21 months. Age-related increase (comparison of A & B) and the drug-induced decrease (comparison of B with C and D) in lipofuscin accumulation are evident.

ALC has antioxidative action (Calvani et al 1999) as evidenced by its antilipidperoxidative and antilipofuscin effects (Kaur et al 2001). The mechanism of its antioxidative action, however, is not clear. The drug does not scavenge hydroxyl radical and its effects on superoxide radicals are doubtful. But it inhibits xanthine oxidase activity (Calvani et al 1999) and this property would favour reduction in oxygen radical formation (Mates et al 1999). Furthermore, the ability of ALC to enhance acetyl coenzyme A levels is also antioxidative since enhanced acetyl coenzyme A levels increase ATP production which would result in lower lactate levels, and that in turn will lower oxidative stress. ALC in fact normalizes tissue levels of pyruvate and lactate and prevents oxidative stress-mediated protein oxidation (Calvani et al 1999). Our own work has further shown that ALC treatment results in the activation of glutathione-s-transferase activity, which may also contribute to its antioxidative capacity. We have also found that the ALC decreases lipid peroxidation and lipofuscin contents and elevates Na^+,K^+-ATPase activity in aged brain regions (Kaur et al 2001). The ALC was also found to augment MUA (Fig.5) in aged rat brain regions indicating the drug's ability to stimulate electrical activity in the ageing brain.It has also been reported to activate EEG activity in elderly human patients (Herrmann et al 1990). Improvement of electrical activity by ALC may in part be due to its effects on lipid peroxidation and the sodium pump activity (Fig.6). Correlational studies showed that the ALC-induced changes in lipid peroxidation, Na^+,K^+-ATPase activity, glutathione-s-transferase activity were correlated with the drug-induced activation of MUA in brain regions (Fig.3). Experimental data have also shown that changes in the membrane Na^+,K^+-ATPase activity are associated with altered membrane electrophysiological activity (Lees

1991). The decreased sodium pump activity may also cause membrane depolarization (Beal 1995). Presynaptic afferent depolarization

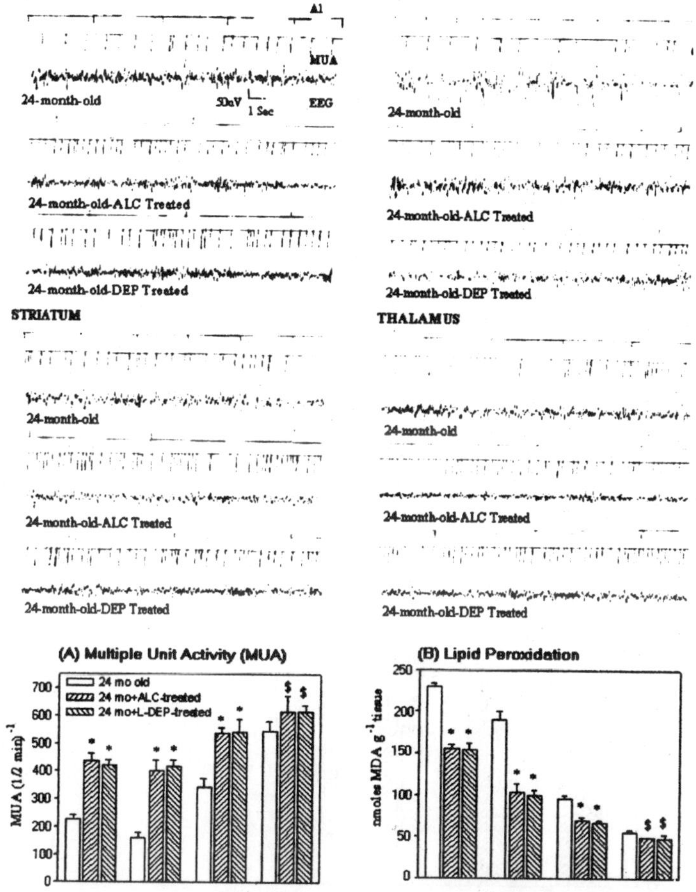

Fig.5 **In upper panels are presented the representative samples of polygraphic recordings of EEG & MUA showing the effect of acetyl-L-carnitine (ALC) and L-deprenyl on MUA in four brain regions of 24 months old rats. Histogram (lower panel) show increase in MUA (A), and decrease in lipid peroxidation (B) after acetyl-L-carnitine (ALC) and deprenyl treatments in four brain regions of 24 months old rats. Drug treatment lasted 3 months beginning at the age of 21 months. Drug-induced increase in MUA and EEG is evident.**

during synaptic neurotransmission results in a lowering of the frequency of action potential generated at the postsynaptic neuron. Since membrane lipid peroxidation significantly influences membrane electrophysiology (Pellmar 1986), shows age-related elevation, and contributes to the ageing of membranes (Mattson 1998), the observed statistical correlations between the age-related decline in MUA and the age-related elevation in lipid peroxidation are indicative of causal relationships between these parameters. The statistical correlation between the age-related inhibition of Na^+,K^+-ATPase activity and the age-related decrease in MUA in various brain regions observed in our study is also reflective of causal relationships.

L-deprenyl

The drug L-deprenyl (a selective and irreversible inhibitor of monoamine oxidase B) is considered a safe drug in the treatment of Parkinson's disease (Tatton and Chalmers-Redman 1996) and works only when taken in combination woth drugs like levodopa. It is found useful in reducing functional decline in Alzheimer's disease (Sano et al 1997). The effects of chronic deprenyl treatment on parameters of normal ageing for assessing its ageing reversal potential are of great interest as the drug is neuroprotective (Bickford et al 1997) and has anti-oxidative effects (Ingram et al 1993,Amenta et al 1994, Thiffault et al 1995, Kaur et al 2003). It is projected as a putative antiageing substance similar to many other agents that may protect cell macromolecules against ageing related oxidative damage (Hinerfeld et al 2004).

We have found (Kaur et al 2003) that L-deprenyl lowers lipid peroxidation and lipofuscin content, and elevates the Na^+,K^+-ATPase and glutathione-s-transferase activities in aged rat brain regions (Fig. 4,5,6). Chronic administration of deprenyl has been found to cause stimulation of acetylcholinesterase and Na+,K+-ATPase recently in other studies also (Carageorgiou et al 2003). It also increases MUA (Fig.5) in brain regions indicating that it reverses the ageing-related decline in brain electrophysiological activity. Deprenyl's antilipidperoxidant and Na^+,K^+-ATPase upregulating effect may in part underlie its augmentation of MUA as the drug-

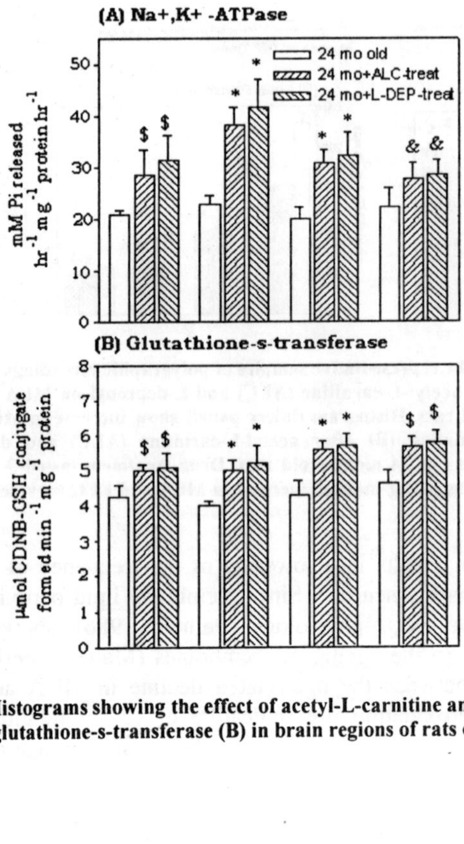

Fig 6 Histograms showing the effect of acetyl-L-carnitine and L-deprenyl on Na^+,K^+ ATPase (A), glutathione-s-transferase (B) in brain regions of rats of 24 months of age.

induced changes in these parameters were correlated with the drug induced elevation in MUA. Deprenyl increases the electrical firing rate of substantia nigra dopaminergic neurons (Huang et al 1997), but it reduces neuronal electrical firing in animal epileptic seizure models (Gupta and Kulkarni 2000). This shows that the drug prevents age-related decline by augmenting electrical activity while it lowers excessive electrophysiological activity associated with epileptiform conditions (Kaur et al 2003). Both ALC and L-deprenyl were found to elevate the basal level of MUA in CA3 subfield of aged hippocampus (Kaur et al 2001,2003). The hippocampal CA3 neuronal network is a crucial region for physiological functions as cells of this gateway network modify the incoming information and transfer the modified message to the CA1 subfield via CA3-CA1 projections (Li et al 1994) for short-term memory formation. Deprenyl has been reported to increase dendritic arborization in CA3 neurons of monkey hippocampus (Lakshmana et al 1998). Deprenyl-induced improvement in the basal level of activity in the aged brain is therefore, of significance as changes in the activity of hippocampal neurons underlie difficulties in memory in normal ageing, and electrophysiological markers correlate with functional alterations in the aged hippocampus (Barnes et al 1997). These two drugs appear also effective against aluminium-induced enhanced ageing (Kaur et al 2003a).

Centrophenoxine

Centrophenoxine has been widely studied for its antiageing effects (Nandy and Bourne 1966, Hasan et al 1974, Roy et al 1983, Patro and Sharma 1984, Patro et al 1992, Sharma et al 1993,Zs-Nagy 2002). It augments the activity of antioxidant enzymes (Roy et al 1983), and inhibits lipofuscin formation (Roy et al 1983) and accumulation (Nandy 1983, Dowson et al 1987, Sharma et al 1993). The drug is a membrane fluidizer (Wood et al 1986), improves synaptic plasticity (Bertoni-Freddari et al 1982), and has a variety of effects on synapses (Giuli et al 1980). We have found that centrophenoxine stimulates MUA in aged rat brain cerebral cortex (Roy et al 1988), and hippocampus (Sharma et al 1993). These data also showed that the MUA declined with age in an inverse relationship with lipid peroxidation indicating that rising lipid peroxidation during ageing may be related to the decline of MUA during ageing.

The brief discussion of the effects of the three drugs on basal MUA in brain regions would clearly show that the drugs prevent ageing-related decline in brain electrical activity and improve aged brain's electrophysiology. The improvement in brain electrophysiology is significantly related to the drug's effect on several aspects of biochemistry such as prevention of oxidative stress and changes in the membrane sodium pump function. Since oxidative stress is a major contributor to the ageing process and to many neurodegenerative diseases, the antioxidative effects of drugs (Hinerfeld et al 2004) are of considerable importance. Although in this article we have focused on the MUA decline as a measure of electrophysiological ageing of the brain, the changes in various other electrophysiologic parameters such as the electroencephalogram, synaptic potentials etc will also be equally important as measures of electrophysiological ageing (Musha et al 2002). Of various regions we studied in this regard the hippocampus and cortex showed higher age-related declines of MUA whereas the thalamus showed the lowest decline (Sharma et al 1993, Kaur et al 1998). This is consistent with the view that the thalamus has a much lower age-sensitivity (Kaur et al 1998) and thus shows a lower level of age-related impairment in MUA. The hippocampus and cortex are known to show greater age-sensitivity (Amenta et al 2001) and, therefore, exhibit a higher impairment of MUA. Our data on age-related decline in MUA derived from cross sectional studies are consistent with the data that have resulted from longitudinal studies on age-related depression in MUA (Malmo and Malmo 1982).

Acknowledgement: The senior author (RS) is supported by a University Grants Commission's Emeritus Fellowship award.

References

Abdulla FA, Abu-Bakra MA, Calaminici MR, Stephenson JD and Sinden JD (1995) Neurobiol Aging **16**: 41-52.
Adams MM, Smith TD, Moga D, Gallaghar M, Wang Y, Wolfe BB, Rapp PR and Morrison JH (2001). J Neurosci 432:230-243.
Almaguer W, Estupinan B, Frey JU and Bergado JA (2002) Neurobiol Aging**23**: 319-324.
Amenta F, Bongrani S, Cadel S, Ferrante F; Valsecchi B and Zeng Y-C (1994) Mech Ageing Dev **75**: 157-167.
Ames BN, Shigenaga MK and Hagen TM (1993) Proc Natl Acad Sci (USA) **90**: 7915-7922.
Apartis E, Poidessous-Jazat F, Epelbaum J and Bassant MH (2000) Brain Res **876**: 37-47.
Arechiga H, Fernandez-Quiroz F, Fernandez de Miguel F and Rodriguez-Sosa L (1993) Chronobiol Int **10**: 1-19.
Aston- Jones G, Rogers J and Shaver RD (1985) Nature (Lond) **318**: 462-464.
Athanassakis I, Zarifi I, Evangeliou A and Vassiliadis S (2002) Brain Res**932**:70-78.
Atkins CM and Sweatt JD (1999) J Neurosci **19**: 7241-7248.
Auerbach JM and Segal M (1997) J Neurosci **17**: 8695-8701.
Avshalumov MV, Chen BT and Rice ME (2000) Brain Res **882**: 86-94
Barnes CA, Suster MS, Shen J and Mc Naughton BL (1997) Nature**388**: 272-275
Barnes CA, Rao G and Mc Naughton BL (1987) J Comp Neurol **259**: 549-558.
Barnes CA, McNaughton BL and O'keefe J (1983). NeurobiolAging **4**: 113-119.
Baudry M, Arst DS and Lynch G (1981) Brain Res**223**: 195-198.
Beal MF (1995) Annals Neurol **38**: 357-366.
Bertoni-Freddari C, Giuli C and Pieri C (1982) Arch Gerontol Geriat**1**: 365-373
Bickford PC, Adams CE, Boyson SJ, Curella P, Gerhardt GA, Heron C, Ivy GO, Lin AMLY, Murphy MP, Poth K, Wallace DR, Young DA, Zahniser NR and Rose GM (1997) Neurobiol Aging **18**: 309-318.
Bickford PC, Gould T, Briederick L, Chadman K, Pollock A, Young D, Shukitt-Hale B and Joseph J (2000) Brain Research **866**: 211-217.
Birren JE and Wall PD (1956) J Comp Neurol **104**: 1-16.
Blair IA (2001) Exp Gerontol**36**:1473-1481.
Bonhaus DW, Perry WB and McNamara JO (1990). Brain Res **532**: 82-86.
Boulianne GL (2001). Mech Ageing Develop**122**: 883-894
Bunting TA and Scott BS (1989) Brain Res **501**:105-115.
Buzsaki G, Bickford RG , Armstrong DM, Ponomareff G, Chen KS, Ruiz R, Thal LJ and Gage FH (1988a) Neuroscience **26**: 735-744.
Buzsaki G, Bickford RG, Ponomareff G, Mandel R and Gage FG (1988b). J Neuroscience**8**: 4007-4026.
Candy C, Evans MS and Brewer GT (2001) Brain Res **921**:1-11.
Calvani M and Arrigoni-Martelli E (1999) Int J Tissue React**21**: 1-6.
Carageorgiou H, Zarros A and Tsakiris S (2003). Pharmacol Res 48: 245-251.
Chatgilialoglu C and O'Neill P (2001) Exp Gerontol**36**: 1459-1471.
Chen BT, Avshalumov MV and Rice ME (2001). J Neurophysiol**85**: 2468-2476.
Chen BT, Avshalumov MV and Rice ME (2002) J Neurophysiol **87**: 1155-1158.
Clark AS, Magnusson KR and Cotman CW (1992) Behav Neurosci **106**:324-335
Davis S, Marcowska AL, Wenk GL and Barnes CA (1993) Neurobiol Aging**14**: 107-115.
Davydov VV, Dobaeva NM and Bozhkov AI (2004) Exp Gerontol 39: 11-16
Deboer T and Tobler L (1995) Brain Res**670**: 153-156.
De Grey ADNJ (2003) Exp Gerontol**38**: 927-934.

Destexhe A and Sejnowski TJ (2003) Physiol Rev **83**: 1401-1453.
Dowson JH (1987) Advances in the Biosciences **64**: 93-100.
Dowson JH, Wilton-Cox, Cairns MR and Ramacci MT (1992) Biol Psychiatry **32**: 179-187.
Duffy FH, McAnulty GB and Albert MS (1996) Neurobiol Aging **17**: 587-599.
Duffy FH, McAnulty GB, and Albert MS (1993) Neurobiol Aging **14**: 73-84.
Franceschi C, Valensin S, Bonafe M, Paolisso G, Yashin AI, Monti D and De Benedictis G (2000). Exp Gerontol **35**: 879-896.
Frankland PW, O'Brien C, Ohno M, Kirkwood A and Silva AJ (2001) Nature **411**: 309-313.
Frick KM, Burlingame LA, Delaney SS and Berger-Sweeney J (2002) Neurobiol Aging **23**:145-158.
Frolkis VV, Tanin SA, Martynenko DA, Bogatskaya LN and Bezrukov VV (1984) Aging of the neurons. In: Physiology of Cell Aging (Frolkis VVand Johnson JE, ds) pp 149-185, Raven Press, New York.
Gilman SC, Bonner MJ and Pellmar TC (1992) Neurosci Lett **140**: 157-160.
Giuli C, Bertoni-Freddari C, Pieri C (1980) Mech Ageing Develop **14**: 265-271.
Gubellini P, Ben-Ari Y and Gaiarasa J-L (2001) Eur J Neurosci **14**: 1937-1946.
Gupta M and Kulkarni SK (2000). Indian J Exp Biol **38**: 332-337.
Harman D (1956) J Gerontol **11**: 289-300.
Hasan M, Glees P and Spoerri P (1974) Tissue Res **150**: 369-375
Hayflick L (2000). Nature **408**: 267-269.
Herrmann WM, Dietrich B and Hiersemenzel R (1990) Int J Clin Pharmacol Res **10**: 81-84.
Hinerfeld D, Traini MD, Weinberger RP, Cochran B, Doctrow SR, Harry J and Melov S (2004) J Neurochem 88: 657-667
Huang C-C, Tsai J-J and K-S (1997). Brain Res **753**: 27-35.
Huttenlocher PP and Rowson MD (1968) Exp Neurol **22**: 118-129.
Ingram DK, Wiener HL, Chachich ME, Long JM, Hengemihle J and Gupta M (1993) Neurobiol Aging **14**: 431-440.
Johnson RD and Murray FT (1992) Brain Res Bull **28**: 61-64.
Jonsson PA, Ernhill K, Anderson PM, Bergemalm, Brannstrom, Gredal O, Nilsson P and Marklund S. (2004) Brain 127 (Part I) : 73-88
Jones RSG and Olpe HR (1984) Neurosci .Lett **50**: 31-36.
Kanungo MS (2003) Proc Indian natn Sci Acad B **69**: 121-140.
Katz PS and Clemens S (2001) Trends Neuroscience **24**; 18-25.
Kaur J, Sharma D and Singh R (1998). Indian J Biochem Biophys **35**: 364-371
Kaur J, Sharma D and Singh R (2001) Neurosci Lett **301**: 1-4.
Kaur J, Singh S, Sharma D and Singh R (2003a) Indian J Biochem Biophys **40**:330-339.
Kaur J, Singh S, Sharma D and Singh R (2003b) Biogerontology **4**: 105-111.
Klann E, Roberson ED, Knapp LT and Sweatt JD (1998) J Biol Chem **273**: 4516-4522.
Knapp LT and Klann E (2002) J Neurosci **22**: 674-683.
Kyriazis M (2003) Biogerontology **4**: 75-90.
Lakshmana MK, Rao BS, Dhingra NK, Ravikumar R, Govindaiah, Ramchandra, Meti BL and Raju TR (1998) Brain Res 796:38-44
Lamour Y, Dutar P and Jobert A (1987) Neuroscience **16**: 835-844.
Landfield PW and Pitler TA (1984). Science **226**:1089-1092.
Landolt H-P and Borbely AA (2001) Clinical Neurophysiol **112**: 369-377.
Lees G and Dougalis A (2004) Brain Res 997:1-14
Leung LS and Yim CC (1991) Brain Res **553**: 261-274.
Lippa AS, Crichett DJ, Ehlert F, Yamamura HI, Enna SJ and Bartus RT (1981) Neurobiol Aging **2**: 3-8.
Lisman JE (1997) Trends Neurosci **20**: 38-43.
Lisman J and Morris GM (2001) Nature **411**: 248-249.
Ludwig M and Pittman QJ (2003) Trends Neurosci **26**: 255-261.
Li X-G, Somogyi P, Ylinen A and Buzsaki G (1994). J Comp Neurol **339**: 181-208.
Maccari F, Arseni A, Chiodi P, Ramacci MT, and Angelucci L (1990 Exp Gerontol **25**:127-134.
Malmo HP and Malmo RB (1977) Electroencephalogr clin Neurophysiol **42**: 501-509.
Malmo HP & Malmo RB (1982) Neurobiol. Aging **3**:43-53.
Martinez M Hernandez AI and Martinez N (2000) Brain Res **855**: 100-106.
Masoro EJ (2000) Exp Gerontol **35**: 299-305.
Masoro EJ (1993. J Amer Geriatr Soc **41**: 994-999.
Mates JM, Perez-Gomez C and de castro IN (1999) Clinical Biochem **32**:595-603.
Mata M, Glorioso JC and Fink DJ (2003) J Gerontol Med. Science 58A: 1111-1118
Mattson MP (1998) Trends Neurosci **21**: 53-57.

Mattson MP, Duan W, Lee J and Guo W (2001). Mech Aging Develop **122**: 757-778.
Mc Evoy LK, Pellouchoud E, Smith ME and Gevins A (2001) Cogn Brain Res **11**: 363-376.
Mizumori SJY, Lavoie AM and Kalyani A (1996) Behav Neurosci **110**: 1006-1016.
Murchison D and Griffith WH (1995) J Neurophysiol **74**: 876-887.
Musha T, Asada T, Yamashita F, Kinoshita T, Chen Z, Matsuda H, Uno M and Shankle WR (2002) Clin Neuophysiol **113**: 1052-1058.
Nandy K (1983) *Aging* 21: 401-413.
Nandy K and bourne GH (1966) Nature (London) **210**: 247-250.
Nodera H, Bostock H, Kuwabara S, Sakamato T, Asanuma K, Jia-Ying S, Ogawara K, Hattori N, Hirayama M, Sobue G and Kaji R (2004) Brain 127 (Part I): 203-211
Olpe HR & Steinmann M (1982) Brain Res **151**:174-176.
Papatheodoropoulos C and Kostopoulos G (1996) Eur J Neurosci **8**: 510-520.
Patro IK and Sharma SP (1984) Exp Gerontol **19**: 241-252.
Patro IK, Patro N, Sharma SP and Chaudhury A (1992) Lipofuscinolysis: An overview. In: Advances in Physiological Sciences (Manchanda SK, Selvamurthy W and Mohan Kumar V eds) pp 441-450, Macmillan India Ltd, Delhi.
Pellmar TC (1986) Brain Res **364**: 377-381.
Pellmar TC (1987) Neuroscience **23**: 447-456.
Pellmar TC (1995) J Neurosci Meth **59**: 93-98.
Pitler TA and Landfield PW (1990) Brain Res **508**: 1-6.
Potier BY, Lamour Y and Dutar P (1993). Neurobiol Aging **14**:17-25
Radeck RJ, Curzon P and Decker W (1994) Brain Res Bull **33**: 183-188.
Raz N (2001) Ageing and the brain. In : Encyclopedia of Life science pp 1-6 Nature Publishing Corporation.
Rice ME, Forman RE, Chen BT, Avshalumov MV, Cragg SJ and Drew KL (2002) Comp Biochem Physiol Part C 133:515-525.
Riddle DR, Gutierrez G, Zheng D, White LE, Richards A and Purves D (1993) J Neurosci **13**: 4193-4213.
Riedel G, Platt B and Micheau J (2003) Behav Brain Res 140: 1-47
Roy D and Singh R (1988) Exp Gerontol **34**: 161-174.
Roy D and Singh R (1983) Exp Gerontol **18**: 185-197.
Salek-Haddadi A, Friston KJ, Lemieux L and Fish DR (2003) Brain Res Reviews 43:110-133.
Sapolsky RM (2003) Nature Reviews Neuroscience **4**: 61-69.
Sata A, Sato Y and Suzuki H (1985) Neurosci Lett **53**:15-20.
Sharma D, Maurya AK and Singh R (1993) Neurobiol Aging **14**: 319-330
Shen JM and Barnes CA (1996). Neurobiol Aging **17**: 439-451.
Seutin V, Scuvee-Moreau J, Masotte L and Dresse A (1995) Brain Res **683**: 275-278.
Shigenaga MK, Hagen TM and Armes BN (1994) Proc Natl Acad Sci (USA) **91**: 10771-10778.
Singh R and Pathak DN (1990). Epilepsia **31**: 15-26.
Siwak CT, Callahan H and Milgram NW (2000) Prog Neuro-Psychopharmacol & Biol Psychiat **24**: 709-726.
Skrandies W (1998) Cognitive Brain Res **6**:173-183.
Smith SJ and Thompson SH (1987) J Physiol **382**: 425-448.
Sowell ER, Peterson BS, Thompson PM and Henkenius AL (2003) Nature Neuroscience **6**: 309-322.
Subba Rao K, Ayyagari S, Raji NS and Murtthy KJR (1996) Curr Sci **71**: 464-469.
Taglialatela G, Caprioli A, Giuliani A and Ghirardi O (1996). Exp Gerontol **31**:577-587.
Taglialatela G, Navarra D, Cruciani R, Ramacci MT, Alema GS and Angelucci L (1994) Exp Gerontol **29**: 55-66.
Tanaka Y and Ando S (1990) Brain Res **506**: 46-52.
Tandon OP and Mahajan AS (1999) Indian J Physiol Pharmacol **43**: 425-434.
Tani Y and Ishihara T (1988) Physiol Behav **44**: 389-392.
Tatton WG and Chalmers-Redman RME (1996) Neurology **47**: S171-S183.
Tavernarakis N and Driscill M (2002) Mech Ageing Develop **123**: 215-229.
Taylor L and Griffith WH (1993) Neurobiol Aging **14**: 509-515.
Thiffault C Aumont N, Quirion R and Poirier J (1995) J Neurochem **65**: 2725-2733
Tretter L and Adam-Vizi V (1996) J Neurochem **66**:2957-2966.
Van Remmen H, Guo Z and Richardson A (2001) The anti-ageing action of dietary restriction. In: *Ageing Vulnerability:Causes and Interventions*, Novartis Foundation Symposium 235,pp 221-233 John Wiley & Sons,LTD, New York.
Verleger R, Heide W, Butt C, Wascher E and Kompf D(1996) *Electroencephalogr Clinical* Neurophysiol **99**: 444-457.

Wood WG, Gorka C, Ambrecht HJ, Williamson LS and Strong R (1986) Life Sci **39**:2089-2095.
Wenk GL and Barnes CA (2000) Brain Res 885: 1-5
Wenk GL, Walker LC, Price DL and Cork LC (1991) Neurobiol Aging **12**:93-98.
Winterer G and Goldman D (2003) Brain Res Reviews 43: 134-163.
Yordanova J, Kolev V, Hohnsbein J and Falkenstein M (2004) Brain 127: 351-362
Zhao H, Xu H and Xu X (2004) Brain Res 996 : 111-116
Zoccarato F, Valente M and Alexandre A (1995) *J* Neurochem **64**: 2555-2558.
Zs.-Nagy I (2002) Arch Gerontol Geriatrics 35 (Suppl 8): 385-398
Zumsteg D and Wieser HG (2002) Clinical Neurophysiol **113**:681-685.

Section D

Brain Diseases

Molecular and Cellular Neurobiology
Editors: M.K. Thakur and S. Prasad

Molecular Biology of Gliomas–the Story so far

Deepak Agrawal and Ashok Kumar Mahapatra

Department of Neurosurgery, Neurosciences center, All India Institute of Medical Sciences
New Delhi-110 029, India

Abstract

Glial tumours are the commonest primary brain tumours and the most frequent cause of solid tumours in young adults. The increasing sophistication of modern therapeutic regimens unfortunately has not been reflected in the outcome of the patients suffering from gliomas. Most gliomas arise from astrocytes and this review will restrict itself to astrocytic tumours that form part of a lineage of progression, with tumours of a lower grade tending to recur as higher grade. Molecular model building studies have basically looked at alterations including overexpression, mutations, loss and rearrangements of known growth factors, oncogenes and tumour suppressor genes corresponding to a defined tumour grade. There is also molecular evidence of two types of GBM — primary (de novo) and secondary (arising from lower grades). Mostly, the utilization of this extensive knowledge of tumour molecular biology for therapeutic purposes is still at the experimental stages, though some drugs are in clinical trials with promising results. This review discusses three targets, the Epidermal Growth Factor Receptor (EGFR), Murine Double Minute 2 (MDM2) and p53 that show promise for intervention in these tumours. It is expected that this knowledge-based approach will lead to more effective and less toxic management strategies for Glial tumours.

Key Words: Glioma, Oncogene, Tumour Suppressor gene, MDM2, p53, EGFR, Genetics, Genomic instability, Molecular targets

List of abbreviations

EGFR, Epidermal growth factor receptor; MDM2, Murine Double Minute 2; GBM, Glioblastoma multiforme; GFAP, Glial fibrillary acidic protein;hMLH1, human mutL homolog; hMSH2, human mutS homolog; IL, interleukin; ILR, interleukin receptor; LOH, Loss of heterozygosity; NO, Nitric oxide; NOS, Nitric oxide synthase; OA, Oligoastrocytoma; RAPD, Random amplification of polymorphic DNA; RAS,Rat sarcoma viral oncogene homolog; Rb, Retinoblastoma; SH2, Src homology 2; TSG, Tumour suppressor gene; TGF, beta-Transforming growth factor-beta; VEGF, Vascular endothelial growth factor.

Introduction

Carcinogenesis is a multistep process in several human cancers. It is now generally acknowledged that morphologic changes that occur in malignant progression reflect the sequential acquisition of genetic alterations. Bridging the gaps between phenotype and genotype has led to an increased understanding of the features that form the basis of the histologic tumor classification (Kleihues et al 2000).

Among tumors of the central nervous system, those of astrocytic origin have been studied most extensively. Astrocytomas constitute a large and rather heterogeneous group of tumors, with wide differences in their clinical incidence, gross and microscopic features and biologic behaviors. Their morphologic diversity is attributed partly to the various structural forms displayed by developing mature and reactive astrocytes, and partly, in diffusely infiltrating examples, to the propensity of neoplastic cells to accommodate their shape and size to the persistence of pre-existing structures in the neural parenchyma.

Progression of gliomas to more malignant phenotypes involves numerous molecular genetic alterations. The genes affected by these alterations, the steps in malignant progression for which they are responsible, their normal function in controlling diverse cellular functions such as differentiation, signal transduction, cell cycle progression and angiogenesis and how they act in concert with other tumor suppressor genes or oncogenes are the focus of intense research around the world as well as in our laboratories at the All India Institute of Medical Sciences.

The activity and regulation of a number of mitogenic signaling pathways is aberrant in astrocytomas, and this is thought to play a crucial role in the development of these tumors. The cascade of events leading to the formation of gliomas are not well understood and have been found to include mutations, amplification and overexpression of various growth factor receptors (EGFR), cell cycle regulatory molecules e.g. MDM2, cyclin dependent kinase 4 and 6, that positively regulate proliferation and cell cycle progression (Srivastava et al 2003, Sarkar et al 2003, Sarkar et al 2000, Misra et al 2001, Misra et al 2000, Jotwani et al 2001, Dil-Afroze et al 1998, Chattopadhyay et al 1997). Inactivating mutations and deletions of signaling and cell cycle progression e.g. p53, retinoblastoma protein (Rb) also participate actively in the development of transformed phenotype. Several mitogenic pathways are also stimulated via the autocrine loop, with astocytoma cells expressing both the receptors and respective cognate ligands.

The potential of immunologically mediated diagnosis and therapy, although greatly enhanced since the advent of monoclonal antibodies has not been fully realized in astrocytomas due to significant problems, the most important being the challenge of identifying antigenic molecules specific to these tumors. With the generation of specific reagents, such as monoclonal antibodies directed to glioma derived antigens or emerging gene therapy, techniques designed to deliver toxins, antibodies or reconstituting gene specifically to the tumor tissue are fast emerging. In recent years, there have been great advances in the understanding of genetic events and the molecular biology of human astrocytomas. The EGFR gene, p53 gene and the MDM2 gene have been found to have a significant role in ongonesesis of glial tumors and the advances made in our understanding of these genes till date.

EGFR Gene

Often associated with chromosome 10 loss is amplification and rearrangement of the EGFR gene, which has been reported almost exclusively in primary in about 50% of cases (Agosti et al 1992, Von-Deimling et al 1992). Moreover, several GBM's display deletions and rearrangement of the EGFR gene in association with amplification. The rearrangements most frequently result in deletion of extracellular portion of the receptor, and its elimination results in a protein that fails to bind the ligand but express constitutive tyrosine kinase activity, thus acting as if the growth factor is always present and stimulating cell proliferation (Ekstrand et al 1993). p53-mdm2 pathway alteration is frequent in human gliomas in which it seems to represent an early event in carcinogenesis (Agosti et al 1992). The p53 mutation is an important step in the progression of astrocytomas that has been demonstrated in a study of three diffuse astrocytomas that reoccurred as high grade astrocytomas

(Sidransky et al 1992). Inactivating mutations in the p53 gene has been described in about one-third of diffuse astrocytomas (Fults et al 1992, Louis et al 1993). The mutation identified in the p53 correlates well with 17p deletion. This 17p deletion occurs in about 17% cases with p53 mutation (Fults et al 1992). Thus, in these tumors both copies of p53 gene are inactivated — one by deletion and another by mutation. In normal cells, p53 protein cannot be detected by IHC technique. Some studies have correlated nucleus staining with p53 antibodies with mutation of p53. The incidence of such staining in diffuse astrocytomas is much higher than the actual incidence of p53 mutation, suggesting that such a correlation is not always present (Newcomb et al 1993). The number of cells with nuclei positive for p53 is rarely over 30% and there is no obvious correlation between the percentage of positive cells in tumor grade. About 30% of tumors with have no detectable p53 mutation (Louis et al 1993). p53 accumulates in the nucleus of cell after DNA damage or during unregulated proliferation. Taken together, the evidence suggests that role of the accumulation of the p53 in the nuclei of some cells in all grades of astrocytomas is currently unclear.

MDM2 gene amplification has been observed in about 10% of primary GBM's but none in secondary GBM's (Reifenberger et al 1993) where its overexpression may constitute a molecular mechanism of the escape from p53 regulated growth control. Amplification of CDK4 has been detected in about 15% of malignant gliomas including primary and secondary GBM's (Biernat et al 1997). The gene is located on chromosome 12q 13 close to the MDM2 gene and in some cases both the genes are co-amplified (Reifenberger et al 1996). Beirnat et al (1997) in an IHC analysis of 29 primary glioblastomas detected strong nuclear expression of MDM2 protein with staining in 15 cases (52%) whereas the positivity was much lower in 27 secondary glioblastomas (3 or 11 % of the cases). In majority of primary glioblastomas, MDM2 overexpression is shown by a significant and higher fraction of positive tumor cells (>50%) while it was less than 5% in the secondary GBM's. Secondly, differential PCR showed an intensity of the 143 BP MDM2 fragment and 113 BP dopamine receptor fragment in all cases. Thirdly, mdm2 amplification was not detected in any of the secondary GBM's. In addition there was no correlation between mdm2 overexpression and p53 mutation.

p53 Tumor Suppressor Gene

p53 made to the publications in 1979 as it was described as a cellular protein found in tight association with the simian virus 40 large T protein (SV4OLT). A lot of interest was generated and it was nominated as the 'molecule of the year' in 1993 by Science magazine. p53 was irrationally incriminated for the development of all cancers after its common association with a variety of malignancies was detected and was believed initially to be an oncogene- a cell cycle accelerator. Now it has been proved to be a potent tumor- suppressor and has been described as the "guardian of genome" by David Lane. p53 is unusual due to its "sensor" function, as it can respond to inappropriate growth-promoting or growth-inhibiting conditions via integration of a multitude of signals from the cell's, internal and external environment.

The p53 network is normally switched 'off' and is only activated when the cells are damaged or stressed. These stressed cells are potentially dangerous to the organism as they are more likely to carry mutations and are prone to escape the normal cell-cycle control, ultimately turning the cells cancerous. The p53 protein acts by shutting off this multiplicative process thus inhibiting the progression of the cell-cycle. It causes apoptosis of the cells in a desperate attempt to control the damage when the cell-cycle control fails. Three pathways are described which can activate the p53 network (Vogelstein et al 2000). The first pathway is mediated by DNA damage, which in turn activates two protein kinases- ATM (Ataxia telangiectasia mutated) and Chk2. The second pathway is turned on by aberrant growth signals e.g. ras and myc oncogenes; and is dependent on the protein p14ARF. The third pathway is induced by different chemotherapeutic agents; UV rays etc. and is mediated by kinases called ATR (Ataxia telangectasia related) and casein kinase II. All these three pathways inhibit the degradation of p53 protein resulting in its stabilization at high concentration. This increased concentration of the protein is covalently modified, binds to particular gene sequences and helps in the transcription of the adjacent genes. The amount of the p53 protein within the cell is determined by the opposing roles of increased rate of synthesis and the rate of degradation, of which the latter seems to be more important. The degradation is dependent on ubiquitin-mediated proteolysis. The ubiquitin molecule flags the p53 molecule for its detection by the degradation machinery. MDM2 forms an important part of this machinery. Increased level of p53 protein is inadequate in itself for its effect and requires conformational changes before it becomes effective as a transcriptional activator. The different genes controlled by p53 have roles in broadly four different areas: a) cell-cycle inhibition- The p53 protein directly stimulates the expression of p21WAF/CIPI that is an inhibitor of cyclin-dependent kinases (CDKs). Through the negative effects on various CDKs, p21 inhibits both the G1 to S and G2 to mitosis transitions (Taylor et al 2001). Other genes e.g. Reprimo, can also arrest the cells in the G2 phase. Topoisomerse II is required during the G2/M transition and its inhibition is sufficient to arrest the cells in G2. p53 has been found to cause downregulation of topoisomerase II by repressing the promoter (21). b) Apoptosis- p53 causes over-expression of mitochondrial proteins like Bax, NOXA and P53AIPI, which lead to apoptosis (Bates et al 1999, Pollack et al 1997, Sheikh et al 2000); c) Genetic stability- p53 causes DNA repair by the induction of 'ribonucleotide reductase'; d) Inhibition of vessel formation- Under normal circumstances, p53 stimulates the expression of genes responsible for the inhibition of vessel formation. Cells in which it is inactivated, there is recruitment of vessel formation. The importance of the p53 protein in the physiological process like development is beyond question (Choi et al 1999, Almog et al 1997).

Though the regulation of p53 can take place by different stresses: DNA damaging agents, decreased oxygen, oncogenic stimuli, cell adhesion, altered ribonuceotide pools and redox states, it is believed that unique pathways are utilized by different stimuli. Importantly, the regulation of p53 takes place predominantly at the protein level. Though the regulation of p53 may take place via the regulation of the mRNA levels during certain physiological processes, amplification of p53 genes, whether wild or mutant type, and the consequent amplification of the mRNA levels are not seen in the human tumors. Rather, regulation of the p53 protein takes place by mutations which either 1) increase the half-life via conformational changes that mitigate the degradation of the protein by the uibiquitin complex or more importantly, 2) decrease the sequence—specific DNA binding and transcriptional activity of the p53 protein. The p53 protein is a short-lived protein whose half-life is increased several fold by damage to the DNA. Many other intracellular regulator proteins e.g. MDM2, HIF (Hypoxia Induced Factor)-la, that prevent the degradation or stabilize the protein, come into play. But it is conjectured that the alteration of the levels of the protein per se is inadequate to explain the regulation of the p53 completely wherein comes in the role of post-translational modifications. These conformational changes can be varied and involves addition and removal of phosphate, acetyl, glycosyl, ribose, ubiquitin or 'sumo' groups (Vogelstein et al 2000, Woods et al 2001).

The p53 protein can be divided into thirds: amino-terminal containing the domain, the central core containing its sequence-specific DNA-binding and the multifunctional carboxy-terminal. The amino terminal is also responsible for the binding of MDM2 protein. The proline-rich linking region between amino- acids 60 and 92 begins the hydrophobic midregion. The critical region lies between the amino acids 100 and 295 and determines the conformation of p53, specific binding to DNA consensus- sequences and sequence-specific transcriptional activity. The positively charged carboxy terminus contains the nuclear localization sequence (amino acids 316-360). The oligomerization domain (amino acids 319-360). The oligomerization of p53 into a dimer or a tetramer is required for DNA binding and transactivation function (Milner et al 1991).

Whereas normal cells express wild-type p53 (wt p53), their cancerous counterparts express mutant forms (Oren et al 1999). The nature of p53 biology predicts that tumors with p53 mutations will be more aggressive than tumors with wild-type p53 genes. Importantly, mutant p53 can drive even the cotranslated wild-type p53 into the mutant conformation (Milner et al 1991). The prevalence of p53 mutations in different cancers varies from 0-60% and is more than 80% in certain subtypes.

Major mutational hotspots occur in four of five domains labeled I-V (I, amino CII, amino acids 13-19; II, acids 117-142; III, amino acids 171-181; IV, amino acids 234-2581; V, amino acids 270-286). Missense mutations are the commonest mutations noted (79%) in the midregion but are uncommon (23%) in the amino and carboxy termini, where nonmissense mutations (nonsense, frameshift, splice and silent) are predominant (Greenblatt et al 1994). The wild-type p53 protein usually resides in the cell nucleus, but is undetectable by IHC, to a short half-life of 5-20 minutes, depending on the cell-type, and the minute amounts in which they are present in the cell normally. In certain tumor lines significant amount of p53 protein resides in the cytoplasm rather than in the nucleus which may be an important of inactivating the function of p53 even in the presence of wild-type genes; MDM2 may play a great role in this preferential localization of the protein. IHC studies lend 75% (36-100%) sensitivity and the predictivity of 63% (8-100%) (Greenblatt et al 1994) to the detection of p53 protein. The false-negative results in the detection of p53 by IHC can have two potential reasons where technical reasons are excluded: a) the underlying lesion can be a gross deletion leading to complete absence of p53 protein rather than a missense mutation, which is the commonest mutation in the p53 gene; b) the point mutation may not stabilize the p53 protein in the cell sufficiently for the detection by IHC. It has been observed that there can be false p53 positivity in malignancies where ras oncogene is involved without any involvement of p53. The p53 gene is responsible for the transcription of a number of genes known as p53 induced genes (PIG), the more important of which are wafl, mdm2 and GADD45 (gene associated with DNA damage. The interesting point is the fact that p53 transcribes genes associated with apparently antagonistic functions, namely cell cycle arrest (e.g. P21wafl/cipl) as well as death (e.g. Bax). To explain the preferential activation of one subset of gene over the other, it has been hypothesized that the extent of DNA damage may regulate the outcome; with low levels of damage the former subset of genes are activated allowing for the DNA damage repair to take place whereas the genes responsible for the activation of apoptotic cascade are activated when the damage is beyond repair (Balint et al 2001). The other cause of this bias could be due to the fact that the mutants activate mostly the cell-cycle arrest while apoptosis is seen with the wild-type p53 (Balint et al 2001).

The p53 mutations in the brain tumors are mostly restricted to astrocytic tumors (Fulci et al 1998) but one of the caveats of these studies is that most of the investigators only analyzed the exons 5-8. The p53 gene is involved in the early stages of astrocytoma tumor genesis. Louis et al (Louis et al 1994) observed that p53 mutation and allelic loss of chromosome 17p are observed in approximately one-third of all grades of adult astrocytomas suggesting that inactivation of p53 is important in the formation of grade II tumor. Moreover, high grade astrocytomas with homogenous p53 mutation evolve clonally from subpopulation of similarly mutated cells present in initially low grade tumors (Sidransky et al 1992). Such studies are complimented by functional studies that have recapitulated the role of p53 inactivation. For instance, cortical astrocytes from mice without functional p53 appear immortalized when in vitro and rapidly acquire transformed phenotype. Cortical astrocyte from mice with one functional copy of p53 behave more like wild type astrocytes and only show signs of immortalization after they have lost one functional p53 copy. Interestingly, those cells without functional p53 become markedly aneuploid, showing that p53 loss result in genomic instability leading to malignant astrocytomas (Van Meyel et al 1994). Thus, the abrogation of astrocytic p53 function appears to facilitate some events integral to neoplastic transformation, setting the stage for further malignant astrocytomas.

Mutations of the p53 gene in particular have been reported to be uncommon in childhood gliomas (Litofsky et al 1994) which contrast with their frequent detection in adult gliomas (Rasheed et al 1992) because p53 dependent apoptosis has been implicated in mediating the toxicity of radiotherapy and chemotherapy (Kerr et al 1994, Lowe et al 1965), it is conceivable that low incidence of TP53 mutation among childhood gliomas may account for their more favorable response to their adjuvant therapy and their generally better overall prognosis than adult tumors. This hypothesis has been validated in other tumor types in which TP53 mutation or nuclear accumulation of p53 (which has been used as an indirect correlate of TP53 mutation) has been associated with poor in vivo chemosensitivity (Rusch et al 1995) and an increased rate of 'disease progression (Esrig et al 1994). Conversely, transfer of a wild type p53 gene to a tumor cell line lacking normal p53 has been associated with facilitation of apoptosis (Shaw et al 1992) and enhancement of chemosentivity in vitro (Fujiwara et al 1994).

Pollack et al (1997) studied 29 cases of pediatric malignant gliomas and found p53 mutations in 11(37.9%) which was higher than previous reports (Litofsky et al 1994) and more in line with frequencies that have been observed in adult gliomas (Rasheed et al 1992). Secondly, TP53 mutations and p53 overexpression were associated with less favorable prognosis and period free survival. Also, it has been shown that even in the absence of TP53 mutations, p53 expression can be increased by cell cycle specific variation, amplification of MDM2, DNA damage and anoxia which can lead to stress induced increases in the expression of wild type p53 (Danks et al 1995). Thirdly, the presence of TP53 mutations occurred independently of histological grade among children with malignant gliomas.

Sirdansky et al (1992) earlier demonstrated that clonal expression of p53 mutant cells in astrocytomas was associated with progression of the tumor. Orellana et al (1998) analyzed 37 pediatric CNS tumor for mutation in exon 5-8 of the TP53 gene, only 2/10 (20%) of astrocytomas showed mutation located in the exons 6-8. Two of 8 astrocytomas and I GBM showed loss of heterozygosity at 17p gene. This study suggested that TP53 may be implicated in at least some of the pediatric high grade astrocytomas or in transition of low to high grade pediatric astrocytomas.

Litofsky et al (1994) analyzed p53 mutation using PCR-SSCP analysis in 35 pediatric astrocytomas and found no mutation in the p53 gene. Five of 21 tumors were homozygous in the region of p53 on the chromosome 17p and no mutation was identified in exons 5-8 in any of these tumors and observed the frequency of p53 mutation in pediatric astrocytomas is significantly less than the frequency of adult tumors.

Sure et al (1997) in a study of 20 pediatric glioblastomas exhibited an overall frequency of p53 mutation in 5 (25%) cases. In contrast to adults, most mutations 4/5 980%) were detected in the group of primary GBMs. In addition, overexpression of EGFR was found in 2/20 GBM. Surprisingly, one of these GBM's also harbored a p53 mutation. Sure et al (1997) also analyzed 10 of GBMs which revealed an accumulation of p53 protein without apparent gene mutation in the evolutionary conserved areas of the gene (exon 5-8). These results support the hypothesis that mechanism other than p53 gene mutation result in stabilization of the p53 protein and its accumulation in the tumor cells.

MDM2 Oncogene

Murine Double Minute 2 (MDM2) was first identified in spontaneously transformed immortalized murine cell line, BALB/c 3T3. The cell line contained 25-30 copies of paired acentric chromatin bodies called double minutes. MDM2 was the second of the two tandem genes cloned together from these amplified sequences.

Tumorogenic potential of this gene was identified in the nude mice experiment, which conferred the label of oncogene to this gene. Later the protein derivative of this gene was identified as a 9OkDa phosphoprotein, which inactivated p53 activity. Increased levels of p53 are found due to the coupled effects of an increased p53 translational rate and decreased p53 degradation. MDM2 is found to be closely related to the p53 degradation rate (Haupt et al 1997). It binds p53 close to the N-terminus of p53 and acts as an ubiquitin ligase. This marks p53 for its degradation by the 26S proteosome (Momand et al 2000), which involves multiple mono-ubiquitination of the p53 molecule. MDM2 constantly shuttles between the nucleus and the nucleolus due to its nuclear localization sequence (NLS) and the nuclear export sequence (NES). Mutations in the NES of MDM2 increase the stability of p53 and thus cause its accumulation in the nucleus. Along with this MDM2 dependent pathway, p53 has an independent pathway of its own for this transport too (Momand et al 2000), which can promote the nuclear export, ubiquitination and the subsequent degradation of p53, even in the absence of direct binding. Besides, the cytoplasmically sequestered wild type p53 protein is resistant to MDM2 mediated degradation in certain tumors (Zaika et al 1999). The cocompartmentlization of MDM2 and p53 is important in the degradative process but the nucleus-to-cytoplasmic transport is not mandatory for the same (Xirodimas et al 2001). The corollary that arises out of the various observations is that the tumor cells with mutant p53 are incapable of transcribing MDM2 protein that can target the mutants for degradation and p53 protein is retained in increased levels (Kubbutat et al 1997).

MDM2 interferes with the activity of p53 by a second mechanism also. It binds to the p53 promoter sequence that prevents the transcription of a variety of genes involved in the cell cycle (Momand et al 2000). It is conjectured that these two are the two steps of the same process rather than two different mechanisms altogether. p53 and MDM2 are involved in a complex negative feedback loop (Alarcon-Vargas et al 2002). In this loop p53 activates MDM2 at a transcriptional level, which in turn downregulates p53. It is believed that MDM2 isoforms may interact with the regulators of p53 rather than the p53 molecule itself and the RING domain of the MDM2 molecule is important in its inhibitory effect on cell proliferation. Phosphorylation of MDM2 might also control the p53-MDM2 complex formation. MDM2 binds molecules that in turn modulate the p53 molecule. One of these molecules is pl 9ARF (alternative reading frame), which has an analogue called pl4ARF in human ARF promotes the degradation of MDM2, and in turn, stabilizes p53 molecule (Zhang et al 1998). p53 upregulates its inhibitor MDM2 by two mechanisms (a) direct transcriptional activation of MDM2; and (b) downregulation of the inhibitor of MDM2, pl4ARF (Momand et al 2000). Few other genes have been identified with structural similarity to p53, namely p73, p63 and p51. p73 is capable of upregulating p53-resposive genes and p73 can be bound and inactivated by MDM2, its transcriptional and apoptotic functions being negated. MDM2 can bind and modulate to other proteins like E2F1 and Rb protein, which are responsible for cell cycle progression. CREB binding protein (CBP) and p300 have p53-binding domain. These proteins are also seen to bind to MDM2. So CBP/p300 can keep the MDM2 away from p53 by competitive inhibition (Momand et al 2000). The cross-talk between the Akt, p53 and MDM2 is believed to have possible implication in the regulation of apoptosis (Gottlieb et al 2002) and this interaction may be a potential therapeutic target.

Conclusions

The outlook for gliomas patients has not changed substantially despite the modern therapeutic modalities, and the median survival time of GBM is still about a year from diagnosis. However, the new knowledge regarding molecular alterations in gliomas will perhaps change the way we look at these tumours. Knowledge of how genomic instability influences tumours will have important implications for our understanding of drug resistance and metastasis, and perhaps lead to pharmacological measures to deal with the same. The identification of small molecules as well as possible antibodies to target specific molecular alterations in gliomas like EGFR, PDGFR, VEGF etc. already show promise and further research will surely lead towards the universal goal of all ligand specific therapies i.e., increased effectiveness and specificity, low side effects and reduced development of drug resistance.

References

Agosti RM, Leuthold M, Gullick WJ, Yasargil MG and Westler OD (1992) Virchows Arch A Pathol Anat Histopathol **420**:321-325.

Alarcon-Vargas D and Ronai Z (2002) Carcinogenesis **23**:541-547.

Almog N and Rotter V (1997) Biochemica et Biophysica Acta **1333**:F1-F27.

Balint E and Vousden KH (2001) Br J Cancer **85**:1813-23.

Bates S and Vousden KH (1999) Cell Mol Life Sci **55**:28-37.

Biernat W, Kleihues P, Yonekawa Y and Ohgaki H (1997) J Neuropathol Exp Neurol **56**:180-185.

Chattopadhyay P, Rathore A, Mathur M, Sarkar C, Mahapatra AK and Sinha S (1997) Oncogene **14**: 871-874.

Choi J and Donehower (1999) Cell Mol Life Sci **55**:38-47.

Danks RA, Chopra G, Gonzales MF, Orian JM and Kaye AH (1995) Neurosurgery **37**:246-254.

Dil-Afroze, Misra A, Sulaiman IM, Sinha S, Sarkar C, Mahapatra AK and Hasnain SE (1998) Gene **206**:45-48.

Ekstrand AJ, Longo N, Collins VP and James CD (1993) J Neuro-Oncol **15**:S8.

Esrig D, Elmajian D, Groshen S, Freeman JA, Stein JP, Chen Sc, Nichols PW, Skinner DG, Jones PA and Cote RJ (1994) N Engl J Med **331**:1259-1264.

Fujiwara T, Grimm EA, Mukhopadhyay T, Zhang WW, Owen-Schaub LB and Roth JA (1994) Cancer Res **54**:2287-2291.

Fulci G, Ishii N and Van Meir EG (1998) Brain Pathol **8**:599-613.

Fults D, Brockmeyer D, Tullous MW, Pedone CA and Cawthon RM (1992) Cancer Res **52**:674-679.

Gottlieb TM, Leal JFM, Segar R, Taya Y and Oren M (2002) Oncogene **21**:1299-1303.

Greenblatt MS, Bennett WP, Hollstein M and Harris CC (1994) Cancer Res **54**:4855-4878.

Haupt Y, Maya R, Kazaz A and Oren M (1997) Nature **387**:296-299.

Jotwani G, Misra A, Chattopadhyay P, Sarkar C, Mahapatra AK and Sinha S (2001). Cancer Genet Cytogenet **125**:41-45.

Kerr JF, Winterford CM and Harmon BV (1994) Cancer **73**:2013-2026.

Kleihues P and Ohgaki H (2000) Toxicol Pathol **6**:164-170.

Kubbutat MHG, Jones SN and Vousden KH (1997) Nature **387**:299-303.

Litofsky NS, Hinton D and Raffel C (1994) Neurosurg **34**:967-973.

Louis DN, Von Deimling A, Chung RY, Rubio MP, Whaley JM, Eibl RH, Ohgaki H, Weistler OD, Thor AD and Seizinger BR (1993) J Neuropathol Exp Neurol **52**:31-38.

Louis DN (1994) J Neuropathol Exp Neurol **53**:11-21.

Lowe SW, Ruley HE, Jacks T and Hammill JF (1965) Arch Neurol **13**:547-554.

Milner J and Medcalf EA (1991) Cell **65**:765-74.

Misra A, Chattopadhyay P, Dinda AK, Sarkar C, Mahapatra AK, Hasnain SE and Sinha S (2000) J Neurooncol **48**:1-12.

Misra A, Chosdol K, Sarkar C, Mahapatra AK and Sinha S (2001) Mutat Res **12**:53-59.

Momand J, Wu H-H and Dasgupta G (2000) Gene **242**:15-29.

Newcomb EW, Madona WJ, Pisharody S, Lang FF, Koslow M and Miller DC (1993) Brain Pathol **3**;229-235.

Orellana C, Hernandez-Marti M, Marinez F, Castell V, Millan JM, Alvarez-Garijo JA, Priepo F and Badia L (1998) Cancer Genet Cytogenet **102**:93-99.

Oren M and Rotter V (1999) Cell Mol Life Sci **55**:9-11.

Pollack IF, Hamilton RL, Finkelstein SD, Campbell JW, Martinez AJ, Sherwin RN, Bozik ME and Gollin SM (1997) Cancer Res **57**:304-309.

Rasheed BK, Fuller GN, Friedman AH, Bigner DD and Bigner SH (1992) Genes Chromosomes Cancer **5**:75-82.

Reifenberger G, Ichimura K, Reifenberger J, Elkahloun AG, Meltzer PS and Collins VP (1996) Cancer Res **56**:514-515.

Reifenberger G, Liu L, Ichimura K, Schmidt EE and Collins VP (1993) Cancer Res **53**:2736-2739.

Rusch V, Klimstra D, Venkatraman E, Oliver J, Martini N, Gralla R, Kris M and Dmitrovsky E (1995) Cancer Res **55**:5038-5042.

Sarkar C, Chattopadhyay P, Ralte AM, Mahapatra AK and Sinha S (2003) Cancer Genet Cytogenet **15**:156-164.

Sarkar C, Rathore A, Chattopadhyaya P, Mahapatra AK and Sinha S (2000) Pathology **32**:84-88.

Shaw P, Bovey R, Tardy S, Sahli R, Sordat B and Costa J (1992) Proc Natl Acad Sci (USA) **89**:4495-4499.

Sheikh MS and Forance AJ Jr (2000) J Cell Physiol **182**:171-181.

Sidransky D, Mikkelsen T, Schwechheimer K, Rosenblum ML, Cavanee W and Vogelstein B (1992) Nature **355**:846-847.
Srivastava T, Chattopadhyay P, Mahapatra AK, Sarkar C and Sinha S (2003) J Neurooncol **66**:51-57.
Sure U, Ruedi D, Tachibana O, Yonekawa Y, Ohgaki H, kleihues P and Hegi ME (1997) J Neuropathol Exp Neurol **56**:782-789.
Taylor WR and Stark GR (2001) Oncogene **20**:1803-1815.
Van Meyel DJ, Ramsay DA, Casson AG, Keeney M, Chambers AF and Cairncross JG (1994) J Natl Cancer Inst **86**:1011-1017.
Vogelstein B, Lane D and Levine AJ (2000) Nature **408**:307-310.
Von-Deimling A, Louis DN, Von Ammon K, Peterson I, Hoell T, Chung RY, Martuza R, Schoenfeld D, Yasargil MG, Weistler OD and Seizinger BR (1992) J Neurosurg 77:295-301.
Woods DB and Vousden KH (2001) Exp Cell Res **264**:56-66.
Xirodimas DP, Stephen CW and Lane DP (2001) Exp Cell Res **270**:66-77.
Zaika A, Marchenko N and Moll UM (1999) J Biol Chem **24**:274-280.
Zhang Y, Xiong Y and Yarbrough WG (1998) Cell **92**:725-734.

Molecular and Cellular Neurobiology
Editors: M.K. Thakur and S. Prasad

Post Traumatic Epilepsy: Iron Model

Varsha Sharma* and P. Prakash Babu**

*School of Life Sciences, Jawaharlal Nehru University, New Delhi-110 067
**Department of Animal Sciences, University of Hyderabad, Hyderabad-500 046, India

Abstract
Intracortical injection of iron ($FeCl_3$ or $FeCl_2$) in the rat brain leads to development of a chronic epileptic focus. This iron-induced focal experimental epilepsy model in rodents is important because it serves as an appropriate model of clinical post-traumatic epilepsy. After its onset in the cortical focus, the epileptiform activity evolves with time, generalizes in the cerebral cortex and even propagates to various subcortical brain structures. It is believed that production of electrographic discharges in iron induced experimental epilepsy is mediated by oxidative stress resulting from iron+ tissue interaction, which leads to the generation of oxygen free radicals. The iron involved in the process comes from hemoglobin liberated from extravasated blood released during intracranial hemorrhages. Lipid derangements such as membrane lipid peroxidation resulting from oxidative stress can alter membrane electrophysiological properties including ion gating causing a state of hyper excitability responsible for the production of epileptiform activity. Tissue *in vivo* possesses natural protective defense system against the adverse effect of oxidants. Several enzymes largely provide the antioxidant defense. Noradrenergic influence originating in locus coeruleus neurons has an inhibitory control over electrographic seizure control. Studies revealed that excitatory amino acids glutamate and aspartate, inhibitory neurotransmitter GABA, various receptors viz. NMDA AMPA, glutamate; calcium and potassium ions too are involved in basic cellular mechanisms. The mechanism underlying the development of post-traumatic epilepsy can be investigated and delineated in this model. The data derived may be helpful in designing suitable therapeutic and prophylactic strategies for posttraumatic epilepsy.

Keywords: Post-traumatic epilepsy, EEG, oxidative stress, lipid peroxidation

Introduction

The most common cause of Traumatic Brain Injury (TBI) is accidents with vehicles, falls, sports, recreational injuries and assaults. Seizures may appear as early as the first week after injury. Although mild cases constitute the vast majority, TBI is the common denominator to the development of PTE (Post traumatic Epilepsy).In case of not severe PTE appears in 5% of all the patients TBI and in 15 – 20% in patients suffering from severe TBI. PTE represents about 2 -3% of the etiologies of epilepsy (Oliveros-Juste et al 2002). It is one of the world's most prevalent non-communicable diseases. Consequently, in the context of large and rapidly increasing populations and accidents in the world, PTE is a significant health and psychological burden requiring urgent attention (The global burden of mental and neurological problems). In order to understand the post-traumatic epilepsy, various forms of head injury have been defined (Adams et al 1997). Following head injury, there are two broad categories of seizures that may occur – early or late. Early seizures are acute seizures that occur as a result of the physical effects of the trauma (i.e., within one to two weeks after injury). Late seizures occur after the patient has recovered from the effects of the injury and can occur weeks, months, or even years after the original injury. These can be single or recurrent events. Recurrent late seizures are considered post-traumatic epilepsy (Langendorf and Pedley, 1997). Post- traumatic epilepsy is defined by the development of chronic seizures following head trauma. However, certain people may develop seizures after only a mild insult, yet why others sustaining even more severe traumatic injury do remain a mystery. Even though the mechanisms are not completely understood, the fact that there is a demonstrable event leading to epileptogenesis allows this type of epilepsy to be studied in animal models and clinical situations.

Basic Mechanism of Epileptogenesis

Recordings from individual neurons in animals with experimental epilepsy and, when possible, from epileptic foci in man, often show a characteristic abnormality. Episodes of membrane depolarization occur (paroxysmal depolarization shifts)-PDSs accompanied by bursts of action potentials. More than one mechanism may be responsible: in PTE depolarization is due to entry of calcium ions in one area and sodium currents are involved in other (Fig.1).

Recording with extracellular microelectrodes shows that burst firing occurs synchronously in groups of neurons. In epileptic foci an increased proportion of neurons exhibit burst firing, the individual units fire less regularly, at slightly greater mean intervals and most significantly, with a much greater incidence of synchronization between neurons (Isokawa-Akesson et al 1987,89; Jensen et al 1997). The mechanism leading to the establishment and maintenance of an epileptogenic focus is currently not yet fully understood (Fig.2). It would be simplistic to view epileptogenesis merely as excessive activity due to an imbalance between excitation and inhibition. Hyper synchrony is probably a crucial feature of epileptogenesis and may involve either excitatory or recurrent inhibitory innervations. Taking into account the numerous electrophysiological investigations, the involvement of thalamocortical circuitry in the generation of abnormal oscillatory rhythms from the particular neuronal network is considered as basic underlying mechanism (Avoli et al 1983, Sharma and Singh1999a). Biochemical mechanisms operative within thalamocortical circuitry during this neuronal

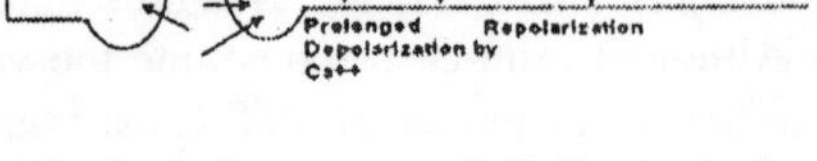

Fig. 1. Ionic movement in Epileptogenesis

oscillation seem to entail phase –locked inhibition by $GABA_B$ alternating with glutamate mediated excitation (Kamphuis et al 1988,92; King et al 1985). The activity of T-type calcium currents in the interaction in excitation and inhibition are

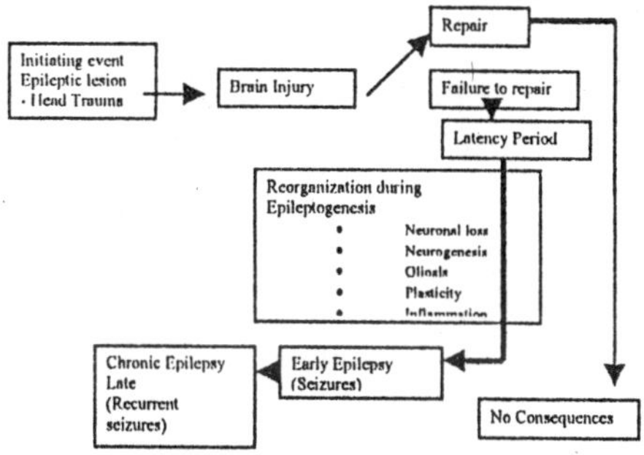

Fig. 2. Mechanisms of Epileptogenesis at Cellular Level (earlier reports)

considered as basic cellular mechanisms Oxidative stress is a result of increased formation of free radicals and/or reduced antioxidative system capacity(Fig.3). Because of presence of unpaired electron, free radicals are highly reactive. Free radicals contribute to degeneration of neurons in epilepsy (Nikushkin et al 1987). High oxygen consumption, lipid content and transition metals are particular risk factors inducing oxidative stress (Halliwell1989, Jesberger and Richardson, 1991).

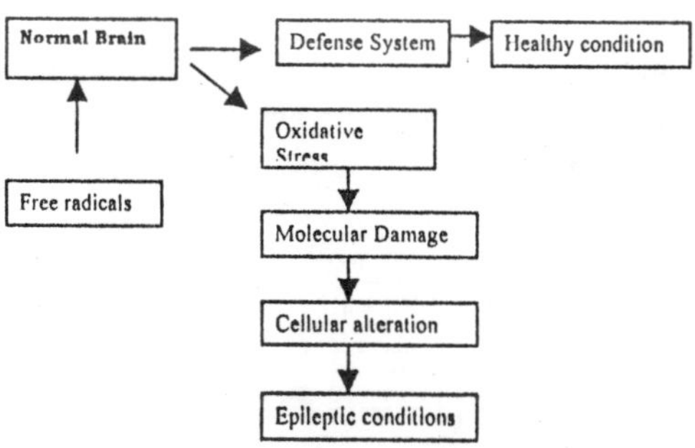

Fig. 3. Oxidative stress

Iron and Copper salts promote the generation of oxidizing free radicals from peroxides for instance release of free metal ions and haemproteins leads to the formation of ·OH, by a process known as Fenton reaction. Only in conditions of iron overload, as stated above there seems to be redox active iron available for free radical production.

$$H_2O_2 + Fe^{++} \ldots\ldots . OH^{\cdot} + OH^{\cdot} + Fe^{+++}$$

Those who absorb significantly more than normal amounts of iron, are at increased risk for posttraumatic epilepsy. In persons who have suffered head trauma, blood extravasations followed by hemolysis results in the deposition of iron within the neuropil and this iron can produce free radicals leading to peroxidation of membrane lipid. That lipid derangements and membrane lipid peroxidation can alter the membrane electrophysiological properties. Since formation of damaging free radicals by blood in the parenchyma of the brain is associated with an increased risk in developing epilepsy, possible mechanisms involving this type of insult have been explored (Singh and Pathak 1990, Sharma and Singh 1999a &b). Both breakdown products have been shown to have physiological effects on synaptic transmission that may lead to epileptogenesis (Yip and Sastry 2000). Indeed, there is an animal model where epilepsy (PTE) is induced by injecting the microlitre quantities of aqueous iron cations into the cortex of rats (Willmore1978; Singh and Pathak 1990, Sharma and Singh 1999 a &b, 2001, Singh et al 2000). If antiperoxidant compounds are given, seizure development can be blocked (Langendorf and Pedley 1997). Another study using chemiluminescence found that iron injections affected both the levels of free radicals and excitatory amino acids (KÜÇÜkkaya et al 1998).

The basic mechanisms in Epilepsy are mediated by

- Alternating inhibitory and excitatory phase regulated by GABA and Glutamate receptors in the thalamocortical circuits.
- Activation of N-methyl-D-aspartate (NMDA) receptors.
- Increase in calcium currents mediated by T calcium channels.
- Decrease in calcium current by L calcium channels

The Spike and wave pattern recorded electrophysiologically seems to be initiated in the thalamus and cortex in iron induced experimental rats (Sharma and Singh, 1999a). Von Krosigk et al (1993) observed that the thalamic reticular nucleus commands oscillatory behaviors in the thalamocortical circuit because of its low threshold for Ca++ dependent spikes mediated by the $GABA_B$ receptors, inhibitory postsynaptic potentials (IPSP), which in turn generate high rate action potentials (Von Krosigk et al 1993)(Fig.4).

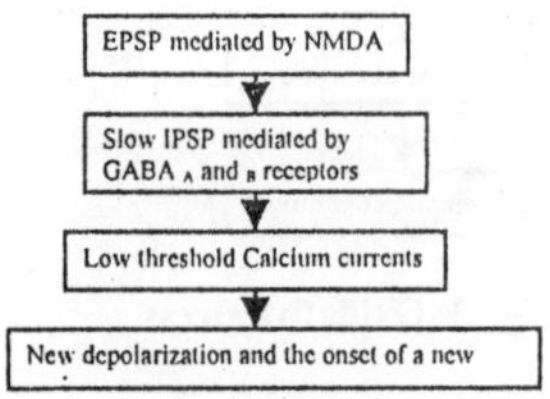

Fig. 4. Initiation and propagation of epilepsy

At the same time, however, it serves as a precursor for the inhibitory neurotransmitter GABA (Gamma Amino Butyric Acid), so its effects on the pathogenesis of epilepsy may be more complicated. Another avenue in the exploration of epilepto-genesis is the role of excitatory amino acids. Studies have found increased levels of glutamate and aspartate in the brains of rats suffering from seizures after brain injury (Singh and Pathak 1990, Nilsson et al 1994). Iron injected into the brain has also been found to affect the release and metabolism of the excitatory neurotransmitter glutamate. Samuelsson et al (2000) found a decrease in glutamate transporter protein (a protein expressed by astrocytes that is involved with glutamate reuptake) after iron injection. Similar results have been found in humans (Ronne Engström et al 1992). Glutamate is widespread in brain as well as in the extracellular fluid. It is involved in many metabolic functions. Glutamate is thought to exert an excitotoxic effect on the brain in large amounts (Evans 1962). Garthwaite et al (1988) and Kraus et al (1998) suggested that the N-methyl-D-aspartate (NMDA) subtype of glutamate receptor participates in the expression of the increased neuronal excitability in the epileptic brain. Calcium conductance is a main factor in the excitability of epileptic neurons (Meyer 1989,Wadman & Connor 1992).

Animal Model of Post-traumatic Epilepsy

A search of the literature reveals that animal models are getting prime importance in the field of research now than was the case a decade or so. Of a number of experimental epilepsy models Dichter and Ayala (1987) have provided the insight into the basic mechanism of various types of human epilepsies. Adams et al (1997) has defined various forms of head injury. When nothing penetrates the skull, it is called closed head injury. Loss of consciousness transiently can be associated with this kind of injury which could do gross damage to the brain depending on the severity of injury. Animal models have been used to examine the consequences of head injury on epileptic activity, which includes fluid percussion injury model, creation of freeze lesion etc. Among these a popular model that is widely used is the placement of metals (iron, cobalt, aluminium) to the brains of animals to produce seizure. The iron-induced epilepsy model is of interest because it may serve as a model of clinical PTE appearing in persons who have suffered head trauma. The animal model of PTE attempts to represent some aspect of the etiology, symptology or treatment of this disorder. It will be particularly useful in the preclinical testing of various anticonvulsant drugs, in studying the functional relevance of GABA neurotransmitter, NMDA, AMPA and glutamate receptors, role of Ca^{++}, K^{+}, Cl^{-} channels implicated in the modulation of epileptic seizures.

Iron as a Focal Epileptogenic Agent

Iron has an advantage over many other substances viz. organic chemicals, is that metals tend not to be absorbed by the body, and not to be changed by metabolism and/or other chemical reaction. Therefore, they tend not to diffuse from the area of injection and to persist at the site of implication or administration for at least several days to weeks. The fact that $FeCl_3$ when injected intracerebrally, produces experimental epilepsy in laboratory rats may be of relevance to human post traumatic epilepsy. It is known that extravasations of blood during head trauma, the development of post-traumatic epilepsy is a frequent sequela of severe head trauma. It has been postulated and confirmed by Magnetic resonance imaging (MRI) that the deposition of iron from hemoglobin and its subsequent sequestration as hemosiderin into brain cells as a consequence of the blood loss may be cause of seizure seen in this condition in humans (Angeleri et al 1999). If this is the case, then iron-induced epilepsy could be an ideal model of posttraumatic epilepsy.

Experimental set up for Iron Model of PTE.

Male Wistar rats (200-250 gms) housed in a constant temperature animal houseroom with a light/dark cycle and were allowed to access food and water adlibitum. The rat is placed in a stereotaxic frame as soon as it is anesthetized (Sodium penta baribitone 35mg/kg). The skull is exposed by using a scalpel and blade blunt dissection. With a dental drill burr holes, 1mm in diameter were made for intracranial implantation of electrodes and for iron application. The coordinates for various brain structures were according to the coordianates given in the rat brain atlas (Paxinos and Watson 1982). For iron application a burr hole was made in the right calvarium over the cerebral cortex 1 mm posterior and 2.5 mm lateral to bregma (Willmore et al 1978 Willmore and Triggs 1986; Singh and Pathak 1990, Sharma and Singh 1999a,b). Through this hole, 5μl freshly prepared aqueous solution containing 100mM $FeCl_3$ /$FeCl_2$was injected for 5 min into the cortex by a microlitre syringe. The age matched control animals received physiologic saline at the same coordinate site. Immediately after injection, bipolar and monopolar (Plastics One Inc.) were implanted in the cortical and subcortical brain areas. These were then fixed with dental cement that clearly isolated the electrodes from skin and muscles. The primary indication of epileptic activity in the animal in awake and calm position is the appearance of repetitive electrocorticographic spiking activity. Electrophysiological recordings from conscious unrestrained experimental and control rats were obtained using a Grass 79D EEG polygraph. Records were made daily from the third post operative day. Connections to the EEG electrodes board were made through low noise wire cables. For the first two days after surgery, the animals were placed in the recording chamber to habituate them to the recording setup before the beginning of the experiments. For recording, animals were in electrical shielded, sound attenuated, ventilated and dimly lit chamber that had a one-way glass window for observation of overt behavioral signs (Roy and Singh 1988). Electroencephalographic recordings were performed for 2 hrs each in both forenoon and afternoon each day to ascertain the presence or absence of epileptic EEG activity. Development and persistence of epileptic EEG activity was followed for 28 days after cortical iron injection (Fig.5). In addition to field potential recordings (electroencephalographically recorded (EEG) potentials (Fig.6), multiple unit potentials (MUA) (Vergenses et al 1987, Silva-Barrat et al 1986, Traub et al 1996 and Babb et al 1987) were also recorded simultaneously from these sites(Fig.7). MUA was amplified, filtered (Grass Modal P511J and 7P511L amplifiers 300Hz to 10 kHz and electronically discriminated using a window discriminator (WPI) and display on a storage oscilloscope. The standard TTL pulses from the window discriminator together with EEG activity were simultaneously recorded on the polygraph.

Advantage of Iron Model

The study of animal models is one of the most promising approaches to understand the mechanism of epileptogenic seizure. When electrophysiological or biochemical methods are used, this model enable us to define exactly which neuronal circuits are involved in the generation of rhythmic activity of the CNS and the molecular mechanism that mediate the oscillatory activity. A major problem, however, is deciding whether or not such a model is really similar to the human counterpart. Advantage in rat model is focal models of epilepsy and epileptogenic focus can be induced at a specific anatomic site and behavioral seizures, as well as EEG spiking, can be reliably produced that resemble the human counterpart, in that they tend to be intermittent and generalized.

Cortex Cx	Hipocampus CA3	Hippocampus CA1	Amygdala AMG	Caudate-puatamen CPU

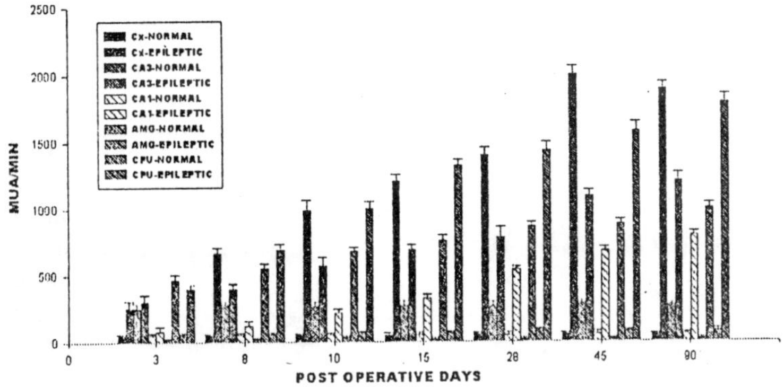

Fig. 5 DEVELOPMENT OF EPILEPTIC ACTIVITY OVERTIME IN CORTICAL AND SUBCORTICAL BRAIN AREAS

Focal Models are Adaptable to variety of studies

Measurement of enzyme activity, transmitter levels, or a large number of other parameters can be measured and related to the behavioral state of the animal, prior to seizure onset, during periods of active seizure or at a time when seizure no longer occurs. A large number of studies have been done that demonstrate the possibilities; a few examples will be indicated here. Nilsson et al 1994, Ronne-Engstrom et al (1992) have shown that increased levels of glutamate and aspartate play a significant role in epileptogenesis. Channel subtypes of NMDA, AMPA, and glutamate have been reported to have convulsive and anticonvulsive properties depending on which channel subtype were being affected (Porter and Meldrum 1998). Calcium induced increase in extracellular potassium too was found to be involved in excitotoxicity (Langerdorf and Pedley 1997). Another area investigated is the inhibitory role of GABA induced Chloride channels (Porter and Meldrum 1998).Montanez et al (2000) have put forward very interesting (although controversial) study that seizure themselves may be a way that the brain copes with injury and recovers. Thus preventing seizures immediately after brain injury (i.e. prophylaxes) may prove more harmful than beneficial. In above experiments a dose of 5 µl $FeCl_3$ /$FeCl_2$ can produce a continuum of seizure activity that showed spontaneous recurrent epileptiform discharges consisting of bursts, spikes and multiple spike wave complexes. The electrographic episodes were spontaneous and recurrent during the observation period the rats were in a state of immobility; i.e. they interrupted cage exploration and became motionless, piloerection and occasional twitching of vibrissae and neck muscles could be observed in some animals. Electroencephalographic discharges continued to persist up to 28 day after $FeCl_3$

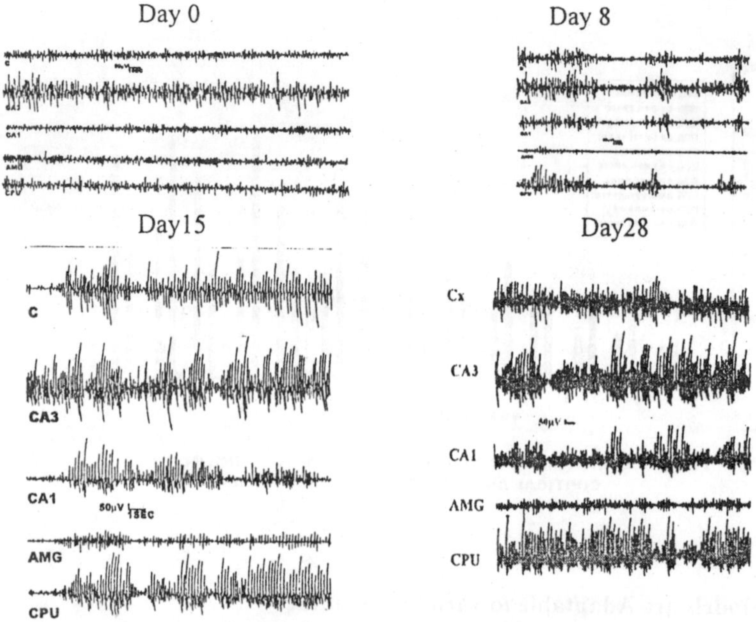

Fig. 6. Simultaneous EEG recordings during epileptic activity in cortical focus (C); Hippocampus (CA3); Hippocampus (CA1); Amygdala (AMG) and Caudate nucleus(CPU) records from experimental animals correspond to day 0,8,15 and 28 after $FeCl_3$ injection

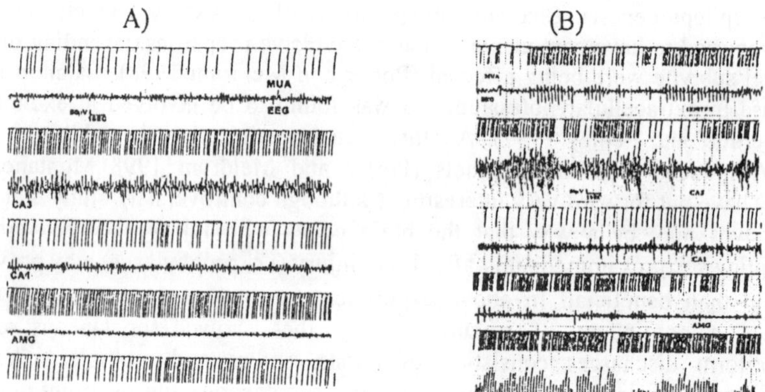

Fig. 7. Simultaneous EEG and MUA recordings during epileptic activity in cortical focus (C); Hippocampus (CA3) ; Hippocampus (CA1); Amygdala (AMG) and Caudate nucleus(CPU) are shown in figure A Control records from the animals injected with acidified saline

solution at day 15 (figure B) records from experimental animals correspond to day 15 after $FeCl_3$ injection.

injection. The iron induced ipsilateral focal epileptiform activity from its site of origin, however, has been found to become generalized to involve almost the entire cerebral cortex of both the hemisphere (Moriwaki et al 1990, 1992). Epileptogenic effects of iron are thought to result from reactivity of cortical neurons in response to iron-induced oxidative stress i.e. neuronal membrane lipid peroxidation caused by active oxygen free radicals generated by iron-mediated reactions in the brain tissue (Willmore and Rubin 1981; Willmore et al 1986; Singh and Pathak 1990, Yokoi et al 1995). That lipid derangement (Nelson and Delgado-Eseueta 1996) and lipid peroxidation can indeed alter membrane electrophysiologic properties including ion gating (Fong and McName 1986). Addition of iron salts to tissue suspension causes formation of several types of oxidants (oxygen free radical and peroxides). Formation of fluorescent products of lipid peroxidation (Triggs and Willmore 1984) and superoxide radical (O_2^-) (Willmore et al 1983), similar to humans during neurotrauma is associated with the head injury (Temkin 2003 Mori et al 1990). In iron-induced epileptic focus, the astrocytes uptake of glutamic acid has also been found to be disturbed, and this appeared to be due to oxidative stress-induced decrease in glial glutamate transporter protein (Samuelsson et al 2003). Thus, higher basal levels of glutamate would also be likely to contribute to epileptogenesis. Increased levels of excitatory aminoacids are known to be associated with head trauma (Annegers and Coan 2000; Bernado 2003). Based on the infliction of traumatic brain injury by fluid percussion injury, a new model of posttraumatic epilepsy has also been recently developed (D'Ambrosio et al 2004). This model is said to more closely reproduce the human traumatic condition of closed head injury, and electrocorticographic epileptic discharges in this model are similar to those found in iron model. Furthermore, the iron-induced seizures are like clinical seizures as they are produced spontaneously by a disordered brain. This is in contrast to seizure models where seizures are triggered 'on demand' by an external stimulus (Levy et al 1990). In the process of generalization of seizure activity, multiple subcortical areas are likely to operate as a network in the elaboration and exacerbation of spike wave seizures after their initiation in the cortical brain site (Lothman 1991).It is of interest to note that occurrence of neuronal epileptic bursting may preclude information transmission among affected structures (Wiest and Nicolele's 2003). Therefore, identification of subcortical brain regions involved is necessary for understanding the process of seizure generalization in this model of epilepsy (Handforth and Ackermann 1995; Baram et al 1997; Brunson et al 1998; Lothman 1985). Stereoencephalographic [depth electroencephalographic (EEG)] study will help to determine the subcortical brain circuits or structures likely to be involved (Moriwaki et al 1992; Baram et al 1997; Medvedev et al 1996). Our initial study of iron epilepsy has shown that the cortical focal epileptic activity spreads to the thalamus, locus coeruleus and substantia nigra (Sharma and Singh 1999a, 1999b). Our subsequent studies (to be published) have indicated the focal epileptiform activity spreads to intracortical areas such as the hippocampus (including its subfields), amygdala and striatum (caudate-putamen) by recording electroencephalographic (EEG) epileptic activity and multiunit action potentials (MUA) simultaneously from these structures.

Iron induced epileptic rats do not show violent motor convulsions. After $FeCl_3$ injection, low voltage fast activity in the cortex buildup progressively into high voltage fast activity with spiking, bursts of polyspiking, or high voltage sharp waves. The paroxysmal seizure activity increased rapidly in duration and complexity after day 8. Seizure patterns consisting of interrupted sequences of continuous and rhythmic high voltage spike and wave complexes were also observed. By day 10-15 after $FeCl_3$ injection showed prolonged bursts of epileptiform activity in all cortical and subcortical structures. 23-28 day after iron injection showed frequent and prolonged epileptiform discharges consisting of high voltage sharp waves. The epileptic activity during day 15-28 can be characterized as frequent status epilepticus like EEG discharges. Simultaneous recording of MUA was done to verify that there was an epileptic activity associated increase in neuronal firing at the recording site (Babb et al 1987). The MUA recordings also ensured that recorded stereo encephalographic epileptic activity represented real epileptogenic EEG signals and cellular activity changes rather than the accidentally recorded field potentials spread (volume conducted) from nearby structures (Vergenes et al 1987 and Verleger et al 1996). Further, the temporal and spatial correlation between epileptic field potential and neuronal action potentials (MUA) provides better localization of focal seizure activity at sites distinct from the initial focal site (Babb et al 1987 and Kandelet al 1996). We also found that the drug ethosuximide abolishes epileptic activity in this model. Ethosuximide has been reported (Pelligrini et al 1989) to desynchronize the hypersynchronizing electrophysiological activity in the reticular-thalamic- cortical net, which occur during the generation of epileptiform activity. This action of the drug may be particularly responsible for its antiabsence effect in humans (Pelligrini et al 1989). Since in the iron-induced cortical focal epilepsy thalamocortical synchronous activity is involved (Sharma and Singh 1999a). Recently, the hypothesis of subcortical origin of absence seizures has been challenged (Manning et al 2004; Timofeev and Steriade 2004), and it appears that a focal seizure initiation site for absence seizure is present in the cerebral cortex rather than in the thalamus. Furthermore, ethosuximide microinfused into the perioral region of the primary somatosensory cortex or rats with genetically determined absence seizures has been found to abolish absence seizures (Manning et al 2004). Ethosuximide's action in iron epilepsy would also be of further interest since iron induction of epilepsy is mediated by oxidative stress-induced lipid peroxidation and the drug ethosuximide being a Ca_2^+ channel antagonist may have anti-lipid-peroxidative activity (Janero and Burghardt 1989, Singh et al 2001) during the generation of epileptiform activity. As the mechanisms behind the development of PTE are discovered, therapies may be developed that can prevent epilepsy after head trauma has occurred.

A class of enzymes known as thiol-disulfide oxidoreductases primarily catalyzes thiol-disulfide interchange reactions. Out of which thioltransferase (glutaredoxin) specifically acts on glutathione-containing mixed disulfides (Gravina and Mieyal 1993, Chrestensen et al 2000). The expression of this enzyme has been reported earlier (Balijepalli et al 1999, 2000, Rajappa et al 2002) in rat and human. We examined the expression (*in situ*) of this enzyme in normal and iron induced experimental epileptic rats (Fig.8) (paper under preparation).

Existing Anti Epileptic Therapy

There has been considerable progress in the pharmacotherapy of epilepsy in the past decade. The selection of Anti Epileptic Drug (AED) is based mainly on its efficacy for specific type of seizures, tolerability and safety. Furthermore, currently there are no drugs available to prevent the development of epilepsy following head trauma

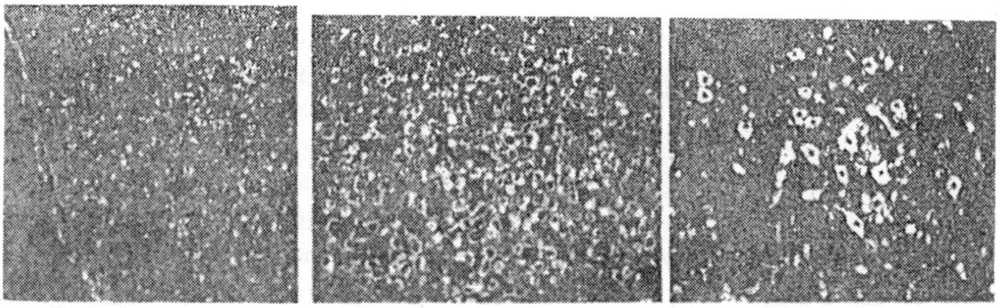

Striatum Hippocampus Reticulum

Fig. 8. Expression of Thioltransferase mRNA in rat CNS after iron induction of epilepsy in different regions

(Löscher 2002). Despite progress in understanding the pathogenesis of seizures and epilepsy the cellular basis of human epilepsy remains a mystery. In the absence of specific etiological understanding, approaches to drug therapy must be directed at control of symptoms, that is the suppression of the seizures. Anticonvulsants have greatly improved the lives of people with epilepsy. Furthermore, there have been significant advances in the medical treatment of epilepsy in recent years (Sabers and Gram 1996; Sharma and Singh 1999b, 2001, Singh et al 2001). Traditionally; pharmacological strategies for the treatment of epilepsy are aimed at suppressing the initiation or the propagation of seizure rather than the process that lead to epilepsy. As a result the currently available AED (Anti Epileptic Drug) cannot prevent the development of epilepsy following brain injury (Harnandez 1997 and Temkin 2001). Furthermore, there is increasing evidence that, despite early treatment and suppression of seizures, AEDs do not affect the progression or underlying natural history of epilepsy (Löscher 1998). Thus in addition to developing drugs that are currently pharmaco-resistant another goal is to develop truly anti-epileptic therapy that prevents either the development or the progression of epilepsy. Better known anticonvulsant drugs generally act:

- By increasing GABAergic action mechanisms, that is predominantly inhibitory;
- By depressing excitatory glutamate neurotransmission;
- By depressing voltage-dependent Ca++ mobilizing mechanisms; or
- By depressing Na+ conductance, this is also voltage dependent.

Phenytoin, CBZ, and Valproate act through blocking Na+ channel, Valproate and Ethosuximide through blocking T calcium channel, Clonazepam by activating GABA channel. Blocking T Ca++ channel afford satisfactory therapy for generalized absence. Action of second-generation drug is remaining controversial and research is yet to establish the mechanism of action. Prophylactic efficacy of other drugs, like lipid peroxidation inhibitors, neuroprotectors (especially antioxidants), glutamate receptor blockers, NMDA receptor blockers, and drugs that modulate apoptosis via caspases inhibition are under investigation.

Conclusion

Epilepsy, one of the most important non-communicable neurological illnesses, is particularly under resourced and under treated in the developing world. Epidemiological studies have made it clear that the magnitude of the problem makes it a public health priority. Large numbers of people are at risk of morbidity and mortality, mainly because of difficulties with treatment infrastructure and the availability of suitable drugs. However, people with epilepsy need more than drug treatment because their local cultural context adds a social and economic burden to the physical burden of their seizures. The education of health workers, patients, and the wider community is therefore essential. Application of iron to cerebral cortex of laboratory animals particularly rodent, leads to the production of a reproducible seizure that leads itself well to a variety of studies. Studies utilize focal epilepsy may significantly increase our knowledge of seizure mechanism and ultimately, our knowledge of human seizure disorder. Since characterized nearly 25 years ago iron induced experimental post traumatic model has become a well established model with which to study epileptogenesis and the mechanisms that maintain an epileptic state. Given the complexity of cellular events that occur with activation of a single/ group of neurons, it is unlikely we will find a single factor that is responsible for the complex network phenomenon that PTE induction represents. Therefore, this model may provide insight into human PTE by allowing us to examine the physiological and behavioral effects of Anti Epileptic Drugs/ other therapeutic treatments; delineate the contribution of neural mechanism involved in mediating progression of seizure; and screen novel compounds etc. As the complexities of cell cycling cell differentiation and cell migration are unrevealed, a better understanding of the mechanism of PTE will be at hand. The techniques of molecular biology will provide new experimental approaches to identify the mode of PTE induction and progression. Although studying an animal model is often rife with difficulties, it remains a necessary means for investigation leading to the understanding, prevention and treatment of neurological disorder/disease.

Acknowledgements

I am thankful to Prof. Alok Bhattacharaya (SLS, JNU) for his consistent encouragement and Prof. Rameshwar Singh (SLS, JNU) for his critical comments. Prof. K. S. Rao (UOH, Hyderabad), Prof. S K Sharma (USIC, JNU), for going through article critically and to DST for financial assistance [Project No SR/WOS-A/ LS-508/2003]

References

Adams R D, Victor M and Ropper A H (1997) Craniocerebral Trauma. In:Principles of Neurology. 6th Edition. pp: 874-889. McGraw-Hill, New York.

Angeler F, Majkowski J, Cacchio G, Sobieszek A, D'acunto S, Gesuita R, Bachleda A, Polonara G, Krolicki L, Signorino M and Salvolini U (1999) Eplepsia **40** : 1222-1230.

Annegers J F; Hauser A Coan S P and Rocca WA (1998) N Engl J Med **338**: 20-24.

Annegers J F, Coan S P (2000) Seizure **9**:453-7.

Avoli M , Gloor P, Kostopoulos G and Gotman J (1983) J Neurophysiol **50** : 819-837.

Babb T L, Wilson C L and Isokawa-Akesson M (1987) Electroencephalographer Clin Neurophysiol **69**: 467.

Balijepalli S, Tirumalai PS, Swamy KV, Boyd MR, Mieyal JJ, Ravindranath V (1999) J Neurochem 72:1170–1178.
Balijepalli S, Boyd MR, Ravindranath V (2000) .Mol Brain Res 85:123–132.
Baram T Z, Gerth A, Schultz L (1997) Brain Res **98**: **265**-70.
Benardo L S (2003) Epilepsia **44**: 27-33.
Brunson K L Schultz L and Baram TZ (1998) Dev Brain Res **111** : 119-128.
Chrestensen CA, Starke DW, Mieyal JJ (2000) J Biol Chem **275**:26556–26565.
D'Ambrosio R, Fairbanks J P, Fender J S, Born D E, Doyle D L, Miller J W. (2004) Brain **127**:304-14.
Dichter M A and Ayala G F (1987) Science **237**: 157-64.
Evans J H (1962) Neurol **12:** 665-674.
Engstrom ER, Hillered L, Flmik R, Kihlstrom L, Lindguist C, Nie JX, Olsson, Y and Silander HC (2001) Epilepsy Res **43**: 135-144.
Fong TM and and Mcnamee GM (1986) Biochemistry **25** : 830-40.
Garthwaite J, Charles SL and Chess-Williams R (1988) Nature **336**:385-387.
Gravina SA, Mieyal JJ (1993) Biochemistry 32:3368–3376.
Handforth A and Ackermann RF (1995) Brain Res. Rev **20**:1-23.
Halliwell B (1989) Acta Neurol Scand **126**: 23-33.
Harnandez TD (1997) Trends Pharmacol Sci **18**: 59-62.
Isokawa-Akesson M, Wilson CL, Babb TL (1987) Epilepsy Res. **1**:17-34.
Isokawa-Akesson M, Wilson C L, Babb TL (1989) Epilepsy Res. **3**: 236-47.
Janero D R and Burghardt B (1989) Biochemical Pharmacol **38**: 4344-4348.
Jensen M S, Azouz R and Yaari Y (1997) J Neurophysiol **71**: 831-835.
Jesberger J A and Richardson J S (1991) Int J Neurosci **57**:1-17.
Kamphuis W, Wadman WJ and Lopes da Silva FH (1988) Brain Res **440**: 205–215.
Kamphuis JA and Gorter FH (1992) Epilepsy and Inhibition, 336–374.
Kandel A, Bragin A Carpu D and Buzsaki G (1996) Epilepsy Res **23** :123.
Kenchappa R S, Diwakar L, Boyd M R, and Ravindranath V (2002) J Neurosci **22**:8402–8410.
King GL, Dingledine R, Giacchino JJ and McNamara JO (1985) J Neurophysiol **54**: 1295–1304.
Kraus JE and McNamara JO (1998) Mol Brain Res **61**: 114–120.
KÜÇÜkkaya B, Aker R, YÜksel M, Onat F and Yal Çin AS (1998) Brain Res **788**: 133-136.
Langendorf F and Pedley T A (1997) Epilepsy: A Comprehensive Textbook. Lippincott-Raven, Philadelphia.
Levy SL, Burnham W M and Hwang PA (1990) Epilepsy Res **1**:12-7.
LÖscher W (1998) Eur J Pharma **342**: **1**-**13**.
LÖscher W (2002) Trends Pharmaco Sci **23**: 113-118.
Lothman E W; Hatelid J M and Zorumski C F (1985) Brain Res **360**: 92-100
Lothman E W 1991 Current Opinion Neurol Neurosurg 4 253-258.
Manning J P, Richards D A, Leresche N, Crunelli V, Bowery N G. (2004) Neuroscience **123**:5-9.
Medvedev A, Mackenzie L, Hiscock J J and Willoughby J O 1996 Electroencephalgr Clin Neuropohysiol **98** : 157-166.
Meyer FB (1989) Brain Res Rev **14**: 227-243.
Moriwaki A Hattori Y Nishida N and Hori Y (1990) Neuroscience Lett 11072-76.
Moriwaki A Hattori Y, Hayashi Y and Hori Y (1992) Electroencephalogr Clin Neurophysiol **83: 281-288.**
Mori A, Hiramatsu M andYokoi L (1990) Pavlonian J Biol Sci **25** : 54-62.
Montañez S, Kline A E, Gasser T A and Hernandez T D(2000) Brain Res. **860**:29-40.
Nelson L and Delgado-Escueta AV (1986) Brain **386** : 32-40.
Nikushkin E V, Kryzhanovski GN, Tupeev I R, Bordyukov MM and Yuzefova S M (1987) Biull Eksp Biol Med **3**: 297-299.
NIH Consensus Development Panel on Rehabilitation of Persons With Traumatic Brain Injury. (1999) JAMA **282**: 974–83.
Nilsson P, Ronne-Engstrom E, Flink R, Ungerstedt U Carlson H and Hillered L (1994) Brain Res **637**: 227-232.
Oliveros-Juste A, Bertol V and Oliveros Cid A (2002) Rec Neurol **34**: 448-459.
Pellegrini A Doss R C, Dal POS F, Emani M, Zanotto L and Testa G (1989) Brain Res **497**: 344-360.
Ronne-Engstrom E, Hillered L, Flink R, Spannare B, Ungerstedt U and Carlson H (1992) J Cereb Blood Flow Meta. **12**: 873-876.
Roy D and Singh R (1988) Exp.Gerontol **23**: 161-74
Paxinos G and Watson C(1982) The rat brain in stereotaxic coordinates, (Academic Press NY)
Porter R J and Meldrum B S (1998) Antiepileptic Drugs. In: Basic and ClinicalPharmacology. 7th Edition Katzung B (ed.) McGraw-Hill, New York. pp. 386-407.

Sabers A and Gram L (1996). Drugs **52**: 483-493.
Samuelsson C, Kumlien E, Elfving A, Lindholm D, Ronne-Engstrom E (2003) Epilepsy Res. **56**:**165-73**.
Samuelsson C, Kumlien E, Flink R, Lindholm D and Ronne-Engstrom E (2000) Neurosçi Lett **289**:**185-188**.
Sharma V and Singh R (1999a) Ind J Exp Biol **37**: 461-467.
Sharma V and Singh R (1999b) Ind J Exp Biol **37**: 468-475.
Sharma V and Singh R (2001) Ann of Neurosci (Abstract) 8:26
Silva Barrat C, Menini Ch, Beyére P and Naquet R (1986) Electroencephalographer Clin Neurophysiol **69**: 455
Singh R and Pathak D N (1990) Epilepsia **31**: 1-8.
Singh R, Bhalla S and Sharma V (2000) Ann of Neurosci (abstract) **7**: 20.21
Timofeev I and Steriade M (2004) Neuroscience **123**:299-336
Temkin N R (2001) Epilepsia **42**: 515-524
Temkin NR (2003) Epilepsia **44**:18-20.
The Global Campaign Against Epilepsy. Geneva, World Health Organization, 2000 (Information Pack for the launch of the Global Campaign's Second Phase, 12–13 February 2001).
The global burden of mental and neurological problems. Geneva, World Health Organization, 1999 (WHO Fact Sheet No. 217).
Traub R D, Borch C Colling S B and Jeffery J G R (1996) Epilepsia **37**: 879
Triggs W J & Willmore L J ,(1984) J Neurochem **42**: 976
Verleger R, Heide W, Butt C, Wascher E and Kompf D (1996) Electroencephalogr Clin Neurophysiol **99**: 44.
Vergenes M, Marescaux C, Depaulis A, Micheletti G and Warter J M (1987) Exp Neurol **96**: 127.
Von Krosigk M, Bal T and McCormick D A (1993) Science **261**: 361-364.
Wadman W J and Connor JA (1992) Neurosci 48: 293–305.
Wiest M C and Nicoleleis MAL (2003) Nature Neurosci **6**:913-914.
Willmore LJ, Hiramatsu M, Kochi H and Mori A (1983) Brain Res **277**: 392-396.
Willmore LJ and Rubin JJ (1981) Neurology (Ny) **31**: 63-69.
Willmore L J, Triggs W J and Gray J D (1986) Brain Res. **382**: 422-426
Willmore LJ, Sypert GW, Munson JB and Hurd RW (1978) Science **200**: 1501-1503.
Yip S and Sastry B R (2000) Brain Res **864**: 1-12.
Yokoi I, Toma J, Liu J, Kabuto H and Mori A (1995) Free Radic Biol Med **19**: 473-479.

Molecular and Cellular Neurobiology
Editors: M.K. Thakur and S. Prasad

The Endothelium and Cerebrovascular Disease

D. Nagaraja and Rita Christopher

National Institute of Mental Health & Neuro Sciences, Bangalore-560 029, India

Abstract
The vascular endothelium is a structurally simple but functionally complex tissue whose integrity is essential for normal vascular function. This dynamic organ releases humoral factors that control blood vessel relaxation and contraction, thrombogenesis and fibrinolysis and platelet activation and inhibition. Thus the endothelium contributes to blood pressure control, blood flow, and vessel wall patency. Functional integrity of the endothelium is crucial for the maintenance of blood flow and prevention of thrombosis. Risk factors for cerebrovascular disease, such as hypertension, diabetes, cigarette smoking and hypercholesterolemia impair endothelial function, which lead to atherosclerosis and thrombosis. Vascular endothelial function is an important and clinically relevant therapeutic target for cerebrovascular disease.

Key words: Endothelial dysfunction, risk factors, stroke

Introduction

Cerebrovascular disease accounts for considerable mortality and morbidity in India. The most common form, atherothrombotic brain infarction, is caused by structural and functional changes in the blood vessel wall. These changes include abnormal vasoconstriction, enhanced interaction of blood cells with the vessel wall, activation of the coagulation mechanisms, and migration and proliferation of the smooth muscle cells (Luscher, 1994). Depending on the stage of the disease, one or more of these factors predominate and have an important role in the pathogenesis of stroke.

The vascular endothelium is a crucial vascular structure that is both a target for and a mediator of cerebrovascular disease. The endothelium is a structurally simple but functionally sophisticated organ that is strategically interposed between the blood and all other tissues in the body, and interacts directly with cells and molecules arriving via the blood or the tissues. Although the endothelium consists of a monolayer of cells, it covers an area of approximately $700m^2$ and weighs approximately 1 to 1.5 kg in a 70 kg man (Cryer, 1983). In the quiescent state the endothelium acts to maintain the blood's fluidity and to resist thrombosis, but thrombotic functions take over when the cells are perturbed. The interaction between neighboring endothelial cells serve to maintain vascular permeability and regulate leukocyte traffic. The vascular endothelium also acts as a selective barrier between the elements in the blood and the extravascular space, and serves to convey signals between the tissues and circulating elements. During infection, or when activated by various cytokines, the endothelium functions to attract leukocytes, allowing their emigration to sites of inflammation. The endothelium also has local regulatory functions, being the key regulator of vascular tone as well as an important modulator of vascular remodeling through its influence on the growth of the underlying vascular smooth muscle cells and by secretion of extracellular matrix components. Towards all these functions, the endothelium synthesizes a diverse armament of hormones and vasoactive mediators, some of which are expressed constitutively and others in response to various stimuli (Trachtenberg and Ryan, 1995).

Endothelial hemostatic functions

The endothelium has both thrombotic and thromboresistant properties. Endothelial cells (ECs) inhibit thrombus formation by interfering with the coagulation cascade, inhibiting platelet adhesion/ aggregation and activating fibrinoytic pathways (Pearson, 1994). The luminal surface of the EC contains anionic heparin-like glycosaminoglycans that are synthesized by the ECs and have the capacity to bind thrombin, the key enzyme of blood coagulation, and the serum protein antithrombin (antithrombin III, AT-III). Antithrombin is a potent inhibitor of the coagulation cascade. It is a non–vitamin K-dependent protease that not only inhibits the action of thrombin but also other procoagulant factors (eg. factor Xa) as well. Antithrombin's activity is markedly potentiated by heparin; potentiation of its activity is the principle mechanism by which both heparin and low–molecular weight heparin produce anticoagulation. In the presence of ECs, thrombin reacts primarily with antithrombin, resulting in rapid inactivation of the enzyme. Thrombin also interacts with thrombomodulin on the EC surface. Thrombomodulin binds both thrombin and protein C, and the thrombin-thrombomodulin–protein C complex markedly accelerates the activation of protein C. Protein C, a major inhibitor of the procoagulant system, is a vitamin K-dependent protein synthesized by the hepatocyte. Activated protein C inhibits coagulation factors VIIIa and Va, the two cofactors necessary for thrombin and factor X activation. Protein C inhibitory activity in degrading activated clotting factors is markedly enhanced by protein S, another vitamin K-dependent glycoprotein. Activated protein C may also promote fibrinolysis by neutralizing plasminogen activator inhibitor-1 (PAI-1) (Bick, 1994).

Deficiencies of antithrombin, protein C or protein S lead to increased risk of venous and arterial thrombosis. Various studies from India have reported deficiencies of these proteins in cerebrovascular disease. In a study of 98 patients with occlusive stroke, Singh et al (2003) have observed a significant depression of the biological as well as immunological activity of antithrombin. Mohanty and coworkers (1999) have found activated protein C resistance (APC-R) to be the most common defect (21.62%) followed by antithrombin deficiency in idiopathic non-embolic cerebrovascular disease. In another study from India, Nagaraja and Christopher (1996) have observed that 8.3% of young patients who presented with ischemic stroke of unknown etiology had deficiencies of the coagulation inhibitor proteins, protein C, protein S or antithrombin. Combined protein C and S deficiency in stroke patients (Jha, 2002) and cerebral venous thrombosis associated with protein C deficiency (Patel et al 2000) has also been reported from India.

The endothelium is also thought to produce a lipoprotein-associated inhibitor of coagulation known as TFPI (tissue factor pathway inhibitor) or LACI (lipoprotein-associated coagulation inhibitor). This factor is a potent inhibitor of coagulation factor Xa and the factor VIIa /TF-complex (Broze, 1992). Although there is no direct evidence of it being made in EC, it is currently hypothesized that the heparin-releasable form is associated with endothelial glycosaminoglycans. Decreased plasma TFPI activity has been reported in patients with ischemic stroke (Abumiya et al 1995). Products released by ECs, which prevent platelet adhesion/aggregation, include prostacyclin (PGI_2), adenosine and nitric oxide. There is synergism between the anti-aggregatory effect of PGI_2 and the effect of nitric oxide at threshold levels (Radomski et al 1987).

Clot dissolution through the enhancement of the fibrinolytic system is another EC anticoagulant defense mechanism. Both tissue-type plasminogen activators (t-PA) and urokinase-type plasminogen (u-PA) activators are synthesized by the EC. The synthesis and release of plasminogen activators by ECs is stimulated by thrombin, activated protein C, epinephrine, vasopressin, and bradykinin. In addition, t-PA appears to be more specific for fresh clot since fibrin but not fibrinogen is required for full enzymatic activity, localizing the fibrinolytic response to areas of ongoing clot formation. Decreased fibrinolytic potential has been reported in South Asian women with ischemic cerebrovascular disease (Kain et al 2001).

In contrast the above anticoagulant mechanisms, the endothelium when perturbed mechanically or exposed to different agents such as inflammatory cytokines, viruses, bacterial endotoxins, thrombin, tumor necrosis factor or interleukin-1, can interact with blood components to promote coagulation. The procoagulant activity is the result of increased production of tissue factor (thromboplastin), vonWillebrand factor (factor VIII_{vw}), plasminogen-activator-inhibitor, platelet activating factor and extracellular matrix components. Tissue factor released by damaged cells functions with factor VII in activating factor X initiating the `extrinsic pathway' of blood coagulation and potentiating the cleavage of factor IX, and thus activating the intrinsic pathway ' as well. In the presence of factor VIII_{vw}, activated factor IX activates factor X. The activated factor X formed by either pathway then converts prothrombin to thrombin in a reaction which requires factor Va. Thrombin converts soluble fibrinogen to insoluble fibrin, activates platelets and inhibits vessel wall fibrinolytic activity. Thrombin also acts on the ECs to stimulate the synthesis and release of various agents, such as inflammatory mediators, growth factors, vasoactive substances and agents to modulate thrombus formation, such as inhibitors and activators of platelet aggregation and the fibrinolytic system (Mantovani et al 1992). Thrombin causes leukocyte adhesion to the endothelium by triggering expression of adhesion molecules on the cell surface and cause disruption of endothelial permeability properties. The majority of the effects of thrombin on the endothelium is mediated by its receptor and requires its lytic activity (Rabiet et al 1994). Factor VIII_{vw} synthesized by ECs is bound by collagen types I, III, IV and V, and helps to mediate platelet adhesion to the subendothelium through a specific glycoprotein receptor on the platelet surface, glycoprotein IIb/IIIa, which also serves as a receptor for fibrinogen. Platelet-activating factor (PAF) is a juxtacrine signaling phospholipid synthesized by activated endothelial cells. PAF stimulates platelet aggregation and smooth muscle contraction via a specific 7-membrane-domain PAF receptor. The endothelium also synthesizes other proteins, which stimulate the coagulation cascade. These include thrombospondin, collagens and fibronectin. Pro-coagulant activity also involves alteration in fibrinolysis through the regulated expression of tissue type plasminogen activator (t-PA) and specific plasminogen activator inhibitors. The activity of t-PA in plasma is based on the amount of t-PA released as well as on the level of plasminogen activator inhibitor (PAI) present (Nilsson et al 1985).

Thus, the hemostatic potential of the endothelium results from a complex balance between active factors, which have opposing biological actions, and the interaction between active factors and specific inhibitors. Upset of this balance could lead to conditions predisposing to both embolic and thrombotic stroke.

Endothelial control of vascular wall function

The vascular endothelium exerts an important influence on cerebral vascular tone through the production and release of diverse group of vasoactive factors (Faraci, 1992). Relaxing factors produced by the endothelium include nitric oxide and prostacyclin. Endothelium-derived contracting factors include cycloxygenase products of arachidonic acid and endothelins. The angiotensin-converting enzyme (ACE) produced by the endothelium has established a role for endothelium in the control of blood pressure. Several pathophysiological conditions are associated with increased formation of endothelium-derived contracting factors or decreased production of relaxation factors. Such endothelial dysfunction in the cerebral circulation may shift the balance of vascular tone towards constriction and may potentially contribute to onset or maintenance of cerebral ischemia.

Endothelium-derived relaxing factors

Nitric oxide (NO), originally called endothelium-derived relaxing factor (EDRF), is formed from L-arginine by a constitutive endothelial NO synthase (eNOS) in response to various stimuli including fluid shear stress and exposure to the neurohumoral factors acetylcholine, bradykinin, serotonin, and substance P (Furchgott and Zawardzki, 1980; Palmer et al 1988). Although short-lived, NO is a potent vasodilator that induces smooth muscle relaxation through the activation of smooth muscle guanylate cyclase. NO-induced endothelium-dependent relaxation can be pharmacologically inhibited by analogues of L-arginine like L-nitroarginine methyl ester (L-NAME) (Nathan and Xie, 1994). NO also inhibits platelet aggregation and leukocyte adhesion to the endothelium. Endothelium-derived NO is now recognized to be an anti-inflammatory and anti-atherosclerotic molecule and indirect evidence suggests that alterations in the NO pathway might be involved in endothelial dysfunction (Davis, 1999). There are at least three major mechanisms in the NO pathway that may lead to endothelial dysfunction and atherogensis; (a) functional abnormalities of the NOS attributable to substrate or cofactor deficiency; (b) increased breakdown of NO; and (c) reduced production of eNOS. All these mechanisms lead to inhibition of the endothelium-dependent relaxation due to decrease in the bioavailability of NO. Mice lacking in the endothelial type NO synthase gene exhibit hypertension and enhanced vascular remodeling in response to injury (Huang et al 1995). Several lines of evidence indicate a significant association between NOS gene and cardiovascular disease (Soma et al 1999; Ichihara et al 1998). Ghilardi and coworkers (2002) have reported that T786C mutation of the eNOS gene is an independent risk factor for moderate to severe internal carotid artery stenosis.

Prostacyclin (PGI_2) is a prostaglandin synthesized from arachidonic acid in the endothelial cells. It acts on smooth muscle prostacyclin receptors to activate adenylate cyclase and inhibit contraction. PGI_2 increases the synthesis of NO by endothelial cells, and in turn, NO increases the activity of PGI2 on smooth muscle relaxation (O'Connell and Genest, 2001). NO and PGI_2 synergistically inhibit platelet aggregation. PGI_2 increases cyclic 3'5'adenosine monophosphate in smooth muscle cells and platelets. The effects of PGI_2 are opposed by thromboxane, a prostanoid produced by activated platelets. Aspirin, an irreversible inhibitor of cyclo-oxygenase, has been shown to be effective in the prevention or treatment of cerebrovascular disease. Its efficacy relates to the ECs ability to synthesize cyclo-oxygenase continuously, whereas the platelet, a non-nucleated cell, cannot. This leads to the cumulative inhibition of thromboxane A_2 formation while the endothelium continues to produce PGI_2 (Trachtenberg and Ryan, 1995).

Endothelium-derived contracting factors

In addition to the vasodilators, the endothelium also releases vasoconstrictor substances in response to a variety of stimuli. Endothelium-derived contracting factors include endothelin-1 (ET-1), vasoconstrictor prostanoids such as thromboxane A_2 and prostaglandin H_2, and components of the renin-angiotensin system such as angiotensin II. Endothelin-1, a linear 21-amino acid peptide secreted by the endothelium, is the most potent vasoconstrictor substance, with a potency of 10 times that of angiotensin II (Yanagisawa et al 1988). Many substances, including thrombin, epinephrine, transforming growth factor-β, increase preproendothelin mRNA as well as the release of vasoactive ET-1 from the endothelial cells. ET-1 binds to specific membrane receptors, leading to the release of calcium from intracellular stores activating protein kinase C. In addition to its long acting pressor and vasoconstrictor action, ET-1 has numerous biologic activities, including mitogenicity for mesangial cells and capillary endothelial cells. Markedly elevated plasma endothelin-1 levels have been reported in acute ischemic stroke (Alioglu et al 2002) and may reflect enhanced synthesis by damaged ECs within the infarcted tissue (Ziv et al 1992). In the brain, local leakage of ET-1 may induce prolonged and severe constriction of collateral vessels, and may significantly worsen the outcome following the ischemic event.

In the cerebral circulation, arachidonic acid, acetylcholine, histamine and serotonin can evoke endothelium-dependent contraction that is mediated by thromboxane A_2 or prostaglandin H_2 (Luscher and Vanhoutte, 1990). Thromboxane and prostaglandin H_2 are derived from the cyclo-oxygenase pathway. Unlike PGI_2 they are potent vasoconstrictors and platelet activators. They activate the thromboxane receptors in vascular smooth muscles and platelets, thereby counteracting the effects of NO and PGI_2 in both types of cell. The ratio of PGI_2 to thromboxane is believed to be important for cardiovascular health (O'Connell and Genest, 2001) In addition, the cyclo-oxygenase pathway is also a source of superoxide anions, which rapidly inactivate NO to form the potent cytotoxic oxidant, peroxynitrite.

The endothelium also regulates the activity of the renin-angiotensin system. Angiotensin-converting enzyme (ACE) is expressed on the endothelial cell membrane. Angiotensin II can activate endothelial angiotensin receptors; these receptors stimulate the production of ET-1 and other mediators such as plasminogen-activator inhibitor-1 (PAI-1) (Vaughan et al 1995). The renin-angiotensin system is also known to play a role in vascular remodeling. The effectiveness of ACE inhibitors in preventing ischemic events may be due to their blunting effects on angiotensin II production and reduction in PAI-1 levels.

Endothelium-leukocyte interactions

Under normal physiologic conditions, leukocytes interact with the endothelium. Over 75% of the granulocytes are adherent to the endothelium, where they remain ready for release by specific stimuli to join the circulating pool. The lymphocytes on the other hand, circulate through the plasma and, as part of normal course, emigrate through specialized post-capillary venules in lymphoid tissues, returning to the plasma after a few hours. Early in the pathogenesis of atherosclerosis, endothelial cells display an increased affinity for leukocytes, which adhere to and migrate through the endothelium ultimately forming the fatty streak. The interaction of the leukocytes with the endothelium is carefully controlled by specific adhesive receptors. Three groups of adhesive receptors/ligands are known to participate in leukocyte adhesion. These include the proteins of the integrin family or Leu-CAMs (leukocyte cellular adhesion molecules), members of the immunoglobulin-related molecule or ICAMs (intercellular adhesion molecules) and selectins or LEC-CAMs (lectin-epidermal growth factor–complement cell adhesion molecule).

Increased endothelial expression of the leukocyte adhesion molecules, vascular cell adhesion molecule-1 (VCAM-1), ICAM-1, and E and P selectins, as well as decreased shear stress and increased leukocyte activation promotes leukocyte adhesion to the endothelium. E and P-selectins are members of the selectin family and are counterparts for L-selectin, which is found in all leukocytes. ICAM-1 and VCAM-1 belong to the immunoglobulin superfamily and are counter receptors for leukocyte integrins. ICAM-1 binds the integrin heterodimer CD11a/CD18 found on all leukocytes and the CD11b/CD18 heterodimer found on monocytes. VCAM-1 binds the integrin heterodimer $\alpha_4\beta_7$ and $\alpha_4\beta_1$. $\alpha_4\beta_1$ is found exclusively on lymphocytes and monocytes, whereas $\alpha_4\beta_7$ is found exclusively on lymphocytes. Circulating leukocyte adhesion molecules can be used as molecular markers for atherosclerosis, particularly ICAM-1 and E-selectin, which correlate strongly with carotid artery atherosclerosis (Hwang et al 1997).

Activation of endothelial cells as a consequence of cerebral ischemia leads to the expression of adhesion molecules such as ICAM-1, VCAM-1, and E-selectin, which can be released into the blood. In experimental animal models of stroke, peripheral blood leukocytes migrate into the brain parenchyma within the first 12 hours after ischemia (Garcia et al 1994). Migration of peripheral blood leukocytes requires prior adhesion to cerebral endothelial cells, which is mediated by adhesion molecules on the surface of cerebral endothelial cells and peripheral blood leukocytes. Upon activation by several stimuli, these cells secrete a multitude of different cytokines, chemokines, and other inflammatory mediators (Kim, 1996). Most of them are thought to induce cell damage and significantly contribute to reperfusion injury. As a result of large inter-individual variability influenced by ischemia-independent factors, soluble adhesion molecules are not reliable candidates as surrogate markers in acute cerebral ischemia (Bitsch et al 1998).

Endothelial dysfunction

Alterations in one or more of the physiological roles of the vascular endothelial cells constitute endothelial dysfunction. It may be the cause or consequence of vascular disease and is a hallmark of known vascular risk factors (Luscher and Barton, 1997). Endothelial dysfunction precedes structural vascular alterations, indicating a protective role of the functionally intact endothelium. Endothelial cell denudation however, occurs only in very late stages of atherosclerosis and plaque rupture. These changes in endothelial cell morphology are most invariably associated with functional alterations and intimal thickening, with accumulation of white blood cells, vascular smooth muscle cells, and fibroblasts and matrix deposition.

Cerebrovascular risk factors and endothelial dysfunction

The major risk factors associated with the etiology of endothelium pathology are smoking, hypertension, diabetes and elevated low-density lipoprotein (LDL) cholesterol. Injury produced by the various risk factors convert the endothelial cells with anticoagulant properties to a lethal arsenal of pro-coagulant molecules.

Cigarette smoking

Cigarette smoking may affect the endothelium in several ways. In animal models, cigarette smoking induces ultrastructural changes, which may be associated with increased endothelial permeability, including blebbing, and an increase in the number and size of plasmalemmal vesicles (Pittilo et al 1982). Exposure to carbon monoxide had been shown to increase arterial endothelial permeability to plasma proteins, whereas nicotine increases their retention in the artery wall (Allen et al 1989). Further, direct exposure to cigarette smoke modifies LDL, in part through a superoxide anion-dependent mechanism, to a form taken up more avidly by macrophages (Yokode et al 1988). Recent evidence has indicated that both active and passive cigarette smoking are associated with dysfunction of normal endothelial physiology, in a dose dependent and potentially reversible way (Puranik and Celermajer, 2003). In a study from India, prolonged exposure of experimental animals to cigarette smoking was observed to unfavorably affect lipoprotein profiles by decreasing high-density lipoprotein and increasing low and very low density lipoproteins (Latha et al 1988). There was increased cholesterogenesis in the heart, lungs and liver, as evidenced by increased activity of HMG-CoA reductase and increased incorporation of labeled acetate into cholesterol.

Hypertension

Hypertension is an important cardiovascular risk factor associated with stroke. Hypertension induces structural and functional abnormalities of the vascular endothelium (Lustier, 1990). In patients with hypertension, basal formation of nitric oxide and the effect of an acetylcholine infusion on the release of nitric oxide were found to be reduced (Panza et al 1990,1993; Taddei et al 1993) whereas flow-induced vasodilatation was maintained (Laurent et al 1990). Such changes in endothelial function could be a cause or a consequence of hypertension. Most experimental data suggest that endothelial dysfunction develops as blood pressure increases and that the degree of dysfunction is related to the blood pressure level. Hence, dysfunction of the L-arginine NO pathway is likely to be a secondary event involved in the maintenance rather than the initiation of hypertension. Most important, however, endothelial dysfunction may contribute to the vascular complications of hypertension, such as myocardial infarction and stroke. Plasma levels of endothelin remain normal in most patients with hypertension except in the presence of renal failure or atherosclerosis. Increased local vascular production of endothelin, however, is likely; because most of the peptide is released abluminally, plasma levels of endothelin do not necessarily reflect local tissue levels (Wagner et al 1992). Vascular endothelin production is reduced in the spontaneously hypertensive rat but increased in angiotensin II-induced hypertension in Wistar-Kyoto rats (Barton et al 1997).

Factors that appear to be important for the development of endothelial dysfunction in hypertension are the duration and severity of hypertension. It is possible that certain groups of hypertensive patients are more likely to have endothelial dysfunction than others because of different pathogenetic mechanisms or dietary and genetic factors. Also, other variables, such as insulin resistance, may be involved. Hypertension and stroke have been observed to occur at a relatively younger age in Asians. Among anti-hypertensive therapy, angiotensin-converting enzyme inhibitors acutely restore a normal coronary response to physiological tests. Calcium-blockers (dihydropyridines) have a permanent vasodilator effect. The restoration of a normal endothelium-dependent vasomotor tone might be one target of antihypertensive therapy.

Diabetes mellitus

Incidence of cerebrovascular disease in diabetic men has been reported to be twice as that of non-diabetics and almost three times greater in diabetic women in the Framingham Study (Kannel and McGee, 1979). Elevated glucose levels in patients with diabetes mellitus cause endothelial dysfunction. The endothelial dysfunctions that have been described include decreased endothelium-dependent vasorelaxation, increased leukocyte-endothelial cell adhesion and vascular permeability, and the altered production of a variety of vasoactive substances, which affect coagulation, extracellular matrix homeostasis, and smooth muscle physiology (Feener and King, 2001). Studies have shown that elevated glucose concentrations increase expression of NO synthase and production of superoxide anion in vitro (Cosentino et al 1997). Vascular dysfunction due to high glucose levels appears to be mediated in vivo by vascular endothelial growth factor via an NO synthase–linked pathway (Tilton et al 1997). In addition, cytokine dysregulation occurs in diabetes and increased production of altered LDL and its immune complex activates monocytes and macrophages, damages endothelium and brings monocytes, platelets and endothelium together in an almost lethal embrace leading to atherogenesis. Activated monocytes also upregulates its tissue factor activity and contributes to secretion of TNF-α, interleukin-1, interleukin-6 and interferons, which by themselves activate the pro-coagulant function of endothelium. Antithrombin levels in diabetes remain unaltered, so also its ability to neutralise many of the activated coagulation factors like factor Xa and thrombin. However antithrombin is an allosteric protein and its ability to neutralise thrombin increases by 100-1000 folds in the presence of trace amount of heparin or heparinoids. Unfortunately due to endothelial cell functional disturbances the output of heparinoids from endothelial cells are compromised, compromising the full potential of the antithrombin molecule.

Hypercholesterolemia

Hypercholesterolemia per se, without atherosclerotic vascular changes, inhibits endothelium-dependent relaxations, which are further reduced in atherosclerosis (Harrison et al 1987). Low-density lipoprotein (LDL) is a major determinant of this phenomenon. Incubation of isolated coronary arteries with oxidized but not native LDL selectively inhibits endothelium-dependent relaxations to serotonin, aggregating platelets, and thrombin (Tanner et al 1991). A similar decrease in response can be achieved by pertussis toxin or an inhibitor of NO formation, suggesting defective activation of the L-arginine NO pathway by G_i protein-coupled receptors (Shimokawa et al 1989). Exogenous L-arginine improves or restores reduced endothelium-dependent relaxation in the presence of oxidized LDL, which suggests that oxidized LDL impairs the activity of NO synthase. The active component of LDL appears to be lysolecithin, which mimics most of the effects of LDL.

An important step in the atherosclerotic process is the attachment of inflammatory cells to arterial endothelium and migration of these cells into the intima. Although lipoprotein deposition has been thought to be a necessary precursor to inflammatory cell recruitment, lipoproteins in their native form do not appear to incite an inflammatory reaction. Minimally oxidized LDL could initiate inflammatory cell recruitment into developing atherosclerotic lesion. In addition, in hyperlipidemia, endothelial cell production of endothelin is increased (Lerman et al 1991), while the expression of endothelin receptors is downregulated (Winkles et al 1993). A likely stimulus for the increased production is LDL, which increases endothelin gene expression and endothelin release from human aortic endothelial cells (Boulanger et al 1992). Plasma endothelin levels are increased and correlate positively with the extent of atherosclerotic lesion formation (Lerman et al 1991). Thus local vascular endothelin may contribute to abnormal vasomotion and vasoconstriction and the proliferation of vascular smooth muscle cells observed in atherosclerosis.

Hyperhomocysteinemia

Homocysteine is a sulfur-containing amino acid formed during the metabolism of dietary methionine. An elevated level of plasma total homocysteine or hyperhomocysteinemia, is emerging as a strong risk factor for cerebrovascular disease (Parnette et al 2004). Hyperhomocysteinemia, can result from an inherited deficiency of enzymes involved in the metabolism of homocysteine, mainly cystathionine β-synthase (CBS) and methylene tetrahydrofolate reductase (MTHFR) or from a deficiency of cofactor for these enzymes, vitamins B_6, B_{12} and folate.

Endothelial dysfunction may be an underlying cause leading to proatherogenic effects associated with hyperhomocysteinemia (Chambers et al 1998). The mechanisms that lead to impaired endothelial function in hyperhomocysteinemia are not fully understood. Studies suggest that homocysteine may interact with physiological mediators of the endothelial matrix. Oxidative mechanisms and decreased biological activity of endothelium-derived nitric oxide (NO) may also contribute to homocysteine-associated endothelial dysfunction (Sydow and Boger, 2001). Recent in vitro and in vivo studies also demonstrate that hyperhomocysteinemia induces endoplasmic reticulum stress, leading to activation of the unfolded protein response (UPR) (Austin et al 2004).

Risk factors and stroke in India

In an analysis of 893 patients with ischemic stroke, Kaul and coworkers (2000) have found the common risk factors, hypertension in 62%, and diabetes in 38% and smoking in 28% of the patients. The frequency of risk factors was similar in patients with lacunar and non-lacunar infarctions. In another study, Nagaraja and coworkers (1998) have observed that hypercholesterolaemia, hyperglycemia and hypertension were more common in elderly stroke patients, while smoking, alcohol and tobacco abuse were more prevalent in younger age group (relative risk 1.2, 1.9 and 1.5 respectively). Hypertension has been reported to be twice more common in stroke patients with Doppler scan and carotid angiographic evidence of extracranial carotid vascular disease compared to the group with normal carotid vessels (Nagaraja et al 1989). The reduction of HDL cholesterol was also more frequent in the group with extracranial carotid vascular disease (63%) and in hypertensive (73%) patients. Other studies (Dalal, 2002; Bharucha et al 1988) have also observed hypertension to be an important risk factor for stroke in the Indian population. The risk for ischemic stroke was considerably higher when there was any combination of hypertension, heart disease and HDL cholesterol level lower than 45 mg/dl (Sridharan, 1992).

In a study of young adults with ischemic stroke, Nayak and coworkers (1997) have reported that hypertension, smoking and hyperlipidemia were significantly more prevalent in the 31-45 year age group compared to the 15-30 year age group. The prevalence of hyperlipidemia has been reported to be 60% in young patients with ischemic stroke of unknown etiology (Bansal et al 1986). However, in another case-control study of ischemic stroke in the young, the serum total cholesterol, HDL, LDL and VLDL-cholesterol, apolipoproteins A-I and B, were not observed to be significantly different in the stroke patients when compared to normal controls (Christopher et al, 1996).

Tests of endothelial function

Diagnostic tests to detect early vascular changes could make possible the identification of patients at risk for the progression of vascular disease and provide a means of monitoring therapy. Tests to assess the function of the L-arginine pathway are still invasive and involve the infusion of nitric oxide inhibitors (such as N^G-monomethyl-L-arginine) (Calver et al 1992) or acetylcholine (which stimulates the release of nitric oxide) (Linder et al 1990). Although vasoconstriction in response to N^G-monomethyl-L-arginine indicates the presence of basal nitric oxide formation, an infusion of acetylcholine allows the assessment of receptor-mediated release of nitric oxide. Vasodilatation in response to acetylcholine in human arteries is endothelium-dependent and is inhibited by N^G-monomethyl-L-arginine, but not by inhibitors of other mediators (Linder et al 1990). Flow-dependent vasodilatation of large arteries is another nitric oxide-dependent response. Tests of endothelial function may detect early vascular abnormalities in patients with cardiovascular risk factors but without clinically evident disease. Mean flow-mediated, endothelium-dependent dilatation as measured by brachial artery diameter response to reactive hyperaemia and sublingual glyceryl trinitrate, has been reported to be impaired in healthy Asian Indians (Chambers et al 1999). In patients with hypercholesterolemia and diabetes, endothelium-dependent vasodilatation in response to acetylcholine has been observed to be reduced in both the peripheral and coronary circulation long before structural vascular changes or clinical symptoms occur (Creager et al 1990; Treasure et al 1992; Zeiher et al 1993). Studies of the coronary circulation suggest that during the natural history of atherosclerosis, impairment of receptor-mediated activation of the endothelium is the first event, followed by flow-induced vasodilatation and later by a reduced response of smooth muscle to nitric oxide (Zeiher et al 1993). Coronary endothelial function testing by intracoronary acetylcholine infusion has also shown that coronary endothelial dysfunction is associated with an increased risk of cerebrovascular events (Targonski et al 2003).

Conclusion

It is obvious that the relation between the endothelium and cerebrovascular disease is complex. In the early stages of vascular disease, alterations in endothelial function are subtle and selective and can affect both basal formation and receptor-mediated release of endothelial mediators. In contrast, the release of mediators in response to vascular shear stresses is affected only later in the course of disease. Not all endothelial mediators are similarly affected unless severe atherosclerotic changes have occurred. In addition, the cerebral and coronary arteries are probably more prone to endothelial dysfunction than other vascular beds. Endothelial dysfunction may occur more readily in patients with hyperlipidemia or diabetes. Finally, even in patients with a particular risk factor, the degree of endothelial dysfunction may differ, as does the incidence of cardiovascular complications at later stages. Hence, we need further studies using sophisticated, noninvasive tests of endothelial function to define more clearly the occurrence, natural history, anatomical distribution, and type of endothelial dysfunction in different populations and individual patients. This may allow more precise risk stratification and eventually more effective therapy.

References

Abumiya T, Yamaguchi T, Terasaki T, Kokawa T, Kario K and Kato H (1995) Thromb Haemost **74**: 1050-1054.

Alioglu Z, Orem A , Bulbul I, Boz C, Ozmenoglu M and Vanizor B (2002) Angiology **53**: 77-82.

Allen DR, Browse NL and Rutt DL (1989) Atherosclerosis **77**: 83.

Austin RC, Lentz SR, Werstuck GH (2004) Cell Death Differ **11** Suppl 1:S56-64.

Bansal BC, Sood AK and Bansal CB (1986) Stroke 17: 1142-1145.

Barton M, Shaw S, d'Uscio LV, Moreau P and Luscher TF (1997) Nephrology **3**: 235-241.

Bharucha NE, Bharucha EP, Bharucha AE, Bhise AV and Schoenberg BS (1988) Neurology **38**: 490-492.

Bick RL (1994) Med Clin N Am **78**: 635-665.

Bitsch A, Klene W, Murtada L, Prange H and Rieckmann P (1998) Stroke **29**: 2129-2135.

Broze GJ (1992) Semin Hematol **58**:159-169.

Calver A, Collier J, Moncada S and Vallance P (1992) J Hypertens **10**: 1025-1031.

Chambers JC, McGregor A, Jean-Marie J and Kooner JS (1998) Lancet **351**: 36-37.

Chambers JC, McGregor A, Jean-Marie J and Kooner JS (1999) Heart **81**: 501-504

Christopher R, Kailasanatha KM, Nagaraja D and Tripathi M (1996) Acta Neuro Scand **94**:127-130.

Cosentino F, Hishikawa K, Katusic ZS and Luscher TF (1997) Circulation **96**: 25-28.

Creager MA, Cooke JP, Mendelsohn ME, Gallagher SJ, Coleman SM, Loscalzo J, Dzau VJ (1990) J Clin Invest **86**: 228-234.

Cryer A (1983) In: Biochemical Interactions of the Endothelium Amsterdam, Elsevier, p1-3.

Dalal PM (2002) Neurol India **50**: S2-S7.

Davis PA (1999) Circulation **99**: 2700-2712.

Faraci FM (1992) Pharmacol Ther **56**: 1-22.

Feener EP and King GL (2001) Heart Fail Monit **1**: 74-82.

Furchgott RF and Zawadzki JV (1980) Nature **288**: 373-376.

Garcia JH, Liu KF, Yoshida Y, Lian J, Chen S and del Zoppo GJ (1994) Am J Pathol **144**: 188–199.

Ghilardi G, Biondi ML, DeMonti M, Bernini M, Turri O, Massaro F, Guagnellini E, Scorza R (2002) Clin Chem **48**: 989-993.

Harrison DG, Freiman PC, Armstrong ML, Marcus ML and Heistad DD (1987) Circ Res **61**: 1174-1180.

Huang PL, Huang Z, Mashimo H, Bloch KD, Moskowitz MA, Bevan JA et al (1995) Nature **377**: 239-242.

Hwang SJ, Ballantyne M, Sharett R et al.(1997) Circulation **96**: 4219-4225.

Ichihara S, Yamada Y, Fujimura T, Nakashima N, Yokota M (1998) Am J Cardiol **81**: 83-86.
Jha SK, Anand AC, Sharma V, Kumar N and Adya CM (2002) High Alt Med Biol Spring **3**: 21-27.
Kain K, Catto AJ, Carter AM, Young J, Bamford J, Bavington J and Grant PJ (2001) Br J Haematol **114**: 155-161.
Kannel WB and McGee (1979) Circulation **59**: 8-13.
Kaul S, Venketswamy P, Meena AK, Sahay R and Murthy JM (2000) Neurol India **48**: 116-119.
Kim JS (1996) J Neurol Sci **137**: 69–78.
Latha MS, Vijayammal PL and Kurup PA (1988) Atherosclerosis **170**: 225-231.
Laurent S, Lacolley P, Brunel P, Laloux B, Pannier B and Safar M (1990) Am J Physiol **258**: H1004-H1111.
Linder L, Kiowski W, Buhler FR and Luscher TF (1990) Circulation **81**: 1762-1767.
Luscher TF (1994) N Engl J Med **330**: 1081-1083.
Luscher TF and Barton M (1997) Clin Cardiol **20**: II-3-II-10.
Mantovani A, Bussolino F and Dejana E (1992) FASEB J **6**: 2591-2599.
Mohanty S, Saxena R and Behari M (1999) Thromb Haemost **8**: 465-466.
Nagaraja D, Taly AB and Rao BS (1989). J Assoc Physicians India **37**: 253-254.
Nagaraja D and Christopher R (1996) In: Cerebrovascular disease: Indian experience Eds. D Nagaraja and AB Taly, IAN Publication, p25-32.
Nagaraja D, Gurmurthy SG, Taly AB, Subbakrishna K and Sridhararama Rao BS (1998) Neurol India **46**: 183-185.
Nagaraja D, Christopher R and Tripathi M (1999). Neurology India **47**: 155-157.
Nagaraja D and Christopher R (2002) Neurol India **50** (Suppl 1): S21-S29.
Nathan C and Xie QW (1994) Cell **78**: 915-918.
Nayak SD, Nair M, Radhakrishnan K and Sarma PS (1997) Natl Med J India **10**: 107-112.
Nilsson IM, Ljunger H, Tengborn L (1985) Br Med J **290**: 1453.
O'Connell BJ and Genest J (2001) Circulation **104**: 1978-1983.
Palmer RMJ, Ashton DS and Moncada S (1988) Nature **333**: 664-666.
Parnetti L, Caso V, Santucci A, Corea F, Lanari A, Floridi A, Conte C and Bottiglieri T (2004) Neuro Sci 25:13-17.
Panza JA, Quyyumi AA, Brush JE Jr, and Epstein SE (1990) N Engl J Med **323**: 22-27.
Panza JA, Casino PR, Badar DM and Quyyumi AA (1993) Circulation **87**: 1475-1481.
Patel A, Jayaram S, Kabra D, Patil A and Jaria R (2000) J Assoc Physicians India **48**: 544.
Pearson JD (1994) Br Med Bull **50**: 776-788.
Pittilo RM, Mackie IJ, Rowles PM et al (1982) Thromb Haemost **48**: 173.
Puranik R, Celermajer DS (2003) Prog Cardiovasc Dis **45**: 443-58.
Rabiet MJ, Plantier L and Dejana E (1994) Br Med Bull **50**: 93-945.
Radomski MW, Palmer RM and Moncada S (1987) Br J Pharmacol **92**: 632-646.
Shimokawa H, Flavahan NA and Vanhoutte PM (1989) Cir Res **65**: 740-753.
Soma M, Nakayama T, Kanmatsuse K (1999) Curr Opin Nephrol Hypertens **1999**: 8: 83-87.
Sridharan R (1992) Neuroepidemiology **11**: 24-30.
Sydow K and Boger RH (2001) Z Kardio l **90**: 1-11.
Taddei S, Virdis A, Mattei P and Salvetti A (1993) Hypertension **21**: 929-933.
Tanner FC, Noll G, Boulanger CM and Luscher TF (1991) Circulation **83**: 2012-2020.
Targonski PV, Bonetti PO, Pumper GM, Higano ST, Holmes DR Jr, Lerman A (2003) Circulation.**107**: 2805-2809.
Tilton RG, Kawarmura T, Chang KC, Ido Y, Bjercke RJ, Stephan CC, Brock TA and Williamson JR (1997) J Clin Invest **99**: 2192-2202.
Trachtenberg JD and Ryan US (1995) In: Brain Ischemia: Basic Concepts and Clinical Relevance (LR Caplan ed) Springer-Verlag, pp173-184.
Treasure CB, Manoukian SV, Klein JL, et al (1992) Circ Res **71**: 776-781.
Vaughan D, Lazos S and Tong K (1995) J Clin Invest **95**: 995-1000.
Wagner OF, Christ G, Wojta J, Vierhappe H, Parzer S, Nowomy PJ, Schneider B, Waldhaus W and Binder BR (1992) J Biol Chem **267**: 16066-16068.
Wong KS (1999) Stroke **30**: 2326-2330.
Yanagisawa M and Masaki T (1989) Trends Pharmacol Sci **10**: 374-378.
Yokode M, Kita T, Arai H et al (1988) Proc Natl Acad Sci USA **85**: 2344.
Zeiher AM, Drexler H, Saurbier B and Just H (1993) J Clin Invest **92**: 652-662.
Ziv I, Fleminger G, Djaldetti R, Achiron A, Melamed E and Sokolosky M (1992) Stroke **23**:1014 –1016.

Molecular and Cellular Neurobiology
Editors: M.K. Thakur and S. Prasad

Molecular Mechanisms of Cerebral Ischemia

Shyam S. Sharma* and Meenakshisundaram Thiyagarajan

Department of Pharmacology and Toxicology, National Institute of Pharmaceutical Education and Research, S.A.S. Nagar Mohali-160 062, India

Abstract

Stroke is a third leading cause of death and the first major cause of chronic disability. More than 80% of the strokes are result of cerebral ischemia. Despite the enormity of the problem, no currently approved therapy reduces stroke size or neurological disability. The pathophysiological mechanisms leading to neurological damage in ischaemic stroke are complex and not yet clearly understood. In this review molecular mechanisms such as glutamate excitotoxicity, calcium overload, free radical generation, poly (ADP-ribose) polymerase overactivation, mitogen activated protein kinases (MAP Kinases) activation, apoptosis, and gene activation, that contribute significantly to the cerebral ischaemic injury are discussed. We also covered the therapeutic interventions that have been explored in preclinical and clinical trials.

Key words: Cerebral ischemia, excitotoxicity, free radicals, poly (ADP-ribose) polymerase, MAP Kinases, gene activation, therapeutics

Introduction

Stroke is an acute neurological injury occurring as a result of interruption of blood supply to brain. More than 80% of the strokes are as a result of cerebral ischemia. Focal cerebral ischemia involves restricted brain regions and occurs in a wide variety of clinical settings but is most commonly a result of cerebral vascular atherosclerosis or occlusion of middle cerebral artery (MCA) by a thrombus or embolus. Global cerebral ischemia involves the entire brain and occurs during cardiac arrest or severe systemic hypotension. Focal ischemia is more frequent than the global ischemia.

Stroke is a third largest cause of death in the western world. In USA, the incidence of stroke is estimated to be approximately 750,000 per annum (www.stroke.org), with a mortality of 150,000 per annum. Community based surveys from many regions of India showed a prevalence rate in the range of 40 to 270/100,000 in rural population as compared to the reported prevalence of 400-800/100,000 in metropolitan cities (Anand et al 2001; Dhamija and Dhamija, 1998). The symptoms of cerebral ischemia ranges from blurred vision to vertigo, dizziness, convulsion, and loss of consciousness depending on the area of the nervous system affected. There are wide ranges of motor and sensorimotor deficits, including tremor, lack of co-ordination and partial or complete paralysis. Currently, the only approved treatment for acute ischemic stroke is tissue plasminogen activator, a thrombolytic agent that limits the ischemic insult by lysing arterial thrombus and restoring blood flow.

* Corresponding author

Thrombolytic (tissue plasminogen activator) therapy is available in India but the success depends on initiating the treatment in the narrow therapeutic time window of 3 h on onset of stroke (Srivastava and Prasad, 2001). Because of narrow therapeutic window very few patients (5-8%) qualify for thrombolytic treatment. There is no approved neuroprotective therapy for reducing neurological damage associated with stroke.

Mechanisms of cerebral ischemia

Reduced blood supply to a territory of brain leads to decrease in oxygen and glucose supply to the affected area. Lack of oxygen supply activates anaerobic metabolism in brain. Anaerobic metabolism cannot completely compensate for the energy needs of neuron, as only a fraction of the needed energy is obtained from the same. The glycogen stores are small and get depleted within minutes result in failure of energy production. Much of the energy in the neuron is spent to maintain the ionic homeostasis. Within minutes of ischemia the ionic pumps fail resulting in depolarization of neurons. Depolarization of the presynaptic neuron releases neurotransmitters like glutamate and dopamine, which are excitatory in nature. Failure of the ionic pumps also produces edema and swelling of neuron. To some extent the early damage is reversible. Continuation of the ischemic condition for prolonged period of time results in the irreversible damage to neurons. Neurons in the severely affected part die immediately due to oxygen starvation. Due to the impaired blood supply the metabolites, which are formed, get accumulate around the ischemic core. This results in the formation of ischemic penumbra where the probability of neuronal loss increases. In ischemic penumbra (outside the stroke's core) where neuronal electrophysiology is not compromised the oxygen supply remains sufficient (20-25% of normal blood flow). If blood flow in penumbra decreases below its critical level and/or energy requirements exceeded, the infarct core inevitably expands.

Despite the enormous scientific efforts that have been made to clarify the pathophysiology of ischemic neuronal injury, the mechanism responsible for neuronal cell death after cerebral ischemia remains unclear (Lo et al 2003). Current understanding of molecular mechanisms such as excitotoxicity, ionic imbalance, free radicals generation, Poly (ADP-ribose) polymerase (PARP) overactivation, mitogen activated protein kinase (MAPK) activation, and alteration in gene expression involved in cerebral ischemic reperfusion (IR) injury (Fig. 1) and strategies developed for neuroprotection are discussed in details in this review.

Glutamate excitotoxicity and calcium overload: Evidences have accumulated implicating glutamate neurotoxicity (excitotoxicity) and increase in intracellular calcium $[Ca^{++}]_i$ as a major contributors of neuronal damage during ischemic insult. Ischemia triggers excessive release of glutamate from presynaptic nerve terminals and astrocytes into the extracellular space, with subsequent overstimulation of glutamate receptors. Glutamate activates N-methyl D-aspartate (NMDA), α-amino–3- hydroxy-5-methyl-4-isoxazole propionic acid (AMPA), Kainate, and G-protein coupled metabotropic glutamate receptors. The NMDA receptor channel is permeable for sodium and calcium ions, whereas AMPA receptors normally gates

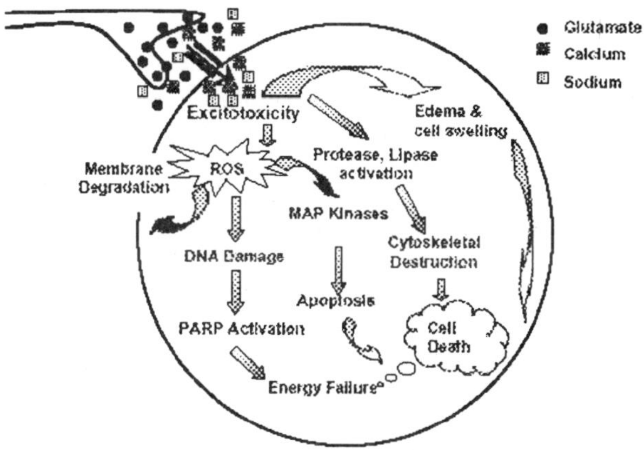

Fig 1. Molecular events involved in neuronal death in cerebral ischemia: Interruption of cerebral blood flow results in decreased energy production, which causes failure of ionic pumps, release of excitatory neurotransmitter such as glutamate from presynaptic terminals and generation of free radicals. These events activate downstream signaling cascade leading to necrotic and apoptotic cell death. ROS-reactive oxygen species, PARP – poly (ADP ribose) polymerase, MAP kinases-mitogen activated protein kinases.

sodium and potassium conductance. Blocking the glutamate-binding site, its modulatory sites such as glycine, polyamine binding site, or the ion channel has been shown to provide neuroprotection in focal ischemia. NMDA antagonists such as dizocilpine (MK-801), dextromethorphan, NPS1506, ACEA 1021, CGS-19755 and CP-101, 606-27 showed neuroprotection in animal models of stroke. NMDA antagonist causes neurobehavioural side effects, which limits their use in therapy of stroke. These adverse events are particularly pronounced with NMDA antagonists such as phencyclidine, ketamine, and dizocilpine. Polyamine site antagonists (eliprodil and ifenprodil) have also been explored for their neuroprotective potential. AMPA/Kainate receptor antagonists are found to be superior to NMDA antagonist in protecting selectively vulnerable hippocampus CA1 pyramidal neurons from delayed death after transient global ischemia (Sheardown et al 1990). DNQX, an AMPA/Kainate receptor antagonist produced neuroprotection and attenuated c-Jun N-terminal kinases-3 (JNK3) activation during cerebral ischemia (Tian et al 2003). Blockade of the NMDA or AMPA/kainate receptor can modulate heat shock protein responses in both neurons and glia within the developing infarct (Pettigrew et al 2003). AMPA antagonist YM872 is under clinical studies in stroke patients. Eight G-protein coupled metabotropic glutamate receptors (mGluR) have been identified and divided into three subgroups. Activation of group I (mGluR1 and mGluR5) metabotropic glutamate receptors is both neurotoxic and neuroprotective (Bruno et al 2001; Kingston et al 1999). Group II (mGluR2 and mGluR3) and III (mGluR4, 6, 7, 8) are presynaptic, and they depress glutamate release (Pin and Duvoisin, 1995). Agonist acting at mGluR2/3 can diminish ischemic damage in animals (Bond et al 2000). Drugs targeted at mGlu receptors have been explored for their potential in preclinical studies.

Growing evidence suggests the involvement of other neurotransmitters (gamma aminobutyric acid and serotonin) and neuromodulators (adenosine) in ischemia. Adenosine may mediate some of the damages caused in stroke and hence blocking Adenosine A2A receptors may spare neurons (Melani et al 2003). Gamma aminobutyric acid (GABA) is the main inhibitory neurotransmitter in the brain and activation of $GABA_A$ receptor has been suggested to have neuroprotective effect in cerebral ischemia (Kristensen et al 2003). One of the GABA agonist – clomethiazole has been evaluated clinically in stroke patients. Serotonin and acetylcholine have also been reported to be neuroprotective. Serotonin agonist (Repinotan) has shown neuroprotection in animals and now being explored in clinical trials.

In CNS normally intracellular calcium $[Ca^{++}]_i$ is regulated by voltage-gated and NMDA receptor operated calcium channels. Excessive and uncontrolled calcium influx into neurons under ischemic conditions is considered to be one of the major mechanisms of neuronal damage. Uncontrolled increase in $[Ca^{++}]_i$ can activate many intracellular events. $[Ca^{++}]_i$ can activate nNOS for generation of nitric oxide (NO) from L-arginine. It can produce rapid and extensive break down of phospholipids, proteins and nucleic acids as a result of activation of calcium dependent phospholipases and proteases. Activation of cytosolic proteases can lead to direct disassembly of the cellular microtubule system and proteolytic degradation of the structural and functional proteins. $[Ca^{++}]_i$ activates calpains, a family of nonlysosomal neutral cysteine proteases. Pathologic calpain activity has been observed in cerebral ischemia and calpain inhibitors have been reported to reduce the ischemic-reperfusion (IR) insult mediated neuronal death (Markgraf et al 1998). Calcium channel antagonists (nimodipine and flunarizine) have been explored for neuroprotective potential in preclinical and clinical studies. These drugs did not show improvement in clinical trials.

***Oxidative stress, nitric oxide and poly (ADP-ribose) polymerase*:** The central nervous system is particularly vulnerable to oxidative damage caused by free radicals (FRs). FRs or reactive oxygen species (ROS) like superoxide ($O2^{\cdot-}$), hydroxyl ($^{\cdot}OH$), peroxynitrite ($ONOO^{-}$) radicals and hydrogen peroxide (H_2O_2) are produced during IR injury. Since ROS are highly reactive, endogenous defense mechanisms are present in the cell that serves to protect the cell against damaging effects of ROS. Superoxide dismutase, glutathione peroxidase and catalase comprise the enzymatic and vitamins and their analogs such as α-tocopherol and ascorbic acid comprises non-enzymatic defense systems against oxidative stress. Levels of endogenous antioxidant enzymes fall during ischaemia whereas FR production increases during ischemia-reperfusion. Exogenous supply of FR scavengers is being explored as a potential strategy for restoring physiological balance and to improve neurological outcome. Mice over expressing superoxide dismutase (SOD) are found to be resistant to neuronal loss after transient cerebral IR insult (Matsuyama, 1997). Antioxidant enzymes (SOD, catalase, glutathione peroxidase) reduced infarct size in animal models of IR insult (Matsuyama, 1997). Tirilazad, a inhibitor of lipid peroxidation and nitrone based free radical scavengers alpha phenyl–N-tert-butyl-nitrone and N-tert-butyl- (2-sulfophenyl)-nitrone have produced significant neuroprotection (Maruki et al 1993). Another study in the cortical neurons indicated that vitamin-E reacts with free radicals thereby preventing apoptosis and necrosis mediated neuronal loss (Tagami et al 1998). Although some of the antioxidants showed efficacy in animal models, most of them did not show beneficial effects in clinical trials.

Nitric oxide (NO) involvement in the pathophysiology of cerebral ischaemia has been postulated by several researchers (Bolanos and Almeida, 1999; Coeroli et al 1998; Read et al 1999; Sharma et al 2000). Nitric oxide is produced from L– arginine by three types of nitric oxide synthases (NOS). Constitutive NOS such as neuronal NOS (nNOS) and endothelial NOS (eNOS) are calcium dependent, however, inducible NOS (iNOS) is calcium independent. During ischemia, the role of eNOS may be beneficial since it has been observed that knockout mice deficient in eNOS developed large infarction after ischemia as compared to wild type control (Huang et al 1996). In contrast nNOS and iNOS appear deleterious in ischemia since several selective inhibitors of these isoform are neuroprotective (Escott et al 1998; Iadecola et al 1996; Sharma et al 2000). NO is a weak radical that can combine with superoxide to form a highly toxic peroxynitrite radical, which induce array of deleterious events (Virag et al 2003). We have demonstrated neuroprotection with peroxynitrite scavengers FeTMPyP and FeTPPS in focal ischemia model in rats (Sharma et al 2002).

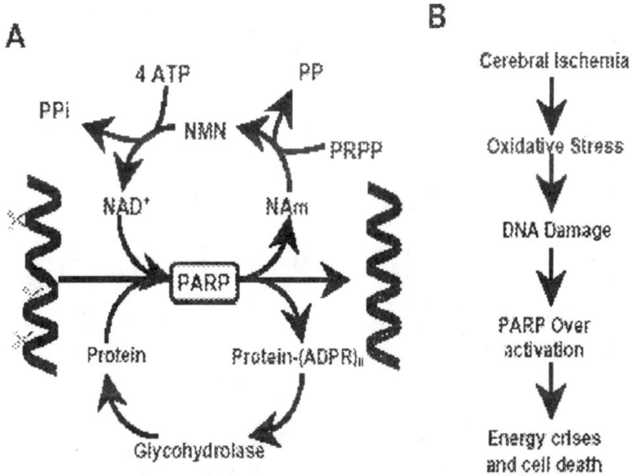

Fig 2. Role of poly (ADP ribose) polymerase (PARP) in DNA repair and cerebral ischemia: A) PARP mediated DNA repair. B) Mediation of cerebral ischemia induced neuronal cell death by PARP. PARP inhibitors such as 3-aminobenzamide and isoquinolinediol have shown neuroprotection in several *in vitro* and *in vivo* studies. (ADPR)n – ADP ribose polymer, NAD^+-nicotinamide adenine dinucleotide, NAm- nicotinamide, NMN-nicotinamide mononucleotide, PP- pyrophosphate, PRPPA.

Oxidative stress produce DNA strand breaks, which activate nuclear enzyme PARP as an attempt to repair strand breaks (Fig. 2). During cerebral ischemia a massive DNA damage occurs as a result of the excessive ROS generation causing overactivation of PARP, which leads to NAD^+ depletion and neuronal death (Bowes et al 1998). Neuroprotective potential of PARP inhibitors have been reported in *in vivo* and *in vitro* model of cerebral ischemia (Virag and Szabo, 2002). PARP inhibitors protects various cell from oxygen-glucose deprivation (OGD), NO, glutamate mediated and ROS mediated cytotoxicity (Thies and Autor, 1991; Virag and Szabo, 2002). Eliasson et al (1997) showed that genetic disruption of the PARP provides profound protection against glutamate and NO-mediated ischemic insults in *in vitro* studies. They also observed major decrease (80%) in the infarct volume in PARP knockout mice (Eliasson et al 1997). We have demonstrated neuroprotective activity of 3-aminobenzamide, nicotinamide, 4-amino-1, 8-napthalimide and isoquinolinediol in stroke model in rats (Gupta et al 2004; Kabra et al 2004; Sharma et al 2004). Novel PARP inhibitors like PJ-34, DPQ, NU1025 and 6(5H)-phenanthridinone are reported to have very good PARP inhibiting activity (Virag and Szabo, 2002) and are under investigation in various *in vivo* models of IR injury.

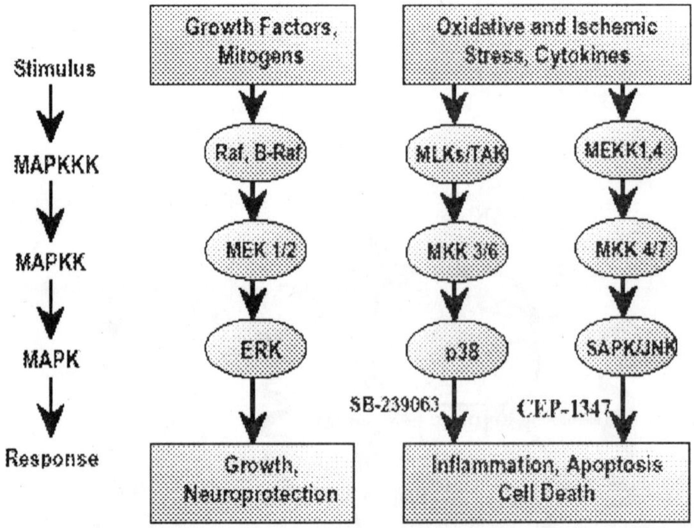

Fig 5. Cascades of Mitogen activated protein kinases (MAPK): Cerebral ischemia induced oxidative stress activate these pathways. Inhibitors of MAP kinases (SB 239063, p38 inhibitor and CEP 1347, JNK inhibitors) have shown neuroprotection in animal studies. ERK – extracellular signal regulatory kinase, JNK – c Jun N-terminal kinase, MAPK – mitogen activated protein kinase, MAPKK or MEK – MAPK kinase, MAPKKK – MAPK kinase kinase, SAPK – stress activated protein kinase, MLK – Mixed lineage kinase.

***MAP Kinase Pathway*:** Mitogen-activated protein kinases (MAPK) pathways are major signaling cascades controlling complex programmes such as embryogenesis, differentiation, and cell death in addition to short-term changes required for homeostasis and hormonal response (Pearson et al 2001). So far three distinct MAPK pathways have been described in mammalian cells (Fig. 3). Extracellular signal regulated protein kinases (ERKs) were the first of these MAPKs to be discovered. Subsequently c-Jun N-terminal kinases (JNKs) and p38 MAPK were identified.

ERKs family protein kinase includes mainly ERK1 and ERK2. Activation of ERK1/2 is important for the several neuronal functions like neuronal differentiation, and survival during development and adaptive response of mature neurons. The cell survival effect is mediated as a result of activation of transcription factors involved in proliferation and differentiation (Marshall, 1995). Phospho ERK1/2 (p-ERK) is upregulated in hippocampal organotypic culture after oxygen glucose deprivation (Irving et al 2000). The role of ERK in IR insult remains controversial. Several inhibitors of MEKs were studied for neuroprotection. PD98059 and U0126 showed neuroprotection and produced inhibition of ERK activation. JNKs mainly convey stress signals to the cell nucleus and its three isoforms of JNK viz JNK1, JNK2 and JNK3 have been identified. JNK1 and JNK2 are found in all tissues, whereas JNK3 is found most abundant in brain, heart and testis. The functional difference of these isoforms yet has not been reported. JNKs can phosphorylate a variety of cytoplasmic and nuclear proteins. The transcription factor c- Jun is the main substrate for JNKs. JNK pathway activation has also been shown to be increased after transient cerebral ischemia (Herdegen and Leah, 1998) and global ischemia (Hu *et al*., 2000). CEP1347 inhibits JNK pathway (Roux et al 2002). Recently more selective JNK inhibitors have been generated, which includes SP600125 and SPC9766.

p38 is widely distributed in mammalian tissue including brain. The p38 pathway is strongly activated by factors such as tumour necrosis factor-α (TNF-α) and interleukin- 1β, which is known to be increased after stroke (Freshney et al 1994). Sustained activation of p38 has been shown to be associated with neuronal cell death/apoptosis and selective p38 inhibitors have been shown to promote cell survival (Kummer et al 1997). p38 MAPK inhibitors (SB-203580 and SB-239063) have been studied in preclinical setting (Legos et al 2002).

Apoptosis

Apoptosis is a programmed cell death process in which cells neatly commits suicide by chopping themselves into membrane-packed bits. The delayed loss of neurons after cerebral ischemia in the penumbral area showed key criterions of apoptosis (Linnik et al 1993; Nitatori et al 1995). It has been observed that moderate glutamate exposure and oxidative stress can lead to apoptotic cell death in neurons.

The Bcl-2 family of proteins is one of the main components regulating apoptosis. It has both cell proliferation and cell death function. In Bcl-2 family, Bcl-xL is an anti-apoptotic protein where as Bax, Bid, Bad are pro-apoptotic in nature. Martinou et al (1994) reported that Bcl-2 protects the cortical neurons in mice from ischemic insult induced by MCA occlusion (Martinou et al 1994). The transgenic mice that overexpressed Bcl-2 in neurons had significantly smaller infarction than wild type mice (Chen et al 1998). Bax is highly expressed in CA1 neurons after global ischemia (Chen et al 1998). Bid deficient mice showed less infarction in which caspases and cytochrome C expression is severly impaired (Yin et al 2002). Caspases (cysteine proteases) are proteolytic enzymes plays a key role in the initiation and execution of the apoptotic program. A dozen of caspases have been discovered till now, amongst caspase 3 plays a major role in apoptosis. Activation of caspase-3 has been demonstrated in IR injury (Chan et al 1998; Du et al 1997). Inhibitors of caspases (z-VAD fmk and z-DEVD) were found to be neuroprotective in cerebral ischemia (Graham and Chen, 2001; Hara et al 1997; Namura et al 1998).

Growth factors

In nervous system growth factors stimulate differentiation (during development), support neuronal survival and generally considered to be neuroprotective. Several growth factors are expressed during cerebral ischemia including brain derived nerve growth factor (BDNF), nerve growth factor (NGF), basic fibroblast derived growth factor (bFGF), platelet derived growth factor (PDGF), transforming growth factor β (TGFβ) and vascular endothelial growth factor (VEGF) (Ren and Finklestein, 1997). Growth factor treatment protected the neuronal cell even if treated long after the insult (Kawamata et al 1997). bFGF treatment also reported to increase the Bcl-2 in IR injury in rat. In clinical studies bFGF treatment did not alter the mortality rate in patients. Even in some cases inherent antigenic property of growth factors resulted in immune reactions (Finklestein and Wada, 2000).

Gene activation and gene therapy

Accumulation of glutamate in the extracellular space and $[Ca^{++}]_i$ overload in neurons activates many downstream processes including gene transcription. More than 100 genes which are denoted as immediate early genes (IEG) have been reported to be expressed within 15 min of the insult (Greenberg et al 1985). These include many proteins such as cytokines, enzymes and transcription factors etc. Transcription factors (*c-fos, fos-B, c-jun* and *junD)* are oncogenes, which are involved, in the transcriptional activation of other genes. Heat shock genes are activated continuously or immediately after the IEG expression to stabilize the stressed cells.

Recent studies have suggested that a number of genes can be potent candidates for the treatment of cerebral infarction. New vectors and gene delivery systems have been identified in recent years. Herpes Simplex is considered to be as one of the common vector systems used to deliver the genes to nervous system. Many proteins (HSP70, calmodulin and Bcl-2) have been experimented to evaluate the neuroprotective potential in *in vitro* and *in vivo* models of IR injury. Gene therapy aimed to upregulate these proteins produced conflicting results. HSP70 over expression failed to produce protection against excitotoxicity induced neuronal loss in primary neuronal cultures (Fink et al 1997). Nevertheless it produced neuroprotection in MCA occlusion induced focal ischemia model in rat. Herpes simplex virus mediated transfection of calmodulin gene resulted in reduction of neuronal loss (Phillips et al 1999). In similar approach many other genes, which are involved in apoptotic neuronal loss, have also been experimented. The safety of the vectors is considered to be an important limiting factor for the gene therapy in human beings. Above mentioned studies suggest the potential of gene therapy for treatment of cerebral ischemia in the future.

Cerebral ischemia research in Indian laboratories

Many academic and industrial organizations in India are working to develop newer neuroprotective approaches for the treatment of cerebral ischemia. Our laboratory at NIPER is involved in identifying new chemical entities which protects brain from cellular and metabolic consequences of IR injury. We are using middle cerebral artery occlusion (MCAO) rat model for focal ischemia and bilateral common carotid artery occlusion gerbil model for global ischemia. Vessel occlusion is routinely confirmed by measuring cerebral blood flow (CBF) using Laser Doppler Blood Flow monitor (Perimed, Sweden). During the occlusion phase blood gases (Nova Biomedical, USA), haemodynamic parameters (Power lab, Australia) and body

Table1. Drugs explored for neuroprotective potential in clinical trials for cerebral ischemia

Drug class	Drug name
Glutamate antagonists	
Competitive NMDA antagonists	CGS 19755 (Selfotel)
NMDA channel blockers	Aptiganel (Cerestat), CP-101606, Dextrorphan, Dextromethorphan, Magnesium, NPS 1506, Remacimide, Memantine, MK-801
Glycine site antagonists	ACEA 1021 (Licostinel), GV 150526
Polyamine site antagonists	Eliprodil (SL 82-0715), Ifenprodil
AMPA antagonists	GYKI 52466, NBQX, YM90K, YM872, ZK-200775
Kainate antagonist	SYM 2081
Modulator of uptake capacity of glutamate transporters and expression of GABA receptors and various astrocytic factors	ONO 2506
Serotonin agonists	Bay x 3702 (Repinotan)
GABA agonists	Clomethiazole
Calcium channel antagonists	Nimodipine, Flunarizine
Calcium chelator	DP b99
Voltage-dependent potassium channel agonists	BMS-204352
Sodium channel antagonists	Fosphenytoin, Lubeluzole, Sipatrigine (BW619C89, 619C89)
Free-radical scavengers	Tirilazad mesylate, Ebselen, NXY 059 (Cerovive)
Ion channel and Nitric oxide inhibitor	Lubeluzole
Phosphatidylcholine Precursor	Citicoline
Growth factors	Fibroblast growth factor
Leukocyte adhesion inhibitor	Anti-ICAM antibody (Enlimomab), Hu23F2G

temperature (Homeothermic blanket system, Harvard, UK) is monitored in animals. Cerebral infarction and edema is calculated from TTC stained brain sections using image analysis software (Leica, Germany). We have demonstrated the neuroprotective activity of antioxidant (curcumin, melatonin, FeTMPyP and FeTPPS) and PARP inhibitors (3-aminobenzamide, nicotinamide, 1, 5-isoquinolinediol, 4-amino-1, 8-napthalimide) in stroke model in rats (Gupta et al 2004; Kabra et al 2004; Sharma et al 2004; Thiyagarajan et al 2004; Thiyagarajan and Sharma, 2004). We are also developing cell culture based system for studying neuroprotective mechanisms.

Dr. Y. K. Gupta laboratory at All India Institute of Medical Sciences at New Delhi has demonstrated the neuroprotective effect of resveratrol, adenosine and combination of melatonin and meloxicam in acute ischemic stroke in rats (Sinha et al 2002). Other research institutes and pharmaceutical industry involve in drug discovery and development of treatment for stroke includes Central Drug Research Institute, Lucknow; Christian Medical College and Hospital, Vellore; Punjabi

University, Patiala; Jamia Hamdard, New Delhi, Torrent Research Center, Ahemdabad and Reliance Life Sciences Pvt. Ltd., Mumbai.

Conclusions

Cerebral ischemia is a complex injury process that occurs when the nutrient blood supply to cerebral structures is reduced below critical levels. Ischemia activates multiple events in the brain that leads to cell death. Mechanisms such as glutamate excitotoxicity, calcium overlaod, free radical generations, poly (ADP ribose) polymerase overactivation, MAP kinase activation, apoptosis, cytokines and gene activations have been clearly documented and pharmacological interventions (Table 1) targeted at these mechanisms have been explored for their neuroprotective potential in various preclinical and clinical studies. Most of the neuroprotective drugs developed to reduce neurological damage have shown significant neuroprotection in animal studies but failed to mimic in stroke patients. Potential reasons for the failure of neuroprotective agents in clinical trials include narrow therapeutic window, suboptimal drug dosage, inappropriate drug levels at the target CNS site, and discrepancy in drug metabolism and their target was one single pathophysiological mechanism. It would also be a candid approach to limit ischaemic injury by using combination of drugs acting at different targets. Therefore, future of stroke treatment likely to result from a combination of therapeutic modalities aimed at different pathophysiological mechanisms and delivering at specific time after cerebral ischemia.

Acknowledgements

Works carried out in the Dr. S. S. Sharma laboratory was supported by grant from National Institute of Pharmaceutical Education and Research (NIPER), Punjab and Young scientist fast track grant (SP/FTP/LS-A-36/2001) from Department of Science and Technology, Government of India.

References

Anand K, Chowdhury D, Singh KB, Pandav CS and Kapoor SK (2001) Neuroepidemiology **20**: 208-11.

Bolanos JP and Almeida A (1999) Biochim Biophys Acta **1411**: 415-36.

Bond A, Jones NM, Hicks CA, Whiffin GM, Ward MA, O'Neill MF, Kingston AE, Monn JA, Ornstein PL, Schoepp DD, Lodge D and O'Neill MJ (2000) J Pharmacol Exp Ther **294**: 800-9.

Bowes J, Ruetten H, Martorana PA, Stockhausen H and Thiemermann C (1998) Eur J Pharmacol **359**: 143-50.

Bruno V, Battaglia G, Copani A, Cespedes VM, Galindo MF, Cena V, Sanchez-Prieto J, Gasparini F, Kuhn R, Flor PJ and Nicoletti F (2001) Eur J Neurosci **13**: 1469-78.

Chan PH, Kawase M, Murakami K, Chen SF, Li Y, Calagui B, Reola L, Carlson E and Epstein CJ (1998) J Neurosci **18**: 8292-9.

Chen J, Nagayama T, Jin K, Stetler RA, Zhu RL, Graham SH and Simon RP (1998) J Neurosci **18**: 4914-28.

Coeroli L, Renolleau S, Arnaud S, Plotkine D, Cachin N, Plotkine M, Ben-Ari Y and Charriaut-Marlangue C (1998) J Neurochem **70**: 2516-25.

Dhamija RK and Dhamija SB (1998) J Assoc Physicians India **46**: 351-4.

Du Y, Bales KR, Dodel RC, Hamilton-Byrd E, Horn JW, Czilli DL, Simmons LK, Ni B and Paul SM (1997) Proc Natl Acad Sci U S A **94**: 11657-62.

Eliasson MJ, Sampei K, Mandir AS, Hurn PD, Traystman RJ, Bao J, Pieper A, Wang ZQ, Dawson TM, Snyder SH and Dawson VL (1997) Nat Med **3**: 1089-95.

Escott KJ, Beech JS, Haga KK, Williams SC, Meldrum BS and Bath PM (1998) J Cereb Blood Flow Metab **18**: 281-7.

Fink SL, Chang LK, Ho DY and Sapolsky RM (1997) J Neurochem **68**: 961-9.
Finklestein SP and Wada K (2000) Update on growth factor treatment for stroke: 2000. In: Pharmacology of Cerebral Ischemia (Krieglstein J and Klumpp S Finklestein SP and Wada K , eds), Medpharm Scientific Publishers.
Freshney NW, Rawlinson L, Guesdon F, Jones E, Cowley S, Hsuan J and Saklatvala J (1994) Cell **78**: 1039-49.
Graham SH and Chen J (2001) J *Cereb Blood Flow Metab* **21**: 99-109.
Greenberg ME, Greene LA and Ziff EB (1985) J Biol Chem **260**: 14101-10.
Gupta S, Kaul CL and Sharma SS (2004) Neurol Res **26**: 103-107.
Hara H, Friedlander RM, Gagliardini V, Ayata C, Fink K, Huang Z, Shimizu-Sasamata M, Yuan J and Moskowitz MA (1997) Proc Natl Acad Sci U S A **94**: 2007-12.
Herdegen T and Leah JD (1998) Brain Res Brain Res Rev **28**: 370-490.
Hu BR, Liu CL and Park DJ (2000) J Cereb Blood Flow Metab **20**: 1089-95.
Huang Z, Huang PL, Ma J, Meng W, Ayata C, Fishman MC and Moskowitz MA (1996) J Cereb Blood Flow Metab **16**: 981-7.
Iadecola C, Zhang F, Casey R, Clark HB and Ross ME (1996) Stroke **27**: 1373-80.
Irving EA, Barone FC, Reith AD, Hadingham SJ and Parsons AA (2000) Brain Res Mol Brain Res **77**: 65-75.
Kabra D, Thiyagarajan M, Kaul CL and Sharma SS (2004) Brain *Res. Bull.* **62**: 425-433.
Kawamata T, Dietrich WD, Schallert T, Gotts JE, Cocke RR, Benowitz LI and Finklestein SP (1997) Proc Natl Acad Sci U S A **94**: 8179-84.
Kingston AE, O'Neill MJ, Bond A, Bruno V, Battaglia G, Nicoletti F, Harris JR, Clark BP, Monn JA, Lodge D and Schoepp DD (1999) Ann N Y Acad Sci **890**: 438-49.
Kristensen BW, Noraberg J and Zimmer J (2003) Brain Res **973**: 303-6.
Kummer JL, Rao PK and Heidenreich KA (1997) J Biol Chem **272**: 20490-4.
Legos JJ, McLaughlin B, Skaper SD, Strijbos PJ, Parsons AA, Aizenman E, Herin GA, Barone FC and Erhardt JA (2002) Eur J Pharmacol **447**: 37-42.
Linnik MD, Zobrist RH and Hatfield MD (1993) Stroke **24**: 2002-8; discussion 2008-9.
Lo EH, Dalkara T and Moskowitz MA (2003) Nat Rev Neurosci **4**: 399-415.
Markgraf CG, Velayo NL, Johnson MP, McCarty DR, Medhi S, Koehl JR, Chmielewski PA and Linnik MD (1998) Stroke **29**: 152-8.
Marshall CJ (1995) Cell **80**: 179-85.
Martinou JC, Dubois-Dauphin M, Staple JK, Rodriguez I, Frankowski H, Missotten M, Albertini P, Talabot D, Catsicas S, Pietra C and et al. (1994) Neuron **13**: 1017-30.
Maruki Y, Koehler RC, Kirsch JR, Blizzard KK and Traystman RJ (1993) Stroke **24**: 724-30.
Matsuyama T (1997) Free radical mediated cerebral damage after hypoxia/Ischaemia and stroke. In: Clinical Pharmacology of Cerebral Ischaemia (Horst G and Korf J Matsuyama T, eds), Humana Press.
Melani A, Pantoni L, Bordoni F, Gianfriddo M, Bianchi L, Vannucchi MG, Bertorelli R, Monopoli A and Pedata F (2003) Brain Res **959**: 243-50.
Namura S, Zhu J, Fink K, Endres M, Srinivasan A, Tomaselli KJ, Yuan J and Moskowitz MA (1998) J Neurosci **18**: 3659-68.
Nitatori T, Sato N, Waguri S, Karasawa Y, Araki H, Shibanai K, Kominami E and Uchiyama Y (1995) J Neurosci **15**: 1001-11.
Pearson G, Robinson F, Beers Gibson T, Xu BE, Karandikar M, Berman K and Cobb MH (2001) Endocr Rev **22**: 153-83.
Pettigrew LC, Holtz ML, Minger SL and Craddock SD (2003) Neurol Res **25**: 201-7.
Phillips RG, Meier TJ, Giuli LC, McLaughlin JR, Ho DY and Sapolsky RM (1999) J Neurochem **73**: 1200-5.
Pin JP and Duvoisin R (1995) Neuropharmacology **34**: 1-26.
Read SJ, Hirano T, Davis SM and Donnan GA (1999) Drugs Aging **14**: 11-39.
Ren JM and Finklestein SP (1997) Eur J Pharmacol **327**: 11-6.
Roux PP, Dorval G, Boudreau M, Angers-Loustau A, Morris SJ, Makkerh J and Barker PA (2002) J Biol Chem **277**: 49473-80.
Sharma SS, Munusamy S, Thiyagarajan M and Kaul CL (2004) J Neurosurg (In Press).
Sharma SS, Shankar M and Kaul CL, Neuroprotective effect of peroxynitrite scavenger in middle cerebral artery occlusion induced focal ischemia in rats. , Presented at XXXV Annual Conference of Indian Pharmacological Society, Gwalior, India, 2002.

Sharma SS, Sinha K and Gupta YK(2000) Protective effect of nNOS inhibitor in middle cerebral artery occlusion induced focal ischemia in rats. International Congress on Frontiers of Pharmacology and Therapeutics in 21st Century, New Delhi, India,.
Sheardown MJ, Nielsen EO, Hansen AJ, Jacobsen P and Honore T (1990) Science **247**: 571-4.
Sinha K, Chaudhary G and Gupta YK (2002) Life Sci **71**: 655-65.
Srivastava AK and Prasad K (2001) Neurol India **49**: 272-6.
Tagami M, Yamagata K, Ikeda K, Nara Y, Fujino H, Kubota A, Numano F and Yamori Y (1998) Lab Invest **78**: 1415-29.
Thies RL and Autor AP (1991) Arch Biochem Biophys **286**: 353-63.
Thiyagarajan M, Kaul CL and Sharma SS (2004) Br J Pharmacol **142**: 899-911.
Thiyagarajan M and Sharma SS (2004) Life Sci **74**: 969-85.
Tian H, Zhang G, Li H and Zhang Q (2003) Neurosci Res **46**: 191-7.
Virag L and Szabo C (2002) Pharmacol Rev **54**: 375-429.
Virag L, Szabo E, Gergely P and Szabo C (2003) Toxicol Lett **140-141**: 113-24.
Yin XM, Luo Y, Cao G, Bai L, Pei W, Kuharsky DK and Chen J (2002) J Biol Chem **277**: 42074-81.

Molecular and Cellular Neurobiology
Editors: M.K. Thakur and S. Prasad

Cerebral Malaria: New Insights from Signal Transduction and Cell Signaling Studies

Kota Arun Kumar[1,3] Varsha Sharma[2] and Phanithi Prakash Babu*[1]

[1]Department of Animal Sciences, School of Life Sciences, University of Hyderabad
Hyderabad-500 46, India
[3]School of Life Sciences, Jawaharlal Nehru University, New Delhi-110 067, India
[3]Micheal Heidelberg Division of Immunology, Dept of Pathology
New York University School of Medicine, NY-10016, USA

Abstract

Cerebral Malaria (CM) is a grave pathological complication of Plasmodium falciparum infection causing neurological disturbances and sequelae. The pathological basis of the disease is attributed to both the parasite infestation in the brain and also the excessive immune responses generated by the host. During circulation through the blood, the erythrocytic stages of Plasmodium adhere to the vascular endothelial cells of various organs, as the deep-seated vessels are preferred sites for the parasite retention due to their relatively low oxygen tension. At these sites the parasites adhere to the endothelial cells, a phenomenon referred to as microvascular sequestration. Microvascular sequestration is a receptor ligand interaction. Both the ligand on the surface of the infected RBC membrane and the receptors on the endothelial cells have been well characterized in various pathological settings of human and murine cerebral malaria. The most well characterized ligands are the members of the var family of antigens like PfEMP-1. On the surface of the endothelial cells various receptors like ICAM-1, VCAM, Thrombospondin, Chondriotin sulphate A and E-selectin have been implicated in interaction with the parasite ligands. One of the most intriguing features of CM is how the parasites residing inside the RBC can send signals to the vascular endothelial cells and bring about profound changes in these cells. Hence any theory that explains the basis of the CM pathology, must invariably explain the factor that is responsible for initiating these pathological events during CM.

Recent evidence suggests that TNF-alpha plays a crucial role in most of the pathological settings of CM. This was clear by the fact that, anti-TNF-alpha therapy abrogated the pathological responses and protected the mice during experimental CM. Moreover the TNF-KO mice never developed CM fatality and finally, TNFR-2 deficient mice did not develop neurological disorders. These observations prove beyond doubt that TNF-α and its receptor, TNFR2, are recruited in signaling during pathogenesis of CM. However, to meet this criterion, relatively high levels of TNF-alpha are a prerequisite to initiate pathological responses. This condition is met with during the rapid multiplication of the erythrocytic stages. The asexual multiplication of the malarial parasites release merozoites that have GPI anchored molecules on their surface that stimulate the macrophages to secrete TNF-alpha. High circulating TNF alpha levels are also known to increase or up regulate the receptors on the endothelial cells, which in turn favor infected RBC (iRBC) retention in blood vessels.

Key words: Cerebral malaria, Plasmodium, apoptosis, necrosis, inflammation

---
* Author for correspondence

Introduction

The adhered iRBC are known to activate the endothelial cells that are manifested as a breakdown of blood brain barrier or the tight junctions of the brain microvasculature. The activation of the endothelial cells involves signaling molecules like kinases that essentially bring about permeability changes in the vascular endothelial further making way to diffusion of the soluble mediators like TNF-alpha into the general brain parenchyma. A second inflammatory response in the brain in the vicinity to the disrupted BBB also produces substantial TNF further deteriorating the pathological responses.

The pathology of CM is thought to be mediated by the above mechanisms, additional evidences suggest that a number of other mechanisms also operate during the host parasite interactions. These include the role of immune elements, like T cells and the NKT cells that fine tune the immune responses, role of transcription factors and, protein kinases that influence gene expression and initiate paracellular signaling pathways in the cells of the CNS. An overview of these mechanisms of CM pathology suggests that multiple pathologic and protective responses are operational during cerebral malaria. Understanding these mechanisms is important not only for scientific interest but also to device strategies and evaluating circumventive interventions. This chapter outlines the current state knowledge pertaining to cell signaling events during CM in humans and discusses the contribution of studies of animal models to elucidate underlying mechanisms.

Overview of Cerebral Malaria Theories

Sequestration Hypothesis: In post mortem brain tissue from CM patients, the small blood vessels are often packed with parasitized red blood cells (PRBC). This phenomenon is called sequestration. Sequestration in brain is significantly more common in CM than in other forms of severe malaria (MacPherson et al 1985; Pongponratn et al 1991; Silmut et al 1999). Early investigators suggested that cerebral vessels clogged with PRBC would be unable to supply sufficient oxygen and nutrients to the local brain tissue (Marchiafava and Begnami 1984). However, there is little evidence of irreversible hypoxic damage to neurons in fatal CM (Medana et al 2001). This "mechanical hypothesis" also fails to explain the lack of long-term neurological problems among the majority of the patients who survive the disease (Brewester et al 1990). PRBC are also sequestered in the brains of non-CM cases (Silmut et al 1999). Although sequestration is necessary for the development of CM, other factors must be involved in initiating and maintaining the process leading to coma.

Sequestration results from adhesion of PRBC to receptors on the endothelial cells (EC) (Turner 1997; Berendt et al 1994). These receptors include adhesion molecules, such as CD36, intracellular adhesion molecule (ICAM-1) and CD31, thrombospondin and chondriotin sulphate A (CSA). PRBC can bind to all these molecules *in vitro* under static and flow conditions. Immunohistochemistry has shown that all these receptors are present on EC in the brain during CM (Silamut et al 1999; Greenwood 1999). ICAM-1 expression is particularly strong on vessels containing many PRBC. Interestingly many of these molecules also act as endothelial receptors for leukocytes during the inflammatory responses and immune surveillance. Leukocyte adhesion is not passive event, the receptors are actively involved in recognition, signaling and functional changes to EC in response to engagement by lymphocyte surface ligand.

The ligands for adhesion on the PRBC are produced by the parasite and expressed on the red blood cell (RBC). The best characterized ligand is called *P. falciparum* erythrocyte membrane protein (PfEMP-1) (Craig and Scherf 2001). The parasite exports and inserts PfEMP-1 and other molecules into the RBC membrane at the adhesive foci called the knobs. Variants of PfEMP-1, expressed by different parasite strains, have different binding affinities for the various endothelial receptors, and this influence where the infected cell sequesters and hence the clinical form of the disease. For example, CSA-binding parasites are implicated in placental malaria (Fried and Duffy 1998; Beeson et al 2001), whereas ICAM-1 binding is important in CM (Turner et al 1974; Smith, et al 2000).

Soluble Mediators Hypothesis: Some authors have tried to explain the pathophysiology of CM by proposing that the cytokines or other soluble mediators could cause the symptoms of coma directly, independent of sequestration (Clark et al 2000). Cytokines such as tumor necrosis factor (TNF-α) and interleukin (IL-1), or neuroactive mediators such as nitric oxide (NO) can be toxic to the central nervous system (CNS) when overproduced. Such short-lived mediators might explain the reversible symptoms of coma in CM. Children with severe malaria and CM have significantly higher plasma levels of TNF-α than those with mild disease (Kwiatkowski et al 1990). Brain tissues from fatal CM cases contain both TNF-α and IL-1β messenger RNA (mRNA) and protein (Brown et al 1999; Udomsangpetch et al 1997). However, the plasma TNF-alpha levels are lower in CM patients than they are in humans infected with *Plasmodium vivax* (Karunaweera et al 1992), which has no neurological symptoms, and in mouse models of non CM. Also, a trial using monoclonal antibody to TNF-α as a treatment for human CM has been unsuccessful (Van Hensbroek et al 1996).

Another potential soluble toxic mediator is NO, which is produced by a group of enzymes called as nitric oxide synthases (NOS). Recent studies found wide spread expression of one of these enzymes, inducible NOS, in endothelium and many other CNS cells during CM (Maneerat et al 2000). Cytokine induced NO could damage neurons or disrupt the nerve function indirectly by disturbing axonal transport or synaptic transmission. However, the results from the studies that measure derivatives of NO in plasma (Taylor et al 1998) and cerebrospinal fluid (CSF) (Dondorp et al 1998) during human CM are inconclusive, and some data suggest that NO could play a protective rather than a pathogenic role (Chiwakata et al 2000).

Factors contributing to the pathology of cerebral malaria

The factors that lead to CM are largely unknown. We are increasingly aware that CM is a heterogeneous condition in humans, even when using strictly defined clinical criteria. Diverse factors may act together, contributing to coma in individual patients with CM. It is probable that both sequestration and release of soluble factors play a role in causing CM. Parasite adhesion may initiate events which can then continue following clearance of infection as a result of an immune response or drug treatment. For instance, infection of a susceptible individual (Bull et al 1998) with a virulent parasite strain, combined with an over vigorous immune response, could predispose that individual to severe malaria and CM. If the host is predisposed to make a vigorous systemic inflammatory response by releasing high levels of TNF-α and up regulating ICAM-1 expression on EC, then this might allow the selection of strongly ICAM-1-binding parasites in cerebral vessels (Adams et al 2000).

The key question in the pathogenesis of CM is how the parasite, developing with in the erythrocytes sequestered in blood vessels, disrupts the function of CNS. Any theory about how CM develops must involve either the transfer of active mediators from the blood into the brain tissue or a parenchymal response to signals coming from the inside the blood vessels. Hence, a crucial interface determining the pathophysiology of the CM will be between the sequestered parasite and the brain, namely the blood-brain-barrier (BBB).

Mechanisms of CM Pathology

Blood Brain Barrier (BBB) Disruption During CM: Signaling through tight Junctions

The BBB is both a structural and functional barrier, which regulates the free passage of molecules and ions into and out of the brain parenchyma (Rubin and Staddon 1999). Together with the blood-CSF barrier, the BBB ensures constant composition of the extracellular fluid in the brain. This is essential for normal neuronal functioning and it also regulates the passage of immune cells into the CNS (Hickey 2001). The physical components of the BBB include cerebral EC and their basement membrane, together with perivascular cells including pericytes, perivascular monocytes, astrocytes and microglia. The specialized features of EC include (i) highly organized tight junctions in between cells; (ii) low rates of fluid phase pinocytosis; and (iii) specific transporters and efflux mechanisms which allow essential molecules to gain access, where as others are returned to the circulation.

Signaling and BBB Integrity

Tight junctions and adherence junctions are both associated with known signaling molecules (eg., the small GTPase proteins rho, rac and cdc42). The proteins in the tight junctions are members of the membrane-associated guanylate kinase family, which are involved in signal transduction. The cytoplasmic domains of junctional proteins undergo enhanced phosphorylation indicative of active signaling, which can alter permeability at these junctions. Hence, junctional proteins between cerebral EC can directly regulate cell signaling process, gene transcription and hence cellular phenotype and paracellular permeability (Staddon et al 1995; Staddon and Rubin 1996).

Cell Adhesion and Junctional Disruption

Much of the data on ICAM-1 mediated signaling comes from *in vitro* studies of T-cell migration across rat endothelial cell lines (Durieu-Trautmann et al 1994; Etienne et al 1998; Adamson et al 1999; Etienne- Manneville et al 2000). The leukocyte adhesion molecule LFA-1 binds to endothelial ICAL-1 and stimulates the phosphorylation of several cytoskeletal components, including cortactin (Durieu-Trautmann et al 1994) rho, focal adhesion kinase (FAK), paxillin and p130cas (Etienne et al 1998; Adamson et al 1999). Recent work has revealed that two separate signaling pathways through ICAM-1 cause these phosphorylation events. In the first pathway (Etienne- Manneville et al 2000), rises in intracellular Ca2+ result in the sequential phosphorylation and activation of PLCγ1, PKC, pp60src and finally, cortactin. In the second pathway, activation of rho results in phosphorylation and activation of p130cas. The rho dependent pathway also stimulates the stress activated protein kinase (SAPK/JNK) pathway, which in turn can regulate gene expression.

The two ICAM-1 signaling pathways appear to reunite at FAK. FAK and paxillin are components of focal adhesions, specialized sites of cell attachment to the extracellular matrix. Although there is evidence that focal adhesion regulate cell junction permeability in some settings, it is not clear whether this also applies to ICAM-1 mediated signaling events. ICAM-1 mediated pathways of adhesion, signaling and changes in junctional permeability could help leukocytes to enter the brain during immune surveillance and inflammation (Perry et al 1997). Since PRBC are known to bind to cerebral EC using many of the same adhesion receptors, especially ICAM-1, it is intriguing to speculate if PRBC adhesion mimic leukocyte adhesion, thus causing signaling and changes in cell functional adhesions (Fig. 1).

Thus the hypothesis one could draw from the above discussion if that PRBC binding to sequestration receptors such as ICAM-1 in cerebral vessels could produce subtle localized and reversible disruption of the BBB. This could involve signaling events, which subvert the pathways used by lymphocytes to cross the BBB. In this model, CM may result when a patient is infected with a parasite that binds to ICAM-1 with high affinity. Engagement of ICAM-1 or other receptors could cause secondary signaling events, influencing the cytoskeleton of the EC and the

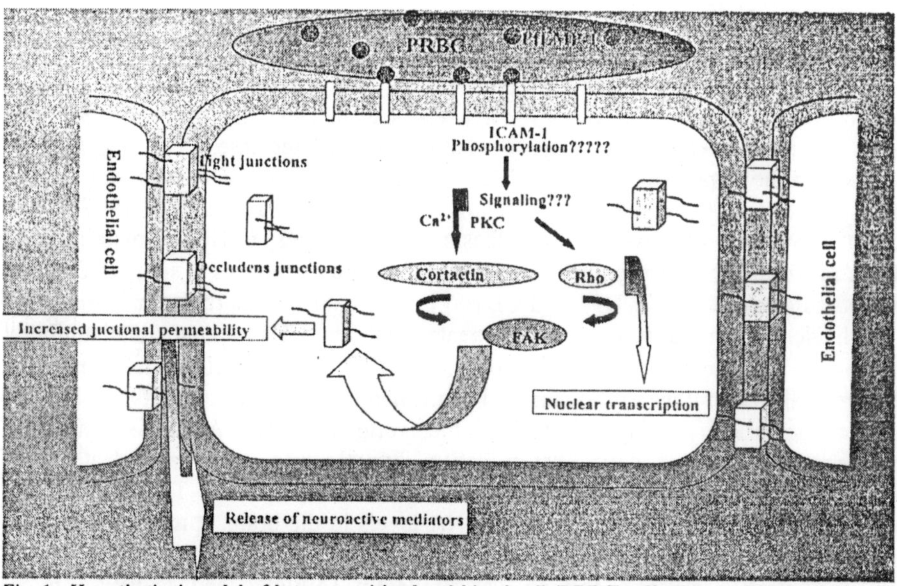

Fig. 1. Hypothetical model of how parasitized red blood cell (PRBC) adhesion could cause changes in BBB permeability (Schematic adapted from: Trends in Parasitology 18: No.8 Aug. 2002).

intercellular junctions. This is a pathway that normally allows unzipping of the tight junctions and allows passage of a leukocyte into the perivascular space. However, the PRBC is not motile, so this signaling and increased junctional permeability may produce a paracellular pathway, through which plasma proteins such as fibrinogen, circulating cytokines or other mediators could pass. On entering the perivascular space, this could cause neuronal dysfunction either directly or indirectly via activation of pericytes and other cells. This model explains some unique characteristics of human CM. The rapid reversibility of symptoms may reflect the repairing of the localized disruption following clearance of parasites. Recovery and the incidence of neurological side-effects would depend on how much secondary neuropathological damage occurred during the phase of sequestration and BBB breakdown.

Evidence for the Role of Tumor Necrosis Factor Receptor 2 (TNFR2) in Cerebral Malaria in Mice

Various studies using antibodies, recombinant cytokines, or knock-out mice have shown that the secretion of tumor necrosis factor (TNF or TNF-α) is important effector of CM (Grau et al 1987; Grau et al 1989; Rudin et al 1997). Evidence for a major role of TNF in human CM is also compelling and includes the presence of TNF in plasma (Kwiatkowski et al 1990) and higher incidence of CM in subjects with high TNF-producing genotypes (Kwiatkowski 2000). TNF and LT-α are ligated to two receptors (TNFR1 and TNFR2), and it has been reported that mortality and coma occur in TNFR1 -/- but not in in TNFR2-/- mice (Lucas et al 1997a), indicating interaction of membrane TNF and TNFR2 is crucial in the development of neurological syndrome in CM.

TNF might be responsible for several of the manifestations of CM, such as increase of the vascular permeability (Abe et al 1990), perturbation of the glucose metabolism at several levels (i.e., glucose transport, hepatic gluconeogenesis and glycolysis) reviewed in previous reports (Calrk et al 1997; Evans et al 1989), and sequestration of cells in the microcirculation. In this later case, TNF might act by increasing the expression of the adhesion molecules CD54 (ICAM-1) and CD106 (VCAM) or CD40 on endothelia (Imhof & Dunon 1995; Lucas et al 1997b), that might contribute to increasing the adhesion of leukocytes, pRBC, or platelets, and there by disturb the microcirculation. This pathogenic hypothesis is supported by the increased expression of CD54 and CD106 in the brain microcirculation during CM and the delayed mortality seen in mice treated with anti CD11a monoclonal antibody (LFA-1, a β2 integrin determinant) (Grau et al 1991) or in CD54 deficient mice (Favre et al 1999). Recently Piguet et al confirmed the essential role of the TNFR2 in cerebral malaria by using a model of *Plasmodium berghei* ANKA strain. They demonstrated that 6-8 days post infection, hypothermia, coma and death were observed in +/+ or TNFR1-/-, but never in TNFR2-/-, mice. TNF production, evaluated by the serum levels or the mRNA levels in the brain, spleen or lung, was similar in +/+, TNFR1-/-, or TNFR2-/- mice. Macrophages or parasitized red blood cell sequestration in brain or lung was similar in TNFR1-/- and TNFR2-/- mice. Accordingly, upregulation of CD54 or CD40 in brain or lung was also similar in TNFR1-/- orTNFR2-/- mice. Platelet loss, manifested by thrombocytopenia and the presence of microparticles in plasma, was similar in TNFR-/- or TNFR2-/- mice.

Breakdown of the blood brain barrier, detected by the diffusion of the tracers, was attenuated in both TNFR1-/- and TNFR2-/-, compared with +/+, mice. Endothelial cells from the brain capillaries, examined by transmission electron microscopy, were similar in TNFR1-/- or TNFR2-/- mice, whereas the basement membrane was enlarged in TNFR1-/-, but not in TNFR2-/-, mice. In addition, infected +/+ and TNFR1-/- mice became insulin resistant, while in contrast TNFR2-/- became extremely insulin sensitive. This study supports the possibility that coma and death are mediated not by cell sequestration or breakdown of vascular permeability, similar in TNFR1-/- or TNFR2-/- mice, but metabolic disturbances selectively mediated by TNFR2 (Pierre et al 2002).

Activation of Intrinsic cell death pathways during cerebral malaria proceeds via mitochondrial Anomalies

Intrinsic pathways in apoptosis are defined as those that initiate cell death in response to release of certain apoptogenic factors like cyt-c, bax and bcl_2 from mitochondria. There exists a direct evidence to indicate that TNF has a profound influence on mitochondrial structural and functional integrity. Since there is an overwhelming induction of TNF-alpha during CM pathology, we studied the structural integrity of mitochondria and its related proteins bax and bcl_2. Our results suggest that mitochondrial anomalies like vacuolation and swelling observed during cerebral malaria (Fig.2) might represent a direct consequence of TNF-alpha. Whether the upstream activation events distal to TNF-alpha utilize procaspase-8 during CM is an intriguing question as most of the TNF mediated events utilizes the procaspse-8 conversion to active caspase-8 form. However, following the mitochondrial anomalies, there is an increase of bax over bcl_2, which presumably might alter the mitochondrial membrane permeability to release cyt-c, that plays a crucial role in setting the stage for apoptosis by activation of caspases (Kumar & Babu 2002) (Fig. 3a). Caspase-3 is also involved in the cascade driven by TNF which is evident from our studies and demonstrates the occurrence of the cleavage product (active fragment/signature fragment) specific only to the infected samples and not to controls.

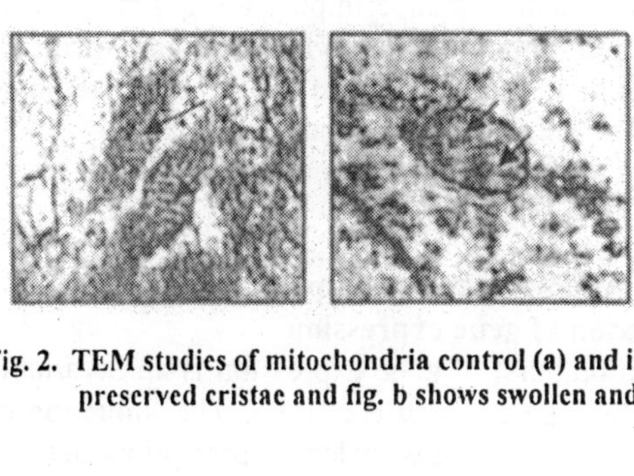

Fig. a shows finely preserved cristae and Fig. b shows swollen and vacuolated Mitochondria (5000x)

Fig. 2. TEM studies of mitochondria control (a) and infected (b) mice brain Fig a shows finely preserved cristae and fig. b shows swollen and vacuolated Mitochondria (5000x).

Ultrastructural and Biochemical changes associated with Nuclei of CM infected brain samples suggest for a mixed mode of cell death

Apart from the overwhelming evidence of apoptosis or programmed cell death during CM, our studies additionally demonstrate an evidence for necrotic mode of cell death during the pathology. This was evident from morphological changes in the nucleus in CM infected brain, which showed nuclear vacuolation and condensed nuclear material (Fig. 4a and b). Further, the TUNEL staining revealed nuclear changes at two distinct stages viz, an early stage of DNA condensation and later stage of DNA fragmentation or Karyorexhesis (Fig 5).

Further the biochemical evidence for the presence of signature fragment of 89 kDa PARP and p53 induction provides an unequivocal evidence for the apoptosis. Our

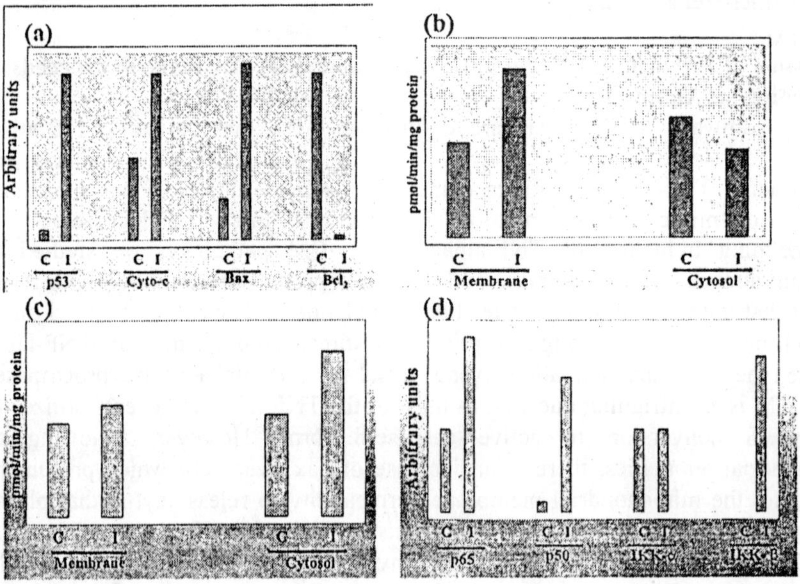

Fig. 3. Graphs reproduced from authors published manuscripts. Neurosci.letts 329 (2000) 319-323 and 340 (2003) 139-142

results suggest that necrosis and apoptosis may not be mutually exclusive events associated with the pathology, rather the heterogeneity that appears in cell death could be an inevitable outcome of cells attempt to superimpose a highly structured apoptosis over a more chaotic necrosis. Further, owing to the local immune mediated responses and inflammatory reactions in the brain, there could be a regional heterogeneity in the mode of cellular damage. These observations pertaining to CM paradigm may prove useful for analyzing the mechanisms that govern cell death under pathological conditions.

Role of Transcription factors in Modulating Host Cell responses during CM Pathology: Implications for the induction of gene expression

In most of the inflammatory models, an outburst of gene expression is an invariable mechanism associated with host cells to cope up with the stress. The induction of gene expression however is associated or influenced by nuclear transcription factors

Fig. 4

a b

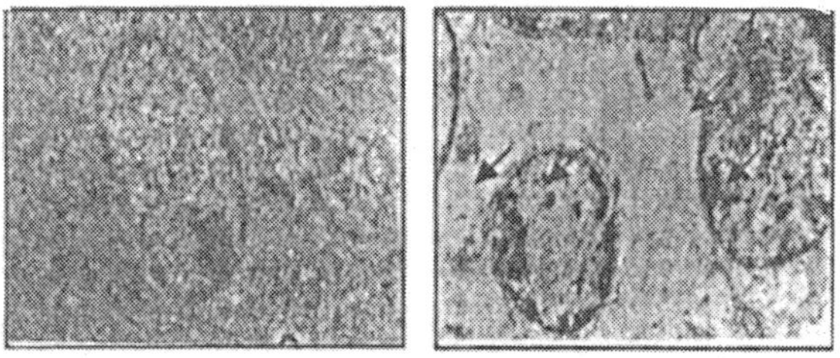

Fig.5

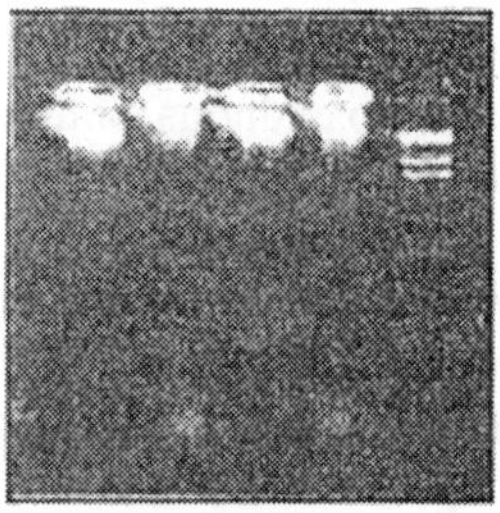

CC IC CCB ICB MW

Fig 4. TEM studies of Nuclei from control (a) and CM infected (b) brain. Fig a shows uniformly distributed granular material but in fig b nuclei shows extensive vacuolation and condensed nuclear material (10,000x)

Fig. 5. Agarose gel electrophoresis showing oligonucleosomal fragmentation of DNA CC: Control Cortex, CCB: Control Cerebellum, IC: Infected Cortex, ICB: InfectedCerebellum, MW: Molecular weight

that predominantly reside in the cytosol in resting state/inactive state, but following an inflammatory stimuli, undergo conformational changes, translocate to nucleus and activate gene expression by binding to the promoter of the cognate DNA sequences. One such factor is the Nuclear transcription factor kappa B. Our studies have implicated a role for this nuclear factor/complex during CM as evident by translocation of two subunits of this complex p50 and p65 from cytosol to nucleus. The apparent decrease in the levels of these subunits in cytosol and a concomitant increase in nuclear fraction not only suggests its translocation but also its probable needs to activate gene expression under conditions of inflammation. Further this activation of the cytosolic NF-κB is under the regulation of a kinase called as IKK complex. Though the IKK complex (inhibitory kappa B complex) is made up of three subunits IKK α,β and γ, overwhelming role of IKK-β and not IKK-α seem to be crucial for NF-κB complex activation as evident by increased levels of IKK-β alone during CM pathology (Kumar & Babu 2003) (Fig 3d).

Modulation of Protein Kinase C and Protein Tyrosine Kinases during CM Pathology

As a perquisite to the pathological or inflammatory settings, cells respond to extra cellular signals with a specific set of mechanisms that regulate or modulate gene expression. Between the signal and the gene, a system of rather different cellular components is assembled to guarantee a specific and successful process of signal transduction. Most of the components of the signal transduction pathways are proteins whose activity is altered by either ligand or secondary messenger binding, by covalent modifications and by subsequent changes in the conformation or subunit number. The majority of the covalent modifications observed are phosphorylations on the tyrosine or serine/threonine residues, and both Tyr and Ser/Thr-specific substrates are present amongst the numerous components of signal transduction. Moreover some of the protein kinases are themselves substrates of other protein kinases or modulate their own activity by (multiple) autophosphorylation reactions.

The PKC and the PTK represent dedicated signal transduction complexes in the cell membrane that are known to collaborately regulate the NF-κB activation in diverse paradigms both infectious (Hyung-Yong K & Yasuko 2002; Matam et al 2001; Souvenir et al 1997; Uwe-Karsten et al 2001) and other cell signaling events (Ching- Chow et al 2001). Our studies implicate a role for these protein kinases in mediating the post TNF signaling events during CM pathology. A general increase in the protein tyrosine kinase activity in the CM infected brain samples was consistent with increased tyrosine phosphorylation of certain proteins in the infected brain samples and decrease in the protein kinase C activity levels in the cytosolic fraction of infected brain (3b &c) coincided with the decreased levels of certain PKC isoforms like PKC alpha and delta (Kumar & Babu 2003). The observed decrease in levels of certain PKC isoforms might also explain for the caspase like activity that could be associated with the CM infected brain samples. On contrary, increased protein kinase C activity in the membrane fraction of CM infected brain without apparent changes in the levels of its isoforms might suggest for some post translational modifications associated with these membrane bound enzymes following TNF signaling (Fig.6).

Role of T cells in Cerebral Malaria Pathology

T cells play a central role in both immunity and pathology of malaria (Omer et al 2000). In mice several lines of evidence point to a major role for T cells in the immunopathology of cerebral malaria. Existing literature suggests that athymic nude mice do not develop CM (Finley et al 1982) and anti-CD4 or anti-CD8 treatment prevent cerebral malaria in C57BL/6-PbK (Curfs et al 1989) and C57BL/6-PbA models. The activation of T cells during cerebral malaria is in response to the schizont rupture, during the erythrocytic stages.

The malaria merozoites that are released following the schizont rupture contain glycosylphosphatidylinositol (GPI)-the anchor molecule, a constituent of merozoite surface protein. These parasite antigens are recognized by the Antigen Presenting Cells (APC) and presented to the T-cells in the context of MHC class II molecules. As a consequence of this antigen presentation to T cells, the T cells are activated which then manifest effector functions like IFN-γ secretion which in turn stimulate the macrophages to secrete TNF-α. GPI toxins are also capable of independently activating macrophages to secrete TNF-α in addition to T-cell mediated TNF-α induction (Brain de Souza & Eleanor 2002) (Fig 7).

Recent studies also highlight the regulation of Murine Cerebral Malaria Pathogenesis by CD1d-restricted NKT cells and Natural Killer complex. NKT cells constitute the non-conventional lymphoid population that co-express NK and T cell receptors (TCR) and upon antigen-specific or polyclonal stimulation through the TCR, CD1d-restricted NKT cells produce large amounts of IL-4 and IFN-γ and thus differentially regulate T_H1/T_H2 polarization, malarial pathogenesis and fatality according to the genetic background of mice (Diana et al).

The role of cytolytic effector mechanisms mediated by CD8 cells in the development of ECM and the involvement of T lymphocytes in brain cell trafficking was elegantly demonstrated by Mazier et al (Josianne et al 2003) Indeed, the trafficking of immunocompetent cells have been highly implicated in the

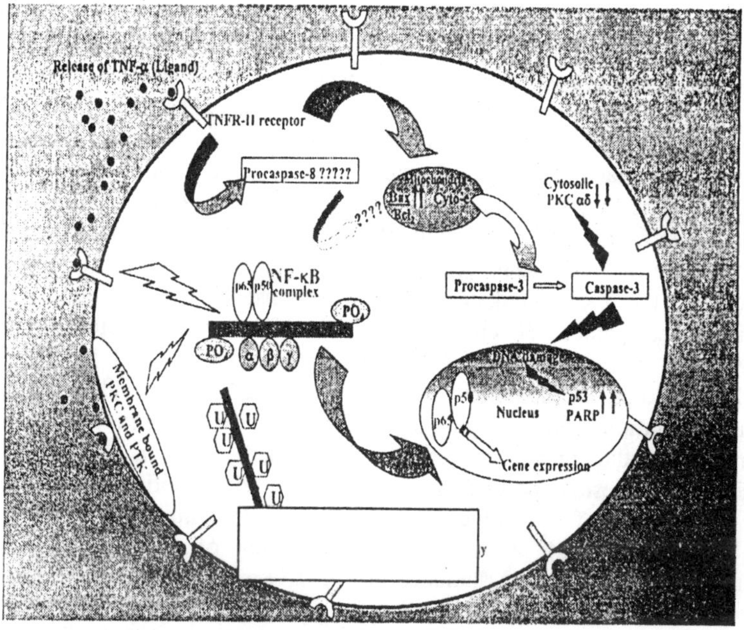

Fig. 6. Hypothetical model showing the activated signaling pathways during CM following receptor mediated binding of TNF-α to $TNFR_2$ (Schematic representation made from the authors published data: Neurosci.lett 329 (2002) 319-323 and 340 (2003) 139-142).

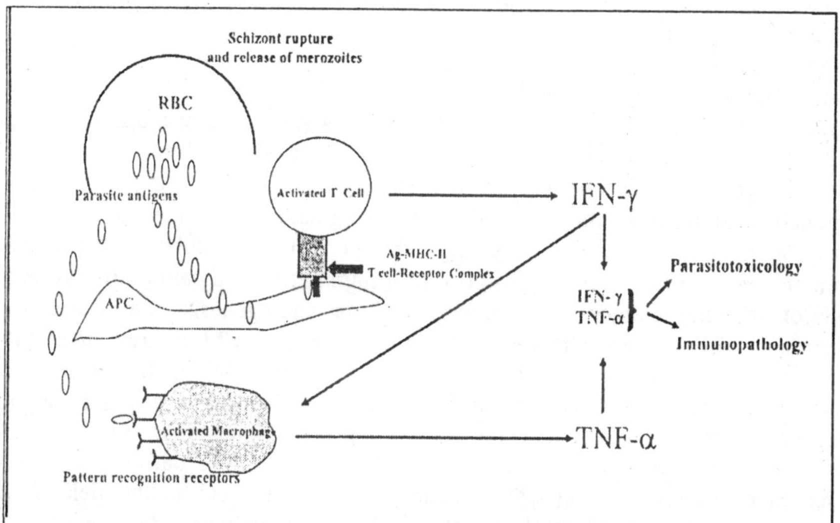

Fig. 7. Sequence of events leading to immune response in malaria Proinflammatory cytokine production by activated T cells and macrophages in response to parasite antigens or toxins (Adapted from: Microbes and infection 4 (2002) 291-300).

establishment of the immune response and disease progression (Gerard & Rollins 2001). Based on chemokine receptor-deficient mice models, chemokines/chemokine receptor have been recently implicated in various pathologies, such as multiple sclerosis in human and murine experimental autoimmune encephalomyelitis.

Direct evidence for the CD8 T lymphocytes involvement in ECM pathogenesis is evident from the fact that mice homozygous for the null allele of the β2-microglobulin gene that lack functional CD8 T cells (Yanez et al 1996) as well as CD8 knockout (KO) mice (Boubou et al 1999) were resistant to ECM. In addition, a selective increase in Vβ8+CD8T cells has been observed during ECM, suggesting a critical role for the T-cell mediated immune response after PbA infection and its neurological consequence. To elucidate the role of cytolytic effector mechanisms mediated by the CD8 cells during ECM, and the involvement of T lymphocytes in brain cell trafficking, ECM models were induced in C57BL/6 wild type (WT) Ipr (Fas-natural mutation-inducing lymphoproliferative disease), and gld (Fas-ligand natural mutation-inducing generalized lymphadenopathy) and in perforin deficient mice (PFP-KO). These experiments precisely lead to identification of cytotoxic molecules involved in effector functions of CD8 cells. In contrast to the WT, Ipr and gld that developed ECM, all groups of mice died of anemia and increased parasitaemia. Further, ECM was induced in PFP-KO mice after adoptive transfer of cytotoxic cells from infected C57BL6 mice, which were directed to the brain of PFP-KO mice. This specific recruitment might involve chemokine/chemokine receptor, since their expression was upregulated on activated CD8 cells. The role of chemokine/chemokine receptors was further illustrated by inducing ECM models in CCR5-KO mice where susceptibility to ECM was delayed. Thus, lymphocyte cytotoxicity and cell trafficking are key players in ECM pathogenesis (Josianne et al 2003).

Conclusion and Future Directions

An overview of the mechanisms of CM pathology suggests that multiple pathologic and protective mechanisms are operational during cerebral malaria. Since susceptibility to experimental cerebral malaria in mice is dependent on the production of proinflammatory and counter regulatory cytokines by T cells and innate immune system, further studies should aim at modulating the excessive immune responses as a strategy to prevent inflammation and pathology. Probing the role of immune elements in cerebral malaria pathogenesis and analyzing the importance of lymphocyte trafficking to the CNS during ECM progression might call for better therapeutic strategy for controlling CM. In addition, the recent insights from signal transduction studies implicating the role of transcription factors to regulate gene expression and the role of protein kinases in modulating the pathological responses during cerebral malaria help us in defining the precise cascades operational during CM infection. Recent advances in the field of laser capture dissection technology (Helen et al 2002) should help us to precisely reach the microenvironment where pRBC cell interacts with the endothelial cells. Analysis of gene expression in these target endothelial cells might give us a cue to the vast number of transcripts that are made during cell-cell interaction that is vital to the loss of integrity of tight junctions and release of parasite toxins and immune mediators in brain parenchyma. For all such studies, murine model is suitable both for determining precise pathogenic features of the cerebral form of the disease and also for evaluating the circumventive interventions.

Although remarkable advances have been made in understanding the pathology of cerebral malaria in past one decade, the number of unsolved fundamental questions still remains large. The mechanistic action of the pathological mediators like TNF-alpha and its ability to induce pleotropic effects by activating multiple signaling pathways and the cross talk between such cascades should be clarified not only because of scientific interest but also to increase the ways of effective treatment. Further it would be particularly important to determine whether the new transgenic technologies (i.e. antisense, conditional knockouts and RNA interference) would allow us to detect a wider range of candidate genes, telling us more about the neurobiological mediators of response to CM. Further accumulation of new knowledge regarding the type of cell death, specific signaling molecules and their receptor specificity involved in their downstream consequences, leading to specific signaling cascade, utilizing the knowledge from new discipline of micro arrays will continue to provide better opportunities to effectively address CM pathology. Finally, there is a very good reason to be optimistic that these types of investigations will lead to novel and efficacious therapies for CM induced pathology.

References

Abe A, Sekiya S, Yamasite T, and Senso F (1990) J Immunol **145**:2902-2907.

Adams S et al (2000) Infect Immun **68**: 264-269.

Adamson P et al (1999) J Immunol **162**: 2964-2973.

Beeson JG et al (2001) Trends Parasitol **17**: 331-337.

Berendt AR et al (1994) Parasitology **108**: 519-528.

Boubou MI, Collette A, Voegtle D, Mazier D, Cazenave PA, and Pied S (1999) Int Immunol **11**:1553.

Brian de Souza J, and Riley EM (2002) Microbes and Infect **4**: 291-300.

Brewester D et al (1990) Lancet **336**: 1039-1043.

Brown H et al (1999) J Infect Dis **180**: 1742-1746.

Bull PC et al (1998) Nat Med **4**: 358-360.

Calrk IA, AI yaman FM, and Jacobson LS (1997) Int J Parasitol **27**: 1237-1249.

Ching-Chow C, Chian-Yin C, Yi-tao S, Wei-Chein H (2001) Cell Signalling **13**: 543-553.

Chiwakata CB et al (2000) Infect Immun **68**: 394-399.

Clark IA et al (1997) Int J Parasitol **27**: 1237-1249

Craig A and Scherf A (2001) Mol Biochem Parasitiol **115**: 129-143.

Curfs JHAJ, Schetters TPM, Hermsen CC, Jerusalem C R, Van Zon AAJC and Eling WMC (1989) Clin Exp Immonol **75**: 136-140.

Diana S H, Mary-Anne Siomos, Lynn Buckingham, Anthony A S, and Louis Schofield, Immunity, **10**. 391-402.

Dondorp AM et al (1998) Am J Trop Med Hyg **59**: 497-502.

Durieu- Trautmann, O et al (1994) J Biol Chem **269**: 12536-12540

Etienne, S et al (1998) J Immunol **161**: 5755-5761.

Etienne-Manneville, S et al (2000) J Immunol **165**: 3375-3383.

Evans RD Argilles JM, and Williamson DH (1989) Clin Sci **77**: 357-364.

Favre N, Williman K, Ryffel B, Weiss N, Imhof BA, Rudin W, Lucas R, and Piguet PF (1999) Microbes Infection **1**:961-968.

Finley RW, Mackey LJ amd Lambert PH (1982) J Immunol. **129**: 165-175.

Fried M and Duffy PE (1998) J Mol Med **76**: 162-171.

Gerard C, and Rollins BJ (2001) Chemokines and disease Nat Immunol **2**:108.

Grau GE, Pointaire P, Piguet PF, Vesin C, Rosen H, Stamenkovic I, Takei F, and Vassalli P (1991) Eur J Immunol **21**: 2265-2267.

Grau GE, Fafardo LF, Piguet PF, Allet B, Lambert PH, and Vasalli P (1987) Science **237**: 1210-1212.

Grau GE, Heremans H, Piguet PF, Pointaire P, lambert PH, Billiau A, and Vasalli P (1989) Proc Natl Acad Sci (USA) **86**: 5572-5574.

Greenwood, B (1999) Bull WHO **77**: 617-618.

Helen J Ball, Brent McParland, Catherine Driussi, Hunt NH (2002) Brain Res Protocols **9**: 206-213
Hickey WF (2001) Glia **36**: 118-124.
Hyung-Yong K, Yasuko R (2002) Infect Immun **70** 4132-4141.
Imhof BA and Dunon D (1995) Adv Immunol **58**:345-416.
Karunaweera ND et al (1992) Proc Natal Acad Sci (USA) **89**: 3200-3203.
Kumar AK and Babu PP (2002) Neurosci Lett **329**: 319-323.
Kumar AK and Babu PP (2003) Neurosci Lett **340**: 139-142
Kwiatkowski D, Hill AVS, Sambou I, Twumasi P, Castracane J, Manogue KR, Cerami A, Brewester DR, and Greenwood BM (1990) Lancet **336**: 1201-1204.
Kwiatkowski D (2000) Curr Opin Gene Dev **10**: 320-324.
Kwiatkowski D et al (1990) Lancet **336**: 1201-1204
Lucas R, Juillard Piroud C, Monso Hinard C, De Kesel T, Buurman WA, Moore MW, Dayer JM, Fiers W, Bluethman H, and Grau GE (1997a) Eur J Imunol **27**, 1719-1725.
Lucas R, Lou JN, Juillard P, Moore M, Bluethman H, and Grau GE (1997b) J Neuroimmunol **72**: 143-148.
MacPherson GG et al (1985) Am J Pathol **119**: 385-401.
Maneerat Y et al (2000) Histopathology **37**: 269-277.
Marchiafava E and Begnami, A (1884) Vol 150 pp 1-234, The New Sydenham Society.
Matam V, Ramachandrana SN and Channe GD (2001) J Biol Chem **276**: 6909-6912.
Medana I et al (2001) Neuropathol Appl Neurobiology **27**: 421-433.
Nitcheu J, Bonduelle O, Combadiere C, Tefit M, Seilhean D, Mazier D and Combadiere B (2003) J of Immunol **170**: 2221-2228.
Omer F M, Kurtzhals J A and Riley E M (2000) Parasitol Today 16 18-23.
Perry VH et al (1997) Mol Med Today **3**: 335-341.
Pierre Francois Piguet, Chen Da Kan and Chriastian Vesin (2002) Lab Invest **82**: 9 (1155-1166).
Pongponratn E et al (1991) Am J Trop Medical Hyg **44**: 168-175.
Rubin LL and Staddon JM (1999) Ann Rev Neurosci **22**: 11-28.
Rudin W, Eugster HP, Bordmann G, Bonato J, Muller M, Yamage M, and Ryffel B (1997) Am J Pathol **150**: 257-266.
Silamut K et al (1999) Am J Pathol **155**: 395-410.
Smith JD et al (2000) Proc Natl Acad Sci (USA) **97**: 1766-1771.
Staddon JM et al (1995) Biochem Soc Trans **23**: 475-479.
Staddon JM and Rubin LL (1996) Curr Opin Neurobiol **6**: 622-627.
Souvenir D T, Peter GS, Ralph S, Novakovic M, McConville L, Schofield (1997) Proc Natl Acad Sci (USA) **94** 4022-4027.
Taylor AM et al (1998) Trans R Soc Trop Med Hyg **92**: 170-175.
Turner G et al (1974) Am J Pathol **145**: 1057-1069
Turner G (1997) Cerebral Malaria Brain Pathol **7**, 569-582.
Udomsangpetch R et al (1997) Am J Trop Med Hyg **57**: 501-506.
Uwe-Karsten H, Macro P, Klemens A et al (2001) Eur J Immunol **31**: 2104 -2115.
Van Hensbroek MB et al (1996) J Infect Dis **174**: 1091-1097.
Yanez DM, Manning DD Cooley AJ, Weidanz WP, and Van der Heyde HC (1996) J Immunol **157**:1620.

Molecular and Cellular Neurobiology
Editors: M.K. Thakur and S. Prasad

Central Response to Peripheral Nerve Injury: Role of Non-Neuronal Cells

I.K. Patro, Seema Pathak and Nisha Patro

Neuroscience Centre, School of Studies in Zoology, Jiwaji University, Gwalior, India

Abstract
It is now established that there is loss of neurons following injury, with age and in neurodegenerative disorders. Possibly all disturbances in the nervous system activates the microglia and astroglia. We now have evidence that activated microglia are able to exert neurotoxic effects. This has attracted attention on microglia research in brain. Understanding the mechanism involved in the process of neuronal death and/or the changes in the neuronal environment as well as availability of novel therapies to speed up neuroregeneration would be beneficial not only to the patients but also to the society by reducing the associated health care costs. Very recently, immunophilin ligands (e.g., FK506) have been observed to exert neuroregenerative and neuroprotective effects via a poorly understood mechanism. They are of two types: immunosuppressant and non-immunosuppressant. These substances are being actively studied for their potentials to act as novel therapies for neuronal loss and neuronal recovery from injury.

Key words: Apoptosis, axotomy, glial activation, immunophilins and immunophilin ligands, neurodegeneration, neuroregeneration

Introduction
Injury to the nervous system immediately needs attention by way of protection and repair. The episode triggers a set of neuroglial and metabolic changes crucial to execute both the physiological processes (Deckert-Schluter et al 1995; Drescher et al 1997; Griffin et al 1992; Kreutzberg, 1996; Raivich et al 1996). Injured neurons are in a state of emergency and thus rapidly attain altered levels of gene expression and also stimulate the microglia and astrocytes in their vicinity for support. Such neuroglial activation is graded and similar to what is readily observed in stroke or ischemia, in neurodegenerative disorders or syndromes during inflammation due to infection or autoimmune disease or may be after axonal injury. Products of proinflammatory cytokine, functional changes in brain vascular endothelial and infiltration of immune system cells into the damage tissue all accompany neuroglial activation. In addition the neuroglial activation steps follow more or less a similar pattern (stereotype) irrespective of the episode that triggers it and thus reflect an evolutionarily conserved programme. It is believed that it plays an important role in protection and repair of the nervous system. Of late people have questioned if an over activation or persistence of chemical changes can negatively influence to non-injured neurons in the vicinity of damaged or recovering neurons (Patro and Patro, 2004). Injury to peripheral nerve induces a highly reproducible but complex sequence of histopathological events both at the site of injury at the cellular level. In this review an effort has been made to evaluate the present status of knowledge on the involvement of non-neuronal cells (astroglia and microglia in particular) in regeneration and degeneration of the neurons facing neuronal injury. This is

expected to open-up newer therapeutic implementations, particularly in relation to immunophilins.

Besides injury, neuron death is also evidenced in several other states of the brain. The mechanism involved in the important phenomenon of neural death in injury, neurodegenerative diseases (like amylotropic lateral sclerosis, Alzheimer's, Parkinsonism) and aging (including senile dementia) remain to be understood. Whatever may be the circumstance, loss of neurons is a life-endangering process and needs to be combated effectively (Patro and Patro, 1992; Deckwerth and Johnson, 1993).

Glial cell activation following peripheral nerve injury and neurodegenerative disorders

Any kind of disturbance, possibly all, in the nervous system activates the microglia and astroglia. Even if the actual injury is at a distance from the CNS and the blood brain barrier is intact, still the glial response becomes prominent. Possibly the mechanism(s) involved in glial cell activity following peripheral nerve injury and during Wallerian degeneration share common elements. What is more important is that activated glial cells influence the progression of neuronal and non-neuronal reactions to the injury. In addition, the glial response determines the outcome of the pathological events.

Following axonal injury neurons mediate their new functional state to astrocytes. The neurons thus share membrane contact with astrocytes. Astrocytes, subsequently, secrete signaling substances those trigger microglial proliferation and early phase of their activation. The activated microglia, in turn, release factors those influence astroglial activities. These astroglial and microglial activities by autocrine mechanisms are amplified by reciprocal signals as well as by neuronal factors (Aldskogius and Kozlova, 1998; Fig. 1).

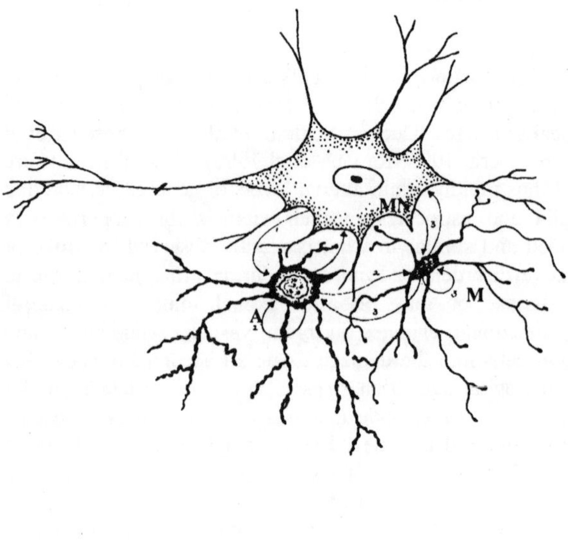

Fig. 1 Neuron-glial and glial-glial autocrine mechanism following motor neuron injury. Diagram illustrating neuron-glial and glial-glial interaction following axon injury in a motor neuron: (1) Injured motor neuron (MN) influence the neighboring astrocytes (A). (2) Astrocytes thus activated along with the injured neuron release microglial activators and stimulate autocrine circuits. (3) Subsequently microglia (M) promote hypertrophy of astrocytes, influence their own status of activation via autocrine loops and, possibly, influence the adjacent neurons.

The pathological activation of glial cells, which forces microglia to transform into immunocompetent cells with cytotoxic capabilities and induces astroglia to upregulate GFAP, amongst other changes, plays an important role in the course and potentially the outcome of neurodegenerative disorders. Activated microglia might induce neuronal apoptosis or *vice versa* the apoptosis of neurons might activate microglia. What is the primary and what is the secondary event remains unclear. Whether and how interference of microglia influences or possibly even slows down the progressive cell death in neurodegenerative diseases like AD, remains to be answered. It is not yet clear, whether the detrimental effects of inflammation outweigh the neuroprotective mechanisms or the opposite holds true. A better understanding of the mechanisms regulating microglial proliferation and modulating its activation is needed to devise more efficient therapies battling neurodegeneration (Box-I and II).

Astrocytes

Astrocytes are most predominant neuroglial cells in the central nervous system. They are of two types (1) the stellar fibrillary astrocytes and (2) protoplasmic astrocytes. The fibrillary astrocytes are present in the white matter. They have slender processes, which stain for glial fibrillary acidic protein (GFAP). The protoplasmic astrocytes are found in the grey matter. They have highly ramified processes with many flappy, membranous extensions. In normal brain most protoplasmic astrocytes are GFAP-negative and it is not possible to study their entire cellular morphology with routine immunohistochemical stainings. Astrocytes, like the mesenchymal cells in peripheral tissue, provide physical support to the neighboring neurons meninges and vasculature. Astrocytes also provide nutrition to highly energy consuming neurons; help in maintaining ionic balance in the extra cellular spaces also remove water and excitotoxic amino acids. In addition, the astrocytes also provide numerous trophic factors particularly following injury. Like microglia, astrocytes also respond to neuronal injury in stages and such activation depends on the extent of trauma. Within 24hrs following axotomy there is a rapid up regulation of GFAP (Chattopadhyay, 1997, Patro et al 1999) this is followed by the appearance of small and slender GFAP-positive processes. After 2-3 days numerous fully stellarized, fibrillary astrocytes appear. Such early morphological transformation of protoplasmic astrocytes to fibrillary is under structural control of cytokine in the neuronal parenchyma.

The reactive fibrillary astrocytes create a physical barrier between damaged and healthy cells. The process is called scarring and the extent of scarring depends on the severity of brain damage. Following 2weeks of axotomy, the reactive fibrillary astrocytes gradually replace microglia from the neuronal surface and enwrap injured neurons with thin flat cytoplasmic processes (Graeber and Kreutzberg, 1988; Guntinas-Lichius et al 1993). The processes of the astrocytic also adhere to each other forming a multi-layered cover of astrocytes lamellae surrounding the injured neuron. Such coverings finally act as small glial scar. On successful regeneration the processes are partially retracted and gradually the synaptic terminals populate on the neuron surface. This evidence that the process of synaptic stripping is a reversible process. Both pro- and anti-inflammatory cytokines have been shown to regulate glial scar formation. Representative photomicrographs of astrocytes at various levels of activation are presented in Fig. 2.

Box-I: Microglial Proliferation

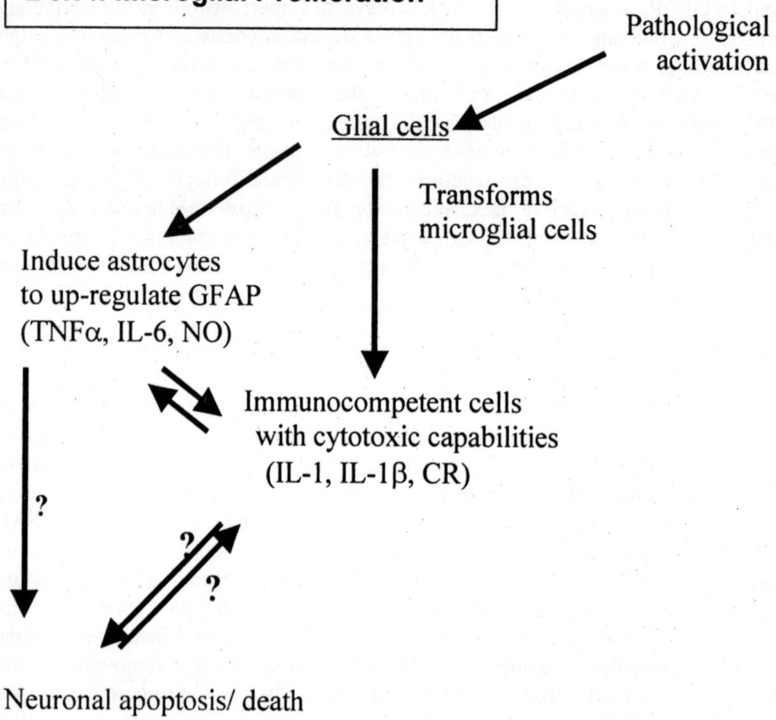

What is unclear?

- Glial activation induces apoptosis or vice versa?
- Glial activation or apoptosis- which is primary?
- Whether or how microglia influence or slow down cell death?
- Whether detrimental effects of inflammation outweigh neuroprotective mechanism or the opposite holds good?

∴ A better understanding of the mechanisms regulating microglial proliferation and modulating its activation is needed to devise more efficient therapy for battling neurodegeneration.

Box-II: Microglial Activation

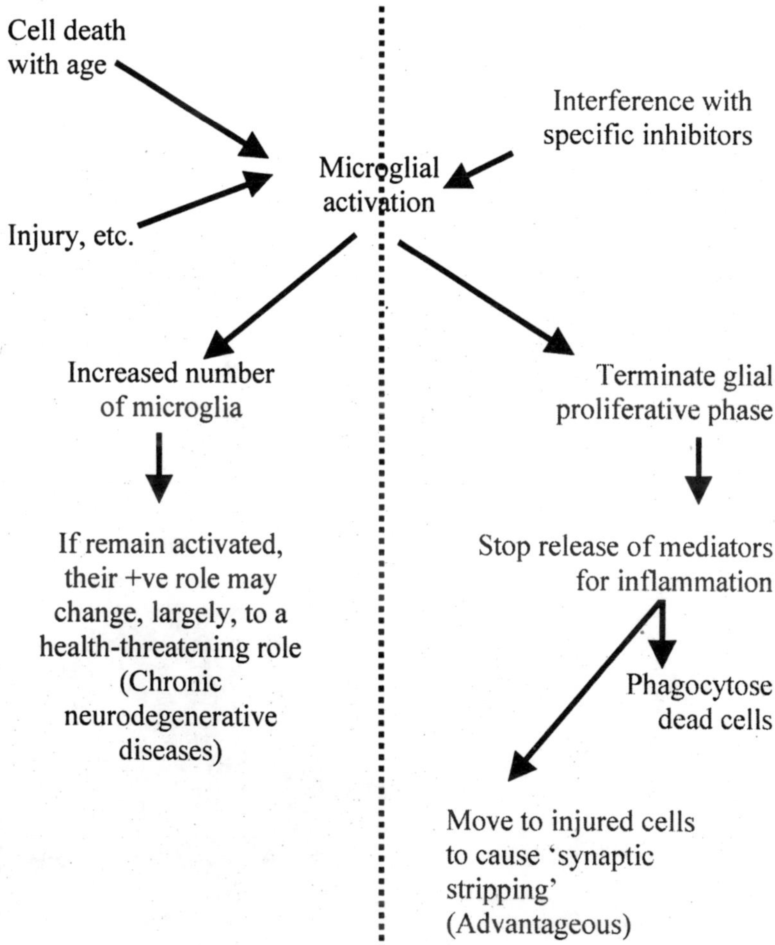

∴ A better understanding of molecular events involved in microglial activation associated neuron death would be essential to develop specific therapeutic strategies.

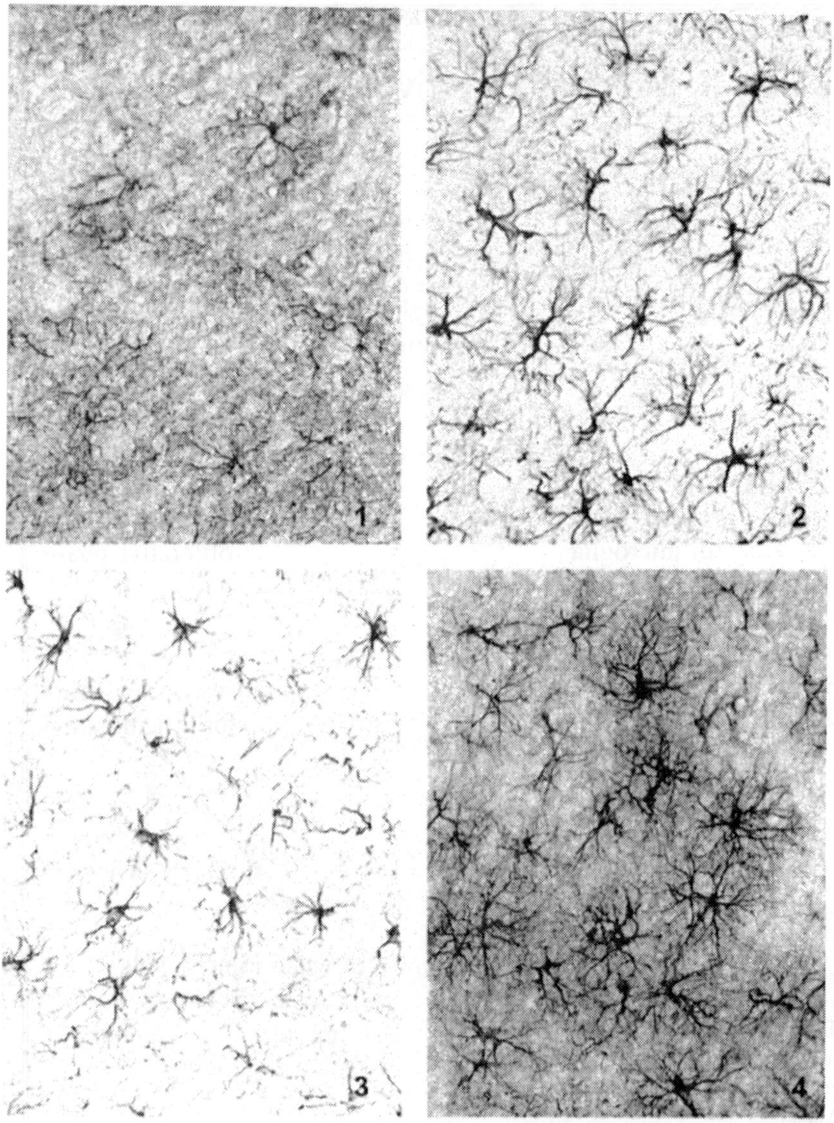

Fig. 2. Morphology of astroglia (GFAP immunostaining) from resting to activated to scar formation in the facial nucleus in rat brain following facial nerve transection.

Immediately after motor axon injury the astrocytes present increased immunoreactivity for connexin 43, a major component of astrocytic gap junctions (Rohlman et al 1994). Within 24 hours the astrocytes present increased immunoreactivity for glial fibrillary acidic protein (GFAP) (Graeber and Kreutzberg, 1986; Ruan et al 1994) and upregulate GFAP mRNA and GFAP protein (Tetzlaff et al 1988). As time advances, increased number of filaments appears in astrocytes, and they develop slender sheet like processes covering non-synaptic areas of neuronal membrane. However, the astrocytes do not proliferate following peripheral axotomy. The astrocytic response to axotomy is in two phases. The first phase tries to normalize within a few days and the second phase of GFAP upregulation recurs after a few weeks of axotomy (Graeber and Kreutzberg, 1988). An upregulation of clustrin parallel to GFAP is evidenced following axotomy (Svensson et al 1995). This protein has multifaceted activities. It is able to inhibit complement mediated cell lysis, involved in cell-adhesion, lipid transport and processes associated with cell survival or cell degeneration (Rosenberg and Silkensen, 1995). However, its role in axotomy-induced astroglial response is unknown.

Microglia

Microglia are distributed randomly throughout the brain with some prevalence in grey matter, cerebral cortex and hippocampus and in lower numbers in dorsal thalamus (Perry and Gordon, 1991; Perry et al 1985). Dying cells probably emit signals, which rapidly activate resting microglia in the neighborhood. Several stages within this activation process can be distinguished, the earliest stage being proliferation. The molecular mechanisms triggering the production and release of microglial growth factors are largely unknown. Hypothetically, such mechanisms might entail disturbance of ion channel properties.

Specific inhibitors capable of switching the cells back to quiescence or turn into a differentiation phase may terminate the proliferation phase. In this phase, microglia release mediators of inflammation and begin to actively phagocytose dying cells and debris. They may also migrate to cell bodies of injured neurons where they reportedly interrupt synaptic contacts ("Synaptic stripping"; Graeber et al 1993). This event may be understood as primarily advantageous to brain tissue functions. Since there is always some ongoing cellular death in the brain during the whole life, one would expect increasing number of microglia with age (Cacabelos et al 1994). If some of the microglia remain activated, their positive role may change, largely, into a health-threatening role as observed in chronic neurodegenerative disorders (Dickson et al 1993). Microglial cells begin to react with their fine ramifications, enlarge and enter mitosis 24 hours after axotomy. Such an activity remains at its peak two days following axon injury and virtually stops after 7 days (Svensson and Aldskogius, 1993, Svensson et al 1994). These microglia migrate towards the injured neuronal perikarya.

There are conflicting interpretations as to how intimate structural relationship the migrating microglia develop with the axotomized neurons. While at lower magnifications these two cells appear to be in direct contact, ultrastructural studies have evidenced that thin astrocytic processes typically separate microglia from the adjacent neuronal membrane (Chen, 1978; Svensson and Aldskogius, 1993). However, Ruan et al (1994) claimed that microglia occupy a perineuronal position before the astrocytic processes begin to extend along non-synaptic areas. It has thus been proposed that microglial cells directly appose the affected neurons just after the injury after which these are replaced by astrocytic processes (Graeber and Kreutzberg, 1988).

Activation of microglia is now known to be the most sensitive molecular sensor in many infections, inflammatory, traumatic, neoplastic, ischemic and degenerative conditions in the CNS and in chronic brain disorders like Alzheimer's disease and other neurodegenerative disorders. The transformation of resting cells into phagocytes happens under strict surveillance *in vivo*. Thus further understanding of mechanisms regulating microglial activation (Box-I) and proliferation (Box-II) *in vivo* may help to develop therapies targeting the potential harmful response in the injured CNS. Dorandeu et al (1998) observed abundance of apoptotic neurons correlated with microglial activation (demonstrated by expression of MHC II) in Fetal Familial Insomnia. Hall et al (1998) investigated the time course of microglial (MHC II antigen reactivity) and astroglial (GFAP immunoreactivity) activation in transgenic mice for ALS. They concluded that the astrocytic activation might exert a trophic influence on the motor neurons. However, the sustained, intense microglial activation may conceivably contribute to the oxidative stress and further damage. If true, then agents, which inhibit microglial activation, may help limit the disease progression.

Microglia are the major producers of IL-1and IL-6 after mechanical injury to rat brain (Giulian et al 1986; Woodroofe et al 1991). The release of IL-1 may stimulate astroglial activation (Giulian and Latchman, 1985) and at this stage delineate the beginning of reactive gliosis and scar formation.

It appears that microglia can be readily activated by inflammatory and infectious agents to generate IL-1. Subsequently, IL-1 induces a whole cascade of activity. These induce the expression of several pro-inflammatory cytokines such as IL-6 and TNF-α (Lee et al 1993b) production of colony stimulating factor (Lee et al 1993c) and induction of inducible form of nitric oxide synthase (iNOS; Lee et al 1993a). This leads to the hypothesis that IL-1 may cause physiological activation of astrocytes in certain inflammatory and infectious conditions of the nervous system, and that this activation may be controlled by microglia through regulation of the production of IL-1.

Activated microglia shows conspicuous alterations, which are associated with the complement system (Svensson et al 1995). The CR_3 receptor is markedly upregulated as also the other complement components like C1 and C1q. In addition C3, C3d, C4 and C9 are also induced. This is thus evident that both the classical and alternative pathways of complement activation are involved. Microglia are the only cellular source for these factors, as the BBB remain intact in such expected conditions.

Even if the target reinnervation is not reestablished the microglial reaction subsides with time. At this stage if the injury is repeated the proliferation and persistent migration of microglia recurs (Summer, 1979).

Graded response of microglia to injury

Ten to 20% of the total glial cells are microglia. Microglia are related to the peripheral tissue macrophages and are derived from hematopoetic or mesodermal cells, which infiltrate into the brain during early development of the central nervous system (Eglitis and Mezey, 1997; Hickey and Kimura, 1988; Hickey and Vass, 1992;

Ling and Wong, 1993). Following injury the microglia present graded response. Raivich et al (1999) in their review on neuroglial activation in injured brain have classified these into five stages, e.g., resting microglia, state of alert, homing, phagosytosis and bystander activation. These stages are described below. Various transformational stages of microglia are presented in Fig. 3.

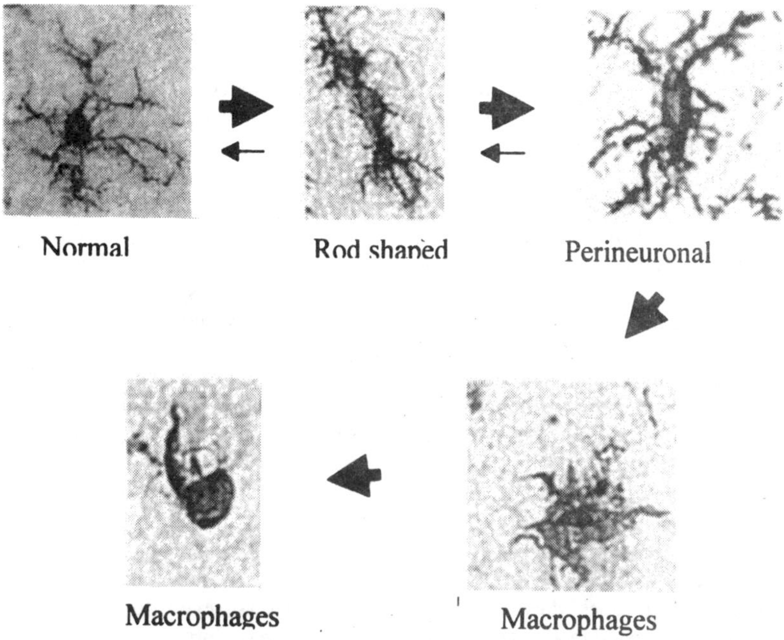

Fig. 3. Transformation of microglia into activated cells following Poly I:C infusion into the brain

Resting microglia: In the normal brain the resting microglia appears to participate in the immune surveillance of the nervous system. The resting microglia have long ramified process. In the white matter they are placed parallel to the nerve fiber. In the grey matter they present a stellate morphology (Compston et al 1997). A highly ramified resting microglia spreads its process over as an area 30-40μm in diameter. While at its peak of ramification they are regarded as performing immune surveillance. An increase in the size of the cell body thinking of proximal process and decrease in ramification of distal branches leads the resting cell to head towards their activated phase. This activation is preceded by molecular events like change in their expression of cell adhesion molecules, cytoskeletal reorganization and antigen presentation

State of alert: The microglia initiates expression of increased amount of the immunological molecules within 24hrs post injury to the neurons. Such microglia are considered to be in a state of alert. Studies have confirmed an increase in microglial IgG immunoreactivity as also increased expression of receptor for complement 3ib fragment. An increase in the important cell surface layout of αMβ2 and αLβ2 including cell adhesive called the ICAM1 is observed. These cells also exhibit rapid expression of the amyloid precursors protein (APP) and these are expected to contribute to the neurodegenerative changes in Alzheimer's patients and other types of dementia.

Homing: In this stage the microglia gets adhered to the damaged structure like lesioned neurons or degenerating neurite terminal. The molecular basis of such activity called 'homing' is under active investigation. The molecules being studied are the chemokines and receptors involved in the process. There is a reduction in ramification (Moller et al 1996). The alert phase αMβ2 and ICAM1are now down regulated (Raivich et al 1998; Werner et al 1998). Now the microglia in fact enter into a biochemically different phase.

These glial cells rapidly come closer to the injured neuronal cell bodies (e.g. following axotomy). Such close apposition of microglia displaces synaptic input and this is a process called synaptic stripping. These microglia have now an increase in expression of receptors for the microglial mitogens, macrophage colony stimulating factor and begin to proliferate with maximal mitotic activity at 2-3 days after axotomy. There is beginning of detection of immunoreactivity for MHC class-I molecules (Raivich et al 1993; Streit et al 1989). Such tight adhesion of microglia to the damaged cells (which are frequently leaky) is helpful as it helps efficient uptake of diffusible molecules and their presentation to infiltrating lymphocytes.

Phagocytosis: Subsequent response of the microglia following homing is dependent upon the presence or absence of cell death. When there is no further damage microglia population goes down, loses activation markers revert to its highly ramified resting phase and moderately express complement receptors.

In presence of neuronal death, microglia further transform into phagocytic cells and remove neuronal debris (Moller et al 1996; Streit and Kreutzberg, 1988). During this phagocytosis stage the microglia express high levels of activation markers already present during state of alert and homing. There is a strong up regulation of MHC class-I molecules (Bohatschek et al 1998; McGeer et al 1993; Simone et al 1995). In the microglia of the grey matter even the MHC class-2 antigen are also present.

There is a difference in phagocytosis in white and grey matter. In grey matter, phagocytosis is faster and microglia remove the cellular debris within a few days and then down regulate their activation markers (Moller et al 1996; Soriano et al 1998). In the other hand, in white matter, phagocytosis and degradation of myelin is an extremely slow process and continue for a longer duration, may be up to years after the initial insult.

Bystander activation: The process of phagocytosis also affects the direct cellular environment of the microglia. Adjacent microglia show an increase in MHC class-I expression, which leads to the formation of blobs of 60-100μm in diameter, with an intensely immunoreactive phagocytic nodule at the center, surrounded by gradually decreasing staining further away from the focal point. What is the function of this bystander activation has attracted attention of neuroscientists. The co-existence of lymphocytes and microglia could be an important immune strategy since a sequential interaction of lymphocytes with phagocytic and non-phagocytic microglia would direct T-cell activation against the antigen present in phagocytic cells.

Sciatic nerve crush lesion: A model for injury and neurodegeneration

The sciatic nerve crush lesion damage has been successfully and repeatedly used as a model for neurodegenerative disorders and neuronal injury (Chattopadhyay, 1997; DeKoning et al 1986; deVrind et al 1993; Haveman et al 1994; Johnson and Deckwerth, 1993; Person and Gatzinsky, 1993). Axonal regeneration ensues following traumatic or mechanically induced axonal degeneration in the peripheral nervous system, resulting in functional recovery (Cajal, 1928). However, the rate of axonal elongation is slow, thus, recovery takes weeks or months depending on the distance between the site of injury and the target tissue (Jacob and Mcquarrie, 1993). Hence, the availability of therapies to speed up regeneration over long distances would be beneficial not only to patients but to society by reducing the associated health care costs.

Experimental lesions have been used widely to induce motoneuron degeneration as a model to test ability of different neurotrophic molecules to prevent lesion associated alterations and/or neurodegenerative disorders. While apoptosis is a prominent feature in neonatal rats, both necrosis and apoptosis occurred in the adult rat following axotomy (Li et al 1998). The craniospinal motoneurons have been extensively used as experimental basis for studying central glial reactions to peripheral nerve injury. These models commonly involve experimental injury to facial, hypoglossal or sciatic nerve. These models are easily standardized and result in highly reproducible central responses.

The pattern and quantum of axotomy associated neuron death is more adverse in immature rats than adults (Tong and Rich, 1997). The apoptotic neuron death of the motor neurons was similar to that induced after trophic factor withdrawal. Ekstrom (1995) have also reported that nerve injury in adult life could lead to nerve death by apoptosis. Sugimoto et al (1998) evidenced apoptotic neuron loss in nerve injury-induced primary neuronal damage in the newborn rats. An age-dependent decline in the ratio of bcl-2 and bax gene products explains the greater susceptibility of immature neurons to apoptosis.

Oliveria et al (1997) demonstrated that spinal motoneurons and sensory dorsal ganglia neurons are lost by apoptosis following sciatic nerve injury. They also found a profound increase in apoptotic interneurons in the deep dorsal horn. It is thus all the more pertinent to explore the search for an agent that could restrict neuron loss by extending survival rate of neurons following axotomy or by restricting the onset of apoptosis.

The current status of drugs for nerve regeneration

Drug development to augment nerve regeneration is an area of intensive research activity. Different types of pharmacologically active substances have been tried but none-of them appear to have attained clinical acceptance without untoward side effects. Several hormones have been tried with variable experimental evidence and limited clinical trials. The widely louded of these are the gangliosides. However, the use of ganglioside has been banned as it develops Guillian-Barre syndrome in some of the patients (Schonhofer, 1991). The other agent isaxomine (Nerfactor) was withdrawn because of hepatotoxicity (Letteron et al 1984). One agent that is in clinical use but needs replacement because of its negative cardiovascular effects, is nimodipine (Murad, 1990) as such it takes almost similar time to fully recover as it may take without a treatment (Angelov et al 1976). We have recently observed the neuroregenerative effects of flunarizine and this needs further investigations (Patro et al 1999). One agent that has attracted attention due to its positive effects is FK506. The claims so far have been impressive and involve the immunophilins. FK506 has also provided basis for searching new compounds, which will have similar, but with non-immunosuppressive properties.

FK506 an immunesuppressant (tacrolimus) is being used for organ transplantation (Hoffman et al 1990). FK506 has been shown to possess two important properties that makes it superior to cyclosporin A (currently most used drug for prevention of allograft rejection). First, FK506 is approximately 10 times more potent immunosuppressant in animals (Todo et al 1987, 1988, 1989) and man thus reducing the instances of transplant rejection (Hoffman et al 1990; McDiarmid et al 1995). Second, initial studies indicate a far less toxicity to humans as compared to that associated with cyclosporin-A (Shapiro et al 1990). Subsequent studies however have indicated its toxic effects depending on the site of transplant (Bronster et al 1995; Klintmalm, 1994; Neuhaus et al 1994; Vincenti et al 1996) in about 3-21% patients. This is being attributed to the tendency to overdose with FK506 (Fung et al 1996). This is how FK506 has become the drug of choice to combat allograft rejection. In addition, FK506 has also been found to possess a variety of neuronal properties, including protection against ischaemic brain injury (Butcher et al 1997; Ide et al 1996; Sharkey and Butcher, 1994; Tokime et al 1996). It is likely that these neuroprotective effects of FK506 are mediated by calcineurin inhibition. FK506 is also expected to inhibit NOS activity thereby reducing formation of NO that has been implicated in mediating neurotoxicity (Dawson et al 1993; Snyder and Sabatini, 1995).

In 1994 FK506 was first reported to have neurotrophism and neuroprotection Subsequently, studies on Immunophilins in the CNS were initiated as FK506 is a prodrug and becomes active once bound to its binding proteins called the FK506 binding proteins (FKBPs). Immunophilins have attracted the attention of neuroscientists because of their presence in higher concentration in the brain as compared to other tissues. We now know of 15 FKBPs. In addition to catalysis of cis-trans confirmation of specific peptidylprolyl bonds (Galat and Riviere, 1998), they also control other cellular activities like gene transcription (e.g. FKBP25, FKBP38), modulation of signaling via the membrane bound epidermal growth factor (FKBP12), intracellular Ca^{2+} release (FKBP12), steroid-hormone-receptor complex formation (FKBP52), and collagen helix formation (cyclosporin A). However, we do not fully understand how the binding of FK506 or its active derivatives alter the function of FKBPs or which FKBPs are involved in the neuroprotective or neuroregenerative effects of FK506.

FK506 and nerve regeneration following injury

Over and above the interesting neuronal properties, FK506 also speeds up nerve regeneration in the peripheral nervous system of the rat with a focal crush lesion of the sciatic nerve (Gold et al 1994, 1995). Under normal circumstances an immunosuppressant is not expected to alter external regeneration since Wallerian degeneration is not an immune mediated event (Griffin et al 1993).

Gold et al (1994, 1995) first reported that systemic administration of FK506 speeds functional recovery following a nerve crush lesion by increasing the rate of axonal regeneration in the sciatic nerve in adult rats. Daily subcutaneous injection of FK506 (1mg/kg) reduced the number of days until the onset of an ability to right the foot and move the toes ("walking"). Wang et al (1997) subsequently worked on the dose dependence of the neuroprotective effects of FK506 and recorded 5mg/kg/i.p./day to be the most effective dose for nerve regeneration. This has initiated studies to understand as to how FK506's neuroprotective action is mediated. Cyclosporin A although shares the immunosuppressive nature of FK506 fails to alter functional recovery or the rate of axonal regeneration (Wang et al 1997). In addition, the neurotrophic effects of cyclosporin A are controversial (Carreau et al 1997; Steiner et al 1997a,b). This leads to an inference that calcineurin inhibition, which is the common mechanism of immunosuppression of both cyclosporin A and FK506, does not increase nerve regeneration. Further the capacity of non-immunosuppressant derivatives of FK506 like V10367 to speed up neuron regeneration without a capacity to inhibit calcineurin ruled out the role of FKBP12 in this process (Snyder and Sabatini, 1995; Cameron et al 1997). This has been answered well in studies with tissues from FKBP12 knockout mice, which retained their responsiveness to the neurite outgrowth promoting property of FK506. Thus FKBP12 is not necessary for FK506 to promote neurite growth.

When FKBP52 were blocked with a mouse monoclonal antibody for FKBP52 in *in vitro* studies this also blocked the ability of FK506 to promote neurite growth for SH-SY54 cells in a concentration dependent manner between 50–100nm (Gold et al 1999). This confirms that FK506's neurite outgrowth promoting factor is totally dependent on its interaction with the immunophilin FKBP52. In the above-mentioned study, surprisingly, FKBP52 antibody itself elicited even longer neurites per unit time as compared to those observed with FK506 (Gold et al 1999). Gold (1999) describes the FKBP52 antibody induced neurite growth as "some of the fastest growing neurites observed to date", and gives a lead that it is perhaps possible to develop immunophilin ligands which can distinguish between FKBP52 or FKBP12 and not both maintaining their ability to increase neurite growth.

GAP-43 (Growth associated protein plays a role in nerve regeneration (Benowitz and Routtenberg, 1997) and daily subcutaneous injection of FK506 to rats has been observed to increase mRNA levels of GAP-43 (Gold et al 1998). However, further studies need to be undertaken. Because the promoter region of the GAP-43 gene contains an AP-1 binding site (Eggen et al 1994), FK506 may increase GAP-43 synthesis via an effect on c-Jun expression and formation of c-Jun homodimers (Herdegen et al 1997). It thus indicates that even a single administration of FK506 alters an important signal transduction pathway regulating nerve regeneration. Herdegen et al (1998, 2000) have reported that FK506 reduced c-Jun expression and apoptosis induced by serum deprivation using cortical neurons in culture. This certainly supports the neuroprotective effects of FK506 (Butcher, 1994; Butcher et al 1997; Dawson et al 1993; Gold, 1999; Ide et al 1996; Kuroda and Siesjo, 1996; Tokime et al 1996; Yagita et al 1996).

Non-immunosuppressant immunophilin ligands as neuroprotectants

Gold et al (1997) demonstrated that FK506-FKBP12 ligands are able to increase nerve regeneration via a calcineurin-independent mechanism. They tested a novel small molecule V10367 that functionally related to the binding domain of FK506, lacks the structural components of the effector domain of FK506 necessary for calcineurin binding. They demonstrated that systemic administration of V10367 speeds functional recovery by accelerating the growth of regenerating axons to the distal musculature following a sciatic nerve crush lesion. They also supported that similar to FK506 there is a reduction in efficacy with high doses. This substance also helps neuron regeneration following oral administration. It is thus possible to separate the nerve regeneration property of FK506 from the immunosuppressant action of FK506-FKBP12 complex. Studies using these novel compounds thereby reveal the potential for developing small molecule FKBP ligands lacking immunosuppressant activity as new therapeutic agents for the treatment of human nerve injuries. FK506 and its FKBP derivatives can be considered as a novel family of neurotrophic and neuroprotective molecules as demonstrated following focal ischaemia (Brecht et al 1998; Sharkey and Butcher, 1994), nerve fiber lesion (Gold et al 1995, 1997, 1998; Lyons et al 1995) and formation or extension of neurites in vitro (Brecht et al 1998; Carreau et al 1997; Steiner et al 1997a,b). It is not known, however, whether FK506 can also protect against axotomy-induced cell death in adult mammalian brain.

The perspective

Axonal regeneration ensues following traumatic or mechanically induced axonal degeneration in the peripheral nervous system resulting in functional recovery. The rate of axonal regeneration being a slow process, recovery takes weeks or months depending on the distance between the site of injury and the target tissues. Hence, the availability of therapies to speed up regeneration over long distances would be beneficial not only to the patients but to the society by reducing the associated health care costs.

The pathological activation of glial cells, which forces microglia to transform into immunocompetent cells with cytotoxic capabilities and induces astroglia to upregulate GFAP amongst other changes, plays an important role in the course and potentially the outcome of neurodegenerative disorders. Activated microglia might induce neuronal apoptosis or *vice versa* the apoptosis of neurons might activate microglia. What is the primary and what is the secondary event, is still unclear. Whether and how interference of microglia influences or possibly even slows down the progressive cell death in neurodegenerative diseases like AD, remains to be answered. It is not yet clear, whether the detrimental effects of inflammation outweigh the neuroprotective mechanisms or the opposite holds true. A better understanding of the mechanisms regulating microglial proliferation and modulating its activation is needed to devise more efficient therapies battling neurodegeneration.

The search for the potential of immunophilin ligands or inhibitors as neuroprotectants and neuroregenerative reagents is a comparatively very recent concept. This is expected to provide novel molecules for protection and regeneration of injury, aging and disease associated degenerating neurons. A recent meeting on Immunophilin in Brain at Frankfurt reveals that FK506 (an immunosuppressant) and its derivative V10367 (a non-immunosuppressant) are both effective in accelerating sciatic nerve regeneration. FK506 enhances the regeneration of central axotomised neurons across the lesion site following spinal cord injury. It is also being evidenced that several other test compounds (Vertex and Gillford compounds) which are FKBP-ligands stimulate neurite outgrowths and suggest the neuroprotection and neuroregenerative potential of novel Vertex compounds (calcineurin-independent) FK506 derivatives.

Only fragmentary data is available on the effects of FK506 or other immunophilin ligands on the non-neural cells in the vicinity of the damaged neurons or the neurons themselves entering the process of apoptosis. The neuroregenerative effects of both immunosuppressant and non-immunosuppressant immunophilin ligands provide a unique opportunity to study the role of astrocytes and microglia in the process of neurodegeneration or neuroprotection following injury. Further information in this subject can offer selective and effective targeting of disorders of the CNS and novel drug therapeutics for neurodegenerative events.

References

Aldskogius H and Kozlova EN (1998) In: Progress on Neurobiology, Elsevier Science Ltd., Great Britain, Vol. 55, pp. 1-26.

Angelov D N, Neiss VF, Streppel M et al (1976) J Neurosci **16**: 1041-1048.

Benowitz L.I and Routtenberg A (1997) Trends Neurosci **20**: 84-91.

Bohatschek M, Jones L.L, Kreutzberg GW and Raivich G (1998) Clin Neuropathol **17**: 286.

Brecht S, Mielke K, Yu MH and Herdegen T (1998) In: Pharmacology of Cerebral Ischemia (ed. Kriegstein, J.), Medpharm, Stuttgart, pp. 151-159.

Bronster DJ, Yonover P, Stein J et al (1995) Transplantation **59**: 1066-1068.

Butcher SP, Henshall DC, Teramura Y et al (1997) J Neurosci **17**: 6939-6946.

Cacabelos R, Alvarez XA, Fernandez-Novoa L et al (1994) Mech. Find Exp Clin Pharmac **16**: 141-151.

Cajal RY (1928) Degeneration and regeneration of the nervous system. Oxford University Press, London.

Cameron AM, Nucifora FC, Fung ET Jr et al (1997) J Biol Chem **272**: 27582-27588.

Carreau A, Gueugnon, J, Benavides J and Vige X (1997) Neuropharmacology **36**: 1755-1762.

Chattopadhyay M (1997) Effect of flunarizine on ageing and injured neurons. Ph.D. Thesis, Jiwaji University, Gwalior, India.

Chen DH (1978) J Comp Neurol **177**: 635-664.

Compston A, Zajicek J, Sussman A et al (1997) J Anat **190**: 161-200.

Dawson TM, Steiner JP, Dawson VL et al (1993) Proc Natl Acad Sci (USA) **90**: 9808-9812.

Deckert-Schluter M, Schluter D, Hof H, Wiestler OD (1995) Neuropathol Appl Neurobiol **21**: 548-551.

Deckwerth TL and Johnson EM (1993) J Cell Biol **123**: 1207-1222.

DeKoning P, Braklee JH and Gispen WH (1986) J Neurosci **74**: 237-246.

deVrind HH, Van den WH, Wondergem J and Haveman J (1993) Int J Radiat Biol **63**: 83-89.

Dickson DW, Lee SC, Mattice LA et al (1993) Glia **7**: 75-83.

Dorandeu A, Wingertsmann L, Chertien F et al (1998) Brain Pathol **8**: 531-537.

Drescher KM, Pease LR, Rodriguez M (1997) Immunol Rev **159**: 177-193.

Eggen BJ.L, Nielander HB, Rensen-de Leeuw MGA et al (1994) Mol Brain Res **23**: 221-234.

Eglitis MA and Mezey E (1997) Proc Natl Acad Sci (USA) **94**: 4080-4085.

Ekstrom PA (1995) Neuroreport **6**: 1029-1032.

Fung JJ, Eliasziw M, Todo S, Jain A et al (1996) J Am Coll Surg **183**: 117-125.

Galat A and Riviere S (1998) Oxford University Press.

Giulian D and Lachman LB (1985) Science **228**: 497-499.

Giulian D, Baker TJ, Shih N and Lachman LB (1986) J Exp Med **164**: 594-604.

Gold BG (1999) Drug Metab Rev **31**: 649-663.

Gold BG, Katoh K and Storm-Dickerson T (1995) J Neurosci **15**: 7505-7516.

Gold BG, Storm-Dickerson T and Austin DR (1994) Restor Neurol Neurosci **6**: 287-296.

Gold BG, Zeleny-Pooley M, Wang MS et al (1997) Exp Neurol **147**: 269-278.

Gold BG, Yew JY and Zeleny-Pooley M (1998) Neuroscience **241**: 25-28.

Gold BG, Densmore W, Shou MM et al (1999) J Pharmacol Exp Therap **289**: 1202-1210.

Graeber MB and Kreutzberg GW (1986) J Neurocytol **15**: 363-373.

Graeber MB and Kreutzberg GW (1988) J Neurocytol **17**: 209-220.
Graeber MB, Bise K and Mehraein P (1993) Acta Neuropathol **86**: 179-181.
Griffin JW, George R and Ho T (1993) J Neuropathol Exp Neurol **52**: 553-560.
Griffin DE, Levine B, Tyor WR and Irani DN (1992) Semin Immunol **4**: 111-119.
Guntinas-Lichius O, Martinez-Portillo F, Lebek J et al (1993) J Neurocytol **26**: 241-248.
Hall ED, Oostveen JA and Gurney ME (1998) Glia **23**: 249-256.
Haveman J, Wondergem J, van Dam WW et al (1994) Neurosci Res Commun **14**: 11-17.
Herdegen T, Claret FX, Kallunki T et al (1998) J Neurosci **18**: 5124-5135.
Herdegen T, Fischer G and Gold BG (2000) TIPS **21**: 3-5.
Herdegen T, Skene P and Bohr M (1997) Trends Neurosci **20**: 227-231.
Hickey WF and Kimura H (1988) Science **239**: 290-292.
Hickey WF and Vass H (1992) J Neuropathol Exp Neurol **51**: 246-256.
Hoffman AL, Makowka L, Banner B et al (1990) Transplantation **49**: 483-490.
Ide T, Morikawa E and Kirino T (1996) Neurosci Lett **204**: 157-160.
Jacob JM and Mcquarrie IG (1993) J Neurobiol **24**: 356-367.
Johnson EM and Deckwerth TL (1993) Annu Rev Neurosci **16**: 31-46.
Klintmalm G (1994) The U.S. N Engl J Med **331**: 1110-1115.
Kreutzberg GW (1996) Trends Neurosci **19**: 312-318.
Kuroda S and Siesjo BK (1996) Neurosci Res Commun **19**: 83-90.
Lee SC, Dickson DW, Liu W and Brosnan CF (1993a) J Neuroimmunol **46**: 19-24.
Lee SC, Liu W, Dickson DW et al (1993b) J Immunol **150**: 2659-2667.
Lee SC, Liu W, Roth P et al (1993c) J Immunol **150**: 594-604.
Letteron P, Fouin-Fortunet H, Tinel M, Danan G et al (1984) J Pharmacol. Exp Ther **229**: 845-850.
Li L, Houenou LJ, Wu W, Lei M, et al (1998) J Comp Neurol **396**: 158-168.
Ling EA and Wong WC (1993) Glia **7**: 9-18.
Lyons WE, Steiner JP, Snyder SH et al (1995) J Neurosci **15**: 2985-2994.
McDiarmid SV, Busuttil RW, Ascher NL et al (1995) Transplantation **59**: 530-536.
McGeer PL, Kawamata T, Walker DG, Akiyama H, Tooyama I And McGeer EG ((1993) Glia **7**: 84-92.
Moller JC, Klein MA, Haas S, Jones LL, Kreutzberg GW and Raivich G (1996) Glia **17**: 121-132.
Murad F (1990) In Goodman and Gilman's: The Pharmacological Basis of Therapeutics (Gilman AG, Rall T, Nies AS and Taylor P eds.) Pergamon Press, New York, pp. 764-783.
Neuhaus P, Pichlmayr R and Williams R (1994) Lancet **344**: 423-428.
Oliveira AL, Risling M, Deckner M et al (1997 Neuroreport **8**: 837-840.
Patro IK and Patro N (1992) Ind Rev Life Sci **12**: 133-146.
Patro N and Patro IK (2004) J Tissue Res **4**: 71-74.
Patro IK, Chattopadhyay M and Patro N (1999) Neurosci. Letters **263**: 97-100.
Perry VH and Gordon S (1991) Int Rev Cytol **125**: 203-244.
Perry VH, Hume DA and Gordon S (1985) Neuroscience **15**: 313-326.
Person HG and Gatzinsky KP (1993) Brain Res **630**: 115-124.
Raivich G, Gehrman J, Graeber MB and Kreutzberg GW (1993) J Histochem Cytochem **41**: 579-592.
Raivich G, Bluethmann H and Kreutzberg GW (1996) Keio J Med **45**: 239-247.
Raivich G, Hass S, Werner A, Klein MA, Kloss CUA and Kreutzberg GW (1998) J Comp Neurol **395**: 342-358.
Raivich G Bohatschek M, Kloss CUA et al (1999) Brain Res Rev **30**: 77-105.
Rohlman A, Laskawi R, Hofer A et al (1994) Neuroreport **12**: 409-412.
Rosenberg ME and Silkensen J (1995) Int J Biochem Cell Biol **27**: 633-645.
Ruan RS, Leong SK and Yeoh KH (1994) Acta Oto-Laryngol **114**: 271-277.
Schonhofer PS (1991) Lancet **338**: 757.
Shapiro R, Fung JJ, Jain AB et al (1990) Transplant Proc **22**: 35-36.
Sharkey J and Butcher SP (1994) Nature **371**: 336-339.
Simone R De Giampaolo A, Giometto B, Gallo P et al (1995) J Neuropathol Exp Neurol **54**: 175-187.
Snyder SH and Sabatini DM (1995) Nature Med **1**: 32-37.
Soriano SG, Wang YF, Lipton SA, Dikkes P et al (1998) Brain Res **780**: 337-341.
Steiner JP, Connolly MA, Valentine HL et al (1997a) Nature Med **3**: 421-428.
Steiner JP, Hamilton GS, Ross DT et al (1997b) Proc Natl Acad (USA) **94**: 2019-2024.
Streit WJ and Kreutzberg GW (1988) J Comp Neurol **268**: 248-263.
Streit WJ Graeber MB and Kreutzberg GW (1989) J Neuroimmunol **21**: 117-123.
Sugimoto T, Xiao C and Ichikawa H (1998) Brain Res **807**: 147-154.
Summer BE (1979) Exp Brain Res **36**: 107-118.

Svensson M and Aldskogius H (1993) Glia **7:** 286-298.
Svensson M, Mattson P and Aldskogius H (1994) J Anat **185**: 537-542.
Svensson M, Liu L, Mattson P et al (1995) Molec Chem Neuropathol **24**: 53-68.
Tetzlaff W, Graeber MB, Bisby MA et al (1988) Glia **1**: 90-95.
Todo S, Podesta L, Chap Chap P et al (1987) Transplant Proc **19**: S64-S67.
Todo S, Ueda Y, Demetris JA et al (1988) Surgery **104**: 239-249.
Todo S, Demetris A, Ueda Y et al (1989) Surgery **106**: 444-451.
Tokime T, Nozaki K and Kikuchi H (1996) Neurosci Lett **206**: 81-84.
Tong JX and Rich KM (1997) J Neurocytol **26**: 339-347.
Torvik A (1972) J Neuropathol. Exp Neurol **31**: 132-146.
Vincenti F, Laskow DA, Neylan JF et al (1996) Transplantation **61**: 1576-1581.
Wang MS, Zeleny-Pooley M. and Gold BG (1997) J Pharmacol Exp Ther **282**: 1084-1093.
Werner A Kloss CUA, Walker J et al (1998) J Neurocytol **27**: 219-232.
Woodroofe MN, Sara GS, Wadhwa M et al (1991) J Neuroimmunol **33**: 227-236.
Yagita Y, Kitagawa K, Matsushita A et al (1996) Life Sci **59**: 1643-1650.

Acknowledgement

Generous gift of Iba1 antibody by Prof. Yoshinori Imai of Japan to the Neuroscience Centre is gratefully acknowledged. We acknowledge the grants from Department of Science & Technology, Govt. of India, New Delhi (to IKP and NP), the Department of Biotechnology, Govt. of India, New Delhi (to IKP) and the Council of Scientific and Industrial Research, New Delhi (to NP) for research in this subject.

Molecular and Cellular Neurobiology
Editors: M.K. Thakur and S. Prasad

Fragile-X-Mental Retardation

Balraj Mittal, Udai Bhan Pandey and Shubha Phadke

Department of Genetics, Sanjay Gandhi Postgraduate Institute of Medical Sciences
Lucknow, India

Abstract
The fragile X syndrome is the commonest cause of inherited mental retardation after Down's syndrome. It is an X-linked semi-dominant condition resulting from hyper-expansion of CGG repeats in the 5' untranslated region of FMR1 gene. Another fragile X condition due to GCC repeat expansion in FMR2 gene is rare. The numbers of normal CGG and GCC repeats in normal individuals vary in different populations and the modal number of 29 and 15 for CGG and GCC repeats respectively in North India is similar to Caucasian populations. The frequencies of fragile X mental retardation among institutional non-specific MR range from 2.5-7.5%. Molecular approaches can help in precise diagnosis, prenatal diagnosis and carrier analysis of fragile X condition. Study of its unique mode of inheritance and unusual features like genetic anticipation serve as a prototype of the triplet repeat disorders. A 'Founder effect' has been proposed to explain the lack of any *de novo* mutations at FMR1 locus. However, two studies from India suggest only weak or insignificant Founder chromosomes in Indian population.

Key Words: triplet repeat disorder, FMR1 gene, founder Effect, CGG and GCC repeat polymorphism

Introduction

Mental retardation is a common problem with major implications for a nation's health, education, and community services affecting 1–3% of the total population (Curry et al 1997). This condition results from failure to develop cognitive abilities and intelligence level appropriate for the age group. The definition of mental retardation (MR) is an intelligence quotient (IQ) of less than 70, with mild MR having IQ between 50-70 and severe MR having IQ less than <50. The prevalence of mild MR is 2-3% and that of severe MR is 0.25-0.8% in the western populations (Vogel and Motulsky 1997). Large diagnostic studies have shown that the etiology remains unknown in 25-50% of the group of severely mentally retarded patients (Van Buggenhout et al 2001). In India, approximately 40% of mentally retarded patients without a known environmental cause are believed to result from genetic defects (ICMR multicentric study 1991). In childhood and adult life the mentally retarded individuals require specialized education or training and supervised care at home or in an institution with attendant personal and economic costs. Caring mentally retarded patients also causes a lot of trauma to the families.

The fragile X syndrome is the commonest cause of inherited mental retardation after Down's syndrome (trisomy 21). About 40% of X linked mental retardation (XLMR) and 4% of all mental retardation cases have been attributed to fragile X syndrome (Frints et al 2002). Although the syndrome mainly manifests in males but up to 30% patients are also females. This is a very subtle dysmorphic syndrome and difficult to diagnose clinically. Long face with prominent mandible, long and mildly dysmorphic ears and macroorchidism are the characteristic features of fragile X syndrome. This phenotype is subtle in young children (Lachiewicz et al 2000) and the features become prominent as the child grows. Fragile X patients have characteristic clinical and behavioral features, hence fragile X syndrome is now placed in X linked syndromic mental retardation.

Fragile X syndrome is the most extensively studied triplet repeat disorder and can be considered as the prototype of the growing number of triplet repeat disorders. CGG and GCC repeats at the FMR1 and FMR2 loci show polymorphism in normal population. Recent surveys of the prevalence of the fragile X syndrome have suggested a incidence of approximately 1 in 4000 maies in general population (Turner et al 1996; Crawford et al 1999). Fragile X mental retardation has been detected in all population and ethnic groups.

The complex inheritance of the fragile X syndrome: the Sherman paradox

The inheritance pattern of fragile X syndrome is complex and atypical for a Mendelian trait. The occurrence of both affected males and females in fragile X families suggests a dominant pattern of inheritance. However, the presence of unaffected males (Fryns et al 1986) who transmit the marker X chromosome to their daughters (also known as normal transmitting males, NTM) point to a mode of inheritance more complicated than a simple X-linked recessive mode (Nielsen et al 1981). Sherman et al (1985) performed large-scale segregation analysis on fragile X syndrome pedigrees and put forward a model of X-linked dominant inheritance with reduced penetrance. Due to absence of de novo cases (with full mutation). The authors argued that each affected male received the mutated gene from his mother. It was proposed that an asymptomatic carrier male is more likely to have grandsons with the disorder than to have brothers with the FXMR. Therefore, the penetrance for the disease increases in succeeding generations of a pedigree (anticipation)—an observation known as the Sherman paradox. The mechanism responsible for the Sherman paradox became clear in 1991 after the cloning of the gene defective in fragile X syndrome.

Cloning and Characterization of the *FMR1* gene

Positional cloning of the gene involved in the fragile X syndrome was accomplished through a combination of linkage studies and the use of somatic cell hybrids (Verkerk et al 1991). The gene was designated *FMR1* (Fragile X Mental Retardation gene 1).

FMR1 gene structure and protein, FMRP

FMR-1 gene is highly conserved orthologs in mammals, chickens, and fruit flies (Wan et al 2000). The gene consists of 17 exons, spanning 38 kb of Xq27.3 (Eichler et al 1994). Full-length fragile X mental retardation protein (FMRP), the protein encoded by *FMR*1, has a molecular weight of 69 kDa, but extensive alternate splicing of exons produces a number of protein isoforms (Eichler et al 1994). Consistent with the primary features of the phenotype, *FMR1* mRNA and protein are highly expressed in testis and in fetal and adult brain, with the majority of signal localized to neurons (Abitbol et al 1993).

In situ hybridization and histochemical staining of human and mouse tissues have indicated that FMRP is widely expressed, but most abundant in neurons (Devys et al 1993). The cellular localization of the FMRP is mainly in the cytoplasm, but some

FMRP has been found in the nucleolus (Willemsen et al 1996). The deduced amino acid sequence of FMRP reveals two K homology (KH) domains and a RGG box.

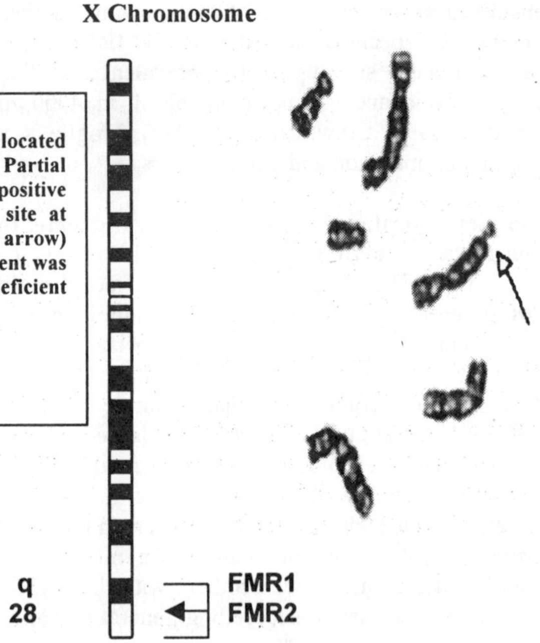

Fig. 1.
FMR1 and FMR2 genes are located at the tip of X chromosome. Partial karyotype of fragile X positive patient showing the fragile site at Xq27.3 position (indicated by arrow) when lymphocytes of the patient was cultured in folic acid deficient medium.

These sequence motifs indicate that FMRP is an RNA-binding protein (Siomi et al 1993). It has been shown that FMRP is able to bind mRNA *in vitro* and that mutations located in one of the two KH domains (I241N or I304N) result in FMRP proteins with impaired RNA binding (Siomi et al 1993). FMRP has been shown to associate with polyribosomes via RNA (Eberhart et al 1996; Mittal and Pandey 2002) indicating a role in the translational machinery. When FMRP found to be absent, the mRNAs normally associated with FMRP-mRNP complexes might be transitionally misregulated, which would lead to impaired synaptic plasticity and cognitive deficit. Recently, many targets of FMRP were identified by micro array analysis.

Disease Mechanism

In the normal population, FMR1 allele sizes varies between 6 and 52 CGG, and in carriers it is in between 52 and 200 CGG (premutation stage). More then 200 CGG repeats (full mutation) are almost always found to be associated with mental retardation (Figure 2). Full mutations are detected in individuals affected by the fragile X syndrome and also in a proportion of unaffected carrier females. There is not yet universal agreement as to the number of repeats that should define the lower limit for the premutation range although most screening studies have used 52 repeats as the lower bound (Hagerman and Hagerman 2002). The association between methylation status of the putative CpG island, located upstream of *FMR*1, and the fragile X syndrome, indicated that suppression of *FMR1* expression could be causing the disease (Vincent et al 1991). Indeed, absence of *FMR1* mRNA transcription in full mutated fragile X males, was demonstrated by Pieretti et al (1991) and methylation status of the CpG island has subsequently been directly correlated to *FMR1* transcription both *in vivo* and *in vitro* (Chiurazzi et al 1998).

Instability of CGG repeats

The premutation alleles of FMR1 gene are unstable. Their instability is seen in virtually every transmission when passing to the offspring, size increases being much more common than rare contractions of the repeat. One important factor that influences expansion is the gender of the parent transmitting the unstable CGG repeat. A premutation expands to a full mutation in offspring with abnormal methylation only when it is transmitted by a female (Snowet al 1993). The larger

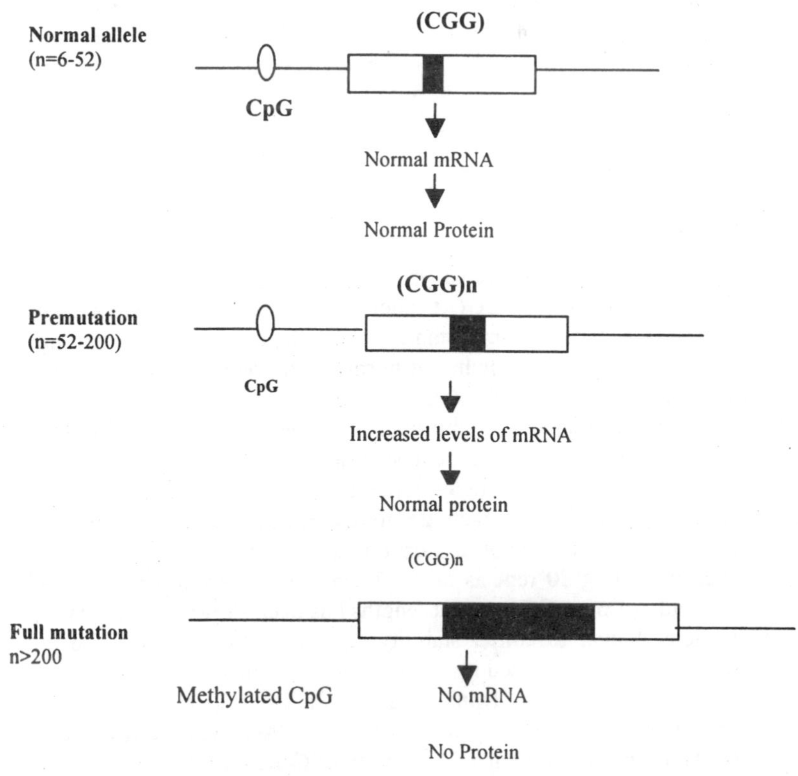

Fig. 2. The molecular mechanism behind the fragile X mutation: In the normal population, the polymorphic (CGG)n trinucleotide repeat region ranges from 5-50 CCG in size; Expansion of the repeat results in highly unstable premutation (50-200 CGG); During inheritance, the premutation expands further to the full mutation (>200 CGG), which leads to hypermethylation of a CpG island located 250 bp 5' to the (CGG)n region. Methylation of the CpG island results in suppression of FMR1 mRNA transcription.

the repeat number in premutation, the more unstable it is upon transmission. Moreover, the risk of transition of a premutation to a full mutation, and resulting risk of having an affected child has been shown to depend strongly on the repeat number of the maternal premutation (Fu et al 1991). In contrast to females, males with premutation have never been reported to have daughters with a full mutation (Yu et al 1992) and only daughters having a premutation have been observed in the offspring of those rare full mutation males who have reproduced (Hori et al 1993). Consistently, it has been shown that full mutation males have only premutations in their sperm (Reyniers et al 1993). Full mutations of the *FMR1* gene are also unstable upon transmission. They usually show further increases in size, reductions of size occurring more rarely (Yu et al 1992). In addition to meiotic instability, full mutations show great somatic instability, so that the CGG repeat lengths vary within and between tissues of an individual. This means that individuals with a full mutation are somatic mosaics as regards repeat size. The size of full mutations can vary greatly in different members of the same sibship, but familial clustering of similar full mutation lengths has also been reported (Nolin et al 1996).

There is no exact definition for the length at which the CGG repeat becomes unstable, because of the considerable overlap between repeat lengths at the high-end range found in the general population and the small premutation alleles detected in fragile X families (Fu et al 1991, Snow et al 1993). The critical region of overlap between normal and premutation alleles appears to be approximately 50 repeats, but in the range of 34-59 repeats, termed the "grey zone", both stable and unstable inheritance of alleles have been reported (Eichler et al 1994, Nolin et al 1996). The unstable alleles in the "grey zone" may represent alleles predisposed to further expansion and/or premutations, which have yet to progress to the full mutation. Further analyses of the structure of the *CGG* trinucleotide repeat region have shown that although the total repeat length is important, it is not the sole factor accounting for instability of the *FMR1* gene. Sequencing of normal and premutation alleles has revealed that the number and position of AGG interruptions within the CGG repeat sequence may significantly influence the stability of the repeat, and in particular, the length of an uninterrupted CGG repeat sequence appears to be an important determinant of instability. Most normal alleles contain one or more regularly spaced AGG units and the uninterrupted CGG tracts do not exceed 30 repeats. In contrast, premutation alleles typically have none or at most one interspersing AGG triplet and a CGG copy number exceeding 30 repeats at the 3' end of the repeat array. There also the highest variability of pure CGG repeat lengths has been shown to take place. The AGG interspersions appear to confer stability and their absence gives rise to longer perfect CGG arrays with increased instability and predisposition to expansion. (Eichler et al 1994, Snow et al 1994, Kunst et al 1997). Nolin et al (1996) have further suggested that in addition to the repeat length and the AGG content, an as yet unidentified familial factor might influence the stability of CGG repeats.

Origin of fragile X chromosomes: Insights into the mutational history of fragile X chromosomes

There is no single case of fragile X mental retardation reported so far where direct transition from normal to full mutation occurred. Therefore, an ancestral mutation has been proposed suggesting that most of the fragile X chromosomes originatedas from small number of founder chromosomes (Macpherson et al 1994). Several groups have used marker-allele analysis as an indirect tool to compare (CGG)n alleles among fragile X patients with alleles in the normal population (Oudet et al 1993; Kunst and Warren 1994; Macpherson et al 1994; Gunter et al 1998; Mingroni-Netto et al 1999). Most of these studies revealed significant association (linkage disequilibrium) between fragile X full mutations and specific polymorphic markers. The most pronounced linkage disequilibrium has been observed in genetically homogenous Finnish population in which 80% of the fragile X full mutations were associated with DXS548-FRAXAC2 haplotype 6-3.

FRAXE syndrome

Another nearby locus FMR2 gene located at Xq28 position, 600 kb distal to the FMR-1 gene has also been found to be responsible for rare type of nonspecific mental retardation which is different from FMR1 mental retardation. Cytogenetic fragile site is also observed in patients with FRAXE syndrome. Molecular characterization revealed that individuals expressing fragile site at Xq28 position (FRAXE) had amplification of a GCC repeat adjacent to a CpG island. This condition of trinucleotide repeat amplification also belongs to the class of dynamic mutations where number of GCC repeats increase in successive generations with decrease in age of onset (Genetic anticipation). Normal individuals show 6-25 GCC repeats while individuals expressing the fragile site have >200 copies and their CpG island is fully methylated. In males the full mutation associated with a variable degree of learning difficulty, whilst females.appear unaffected. Sequences adjacent to the repeat are highly conserved across animal species and showed mRNA transcripts on Northern blot hybridization. The FMR-2 gene encodes a protein of 1311 amino acids, which probably functions as transcription transactivator (Gecz et al 1997).

The frequency of FMR-2 mutations may be very low because large screening programs in different candidate populations of mentally retarded boys detected none (Allingham-Hawkins et al., 1995; Syrrou et al 1998) or only a few cases of FRAXE MR (Holden et al 1996). Based on these studies, the prevalence of FRAXE was estimated to be approximately 1:50000 males (Brown et al 1996).

Fragile X mental retardation in India

India shares large burden of mental retardation but a large number of such individuals are not characterized. In a multicentric large study on genetic causes of mental retardation in India conducted by ICMR to determine the extent and pattern of genetic causes of mental retardation among children in different parts of India, 42.3 per cent patients had mild, 25.3 per cent moderate, 19.2 per cent severe and 13.1 per cent profound mental retardation (ICMR multicentric study. 1991). There are several reports from India about the screening of fragile X mental retardation mainly performed on mentally retarded individuals from institutions (Baskaran et al 1998; Jain et al 1998; Sharma et al 2001; Saha et al 2001; Pandey et al 2002, Pandey et al, 2004). Earlier study based on cytogenetic analysis, conducted among 300 MR patients from western India estimated the frequency to be approximately 2% (Murthy and Kar 1991).

Molecular study performed at a genetic referral center in New Delhi (north India) reported a frequency of fragile X syndrome 5.3% among children with unexplained mental retardation (Jain et al 1998). Baskaran et al (1998) found a prevalence of 6.8 % of fragile X in south Indian population, while in a study performed in West Bengal (eastern India) 7% frequency was observed (Saha et al 2001). A recent study done from New Delhi suggests that 7.7 % of mentally retarded patients have fragile X syndrome (Sharma et al 2001). We screened 160 individuals (128 males and 37 females) with unexplained mental retardation for the CGG/CCG expansion in the *FMR1* and *FMR2* genes. This group was identified from 368 mentally retarded cases after clinical examination. Cases with obvious causes of MR (like Down syndrome) were excluded from the study. Using PCR and Southern hybridization, three males showed full mutation at FMR-1 locus but none of females were positive. Thus, the prevalence of Fragile X mental retardation in mentally retarded males without obvious cause in our study is 2.4%. One female was identified with approximately 60 CGG repeats at her one of X chromosomes and other X chromosome was found to be normal on Southern blot analysis (Hind III+ Eag I digest).

Our study at Lucknow indicates significantly lower prevalence of fragile X in Indian patients with nonspecific MR. The difference may be due to different selection criteria of patients used in various studies. Preselection criteria based on clinical scoring can influence the frequency of fragile X mental retardation. Therefore, it is difficult to comment whether there is really any difference in the prevalence of fragile X syndrome in different regions of India. One of the main limitations of all these studies in India is that many of the patients particularly with mild to moderate mental retardation do not attend any special school for economic or social reasons. However, large surveys all over world generally indicate 2-5% prevalence of fragile X mental retardation (Youings et al 2000).

There are only two studies where FMR2 gene was analysed for GCC hyperexpansion among patients with unexplained mental retardation. However, not a single case of FRAXE was found to be positive in two published studies from India (Sharma et al., 2001; Pandey et al 2002), which reiterated that this syndrome is very rare among individuals with unexplained MR (Holden et al 1996).

CGG AND GCC Repeat polymorphism in Indian population

The trinucleotide repeat numbers of the FMR1 and FMR2 genes are polymorphic in humans. The CGG repeat sequence involved in the manifestation of Fragile X syndrome shows marked variation in normal individuals and in fragile X patients.

There are four different published reports from different parts of India showing different model alleles at FMR1 locus (Table 1). A study done in South Indian population revealed 31 CGG repeats as the most common FMR1 allele (Baskaran et al 1998). In order to assess the extent of variation of the CGG repeat in the population from the eastern region of India, Saha et al (2001) identified 21 distinct alleles ranging in size from 8 to 44 CGG repeats and a repeat size of 28 was the most frequent. Another study from northern part of India (New Delhi) found a total of 26 distinct alleles ranging from 19 to 50 CGG repeats and the most frequent allele size in this population had 29 CGG repeats (in 35.3% of the X chromosomes), with 28 repeats (27.9%) being the next frequent. We have observed nineteen different alleles at the FMR1 locus that ranged from 9-37 repeats in our study done in at Lucknow,

Model repeat number was 29 CGG repeats in our studied samples (Pandey et al 2002).

There are only two published studies on the GCC repeat distribution pattern at FMR2 locus. New Delhi study revealed a total of 20 different alleles in the range of 3–27 GCC and the modal allele was 15 GCC repeats (in 30.1% of the X chromosomes) with 18 repeats (18.7%) being the next most frequent. We have analyzed 330 X-chromosomes from normal individuals and 38% chromosomes had 15 GCC repeats. The second most common allele was 14 GCC repeats (11.34%). There were 18 different alleles ranging in size from 8 – 26 repeats at FMR2 locus (Pandey et al 2002). Most of the studies in Caucasian population reported 15 GCC repeats as the model repeat number at FMR2 locus.

The FMR1 allelic frequency in Indian population is overall similar to other populations, with minor differences compared to the other Asian populations. There are some differences in repeat number in three regions (North, south and eastern India). The modal repeat number in several populations of European ancestry was reported as 30 (Chiurazzi et al 1996a) or 29 (Fu et al 1991; Jacobs et al 1993). CGG distribution pattern at the FMR1 gene in African population clearly showed that the African distribution had two modes, with substantial proportion of alleles at 30 repeats (27%) and 29 repeats (27%), coinciding with the European and Asian modes, respectively (Chiurazzi et al 1996b).

Founder father effect in Indian population:

As mentioned earlier, due to absence of any de novo mutation among fragile X families, it was hypothesized that the mutations existing in the present fragile X population are derived from a few founder mutations many generations ago (so called founder-effect). The hypothesis has been tested by studying the linkage disequilibrium of flanking polymorphic loci between normal and fragile X individuals in a population. There are only two studies (Sharma et al 2002; unpublished data) carried out so far to find out the founder father effects in genetically complex Indian population. Microsatellite markers flanking the FMR1 gene, FRAXAC1, DXS548 and ATL1 have been studied in normal individuals and fragile X positive patients. Both of the studies have been done in north India.

Four alleles of FRAXAC1 were observed in Indian samples (in both studies) as compared to five in the many Caucasians (Chiurazzi et al 1996a) and only two in Chinese populations (Jara et al. 1998). Comparative analysis of FRAXAC1 marker distribution pattern in normal world population was similar to Italian and White American populations. However, substantial differences in FRAXAC1 distribution pattern were observed as compared to normal Chinese and Japanese populations. FRAXAC1-C allele was the most common in studies done at two different northern Indian samples (Sharma et al., 2003) and our study) in normal subjects as well as in fragile X patients. The FRAXAC1-A allele and the DXS548-2 allele were frequently found on FMR1 founder chromosomes in Europeans (Macpherson et al 1994; Chiurazzi et al 1996a). However, there was no significant difference in the distribution pattern of these markers in normal and fragile X chromosomes in India.

DXS548-7 was the most frequently observed allele accounting for more than 45% of the fragile X as well as normal chromosomes in mainly coming from northern

Table 1. Comparison of (CGG)n repeat characteristics in FMR1 among few populations(A) Fu et al., 1991; (B) Pergolizzi et al., 1992; (C) Brown et al., 1993; (D) Kremer et al., 1991.

Country	X Chromosomes	Repeat Range	Allelic variants	Mode	Reference	Method
U S (mixed)	492	6-54	31	29	Fu et al 1991	A
U K	309	13-61	30	29	Jacobs et al 1993	A
U S (white)	444	13-61	33	30	Snow et al 1993	B
Japan	125	13-39	15	28	Richards et al 1994	D
Central China	123	21-43	12	29	Zhong et al 1994a	C
Italy	141	18-47	26	30	Chiurazzi et al 1996b	A
South China	117	17-37	16	29	Chen et al 1997	A
Chile	192	19-44	19	30	Jara et al 1998	A
S. India	161	8-44	17	31	Baskaran etal 1998	A
Cyprus	600	12-59	34	29.9	Patsalis et al 1999b	C
China {Hong kong}	1404	19-54	34	29	Pang et al. 1999	C
N. India	368	19-50	26	29	Sharma et al 2001	A
N. India	327	9-37	26	29	Pandey et al 2002	A

India. This allele was the most common in different world population studied so far. Intermediate alleles DXS548-6.5 was rare in our samples as well as in European population but more of these alleles have been observed in normal black African population (Chiurazzi et al 1996b). In contrast to the findings of Sharma et al (2003), we observed a rare DXS548-6.5 allele in our sample population. These alleles are generated by single base polymorphism of the distal cytosine residues in the $(C)_4G(C)_{11}$ tract (Chiurazzi et al 1996a). Our study does not support the hypothesis of Sharma et al (2003) that these alleles are rare in ethnically younger populations like ours and only restricted to genetically ancient groups such as Africans. No clear evidence of linkage disequilibrium with fragile X chromosomes at DXS548 allele has been observed among five different studies (Buyle et al 1993; Oudet et al 1993; Malmgren et al 1994; Zhong et al 1994a; Chiurazzi et al 1996a, b) as well as in our studies from India.

ATL1- G allele has been shown to be an ancestral allele as it is present in 96% of the isolated Africans (Gunter et al 1998). In Caucasians, the G allele is over-represented in fragile X syndrome. The ATL1-G allele was more frequent than ATL1-A in our population in fragile X and normal individuals, and no significant disequilibrium could be ascertained.

By combining data from individual markers, various haplotypes could be constructed. DXS548-FRAXAC1 is the most frequently studied haplotype with an observed near universal predominance of the 7-C haplotype among control X chromosomes from white Caucasians and African-Americans (Macpherson et al 1994; Rousseau et al 1995; Crawford et al 2000). The 7-C is also modal haplotype in India (Sharma et al 2003; our data). However, 7-D haplotypes are most prevalent among normal X chromosomes in Pacific Asian populations of China and Thailand (Zhong et al 1999; Limprasert et al 2001).

Haplotypes 2-A was observed in 11.8% of our fragile X samples as compared to 4.8% in normal X chromosomes (Table 2). Earlier 2-A haplotype was found to be associated with Caucasian fragile X chromosomes (Chiurazzi et al., 1996a) but neither our study nor study done by Sharma et al (2003) found any linkage disequilibrium of 2-A haplotype in between fragile X and controls. Sharma et al (2003) reported positive association of 2-C haplotype of DXS548-FRAXAC1 with fragile X chromosomes (2-C haplotype was present in 2.5% of fragile X positive chromosomes but none of normal chromosomes). In contrast, distribution of 2-C haplotype was almost same in our fragile X (7.1%) and normal subjects (8.8%). Hence, in our samples this haplotype was not associated with fragile X positive chromosomes (Table 2). The findings from India apparently contrast with many other reports on fragile X founder father effects in different ethnic groups (Churazzi et al 1996a,b). There are also reports (Pesso et al 1997; Limprasert et al 2001) showing no founder effects in a specific population, that of Ashkenazi Jews and Thai population. The study from Delhi (Sharma et al 2003) showed weak founder father effects in fragile X patients while our results do not suggest any founder effect.

From the above results it is difficult to conclusively prove whether founder effect has played any role in the origin of fragile X chromosomes in the Indian population. The population heterogeneity and relatively smaller sample size of fragile X chromosomes can affect the outcome of our study. In fact, there has been no protective or at risk haplotypes identified in any Asian population which may also

Table 2. **Comparison of selective DXS548-FRAXAC1 haplotypes in different fragile X and normal populations**

Population/ No. of Fragile X (normal) individuals	DXS548- FRAXAC1 haplotypes given as percent in fragile X (normal) chromosomes						
	7-D	7- C	6- D	6-C	2-A	2-C	7-A
India, Lucknow 28 (122)	7.2 (3.2)	50.0 (44.0)	4.8 (7.2)	10.7 (4.8)	11.2 (4.8)	7.1 (8.8)	1.2 (1.6)
India, Delhi 40 (262)	12.5 (2.3)	30.4 (33.2)	5.0 (3.0)	11.0 (6.1)	12.5 (4.0)	2.5 0.0	2.5 0.0
Chinese 24 (206)	20.80 (54.9)	4.2 (26.2)	62.5 (9.7)	8.3 (2.4)			
Thai 25 (125)	64.0 (60.0)	24.0 (29.6)	0.0 (4.0)	8.0 (1.6)			
White Americans 152 (564)	2.6 (6.9)	20.4 (62.2)	32.9 (7.1)	4.6 (6.1)	23.0 (3.9)	3.2 (1.9)	3.2 (1.4)
Finn 36 (54)	0.0 (13:0)	0.0 (55.6)	11.1 (13.0)	83.3 (3.7)	5.6 (5.6)		
Italian 125 (202)	4.0 (9.9)	20.8 (63.9)	16.8 (7.4)	1.6 (5.4)	24.0 (2.5)		
Czech 35 (33)	0 (3.0)	8.6 (66.7)	17.1 (0)	11.4 (6.1)	22.9 (6.1)		

reflect paucity of clinical cases in less affluent Asian nations or microsatellite alleles with low polymorphism information contents. As haplotype associations are the molecular 'remnants' of founder effects in a population, it is expected that 'at risk haplotype' would be determined in Indian population once sufficient polymorphic markers and clinical cases are identified.

References

Abitbol M, Menini C, Delezoide AL, Rhyner T, Vekemans M and Mallet J (1993) Nat Genet **4**: 147-153.

Allingham-Hawkins DJ and Ray PN (1995) Am J Hum Genet **57**: 72-76

Baskaran S, Naseerullah MK, Manjunatha KR, Chetan GK, Arthi R, Rao GV, Girimaji SR, Srinath S, Sheshadri S, Devi RR and Brahmachari V. (1998) J Med Res **107**: 29-36.

Brown WT, Houck G, Jeziorowska A, Levinson FN, Ding X, Dobkin Cet al (1993) JAMA **270**: 1569-75

Brown WT (1996) Am J Hum Genet **58**: 903

Buyle S, Reyniers E, Vits L, De Boulle K, Handig I, Wuyts FL, Deelen W, Halley DJ, Oostra BA and Willems PJ (1993) Hum Genet **92**: 269-272.

Chiurazzi P, Destro-Bisol G, Genuardi M, Oostra BA, Spedini G and Neri G. (1996a) Am J Med Genet **64**: 216-219

Chiurazzi P, Genuardi M, Kozak L, Giovannucci-Uzielli ML, Bussani C, Dagna-Bricarelli F, Grasso M, Perroni L, Sebastio G, Sperandeo MP, Oostra BA and Neri G (1996b) Am J Med Genet **64**: 209-215.

Chiurazzi P, Pomponi MG, Willemsen R, Oostra BA and Neri G. (1998) Hum Mol Genet **7**: 109-113.

Crawford DC, Meadows KL, Newman JL, Taft LF, Pettay DL, Gold LB, Hersey SJ, Hinkle EF, Stanfield ML, Holmgreen P, Yeargin-Allsopp M, Boyle C and Sherman SL (1999) Am J Hum Genet **64**: 495-507

Crawford DC, Wilson B and Sherman SL (2000) Hum Mol Genet **9**: 2909-2918.

Curry CJ, Stevenson RE, Aughton D, Byrne J, Carey JC, Cassidy S, Cunniff C, Graham JM Jr, Jones MC, Kaback MM, Moeschler J, Schaefer GB, Schwartz S, Tarleton J & Opitz J (1997) Am J Med Genet **72**: 468.

Devys D, Lutz Y, Rouyer N, Bellocq JP and Mandel JL (1993) Nat Genet**4**: 335-340.

Eberhart DE, Malter HE, Feng Y and Warren ST (1996) Hum Mol Genet**5**: 1083-1091.

Eichler EE, Richards S, Gibbs RA and Nelson DL (1994) Hum Mol Genet **3**: 684-685.

Frints SG, Froyen G, Marynen P, Fryns JP (2002) Clin Genet**62**: 423-32.

Fryns JP, Kleczkowska A, Dereymaeker A, Hoefnagels M, Heremans G, Marien J and van den Berghe H (1986) Clin Genet **30**: 315-323.

Fu YH, Kuhl DP, Pizzuti A, Pieretti M, Sutcliffe JS, Richards S, Verkerk AJ, Holden JJ, Fenwick RG Jr, Warren ST, et al (1991) Cell **67**:1047-1058.

Gecz J, Oostra BA, Hockey A, Carbonell P, Turner G, Haan EA, Sutherland GR and Mulley JC (1997) Hum Mol Genet **6**: 435-441

Gunter C, Paradee W, Crawford DC, Meadows KA, Newman J, Kunst CB, Nelson DL, Schwartz C, Murray A, Macpherson JN, Sherman SL and Warren ST (1998) Hum Mol Genet**7**: 1935-1946

Holden JJA, Julien-Inalsingh C, Chalifoux M, Wing M, Scott E, Fidler K, Swift I, Maidment B, Knight SJ, Davies KE and White BN (1996) Am J Med Genet**64**: 420-423.

Hori T, Yamauchi M, Seki N, Tsuji S, Kondo I (1993).Clin Genet**43**:34-8

ICMR multicentric study.(1991) Indian J Med Res**94**:161-9

Jacobs PA, Bullman H, Macpherson J, et al (1993) J Med Genet **30**: 454-459

Jain U, Verma IC and Kapoor AK (1998) Indian J Med Res **108**: 12-16

Jara L, Aspillaga M, Avendano I, Obreque V, Blanco R and Valenzuela CY. (1998) Am J Med Genet **75**: 277-282.

Kremer, EJ, Yu S, Pritchard M, Nagaraja R, Heitz D, Lynch M, et al (1991) Am J Hum Genet **49**: 656-661.

Kunst CB and Warren ST (1994) Cell **77**: 853-861.

Kunst CB, Leeflang EP, Iber JC, Arnheim N and Warren ST (1997) J Med Genet**34**: 627-631.

Lachiewicz AM, Dawson VV, Spiridigliozzi UA (2000) Am J Med Genet**92**: 229-36.

Limprasert P, Saechan V, Ruangdaraganon N, Sura T, Vasiknanote P, Jaruratanasirikul S and Brown WT (2001) Am J Med Genet **98**: 224-229

Macpherson JN, Bullman H, Youings SA and Jacobs PA (1994) Hum Mol Genet**3**: 399-405.

Malmgren H, Gustavson KH, Oudet C, Holmgren G, Pettersson U and Dahl N (1994) Eur J Hum Genet **2**:103-109.

Mingroni-Netto RC, Costa SS, Angeli CB and Vianna-Morgante AM (1999) Am J Med Genet **84**:204-207.

Mittal B and Pandey UB (2002) Clinical Genetics **62**: 191-195.

Murthy SK and Kar B (1991) Am J Hum Genet **42** (suppl 49):270

Nielsen KB, Tommerup N, Poulsen H, Mikkelsen M (1981)Hum Genet**59**: 23-5.

Nolin SL, Lewis FA 3rd, Ye LL, Houck GE Jr, Glicksman AE, Limprasert P, Li SY, Zhong N, Ashley AE, Feingold E, Sherman SL and Brown WT (1996) Am J Hum Genet**59**: 252-1261.

Oudet C, von Koskull H, Nordstrom AM, Peippo M and Mandel JL (1993) Eur J Hum Genet**1**:181-189

Pandey UB, Phadke SR and Mittal B (2002) Genet Test **6**: 335-339.

Pandey UB, Phadke SR and Mittal B (2003) Neurology Ind**51**: 367-369.

Pergolozzi RG, Erster SH, Goonewardena P and Brown WT (1992) Lancet**339**: 271-272.

Pesso R, Barkai G, Ravia Y, Gak E, Frydman M, Goldman B and Friedman E (1997) Hum Genet **101**: 186-189

Pieretti M, Zhang FP, Fu YH, Warren ST, Oostra BA, Caskey CT and Nelson DL (1991) Cell **66**: 817-822.

Reyniers E, Vits L, De Boulle K, Van Roy B, Van Velzen D, de Graaff E, Verkerk AJ, Jorens HZ, Darby JK, Oostra B et al (1993) Nat Genet **4**: 143-146.

Richards RI, Kondo I, Holman K, Yamauchi M, Seki N, Kishi K, Staples A, Sutherland GR and Hori T (1994) Am J Med Genet **51**: 412-416.

Rousseau F, Rouillard P, Morel ML, Khandjian EW and Morgan K (1995) Am J Hum Genet **57**:1006-1018

Saha S, Karmakar P, Chatterjee C, Banerjee D, Das S and Dasgupta UB (2001) Ann Clin Biochem **38**:264-271
Sharma D, Gupta M and Thelma BK (2001) Genet Epidemiol **20**: 129-144
Sharma D, Gupta M and Thelma BK (2003) Hum Genet **112**: 262-271.
Sherman SL, Jacobs P, Morton N, Froster-Iskenius U, Howard-Peebles P, Nielsen K, Partington N & Sutherland G (1985) Hum Genet **69**: 289-299.
Siomi H, Siomi MC, Nussbaum RL and Dreyfuss G (1993) Cell **74**: 291-298.
Snow K, Doud LK, Hagerman R, Pergolizzi RG, Erster SH and Thibodeau SN (1993) Am J Hum Genet **53**: 1217-1228
Syrrou M, Georgiou I, Grigoriadou M, Petersen MB, Kitsiou S, Pagoulatos G and Patsalis PC (1998) Genet Epidemiol **15**:103-109.
Turner G, Daniel A and Frost M (1980) J Pediatr **96**: 837-841.
Turner G, Webb T, Wake S and Robinson H (1996) Am J Med Genet **12**; 64: 196-197
Van Buggenhout GJ, Trommelen JC, Brunner HG, Hamel BC and Fryns J (2001) AnnGenet **44**: 47-55.
Verkerk AJ, Pieretti M, Sutcliffe JS, Fu YH, Kuhl DP, Pizzuti A, Reiner O, Richards S, Victoria MF, Zhang FP, et al (1991) Cell **65**: 905-914.
Vogel F and Motulsky AG (1997) Human Genetics, problems and approaches, 3rd ed, Springer, Berlin
Wan L, Dockendorff TC, Jongens TA and Dreyfuss G (2000) Mol Cell Biol **20**: 8536-8547.
Willemsen R, Oosterwijk JC, Los FJ, Galjaard H and Oostra BA (1996) Lancet **348**: 967-968.
Youings SA, Murray A, Dennis N, Ennis S, Lewis C, McKechnie N, Pound M, Sharrock A and Jacobs P (2000) J Med Genet **37**: 415-421.
Yu S, Pritchard M, Kremer E, Lynch M, Nancarrow J, Baker E, Holman K, Mulley JC, Warren ST, Schlessinger D, et al (1991) Science **252**: 1179-1181.
Zhong N, Ju W, Xu W, Ye L, Shen Y, Wu G, Chen SH, Jin R, Hu XF, Yang A, Liu X, Poon P, Pang C, Zheng Y, Song L, Zhao P, Fu B, Gu H and Brown WT (1999) Am J Med Genet **84**: 191-194.
Zhong N, Liu X, Gou S, Houck GE Jr, Li S, Dobkin C and Brown WT (1994) Am J Med Genet **51**: 417-422

Molecular and Cellular Neurobiology
Editors: M.K. Thakur and S. Prasad

Genetic Variability Leading to Narcotic Abuse: Potential Implication of Single Nucleotide Polymorphisms

Ishani Deb and Sumantra Das

Neurobiology Division, Indian Institute of Chemical Biology, 4 Raja S.C. Mullick Road
Kolkata-700 032, India

Abstract
Addiction to narcotics is an important social hazard and stems from a physiological disorder in the human brain that causes the addicted individual to crave and continue to use the substance of abuse despite the risk of physical or psychological harm. Although the biochemical factors that contribute towards narcotic abuse are still poorly understood, evidences indicate a significant role of heredity in narcotic addiction in addition to social and environmental influences. Such a genetic predisposition in narcotic abuse disorders has prompted examination of the various candidate genes for association with the substance of abuse with a view to better understand their pathogenesis with the ultimate objective of improvement of preventive strategies, diagnostic tools and treatment. Single nucleotide polymorphisms (SNPs) are being used to examine the chromosomal mapping and eventual cloning of genetic determinants of complex quantitative phenotypes. In this chapter we have reviewed the various studies on the possible correlation between SNPs of candidate genes with narcotic tolerance and dependence for a better understanding of this complex disorder of the brain.

Keywords: Narcotic addiction, mesolimbic, opioid, gene, SNP

Introduction

Opioid drugs, also called opiates or narcotics, have been used since the beginning of recorded history. Opium, from which most of these opioids were originally discovered, is derived from a white liquid produced by the poppy plant, first cultivated in the Middle East and Asia. It was used therapeutically in ancient cultures to induce calm and to relieve pain, and recreationally to induce euphoric dream states. Its main alkaloid, morphine is the most widely prescribed narcotics for pain relief in addition to other natural compounds like codeine, methadone, meperidine etc. The acetylated form of morphine, known as heroin, is a synthetic product and found to be ten times more potent than morphine.

In spite of its beneficial effects, repeated use of narcotics causes development of tolerance towards the drug as well physical dependence. Additionally the drugs elicit compulsive drug-seeking behaviours upon repeated administration, which ultimately leads to the phenomenon of narcotic addiction or abuse in many of the users. Narcotic abuse is one of the most important public health issues worldwide. While there is no authentic statistic on the number of drug users in India, a survey conducted in the USA show that approximately 1% of the adult population in the United States abuses narcotics of which about one-half million people are addicted to heroin. This costs the nation $10 billion a year in treatment, care, and lost productivity; this does not include the cost of treating use-related diseases, like AIDS.

Such uncontrolled abuse of narcotics, worldwide, has triggered widespread search for potent analgesics. However, for the present, the opioids still retain their special place in the treatment of pain, particularly in cases of terminal pain, pain due to burns, post-operative pain etc., thereby necessitating concerted and intensive efforts towards management of the inevitable abuse of the drugs.

Mechanism of action of opioids

Opioids bind to their receptors in the brain. There are mainly three types of opioid receptors found in the human brain, μ, δ and κ. Morphine, preferentially binds to μ-opioid receptors in the nervous system. Cerebral μ receptors mediate analgesia, sedation, respiratory depression, catalepsy, inhibition of pituitary hormone secretion, and euphoria. In addition, peripheral effects of μ-receptor stimulation results in constipation. κ-Opioid receptors also mediate analgesia, but unlike μ-opioid receptors lead to dysphoria (Pfeiffer et al 1986). Several lines of evidence indicate that the δ-opioid receptor (DOR) is essential for morphine tolerance. Initial experiments showed that blockade of DOR with antagonists (Abdelhamid et al 1991) or DOR down regulation using antisense probes (Kest et al 1996) disrupted the development of tolerance. The opioid receptors are activated physiologically by endogenous neuropeptides called β-endorphin, leu-enkephalin, met-enkephalin, and dynorphin, respectively. These peptides are not selective for a certain type of opioid receptor. Nevertheless, β-endorphin binds preferentially to μ receptors, the enkephalins bind preferentially to δ, and dynorphin binds preferentially to κ receptors. The enkephalins, dynorphin, and β-endorphin are derived from three distinct precursor molecules, proenkephalin, prodynorphin, and proopiomelanocortin (POMC), respectively, each of which is encoded by a different gene. The POMC gene product not only contains the sequence for β-endorphin, but also for adrenocorticotrophin (ACTH) and α-, β-, and γ-melanocyte-stimulating hormone (MSH).

Neurobiology of addiction

Considerable efforts have been directed towards delineating the mechanism of long term effect of opioids leading to addiction. Although drugs of abuse are chemically divergent molecules acting via different pathways, the resultant addictions share some common features (Nestler, 2001). There is a general characteristic of the activation of the mesolimbic dopamine system whereby there is increased firing of dopamine neurons in the ventral tegmental area (VTA) of the midbrain and a subsequent increase of dopamine released into the nucleus accumbens (NAc) and other regions of the limbic forebrain like the prefrontal cortex. Several drugs of abuse also activate the mesolimbic dopamine system by mimicking (opiates) or activating (alcohol, nicotine) endogenous opioid pathways that innervate the VTA and NAc. Others act directly in the NAc (cannabinoids and phencyclidine). These various actions seem to produce some similar net effects (generally inhibition) on medium spiny neurons of the NAc. It is considered that these various mechanisms are central in mediating the acute reinforcing properties that are shared by all drugs of abuse (Robbins and Everitt, 1996; Koob et al 1998; Wise, 1998; Robinson and

Berridge, 2000). The mesolimbic dopamine system and its forebrain targets are part of the motivational system that regulates responses to natural reinforcers such as food, drink, sex and social interaction. Drugs of abuse affect this pathway (Fig. 1) with a strength and persistence of much greater magnitude than those produced by natural reinforcers. It is postulated that repeated exposure of drugs of abuse cause strong stimulation of the target neurons which undergo changes in a way that leads to marked alterations in reinforcement mechanisms and motivational state.

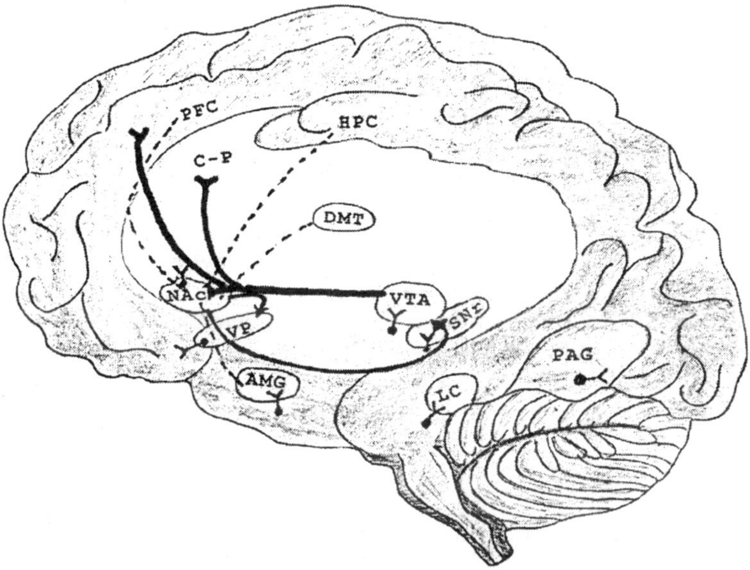

Fig. 1. Neural circuits of addiction. Dotted lines indicate limbic afferents to the nucleus accumbens (NAc). Lines with arrowheads represent efferents from the NAc thought to be involved in drug reward. Deep lines indicate projections of the mesolimbic dopamine system thought to be a critical substrate for drug reward. AMG, amygdala; C-P,caudate-putamen; DMT, dorsomedial thalamus; HPC, hippocampus; LC, locus coeruleus; PAG, periaqueductal grey; PFC, prefontral cortex; SNr, substantia nigra pars reticulata; VP, ventral pallidum; VTA, ventral tegmental area. •–< indicate presence of opioid receptor.

Apart from the VTA and NAc, many other brain structures undergo changes during addiction. In addition to the limbic system, the brainstem respiratory nuclei, the cerebral cortex, and the Locus Coeruleus (LC) are some of the centers where the drugs of abuse have clear influence. LC is an important brain nucleus having homogeneous set of noradrenergic neurons which projects to many brain regions. Due to a high density of opiate receptors and the presence of endorphins, the LC has a major involvement for development of opiate dependence (Rasmussen et al 1990). It is worth noting that as LC projects to several regions and receives axons from many brain nuclei, changes in its behavior may be the synergetic sum of afferent signals, including the feedback, through some pathways, of its own action. Addiction would then be the result of a complex interaction between brain sites imbalanced by excessive opiate exposure. Based on the action of morphine on the various regions of the brain involved in narcotic addiction, Simonato (1996) proposed a hypothesis as to how morphine induces changes in specific brain regions that account for important aspects of addiction. First, in the VTA, morphine inhibits GABA neurons via μ-opioid receptors, thus disinhibiting dopaminergic neurons and increasing dopamine input in the NAc and in other areas. The phenomenon may be associated in the mechanism of reward, a positive reinforcer to opioid addiction. Second, the firing rate of the LC is acutely inhibited by morphine, returns to baseline level with tolerance, and rises dramatically during withdrawal. This phenomenon, on the other hand, may be involved in the mechanism of dependence, the negative reinforcer to addiction. Since the dopaminergic and noradrenergic neurons of VTA and LC respectively are innervated diffusely to various target regions of the brain, additionally morphine may indirectly affect these neurons, which may possess other neurotransmitter systems. A combination of both these effects may determine the final response of specific neuronal population in specified areas. Some of the important neurotransmitter systems, which play a role in drug dependence, are GABAergic neurons of the VTA, κ-opioid, GABA and cholinergic systems of the cerebral cortex.

How does chronic narcotic use result in stable changes in the target neurons? Regulation of gene expression is one mechanism whereby repeated exposure to a drug of abuse would eventually lead to changes in nuclear function and to altered rates of transcription of particular target genes by causing repeated perturbation of intracellular signal transduction pathways. Altered expression of these genes would lead to altered activity of the neurons in which those changes occur and, ultimately, to changes in the neural circuits in which those neurons operate. The result would be stable changes in behaviour. Whereas neural genes are probably regulated by hundreds of distinct types of transcription factor, two transcription factors in particular have so far been implicated in addiction: the cyclic-AMP response-element-binding protein (CREB) and FosB (Nestler, 2001).

Hereditary trait in addiction

Evidences indicate that addiction to substances of abuse has a genetic predisposition. The co-occurrence of various drug use with alcohol addiction is well documented. Alcoholic patients developed another substance abuse six times more frequently than other subjects with mental disorders and drug abusers developed alcoholism four times more frequently than other subjects (Regier et al 1990). In another study, the odds ratio between alcohol abuse and drug abuse was 13.7 (Vandenbergh et al 1997). Additionally, familial, twin and adoption studies corraborate this hypothesis. Family studies showed an increased rate of drug disorders among relatives of drug addicts compared to the general population (Chaudry et al 1991; Mirin et al 1991; Rounsaville et al 1991; Luthar et al 1992; Merikangas et al 1998;). Although these studies supported a substantial degree of genetic vulnerability for substance abuse, these failed to dissociate the contribution of familial environment from purely genetic factors. Twin studies afforded better means of determining the role of genetic elements in susceptibility to drug abuse. Tsuang et al (1996) demonstrated a significant genetic component in drug disorder on the basis of a large population of twins in the Vietnam-era twin-pair registrants. Many other twin data studies support the above (Grove et al 1990; Pickens and Svikis, 1991; Tsuang et al 1998; Kendler et al 1999; Karkowski et al 2000). It was observed that genetic components appear to be more prominent in the more severe abusers and in some categories of drugs like heroin. A few adoption studies have also been carried out which demonstrate that both inherited and environmental factors operate in the etiology of substance abuse. Cadoret et al (1986, 1995) undertook cross-fostering studies in adoptees, who had been separated at birth from biologic parents, and found significant correlation between drug disorder in the adoptee and alcohol problems in the biological parent.

Single nucleotide polymorphisms and their implications in narcotic abuse

Such a significant role of heredity in contributing to drug abuse disorders has prompted examination of the various candidate genes for association with the substance abuse with a view to better understand their pathogenesis with the ultimate objective of improvement of preventive strategies, diagnostic tools and treatment. Considerable efforts have been directed in recent years to detect the genetic variants contributing to substance abuse. Association and linkage studies are the two dominant strategies being used to find the disease predisposition alleles. Genetic variants—or polymorphisms—arise from new mutations. The simplest type of polymorphism is a single base mutation, which substitutes one nucleotide for another, referred to as a single nucleotide polymorphism (SNP). More SNPs are thought to exist in the human genome than any other type of polymorphism (Wang et al 1998). To date, more than 2.5 million non-redundant human SNPs have been reported in the public domain, of which approximately 100,000 have been validated by either independent investigators or by independent methods. SNPs may not necessarily have any relevance to disease or outcome; they can be anonymous variants within or between genes (i.e., uncharacterised with respect to protein coding or gene function), or could be functional, causal mutations. Nevertheless, SNPs have heralded as key to understanding the genetic factors that determine susceptibility and predisposition to such common diseases as diabetes, hypertension, and schizophrenia (Risch and Merikangas, 1996; Lander, 1996; Collins et al 1997), or predict variability to drug response (Roses, 2001). However, the number of instances in which certain alleles could be definitely associated with a certain phenotype remains small. Among the more well known examples are the association of the apolipoprotein E type 4 allele with late-onset familial Alzheimer disease (Strittmatter et al 1993), the protection against HIV-1 infection in individuals homozygous for the CCR 5- Δ32 allele (Carrington et al 1999), the association of the common Pro12Ala polymorphism in peroxisome proliferator-activated receptor-γ with type 2 diabetes (Altshuler et al 2000), and the APC T3920A (Laken et al 1997) and CHEK 2 1100delC (Meijers-Heijboer et al 2002) variants, that are responsible for an increased susceptibility to colorectal and breast cancer, respectively. In this review, we restrict our attention to studies on the possible correlation between SNPs of candidate genes with narcotic tolerance and dependence.

Review of the literature

The importance of dopamine activity in mesolimbic/mesocortical circuits for behavioral reward in the pathogenesis of narcotic abuse has been described earlier. This has prompted several studies in investigating the importance of specific alleles of dopamine receptor subtypes form vulnerability towards narcotic addiction. Kotler et al (1997) analysed the exon III VNTR of D4 receptor (DRD4) gene in opiate-dependent subjects from Israel, and found a significant excess of the 7-repeat allele. Using a Han Chinese case-control sample of heroin-dependent subjects and normal controls, Li et al (1997) found two 7-repeat alleles which occurred exclusively in the patient group. When the D4 VNTR was divided into 'long' (5-7 repeats) and 'short' (2-4 repeats), a significant excess of long alleles was observed in the patient group. On the other hand, employing independent case-control and family-based association samples of large population, Franke et al (2000) found long alleles of the DRD4 exon III VNTR in similar frequency among heroin addicts and controls. Furthermore, no preferential transmission of long alleles to affected offspring was observed in a sample of 111 patients and their parents thereby refuting the hypothesis that alleles of the DRD4 exon III VNTR are susceptibility factors for opioid dependence in man.

Attempts to correlate an association of Bal I polymorphism at the DRD3 gene in opiate addicts (Duaux et al 1998) showed that there were no marked differences in genotypes between patients as a whole and controls. However, it was observed that patients with a high sensation-seeking score were more frequently homozygotes for both alleles than patients with a low sensation-seeking score suggesting that the DRD3 gene may have a role in drug dependence susceptibility in individuals with high sensation-seeking scores. Studies on the frequency of polymorphism of Ser9Gly in the DRD3 gene in Chinese heroin addicts (Li et al 2002) did not show any significant difference from controls.

The distribution of the D2 dopamine receptor (DRD2) gene variant marked by TaqI restriction fragment length polymorphisms has been investigated in a number of studies. Unlike polysubstance users, abusers with more prominent opiate use, the distribution of TaqI A1 allele were similar to control genotypes (Persico et al 1996). Lawford et al (2000) an interesting study in Caucasian opioid-dependent patients who were followed over a one-year period in an outpatient methadone treatment program. The patients were divided into three groups based on their response to treatment. One group of patients dropped out of the methadone program (Group A), another group had successful treatment (Group B) while the remaining group had poor treatment (Group C) outcome. The frequency of the A(1) allele was highest in Group C (42.1%), followed by Group A (22.7%) and was lowest in Group B (9.3%). The results indicate that DRD2 variants are predictors of heroin use and subsequent methadone treatment outcome and suggest a pharmacogenetic approach to the treatment of opioid dependence. Li et al (2002), on the other hand, did not find any association of the three polymorphisms (promoter - 141DeltaC, Ser311Cys, and TaqI) in the DRD2 gene in Chinese heroin addicts.

Catechol-O-methyltransferase (COMT) also plays a prominent role in dopaminergic circuits important for drug reward, and COMT alleles encode enzymes whose activities vary from three- to four-fold. A high degree of association between the high activity COMT polymorphism and polysubstance abuse in a group of North American subjects has been reported (Vandenbergh et al 1997). This has also been confirmed by genotyping Israeli heroin addicts and both parents using a robust family-based haplotype relative risk (HRR) strategy (Horowitz et al 2000). An excess of the val COMT allele and a trend for an excess of the val/val COMT in the heroin addicts compared to the HRR control group was observed. Cao et al (2002) undertook association studies between heroin-dependence and -287 A/G polymorphism of COMT gene and observed a weak but significant difference in genotype of -287 A/G polymorphism of COMT gene between heroin-dependent subjects and controls.

Other non-opioid candidate genes, which have been analyzed for association with heroin addicts, include GABA (A) gamma2 (Li et al 2002) and the serotonin transporter HTR2A (Kotler et al 1999; Li et al 2002) genes and do not indicate any likelihood of these genes to be a major risk factor in heroin addiction.

Human μ-opioid receptor (OPRM1) is the primary site for the analgesic action of most opioid drugs such as morphine, methadone and heroin and pharmacological and electrophysiological evidence suggest that these receptors are involved in the mechanism of narcotic dependence. Thus, μ-opioid receptors are appropriate candidate genes for case-control association studies of narcotic dependence. Table - 1 shows the distribution of the various polymorphisms in the OPRM1 gene.

Bond et al (1998) identified five different SNPs in the coding region of the μ opioid receptor gene of which nucleotide substitution at position 118 (A118G), predicting an amino acid change at a putative N-glycosylation site appeared to be most prevalent. The variant receptor resulting from the A118G SNP did not show altered binding affinities for most opioid peptides and alkaloids tested except that it binds beta-endorphin, an endogenous opioid that activates the μ opioid receptor, approximately three times more tightly than the most common allelic form of the receptor. Results suggested that SNPs in the μ-opioid receptor gene can alter binding and signal transduction in the resulting receptor and may have implications for normal physiology, therapeutics, and vulnerability to develop or protection from diverse diseases including the addictive diseases. Studies undertaken to delineate allele and haplotype frequencies of Ala6Val and Asp40Asn variants in exon 1 of OPRM1 gene, in control and substance-dependent populations of African American (AA), European American (EA) and Hispanic origin, and in a series of populations differing in geographic origin (Japanese, Ethiopians, Bedouins, and Ashkenazi Jews), showed that allele frequencies for both polymorphic systems were significantly different between AA and EA subjects, and there was significant heterogeneity among the more extensive set of populations (Gelernter et al 1999). However, there were no significant differences in allele frequency by diagnosis; that is, neither polymorphism appears to be a direct risk factor for substance dependence. Recently efforts have been made to screen the complete coding region of the OPRM1 gene for SNPs in Han-Chinese heroin addicts and normal control (Shi et al 2002). While three novel SNPs were detected, one in exon3, one in intron3 and one in the 3' untranslated region, it was observed that the SNP A118G had no statistically significant correlation with heroin addiction in Han Chinese. However, addicted subjects with the SNP in intron2 (IVS2 +G31A) tended to show much higher heroin intake dosages than those without this SNP. Moreover, individuals carrying both SNP A118G and IVS2 +G31A consumed relatively more drugs compared to other addicts. The study emphasized the importance of studying the various regions of the μ– opioid receptor gene, coding as well as non-coding, for genetic markers that may be linked to, or directly contribute to opioid drug-seeking behavior. Very recently, an extensive study on the distribution of three polymorphisms of OPRM1 gene was investigated in four different Asian populations, and in heroin-dependent subjects deriving from three of these populations (Tan et al 2003). Although a significant differences in allele and genotype frequencies between different ethnic groups for A118G polymorphism was observed, a highly significant association with heroin dependence in Indians for both genotype distribution and allele frequency was demonstrable. For the C17T

Table 1. Distribution of various polymorphic alleles in human μ-opioid receptor (OPRM1) gene.

Regions	Size (bp)	SNPs	References
5'- upstream	34	G-54T G-38T	PNAS, USA **96**: 7752, 1999. Addict Biol **2**: 303, 1997.
Exon-1	492	C12G C17T G24A A118G	PNAS USA **96**: 7752, 1999. Addict Biol **2**: 303, 1997. PNAS USA **95**: 9608, 1998. Mol. Psychiatry **2**: 490, 1997.
Intron-1	49,993	---------	
Exon-2	352	C440G A454G	Mol. Psychiatry **2**: 490, 1997. PNAS USA **96**: 7752, 1999. Eur J Pharm **410**: 249, 2000.
Intron II	773	IVS2G31A IVS2G51A IVS2C691G	Hum Mutat **19**: 459, 2002. J Mol Med **76**: 525, 1998. J Mol Med **70**: 525, 1998. Molec Psychiatry **2**: 490, 1997. Addict Biol **5**: 181, 2000.
Exon-3	520	G779A G794A T802C G942A G877A	PNAS USA **95**: 9608, 1998. Eur J Pharm **410**: 249, 2000. Eur J Pharm **410**: 249, 2000. PNAS USA **95**: 9608, 1998. Hum Mutat **19**: 459, 2002.
Intron III	27,171	IVS3G-77A (77 bp upstream of exon 4)	Hum Mutat **19**: 459, 2002.
Exon-4	815	-------	
3' - UTR		C9T (9 bp downstream of stop codon)	Hum Mutat **19**: 459, 2002.

polymorphism, the minor allele was documented for the first time in Chinese and Malays. Molecular haplotyping revealed complete linkage disequilibrium between the A118G and C17T polymorphisms and almost complete between the A118G and C1031G polymorphisms. Studies have been undertaken in our laboratory to investigate the possible association of six SNPs of exon-1 of OPRM1 gene in narcotic addicts of Kolkata. We have investigated in three groups: non-addicts, alcohol addicts and narcotic addicts. Unlike the observations of Tan et al (2003), our unpublished observations suggest that there is no significant correlation between A118G polymorphism and addiction.

The importance of δ-opioid receptor in morphine tolerance has been described earlier. It has been shown that in mice with a mutation in the DOR-1 gene, analgesic tolerance to morphine after daily injection of a fixed morphine dose was abolished (Zhu et al 1999). Recent studies on mice lacking a functional gene for the endogenous ligand preproenkephalin (ppENK), the endogenous agonist for the δ-opioid receptor also show a similar tolerance deficit (Nitsche et al 2002). These studies indicate that morphine tolerance can be genetically dissociated from physical dependence. Comings et al (1999) tested the hypothesis that the alleles at the (CA)n repeat of the proenkephalin gene (PENK) might be associated with opioid addiction in non-Hispanic Caucasian subjects with opioid dependence (heroin) and observed that 66% of the subjects with opioid dependence carried the > or = 81 bp allele compared with 40% of subjects with other types of substance abuse and 49% of ethnically matched controls. These results are consistent with a role of the PENK gene in opioid dependence. Franke et al (1999), on the other hand, investigated the contribution of a silent T to C change in the coding region of the human delta-opioid receptor gene to the development of heroin and alcohol dependence using large case-control and family-based association samples and observed no statistically significant differences between patients and controls. They also did not find preferential transmission of the C allele from parents to affected offspring which do not support an association between genetic variation of the delta-opioid receptor and addictive behavior in man. In another study, two single nucleotide polymorphisms (SNPs), OPRD1 921T > C and 80G > T, of the human delta opioid receptor gene (OPRD1) were used in population-based studies of Chinese heroin dependent patients. Genotype and allele frequencies at OPRD1 921T > C were not significantly different, and the OPRD1 80G was absent from both Chinese opioid dependence patients and controls thereby ruling out any association between the OPRD1 gene and heroin dependence in the Chinese population (Xu et al 2002).

Conclusion

With the advent of the postgenomic era in the new millennium, molecular genetics has come a long way in the unraveling of the etiologies of most of the single-gene disorders that lead to neurodegenerative diseases. Genes considered as risk factors like the prion gene in iatrogenic and idiopathic Creuzfeldt–Jakob disease, the apolipoprotein E gene in Alzheimer's disease, the tau gene in progressive supranuclear palsy etc. have been identified. However dissection of the more complex genetics of disorders like substance abuse as well as most neurodegenerative diseases, which do not show simple patterns of inheritance, has just begun. Given the huge socioeconomic impact of narcotic abuse it will be a challenge for molecular geneticists in future years to facilitate this endeavor. With the rapid development of detection technologies and scoring of SNPs and extensive cataloguing of SNPs across the genome, SNPs are being increasingly used as means of investigation of the genetic causes of complex human disease. Nevertheless, reports of successful applications of SNPs to genome-wide linkage analysis of both mono- and polygenic traits are rare. Newer insights on recruitment strategies to avoid population stratification, increasing of sample size and shifting away from case-control and cohort studies towards family-based association designs are some of the current approaches used to overcome these problems. Progress in the analysis of SNPs also needs to be complemented by methods that allow the systematic elucidation of both primary and secondary phenotypes of genes.

References

Abdelhamid EE, Sultana M, Portoghese PS, Takemori AE (1991) J Pharmacol Exp Ther **258**: 299–303.

Altshuler D, Hirschhorn JN, Klannemark M, Lindgren CM, Vohl MC, Nemesh J, Lane CR, Schaffner SF, Bolk S, Brewer C, Tuomi T, Gaudet D, Hudson TJ, Daly M, Groop L, Lander ES (2000) Nature Genet **26**: 76-80.

Bond C, LaForge KS, Tian M, Melia D, Zhang S, Borg L, Gong J, Schluger J, Strong JA, Leal SM, Tischfield JA, Kreek MJ, Yu L (1998) Proc Natl Acad Sci (USA) **95**: 9608-9613.

Cao L, Li T, Xu K, Liu X (2002) Zhonghua Yi Xue Yi Chuan Xue Za Zhi **19**: 499-501.

Cadoret RJ, Troughton E, O'Gorman TW, Heywood E (1986) Arch Gen Psychiatry **43**: 1131-1136.
Cadoret RJ, Yates WR, Troughton E, Woodworth G, Stewart MA (1995) Arch Gen Psychiatry **52**: 42-52.
Carrington M, Dean M, Martin MP, O'Brien SJ (1999) Hum Mol Genet **8**: 1939-1945.
Chaudry H, Arria A, Tarter R, Chaudhry S, Chaudhry N (1991) Br J Addicts **86**: 785-788.
Collins FS, Guyer MS, Chakravarti A (1997) Science **278**: 1580-1581.
Comings DE, Blake H, Dietz G, Gade-Andavolu R, Legro RS, Saucier G, Johnson P, Verde R, MacMurray JP (1999) Neuroreport **10**: 1133-1135.
Duaux E, Gorwood P, Griffon N, Bourdel MC, Sautel F, Sokoloff P, Schwartz JC, Ades J, Loo H, Poirier MF (1998) Mol Psychiatry **3**: 333-336.
Franke P, Nothen MM, Wang T, Neidt H, Knapp M, Lichtermann D, Weiffenbach O, Mayer P, Hollt V, Propping P, Maier W (1999) Am J Med Genet **88**: 462-464.
Franke P, Nothen MM, Wang T, Knapp M, Lichtermann D, Neidt H, Sander T, Propping P, Maier W (2000) Mol Psychiatry **5**: 101-104.
Gelernter J, Kranzler H, Cubells J (1999) Mol Psychiatry **4**: 476-483.
Grove WM, Eckert ED, Heston L, Bouchard TJ, Segal N, Lykken DT (1990) Biol Psychiatry **27**: 1293-1304.
Horowitz R, Kotler M, Shufman E, Aharoni S, Kremer I, Cohen H, Ebstein RP (2000) Am J Med Genet **96**: 599-603.
Karkowski LM, Prescott CA, Kendler KS (2000) Am J Med Genet **96**: 665–670.
Kendler KS, Karkowski L, Prescott CA (1999) Acta Psychiatr Scand **5**: 368-376.
Kest B, Lee CE, McLemore GL, Inturrisi CE (1996) Brain Res Bull **39**: 185–188.
Koob GF, Sanna PP, Bloom FE (1998) Neuron **21**: 467–476.
Kotler M, Cohen H, Segman R, Gritsenko I, Nemanov L, Lerer B, Kramer I, Zer-Zion M, Kletz I, Ebstein RP (1997) Mol Psychiatry **2**: 251– 254.
Kotler M, Cohen H, Kremer I, Mel H, Horowitz R, Ohel N, Gritsenko I, Nemanov L, Katz M, Ebstein RP (1999) Mol Psychiatry **4**: 313-3144.
Lander ES (1996) Science **274**: 536-539.
Laken SJ, Petersen GM, Gruber SB, Oddoux C, Ostrer H, Giardiello FM, Hamilton SR, Hampel H, Markowitz A, Klimstra D, Jhanwar S, Winawer S, Offit K, Luce MC, Winzeler KW, Vogelstein B (1997) Nature Genet **17**: 79-83.
Lawford BR, Young RM, Noble EP, Sargent J, Rowell J, Shadforth S, Zhang X, Ritchie T (2000) Am J Med Genet **96**:592-598.
Li T, Liu X, Zhao J, Hu X, Ball DM, Loh el-W, Sham PC, Collier DA (2002) Am J Med Genet **114**: 329-335.
Li T, Xu K, Deng H, Cai G, Liu J, Liu X, Wang R, Xiang X, Zhao J, Murray RM, Sham PC, Collier DA (1997) Mol Psychiatry **2**: 413-416.
Luthar SS, Anton SF, Merikangas KR, Rounsaville BJ (1992) J Nerv Ment Dis **180**: 153-161.
Meijers-Heijboer H, van den Ouweland A, Klijn J, Wasielewski M, de Snoo A, Oldenburg R, Hollestelle A, Houben H, Crepin E, van Veghel-Plandsoen M, Elstrodt F, van Duijn C, Bartels C, Meijers C, Schutte M, McGuffog L, Thompson D, Easton D, Sodha N, Seal S, Barfoot R, Mangion J, Chang-Claude J, Eccles D, Eeles R, Evans DG, Houlston R, Murday V, Narod S, Peretz T, Peto J, Phelan C, Zhang HX, Szabo C, Devilee P, Goldgar D, Futreal PA, Nathanson KL, Weber B, Rahman N, Stratton MR (2002) Nature Genet **31**: 55-59.
Merikangas KR, Stolar M, Stevens DE, Goulet J, Preisig MA, Fenton B, Zhang H, O'Malley SS, Rounsaville BJ (1998) Arch Gen Psychiatry **55**: 973-979.
Mirin SM, Weiss RD, Griffin ML, Michael JL (1991) Compr Psychiatry **32**: 36-51.
Nestler EJ (2001) Nature Reviews **2**: 119-128.
Nitsche JF, Schuller AG, King MA, Zengh M, Pasternak GW, Pintar JE (2002) J Neurosci **22**: 10906-10913.
Persico AM, Bird G, Gabbay FH, Uhl GR (1996) Biol Psychiatry **40**: 776-784.
Pfeiffer A, Brantl V, Herz A, Emrich HM (1986) Science **233**: 774– 776.
Pickens RW, Svikis DS (1991) Arch Gen Psychiatry **10**: 202-213.
Rasmussen K, Beitner-Johnson D, Krystal JH, Aghajanian GK, Nestler EJ (1990) J Neurosci **10**: 2308-2317.
Regier DA, Farmer ME, Rae DS, Locke BZ, Keith SJ, Judd LL, Goodwin FK (1990) JAMA **264**: 2511-2518.
Risch N, Merikangas K (1996) Science **273**: 1516-1517.
Roses AD (2001) Hum Mol Genet **10**: 2261-2267.
Robbins TW, Everitt BJ (1996) Curr Opin Neurobiol **6**: 228–236.
Robinson TE, Berridge KC (2000) Addiction **95**: S91–S117.

Rounsaville BJ, Kosten TR, Weissman MMP, Pauls D, Anton SF, Merikangas K (1991) Arch Gen Psychiatry **488**: 33-42.
Shi J, Hui L, Xu Y, Wang F, Huang W, Hu G (2002) Hum Mutat **19**: 459-460.
Simonato M (1996) Trends Pharmacol Sci **17**: 410-415.
Strittmatter WJ, Saunders AM, Schmechel D, Pericak-Vance M, Enghild J, Salvesen GS, Roses AD (1993) Proc Natl Acad Sci (USA) **90**: 9649-9653.
Tan EC, Tan CH, Karupathivan U, Yap EP (2003) Neuroreport **14**: 569-572.
Tsuang MT, Lyons MJ, Eisen SA, Goldberg J, True W, Lin N, Meyer JM, Toomey R, Faraone SV, Eaves L (1996) Am J Med Genet **67**: 473-477.
Tsuang MT, Lyons MJ, Meyer JM, Doyle T, Eisen SA, Goldberg J, True W, Lin N, Toomey R, Eaves L (1998) Arch Gen Psychiatry **55**: 967-972.
Vandenbergh DJ, Rodriguez LA, Miller IT, Uhl GR, Lachman HM (1997) Am J Med Genet **74**: 439-442.
Vandenbergh DJ, Zonderman AB, Wang J, Uhl GR, Costa PT (1997) Mol Psychiatry **2**: 417-419.
Wang DG, Fan JB, Siao CJ, Berno A, Young P, Sapolsky R, Ghandour G, Perkins N, Winchester E, Spencer J, Kruglyak L, Stein L, Hsie L, Topaloglou T, Hubbell E, Robinson E, Mittmann M, Morris MS, Shen N, Kilburn D, Rioux J, Nusbaum C, Rozen S, Hudson TJ, Lipshutz R, Chee M, Lander ES (1998) Science **280**: 1077–1082.
Wise RA (1998) Drug Alcohol Dependence **51**: 13–22.
Xu K, Liu XH, Nagarajan S, Gu XY, Goldman D (2002) Am J Med Genet **110**: 45-50.
Zhu Y, King MA, Schuller AG, Nitsche JF, Reidl M, Elde RP, Unterwald E, Pasternak GW, Pintar JE (1999) Neuron **24**: 243–252.

Section E

Neurodegenerative Diseases

Molecular and Cellular Neurobiology
Editors: M.K. Thakur and S. Prasad

Neurodegenerative Diseases Involving Trinucleotide Repeat Expansion

Nihar Ranjan Jana

National Brain Research Centre, Manesar, Gurgaon-122 050, India

Summary

A growing number of neurodegenerative diseases have been found due to the expansion of an unstable trineucleotide repeats. Over the past several years, great progress has been made in elucidating the molecular mechanisms by which the expanded polyglutamine tract causes disease pathogenesis. The purpose of this review is to provide a brief description about the polyglutamine diseases and the recent advances in understanding the disease pathogenic mechanism.

Keywords: Neurodegenerative diseases, Trinucleotide repeats, Chaperones, Ubiquitin-proteasome pathways.

Introduction

In 1991, a new mutational mechanism for disease was discovered: the expansion of unstable trinucleotide repeats. The pathological expansion of unstable triplet repeat currently is known to cause 15 neurological diseases, nine of which are neurodegenerative diseases resulting from expansion of CAG repeats coding for polyglutamine tracts. Several defining features are common amongst disorders caused by trinucleotide repeat expansions. First, the mutant (expanded) repeats show both somatic and germline instability and, more frequently, they expand rather than contract in successive transmissions. Secondly, an earlier age of onset and increasing severity of phenotype in subsequent generations (anticipation) generally are correlated with larger repeat length. Finally, the parental origin of the disease allele can often influence anticipation, with paternal transmissions carrying a greater risk of expansion for many of these disorders.

Triplet repeats diseases can be divided into two groups based on the relative location of the trinucleotide repeat to a gene. The first group, presently consisting of six diseases, has its repeats in non-coding sequences (see Table 1 for details) whereas the second group is characterized by the expansion of CAG repeats in the coding region of the target gene. The later group is referred to collectively as "polyglutamine diseases" (Table 2). The non-coding trinucleotide repeats diseases typically are characterized by large and variable repeat expansions that result in multiple tissue dysfunction or degeneration. Phenotypic manifestations within a disease are variable, perhaps due to somatic heterogeneity. The larger mutations often are transmitted from a small pool of clinically intermediate size expansions, termed as pre-mutations. In spite of these similarities, the triplet repeat sequences vary in this subclass (CGG, GCC, GAA, CTG and CAG), and it is clear that the particular triplet sequence, as well as its location with respect to a gene, are important defining factors in dictating the unique mechanism of pathogenesis for each disease. The pathogenic mechanism will also vary disease to disease; depending on the consequences of the lost function of the respective proteins or, in some cases

Table 1. Diseases caused by expansion of non-coding trinucleotide repeats

Disease	Gene/locus	protein	Schematic representation	Proposed mechanism
Fragile X syndrome	FMR1(FRAXA) Xq27.3	FMR1 protein		Loss of function: Loss of FMR1:abnormal RNA metabolism
Fragile XE syndrome	FMR2(FRAXE) Xq28	FMR2 protein		Loss of function: Loss of FMR2:disruption of neuronal gene regulation?
Friedreich ataxia	X25 9q13-21.1	Frataxin	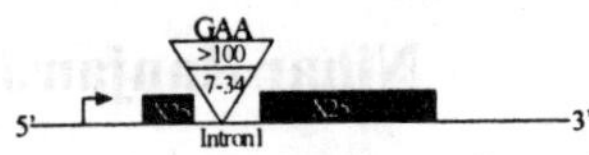	Loss of function: Reduced frataxin causing altered iron homeostasis and mitochondrial dysfunction
Myotonic dystrophy	DMPK 19q13	Myotonic dystrophy protein kinase(DMPK)	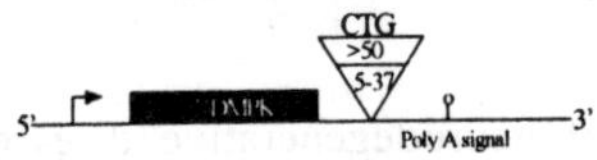	Loss and/or gain of Function: Reduced DMPK, disruption of kinase activity
SCA8	SCA8 13q21	None	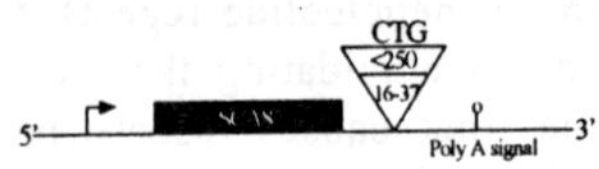	Loss of Function? Abnormal RNA regulation?
SCA12	SCA12 5q31-33	PP2A-PR55β		Loss of Function? Disruption of phosphatase activity?

acquired function of a toxic transcript (Zoghbi and Orr, 2000, Cummings and Zoghbi, 2000).

Polyglutamine neurodegenerative diseases

In comparison with the non-coding triplet repeat disorders, polyglutamine diseases have repeat expansions that are much smaller in size and variation. So far nine polyglutamine diseases have been known (Zoghbi and Orr, 2000). Those are Huntington's disease (HD), dentatorubral pallidoluysian atrophy (DRPLA), X-linked spinal bulbar muscular atrophy (SBMA), and several spinocerebellar ataxias (SCA1, SCA2, SCA3, SCA6, SCA7 and SCA17). Each of these diseases has unique pathophysiologies, which is described below.

Clinical features

HD: Chorea, the most common motor symptom in adult patients usually presents along with memory deficits, affective disturbance and changes in personality. Other forms of motor dysfunction such as Parkinsonism, dystonia and involuntary motor impairments may all be present. Juvenile onset patients show bradykinesia, rigidity, epilepsy, severe dementia and a more rapidly progressing disease. Caudate atrophy is visible in CT and MRI scans. Diffuse, severe atrophy of the neostriatum is the pathologic hallmark of HD. Medium spiny striatal neurons containing GABA are more vulnerable than neurons positive for NADPH-diaphorase. Early onset cases of HD most often show loss of cerebellar Parkinje cells.

***SBMA or Kenedy disease*:** This is unique among the polyglutamine diseases in showing an X-linked recessive pattern of inheritance. Affected males may suffer muscle crams for years before presenting proximal muscle weakness in the fourth or fifth decade. Tendon reflexes are usually absent and sensation

Table 2. Molecular characteristics of polyglutamine neurodegenerative diseases

Disease	Gene locus	Gene' product	Normal CAG(n)	Expanded CAG (n)	Protein localization	Brain regions most affected
HD	4p16.3	Huntingtin	6-34	36-121	Cytoplasmic	Striatum, cerebral cortex
SBMA	Xq11-12	Androgen Receptor	9-36	38-62	Nuclear and Cytoplasmic	Anterior horn and bulbar neurons, dorsal root ganglia
DRPLA	12q	Artophin-1	6-36	49-84	Cytoplasmic	Cerebellum, cerebral cortex, basal ganglia
SCA1	6p22-23	Ataxin-1	6-44	39-82	Nuclear	Cerebellar Purkinje cells, dentate nucleus brain stem
SCA2	12q23-24	Ataxin-2	15-31	36-63	Cytoplasmic	Cerebellar Purkinje cells, brainstem, fronto-temporal lobe
SCA3	14q24.3-31	Ataxin-3	12-41	62-84	Cytoplasmic	Cerebellar dentate neurons, basal ganglia, brain stem, spinal cord
SCA6	19p13	P/Q Ca++ channel	4-18	21-33	Cell membrane	Cerebellar Purkinje cells, dentate nucleus, inferior olive
SCA7	3p12-p21.1	Ataxin-7	4-35	37-306	Nuclear	Cerebellum, brain stem, macula, visual cortex
SCA17	6q27	TATA-binding protein	29-42	47-55	Nuclear	Cerebellum, basal ganglia

is normal. Weakness, wasting and fasciculation, gynecomastia is common. Hypo-gonadism, loss of libido and late sterility has been reported. Pathologically, SBMA causes anterior horn cell, bulbar neuron and dorsal root ganglion cell degeneration. Muscle biopsies show evidence of chronic denervation often with collateral re-enervation.

***DRPLA*:** DRPLA is distinct from most SCAs in its characteristics choreoathetosis, myoclonic epilepsy and dementia. Juvenile-onset individuals may also suffers mental retardation. Marked neuronal loss has been observed in the cerebral cortex, globus pallidus, striatum and cerebellar cortex and in the subthalamic red and dentate nuclei.

***SCAs*:** The dominantly inherited spinocerebellar ataxias are a heterogeneous group of neurologic disorders characterized by variable degree of degeneration of the cerebellum, spinal tracts and brain stem. Some clinical features are common to all SCAs like ataxia and dysarthria due to the cerebellar degeneration. However, there are variable features seen in some subtypes that helps to distinguish certain ataxias. Hypo-or areflexia and extremely slow saccades suggest SCA2. Bulging eyes, early faciolingual fasciculation and Parkinsonism suggest SCA3. Slowly progressive or episodic ataxia, complete absence of extrapyramidal signs, spasticity, cognitive impairment and visual deficit can suggest SCA6. SCA7 patient may first present with cerebellar ataxia or visual deficit while SCA17 patient exhibits cerebellar ataxia associated with dementia, pyramidal tract and basal ganglia symptoms.

Common characteristics features of polyglutamine diseases

The polyglutamine diseases have several similar features and probably share common mechanisms of pathogenesis although, the mutant proteins do not share any homology outside the polyglutamine tract (Zoghbi and Orr, 2000). All nine known polyglutamine disorders are characterized by progressive neuronal dysfunction that typically begins in mid-life and results in severe neurodegeneration. Despite the ubiquitous expression of all nine genes, only a certain subset of neurons is vulnerable to degeneration. Studies of cell cultured and animal models clearly demonstrated that these diseases are caused by gain-of-function mechanism and that the expanded polyglutamine tract are at the core of pathogenesis. Although expanded polyglutamine tracts are extremely toxic in *in vitro* cell culture, the selective neuronal loss in particular to each disease occurs only when the full-length protein harboring an expanded polyglutamine tract is expressed.

***Inverse relationship of repeat size and disease onset*:** Analysis of repeat size and symptomatology in polyglutamine disease patients has clearly demonstrated an inverse relationship between the size of the repeat and the age of onset, though the contribution of the repeat size to age of onset varies according to the protein context.

***Phenotypic overlap and loss of cell specificity*:** It is interesting to note that when the expansion are large, leading to severe juvenile-onset disease, there is significant overlap in the phenotypes of these disorders. Juveniles with SCA1 manifest not only the characteristics ataxia and brain stem dysfunction but also some cognitive impairment and dystonic features. Juvenile onset Huntington's disease patients develop dystonia and seizures in addition to the classical phenotype of chorea and dementia seen in adult patient. This loss of cell specificity with large expansions is important and hints that toxicity is probably much more widespread through neuronal subtypes that are normally spared when the repeat sizes are in moderate ranges.

***Protein aggregation*:** One aspect of subcellular distribution of polyglutamine proteins in SCA1, SCA3, SCA7, HD, SBMA and DRPLA is the localization of mutant proteins to ubiquitinated microscopic nuclear inclusions or aggregates within neuronal nuclei (see Figure 1). In human disease tissue, transgenic animal models, and transfected cells, expanded polyglutamine proteins have been shown to undergo intracellular aggregation, in most cases forming neuronal intranuclear inclusions (NIIs) (Paulson, 1999). The association of aberrant deposits of polyglutamine protein with neurological disease was first reported in 1997, with reports of NIIs of mutant huntingtin fragments in the brain of HD transgenic mice and within the brain of HD patients. Subsequently, NIIs have been detected in the brains of other polyglutamine diseases like SCA1, MJD/SCA3, SCA7, SBMA and DRPLA (Zoghbi and Orr, 2000). The NIIs cannot be observed in SCA2 or SCA6 brain samples, however, cytoplasmic aggregation of the mutant proteins can be seen in the SCA6 and SCA2 brain samples. The time course for the formation of these aggregates either in animal or cellular models of pathology is consistent with them being causative. However, a number of studies argue that it is the soluble polyglutamine protein that is critical for triggering pathogenesis (Saudou et al 1998; Klement et al 1998). Formation of aggregates may be a protective strategy of the cell by reducing its ability to associate with other cellular components.

Pathogenic mechanisms of polyglutamine diseases

In the past decade, enormous progress has been made in elucidating the molecular mechanisms of polyglutamine disease pathogenesis. Mouse, Drosophila, Caenorhabditis elegans, yeast and tissue cultured models have provided tremendous insights into the disease pathogenic mechanisms.

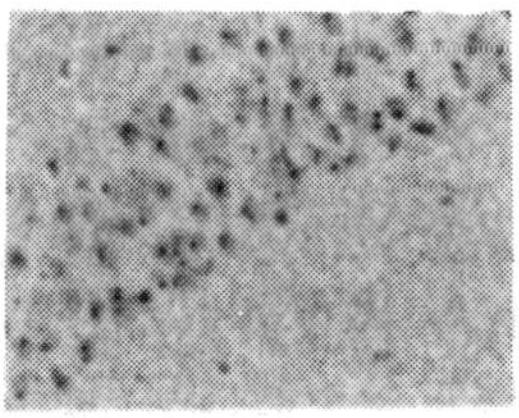

Figure 1. Typical nuclear aggregates (arrow heads) in the brain of HD exon1 transgenic mice. Brain sections were processed for immunohistochemical staining using antibody against ubiquitin.

Misfolding and altered turnover of polyglutamine proteins**:** Polar zipper formation (Perutz et al 1994) and misfolding are the two primary factors that are known to cause the rapid aggregation of the polyglutamine proteins. However, there are several other factors that might also enhance the aggregation process, such as transglutaminase-catalyzed cross-linking of polyglutamine protein, aberrant interaction with other proteins dependent on polyglutamine length, or proteolytic processing of the full-length disease protein. According to the polar zipper hypothesis, polyglutamine repeats that are longer than certain critical length might be able to form highly stable intra-molecular β–sheet structure by creation of hairpin loop and many such molecules can join together by forming inter-molecular β–sheet and thereby causing rapid aggregation of the proteins. The formation of intra-molecular β-sheet structure most likely cause the misfolding of the polyglutamine proteins and the increase in length of β-sheet structure due to increasing number of glutamine repeats increases the misfolding propensity.

Correct protein folding is an essential biological process and for that many proteins interact with molecular chaperones in the cellular environment. The heat shock protein 70 (Hsp70) class of molecular chaperone binds with the misfolded substrate early in the folding process and maintain the polypeptide in the soluble conformation. Both Hsp70 and Hsp40 family of chaperones have been shown to interact with the expanded polyglutamine proteins and have been co-localized with the polyglutamine aggregates in the cellular and transgenic mice models of polyglutamine diseases and also in patient tissue samples (Cummings et al 1998, Jana et al 2000). Furthermore, it has been demonstrated that chaperone over expression suppresses the polyglutamine aggregation and cell death in the mammalian cultured cells (Cummings et al 1998, Jana et al 2000). Overexpression of Hsp70 chaperones also suppresses the polyglutamine-induced neuropathology and delays the progression of the disease in the polyglutamine disease model of Drosophila and SCA1 transgenic mice (Cummings et al 2001).

The cell has two main protective strategies to handle unfolded polypeptides. At first it tries to repair the misfolded proteins and, failure to refold next leads to their selective degradation by the proteasome system. The Hsp70/Hsp40 chaperone pair not only plays an essential role in the refolding of misfolded proteins and preventing aggregation, but is also important for the rapid ubiquitination and degradation of many abnormal proteins by proteasomes. Degradation of a protein by ubiquitin-proteasome pathway involves two distinct and successive steps: (1) covalent attachment of multiple ubiquitin molecules to the target protein and (2) degradation of the target protein by 26S proteasome. In the polyglutamine diseases, the aggregates are not only associate with the Hsp70/Hsp40 chaperones, but also highly ubiquitinates and associates with several proteasome components (Cummings et al 1998, Jana et al 2001). The recruitment of proteasome with the polyglutamine aggregates possibly indicates the unsuccessful attempt by the cells to remove expanded polyglutamine proteins by the proteasome.

Impairment of the ubiquitin-proteasome pathway: Controlled proteolysis of the substrate by ubiquitin-proteasome pathway is crucial in controlling the intracellular levels of a variety of short-lived proteins and for maintaining cellular growth and metabolism. Pharmacological inhibition of the proteasomal function induces apoptosis (Glickman and Ciechanover, 2001). Dysfunction of the ubiquitin-proteasome pathway has been implicated in the pathogenesis of a number of diseases including various neurodegenerative diseases (Glickman and Ciechanover, 2001). Recent studies indicate that the proteasome activity is inhibited in the brains of Alzheimer's disease, Parkinson's disease and also in the ageing brain. In the cellular model of polyglutamine disease, the expression of truncated N-terminal huntingtin causes inhibition of proteasome activity over time (Jana et al 2001; Bence et al 2001). Similar proteasomal malfunction has been reported after transient over-expression of the mutant α-synuclein, which is responsible for autosomal dominant Parkinson disease (Tanaka et al 2001). It seems that impairment of the ubiquitin-proteasome pathway is a common phenomenon that can be caused by any protein aggregation (Bence et al 2001). But questions remain is how an aggregate causes proteasomal malfunction?

Oxidative stress and mitochondrial defect: The generation of free radicals and oxidative damage have long been implicated in several neurodegenerative diseases including polyglutamine diseases (Beal, 2000). Patient with HD showed decreased levels of mitochondrial electron transport enzymes, elevated levels of lactate in the cerebral cortex and basal ganglia, decreased levels of phosphocreatine in the resting muscles. Mitochondrial toxins produces selective damage in the striatum of animals, which closely resembles the pathology of HD. Studies also show increased susceptibility of mitochondria to depolarization in HD lymphoblasts and fibroblasts. The HD transgenic mice show increased vulnerability to the mitochondrial toxin 3-nitropripionic acid. This may be a consequence of mitochondrial dysfunction. Furthermore, dietary creatine supplementation significantly improved survival, slowed the development of brain atrophy in the HD transgenic mice.

***Caspase cleavage, toxic fragments and apoptotic cell death*:** Apoptosis is a highly regulated cell death pathway that has been shown to have a role in neurodegenerative diseases like Alzheimer's disease, amyotrophic lateral sclerosis and polyglutamine diseases. Execution of the apoptotic program involves the activation of a cascade of proteases known as caspases. Evidences for the activation of caspases have been observed in HD brain and also in the brains of HD transgenic mice (Ona et al 1999; Sanchez et al 1999). The expression of expanded polyglutamine in animal cell culture also promotes apoptosis (Sanchez et al 1999). A number of cell culture experiments also demonstrate that the expanded polyglutamine tracts within truncated proteins are more toxic than expanded tract within full-length proteins.

***Aberrant protein interactions and altered gene expression*:** The functional significance of expanded polyglutamine proteins is unknown. One of the possibilities is that the expanded polyglutamine proteins might aberrantly interact with various others cellular proteins and thereby disturb their normal cellular functions. By using yeast two-hybrid screens, various proteins have been identified that interact with polyglutamine proteins and some of these interactions are sensitive to polyglutamine-tract length. A growing body of evidence indicates that the expanded polyglutamine tract interacts with several transcription and co-transcription factors like TATA-binding protein, CREB-binding protein, Nuclear receptor co-repressor, $TAF_{II}130$, Spl and p53 (Cha et al 2000). The aberrant interaction of these factors with the expanded polyglutamine tract or their trapping with the polyglutamine aggregates leads to their unavailability to perform normal cellular function. Based on the aberrant interaction between polyglutamine-expanded proteins and several transcription or co-transcription factors, it is not surprising that the level of expression of various genes will be altered in polyglutamine diseases. The transcriptional alteration has been assessed by microarray technology and other comprehensive approaches. In both HD and SCA1 mouse models, mRNA encoding proteins involved in neural signal transduction and calcium homeostasis were preferentially decreased (Luthi-Carter et al 2000). In HD models mRNA encoding heat shock proteins, proteasome subunits and other stress related proteins are increased.

Treatment and prevention of polyglutamine diseases

The understanding of the molecular pathology of polyglutamine diseases and cellular responses to expanded polyglutamine protein expression has matured enough to develop rational strategy for drug discovery. Most interesting approach is high-throughput screening for small molecules that affect specific molecular target. Potential targets include chaperones, caspases, the ubiquitin-proteasome system, transcription factors, mitochondria and offending the polyglutamine-proteins themselves. The success with the caspase inhibitors and minocycline warrants further study of caspase inhibitors. Because HSP70 over-expression abolishes toxicity in several models, agents that enhance HSP70 expression might also be useful. Much attention also has been focused on screens for drugs which prevent aggregation of polyglutamine proteins. Antioxidants and mitochondrial support agents like creatine and coenzyme Q10 also could be potential therapeutic molecule, since oxidative stress and mitochondrial disturbances has been implicated in the polyglutamine diseases. Although these treatment strategies look exciting and promising, much work remains to be done to either prevent or cure of polyglutamine diseases.

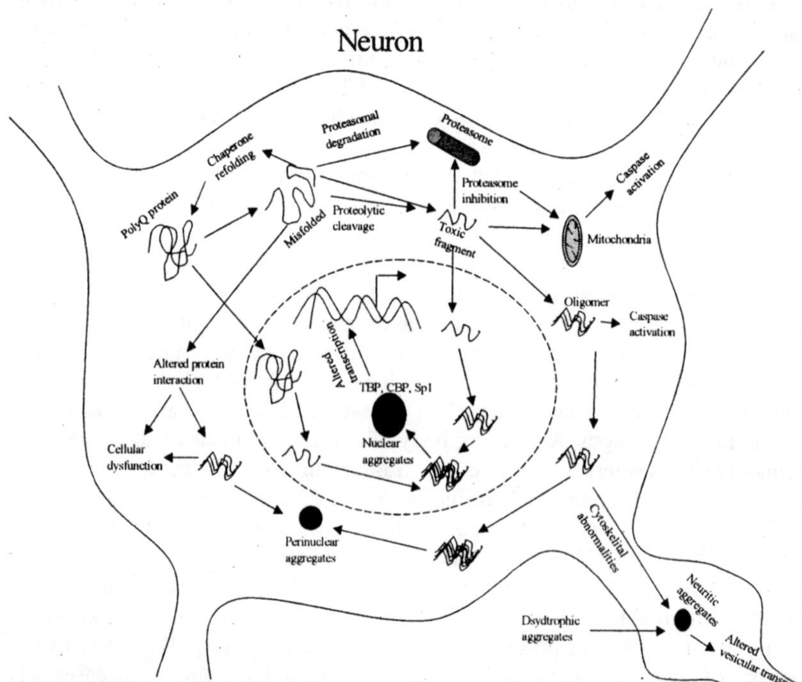

Figure 2. Model for polyglutamine disease pathogenesis.

References

Beal MF (2000) Trends Neurosci **23**: 298-304.

Bence NF, Sampat RM and Kopito R (2001) Science **292**: 1552-1555.

Cha JJ (2000) Trends Neurosci **23**: 387-392.

Cummings CJ, Mancini MA, Antalffy B, DeFranco DB, Orr HT and Zoghbi H Y (1998) Nature Genet **19**: 148-154.

Cummings CJ, Sun Y, Opal P, Antalffy B, Mestril R, Orr HT, Dillmann WH and Zoghbi HY (2001) Hum Mol Genet **10**: 1511-1518.

Cummings CJ and Zoghbi HY (2000) Hum Mol Genet **9**: 909-916.

Glickman MH and Ciechanover A (2001) Physiol Rev **82**: 373-428.

Jana N R, Tanaka M, Wang G and Nukina N (2000) Hum Mol Genet **9**: 2009-2018.

Jana NR, Zemskov EA, Wang G and Nukina N (2001) Hum Mol Genet **10**: 1049-1059.

Klement IA, Skinner PJ, Kaytor MD, Yi H, Hersch SM, Clark HB, Zoghbi HY and Orr HT (1998) Cell **95**: 41-53.

Luthi-Carter R, Strand A, Peters NL, Solano SM, Hollingsworth ZR, Menon A S, Frey AS, Spektor BS, Penney EB, Schilling G, Ross CA, Borchelt DR, Tapscott SJ, Young AB, Cha JH and Olson JM (2000) Hum Mol Genet **9**: 1259-1271.

Ona VO, Li M, Vonsattel JP G, Andrews LJ, Khan SQ, Chung WM, Fre AS, Menon A, Li X.-J, Stieg PE, Yuan J, Penne JB, Young AB, Cha J-H J and Friedlander RM (1999) Nature **399**: 263-267.

Paulson HL (1999) Am J Hum Genet **64**: 339-345.

Perutz MF, Johnson T, Suzuki M and Finch JT (1994) Proc Natl Acad Sci USA **91**: 5355-5358.

Sanchez I, Xu C-J, Juo P, Kakizuka A, Blenis J and Yuan J (1999) Neuron **22**: 623-633.

Saudou F, Finkbeiner S, Devys D and Greenberg M E (1998) Cell **95**: 55-66.

Tanaka Y, Engelender S, Igarashi S, Rao RK, Wanner T, Tanzi RE, Sawa A L, Dawson V, Dawson TM and Ross CA (2001) Hum Mol Genet **10**: 919-926.

Zoghbi HY and Orr HT (2000) Ann Rev Neurosci **23**: 217-247.

Molecular and Cellular Neurobiology
Editors: M.K. Thakur and S. Prasad

Molecular Genetics and Biology of Polyglutamine Expansion Diseases

Biswanath Chattopadhyay, Kanad Bakshi
Pritha Majumdar and Nitai P. Bhattacharyya

Crystallography and Molecular Biology Division, Saha Institute of Nuclear Physics
Bidhan Nagar, Kolkata-700 064, India

Abstract
To gather indirect information on the prevalence of the neurological diseases caused by the expansion of normally polymorphic CAG/CTG repeats, we have been studying variation of CAG/CTG repeats in normal populations and clinically diagnosed individuals at Huntingtin (Htt), SCA1, SCA2, SCA3, SCA6, SCA7, SCA8, SCA12, SCA17 and DRPLA loci since 1995. Haplotype analysis using linked markers in a number of loci has identified the reservoir alleles that are prone for expansion to the pathological state. Further, investigations on single nucleotide polymorphism and repeat variations in genes thought to involve in apoptosis and proteins that are known to interact with poly-Q (coded by CAG repeats) and form aggregation indicated that variations in these genes could modulate the age at onset in HD and SCA2. Generation of cell model with respect to accelerated apoptosis induction by expressing expanded poly-Q, proteomic studies using this cell model and 3-D structure determination of Htt interacting proteins would expect to throw light upon the pathology of the diseases and better intervention of the diseases

Keywords: ethnic populations, genetic epidemiology, eastern India, biology of polyglutamine expansion diseases

Introduction

Triplet repeats expansion disease constitutes a group of neurodegenerative/neuromuscular diseases caused by a novel mechanism of mutation: increasing the number of copies of the triplet repeats in the respective genes. Beginning with the identification of an expanded CCG repeat as the mutation causing Fragile-X mental retardation in 1991, till date there are at least 16 genetic diseases caused by the expansion of tri nucleotide repeats. These include Fragile-X mental retardation A (FRAX-A), E (FRAX-E), F (FRAX-F), myotonic dystrophy (DM), SCA8, SCA12 where the positions of the repeats are in the untranslated region of the genes. In spinobulbar mascular atrophy (SMBA), Huntington's diseases (HD), autosomal dominant spinocerebellar ataxia subtypes SCA1, SCA2, SCA3, SCA6, SCA7, SCA17, and dentatorubral-pallidoluysian atrophy (DRPLA), CAG repeats (coding glutamine, Q repeats) in coding sequences are expanded. Besides, in autosomal dominanat spinocerebellar ataxia subtype 10 (SCA10) where a pentanucleotide repeat (ATTCT) and in autososmal recessive disease Friedreich's ataxia (FRDA), GAA repeats are expanded in the intron of the genes. Summary of these diseases, locations of the genes on chromosomes, type and location of the repeats are shown in the Table 1. Short GCG expansion of the order of 8-13 repeats in the PABP2 gene causes oculopharyngeal muscular dystrophy (Brais et al 1998). In progressive myoclonous epelipsy, expansion of 12 nucleotides motif at the promoter region of Cystain B gene has been observed (Lalioti et al 1997).

There are two common features of the inheritance of these diseases and make them distinct from classical mode of genetic inheritance. Firstly, "anticipation", which is characterized by an increase in the severity of the disease and/ or a decrease in the age at onset as the disease is transmitted through generations, is often seen. Anticipation is associated with progressive increase in the repeats from pre-mutation to full mutation range, and further expansion from then onwards, through successive generations. The sole exception is spinocerebellar ataxia type 6 (SCA6), where "anticipation" is not apparent, possibly because expanded SCA6 alleles are usually transmitted stably (Zhuchenko et al., 1997). Secondly, some of these diseases can be associated with a parental sex bias on transmission (Telenius et al., 1993). In the present review, we shall concentrate only on the diseases where the expansion takes place in the coding sequences of the genes; increasing the number of glutamines (Q) in the respective proteins. This group of diseases is often referred to as "poly-Q" expansion diseases. Normally polymorphic repeats, in general, are believed to expand in two steps. In the first step, from a small number of repeats (normal repeat range) there is expansion to an intermediate number of repeats (pre-mutation). In the second step, there is a further expansion of the intermediate size to a larger number of repeats leading to the disease manifestation; premutation to full-mutation. Within families, carrier individuals carry the premutation, affected individuals carry the full mutation and unaffected individuals have alleles in the normal range. Individuals with alleles in the pre-mutation range show very little or no phenotypic expression of the disease but the alleles are genetically unstable. Prevalence of these diseases varies considerably worldwide. The prevalence of HD is 5-10 per 100,000 individuals in the most Western Europe, U.S.A. and Canada, while in Japan and China the prevalence is very low (Hayden et al 1980, Hayden 1981, Kanazawa 1983, Harper, 1991). Relative frequency of SCA3 has been the most common among all ADCA subtypes (see Table 2 for details). In recent studies on DM, HD and SCAs, the frequency distribution of large CTG/CAG repeat alleles in the normal range (large ANs) in populations have been shown to positively correlate with disease.

Frequency of CAG repeats greater than 20 in HD locus was correlated with the prevalence (Rubinsztein et al., 1994, Squiteiri et al., 1994). Similarly, presence of large ANs at SCA6 locus in Japanese populations has been correlated with the higher frequency of SCA6. On the contrary, presence of large ANs at SCA1 locus was higher in Caucasian populations compared to the Japanese. The frequency of SCA1 was also higher among Caucasian patients, compared to the Japanese patients (Takano et al 1998). These observations led to the postulation that alleles with large ANs are relatively unstable and undergo expansion to reach an intermediate range from where further expansion to the disease range takes place. Thus, investigations on the prevalence of large ANs in populations would reflect upon the prevalence of the diseases among populations.

We have been studying CAG repeats in HD, HDL2, SCA1, SCA2, SCA3, SCA6, SCA7, SCA17 (TBP) and DRPLA loci among clinically diagnosed patients and normal individuals originated from the same geographical region from where the patients originated (mainly from eastern India). Due to lack of well documented epidemiological data, major focus has been provided on the study of distribution of

CAG repeat variations at 5 of these disease loci among normal individuals of different ethnic groups of eastern India, the geographical region from where the patients originated.

Table 1. Trinucleotide repeats expansion disorders, chromosomal locations, repeat motif, position of the repeats in the genes and mutations

Disease	Gene	Chromosomal Location	Repeat Motif	Coding	Normal Range	Expanded Range
FRAXA	FMR1	Xq27.3	CGG	5'UTR	6-52	230->2000
FRAXE	FMR2	Xq28	GCC	5'UTR	6-35	200->1000
DM1	DMPK	19q13.1-q13.3	CTG	3'UTR	5-37	50->2000
SCA8	?	13q21	CTG	3'UTR	16-91	107-127
SCA12	PPP2R2B	5q31-q33	CAG	5'UTR	7-28	66-78
DM2	ZNF9	3q21	CCTG	Intron	?	75-11000
FRDA	Frataxin	9q13	GAA	Intron	7-22	120-1700
SCA10	?	22q13	ATTCT	Intron	10-22	280->4500
SBMA	AR	Xq11-q12	CAG	Poly-Q	7-34	36-68
HD	Huntingtin	4p16.3	CAG	Poly-Q	9-37	38-121
HDL2	Juncto -Philin-3	16q24. 3	CTG	Exon	7-26	44-57
SCA1	Ataxin1	6p22-p23	CAG	Poly-Q	6-44	40-83
SCA2	Ataxin2	12q	CAG	Poly-Q	11-31	35-59
SCA3	Ataxin 3/MJD1	4q32.1	CAG	Poly-Q	12-41	60-84
SCA6	CACNL1A	19p13	CAG	Poly-Q	4-17	20-30
SCA7	Ataxin7	13p12-13	CAG	Poly-Q	7-17	38-130
SCA17	TBP	6q27	CAG	Poly-Q	29-42	47-63
DRPLA	Atrophin-1	12p13.31	CAG	Poly-Q	3-35	49-88

In collaboration with the neurologists from Bangur Institute of Neurology and Calcutta Medical College, Kolkata; Advanced Diagnostic Center, Ranchi and Bombay Hospital, Mumbai and other neurologists in and around Kolkata, we have been diagnosing various neurological disorders caused by the expansion of triplet repeats. Summary of the data is presented in the Table 3. We did not find any expansion in HDL2 and DRPLA loci among the patients, which were found negative for HD. Besides, CAG repeats at DRPLA locus was also tested among SCA patients, which were negative for known SCAs tested (see Table 3). Patients tested for DM (68 patients from 76 families) and FRDA (8 patient from 6 families) were not included in the table. Expansion at SCA8 and SCA17 loci were not detected in any one of 190 families.

Variation in frequencies of large normal alleles at various loci: comparison with global populations

Various ethnic population-specific frequency distributions of CAG repeat alleles at HD, SCA1, SCA2, SCA3 and SCA6 loci have been published (Basu et al 2000a, Pramanik et al., 2000, Chattopadhyay et al 2003b). Summary of the result is shown in the Table 4.

Table 2. Relative prevalence of dominant spinocerebellar ataxias in patients of diverse ethnicity

POPULATION	SCA1	SCA2	SCA3	SCA6	DRPLA	References
CAUCASIANS (mixed)	15%	14%	30%	5%	0%	Takano et al. (1998), Am J Hum Genet 63: 1060-1066.
JAPANESE (Niigata Univ.)	3%	5%	43%	11%	40%	Takano et al. (1998), Am J Hum Genet 63: 1060-1066.
JAPANESE (central Hoshu island)	0%	5.9%	33.7%	5.9%	20%	Watanabe et al. (1998) Clin Genet 53:13-19.
JAPANESE (Hokkaido area)	10.5%	4.4%	24.6%	11.8%	1%	Sasaki et al. (1999) Nippon Rinsho 57:787-791
PORTUGUESE	0%	4%	74%	0%	0%	Silveira et al. (1998) Am J Med Genet 81:134-138
KOREAN	0%	12.6%	4.6%	6.9%	3%	Jin et al. (1999) J Neurol. 246:207-210
SPANISH	6%	15%	15%	1%	1%	Pujana et al. (1999) Hum Genet. 104:516-522.
UK	1.4%	4.1%	0.7%	4.8%	2%	Leggo et al. (1997) J Med Genet 34:982-985
INDIAN (North)	7.7%	25%	5%	0%	0%	Saleem et al. (2000) Hum Genet 106:179-187
INDIAN (East)	10.5	17.5%	7%	1.8%	0%	Basu et al. (2000a) Hum Genet 106:597-604
COMBINED	6.2%	8.7%	23.5%	6.4%	5.4%	

Several investigators noted a positive correlation between frequency of large alleles (>20 repeats) at the HD locus and prevalence (Rubinsztein et al 1994, Squiteiri et al 1994). These large normal alleles are prone to expand to the disease range and are possibly a reservoir for expanded alleles. Squiteiri et al (1994) have found lesser number of chromosomes with >20 CAG repeats in populations (for example, native African, Chinese, Japanese and Finnish) with lower prevalence of HD, compared to western populations. In Indian populations 6.9% (47 alleles) of the pool of 672 chromosomes is in the >20 CAG repeat range, even though there is variation across ethnic groups (Pramanik et al 2000). Across ethnic groups the frequency of alleles >20 repeats varies from absence in the Santals (Austro-Asiatic speaking tribal populations) to 20% in the Agharias. When a pool of the tribal populations were considered (N=340), they were found to contain significantly less large ANs >20

Table 3. Molecular diagnosis of neuro-genetic diseases caused by the expansion of triplet repastat SINP

Disease	Patient families tested	Independent families found positive (total)	Repeat range	Frequency of SCAs
HD	115	81 (92)	41-78	-
SCAs	**190**			
SCA1		15 (18)	44-53	0.08
SCA2		29 (41)	36-45	0.15
SCA3		8 (10)	61-69	0.04*
SCA6		1	21	0.005
SCA7		2	70-72	0.01
SCA12		6 (11)	53-67	0.03**

*Seven families were from West Bengal only ** Five families were from a specific community (decided from the family name). Dr. K.K. Sinha, Ranchi, sent to us all the families

repeats (χ^2= 8.14; P<0.01) compared to the caste populations (N=332). It is interesting to note that we have earlier shown at DMPK locus, one of the two Austro-Asiatic speaking tribal populations (the Santal) had the lowest mean CAG repeat size and does not have large normal alleles greater than 18 CAG repeats as observed for African populations (Basu et al 2001). In HD locus also the low frequency present in the tribal populations is similar to what has been reported for the native Africans (Squiteiri et al 1994). At the SCA1 locus, the frequency of $(CAG)_{>31}$ alleles, and at the SCA2 locus, the frequency of $(CAG)_{>22}$ alleles in any population are of epidemiological interest and correlates well with the prevalence of SCA1 and SCA2, respectively (Takano et al 1998). Globally, the Caucasian whites have both high frequencies of $(CAG)_{>31}$ repeats and also prevalence of SCA1 compared to the Japanese, who have both low prevalence of SCA1 and low frequency of large ANs. A similar observation has also been made with SCA2: ANs >22 and prevalence of SCA2 being higher in Caucasians than Japanese. In contrast, both large-size normal alleles and prevalence of SCA6 are higher in the Japanese population compared to Caucasians (Takano et al 1998). In Indian populations, the $(CAG)_{>31}$ frequency varies from 3.8% (Agharia) to 43.5% (Lodha), among the study populations (Basu et al., 2000a). Considering anthropological groups of a pool of castes (N=366) against tribals (N=368), it was observed that the frequency of large ANs >31 repeats was significantly higher in the tribals than the castes (χ^2 = 13.77; P<0.001). The $(CAG)_{>31}$ frequency of the pooled Indian data (N=734) is 23.3%. When this was compared with other world populations, the frequency was significantly higher than the Mongoloids (Japanese and Chinese; P<0.001), Dogrib,

Pehuenche, Cabecar, American and West Samoan, and was significantly less than only the New Guinea Highlander. However, there was no significant difference with Brazilian Blacks, Nigerians or the Caucasians (German, CEPH, Brazilian White).

At the SCA2 locus, the frequency of large ANs >22 repeats is 4.7% in the population pool out of 830 chromosomes. There were interesting patterns of variation of large ANs among ethnic groups. First, the maximum frequency at this range was found in the Kurumba, a tribal population of south India (15.4%), while alleles in this range were same in two other tribal populations; the Lodha of east India and the Kota of south India. This interesting variation among Kota and Kurumba was also seen in the DMPK and HD loci. Second, while it was expected that the Brahmins of UP, who are considered to have Caucasian affinities, should have a high percentage of large ANs but in reality they were found to be exclusively monomorphic for the 22 repeats allele. However, the Brahmins of West Bengal, another upper caste group from eastern India, did have the second highest frequency of large ANs among the study populations. Third, unlike most of the other loci, frequency of large ANs was not significantly different between pooled tribals (N=378) and pooled castes (N=452). The Indian pool (N=830) was further compared with the Japanese and Caucasian population presented in the study by Takano et al. The percentages of alleles >22 repeat in Indian, Japanese and Caucasian populations are 4.7%, 1% and 12%, respectively. Frequency of large ANs was significantly higher in the Indian population (pooled) compared to the Japanese population (χ^2 =17.96, P<0.0001). Similar comparison revealed that frequency of large ANs was significantly lower in the Indian population than in the Caucasian population (χ^2 =25.13, P<0.0001). The frequency of large ANs >22 repeats in our population is in the intermediate range between the two populations. However, the prevalence of SCA2 in our population appears to be highest among the SCA subtypes (Table 3) and is even higher than the Caucasians. This could either be due to small patient sample size, or this could also be due to the fact that the total number of large ANs in our pooled population is underrepresented, since at least four populations, the Lodha, Kota, Brahmin of UP and the Gaud, do not contribute any large ANs to the pooled data. At the SCA6 locus, the frequency of large ANs >13 repeats is 0.059 for the overall Indian populations. The maximum frequency of alleles in this range is found among the Kota (21.4%) while the Brahmin of WB, Santal, Tanti, Agharia, Bagdi and Rajput do not have any alleles in this range. The frequencies of alleles >13 repeat in the Indian, Japanese and Caucasian populations are 0.059, 0.2, and 0.04, respectively. The frequency of large ANs (>13 repeats) was significantly lower in the Indian population compared to the Japanese population (χ^2 =57.71, P<0.0001). No

Table 4. Distribution of (CAG)n repeats at the HD, SCA1, SCA2, SCA3, SCA6 loci in the pooled populations

Locus (2n)	Repeat range	Total no of alleles	Heterozygocity	CAG repeat sizes (mean ± S.D.)	Large normal alleles (frequency)
HD (672)	11-31	17	0.68	16.8 ± 2.0	>21 (0.069)
SCA1(734)	23-41	17	0.84	30.5 ± 2.5	>31 (0.23)
SCA2 (830)	14-31	12	0.14	22 ± 1.0	>22 (0.047)
SCA3 (824)	13-39	23	0.86	22.3 ±5.1	>33 (0.015)
SCA6 (638)	7-16	9	0.70	11.3 ± 1.9	>13 (0.059)

significant difference was observed between the Indian and Caucasian population. However, compared to a prevalence of SCA6 of 11% and 5% respectively among Japanese and Caucasian SCA patients, our data show a lower prevalence of SCA6 among SCA patients. This observation is consistent with the fact that 6 of the 14 ethnic populations studied do not have any large ANs greater than 13 repeats at this locus. It is interesting to note that the Tripuris, who are a Tibeto-Burman speaking tribal group, and the only Mongoloid population studied (they are a part of the Bodo ethnic stock who migrated from Tibet several centuries ago), stands out from the rest at this locus. They have the highest heterozygosity and the maximum number of alleles among the study populations, and unlike the other populations, have the 13-repeat allele as the most frequent allele. Detail analysis of CAG repeats at DRPLA locus in various ethnic populations of India is shown in the Table 5. Like the other loci, a similar positive correlation between frequency of large alleles (>17 repeats) at the DRPLA locus and prevalence has been noted (Takano et al 1998). In the Indian pool, 6.4% out of the 578 chromosomes is in the >17 CAG repeat range, even though there is variation across ethnic groups. Across ethnic groups the frequency varies from absence in the Kota, Tripuri and Chamar to 17.2% in the Mahishya. There was no significant difference between pooled tribal (N=218) and pooled caste populations (N=360).

The information gathered from the above studies in Indian populations, especially from eastern part of India throw some light on the prevalence of the diseases (Basu et al 2000a, 2001, Pramanik et al 2000, Chattopadhyay et al 2003b). This information together with information from other populations is summarized in the Table 6. Haplotype analysis using linked markers in and around DM locus, has shown that majority of the DM patients had the same haplotype indicating a common origin of the DM mutation worldwide. However, about 10 % of the Indian DM patients have different haplotype than that observed in other populations. This specific haplotype is also present in normal populations indicating that this haplotype is the characteristic of Indian populations. Further, the majority of large normal alleles (> 18 CTG repeats), had the same haplotype as observed in the patients, indicating that these large normal alleles act as the reservoir of the expanded alleles (Basu et al 2000b, 2001). In SCA3 locus, we have observed that in all the SCA3 patients, the expansion took place on the same haplotype back ground as observed worldwide. Besides, similar to what we observed in DM, majority of the large normal alleles (>33 CAG repeats) harbor the same haplotype as observed in the expanded chromosomes supporting the notion that the large normal alleles act as the reservoir of the expanded repeats (Chattopadhyay et al 2003b). Similar result was also obtained in HD locus (Unpublished observations).

Molecular pathology in Huntington's disease

A large number of studies using animal model (knock in as well knock out) and cell model (for detail of the cell model see Table 7) was developed in last 10 years to understand the pathology of triplet repeat expansion disorders. In the present section, we shall only concentrate on Huntington's disease due to limitation in space. Aggregation of expanded poly-Q, popularly known as neuronal intranuclear inclusions (NII), possibly due to conformation changes of the protein, is the hallmark of this group of diseases (see Table 7, reviewed by Hoffner and Dijian, 2002). Subsequently, over the years, it has been found that more than 60 proteins interact with Htt and are co-localized with such aggregation (reviewed by Harjes and Wanker, 2003). These interacting proteins involved in protein trafficking (e.g. HIP-1), cell structure maintenance (e.g. tubulin), cell signaling (e.g. Grb-2), transcription

Table 5. Analysis of CAG repeats at the DRPLA locus in ethnic populations of India

Population (2n)	CAG repeat frequency at different class intervals				Allele range	Total no. of alleles	Heterozygosity	Mean±SD
	6-9	10-13	14-17	>17				
BRA-WB (28)	0.179	0.071	0.679	0.071	7-19	8	0.81	13.6±3.4
SAN (38)	0.105	0.158	0.632	0.105	7-23	11	0.85	14.5±3.0
LOD (50)	0.26	0.12	0.56	0.06	6-21	12	0.88	13.2±3.7
TRI (42)	0.238	0.143	0.619	0	7-17	10	0.84	13.3±3.4
MAH (58)	0.207	0.103	0.517	0.173	7-25	15	0.88	14±4.0
AGH (42)	0.143	0.167	0.619	0.071	6-18	9	0.83	13.5±3.3
GAU (20)	0.15	0.05	0.70	0.10	7-20	10	0.80	14.1±3.1
TAN (30)	0.167	0.167	0.566	0.1	7-35	9	0.84	14.7±6.2
BAG (62)	0.161	0.065	0.709	0.065	7-25	12	0.78	13.9±3.5
BRA-UP (26)	0.346	0.077	0.538	0.039	7-19	9	0.85	12.5±3.8
RAJ (64)	0.156	0.203	0.625	0.016	7-18	11	0.80	13.1±2.6
CHA (30)	0.366	0.234	0.4	0.0	7-17	9	0.85	11.5±3.3
KOT (54)	0.13	0.129	0.741	0.0	7-17	7	0.53	13.0±2.4
KUR (34)	0.118	0.088	0.676	0.118	8-23	11	0.81	14.9±3.6
POOLED (578)	**0.189**	**0.129**	**0.618**	**0.064**	**6-35**	**21**	**0.86**	**13.5±3.6**

Abbreviations used for ethnic populations:
BRA-WB: Brahmin of West Bengal; SAN: Santhal; LOD: Lodha; TRI: Tripuri; MAH: Mahishya;AGH : Agharia; GAU: Gaud; TAN: Tanti; BAG: Bagdi; BRA-UP : Brahmin of Uttar Pradesh; RAJ: Rajpu · CHA: Chamar; KOT: Kota; KUR: Kurumba. (For detail of the populations etc. see Basu et al., 2001)

modulator (e.g. SP1, p53), receptor (e.g. N-CoR) and apoptosis (e.g. caspase 8 and caspase 10). In some cases, it has been shown that such interaction depends on the number of reiterating Q-residues in Htt protein. On the basis of large number of experiments, it has been proposed that these aggregations of proteins stimulate the apoptotic pathway. Initial report of involvement of apoptosis in HD (Dragunow et al. in 1995) was subsequently supported by a large number of observations in animal and cell model. Release of Cytochrome-c from the mitochondria, an indication of apoptosis and activation of caspase 9 have been reported very recently in HD patients and in the mouse model of HD (Kiechle et al 2002). Role of caspase1 is predicted from the observations that in the transgenic mice model of HD, where caspase 1 is also absent, showed delayed progression of HD compared to the mice containing expanded poly glutamine and caspase 1 (Ona et al 1999). It has also been observed that inhibitor of caspase 1 and caspase 3 delayed the mortality in HD mouse model (Chen et al., 2000). Subsequently, it has been shown that Caspase 1 as well as caspase 3 expressions is increased in the brain of HD mouse and HD patients. Caspase 8 is also required for cell death induced by poly glutamine repeats (Sanchez et al 2001). Caspase-12 has also been activated by poly-Q aggregates via a pathway independent of caspase-8 and caspase-3 activation, and caspase-12 activation is closely associated with poly-Q aggregate-mediated cell death (Kouroku et al 2002).

Table 6. Summary of the information on the prevalence of the diseases from our studies in comparison with other published data

Disease	Populations	Prevalence of the disease	Frequency of large Normal alleles (large- ANS)	Comments
HD	German	High	0.21	
	Japanese	Low	0.05	
	India	Not available	0.07	Intermediate in India
SCA1	German	0.15*	0.18	
	Japanese	0.02	0.03	
	India	0. 08	0.23	????
SCA2	German	0.14	~0.12	
	Japanese	0.05	~0.03	
	India	0.15	0.05	Similar to Germany
SCA3	German	0.30	0.02	
	Japanese	0.43	0.05	
	India	0.04	0.01	Lowest in India
SCA6	German	0.05	0.04	
	Japanese	0.11	0.16	
	India	0.005	0.05	Rare in India
SCA7	German	0.12	-	
	Japanese	Not available	Not available	
	India	0.01	0.01	Rare in India
DRPLA	German	<0.05	0.08	
	Japanese	0.20	0.17	
	India	<0.008	0.06	Absent in India

* Prevalence of SCAs was determined among the autosomal dominant cerebellar ataxias tested in each communication. The references are not included.

Neurodegenaration and cancer

From the above discussion it is apparent that apoptotic pathway, especially caspases are involved in neuronal cell death and likely to be enhanced in HD. On the contrary, in cancer it is known that the apoptotic pathway is reduced generally (reviewed by Johnstone et al 2002). Thus, the proteins involved in apoptosis may play an opposite role in these two diseases: in HD there is enhanced apoptosis, while in cancer the apoptosis is reduced in general. It has been shown in a large collection of HD families, that prevalence of cancer is significantly lower among HD patients than that observed in general populations and proposed that expanded polyglutamine stretch in Htt protein protects against cancer by inducing or increasing apoptosis (Sorensen et al 1999). Further, the indication of overlapping pathway between the cancer and HD came from the observations that Htt interacting protein HIP-1 has been found to over express in a number of cancer (Rao et al 2002) and fusion of HIP-1 to platelet-derived growth factor beta receptor has been implicated in leukemia. It has recently

Table 7. Poly glutamine expansion, aggregation and apoptosis in diseases

Disease	Anticipation	Gene	Coding	Expanded range	Aggregation in cell model(C), animal model (A) or human brain (H)	Apoptosis
SBMA	No	AR	PolyQ	36-68	C[1], A[2]	
HD1	Yes	Huntingtin	PolyQ	38-121	C[3], A[4], H[5]	yes
HDL2	NR	Juncto-Philin-3	?	44-57	NR	NR
SCA1	Yes	Ataxin1	PolyQ	40-83	C[6], H[7], A[7]	
SCA2	Yes	Ataxin2	PolyQ	35-59	C[8] (cytoplasmic micro-aggregate)	
SCA3	yes	Ataxin3/ MJD	PolyQ	60-84	C[8], H[10], A[9]	Yes
SCA6	Yes/No	CACNL1A	PolyQ	20-30	NR	
SCA7	Yes	Ataxin7	PolyQ	38-130	C[11], A[12], H[13]	
SCA17	NR	TBP	PolyQ	47-63	C[14], H[15]	
DRPLA	Yes	Atrophin-1	PolyQ	49-88	C[16], H[17]	

1. Merry *et al.* Hum Molec Genet 7: 693-701, 1998; 2. Ross CA. Neuron 35: 819-822, 2002; 3. Hackam *et al* J Cell Biol 141: 1097-1105, 1998; 4. Davies *et al* Cell 90:537-548, 1997 ; 5. DiFiglia *et al* Science, 277:1990-1993, 1997; 6. Matilla *et al.* Nature 389: 974-978, 1997; 7. Skinner *et al.* Nature 389:971-974, 1997; 8. Huynh *et al* Nature Genet, 26: 44-50, 2000; 9. Ikeda *et al.* Nature Genet 13:196-202, 1996; 10. Paulson *et al* Neuron 19:333-344, 1997; 11. Brice *et al* Hum Molec Genet 10: 2569-2579, 2001; 12. Yoo SY *et al* Neuron 37: 383-401, 2003; 13. Holmberg *et al* Hum Molec Genet 7: 913-918, 1998; 14. Reid *et al* Neurobiol Disease 13:37-45, 2003; 15. Nakamura *et al* Hum Molec Genet 14:1441-1448, 2001; 16. Shimohata *et al* Neurosci Lett 323: 215-218, 2002; 17. Hayashi *et al* Acta Neuropath 96:547-552, 1998

been shown that poly glutamine toxicity in drosophila model system for HD is suppressed by a drosophila homologue of myeloid leukemia factor 1 (MLF1). Besides, MLF1 has been co-localized with the poly-Q aggregates (Kazemi-Esfarjani and Benzer, 2002). MLF1 is involved in normal hematopoietic differentiation as well as in erythroid/myeloid lineage switching (by similarity) and is localized in cytoplasm. This protein is the most abundant in testis, ovary, skeletal muscle, heart, kidney and colon and low expression in spleen, thymus and peripheral blood leukocytes. These set of data together with a number of publications showing involvement of signaling proteins in HD indicate that the pathway of neurodegeneration and cancer may overlap.

Alterations of gene expressions by poly-Q expansion

Modulation of gene expression by the expansion of poly-Q stretches in cell and animal model has been reported in a number of cases and has been reviewed recently. Analyzing a number of data derived from the microarray, it has been proposed that disruption of transcriptional pathways plays an important role in poly-Q expansion diseases (Sugars and Rubinsztein, 2003). At least seven transcription factors have been shown to interact with the poly-Q aggregates. For example, p53 protein, popularly known as the "guardian of the genome" and mutated in 50% of all type of cancer, has been identified with the aggregates (Steffan et al., 2000). Similarly, other transcription factors like CBP, SP1, P300 and transcriptional co-repressor mSin3a have also been identified with the aggregates and likely to alter the transcription in the diseases (Sugars and Rubinsztein, 2003).

Studies of single nucleotide polymorphisms in proteins involved in apoptosis and interact with aggregation: possible role of these proteins in modulation of HD

We have earlier shown that variations in GluR6 gene and CCG repeats in htt gene may explain 6.2% and 3.1%, respectively, of the unexplained variability in the age at onset in HD. Similar analysis with SCA2 patients indicated that variations in RAI1 gene may explain about 13% of the unexplained variability in the age at onset. Specific alleles in GluR6 and CA150 locus were only observed in HD patients (Chattopadhyay et al 2003a). We have further studied single nucleotide polymorphisms in apoptotic genes (human caspase activated DNase: hCAD, survivin, caspase 9, caspase 8), Htt interacting protein p53 and the candidate gene for aggregation formation saitohin among HD, SCA and cancer patients as well as normal individuals. Preliminary result indicated that different alleles at hCAD genes are over represented in HD and cancer patients in comparison with that found in normal individuals. This result supports the notion that different alleles in hCAD gene act oppositely in HD and cancer. Besides, variations in an anti-apoptotic gene survivin and pro-apoptotic gene caspase 9 and hCAD modulate the age at onset in HD patients. Similar result has also been obtained with variations in p53 gene and saitohin gene (data not shown). This set of data indicates that apoptotic genes as well as other genes involved in aggregation formation might play a role in the pathology of HD.

Future Perspectives

Complete descriptions of the clinical and familial features of HD have been first described as early as 1872 by George Huntington. After more than 100 years the gene has been localized on the short arm of chromosome 4 (4p16.3) in 1983. Finally the gene, initially called IT15 (Interesting Transcript # 15) and now known as huntingtin has been cloned and characterized. The gene consists of 67 exons spanning about 180 kb genomic region and codes for 3,144 amino acids with unknown function. However, presently no treatment is available for this devastating disease. Researches in the last 10 years indicated that the mutation to this gene alter the apoptotic pathway by interacting with a large number of other proteins, among which several transcription factors attract more attention. Based on the cell and animal model studies several inhibitors of apoptosis (like minocyclin, which actually inhibits caspases) are now being tried for the treatment of the disease. Recently, a DNA based vaccine has been proposed for the cure (Miller et al 2003). Further studies involving the variations in the genes involved in apoptosis and huntingtin interacting proteins as well as 3-D structures of the interacting proteins are necessary to understand the pathology of the disease for the better intervention.

Acknowledgements

We acknowledge Dr. Priyadarshi Basu and Dr. S. Pramanick, for carrying out some of the experiments in our laboratory. One of us (NPB) acknowledges partial financial support from the Department of Biotechnology, Govt. of India for some of the work presented here. We are thankful to Dr. T. Roy, Dr. P. K. Ganguly, Dr. S. Das, Bangur Institute of Neurology; Dr. S.C. Mukherjee, Calcutta Medical College Hospital, Dr. K.K. Sinha, Advanced Diagnostic Center, Ranchi, Dr. B. S. Singhal, Bombay Hospital, Mumbai and other neurologists who provided blood samples of the patients. At last but not the least the anonymous patients and normal individuals who voluntarily participated in the study are being acknowledged. Due to space limitation all the papers in the area were not referred.

References

Basu P, Chattopadhyay B, Gangopadhaya PK, Mukherjee SC, Sinha KK, Das SK, Roychoudhury S, Majumder PP, Bhattacharyya NP (2000a) Human Genet **106**: 597-604.

Basu P, Gangopadhaya PK, Mukherjee SC, Das SK, Sinha KK, and Bhattacharyya NP (2000b) Human Mutaion, Mutaion in Brief 369 online.

Basu P, Majumder PP, Roychoudhury S and Bhattacharyya NP (2001) Human Genet **109**: 310-317

Brais B, Bouchard JP, Xie YG, Rochefort DL, Chretien N, Tome FM, Lafreniere RG, Rommens JM, Uyama E, Nohira O, Blumen S, Korczyn AD, Heutink P, Mathieu J, Duranceau A, Codere F, Fardeau M, Rouleau GA, Korcyn AD (1998) Nature Genet **18**: 164-167.

Chattopadhyay B, Gangopadhyay PK, Das SK, Roy T, Sinha SK, Jha DK, Mukherjee SC, Chakraborty, A, Singhal BS, Bhattacharya, AK and Bhattacharyya NP (2003a) Neuroscience Letters **345**: 93-96

Chattopadhyay B, Basu P, Gangopadhyay PK, Roy T, Mukherjee SC, Sinha KK, Chakraborty A, Roychoudhury S, Majumder PP and Bhattacharyya NP (2003b) Acta Neurologica Scandinavica **108**: 407-414.

Chen M, Ona VO, Li M, Ferrante RJ, Fink KB, Zhu S, Bian J, Guo L, Farrell LA, Hersch SM, Hobbs W, Vonsattel JP, Cha JH and Friedlander RM (2000) Nature Med **6**: 797-801.

Davies SW, Turmaine M, Cozens BA, DiFiglia M, Sharp AH, Ross CA, Scherzinger E, Wanker EE, Mangiarini L and Bates PG (1997) Cell **10**:537-548.

Dragunow M, Faull RL, Lawlor P, Beilharz EJ, Singleton K, Walker EB and Mee E (1995) Neuroreport **6**: 1053-1057

Harjes P and Wanker EE (2003) Trends in Biochem Sciences **28**: 425-433.

Harper PS. (1991) Huntington's disease. WB Saunders Co., London.

Hayden MR, Macgregor J and Beighton PH (1980) S Afr Medical J **58**: 193-196.

Hayden MR (1981) Huntington's chorea. New York: Springer-Verlag

Hoffner G and Djian P (2002) Biochimie **84**: 273-278.

Jin DK, Oh MR, Song SM, Koh SW, Lee M, Kim GM, Lee WY, Chung CS, Lee KH, Im JH, Lee MJ, Kim JW and Lee MS (1999) J Neurol **246**: 207-210.

Johnstone RW, Ruefli AA and Lowe SW (2002) Cell **108**: 153-164.

Kanazawa I (1983) Ministry of Health & Welfare of Japan 151-156.

Kazemi-Esfarjani P and Benzer S (2002) Hum Mol Genet **15**: 2657-2672.

Kiechle T, Dedeoglu A, Kubilus J, Kowall NW, Beal MF, Friedlander R, Hersch SM and Ferrante RJ (2002) Neuromolecular Med **1**: 183-195.

Kouroku Y, Fujita E, Jimbo A, Kikuchi T, Yamagata T, Momoi MY, Kominami E, Kuida K, Sakamaki K, Yonehara S, Momoi T (2002) Hum Mol Genet **11**: 1505-1515.

Lalioti, MD, Scott HS, Buresi C, Rossier C, Bottani A, Morris MA, Malafosse, A and Antonarakis SE (1997) Nature **386**: 847-851.

Leggo J, Dalton A, Morrison PJ, Dodge A, Connarty M, Kotze MJ and Rubinztein DC (1997) J Med Genet **34**: 982-985.

Miller TW, Shirley TL, Wolfgang WJ, Kang X and Messer A (2003) Mol Ther **7**: 572 579.

Ona VO, Li M, Vonsattel JP, Andrews LJ, Khan SQ, Chung WM, Frey AS, Menon AS, Li XJ, Stieg PE, Yuan J, Penney JB, Young AB, Cha JH, Friedlander RM (1999) Nature **399**: 263-267.

Pramanik S, Basu P, Gangopadhaya PK, Sinha KK, Jha DK, Sinha S, Das SK, Maiti BK, Mukherjee SC, Roychoudhury S, Majumder PP and Bhattacharyya NP (2000) Eur J Hum Genet **8**: 678-682.

Pujana MA, Corral J, Gratacòs M, Combarros O, Berciano J, Genís D, Banchs I, Estivill X and Volpini V (1999) Hum Genet **104**: 516-522.

Rao DS, Hyun TS, Kumar PD, Mizukami IF, Rubin MA, Lucas PC, Sanda MG and Ross TS (2002) J Clin Invest **110**: 351-360.

Rubinsztein DC, Amos W, Leggo J, Goodburn S, Ramesar RS, Old J, Bontrop R, McMahon R, Barton DE, Ferguson-Smith MA (1994) Nat Genet **7**: 525-530.

Saleem Q, Choudhury S, Mukherji M, Bashyam L, Padma MV, Maheshwari MC, Jain S and Brahmachari SK (2000) Hum Genet **106**: 179-187.

Sanchez Mejia RO and Friedlander RM (2001) Neuroscientist **7**: 480-489.

Sasaki H and Tashiro K (1999) Nippon Rinsho **57**: 787-91.

Silveira I, Coutinho P, Maciel P, Gasper C, Hayes S, Dias A, Guimaraes J, Loureiro L, Sequeiros J and Rouleau GA (1998) Am J Med Genet **81**: 134-138.

Sorensen SA Fenger K and Olsen JH (1999) Cancer **86**: 1342-1346.
Squitieri F, Andrew SE, Goldberg YP, Kremer B, Spence N, Zeisler J, Nichol K, Theilmann J, Greenberg, J, Goto J, Kanazawa I, Vesa J, Peltonen L, Almqvist E, Anvert M, Telenius H, Lin B, Napolitano G, Morgan K and Hayden MR (1994) Hum Molec Genet **3**: 2103- 2114.
Sugars KL and Rubinsztein DC (2003) Trends in Genet **19**: 233-238.
Takano H, Cancel G, Ikeuchi T, Lorenzetti D, Mawad R, Stevanin G, Didierjean O, Durr A, Oyake M, Shimohata T, Sasaki R, Koide R, Igarashi S, Hayashi S, Takiyama Y, Nishizawa M, Tanaka H, Zoghbi H, Brice A and Tsuji S. (1998) Am J Hum Genet **63**: 1060-1066.
Telenius H, Kremer HP, Theilmann J, Andrew SE, Almqvist E, Anvret M, Greenberg C, Greenberg J, Lucotte G, Squitieri F, Starr E, Goldberg YP and Hayden MR (1993) Hum Mol Genet **2**: 1535-1540.
Watanabe H, Tanaka F, Matsumuto M, Doyu M, Ando T, Mitsuma T and Sobue G (1998) Clin Genet **53**: 13-19.
Zhuchenko O, Bailey J, Bonnen P, Ashizawa T, Stockton DW, Amos C, Dobyns WB, Subramony SH, Zoghbi HY and Lee CC (1997) Nature Genet **15**: 62-69.

Molecular and Cellular Neurobiology
Editors: M.K. Thakur and S. Prasad

Huntington's Disease, an Excitotoxic Neurodegenerative Disorder Engineered Genetically

Shripad B. Deshpande

Department of Physiology, Institute of Medical Sciences
Banaras Hindu University, Varanasi, India

Summary

Huntington's disease (HD) is an autosomal dominant disorder of basal ganglia affecting the elderly. More than normal numbers of CAG repeats are seen on a gene in chromosome 4 which expresses polyglutamine in *huntingtin* protein. This protein is distributed widely in all the neurons and its function is yet not known. Since this protein is present in the entire nervous system, the reasons for selective loss of striatal cells are not known. The possible mechanisms underlying the selective loss striatal neurons are attributed to glutamate excitotoxicity, anatomical vulnerability of lateral striatal artery, dopamine autotoxicity, astrocytic dysfunction, endothelial dysfunctions, hormonal influences and metabolic factors. The primary abnormality in astrocytes has also been implicated for the neuronal loss. Taken together, each one of them certainly influences the neuronal functions but when combined with the other factors, advance the process of neurodegeneration. Thus, HD is produced by complex series of pathological changes in a genetically weak animal.

Keywords: Astrocytes, 3-Nitropropionic acid, Excitotoxicity, Dopamine autotoxicity, Blood-brain barrier

Introduction

The neurons, cells of nervous tissue, continuously work to regulate the bodily functions in man and animals. These cells are subjected to toxic, metabolic, ischemic, and/or infectious insults. Genetically abnormal individuals express different types of proteins so that even the innocuous insults can produce disintegration or degeneration of neurons. The selective degeneration of neurons with ageing can occur in three different types of cells in the central nervous system, causing three different, progressive, crippling and eventually fatal disorders. Degeneration of cholinergic neurons in the basal forebrain area leads to Alzheimer's disease (Kelleher and Growdon, 2002); degeneration of dopaminergic neurons in the nigro-striatal region result in Parkinson's/ Huntington's disease (Aminoff, 2001); and degeneration of cholinergic motoneurons in the brain stem and spinal cord is associated with amyotrophic lateral sclerosis (Brown, 2001). The abnormalities in chromosome 14, 1 and 21 are implicated in Alzheimer's disease (Kelleher and Growdon, 2001). The mutation of Cu/Zn superoxide dismutase (SOD-1) on chromosome 21, which results in the accumulation of free radicals to kill the neurons, as, happens in amyotrophic lateral sclerosis (Brown, 2001). The spinocerebellar ataxias are also included in the class of neurodegenerative disorders.

The neurodegenerative disorders in general are shown to be resultant of excitotoxicity of varying degree in different regions of the brain. Thus, they have some common pathological features apart form the dementia and genetic linkage. Further, they are thought to be a variant of same disease complex affecting different regions of the brain. Thus, understanding one disease may benefit the others. In this article the pathophysiology of Huntington's disease is described as it may throw light on other disease processes.

Huntington's disease

Huntington's disease (HD) is a neurodegenerative disorder involving basal ganglia with the loss of striatal dopaminergic neurons. HD causes progressive dementia in addition to chorea, indicating the involvement of neocortical structures. No single factor is still held responsible for neuronal loss. Its pathological characteristic is the selective degeneration of subsets of neurons, primarily those in striatum and neocortex (Aminoff, 2001). Medium spiny neurons in the striatum with the preservation of subpopulation of neurons containing acetylcholinesterase have been shown.

Genetic link of HD

HD is an autosomal dominant genetic disorder and the gene is present in chromosome 4P. Presence of this gene results in expansion of CAG repeats (normal repeats = 15-20) more than 45. The severity of the disease can be directly correlated with greater number of CAG expansions. These expansions lead to the expression of polyglutamine in a protein of unknown function known as *huntingtin* (Behrens, 2000). The greater number of polyglutamine residues result in altered protein binding reaction which in turn interfere with the mitochondrial function. This results in protein fragmentation and neuronal death. Recent studies in transgenic animals expressing a construct of *huntingtin* protein as well as examination of autopsy material from HD patients have revealed the presence of intranuclear inclusions of ubiquinated *huntingtin* peptide containing pyroglutamine repeats (Behrens, 2000). The presence of genetic trait may not alone be responsible for the aetiopathogenesis of the disorder, as it can not answer all the questions for the selective loss of the striatal neurons. Hence, several other factors are also proposed for the aetiopathogenesis of the disorder. They are, glutamate excitotoxicity, dopamine autotoxicity, metabolic, hormonal, nutritional, infectious, anatomical and other factors. In order to examine the disease process appropriate animal models are required.

Animal models of HD

Several experimental models of HD have been described. Transgenic mice are genetically engineered animals and have features of both Parkinson's disease and HD. Further, they mostly resemble the genetically vulnerable individuals but the occurrence of disease in otherwise normal individuals cannot be explained or examined. The 1-methyl-4-phenyl-1,2,3,6 tetrahydropyridine (MPTP)- toxicity is another model and resemble Parkinson's disease rather than the HD. In animals, HD can also be produced by injecting 6-hydroxydopamine (6-OHDA) locally to destroy monaminergic neurons. Injection thus should be restricted to the area, if the spread of 6-OHDA is not restricted then the surrounding monoaminergic neurons are also destroyed. Injection of quinolonic acid (QA) locally also destroys the striatal neurons. This produces the behavioral changes that mimic more of the earlier symptoms of HD (Dunbar et al 2000). This model has also the limitation as described for 6-OHDA. The most well known animal model of HD is that produced by the administration of 3-nitropropionic acid (3-NPA; Beal et al 1993; Nishino et al., 1995). 3-NPA is a fungal toxin from the mildewed sugarcane and inhibits the mitochondrial succinate dehydrogenase irreversibly (Alston et al 1977). The 3-NPA *in vivo* produces selective loss of striatal neurons (Nishino et al 1995, 1997, 1998).

3-NPA toxicity

The mitochondrial toxin, 3-NPA has received substantial attention, as it produces pathological and behavioral deficits observed in HD. In mid 1970s, the 3-NPA was isolated as a toxic substance for the wide spread poisoning of the domestic livestock (Dunbar et al 2000). However, its significance as potential tool for gaining a better understanding of HD was catapulted by reports of its effects on children in China who were exposed to it after ingesting the mouldy sugarcane (Ludolph et al 1991; Lui et al 1992). Some of these children displayed dystonia or severe plastic rigidity, including tonic flexion of the arm and legs. The computed tomography (CT) scans of these children revealed bilateral basal ganglia hypodensity (Ludolph et al 1991). The major neurochemical action of 3-NPA is irreversible inhibition of the mitochondrial enzyme succinate dehydrogenase (SDH) present in the tricarboxylic acid cycle and component of complex II of the electron transport chain.

Despite knowing the primary action of 3-NPA, it is unclear why repeated systemic administration results in preferential damage to striatum. The striatal vulnerability is not the result of enrichment of SDH at this site (Pang et al 1996) and systemic administration of 3-NPA does not result in greater impairment of SDH activity in the striatum as compared to other regions of the brain (Gould et al 1985; Brouillet et al 1998). The simple explanation could be a greater sensitivity of striatal neurons to 3-NPA. However, cultured striatal hippocampal, septal and hypothalamic neurons are equally sensitive to 3-NPA (Fink et al 1996). In addition, hippocampus is as sensitive as striatum to the effects of malonate, a reversible inhibitor of SDH (Pang et al 1996). Thus it was proposed that the selective striatal vulnerability to 3-NPA is due to the disruption of blood-brain barrier (BBB) at striatal area (Hamilton and Gould, 1987; Nishino et al 1995).

Glutamate excitotoxicity

The 3-NPA toxicity resembled that of kainic acid and was proposed to be due to the excitotoxic mechanism (Hamilton and Gould, 1987). Excitotoxicity refers to the over stimulation of neurons with glutamate, or toxins such as kainate resulting in neuronal death (Olney, 1990). Excitotoxicity is largely mediated by ionotrophic glutamate receptors, with the activation of NMDA receptors resulting in more rapid neuron death than with non-NMDA receptors (Choi, 1992; Geddes et al 2000). In cultures, the neuron death involves rapid swelling, mediated by the influx of Na^+ and Cl^- and delayed component resulting from Ca^{2+} influx (Choi, 1992). Although excitotoxin-induced neuron death is typically necrotic, recent studies indicate the apopototic component (Ankarcrona et al 1995; Ferrer et al 1995). Excitotoxic neuronal damage should be associated with high levels of extracellular glutamate (Siesjo and Bengtsson, 1989) but no such increase in extracellular glutamate was seen. Hence, an indirect excitotoxic action was proposed (Novelli et al 1988). According to this, impairment of energy metabolism results in the lowering of ATP. This leads to the failure of ion pumps (Na^+-K^+ ATPase) that maintain membrane polarity. The depolarization of the membrane overcomes the voltage-dependent block exerted by Mg^{2+} on the NMDA receptor. This allows glutamate to chronically activate the NMDA receptor and results in excitotoxic neuronal damage. Subsequent studies demonstrate that neuronal damage following metabolic impairment is mediated via NMDA receptor activation and excitotoxic mechanisms (Zeevalk and Nicklas, 1991, 1992).

The systemic administration of 3-NPA does not result in an elevation of extracellular glutamate levels in the CNS (Beal et al 1993, Fu et al 1995) yet much of the neuronal damage is thought to occur by excitotoxic mechanisms as evidenced by the relative sparing of NADPH-diaphorase neurons (Beal et al 1993) and protection against striatal toxicity by prior decortication (Beal et al 1993, Fu et al 1995). Moreover, systemic administration of 3-NPA exacerbates the toxicity of intra-striatal NMDA administration (Simpson and Issacson, 1993). The indirect excitotoxic hypothesis helps to explain the similarity between neuronal damage induced by repeated 3-NPA administration and acute excitotoxic insults such as kainic acid toxicity or ischemia (Hamilton and Gould, 1987).

Dopamine autotoxicity

As outlined before, 3-NPA administration can result in rapid necrotic or slow apoptotic death, both *in vivo* and *in vitro.* Ionotrophic glutamate receptors, particularly NMDA receptors, mediate the necrotic but not the apoptotic death. However, these results do not explain the preferential vulnerability of striatum to neuronal damage following systemic 3-NPA administration. The striatum is densely innervated by dopaminergic fibres that originate in substantia nigra. Several lines of evidences indicate that dopamine may exacerbate the excitotoxicity in the striatum. Dopamine (DA) levels are increased in the striatum after forebrain ischemia (Globus et al 1987; Slivka et al 1988). Moreover, dopamine depletion attenuates striatal damage induced by both NMDA and non-NMDA receptor agonist (Buisson et al 1991; Chapman et al 1989; Filloux and Wamsley, 1991).

Recently, a moderate increase in DA and DOPAC concentrations in the striatal dialysate was demonstrated after a single administration of 3-NPA in rats with no behavioral disturbances (Nishino et al 1997). However, the second administration of 3-NPA manifested with increase in DA and DOPAC for more than 2 fold and maintained for more than 6 hours with concurrent development of motor symptoms in the animals (Nishino et al 1997). Thus, the increase in DA and its turnover closely paralleled with appearance of behavioral disturbances. Further DA also increases the Ca^{2+} level in cultured astrocytes. Thus, under normal physiological conditions astrocytes might be involved in the regulation of striatal neuronal activity by modulating the intracellular Ca^{2+}. DA is easily oxidized to yield reactive oxygen intermediates (Ben Sacher et al 1995). Thus, excessive turnover/release of DA might lead to increased production of free radicals which might occur either directly (by oxidation of DA) or indirectly by Ca^{2+} overload in the striatal cells. Now it can be proposed that the sustained increase in intracellular Ca^{2+} brought about by 3-NPA intoxication and increased levels of DA trigger the non-apoptotic astrocytic death.

Dopamine depletion by 6-OHDA treatment or attenuation of DA release by quinipirole (D2 receptor agonist) decreased the incidence of motor symptoms as well as IgG extravasation. On the other hand spiperone, D2 receptor antagonist, enhanced both motor symptoms as well as striatal damage. Thus, greater striatal DA release would enhance the 3-NPA toxicity (Nishino et al 2000).

Role of astrocytes for the neuronal loss

The 3-NPA toxicity of rats was manifested with the loss of GFAP immunoreactive areas in the striatum with abundant GFAP-labeled astrocytes in the periphery (Nishino et al 1995). Further, the terminal depolarization of the neurons was caused by failure of ion exchange due to the large depolarization of the glia (Reipe et al 1992). Further, 3-NPA increased the influx of Ca^{2+} in the astrocytes involving L-type of Ca^{2+} channels partially. But, the increase of intracellular Ca^{2+} was completely blocked by the blockers of Na^{+}-Ca^{2+} exchanger system. The increase of Ca^{2+} did not involve the glutamate receptors (Deshpande et al 1997). Moreover, astrocytes are also necessary for the maintenance of blood-brain barrier (BBB) and in turn protect the neurons. This was further supported by the fact that in a mixed culture (neurons and astrocytes) the increase in the intracellular Ca^{2+} was much greater in astrocytes than in neurons. The astrocytes further exhibited blebbing and bursting after prolonged exposure to 3-NPA but not the neurons (Fukuda et al 1998). These evidences indicate that astrocytes play a vital or primary role for the neuronal survival (Fukuda et al 1998). Basic fibroblast growth factor and thrombin protected the astrocytes in culture against 3-NPA toxicity (Deshpande and Nishino, 1998). Both these agents are neurotrophic factors synthesized by the astrocytes. Neurotrophic factors thus partly prevent the neurodegenerative process.

Estrogen protects against the 3-NPA toxicity

The lower incidence of neruodegenerative diseases in females in the reproductive age group as compared to males indicates the protective effects of estrogen (Deshpande, 2000). Recently, it was demonstrated that 3-NPA toxicity was absent or lesser in female rats than in males (Nishino et al 1997). Further, they found that estrogen protected and testosterone exacerbated the 3-NPA toxicity. The 3-NPA toxicity was lesser in castrated males treated with estrogen. Conversely it was more in ovariectomized rats. But the toxicity was lesser in the ovariectomized rats treated with estrogen. Tamoxifen, an estrogen receptor antagonist, reversed the toxicity of 3-NPA in female or "ovariectomized + estrogen treated" rats to the level of males. Thus, they concluded that the estrogen or estrogen receptors protect against the 3-NPA toxicity (Nishino et al 1997). The neuroprotective actions of estrogen are proposed to be due to the increased tyrosine kinase activity (Gibbs, 1994). Estrogen receptors not only distributed on the neurons but also on astrocytes, endothelial and epithelial cells in the brain (Langub and Watson, 1992). Estrogen treatment increased the number glial processes, enhanced the release of endothelium derived relaxing factor and also lowered the incidence of atheroma formation (Tranque et al 1987; Gisclard et al 1988; Ramwell and Foegh, 1992). These observations suggest that estrogen provide protection to neurons, astrocytes, endothelial cells and others.

Role of endothelial cells and NO

Within hours following the injection of 3-NPA, specially the second dose, histological damages showing the disturbances in blood-brain barrier (BBB) are seen and mostly so in the centrolateral striatum. The BBB is composed of a sheet of endothelial cells with tight junctions and the astrocytic end feet covering the basal lamina of the capillaries (Barres 1993; Cancilla and Berliner, 1993). In the earlier studies elsewhere, 3-NPA concentration was higher in the blood than in the brain tissue after the first injection (Nishino et al 1995). This indicates that 3-NPA does not cross the BBB. However, the second exposure breaks down the BBB as evidenced by the IgG extravasation associated with accumulation of H^3 labeled 3-NPA in the striatal regions in contrast to other regions of the brain. This further suggests the presence of high affinity binding sites for 3-NPA at the striatum.

Systemically administered 3-NPA would be distributed uniformly to all the regions of brain and other tissue. However dysfunction of BBB was observed in centrolateral part of the striatum supplied by lateral striatal artery. Thus, it is suggested that endothelial cells of the lateral striatal artery may be more susceptible to the effects of 3-NPA than the other areas. The endothelial cells were the first target cells that confront with the intraluminal toxins, metabolites, stress or insufficiency (Nishino et al 1997, 1998). The damage to the endothelial cells initiates the interaction with other cells (such as, platelets/leukocytes) and other blood-borne molecules (Ware and Heistad, 1993; Henn et al 1998). This would damage the cytoskeletal structures of the endothelium so as to result in extravasation of albumin, IgG, complement factors, and even the leukocytes (Nishino et al 2000). Further, our recent data provide evidences for the involvement of endothelial cells via nitric oxide (NO) synthatase activity which may be responsible for HD or other neurodegenerative disorders.

Morphological vulnerability of striatal artery

The interesting feature of 3-NPA toxicity is that there is dysfunction of BBB in the centrolateral part of the striatum, which is supplied by the lateral striatal artery. The dysfunction of BBB implies either the dysfunction of endothelial cells or the astrocytes. Since the endothelial cells are the first cells to be exposed to 3-NPA, hence must be vulnerable to the effects of the toxin. The underlying mechanisms for greater vulnerability still remains unclear but the morphological evidences of arterial branches at these regions have specific properties. These are perforate arteries, having higher intraluminal pressure. They have higher diameter to thickness ratio. Thus even in physiological states, the wall of the lateral striatal artery is having greater tension (Laplace law). Further, these arteries have acute bifurcations or hairpin bendings producing a turbulent flow, which activates the NO generation. This condition when compounded with other conditions such as toxins, ischemia, hypertension, infection, etc., damages the endothelial cells. The damage to the endothelial cells disrupts the blood-brain barrier and allows the toxic and other metabolic substances to pass into the extracellular space of the nervous system. The astrocytic end-feet, the next barrier that covers the basement of the capillaries are equipped with glutamate transporters, to transport the glutamate or similar molecules. 3-NPA resembles the structure of glutamate and is also transported through the glutamate transporters. Thus, the intracellular 3-NPA concentration in astrocytes will increase and injure them. This leads to subsequent injury and destruction of astrocyte-dependent cells (neurons).

Metabolic

HD patients have high calorie needs beyond that can be explained by their movement disorder. These patients show marked weight loss during the active phase. Increase in lactate concentrations in certain areas of brain with the decreased mitochondrial enzyme activities in HD patients has been demonstrated.

Mitochondria serve as the efficient source of ATP for cell function. The cells from the critical/vital organs such as the heart and brain are heavily dependent on the energy provided by mitochondrial oxidative metabolism. Consistent with their ancestry, the mitochondria have their own DNA that encodes many but not all their proteins. The mitochondria and their genes come from the mother via the ovum since sperm head do not possess the mitochondria (Coyle, 2000). This extranuclear form of inheritance derived exclusively from the females proven to be a powerful protective tool of the females.

More than 80 % of free radicals generated have mitochondrial origin. The reactive oxygen species were shown to be the mediators of delayed neuronal degeneration caused by activation of ionotropic receptors. Oxidative stress is also shown to precipitate the apoptosis. Ca^{2+} increase, as seen with 3-NPA also causes activation of glutamate ionotrophic receptors thus further enhance the apoptotic or necrotic process. This may the reason for the wide spread degeneration of neurons resulting from mitochondrial poisons such as 3-NPA, cyanide or CO (Greene and Greenmyre, 1996; Dawson et al 1995).

Conclusions

The findings so far indicate that the striatal artery has inherent anatomical predisposition for the stressful conditions. This makes the area (basal ganglia) supplied by it more vulnerable for the ischemic or metabolic insults. Presence of genetic or other linkages makes the damages overt. Further, the astrocytes, and their greater sensitivity to toxins like 3-NPA opens up additional parameters such as break down of BBB, accumulation of transmitters, mitochondrial dysfunction and generation of free radicals. All of them are complimentary to the process of degeneration of astrocyte-dependent cells (neurons). Thus, the future strategy may be to protect the astrocytes to the injurious effects of toxins or insults or replace their functions so as to prevent the neuronal damage.

References

Alston TA, Mela L and Bright HJ (1977) Proc Natl Acad Sci (*USA*) **74**: 3767-3771.

Aminoff, MJ (2001) In: Harrison's Principles of Internal Medicine, Eds. Braunwald E, Fauci AS, Rasper DL, Hauser SL, Longo DL and Jameson JL, 15th Ed, Vol-2, pp, 2399-2406, McGraw-Hill, New York.

Ankarcrona M, Dypbukt JM and Bonfoco E, Zhivotovsky B, Orrenius S, Lipton SA and Nicotera P (1995) Neuron **15**: 961-973.

Barres BA (1993) J Neurosci **11**: 3685-3694.

Beal MF, Brouillet E, Jenkins BG, Ferrante RJ, Kowall NW, Miller JM, Storey E, Srivastava R, Rosen BR and Hyman BT (1993) J Neurosci **13**: 4181-4192.

Ben Sacher D, Zuk R and Glinka Y (1995) J Neurochem **64**: 718-723.

Behrens MI (2000) In: Mitochondrial inhibitors and Neurodegenerative disorders, Eds. Sanberg PR, Nishino H and Borolongan CV, pp.167-176, Humane press, Totowa, NJ, USA.

Brouillet E, Guoyt MC, Mittoux V, Altairac S, Conde F, Palfi S and Hamtraye P (1998) J Neurochem **70**: 794-805.

Brown Jr RH (2001 In: Harrison's Principles of Internal Medicine, Eds. Braunwald E, Fauci AS, Rasper DL, Hauser SL, Longo DL and Jameson JL, 15th Edition, Vol-2, pp, 2412-16, McGraw-Hill, New York.

Buisson A, Plateu V, Plotkine M and Boulu RG (1991) Neurosci Lett **131**: 257-259.

Cancilla PA and Berliner J (1993) In: The Blood-brain barrier, Ed. Pardrid WM , pp 25-46, Raven Press New York.

Chapman AG, Durmuller N, Lees GJ and Meldrum BS (1989) Nerosci Lett **107**: 256-260.

Choi, DW (1992) J Neurobiol **23**: 1261-1276.

Coyle JT (2000) In: Mitochondrial inhibitors and Neurodegenerative disorders, Eds, Sanberg PR, Nishino H and Borolongan CV, pp. vii-viii, Humana press, Totowa, NJ, USA.

Dawson R Jr, Beal MF, Bondy SC, Di Monte DA and Isom GE (1995) Toxicol Appl Pharmacol **134**: 1-37.

Deshpande SB and Nishino H (1998) Brain Research **783**: 28-36.

Deshpande SB, Fukuda A and Nishino H (1997) Exp Neurol **145**: 38-45.

Deshpande SB (2000) Ind J Physiol Pharmacol **44**: 43-49.

Dunbar GL, Shear DA, Dong J and Haik-Creguer KL (2000) In: Mitochondrial Inhibitors and Neurodegenerative Disorders, Eds, Sanberg PR, Nishino H and Borolongan CV, pp 73-92, Humana Press, Totowa, NJ.

Ferrer I, Maritn F and Serrano T, Reiriz J, Perez-Novarro E, Albrech J and Macaya A (1995) Acta Neuropathol **90**: 504-510.

Fink SL, Ho DY and Saplosky RM (1996) Exp Neurol **138**: 298-304.

Filloux F and Wamsley JK (1991) Synapse **8**: 281-288.

Fu Y, He F, Zhang S, Huang J, Zhang J and Jio X (1995) Neurotoxicol Teratol **17**: 333-339.

Fukuda A, Deshpande SB, Shimano Y and Nishino H (1998) Neuroscience **87**: 497-507.

Geddes JW, Bondada V and Pang Z (2000) Eds. Sanberg PR, Nishino H and Borolongan CV, pp 107-120, Humana Press, Totowa, NJ.

Gibbs RT (1994) Ann N Y Acad Sci **743**: 165-196.

Gisclard V, Miller VM and Vanhoutte PM (1988) J Pharmacol Exp Therap **244**: 19-22.

Globus MY, Ginsberg MD, Harik SI, Busto R and Dietrich WD (1987) Neurology **37**: 1712-1719.

Gould DH, Wilson MP and Hamar DW (1985) Toxicol Lett **27**: 83-89.

Greene JG and Greenmyre JT (1996) In: Olanow CW, Jenner P and Youdim M. Eds, Neurodegeneration and Neuroprotection in Parkinson's disease, Academic Press, London, pp 125-142.

Hamilton BF and Gould DH (1987) Acta Neuropathol **74**: 67-74.

Henn V, Slupsky JR and Grafe M (1998) Nature **391**:591-594.

Kelleher RJ and Growdon JH (2002) In: Diseases of the Nervous system, Eds. Ashbury AK, McKhann GM, McDonald WI, Goadsby PJ and McArthur JC Vol. 1, pp, 252-266, Cambridge University press, Cambridge.

Langub MC and Watson RE Jr (1992) Endocrinology **130**: 364-372.

Ludolph AC, He F, Spencer PS, Hammersted J and Sabri M (1991) Can J Neurol Sci **18**: 492-498.

Lui X, Luo X and Hu W (1992) Biomed Environ Sci **5**: 161-177.

Nishino H, Shimano Y, Kumazaki, M, and Sakurai T (1995) Neurosci Lett **186**: 161-164.

Nishino H, Kumazaki M, Fukuda A, Fujimoto I, Shimano Y, Hida H, Sakurai T, Deshpande SB, Shimizu H, Morikawa S and Inubushi T (1997) Neurosi Res **27**: 343-355.

Nishino H, Fukuda A, Nakajima K, Kumazaki M, Muramatsu K, Shimano Y, Sakurai T, Deshpande SB, Inubushi T, Morikawa S and Sanberg PR (1998) Neurosci Res **30**: 303-312.

Nishino H, Hida H, Kamazaki M, Shimano Y, Nakajima K, Shimizu H, Ooiwa T and Baba H (2000) J Neurotrauma **17**:251-260.

Novelli A, Reilly JA, Lysko PG and Henneberry RC (1988) Brain Res **451**:205-212.

Olney JW (1990) Prog Brain Res **86**: 37-51.

Pang Z, Umberger GH and Geddes JW (1996) J Neurochem **66**: 474-484.

Ramwell PW and Foegh ML (1992) J Heart Lung Transplant **11**: 1188-1193.

Seisjo BK and Bengtsson F (1989) J Cerb Blood Flow Metab **9**: 127-140.

Simpson JR and Isacasson O (1993) Expt Neurol **121**: 57-64.

Slivka A, Brannan TS, Weinberger J, Knott PJ and Cohen G (1988) J Neurochem **50**: 1714-1718.

Reipe M, Hori N, Ludolph AC, Carpenter DO, Spencer PJ and Allen CN (1992) Brain Res **586**: 61-66.

Tranque PA, Suarez I, Olmos G, Fernandez B and Garcia-Segura LM (1987) Brain Res **406**: 348-351.

Ware JA and Heistad DD (1993) New Eng J Med **328**: 628-635.

Zeevalk GD and Nicklas WJ (1991) J Pharmacol Exp Ther **257**: 870-878.

Zeevalk GD and Nicklas WJ (1992) J Neurochem **59**: 1211-1220.

Molecular and Cellular Neurobiology
Editors: M.K. Thakur and S. Prasad

Mutation in Cu/Zn Superoxide Dismutase and Familial Amyotrophic Lateral Sclerosis

Anand Goswami and Nihar Ranjan Jana

National Brain Research Centre, Manesar, Gurgaon-122 050, India

Summary

Amyotrophic lateral sclerosis (ALS) is a progressive degenerative disease of motor neurons. For the vast majority of ALS cases, the factor triggering focal initiation and then spreading of motor neuron degeneration in sporadic ALS remains unknown. With the help of modern genetics, scientists have identified mutations in superoxide dismutase 1 (SOD1) and ALS2 as primary cause of the disease and have implicated others as potential contributors. This review will provide a brief description about ALS and the probable mechanisms by which mutations in SOD1 cause familial ALS.

Keywords: Amyotrophic lateral sclerosis, Cu/Zn superoxide dismutase, Transgenic mice, Apoptosis.

Introduction

ALS is a common neurodegenerative disease causing cell death of motor neurons, and progressive muscle weakness resulting in paralysis. It is the most common adult onset neurodegenerative disease having a prevalence of approx 5 in 100,000 individuals. ALS was first described in scientific literature in 1869 by the French neurologist Jean-Martin Charcot. ALS is commonly known in the US as Lou Gehrig's disease. In England it is known as motor neuron disease. The French refer to ALS as maladie de charcot. It is characterized by selective degeneration and death of major neuronal component of the motor system, the upper motor neurons in the motor cortex and lower motor neurons of the brain stem and the spinal cord (see Fig.1 for details). The loss of motor neurons leads to the progressive decline in muscular function resulting in eventual paralysis and death due to respiratory failure. Majority of the ALS patients die within 2-5 years of clinical onset.

About 10% of ALS patients are familial (inherited) cases and apparently sporadic in 90% of the cases with no known genetic component. While current evidence suggests that multiple genetic and environmental factors may be implicated in ALS pathogenesis (see Table 1). Both sporadic and familial ALS share common pathological features such as presence of abnormal neurofilament accumulations in the degenerating motor neurons.

Clinical features

Symptoms of upper motor neuron damage include spasticity, hyperreflexia, and pathological reflexes (Babinski's sign - the one where they run an object from the heel to the big toe), weakness, pseudobulbar palsy, dysarthria, swallowing difficulty, hyperactive reflexes, pathologic laughing and crying, fatigue etc. Lower motor neuron signs are listed as muscle weakness, including truncal muscles, muscle atrophy, fasciculation's, hypotonicity or flaccidity, and muscle cramps. Bulbar signs are listed as dysarthria (speech problems), dysphagia (swallowing problems), and sialorrhea (drooling). Although ALS can strike anyone at any age, generally the onset of the disease is in the fifth or sixth decade of life. As the disease progresses, more muscle groups are affected and patients become progressively incapacitated.

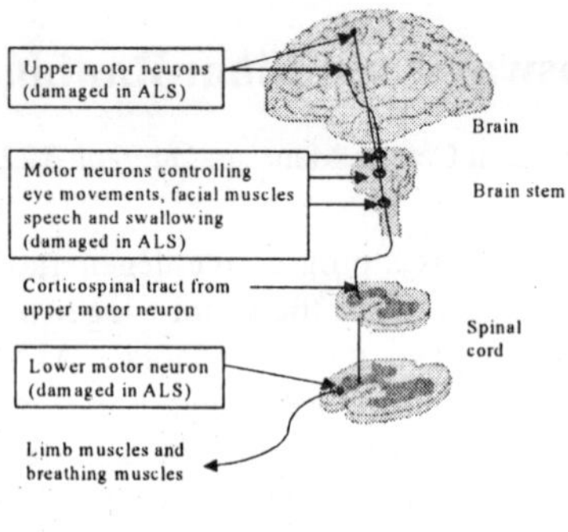

Fig.1. Human motor system

Genes associated with ALS

Very little is known about the genetic defects that cause or predispose to ALS. Till date the only proven cause of ALS are missense mutations in the SOD1 gene. However, recently several genes have been shown to be associated with either familial or sporadic ALS.

Sporadic ALS

Apurinic/apyrimidine endonuclease (APEX): Advance study on linkage analysis has shown that apex gene locus is at 14q 11.2-12, which is involved in DNA repair and protection against oxidative stress. Missense mutation in this gene can lead to a very small number of sporadic ALS. Still there is a lot of uncertainty regarding the mechanism of action of the protein.

Excitatory amino acid transporter 2 (EAAT2) : It is found at the chromosomal locus 11p13-p12. Mutation in this gene was previously thought to be the major cause of sporadic ALS. It codes for a glutamate channel. The splice variants of this gene are found in 61% of sporadic ALS patients (Aoki et al 1998).

Neuronal apoptosis inhibitory polypeptide (NAIP): Found at chromosomal locus 5q12.2-q13.2 and mutation is very rare and till date found in 1 sporadic ALS patient.

Apolipoprotein E4 (APOE4): Found at chromosomal locus 19q13.2 and it is linked to small number of sporadic ALS patients. Current studies have shown that in a small number of sporadic ALS patient (~1% of cases), codon deletion or insertion of KSP

Table 1. Genes associated with ALS

Disease	Inheritance	Linkage	Gene/protein	Onset/course	Features
Sporadic ALS	None	None	Largely unknown	Adult	Includes >90% of incidence of ALS
Familial ALS					
ALS	Dominant	21q22.1	SOD1	Adult	Comprises 20% of inherited cases, more than 100 mutantsknown all but one of which is dominant
ALS with fronto-temporal dimentiia	Dominant	9q21-22	?	Adult	Dementia
ALS with dementia/ Parkinsonism	Dominant	17q21	TAU	Adult	Dementia, parkinsonism, Amyotrophy
ALS	X-linked	Xp11-Xq12	?	Adult	
Juvenile type 1	Recessive	15q15-22	?	Adolescence	Slowly progressing
Juvenile type 3	Recessive	2q33	ALS2	Adolescence	Slowly progressing, mutant gene product seems to be a guanine exchange factor
Juvenile	Dominant	9q34	?	Before 25	Slowly progressing

repeat motif of the neurofilament NF-H gene has been identified, but no such NF-H mutants has been detected in over 1000 control DNA samples. The combined data suggests that NF-H variants may represents risk factor for ALS (Al-Chalabi 1999). Besides abnormalities in neurofilament heavy subunit, apolipoprotein E4 genotype and the genes linked to spinal muscular atrophy, no gene mutation has been found out in the 80% cases of sporadic ALS cases and the field has been keenly awaiting a new ALS gene, with all the attendant opportunities for elucidating key pathway for motor neuron injury.

Familial ALS

Cu/Zn superoxide dismutase I (SOD1): Although CHARCOT first described the neurodegeneration in ALS in 1869, the major breakthrough came in 1993 when researcher discovered the genetic basis of approximately 20% causes of autosomal dominant familial ALS. This turned out to be on chromosome 21, encoding Cu/Zn superoxide dismutase SOD1, and a well-characterized enzyme that protects the cell from damage by free radicles. Over 100 mutations have been discovered spanning all exon of SOD1 gene, with a smattering of other type of genetic lesion. Recently a genetic linkage study has identified another locus, for a subtype of familial ALS with frontotemporal dementia (FTD), that map to chromosome, q21-q22. Three loci for juvenile onset ALS have also been mapped. Two recessively inherited loci map to chromosome 2q33 and 15q15-22 whereas one dominant juvenile onset locus map to chromosome 9q34.

SOD1 associated familial ALS

The Cu/Zn SOD protein, the product of SOD1 gene, is an endogenously produced intracellular enzyme present in essentially every cell in the body. Cellular SOD is actually represented by a group of metallo-enzymes with various prosthetic groups. The prevalent enzyme is cupro-zinc (Cu/Zn) SOD, which is a stable dimeric protein of molecular weight (32 KDa). SOD appears in three forms: (1) Cu/Zn SOD in the cytoplasm with two subunits, and (2) Mn SOD in the mitochondrion. A third extra cellular SOD which has been described recently, contains Copper (Cu SOD). Cu/Zn SOD is considered fundamental in the process of eliminating reactive oxygen intermediate by reducing (adding an electron to) superoxide to form hydrogen peroxide (H_2O_2). Catalase and the selenium-dependent glutathione peroxidase are responsible for reducing H_2O_2 to H_2O. The respective enzymes that interact with superoxide and H_2O_2 are tightly regulated through a feedback system. Excessive superoxide inhibits glutathione peroxidase and catalase to modulate the equation from H_2O_2 to H_2O. Likewise, increased H_2O_2 slowly inactivates Cu/Zn SOD. Meanwhile, catalases and glutathione peroxidase, by reducing H_2O_2, conserve Cu/ZnSOD and Cu/Zn SOD, by reducing superoxide, conserves catalases and glutathione peroxidase. Through this feedback system, steady low levels of Cu/Zn SOD, glutathione peroxidase, and catalase, as well as superoxide and H_2O_2 are maintained, which keeps the entire system in a fully functional state (Fig.2). Cu/Zn SOD also exhibits antioxidant activity by reducing $.O_2^-$ that would otherwise lead to the reduction of Fe^{3+} to Fe^{2+} and thereby promote ·OH formation.

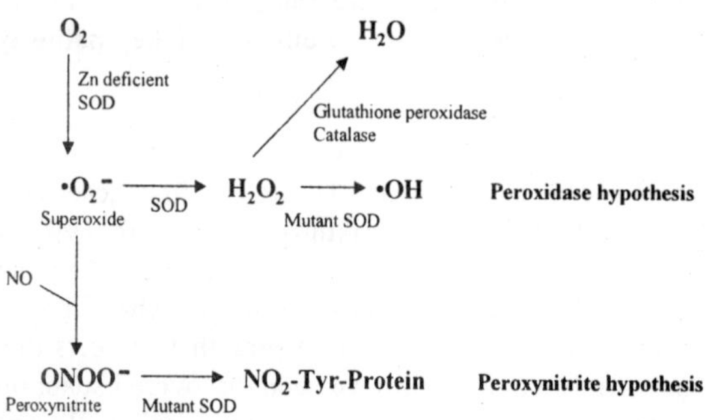

Fig.2. Proposed Copper-mediated oxidative mechanism to explain the toxicity of mutant SOD1.

When the catalase activity is insufficient to metabolize the H_2O_2 produced, Cu/Zn SOD will increase the tissue oxidant activity. Hence, it was found that the antioxidant enzymes function as a tightly balanced system, any disruption of this system would lead to promotion of oxidation.

To date approximately 100 different point mutations in SOD1 gene have been identified in ALS families. Many of these mutations lead to the substitution of an amino acids within regions of enzyme with distinct structural and functional role. In addition to individual amino acid substitutions, there are also a smaller group of mutations resulting in amino acids deletion and truncation. SOD1 associated familial ALS has been demonstrated to occur by a gain-of-function mechanism. Transgenic mice expressing the human mutant SOD1, in addition to their normal wild-type SOD1, develop motor neuron disease similar to human ALS; whereas transgenic mice expressing the human wild-type SOD1 transgene, in addition to their normal mouse wild-type SOD1 gene, do not. In both cases, the transgenic mice have elevated levels of Cu/Zn SOD activity relative to the nontransgenics, indicating that the disease is not caused by lowering of Cu/Zn SOD activity, but the presence of the mutant Cu/Zn SOD polypeptide itself is toxic to motor neurons. Despite the apparent toxicity of superoxide, SOD1 knockout mice have a surprisingly mild phenotype. They develop relatively normally but display subtle motor symptoms by six months of age, as well as impaired fertility. They do not develop ALS, an observation that further support the theory that ALS mutant Cu/Zn SOD protein are inherently toxic.

Pathogenic mechanisms of SOD1 associated familial ALS

Despite the explosion of ALS research, the actual nature of the gained function, by which mutant Cu/Zn SOD kills motor neurons in ALS remains elusive. Two hypothesis have dominated to explain the toxicity: the oligomerization hypothesis and the oxidative stress hypothesis (Cleveland and Rothstein, 2001). The oligomerization hypothesis maintains that mutant Cu/Zn SOD proteins are misfolded and consequently oligomerize into increasingly high-molecular weight species that ultimately aggregate and end up in proteinaceous inclusions. The oligomerized or aggregated proteins, at some stage in their formation, are selectively toxic to the motor neurons. The oxidative damage hypothesis maintains that mutant Cu/Zn SOD proteins catalyze reactions with hydrogen peroxide or peroxynutrite that damage exogenous substrates critical for viability of the affected cells.

***Transgenic mouse model*:** Several line of transgenic mouse expressing various SOD1 mutants have been generated, showing various pathological changes similar to human ALS. The G37R SOD1 transgenic mice provide an excellent illustration of these models. G37R SOD1, which retains full Cu/Zn SOD activity, accumulate 3 to 12 times the endogenous level in the spinal cord and the level of the mutant protein influences the age of onset. Toxic Cu/Zn SOD is transported into axon, and early on, it accumulates in exon where it is associated with structural pathology. Approx 2-3 months before the appearance of clinical symptoms, Cu/Zn SOD accumulates in irregular, swollen, intraparenchymal portion of motor axon and the axonal cytoskeleton. The first neuropathological changes seen in the transgenic mutant SOD1 mice are perikarya, axonal dendritic vacuoles in motor neurons with undetectable neuronal loss or gliosis. By the time the first symptom (fine limb tremor) arises, vacuolizations is prominent, and some neuronal loss, especially of large motor neurons, is observed in the spinal cord. At the end stage, paralysis due to the dramatic loss of motor neurons and marked gliosis is observed. During this time, the motor neuron shows Cu/Zn SOD, ubiquitin and phosphorylated NF-H immunoreactive inclusion. To test the role of neurofilament in disease NF-L null mice were mated to transgenic mouse expressing mutant SOD1 (G85R-SODI). G85R mutant mice without NF-L show delay in onset of disease and extension of survival. Thus assembled neurofilaments are not required for mutant Cu/Zn SOD mediated disease (Cleveland and Julian 1998). To further examine disease mechanism, G85R SODI transgenic mice were mated to the mice expressing wild type SODI mice or SODI null mice. Elevation or elimination of wild Cu/Zn SOD did not influence the SODI mutant associated disease. In addition, SOD1 knockout mice did not develop motor neuron disease (Reaume et al 1996). Therefore the conclusion from the combined result was that the mutation in SODI provoked a gain of toxic property, derived from superoxide associated oxidative stress.

***Toxicity and aberrant enzyme catalysis*:** Cu/Zn SOD is an abundant and ubiquitously expressed protein. In view of its normal function in catalysing the conversion of superoxide anions to H_2O_2, it was initially hypothesised that the toxicity of different Cu/Zn SOD mutants could result from reduced activity of free-radical scavenging enzyme, however it was not of relevance against the measurement of the enzymatic studies and by transgenic mouse studies. There is a remarkable degree of variation in Cu/Zn SOD mutation in terms of enzymatic activity, polypeptide, half life and resistance to proteolysis and this variable did not correlate with the age of onset or rapidity of human disease progression.

***Copper toxicity*:** SOD1 is a homodimer protein and each subunit binds one Zinc and one Copper atom, improper metal binding has been proposed as a mechanism by which mutations could excert their toxic gain of function. A variety of studies reveal that cell death could be initiated by aberrant oxidative chemistry catalysed by the Copper atom bound to the mutant Cu/Zn SOD especially after loss of Zinc binding to Cu/Zn SOD. But it's a subject of controversy because the mutant Cu/Zn SOD subunit could fail to bind the Zinc atom allowing rapid reaction of mutant SOD1 of the Cu^{++} form. This reduced form of enzyme could catalyse the backward reaction converting oxygen to superoxide; which could combine with nitric oxide to produce peroxynitrite. However there is no *in vivo* proof for this aberrant Copper chemistry, proposed to be central to the pathogenesis of the familial ALS. In fact recent studies showed that onset and progression of the disease in the G85R mutant mice is independent of the level of wild type Cu/Zn SOD activity (Reaume et al 1996). Results interpreted as inconsistent with the peroxynitrite hypothesis because the absence of the wild type Cu/Zn SOD would expect to exacerbate toxicity, whereas increasing Cu/Zn SOD would slow disease progression (Bruijin et al 1998). The toxicity of Zinc depletion in Cu/Zn SOD was demonstrated by the introduction into a liposome fusion of purified Zinc depleted Cu/Zn SOD into cultured motor neurons. In this experimental set-up both the wild type Cu/Zn SOD and mutant Cu/Zn SOD levelling Zinc were equally toxic to cultured neurons, however it was questioned due to result of transgenic mouse. Because in a similar condition, when the model is right, a disease in nitric oxide production would attenuate the disease. Yet the destruction of neuronal nitric oxide synthase did not increase survival of SODI G93A transgenic mouse. Although the copper mediated toxicity model remains controversial, the discovery of Cu-chaperones (CCS) now provides opportunity to test directly whether Copper within mutant Cu/Zn SOD mediates motor neuron degeneration in familial ALS.

***Toxic protein aggregation*:** Inclusion body formation is a particularly common finding in neurodegenerative disorders. Cytosolic protein aggregates containing Cu/Zn SOD also appear to be common in both sporadic and familial ALS as well as in transgenic mouse models expressing mutant human Cu/Zn SOD (Julian 2001, Valentine and Hart 2003). It was reported that the lewy body-like hyaline inclusions were present in the affected neurons during the clinical course of familial ALS in transgenic mice model (Johnston. et al 2002). The protein aggregates containing Cu/Zn SOD are common to the disease caused by different mutants, implying that co-aggregation of Cu/Zn SOD mutants with unidentified components or aberrant catalysis by misfolded mutants underlies mutant-mediated toxicity. However, whether the aggregates of Cu/Zn SOD are toxic to the cells is still under debate. There is a good reason to believe that they are the ultimate product of a type of cellular defence mechanism that occurs generally in the cells when the burden of damaged or misfolded protein exceeds the capacity of the protein degradation machinery to eliminate them (Sharman and Goldberg 2001). In mammalian cells, this mechanism has been shown to involve transport of smaller protein aggregates along the microtubules to the microtubules organizing center, where they are incorporated into larger structures termed as aggresomes. These cytoplasmic entities are structures in which aggregated, multiubiquitinated misfolded proteins are sequestered, where they await degradation by the proteasome or packing into larger inclusion bodies. Aspects of toxicity can arise either through aberrant chemistry, mediated by the misfolded aggregated Cu/Zn SOD or through loss or sequestration of essential cellular components like the protein-folding chaperones or proteasomes. The aggregates are highly immunoreactive with antibodies to ubiquitin and the components of proteasome. Partial inhibition of proteasome is sufficient to increase the aggregation of mutant Cu/Zn SOD leading to the proposal that proteasome activity could be limiting by combating such aggregates. Thc involvement of proteasome in the mechanism of mutant Cu/Zn SOD associated ALS could explain the late onset of the disease because proteasome function is known to decline with age.

***Exitotoxicity*:** In ALS, motor neuron injury, a potential role for glutamate excitotoxicity has been hypothesized, since glutamate was first demonstrated to induce neuronal injury in both *in vitro* and *in vivo* experimental models. Since that time numerous studies have supported that glutamate induced excitotoxicity is a potential contributor to ALS pathogenesis. Depolarisation of the neuronal membrane after activation of neuronal glutamate receptors activates voltage-dependent Ca^{2+} channels, allowing Ca^{2+} entry into the cell. Thus, excess activation of neuronal glutamate receptors can cause cell death via alterations in cytosolic free Ca^{2+} homeostasis. For spinal motor neurons, the glutamate transporters EAAT2 present in astrocytes accomplish rapid recovery of synaptic glutamate. Loss of EAAT2 transporters could lead to increased extracellular concentrations of glutamate and excitotoxic degeneration of motor neurons. The excitotoxicity hypothesis is supported by the observation that the majority of sporadic ALS cases have a reduction in astroglia glutamate transporters EAAT2 in motor cortex and spinal cord. The decreased EAAT2 levels in ALS may be due at least in part to abundance of aberrant EAAT2 mRNA species resulting from RNA processing error (Lin et al 1998). Proteins translated from these mRNAs could be rapidly degraded and/or produce dominant negative effects, resulting in loss of EAAT2 protein and activity. The glutamate -induced excitotoxicity resulting from the loss of astroglia's EAAT2 is most likely a secondary effect of the disease process in ALS.

Mitochondrial dysfunction: Much of the studies in mice showed that the level of mutant Cu/Zn SOD expression modulates the age of onset of symptoms but has little effect in the rate of progression of disease after onset. Interestingly, the onset of disease in a mutant SOD1 mice correlates with a sudden increase in the number of vacuoles, representing dilated mitochondria. This leads to the possibility that the toxicity of mutant Cu/Zn SOD may be mediated by damage of mitochondria in motor neurons. Mitochondrial pathology is also present in central nervous system tissue from human ALS cases. How the aggregates of Cu/Zn SOD disrupt the mitochondrial function is unclear. Inhibition of components of the mitochondrial electron transport chain has been demonstrated in both cell culture and transgenic mice models of the disease. In the transgenic mice the mitochondrial enzyme activity was progressively decreased in spinal cord, particularly in the ventral horn, but not at all decreased in the cerebellum (Jung et al 2002). Mitochondrial dysfunction can produce excess superoxide ion and release cytochrome 'c' into the cytoplasm, which further activates caspase-3, and affect Ca++ homeostasis. Oxidative damage to mitochondrial DNA leading to the accumulation of mitochondrial DNA mutations as well as other mitochondrial damage could be other important mechanisms, contributing to selective loss of motor neuron is ALS.

Apoptotic death: Recent studies and evidence suggests that the mechanism of the neuronal death in human ALS proceeds by programmed cell death (Guegan and Przedborski 2003). Many evidence points to apoptotic pathways playing a role in human and mouse model of the disease. This is supported by the generation of DNA fragmentation as determined by *in situ* and labelling of DNA strand breaks and increased immunoreactivity for proapoptotic Bax protein in selectively vulnerable CNS regions of sporadic cases of ALS. There is also compelling evidence that motor neuron death involves apoptosis in disease caused by SODI mutations. Transfection studies with cultured neuronal cells demonstrated the pro-apoptotic effect of mutant Cu/Zn SOD. Moreover hallmark of apoptotic death, i.e. DNA fragmentation; caspase activations and altered expression of Bcl2 were found in the spinal cord of SODI transgenic mice. The members of the Bcl2 family have been widely implicated in the regulation of cell death and have been examined in transgenic SODI G93A mice. Additional support for the role of Bcl2 in ALS is provided from the report that over expression of Bcl2 transgene delays onset of disease and motility in SOD1 mice by over one month. Recent studies have revealed that mutant Cu/Zn SOD could also activate P53 a nuclear phosphoprotein that may play causative role in apoptosis. The important apoptotic pathways in the pathogenesis of ALS also supported by the neuro-protective effect of dominant negative caspase-1 inhibitor transgene in transgenic mutant SOD1 G93A mice (Kostic et al 1997). Caspase-1 mRNA also has been shown to be up regulated in this transgenic mice. Because of the extreme

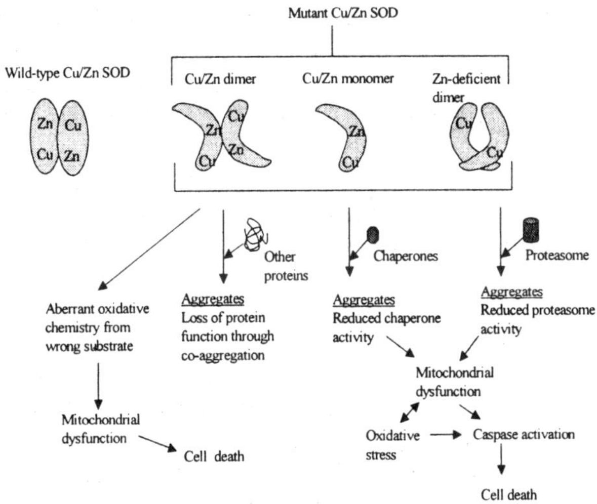

Fig. 3. Models for mutant Cu/Zn SOD mediated motor neuron death.

similarity in the behaviour, histology and molecular mechanism between the mutant SOD1 G93A transgenic mice and human with ALS indicates that caspases play a role not only in the end stage of ALS but also in early progression of the disease.

Therapeutic intervention, treatment and cure

Till date there is no drug to cure motor neuron degeneration in human ALS. In the past several years, various approaches to therapy have been tested in mutant SOD1 mice and have been found to slow the disease progression (Table 2). Riluzole, an inhibitor of synaptic glutamate release, is so far the only approved drug for treatment of human ALS. The best pharmacological treatments in transgenic mice to date have been with zVAD-fmk (a broad range caspase inhibitor) and with the addition of creatine (mitochondrial stabilizer) to drinking water (Klivenyi et al 1999). Current researches further demonstrate that minocycline (down regulate caspase 1 expression and inhibit cytochrome c release) antioxidants, anti-inflammatory agents, metal ion regulators, various growth factors and chaperones could be potential therapeutic molecules (Guegan and Przedborski 2003).

Advance research on molecular genetics of motor neuron degeneration has revealed that gene therapy approaches, involving the use of recombinant virus, offer a promising strategy for the delivery of genes enhancing motor neuron survival and repair. Further insights into the cause and molecular mechanisms of ALS pathogenesis will help to develop effective therapeutic strategies to cure or prevent this devastative neurodegenerative disease.

Table 2. Therapeutic interventions in SOD1 transgenic mice

Strategies	Transgenic mouse	Increased life span
Treatments		
Riluzole (inhibits glutamate release)	SOD1G93A	10-15 days
Creatine (enhances energy reserve)	SOD1G93A	30 days
zVAD-fmk (inhibits caspases)	SOD1G93A	27 days
Transgenic overexpression		
Bcl2	SOD1G93A	30-35 days
Dominant inhibitor of caspase1	SOD1G93A	27 days
Gene knockout		
nNOS	SOD1G93A	No change
P53	SOD1G93A	No change
Gene therapy		
AAV/Bcl2	SOD1G93A	No change

References

Al-Chalabi A, Andersen PM, Nilsson P, Chioza B, Andersson JL, Russ C, Shaw CE, Powell JF, and Leigh, PN (1999) Hum Mol Gen **8**: 157-164.

Aoki M, Lin CL, Rothstein JD, Geller BA, Hosler BA, Munsat, Dupuis L, de Tapia M, Rene F, Lutz-Bucher B, Gordon JW, Horvitz HR, and Brown RH Jr (1998) Ann Neurol **43**: 645- 653.

Bruijin LI, Houseweart MK, Kato S, Anderson KL Anderson SD, Ohama E, Reaume AG, Scott RW and Cleveland DW (1998) Science **281**: 1851-1854.

Cleveland DW (1999) Neuron **24**: 515-520.

Cleveland D W and Rothstein J D (2001) Nature Rev **2**: 806-819.

Cleveland DW and Julian JP (1998) Proc Natl Acad Sci, USA **95**: 9626–9630.

Dal Canto MC and Gurney ME (1994) Am J Pathol **145**: 1271-1279.

Ferrante RJ, Browne SE, Shinobu LA, Bowling AC, Baik MJ, MacGarvey U, Kowall NW, Brown RH Jr, and Beal MF (1997) J Neurochem **69**: 2064-2074.

Guegan C and Przedborski S (2003) J Clin Invest **111**: 153-161.

Johnston JA, Dalton MF, Gurney ME and Kopito RR (2002) Proc Natl Acad Sci USA **79**: 12571-12576.

Julian JP (2001) Cell **104** : 581-591.

Jung C, Higgins CM and Xu Z (2002) J Neurochem **83**: 535-545.

Kostic V, Jackson-Lewis V, de Bilbao F, Dubois-Dauphin M and Przedborski S (1997) Science **277**: 559-562.

Lin CL, Bristol LA, Jin L, Dykes-Hoberg M, Crawford T, Clawson L, and Rothstein JD (1998) Neuron **20**: 589-602.

Reaume AG, Elliott JL, Hoffman EK, Kowall NW, Ferrante RJ, Siwek DF, Wilcox HM, Flood DG, Beal MF Brown RH Jr, Scott RW and Snider WD (1996) Nat Genet **13**: 43-47.

Sherman MY and Goldberg AL (2001) Neuron **29**: 15-32.

Valentine JS and Hart JP (2003) Proc Natl Acad Sci, USA **100**: 3617-3622.

Wong PC, Waggoner D, Subramaniam JR, Tessarollo L, Bartnikas TB, Culotta VC, Price DL, Rothstein J and Gitlin JD (2000) Proc Natl Acad Sci USA, **97**: 2886-28891.

Molecular and Cellular Neurobiology
Editors: M.K. Thakur and S. Prasad

Protective Role of Sodium Pyruvate in Hydrogen Peroxide Induced Apoptosis using Human Neuroblastoma Cell Line as a Model

***Padma Shastry, Jayashree C. Jagtap**
Anmol Chandele and Vandna Prasad

*National Centre for Cell Science, NCCS Complex, Ganeshkhind, Pune-411 007, India

Abstract

Nerve cell death is the central feature of many neurodegenerative diseases and is believed to occur by apoptosis initiated by oxidative stress. Hydrogen peroxide is found to be a useful agent for inducing oxidative stress in cells because it has low reactivity and easy penetrability to cellular membrane. This article summarizes the current literature on molecules and the pathways in apoptosis in general and their role in neurodegeneration. The effectiveness of sodium pyruvate in protection is discussed in context with experimental models using H2O2 as an apoptotic inducer in neuroblastoma cell lines. The prospects of studies for prevention of neurodegeneration and in design of compounds/peptides for protection of functional neurons induced to undergo apoptosis; in post genomic era is high lighted.

Keywords: Apoptosis, Neurodegeneration, Neuroprotection, Sodium pyruvate, Hydrogen peroxide

Apoptosis- Programmed Cell Death

Over the course of an organism's life, cells divide, grow, differentiate and die. Cell death is an invariable part of animal development. Millions of cells die every minute in the animal body but the size remains the same and this is because of a poorly understood mechanism, where by cell division exactly balances cell death (Raff,1998). The two major pathways leading to cell death are – necrosis and apoptosis. Necrosis results following exposure to injurious agents and is a feature more characteristic during pathological conditions. It does not require the input of energy or the synthesis of either proteins or nucleic acids and is characterized by changes like cell swelling, leakage of cellular contents or rupture of nuclear organelle and plasma membrane leading to inflammation of surrounding tissues. Apoptosis, on the other hand is a programmed suicide of cells and plays a major role during development, homeostasis, and in many neurodegenerative diseases. Apoptosis is characterized by changes such as shrinking of cells, blebbing, break down of mitochondrial membrane potential with the release of cytochrome c, degradation of the chromatin in the nucleus and eventually engulfment by phagocytic cells that does not lead to an inflammatory response. (Schweichel and Merke, 1973).

Apoptosis as a process can be divided into 4 sequential phases (Vaux and Strasser, 1996) - 1) The earliest phase is the stimulus that provokes the apoptotic response. This may be an external signal delivered through surface receptors or may originate inside the cell from the action of a drug, toxin or radiation. 2) The second phase includes detection of this signal or metabolic state and transduction of the signal. Signal transduction pathways send this message to the cell death effector machinery. 3) The effector phase is the third phase of the cell death mechanism and includes the proteases that are activated during apoptosis, as well as their positive and negative regulators, and 4) The phase of cell death is the postmortem phase, in which the cells chromatin condenses and its DNA is degraded.

Molecules involved in apoptosis

Molecules involved in apoptosis, both pro and antiapoptotic, range from proteases, kinases, adaptor molecules, transcription factors and many such proteins. During apoptosis, not only are new proteins synthesized, but other molecules already present in the cells in non-active form are converted to active forms on induction of apoptosis to initiate a self destruction program (Amarante-Mendes and Green 1999).

The Bcl-2 family members

The mitochondrial–dependent pathway for apoptosis is governed by Bcl-2 family proteins. The Bcl-2 gene was first identified as a proto-oncogene in follicular B-cell lymphoma. Unlike the other oncogenes studied previously Bcl-2 functioned in preventing cell death rather than promoting proliferation, thus establishing a new class of oncogenes. The Bcl-2 family has ever since expanded significantly and includes both pro and antiapoptotic molecules. It is believed that the ratio between the two subsets in part determines the susceptibility of the cells to a death signal. The Bcl-2 family members possess up to four conserved homology domains designated as BH1, BH2, BH3 and BH4 that correspond to the ά helical segment. Many antiapoptotic proteins display sequence conservation in these domains (Hengartner, 2000). Another characteristic feature of the Bcl-2 family is the ability to form homo and hetero dimers, suggesting neutralization competition between these proteins. Considerable portions of proapoptosis versus antiapoptotic molecules localize to separate cellular compartments in the absence of cell death. Antiapoptotic proteins are initially found in the mitochondria, endoplasmic reticulum and the nuclear membrane. In contrast, substantial fraction of the pro apoptotic members localize to the cytosol or the cytoskeleton prior to the death signal (Gross et al 1999, Kromer, 1997).

Caspases

The caspases (Cysteine-Dependent Aspartate Directed Proteases) are a family of cysteine proteases that are activated during programmed cell death (Kaufmann and Earshaw 2000). These proteases are mainly synthesized as proenzymes (or zymogens) and remain inactive in most healthy cells. Caspases can be categorized according to their substrate specificity. Caspases 6, 8 and 9 are specific for the V/LEXD site whereas 2, 3 and 7 are specific for DEXD site. Caspases are also at times classified according to their interaction module – the death effector domain (DED) or the Caspase Recruitment domain (CARD). Caspases play a critical role in apoptosis either as "initiator" caspases involved in transduction of the apoptotic signal or as "effector" caspases involved in execution of cell death (Bamford et al 2000). In case of both, receptor mediated (Caspase 8 activation) and mitochondrial mediated (Caspase 9 activation) apoptosis, caspase 3 and 7 get activated as a down stream effector caspase. These are largely responsible for the morphological and biochemical changes with in the cell. Caspase 3 targets proteins such as lamins, gelsolin, fodrin, DFF-45, and Poly (ADP-ribose) polymerase. Caspase 3 is also suspected to be involved in the cleavage of proteins such as Bcl-2 and Bcl_{xl} that regulate the survival function (Kirsch DG et al 1999).

The Poly (ADP-ribose) Polymerase and its role in cell death

Poly (ADP-ribose) Polymerase (PARP) plays a pivotal role in maintaining genome integrity through modulation of multiple cellular responses (including base excision repair, necrosis and apoptosis) in answer to genotoxic stress. Poly (ADP-ribosyl)ation involves the cleavage of NAD+, the attachment of ADP ribose moiety to the acceptor protein, the release of nicotinamide and the addition of further ADP-ribose moieties leading to the generation of poly (ADP-ribose) consisting of more than a hundred units. The immediate goal of poly (ADP-ribosyl)ation following the occurrence of DNA lesions is to avoid transcription of damaged genes both by preventing initiation and speeding up ongoing transcription. Modification of transcription factors by Poly (ADP-ribosyl)ation prevents them from binding to promoters. After the silencing of transcription and initiation and facilitating DNA repair, the Poly (ADP-ribose) is removed by Poly (ADP-ribose) glycohydrolase (Ziegler and Oei 2001). PARP has been shown as a major player in both apoptosis and necrosis. Infact PARP was the first cellular protein to be identified as specifically being cleaved during apoptosis and its cleavage was subsequently shown to be a universal phenomena occurring during apoptosis (Lix and Darzykiewicz 2000). Specific proteolysis of PARP occurs in the DNA binding domains at the conserved 211 DEVD 214 site between Asp 214 and Gly 215, as a result of which the 116-kDa molecule gets cleaved to an 89 and 24 kDa subunits. Caspase 3 and caspase 7 are activated during apoptosis and are considered as the most efficient proteases for PARP cleavage (Herceg and Wang 2001). An important aspect of PARP cleavage is the channelization of the cell towards a particular type of cell death – apoptosis or necrosis. Apoptosis is an energy dependent process and the functions of PARP expend a lot of energy. Therefore during apoptosis PARP cleavage occurs so as to prevent the depletion of energy pools necessary for later stages of apoptosis.

Apoptosis In Neurodegeneration

Apoptosis is crucial to embyrogenesis, clonal selection in the immune system, tissue homeostasis and prevention of disease including carcinogenesis and viral attacks by removing damaged or infected cells. During development, cell death helps sculpt parts of the body, caring out cavities or separating digits. It also eliminates structures that once served a function, but are no longer needed, such as the tail of the tadpole when the tadpole becomes a frog.(Meier et al 2000). Nerve cell death is the central feature of many neurodegenerative diseases. Previously it was thought that nerve cell death in these disorders occurs by necrosis. Recent evidences now indicate that it is the process of apoptosis rather than the necrosis that primarily contributes to the nerve cell death in neurodegeneration (Talton and Olanow, 1999) and has been associated with both acute and chronic neurological diseases such as spinal chord injury, stroke, Alzheimer's disease, amyotropic lateral sclerosis, Parkinson's disease and Huntington's disease (Thompson, 1995).

Reactive oxygen species (ROS) and neurodegeneration

Reactive oxygen species (ROS) is a collective term including free radicals containing oxygen atoms and highly reactive non-radical oxygenated compounds such as hydrogen peroxide (H_2O_2) and peroxynitrite ($ONOO^-$). A characteristic of ROS is that some of them, such as superoxide (O_2^-) and hydrogen peroxide are normally produced in living cells and may have various biological effects, including the generation of novel harmful radicals. A free radical is defined as any atom or molecule that possesses an unpaired electron. It can be anionic, cationic or neutral. In biological and related fields, the major free radical species of interest have been those of oxygen i.e. oxygen free radicals (OFRs). The OFRs are part of a greater group of molecules often called reactive oxygen species (ROS) that are all more strongly oxidizing than molecular oxygen itself. These include hydrogen peroxide (H_2O_2), lipid peroxide (LOOH), singlet oxygen ($'O_2$), hypochlorus acid (HOCl) and other N-cholarmine compunds. Free radicals are also generated by electron transport chain reactions and play a part in many endogenous synthetic pathways. In the normal metabolic reduction of oxygen to water there is the potential for the generation of free radical species. The OFRs are potentially very toxic to cells. Due to their highly reactive nature they can readily combine with other molecules such as enzymes, receptors and ion pumps causing oxidation directly, and inactivating or inhibiting their normal function. Some of the products of OFR attack other molecules and can interfere with nucleic acid function, generating alterations in the base sequence with the potential for mutations. Such changes lead to extreme pathological situations such as cancers or degenerative diseases. The brain is acutely dependent on energy supplies for normal functioning. The mitochondria are intracellular founts of the brain's energy supplies. Mitochondrial derived pathological free radicals have been implicated in numerous diseases and the aging process itself. The human CNS is relatively deficient in oxidative defenses, rendering it more susceptible to ROS induced damage. Oxidative stress has been implicated in neurodegenerative diseases such as Alzheimer's disease, Parkinson's disease and Amyotrophic lateral sclerosis (Beal, 1996). Cell death in these conditions has also been correlated with increase in oxidative stress (Jiang et al 2001). Oxidative damage to membranes due to mitochondrial production of ROS may account for the loss in membrane fluidity with aging, which parallels the increase in mitochondrial peroxides. Oxidative damage to proteins may also accumulate with aging progressively impairing electron transport chain activity, causing further ROS production and oxidative damage in a vicious cycle. Progressive loss of ATP may ultimately lead to loss of Na/K- ATPase activity with consequent plasma membrane depolarization, loss of cellular homeostasis and activation of the apoptotic cascade.

Hydrogen Peroxide (H_2o_2) As An Apoptotic Inducer

Hydrogen peroxide is a normal product of cell metabolism. Hydrogen peroxide is produced endogenously by several physiological processes, e.g. during oxidative phosphorylation and by the inflammatory cell respiratory burst. Because it is freely diffusible, H_2O_2 can potentially reach the nucleus to interact with DNA (Uberti et al 1999). It also enhances the damaging effect of ultraviolet radiation, usually the damage seen in the presence of both agents is much greater than done by each agent alone (a synergistic effect). Hydrogen peroxide production, probably, mainly via O_2^-, has been frequently observed from mitochondria and microsomes. Hydrogen peroxide is found to be an useful agent for inducing oxidative stress in mammalian and plant cells because it has low reactivity and it easily penetrates cellular membrane causing further ROS production within the cells. Hydrogen peroxide has been widely used as an inducer for apoptosis and necrosis in different cell types at different concentrations depending on the sensitivity of the cells (Hampton and Orrenius, 1997, Jiang et al 1996) .

Neuroprotective Agents

Antioxidants

It has been well documented that H_2O_2, a free radical generator, induces neuronal loss in neurodegenerative diseases like AD through apoptosis. Naturally occurring and axotomy induced death of developing neurons seem to involve the active participation of oxidative events explaining the protective effects of antioxidants. To protect cells from active oxygen species, organisms have developed enzymatic defenses and non-enzymatic defenses. Another important antioxidant defense mechanism is the binding of iron and copper ions to protein whenever possible, so that generation of OH^- by Fenton reaction is largely prevented (Gabriele et al 2000). Brain is more vulnerable than other tissues because it contains large quantities of oxidizable lipids and metals and has comparatively less antioxidant mechanisms. The brain may efficiently metabolize superoxide but may have difficulties in eliminating the H_2O_2 as it has less catalases than superoxide dismutase and glutathione peroxidase. Neurons contains various molecules such as vitamin C and E which scavenge free radicals (Castagne et al 1999). Cells can tolerate mild oxidant stress and often respond by increasing synthesis of antioxidant defense enzymes and other protective proteins. Any imbalance to this system leads to cell death. There are many external sources, which protect the cells from free radicals for e.g. flavonoids, carotenoids such as β carotene, melatonin, vitamins and sodium pyruvate.

Flavonoids

The biological significance of flavonoids is their anti-oxidative effect. The anti-oxidant activity of flavonoids is due to their ability to chelate metal ions. Flavonoids have been shown to be capable of both directly scavenging free radicals and inhibiting xanthine oxidase competitively. The anti-oxidant property of flavonoids has been also found to enhance the capacity by modulating the level of antioxidant proteins. This property of flavonoids results in an additional anti-inflammatory activity by neutralizing the effect of inflammatory molecules like histamines, prostaglandins and leukotrienes.

Carotenoids

Carotenoids are phytonutrients, the nutritional elements that occur naturally in fruits and vegetables. They contain a conjugated backbone composed of isoprene units, which is usually inverted at the center of the molecule, imparting symmetry. Carotenoids have diverse biological functions and play different roles. Certain carotenoids are precursors of vitamin A and can be metabolically converted into vitamin A in the body. They function as chain breaking antioxidants, protecting cells and other body components from free radical attacks. It quenches singlet oxygen, induces gap junction communications and inhibits lipid peroxidation. Oxidative damage has been linked to the onset of degenerative diseases. Recent scientific data suggest that optimal intake of carotenoids and other antioxidants can help delay the onset of cancer, atherosclerosis, cataracts, macular degeneration and other major degenerative disorders (Weisburger, 1991).

Sodium Pyruvate

Pyruvate is a three-carbon compound and found naturally in the body as a product of glucose metabolism. It is involved in various metabolic pathways of energy production, and represents an important regulatory branch point. Sodium pyruvate acts as an intermediate in sugar metabolism and enzymatic carbohydrate degradation (alcohol fermentation) where it is converted to acetaldehyde and carbon dioxide by carboxylase. Thus, the addition of pyruvate to cell culture provides both an energy source and a carbon skeleton for anabolic processes. Certain specialized cells may require sodium pyruvate due to higher energy needs. Pyruvate is available in supplement form as a pyruvate salt. A pyruvate salt is formed by combining pyruvate with sodium or calcium. Extensive research has indicated that pyruvate supplementation is a potent, natural weight and fat reduction aid, making it popular with that seeking weight loss. Research has also shown that pyruvate acts as an antioxidant, protecting cells and tissues by scavenging harmful oxygen free radicals and preventing the formation of new free radicals (Borle and Stanko 1996).

Various experimental models have been used to examine the protective effects of agents against apoptosis induced in neuronal cells. Neuroblastoma cell lines have the ability to proliferate as well as differentiate as they arise due to arrest in differentiation program of neural crest progenitor cells. They have been widely used as a model in cancer research, in studies for susceptibility to neurotrophic viruses, neuronal differentiation and apoptosis (Shastry et al 2001). In our laboratory, we demonstrated the protective effects of sodium pyruvate against H_2O_2 induced apoptosis using human neuroblastoma cell line SK-N-MC cell line as an *in vitro* model. In this study the hallmark changes in the apoptotic cascade like the caspase 3 activity, cleavage of PARP and alterations in mitochondrial membrane potential were determined during apoptosis induced by H_2O_2 (Jagtap et al 2003).

Caspases are family of cysteine proteases that play an important role during apoptosis. Cleavage of PARP is one of the earliest detectable proteolytic events that occur following high molecular fragmentation of chromatin DNA (Lazebnik et al 1994). The sequential events in the apoptotic pathway suggest a defined cascade of proteolytic events. Secondly, ROI mediated DNA damage has been shown to cause the activation of poly-ADP–ribose transferase, a step important in apoptosis (Schwartzman and Cidlowski, 1993). Our study showed clearly that addition of sodium pyruvate along with H_2O_2 decreased the caspase 3 activity by 2 fold. These findings are in line with data on the kinetics of the enzyme activity reported with Jurkat T lymphocytes treated with 50 μM H_2O_2. Sodium pyruvate inhibited caspase 3 activity resulting in no cleavage of the 116 kD PARP. The data with H_2O_2 corroborated the finding demonstrating increased caspase 3 activity and cleavage of PARP in a similar time dependent manner in HL-60 cells treated with comparable concentration of H_2O_2 (DiPietrantonio et al 1999).

A potentially central factor in cell death in neurodegeneration, is the mitochondrial transition pore (MTP). There is increasing evidence for the role of mitochondria in control of apoptosis induced by different stimuli. It acts as a sensor and an integrator to receive as well as amplify signals from diverse upstream signaling pathways to initiate down stream execution steps (Ferri and Kroemer 2001). The MTP is a nonselective high conductance channel that spans the inner and outer mitochondrial membrane. The integrity of the impermeable inner membrane of mitochondria is essential for maintaining the electrochemical potential ($\Delta\psi M$), which is required for oxidative phosphorylation. It is now well established that in cells undergoing apoptosis, events related to formation of apoptosome involving Apaf 1 and caspase 9 are initiated by release of cytochrome C, which is the result of increased permeability in mitochondrial membrane. We observed an increase in membrane potential was seen as early as 5 hr on exposure to H_2O_2. Further more, there was a decrease in membrane potential SK-N-MC cells treated with H_2O_2 in combination with sodium pyruvate compared to the cells treated with H_2O_2 alone. These findings assume importance in the context that mitochondrial derived pathological free radicals have been implicated in numerous diseases and the aging process itself. The human CNS is relatively deficient in oxidative defenses rendering it more susceptible to the reactive oxygen species induced damage (Martilla, 1988). Recently, the significance of the permeability transition and pore formation in apoptosis triggered by H_2O_2 was demonstrated in L929 cells (Takeyama et al 2002). The findings from our study using the SK-N-MC cells, indicated that mitochondria is indeed the target for the action of H_2O_2, which then initiates a cascade of events leading to DNA fragmentation and apoptosis.

As, flavonoids are reported to exhibit biological effects such as antioxidant and free radical scavenger properties; we investigated whether flavonoids such as β-carotene, rutin and quercetin had any protective effect on apoptosis induced by H_2O_2 in SK-N-MC cells. Interestingly, none of these compounds had any protective effect on apoptosis induced in these cells though other studies have shown with protective effects of quercetin and rutin (upto 200μM) Caco-2 and Hep G2 cells (Aherne and O'Brien, 1999). In sum, our data suggested that it may be worth while to evaluate the neuroprotective property of sodium pyruvate in a broader perspective and explore it's use in in vivo experimental models of neurodegenerative diseases.

Prospective Studies In Post Genomic Era

One of the most significant achievements in the modern biology has been the unraveling of the human genome sequences in 2003 and this has generated a lot of excitement in the field of neuroscience. One of the important aspects that can be explored in light of the voluminous data provided by human genome project is the studies on genetic basis of neurodegenerative diseases. The information from the data bank will aid in identification of genes that are involved in different neurodegenerative diseases to identify therapeutic genes or targets for gene therapy (Plomin and McGuffin 2003). Studies can now be focused to clarify role of new genes in neuronal functions from huge repertoire of anonymous genes of which only few DNA sequences are identified. Techniques such high through put DNA microarrays can be used for gene profiling and screening for up- and down regulation of genes (Young, 2000) in varied neurodegenerative diseases with different etiologies. The alterations in gene profiles can be extrapolated to protein profiles and their functions to correlate them with partial or total loss of neuronal functions. Further more, elucidation of molecular mechanisms in apoptosis, with expression of specific pro and antiapoptotic molecules in neurodegeneration in various neuropathologies, in the aging nervous systems may be of invaluable help to develop novel therapeutic approaches and in development of peptides and drugs/compounds which can inhibit the proapoptotic proteins.

References

Aherne SA and O'Brien NM (1999) Nutr Cancer **34**: 160-166.
Amarante-Mendes GP and Green DR (1999) Braz J Med Biol Res **32**: 1053-1061.
Bamford M, Walkinshaw G and Brown R (2000) Exp Cell Res **256**: 1-11.
Beal MF (1996) Curr Opin Neurobiol **6**: 661-666.
Borle and Stanko (1996) Am J Physiol **270**: 535-540.
Castagne V, Gautschi M, Lefevre K, Posada A, Clarke PG and Castagne V (1999) Prog Neurobiol **59**: 397-423.
DiPietrantonio AM, Hsieh T and Wu JM (1999) Biochem Biophys Res Commun **255**: 477-482.
Ferri KF and Kroemer G (2001) Bio Essays **23**: 111-115.
Gabriele S, Alberto P, Sergio G, Fernanda F and Marco MC (2000) Toxicology **155**, 1-15.
Gross A, Mc Donnell JM and Korsmeyer SJ (1999) Genes Dev **13**: 1899-1911.
Hampton MB and Orrenius S (1997) FEBS Lett **414**: 552-556.
Hengartner MO (2000) Nature **407**: 770-776.
Herceg Z and Wang ZQ (2001) Mutat Res **477**: 97-110.
Jagtap JC, Chandele A, Chopde BA and Shastry P (2003) J Chem Neuroanat **26**:109-18
Jiang D, Jha N, Boonplueang R and Andersen JK (2001) J Neurochem **76**: 1745-1755.
Jiang MC, Yang YHF, Lin JK and Yen JJ (1996) Oncogene **13**: 609-616.
Kaufmann SH and Earshaw WC (2000) Exp Cell Res **256**: 42-49.
Kirsch DG, Doseff A, Chau BN, Lim DS, de Souza-Pinto NC, Hansford R, Kastan MB, Lazebnik YA and Hardwick JM (1999) J Biol Cell **274**: 21155-21161.
Kromer G (1997) Nature Medicine **3**: 614-620.
Lazebnik YA, Kaufmann SH, Desnoyers, Poirier GG and Earnshaw WC (1994) Nature **37**: 346-347.
Lix and Darzykiewicz Z (2000) Exp Cell Res **256**: 19-26.
Martilla RJ, Roytta M, Lorentz H and Rinne UK (1988) J Neural Trans **74**: 87-95.
Meier P, Finch A and Evan G (2000) Nature **407**: 796-801.
Plomin R and McGuffin P (2003) Ann Rev Psychol **54**:205-28.
Raff M (1998) Nature **396**: 119-122.
Schwartzman RA and Cidlowski JA (1993) Apoptosis Endocr Rev **14**: 133-151.
Schweichel JU and Merker HJ (1973) Teratology **7**: 253-266.
Shastry P, Basu A and Rajadhyaksha MS (2001) Int J Neurosci **108**: 109-126.
Takeyama N, Miki S, Hirakawa A and Tanaka T (2002) Exp Cell Res **274**: 16-24.
Talton WG and Olanow CW (1999) BBA/Bioenergetics **1410**: 195-213.
Thompson CB (1995) Science **267**: 1456-1462.
Uberti D, Yavin E, Gil S, Ayasola K, Goldfinger N and Rotter V (1999) Mol Brain Res **65**: 167-175.
Vaux DL and Strasser A (1996) Proc Natl Acad Sci **93**: 2239-2244.
Weisburger JH (1991) Am J Clin Nutr **53**: 226S-237S.
Young RA (2000) Cell **102**: 9-15.
Ziegler M and Oei SL (2001) Bio Essay **23**: 543-548.

Molecular and Cellular Neurobiology
Editors: M.K. Thakur and S. Prasad

Therapeutic Strategies for Neurodegenerative Disorders: Leads from Herbal Drugs

Maheep Bhatnagar

Department of Zoology, University College of Science
Mohan Lal Sukhadia University, Udaipur, India

Abstract

Aim of the present communication is to summaries and describe in short, some salient features of information available and ongoing research for diverse types of neurodegenerative disorders and possibilities for identifying new pharmacological targets or leads for such disorders from plant sources. Efforts made in other laboratories and in our laboratory using existing knowledge on several popular herbal extracts indicated a possible lead from herbs or their products. A few promising leads identified are worth exploiting..

Keywords: Neurodegeneration, Neuroprotection, Herbal extracts, Alzheimer's, Parkinson's

Introduction

Traditional herbal formulations existed in form of household medicines before application of the modern scientific methods to health care. Even today, majority of the rural and urban population practices herbal health care. In fact plant based natural products have played a major role in development of large number of synthetic drugs now in common use for the treatment of common cold and cough to the cancer and even diseases related to central nervous system. They are the most widely used medicines because of their easy availability, cheaper, effective and also because of general notion that they are always safe than synthetic drugs (although which is not always true in many cases). Written records about the medicinal plants and their uses date back at least 5000 years to Sumerians (Swardlow, 2000). Archeological records suggest even earlier use of medicinal plants.

The use of plant based drugs or products are increasing world over not only as alternative therapy but also as preparations alternative to caffeinated beverages, food supplements, cosmetics etc. According to an estimate herbal products now constitute major share of international pharmaceutical industry. About 1500-1800 herbal products are sold in the US markets only as dietary supplements or ethnic medicines besides other preparations (Seth and Kakkar, 2002). The historic bond between plants and human health began to establish in 1897, when Friedrich Bayer and Co introduced synthetic acetylsalicylic acid (Aspirin). Aspirin is a safer synthetic analogue of the salicylic acid, which is an active ingredient of a willow bark.

Knowledge gained from the long history of use of medicinal herbs or isolation and charaterization of their bioactive molecules has served as the foundation for much of our modern pharmacology as well. Many drugs of common use in allopathic practices such as aspirin, digitalis, morphine, quinine are all originally isolated from plants. Simultaneous advancement in the field of chemistry, biochemistry and development of new and sophisticated techniques of isolation, separation, biosynthesis, various active compounds were isolated and screened for their bio-activity. With the advancement of organic biochemistry most of the active constituents from plants have been synthesized. However, in spite of such phenomenal progress in the area of development of new drugs from synthetic sources, plants still continue to provide basic raw material for some of the most important drugs, some of them which are well accepted for their therapeutic use are *Achyranthus aspera* (Diuretic), *Acorus calamus*, (Tranquilizer), *Artemisis ulgaris* (Cardiac tónic), *Butea frondosa* (Antihelminthic), *Baccopa monnieriae* (Memory), *Boerhavia diffusa* (Anti inflammatory), *Cassia fistula* (Cathartic), *Centella asiatica* (Intelligence), *Eugenia jambolana* (Hypoglycemic), *Euphorbia thermofolia* (Antihistatic), *Podophylum hexandrum* and *P. emodi* (Anti cancer), *Catharanthus roseus* (Anti cancer), *Atemisia annua* (Anti malarial), *Commifora mukul* (Hypolipidaemic), *Theobroma cocoa* L (CNS stimulant) and *Caffea arabica* (CNS stimulant and diuretic). These plants and their products have been approved by FDA in USA. In addition to the pure constituents, crude extracts of Belladona, Ipec, Opium, Glycyrrihza, Rhuarb, Valarian, Capsicum, Digitalis, Aloe are also used in modern medicinal system. Essential oil of Japanese mint, pepper mint, Eucalyptus, Clove, Cinnamon, Lemon grass and Camphor are not only commonly practiced in modern medicine but also in cosmatic preparations. Several medicinal plants have been also investigated for their CNS effects by researchers. Satyavati (1995), Dandiya (1991), Dhawan (1995) published a comprehensive list of various Indian medicinal plants with proven CNS activity. Details of some of these plants is summarized as Table.1

Table 1. Medicinal herbs with CNS active activity (Satyawati, 1995, Dandiya, 1995, Dhawan, 1995).

Botanical name	Common name	Acitivity
Lactuca sativa	Garden lettuce	Stimulant
Piper longum	Black pepper	‘’
Acorus calamus	Vacha	Depressant
Baccopa monniera	Brahmi	‘’
Cassia fistula	Indian laburnum	‘’
Celestrus peniculatus	Jyotishmati	‘’
Curcuma aromatica	Wild turmeric	‘’
Cymbopogon nardus	Citronella	‘’
Cyprus rotundus	sedge plant	‘’
Emblica officianalis	Indian gooseberry	‘’
Euphorbia firucalli	Milk brush	‘’
Glycyrihza glabra	Liqorice	
Hibiscus rosasinensis	China rose	
Nymphea stellata	Water lilly	
Oscimum sanctum	Tulsi	

It is true that a lot of scientific studies has been done on a large number of medicinal plants in India as well as abroad. In spite of this, traditionally proven herbal formulations have not been yet standardized with modern standards of drug testing, safety and efficacy. Though opportunities for developing these traditional drugs with international acceptance are enormous, there is hardly any concerted effort in this direction at the government or industry level in our country. Result is the declining enthusiasm for studying plant extracts for their therapeutic value. Many pharmaceutical industries are on the verge of closing. but one has to believe that 250,000 living plant species world over contain a much greater diversity of bio-active compounds than any chemical library made by a man. Although this is also true that extracts prepared from medicinal plants contain a variety of molecules with potent biological activities but unfortunately, it is often difficult to analyze their biological activities because of their complex nature and possible synergistic actions.

In modern therapeutic system, therapies for numerous central nervous system disorders are at present not available. Drugs available for treatments of anxiety, depression and mental health conditions are not satisfactory. Majority of currently available CNS drugs are of synthetic origin and most of them are derived from yet other synthetic molecules. For example, Chlorpromazine and Reserpine, tricyclic and MAO inhibiting antidepressants, Benzodiazines, Meprobamate etc for treatment of pscyhotic patients, Bicuculline, Pentylenetetrazol, Amphetamine, Methylphenidates etc as CNS stimulants, Barbiturates, Hydantoin, Oxazolidone, Succinimides, Acetyl urea as antileptics, Bromocriptine, Apomorphine, levodopa, Amatadine, Trihexiphenydyl, Procyclidines for the treatment of Parkinson's. Most of these drugs have severe side effects and also develop dependency.

Herbal remedies for such conditions have been known since time immortal and efforts made during the past few decades reconfirm that several of herbs are indeed therapeutically useful for treatment of diverse CNS disorders (Table.2).
Although all active principles and the modes of actions these herbal drugs/extracts are not yet precisely defined, but available data on their neuropharmacological activity profiles as well as bioactive components do indicate that they could be useful sources for developing better, effective, safe drugs with novel modes of action. In the past few years, number of reports on properly controlled clinical trials on standardized herbal extracts dealing with neuronal disorders has been increasing. Although initially these reports came from research laboratories in Asia and Europe only but in recent years researchers around the globe are reporting their observations based on strict experimental and clinical studies. No doubt Consequences of these efforts will result in a better-standardized herbal extracts with acceptance by patients and industry as therapies for mild to severe central nervous system disorders.

Ayurveda and CNS active herbal drugs

Natural products or plant based products are main source of the medicines in traditional Indian medicinal system In Ayurveda, these are used as either alcoholic extracts or decoction. Charak Samhita (900 BC) is the first recorded treatise of the Ayurveda. It consist of eight sections with descriptions of more than 341 plants used in medicine. In relation to neurological diseases the drugs, which are acting on central nervous system, are broadly classified into ten categories

1. Medya (Intelligence and memory
2. Samjnasthapna (resuscitative)
3. Madakari (Narcotic/addictive)
4. Nidrajanana (hypnotic/sedative)
5. Vedanasthapana (Analgesic)

6. Apasmarahara (Antileptic)
7. Unmadahara (anti-insanity)
8. Jwarahara (antipyretic)
9. Chardi -nigrahana) (antiematic)
10. Vamanopza (emetic)

Table 2. Composition of certain fomulations in practice for treatment of neurodegenerative disorders.

S.No.	Formulation	Composition	Theraputic implications
1	-	*Hyocyamus reticulata* *Mucuna pruriens* *Sida cordifolia* *Withania somnifera*	Anti-parkinson's
2	Trasina	*Banisteria cappi* *Claviceps purpura* *Datura stramonium* *Mucuna pruriens* *Vicia faba*	Anti Alzheimer's
3	BR-16A	*Acorus calamus* *Bacopa monnierie* *Celestrus peniculatus* *Centella asiatica* *Evolvulus alsenoides* *Mukta pisthi* *Nardostaschys jatamansi* *Orchis mascula* *Prunus amygadalus* *Syzygium aromaticum* *Valerana wallichii* *Withania somnifera* *Zinziber officinale*	Anti-parkinson's
4.	Galentamine	*Galanthus wornorii*	Anti Alzheimer's
5.	Egb761	*Ginkgo bioloba*	Dementia
6.	Egb761`	*Ginkgo bioloba*	Dementia of Alzheimer's type and Multi infarct dementia
7.	Brahmi	*Bacopa monniera* extract	Cognitive functions
8	GB8	Swiss *Ginkgo bioloba*	Cognitive disorders

The term medya in Ayurveda refers to all substances /agents that promote or are beneficial to the medha (i.e. intellect), dhriti (concentration and retention) and smiriti (memory). In the modern pharmacology medhya is called as nootropic agents. Both animal and human experiments have shown that medhya are important for general medicine and psychiatry (Satyawati, 1995). A number of Ayurvedic drugs have been grouped under the general class of Rasayana (Rejuvenators). According to Ayurvedic physicians, rasayana increase the life span and delay aging by counteracting the degenerative changes associated with aging. Modern medicines have also recognized the rasayana group of drugs as anti-stress / adaptogens and could be useful in treating various CNS disorders.

Neurodegenerative diseases

Unlike other categories of diseases such as infections or trauma these diseases share etiology or some clinico-pathologic features. There are two major common features of the degenerative diseases:

1. They are disease of neurons and they selectively affect one or more functional systems of neurons for example there is selective degeneration of the striato-nigral dopaminergic neurons
2. They are marked generally by symmetric and progressive involvement of the CNS.

but they also differ among them selves quite sharply. Some have pattern of heritability, others are sporadic, while others show atrophy and loss of affected neurons with or without specific features, for example, disease of the cortex may manifest as dementia and similarly, disorders of the basal ganglia as extra-pyramidal movement disorders. Thus specific diagnosis is made only by correlating the clinical and pathological findings. For ease of description these diseases are grouped according to the part or parts of the brain that are principally involved. An abbreviated list is given below (Table- 3).

Alzheimer's disease

Alzheimer's disease (Walton, 1996) is the most common cause of senile dementia. Onset of AD before age 50year is rare. The most frequent and initial complaints are impairment of concentration, memory, and other higher intellectual functions. Over the course of five to ten years, progressive memory loss, disorientation and language dysfunction often leads to a mute, immobile state. Death usually results from infection. Although at-least 10% of cases are clearly familial, most are sporadic. Pathologically, disease is characterized by widening of cerebral sulci. Microscopic features are presence of neurofibrillary tangles and senile plaques. Senile plaques are found most frequently in the cerebral cortex and limbic structures. They range from 20 to 150μm in diameter, they are composed of focal collections of dilated, tortuous silver stained neurites that often contain neuro fibrillary tangles, arranged around a central amyloid core. This congo red positive core composed of a β –amyloid protein . In addition to these two characteristics, AD is also charaterized by deficiencies of acetylcholine (ACh) and associated enzymes acetylcholinesterase (AChE)and choline actyl transferase (ChAT) in cerebral cortex, hippocampus and amygdala.. The gene for this β-APP protein is located on chromosome21.

Huntington's disease

The disease usually appears between 20 to 50 years of age. Extra-pyramidal movements combined with progressive dementia characterize it. It is an autosomal dominant condition. Those who inherit it from their fathers usually manifest it much earlier in life than those who inherit it from their mothers. The defective gene t is assigned to chromosome 4. In Huntigton's the brain is comparatively small (less than 1000 gm) and demonstrate marked atrophy in Caudate and Putman (with conspicuous dilatation of the frontal poles of the lateral ventricles There may be secondary atrophy of the cortex and Globus palidus as well . Microscopically, there is severe loss of both large and small neurons, in the dorsal regions of these nuclei. There is also marked fibrillary gliosis.

Table. 3. Various Neurodegenerative diseases

Areas of brain affected	Name of the disease
Basal ganglia and brain stem	1. Alzheimer's disease (AD) 2. Pick's disease
	1. Huntington's disease 2. Ideopathic Parkinson's disease 3. Postencephalitic parkinson's 4. Striatonigral degeneration 5. Progressive supranuclear plasy 6. Shy-Drager syndrome
Spinocerebellar area	1. Olivopontocerebellar degeneration 2. Friedreich's ataxia 3. Ataxia telengiectasia
Motor neurons	1. Amyotrophic lateral sclerosis 2. Werdnig Hoffman disease 3. Kugelberg Welander syndrome

Parkinson's disease

Parkinson's is known as a disturbance of motor function characterized by expression less face, slow voluntary movements andProgressively shortened, accelerated step), rigidity and in most cases characteristic tremors. Such motor disturbance are seen in a number of other disease states, which show damage to the nigrostriatal dopaminergic system. Disease may be produced by drugs (dopamine agonists and toxins), pesticides, stress etc. Ideopathic Parkinson's or Paralysis agitans is sporadic, progressive disorder and is the most common form of the disease. It has onset between 50 to 80 years of age. Grossly, disease shows de-pigmentation of substantia nigra and locus ceruleus. Microscopically, there is loss of melanin containing neurons from these regions, with extra-neuronal pigment diposition and presence of lewy bodies in remaining neurons. Lewy bodies are intra-cytoplasmic, eosinophilic, round or elongated inclusions that consist of dense core surrounded by a paler rim. The loss of the dopamine containing neurons from substantia nigra results in dopamine depletion in the striatum because this is the principal site of their axonal projections. The severity of the Parkinson's syndrome is proportional to the severity of the dopamine deficiency. This deficiency at least in part, can be compensated for by replacement therapy with L-dopa.

Motor neurone disease (Amyotrophic lateral sclerosis complex)

In ALS neuronal degeneration is mainly concentrated in upper and lower motor neurons of the pyramidal motor system. The upper motor neurons (UMNs) are in motor cortex and their axons traverse the internal capsule and corticospinal tract to synapse on the lower motor neurons (LMNs) in the cranial motor nerve nuclei and the motor neurons in the anterior horn of the spinal cord. Pathologically, degeneration of UMNs produces axonal loss, atrophy of the corticospinal tracts in the lateral columns of the spinal cord. The neurological symptoms exhibited by the patients include spasticity, hyperflexia in UMN and muscle atrophy and weakness in LMN. Disese is also sporadic in most cases, but about 2:1 male predominance was noted. Onset is typically in late middle age, with usualy fatal outcome in 2 to 6 years. There is no known treatment as well.

Potential herbal drugs

***Hypericum perforatum*:** *Hypericum perforatum* (St. John's wort) has been in use all over the world as an herbal medication for depression and related disorders. An intensive investigation showed that the plant extracts are significantly useful in treating the depression related disorders (Linde et al 1996). Bhattacharya et al 1998). Hypericin activates dopamine receptors but inhibits serotonin receptor expression (Muller, 1994). Chatterji et al (1998) reported that it is not hypericin but hyperfoin is responsible for the anti-depressent activity. Studies in 3250 patients have demonstrated improvement in 50% of cases.

***Ginkgo biloba*:** Also called as "living fossil". Ginkgo is one of the oldest living tree species. Almost extinct following ice age, the tree survived only in Asia and later in 18th century brought to Europe and America as ornamental tree. Extract obtained from its green leaves was identified as therapeutically useful for the treatment of peripheral circulatory disturbances as it is a vasodilator. Subsequent observations made in pateints and experimental animals revealed that this extract refereed to as Egb761 is not a vasodilator in a classical sense but CNS function modulating and neuroprotective agent suitable for the therapy of patients with cerebro vascular disorders or cerebral insufficiencies. Ginko is widely used in Europe for treating dementia. It improves blood flow in the brain and contains flavnoids (plant substances) that act as antioxidants. Although, many of the clinical trials have been scientifically flawed, the evidence that ginkgo may improve thinking, learning and memory and in people with AD has been higly promising (Watnabe et al 2001).

More than 40 components of *Ginkgo biloba* have been identified. Two of the most important groups of active chemicals are flavonoids (Quercetin, Kaempferol, isorhemnetin) and terpenes lactons or terpenoids which include bilobalide and several ginkgolides (A, B, C, J and M). Individual constituents have been studied in hundreds of in vitro, animals and human pharmacological experiments (Chavez and Chavez; 1998; van Beek et al 1998) Recent research concentrated on effect of Ginkgo on dementia of Alzheimer's type. and closely related multi-infarct dementia. These double blind, randomized control study trials in patients have shown that EGB761 obtained from Ginkgo is efficacious in delaying the clinical deterioration of patients with dementia (Alzheimer's and multi-infarct type) or in brining about symptomatic improvement. (Le Bars et al 1997; Oken et al 1998; Ernst and Pittler, 1999).The mechanism of ginkgo's theraputic effects are not fully understood but attributed to the synergistic effects of its constituents (Behl et al 1994; Maitra et al 1995).These act to varying degrees as scavengers of free radicals, chemicals implicated in the pathophysiology of Alzheimer's disease. The dose is 40mg /3 times per day of an extract standardized to 24% flavonoid glycoside and 6% terpenoids.

***Bacopa monniera*:** Commonly known as Brahmi. The plant is found in wet marshy and damp places throughout India thus also known as Jalbrahmi or Nirbrahmi. It is succulent, creeping herb with 10-20 long stem rooting at nodes. It produces numerous branches. Leaves are sessile, decussate, o.6-2.5 cm long . The drug contains alkaloids brahmine and herpestine, D-mannitol and saponins, bacosides A and B. The plant is used as nervine tonic, diuretic and commonly used to treat asthma, epilepsy, insanity and hoarseness. It is a major constituent of Medhya rasayana formulations, which facilitates learning, and improve memory. Initial studies using 50% ethanolic extract of the whole plant with out roots demonstrated its effects on short and long term memory retention. Rastogi and Dhar (1960) and Chatterji et al (1963) isolated and characterized two Saponins later called Bacoside A and Bacoside B. Subsequently, the bacogenins were isolated and characterized (Kawai et al 1973, Rastogi and Kulshrestha, 1974,75,93). More recently they isolated another minor saponin BacosideA1 and two triterpinoid saponins. Garai and his colleagues (1996) isolated 3 new type triterpinod saponins bacosaponins A, B and C and pseudojujubogenin glucoside bacosaponin D. Bhattacharya et al (1999) studied the effects of *Bacopa monnniera* extract on rat model of Alzheimers disease. Oral administration of the extract in doses of 5 -10mg/kg markedly reduced the magnitude of memory deficit caused by the neurotoxins. The treatment also reversed the reduction in acetylcholine concentration, Choline acetylase activity and muscarinic receptor binding in hippocampus and frontal cortex. Bhattacharaya et al (1999) have reported that effects of *Baccopa monniera* extract on rat model of Alzheimer's disease. The lession was produced in nucleus basilis magnocellularis. . produced marked deficit in retention of acute avoidance learning in 7-14 days time. Oral administration of the extract in doses of 5-10mg/kg markedly reduced the magnitude of memory deficit caused by the neurotoxins. Singh et al (1979) had earlier reported an increase in acetylcholine content in frontal cortex of normal rats given 40mg/Kg *Baccopa monniera* extract orally once daily for 15 days.

Mucuna pruriens: English name is "cowage" plant. It is also called as Atmagupta. It is a twiner with trifoliate leaves, purple flowers and turgid S shaped pods covered with hairs which cause intense itching on contact with the skin. The plant belongs to family leguminosae and the beens have been used as neutricious and energy giving food in India. *Mucuna pruriens* have been described as useful therapeutic agents in various diseases of reproductive and the nervous systems (Naddkarni, 1908, Vaidya, 1925; Dutt,1980) in Ayurveda. An Ayurvedic formulation named Masabaladfi Pacana contains Mucuna beens is useful in Kampavata. L dopa was isolated from Mucuna seeds by Indian workers Damodaran and Ramasawmy in 1937. HPLC studies of Mucuna seeds showed 4.02% content of L-Dopa in whole seeds. Vaidya et al (1978) successfully treated patients of Parkinson's disease with a powder made from whole been of *Mucuna. Manyam* and Bhatt (2001) have also discussed that Levo dopa is the major active compound in *Mucuna pruriens*.

Tinospora Cordifolia: *Tinospora cordifolia* has been extensively studied and reported to have potent immuno-stimulant (Agarwal et al 2002) and antistress effects. Ayurvedic literature recommends a rejuvenating recipe where *Tinospora cordifolia* being an important –constituent to enhance the memory function. However, there are experimental evidence to support memory enhancing property of *Tinospora cordifolia.*

***Centella asiatica*:** Also called Indian pennywort or Mandooki. It is found through out India in the marshy areas up to 2000 m, also in Pakistan, Sri Lanka and Madagaskar. It is herbaceous creeping herb with long prostrate reddish stem possessing long internodes and rooting at the nodes. Leaves arise from each node of the stem. The drug contains triterpenoid saponin glycosides called asiaticoside, oxyasiaticoside, brahmoside, brahminosides and thankuniside. Hydrolysis of the asiaticoside affords to two molecules of fucose, one rhamnose and aglycone asiatic acid. In addition to these the drug contains alkaloids, sterols (β-stitosterol, stigmasterol), tannins, amino acids, inorganic salts. The plant has been commonly used in diseases of skin, nerves and blood. Leaves are taken as tonic for improving memory.

***Galanthus wornorii*:** *Glanthus wornorii* is a Caucasian snow drop Plant. Galentamine is a pure unalterated extract of Galanthus (Venturi et al 1965). A recent study (Willcock et al 2000) has shown that galantamine appears to slow the progression of the neurodegenerative condition. It also reversibly and competitively inhibits acetylcholinesterase and enhances the response of nicotinic receptors to acetylcholine. Thus Galentamine inhibits the production of AChE and enhances the response of brain to ACh. In a study 653 Alzheimer patients diagnosed mild to moderate AD galantamine was shown to slow down the decline of the functional abilities as well as cognition.

Huperzine serrata

Huperzin A is a substance first found in the chinese medicnal herb *Huperzine serrata*. In a well designed, placebo controlled trial, 58% of the patients with Alzheimer's disease had significant improvement in memory and cognitive and behavioral functions after taking 200mg of the drug. A twice per day of eight weeks showed significant improvement. Another double blind report using injected Hyperzine A confirmed a positive effect in patients with dementia (Grant, 1997).

***Withania somnifera*:** *Withania somnifera* Dunal commonly known as "Ashwagandha is used as Rasayana in indian traditional medicinal system. It is an erect shrub growing through out the dry and subtropical parts of India. i.e. MP, UP, Punjab, Gujrat and Rajasthan. It is also cultivated in Congo, South Egypt, Morroco, Jordon and Pakistan. Flowers are greenish, yellow and Berry is red and smooth. Though primarily roots but seeds and leaves are also used for medicinal purposes. In Ayurvedic preparations it is used as powder, decoction, medicated wine etc. The main pharmacologically active constituents are alkaloids and steroid lactones. Among alkaloids withaine, cuscohygraine, tropane, anahygraine, somniferrin,, anaferine, withaminine, withaninine are main while among lactones are withanoloids. The total alkaloid content of roots very between 0.1 to 0.3% although in some cases higher yield is also reported. Karnic (1991) suggested that ashwagandha should be considered as the premier herb for all negative conditions associated with aging. Evidences show that drug is useful in preventing senile dementia and Alzheimer's disease (Nadkarni et al 1976). Alzheimer's develop as an progressive loss of memory and gradual loss of the ability to function and take care themselves. Drug helps in slowdown the progression of the disease (Bhattacharya et al 1995). Glycowithanolides (WSG) were also found to reverse both cognitive deficits and perturbed central cholinergic markers induced as a result of neurodegeneration produced by neurotoxins. Bhattacharya et al (1997, 2001) assaying foot shock induced changes in the rat brain showed that Ashwagandha also normalizes SOD and LPO activity and enhances CAT and GPX activity.

Anti-oxidants and neurodegenerative diseases

Increased generation of oxidative free radicals or impaired antioxidant defense mechanisms have been identified in aging process, neurodegenerative conditions, including Parkinson's and Alzheimer's disease and in chronic stress induced perturbed state (Mexwell, 1995, Jain et al 1998). Major oxidative free radical. scavenging enzymes are super oxide dismutase (SOD), catalase (CAT), and glutathione peroxidase (GPX) deficient functioning of these enzymes leads to accumulation of toxic oxidative free radicals and consequent degenerative changes. Potential antioxidant therapy for CNS disorders should therefore include either natural antioxidant enzymes or agents that are capable of augmenting the function of these oxidative free radical scavenging enzymes (Bast et al 1991). By virtue of their proposed properties and clinical uses in Ayurveda rasayana may provide potential intervention against oxidative threats both in health and disease. (Ghosal, 1991). Bhattacharya et al (1997) have shown that active principal WSG of *W. somnifera* increases cortical and striatal concentration of antioxidant enzymes. Shukla et al (2001) also reported the effects of S. anacardium and *W. somnifera* on SOD, Cat, GPX and LPO and demonstrated significant increase in SOD and LPO. Panda and Kar (1997) investigated the effects of Ashwagandha root powder administered for 15 and 30 days on lipid peroxidation, SOD and CAT in mice. 30 days treatment with root powder produced a significant decrease in LPO and increase in both SOD and CAT.

Dai et al (1987) studied that some medicinal plants are protective and efficient to certain extent in delaying aging process, which might be associated with the generation of free radicals. They also reported that SOD scavenge the free radicals. Tripathi et al (1996) examined the action of *Rubia cordifolia, Strychnos-nux vomica, Moringo oleifora, Bacopa monneira, Nardostachys jatamanasi , Muccuna pruriens* and Tamra bhasma on free radical induction and lipid peroxidation. All these plants exhibited significant anti-lipid peroxidative property but mechanism of action was different for different plants. Bhattacharya et al (2000) observed that *Baccopa monniera* like deprenyl also exhibit significant antioxidant activity after sub chronic administration.

Conclusion

Use of herbal drugs for many CNS disorders are known since long and efforts made during past few decades confirm that several of these herbs are indeed therapeutically potent not only as CNS active agents but also as neuroprotective agents. Although all active principles and mode of action of many of these are not yet precisely definable available data on their neuropharmacological activity indicate that they could be sources for better therapy in CNS related disorders including neurodegenerative diseases. Extracts prepared from medicinal plants and other natural sources although contain a variety of molecules with potent biological activities. Unfortunately it is extremely difficult to analyze them because of the

Table. 3. Various Neurodegenerative diseases

Areas of brain affected	Diseases
Diseases of the basal ganglia and brain stem	3. Alzheimer's disease (AD) 4. Pick's disease
	7. Huntington's disease 8. Ideopathic Parkinson's disease 9. Postencephalitic parkinson's 10. Striatonigral degeneration 11. Progressive supranuclear plasy 12. Shy-Drager syndrome
Spinocerebellar degeneration	4. Olivopontocerebellar degeneration 5. Friedreich's ataxia 6. Ataxia telengiectasia
Motor neurons	4. Amyotrophic lateral sclerosis 5. Werdnig Hoffman disease 6. Kugelberg Welander syndrome

complex nature and possible synergistic effects of the components. In spite of these problems, bioactive substances of plant origin provide safer and holistic approach to disease treatment and prevention of the disease.

References

Agarwal A, Malini S, Bairy KL and Rao MS (2003) Ind J Pharmacol **34**: 339-349.

Bast A, Haenen GR, and Doleman GJ (1991) Am J Medical **91**: 2S Cited in: Bhattacharya SK, Satyan KS Ghosal S (1997) 35: 236-239.

Bhattacharya SK, Kumar A and Ghosal S (1995) Phyto Ther Res **9**: 110-113. In: Bhattacharya SK, Satyan KS and Ghosal S (1997) **35**: 236-239.

Bhattacharya SK, Chakrabarti A and Chaterjee SS (1998) Pharmacopsychiatry **31**: 22-29.

Bhattacharya SK, Kumar A and Ghosal S (1999) Pharmacol Toxicol **4**:11-12.

Bhattacharya SK, Bhattacharya A, Kumar A and Ghosal S (2000) Antioxidant activity of *B. monniera* in rat frontal cortex, striatum and hippocampus. Phyto Ther Res **14**:174-179.

Bhattacharya A, Ghosal S and Bhattachrya SK (2001) J Ethanopharmacol **74**: 1-6.

Behl C, Davis J, Leslie R, and Schubert D (1994) Cell **77**: 817-827.

Chatterji N, Rastogi RP and Dhar ML (1963) Ind J Chem **1**: 212-215.

Chatterji N, Rastogi RP and Dhar ML (1965) Ind J Chem **3**:24 -29.

Chatterjee SS, Bhattacharya SK Wonneman M Singer A and Miller WE (1998) Life Sci **63**: 499-510.

Chavez ML and Chavez PI (1998) Ginkgo (Part I): Hosp Pharm **33**: 658-672.

Damodaran M and Ramaswami R (1937) Biochem **31**: 2149.

Dai YR, Gao CM, Tian QL and Yin Y (1987) Plant Med **53**: 309-310.

Dhawan BN (1995) Koslow SH, Murthy RS, and Coelho GV Eds. pp 197-202, National Inst Health USA.

Dhawan BN and Sing HK (2002) Annl Neurosci **9**: 34- 40.

Dutt VC Materia (1980) 148-149.

Ernst E and Pittler MH (1999). Clin Drug Invest **17**: 301-308.

Fransworth NR and Morris RW (1976) Am J Pharm Educ **148**: 46-52.

Garai.S, Mahto SB, Ohtani.K and Yamasaki K (1996) Phytochem **42**: 815-820.

Ghosal S (1991) Ind J Indg Med **8**:1 Cited in: Bhattacharya. SK, Satyan KS Ghosal S **35**: 236-239, (1997).

Grant WB (1997) Alz Dis Rev **2**: 42-55.

Jain S, Shukla SD, Sharma K and Bhatnagar, M (1998) Ind J Gerontol **12**: 60-69.

Karnick CR (1991) Ind Med **3**: 1-5.

Kawai K, Istaka Y, Shibata.S, Kulshreshta DK and Rastogi RP (1973) Acta Cryst **29B**: 2497-2953.

LaFrance WC Jr, Lautebach EC, Coffey CE, Salloway SP, Kaufer DI, Reeve A, Royall DR, Aylward E, Rummans TA and Lovell MR (2000) J Neuropsychiat Clin Neurosci **12**: 177-192.

Le Bars PL, Katz MM, Berman N, Itil TM, Fredman AM and Saltzberg AFA (1997) JAMA **278**: 1327-1332.

Limig E, Huazhen P and Waiman L (1987) Clin J Intg Trad Med **765**: 288-290.

Linde K, Ramirez G, Murlow CD, Pauls A, Weildenhammer W, and Melchart D (1996) BMJ **313**: 253-258.

Linde K, Riet GT, Hondras M, Vickers A, Saller R and Melchart D (2001) Comp Alt Medicine **1**: 5.

Maitra I, Marcocci L, Droy-Lefais M and Packer L (1995) Biochem Pharmacol **49**: 1649-1655.

Manyam, BV and Bhatt NV (2001) Ayu International. 1: 6-8

Manyam BV and Parikh KM (2002) Annals Neurosci **9**: 40-46.

Maxwell, SRJ (1995) Drugs **49**: 345-361.

Muller WE and Kasper S (1997) Pharmacopsychiatry **30**: 71-134.

Nadkarni, KM (1908) Indian Plants and Drugs. pp242 Norton and Co Madras.

Nadkarni, KM (1976) Ind Metiria Medica **I**: 1292-1294, Popular Prakashan, Bombay.

Ogern SO (2002) Annls Neurosci **9**: 46-53.

Oken BS, Storzbach DM and Kaye JA (1998) Arch Neurol **55**:1409-1415.

Panda, S and Kar A (1997) Ind J Physiol Pharmacol **41**:424-426.

Rastogi RP and Dhar ML (1960) J Sci Ind Res **19B**:455-456.

Rastogi S, Pal R and Kulshrestha DK (1993) Bacoside A5:. Id. J Heterocycl Chem **2**: 149-150.

Rastogi S and Kulshrestha DK (1999) Ind J Chem **38 B**: 353-356.

Satyawati, GV (1995) Koslow SH, Murthy RS and Coelho GV Eds. Pp 185-189, Nat Inst of Heath. USA.

Seth PK and Kakkar, P (2002) Annl Neurosci **9**: 17- 26.

Singh.RH, Singh.L and Sen SP (1979) J Res In Aurved Siddha **1**: 133-148.

Shukla SD (2001) Ph.D thesis submitted to MLS University, Udaipur.

Swardlow J (2000) In: Raskin I, Ribnicky DM, Komarnytsky S, Ilic N, Poulev A, Borisjuk N, BrinkerA B, Moreno DA, Ripoll C, Yakody N, O'Neal JM, Cornwell T, Pastor I, Fridlender B. (2002) Plants and human health in twentty-first century, Trends in Biocem **20**: 522-531.
Tripathi YB, Chaurasia S, Tripathi E, Upadhyaya A and Dubey GP (1996) Ind J Exp Biol **34**: 521-526.
Vaidya ND (1925) pp100, Vaidya Kalptaru, Ahmedabad.
Vaidya AB, Rajgopalan TG, Mankodi NA, Antarkar DS, Tatehd PS, Purohit AV and Wadia NH (1978) Neurol India **26**:171-176.
Van Beek TA, Bombardelli E, Morazzoni P and Peterlongo F (1998) *Ginkgo biloba*L. Fitoterapia **69**: 195-244.
Venturi VM, Piccinin GL and Taddei I (1965) Boll Soc Ital Biol Sper **11**:593-597.
Watnabe CMH, Wolffram S, Ader P, Rimbach G, Pecker L, Maguire JJ, Schultz PG and Gohil KC (2001) Proc Nat Acad Sci **98**: 6577-6580.
Walton J (1998) 9th edition, ELBS, Oxford Univ Press.
Wilcock GK, Lilienfeld S and Gaens E (2000) BMJ **7274**: 1445-1449.

DECAY HEAT REMOVAL AND NATURAL CONVECTION IN FAST BREEDER REACTORS

DECAY HEAT REMOVAL AND NATURAL CONVECTION IN FAST BREEDER REACTORS

Edited by
Ashok K. Agrawal
and
James G. Guppy
both at Brookhaven National Laboratory

HEMISPHERE PUBLISHING CORPORATION
Washington New York London

DISTRIBUTION OUTSIDE THE UNITED STATES
McGRAW-HILL INTERNATIONAL BOOK COMPANY
Auckland Bogotá Guatemala Hamburg Johannesburg
Lisbon London Madrid Mexico Montreal New Delhi
Panama Paris San Juan São Paulo Singapore Sydney
Tokyo Toronto

1981

DECAY HEAT REMOVAL AND NATURAL CONVECTION IN FAST BREEDER REACTORS

1 2 3 4 5 6 7 8 9 B C B C 8 9 8 7 6 5 4 3 2 1

Library of Congress Cataloging in Publication Data

Main entry under title:

Decay heat and natural convection in fast breeder reactors.

Papers of a meeting held in 1980, sponsored by the American Nuclear Society and the Brookhaven National Laboratory.

Bibliography: p.

Includes index.

1. Breeder reactors—Cooling—Congress. 2. Breeder reactors—Shutdown—Congresses. I. Agrawal, Ashok K., date. II. Guppy, James G., date.

TK9203.B7D43 621.48′35 81-329

ISBN 0-89116-196-1 AACR1

Contents

[illegible]

DECAY HEAT REMOVAL IN GCFR

PANEL DISCUSSION: FUTURE RESEARCH AND DEVELOPMENT — WHAT NEEDS TO BE DONE?

Contributors

S. L. ADDITON
Hanford Engineering Development Laboratory, Richland, Washington, USA (227, 241, 409)*

A. K. AGRAWAL
Department of Nuclear Energy, Brookhaven National Laboratory, Upton, New York, USA (69,335, 409)

R. ANDERSON
Fast Reactor Technology Group, UKAEA Dounreay Nuclear Power Development Establishment, Dounreay, Thurso, Caithness, Scotland (293)

J. C. ASTEGIANO
Commissariat a l'Energie Atomique, C.E.N. Grenoble, 38041 Grenoble, Cedex (Isere), France (279)

G. BASQUE
Commissariat a l'Energie Atomique, C.E.N. Cadarache, Boite Postal No. 1, St. Paul-Lez-Durance, France 13115 (83)

C. B. BAXI
General Atomic Company, San Diego, California, USA (379)

T. R. BEAVER
Hanford Engineering Development Laboratory, Richland, Washington, USA (227)

W. A. BEZELLA
Reactor Analysis and Safety Division, Argonne National Laboratory, Argonne, Illinois, USA (37)

D. BLANC
Commissariat a l'Energie Atomique, C.E.N. Cadarache, Boite Postal No. 1, St. Paul-Lez-Durance, France 13115 (83)

G. D. BOUCHEY
FFTF Project, U.S. Department of Energy, Richland, Washington, USA (241)

J. D. BRADLEY
Westinghouse Electric Corporation, Advanced Reactors Division, Madison, Pennsylvania, USA (163)

M. J. CHEN
Components Technology Division, Argonne National Laboratory, Argonne, Illinois, USA (53)

H. CHI
General Atomic Company, San Diego, California, USA (391)

H. S. CHUNG
General Atomic Company, San Diego, California, USA (391)

K. S. CHUNG
Reactor Analysis and Safety Division, Argonne National Laboratory, Argonne, Illinois, USA (37)

R. D. COFFIELD, JR.
Westinghouse Electric Corporation, Advanced Reactors Division, Madison, Pennsylvania, USA (179)

C. W. DAWSON
Fast Reactor Technology Group, UKAEA Dounreay Nuclear Power Development Establishment, Dounreay, Thurso, Caithness, Scotland (307)

E. M. DEAN
EBR-II Project, Argonne National Laboratory, Idaho Falls, Idaho, USA (195)

M. DEBRU
NOVATOME, 20 Avenue E. Herriot, 92350 Le Plessis Robinson, France, (267)

J. DECKERT
NERSA, 20 Avenue E. Herriot, 92350 Le Plessis Robinson, France, (267)

G. M. d'HOSPITAL
General Atomic Company, San Diego, California, USA (379)

H. M. DOMANUS
Components Technology Division, Argonne National Laboratory, Argonne, Illinois, USA (53)

G. S. DRUCKER
Energy Systems Group, Rockwell International, Atomics International Division, USA (147)

M. DUEWEKE
International Atomreaktorbau GmbH, Federal Republic of Germany (257)

F. E. FELDMAN
EBR-II Project, Argonne National Laboratory, Argonne, Illinois, USA (209)

M. J. GABLER
Energy Systems Group, Rockwell International, Atomics International Division, California, USA (351)

J. H. GERMER
Advanced Reactor Systems Department, General Electric Company, Sunnyvale, California, USA (321)

E. GESI
Commissariat a l'Energie Atomique, C.E.N. Cadarache, 13115 St. Paul-Lez-Durance, France (279)

J. L. GILLETTE
EBR-II Project, Argonne National Laboratory, Argonne, Illinois, USA (195)

C. V. GREGORY
Fast Reactor Technology Group, UKAEA Dounreay Nuclear Power Development Establishment, Dounreay, Thurso, Caithness, Scotland (293)

N. GRIMM
Westinghouse Electric Corporation, Advanced Reactors Division, Madison, Pennsylvania, USA (410)

J. G. GUPPY
Department of Nuclear Energy, Brookhaven National Laboratory, Upton, New York, USA (335, 407)

M. P. HEISLER
Energy Systems Group, Rockwell International, Atomics International Division, California, USA (115)

P. HUEBOTTER
Components Technology Division, Argonne National Laboratory, Argonne, Illinois, USA (410)

C. N. KELBER
Division of Reactor Safety Research, U. S. Nuclear Regulatory Commission, Washington, D.C., USA (3)

K. KESAVAN
Westinghouse Electric Corporation, Advanced Reactors Division, Madison, Pennsylvania, USA (163)

M. KHATIB-RAHBAR
Department of Nuclear Energy, Brookhaven National Laboratory, Upton, New York, USA (69, 335)

J. S. KILLIMAYER
Westinghouse Electric Corporation, Advanced Reactors Division, Madison, Pennsylvania, USA (179)

H. KINJO
Advanced Reactors Engineering Department, Nuclear Development Center, Mitsubishi Atomic Power Industries, Inc., Omiya, Saitama, Japan (131)

T. M. KUZAY
Components Technology Division, Argonne National Laboratory, Argonne, Illinois, USA (53)

R. T. LANCET
Energy Systems Group, Rockwell International, Atomics International Division, California, USA (351)

P. LAURET
NOVATOME, 20 Avenue E. Herriot, 92350 Le Plessis-Robinson, France, (267)

D. LETEINTURIER
Commissariat a l'Energie Atomique C.E.N., Cadarache, Boite Postale No. 1, 13 St. Paul-Lez-Durance, France 13115, (83, 411)

D. J. LORD
Fast Reactor Technology Group, UKAEA Dounreay Nuclear Power Development Establishment, Dounreay, Thurso, Caithness, Scotland (293)

I. K. MADNI
Department of Nuclear Energy, Brookhaven National Laboratory, Upton, New York, USA (69)

R. MAKIURA
Toshiba Nuclear Engineering Laboratory, Toshiba Corporation, Kawasaki-ku, Kawasaki, Japan (25)

R. A. MARKLEY
Westinghouse Electric Corporation, Advanced Reactors Division, Madison, Pennsylvania, USA (179)

R. MARTIN
Commissariat a l'Energie Atomique, Service des Tranferts Thermiques, C.E.N. Grenoble, 38041 Grenoble Cedex (Isere), Grenoble, France (279)

K. MAWATARI
Toshiba Nuclear Engineering Laboratory, Toshiba Corporation, Kawasaki-ku, Kawasaki, Japan (25)

W. A. MEDWID
General Atomic Company, San Diego, California, USA (365)

B. MENANT
Commissariat a l'Energie Atomique, C.E.N. Cadarache, Boite Postal No. 1, 13 St. Paul-Lez-Durance, France 13115 (279)

J. C. MILLS
Energy Systems Group, Rockwell International, Atomics International Division, California, USA (351)

D. MOHR
EBR-II Project, Argonne National Laboratory, Argonne, Illinois, USA (195, 209)

F. NAMEKAWA
Toshiba Nuclear Engineering Laboratory, Toshiba Corporation, Kawasaki-ku, Kawasaki, Japan (25)

D. E. NOE
Westinghouse Electric Corporation, Advanced Reactors Division, Madison, Pennsylvania, USA (163)

W. T. NUTT
Hanford Engineering Development Laboratory, Richland, Washington, USA (241)

Y. OKUBO
Advanced Reactors Engineering Department, Nuclear Development Center, Mitsubishi Atomic Power Industries, Inc., Omiya, Saitama, Japan (313)

L. L. PARME
General Atomic Company, San Diego, California, USA (365)

R. J. RIBANDO
Oak Ridge National Laboratory, Oak Ridge, Tennessee, USA (9)

F. ROESGEN
International Atomreaktorbau GmbH, Federal Republic of Germany (257)

W. M. ROHSENOW
Mechanical Engineering Department, Massachusetts Institute of Technology, Cambridge, Massachusetts, USA (97)

T. SAWADA
Advanced Reactors Engineering Department, Nuclear Development Center, Mitsubishi Atomic Power Industries, Inc., Omiya, Saitama, Japan (131)

J-C. SCHNEIDER
NERSA, 20 Avenue E. Herriot, 92350 Le Plessis-Robinson, France (267)

F. W. SCIACCA
Advanced Reactor Systems Department, General Electric Company, Sunnyvale, California, USA (321)

W. T. SHA
Components Technology Division, Argonne National Laboratory, Argonne, Illinois, USA (53)

A. SHENOY
General Atomic Company, San Diego, California, USA (365, 391)

R. M. SINGER
EBR-II Project, Argonne National Laboratory, Argonne, Illinois, USA (115, 195)

J. E. SULLIVAN
EBR-II Project, Argonne National Laboratory, Argonne, Illinois, USA (195)

T. TAMAOKI
Toshiba Nuclear Engineering Laboratory, Toshiba Corporation, Kawasaki-ku, Kawasaki, Japan (25)

Y. S. TANG
Westinghouse Electric Corporation, Advanced Reactors Division, Madison, Pennsylvania, USA (179)

J. H. TESSIER
Reactor Analyses and Safety Division, Argonne National Laboratory, Argonne, Illinois, USA (37)

D. H. THOMPSON
Reactor Analysis and Safety Division, Argonne National Laboratory, Argonne, Illinois, USA (37)

F. TIMMERMANN
International Atomreaktorbau GmbH, Federal Republic of Germany (257)

N. E. TODREAS
Department of Nuclear Engineering, Massachusetts Institute of Technology, Cambridge, Massachusetts, USA (97, 411)

J. V. TOKAR
EBR-II Project, Argonne National Laboratory, Argonne, Illinois, USA (195)

D. M. TURNER
Hanford Engineering Development Laboratory, Richland, Washington, USA (227)

H. VOSSEBRECKER
International Atomreaktorbau GmbH, Federal Republic of Germany (257)

H. WADA
Advanced Reactors Engineering Department, Nuclear Development Center, Mitsubishi Atomic Power Industries, Inc., Omiya, Saitama, Japan (131)

S-F. WANG
Department of Nuclear Engineering, Massachusetts Institute of Technology, Cambridge, Massachusetts, USA (97)

R. WEBSTER
Fast Reactor Technology Group, UKAEA Dounreay Nuclear Power Development Establishment, Dounreay, Thurso, Caithness, Scotland (293)

H. R. ZWEIG
Energy Technology Engineering Center, Energy Systems Group, Rockwell International, California, USA (351)

*Numbers in parentheses indicate the pages on which the authors' contributions begin.

Preface

The shutdown heat removal from nuclear power plants is a subject of great importance, as evidenced by the March 28, 1979 incident in the Three Mile Island Unit 2 plant at Harrisburg, Pennsylvania. Adequate heat removal is related to public safety and to the overall plant design. The system needed (active or passive) to do the job must be incorporated into the design from the very beginning—it should not be an added, patched affair. Careful analysis substantiated by experimental data, therefore, must be undertaken to produce an acceptable model for decay heat dissipation for all foreseeable conditions.

In fast breeder reactors, as in other types of nuclear reactors, the shutdown heat is generally removed via forced flow of the primary coolant through the reactor core. The electric power required is usually provided by at least two independent and guaranteed sources. But, in the event of a total loss of all electric current to pumps, it has been argued that the buoyancy-driven natural circulation can be adequate, at least, in a properly designed liquid-metal-cooled fast breeder reactor. Such arguments have been carefully scrutinized and substantiated in the British Prototype Fast Reactor, French PHENIX, and American EBR-II. Also, similar tests in the recently completed Fast Flux Test Facility are being planned for 1981. In fact, based on the recent PFR tests, for example, Cliff Blumfield, director of the Dounreay Nuclear Power Development Establishment, UKAEA, has claimed "walk away from it" safety.

In 1978—long before the TMI-2 incident—one of us (A.K.A.) proposed to the American Nuclear Society a specialists' meeting on the subject of decay heat removal from fast breeder reactors. The idea generated from the fact that, at least in the U.S., most of the fast reactor safety research and development had been concentrated in the area of core disruptive accidents where failure to scram is a prerequisite and that little work had been done toward systematic accident delineation. The shutdown heat removal is a key event that, although it takes full credit for the reactor scram, can lead to core disruption unless a reliable model for decay heat dissipation exists.

This specialists' meeting, sponsored by the Nuclear Reactor Safety Division—Long Island Section, the Technical Group for Thermal Hydraulics of the American Nuclear Society, and Brookhaven National Laboratory, was the first of its kind devoted exclusively to the issues related to the shutdown heat removal from fast breeder reactors. It provided an international forum for the state-of-the-art presentations on deterministic analyses and their verifications through experimental measurements. Papers were presented on data from tests performed in the EBR-II, PHENIX, and PFR reactors. Various models and analytical predictions were also discussed and applied to these reactors as well as other reactors in advanced stages of construction. These reactors included FFTF (criticality February 8, 1980), SUPER PHENIX, SNR-300, and other commercial designs. In addition, future areas of concern were presented in a panel discussion.

These proceedings of the specialists' meeting are structured in six parts: The first part is an invited overview given by a leading expert of the fast breeder reactor physics and safety. The contributed papers are grouped in the next four parts: Out-of-Pile Experiments and Analyses, Analytical Methods and Modeling, Experiments and Analyses, and Decay Heat Removal in GCFR. The last part is a transcribed version by one of us (J. G. G.) of the panel discussion, Future Research and Development—What Needs to be Done?

There was a general agreement that many system codes are now available but that there still is a need for benchmark three-dimensional thermohydraulic work. On the design side, a number of options are still available, but some form of independent in-vessel cooling system may become mandatory. In addition, there appears to be a need for greater international cooperation and exchange of data from whole-plant tests.

Many people contributed to the organization of the meeting, A. K. Agrawal served as the General Chairman, and J. G. Guppy and R. M. Singer (Argonne National Laboratory) were the Technical Program Co-Chairmen. Other members of the Technical Program Committee were S. L. Additon (HEDL), G. D. Bouchey (U.S. Department of Energy), R. T. Curtis (U.S. Nuclear Regulatory Commission), D. L. Hetrick (University of Arizona), R. A. Markley (Westinghouse Advanced Reactors Division), and A. Torri and A. Shenoy (General Atomic Company). This committee was responsible for reviewing the contributed abstracts. Completed papers were not reviewed, hence they represent the opinion of their respective authors. All the papers except that by C. Dawson were presented and all appear here in approximately the same order as their presentation.

In addition to the members of the program committee, a number of others actively participated in making this timely meeting a great success. We wish to thank the committee members, most of whom doubled as session chairmen, for their time and contributions. We also thank W. T. Sha and N. E. Todreas, who also served as session chairmen. We particularly wish to thank all authors who presented their work so well. Thanks are also due to the panelists not mentioned thus far—N. Grimm of Westinghouse and D. Leteinturier of C.E.N., Cadarache.

It is our pleasure to thank D. Clay (now Mrs. D. Cady) and P. Siemen for their extraordinary effort in making all meeting arrangements. Finally, we express our appreciation to W. Y. Kato of Brookhaven National Laboratory and R. T. Curtis, C. N. Kelber, and P. M. Wood of the U. S. Nuclear Regulatory Commission for their continued encouragement and support.

Ashok K. Agrawal
James G. Guppy

OVERVIEW

Current Trends and Options in Liquid-Metal Cooled Fast Reactor Safety*

CHARLES N. KELBER
Advanced Reactor Safety Research
Division of Reactor Safety Research
Nuclear Regulatory Commission

ABSTRACT

A review of current trends and options in fast reactor safety indicates a high degree of significance attached to assurance of decay heat removal via a passive alternate system. Research is required to assure proper operation of such a system over the wide range of conditions that include station blackout; conditions involving severely degraded cores or perturbed flow patterns.

At the other end of the spectrum, recent trends indicate some areas of similarity between core melt accidents in LWR's and LMFBR's, with these differences: (a) vaporized fuel in LMFBR; (b) Pu aerosols, (c) possibly damaged primary system integrity, and (d) lots of sub-cooled sodium around.

As I write this there are no trends and fewer options for fast reactor safety research in the United States. While I am confident that such work will be needed in the future, even in the United States, it appears to me wisest to weight foreign views - as much as I know them - most heavily.

It has long been realized that in principle a fast reactor can be so designed that even in the event of a complete loss of all AC power, provided it scrams, the reactor can be safely shutdown and maintained that way. One can "walk away from it." I so testified to a sub-committee of the ACRS several years ago. Now we find that the time has come to put that idea into practice. Dieter Smidt described this problem in his talk at the Seattle Meeting (1) and pointed out that, using simple design criteria, a reliable, largely passive system can be used. This is a very real advance in safety and one closely tied to the aims of this meeting. There is a trend in thinking about LMFBR operation that indicates favorable consideration of the operation in the natural convection mode as a normal mode of shutdown. A problem to be solved by the study of natural convection and related matters is what design and operating criteria are needed in detail for such a mode of shutdown. Particular attention needs to be given to the problems associated with

*Work performed under the auspices of the U.S. Nuclear Regulatory Commission

the creep-fatigue range since low flow rates in sodium may give rise to temperature fluctuations in the boundary layers of mixing tees and similar elements.

So far as materials are concerned it is by no means clear that the incentive to go to operation in the creep-fatigue range is justified. If the additional cost is justified, then the current set of standards has to be markedly improved. Certainly there has to be an increased effort related to long-term behavior of materials in these ranges, and as important, their inspection. Although there has been a significant advance in understanding. Particular attention needs to be paid over the entire operating range to assessment of seismic response.

Earthquakes are of concern at many sites. An LMFBR with thin walled piping does not respond like a PWR with thick walled pipe and it does seem out of place to assume that it does. Since the problem is readily solved it seems a pity not to do so.

As you can see, I have been summarizing safety trends with respect to operating transients. The stopping place for such a summary is the control rods. Everyone at Seattle was agreed that a reliable safety and control rod system was a good thing, and I guess I do, too. I do think that you can have too much of a good thing. A "Rasmussen Report" for an LMFBR should establish the point of diminishing returns for such systems, with these caveats: The accuracy of the estimate may not be good enough to make a decision; the report has to be well-prepared: beware of cheap imitations; even though a complex system may be estimated to be highly reliable, establishing the validity of the estimate may well be impossible in a single man's lifetime. Finally, one should ask if that reliability will hold up under accident conditions, say earthquake, fire, or similar naturally occurring disaster, as well as under conditions of sabotage or terror. When you deal with low probability events you have to consider these things.

There is a possible way to simplify this problem. It may well be that cores can be built that have very large temperature coefficients so that under a loss of flow transient even without scram there is an extensive period of time (seven to thirty minutes, depending on the estimator) during which the power to flow ratio remains within tolerable bounds even though flow is decreasing faster than power. This is the estimate advanced by the Super-Phenix group (2) and there is no reason to expect their estimate to be in serious error. With such times at operator's disposal thought can be given to the use of slow but sure methods of shutdown: e.g. lithium injection. The transient flow distribution under these conditions, and especially under conditions other than precisely normal, may will dictate the possible success of such an approach. I hope such topics will receive proper attention at this meeting.

If next we suppose that we consider accidents where there is no scram we find that we are confronted with the core disruptive accident. Never mind how we get there: there are plenty of sources of positive feedback in the core once fuel gets to the melting point.

There are two schools of thought about the core disruptive accident: legislate it out of consideration, or think deeply about it. We chose, over considerable opposition, to do the latter. The result has been the well-known SIMMER (3) code, and that code has now been distributed to a large number of users including FRG, UK, ISPRA and Japan. I shall now use a certain prerogative and leap to a conclusion, even though the work in this area is still far from complete. (This prereogative is usually reserved for Swans of advanced age.)

It is my current conclusion based on SIMMER work to date that the progression of a whole core accident can be summarized in a series of several, but a small number of events.

This summary is given in Table 1. There are some key points to be gathered from this summary:

- damage to the primary system is unlikely
- it is important to determine if and how a badly damaged core can be cooled
- if cooling doesn't succeed, the accident progression becomes a race as the chaotically melting core either finds a pathway to dispersal in the lower plenum or else is forced by fuel vapor to disperse all over the primary system. During this race minor excursions can also occur, but they do not affect the result.

Note that the cooling of badly damaged cores is within the purview of this meeting. We know such cores can be cooled: witness TMI-2.

A key function of large in-pile safety tests is to simulate such extensive core damage and determine the primary circuit flow characteristics that permit such cooling, including an eventual successful transition to cooling by natural convection. No facility currently exists to conduct such tests, but I view them as needed to establish design criteria for the primary circuit.

In any event, the progression of such events in LMFBR's now bears at least a family resemblence to similar sequences (such as TMLB) in PWR's. By thinking about the problem seriously we have managed to find a way to present it in reasonable, logical form suitable for resolution.

But, as you all know, such is usually the case. I hope this meeting sets you all thinking seriously about key safety issues and how to resolve them. Good luck!

REFERENCES

1. D. Smidt, "Targets for the Development of the Safety of Liquid Metal Fast Reactors," Proc. Int. Meeting on Fast Reactor Safety, Seattle, August 19-23, 1979.

2. Y. Ballofet, R. de Fremont, F. Josso, D. DeLapparent and R. Rossignol, "Calculations of the Loss of Flow Accident in Large LMFBR: Influence of Core Parameters," Proc. Int. Meeting on Fast Reactor Safety, Seattle, August 19-23, 1979.

3. La Ron Smith, Editor, "SIMMER-II: A Computer Program for LMFBR Disrupted Core Analysis," NUREG/CR-0453; LA-7515-M, (October, 1978).

TABLE 1. WHOLE CORE ACCIDENT EVENT SEQUENCE

EVENT	OUTCOMES (RANGING FROM MOST TO LEAST LIKELY)
INITIATION	A. EXTENDED DEBRIS BED FORMED ABOVE, BELOW AND WITHIN CORE REGION
	B. DEBRIS BED IN LOWER PLENUM ONLY
	C. DAMAGE TO PRIMARY SYSTEM INCLUDING FUEL VAPOR RELEASE
FLOW RE-ESTABLISHED	A. CRITICALITY CONTROLLED BY ROD OR LITHIUM INSERTION, PROBLEM IS TO MAINTAIN COOLING OF DAMAGED CORE.
	B. EFFECTIVE COOLING OR CRITICALITY CONTROL NOT ES-TABLISHED.
DEBRIS BED ATTEMPTS	A. AN SHOWN BY D-SERIES EXPERIMENTS, BEDS CAN BE COOLED.
	B. BED NOT COOLED, BOILS DRY AND MELTS.
HOMOGENEOUS MASS FORMS FROM CHAOTIC MELTING	A. MOLTEN FUEL MELTS CONTROL ROD GUIDE TUBE AND DRAINS INTO LOWER PLENUM, FORMING A DEBRIS BED
	B. RECRITICALITY EVENT (CA. $70/SEC.) DISPERSES 40% OF CORE, EXTENDED, SUBCRITICAL DEBRIS BED FORMED.

OUT-OF-PILE EXPERIMENTS AND ANALYSES

Comparison of Numerical Results with Experimental Data for Single-Phase Natural Convection in an Experimental Sodium Loop

R. J. RIBANDO
Oak Ridge National Laboratory
Oak Ridge, Tennessee 37830

ABSTRACT

A comparison is made between computed results and experimental data for a single-phase natural convection test in an experimental sodium loop. The test was conducted in the Thermal-Hydraulic Out-of-Reactor Safety (THORS) facility, an engineering-scale high temperature sodium loop at the Oak Ridge National Laboratory (ORNL) used for thermal-hydraulic testing of simulated Liquid Metal Fast Breeder Reactor (LMFBR) subassemblies at normal and off-normal operating conditions. Heat generation in the 19-pin assembly during the test was typical of decay heat levels. The test chosen for analysis in this paper was one of seven natural convection runs conducted in the facility using a variety of initial conditions and testing parameters. Specifically, in this test the bypass line was open to simulate a parallel heated assembly and the test was begun with a pump coastdown from a small initial forced flow. The computer program used to analyze the test, LONAC (LOw flow and NAtural Convection) is an ORNL-developed, fast-running, one-dimensional, single-phase, finite-difference model used for simulating forced and free convection transients in the THORS loop. In addition to simulating this particular run, the code has been used to predict results for other runs and for some hypothetical cases involving higher heat generation rates, less heat transfer through the test section walls, etc.

Comparison of LONAC results with experimental data shows fairly good agreement early in the transient, but shows the need for better thermal models of the heat dump, pump and other loop components if long transients are to be simulated. While the THORS loop is only mildly prototypic of an LMFBR primary loop, this work has provided some useful insights into the problem of confirming natural convection as a viable means of removing decay heat in the event of loss of forced flow in a loop-type LMFBR.

1. INTRODUCTION

At full-power conditions, buoyancy is of little importance to the operation of Liquid Metal Fast Breeder Reactors (LMFBRs). Under decay heat conditions, however, buoyancy will increase the flow over what would be produced by pony motor operation alone. In fact, in current designs buoyancy-driven flow should cool a shut down reactor enough to give a sufficient margin of safety in the unlikely event of a complete loss of forced flow. A state-of-the-art review of work in the area of natural convection in loop-type LMFBRs in general and in the Fast Flux Test Facility (FFTF) in particular has been given by Additon and Parziale.[1] Singer et al.[2] have made an analogous review for pool-type reactors.

Since no loop-type reactors are presently operating in the United States and loop natural convection data are extremely scarce, most current work is aimed at analytically and experimentally verifying numerical models of components used in system codes such as DEMO[3] for predicting plant operation when making the transition to and operating in the natural convection mode.[4] The test discussed and analyzed in this paper was selected from a series of seven which were conducted in the Thermal-Hydraulic Out-of-Reactor Safety (THORS) facility at the Oak Ridge National Laboratory (ORNL). This loop is used for thermal-hydraulic testing of simulated LMFBR subassemblies at normal and off-normal operating conditions. While not considered a valid surrogate for a reactor primary loop, a comparison of THORS data has been made with results obtained using LONAC (LOw flow and NAtural Convection),[5] an ORNL-developed one-dimensional system-type, single-phase, finite-difference numerical model used for simulating THORS operation. The exercise has given valuable insights into the problems of establishing natural convection as a viable means of removing decay heat in the event of loss of forced flow in a loop-type LMFBR.

2. DESCRIPTION

2.1 Test Facility

The test discussed in this paper was conducted in the THORS [formerly the Fuel Failure Mockup (FFM)] facility. An isometric drawing of primary piping and major components as they were at the time of the test is given in Fig. 2.1. Forced flow is provided by a dc-motor-driven vertical-shaft centrifugal pump capable of delivering 40 ℓ/s (600 gpm) at a pressure difference of 960 kPa (140

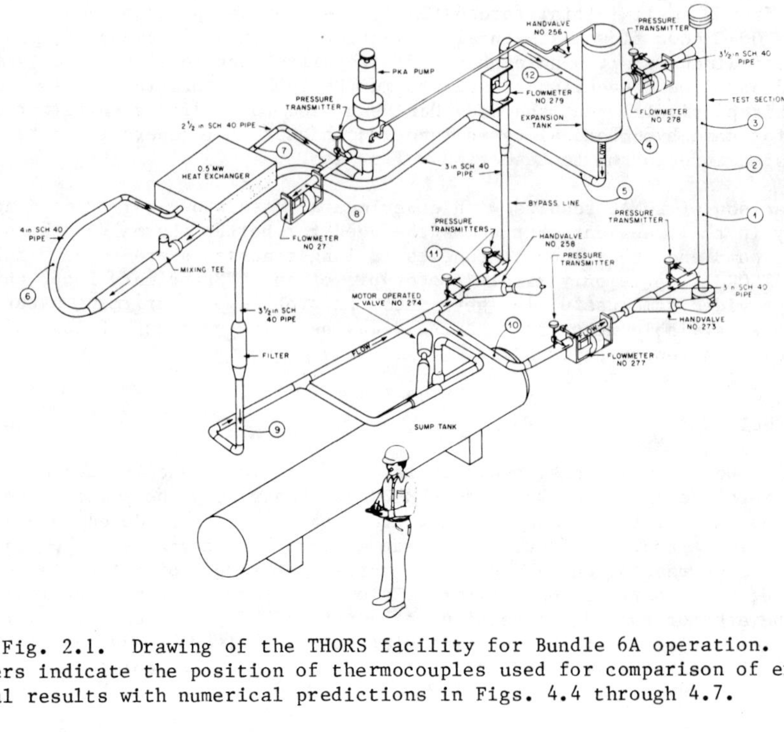

Fig. 2.1. Drawing of the THORS facility for Bundle 6A operation. Circled numbers indicate the position of thermocouples used for comparison of experimental results with numerical predictions in Figs. 4.4 through 4.7.

psi). Heat generation in the simulated LMFBR subassemblies is provided by electrically heated fuel pin simulator (FPS) units. Pressure transmitters, flow meters, and thermocouples are installed to measure the system operating parameters. Electrical trace heaters are installed on all system lines to preheat the system and minimize heat losses during operation.

A test-section inlet valve is available for simulating the pressure drop of the reactor inlet flow paths but was kept fully open in this test. A bypass line with a control valve which can be adjusted to give hydrodynamic simulation of other assemblies in a reactor core is in parallel with the test section. The hot sodium from the test section and sodium from the bypass line mix in the expansion tank. The expansion tank and the piping to it were designed to simulate the hydraulic characteristics downstream of a reactor core, and the expansion tank serves to mix hot and cold sodium, thus protecting the remainder of the loop from excessive temperatures. The ratio of bypass cross-sectional area to that in the test section is approximately 13:1. Programmable power control and pump-speed control capabilities are provided so that prescribed and reproducible flow and/or power transients can be performed.

The pressure of the cover gas above the free surface in the pump bowl is normally regulated at a fixed level [34 kPa (5 psig)]. The pressure of the cover gas above the free surface in the expansion tank may be either fixed at the same level (by opening the cover gas equalizing valve HV-256) or allowed to float (HV-256 closed).

2.2 Bundle Description

The test section (referred to as Bundle 6A in this discussion) is a full-length simulated LMFBR fuel subassembly. It consists of 19 electrically-heated pins 5.84 mm (0.230 in.) in diameter spaced by 1.42-mm-diam (0.056-in.) helical wire-wrap spacers on a 305.-mm (12.0-in.) helical pitch. The gap between the bundle and its containing hexcan is 0.71 mm (0.028 in.), one-half the nominal pin-to-pin spacing. This configuration is used to flatten the sodium temperature across the bundle. In order to give a better simulation of the central region of a full reactor subassembly, appreciable effort was expended in designing and fabricating a low thermal inertia bundle housing. The bundle hexcan is 0.51 mm (0.020 in.) thick type 316 stainless steel backed by ~20 mm (~1 in.) of calcium silicate block insulation contained in a stainless steel jacket (Fig. 2.2). Ear-

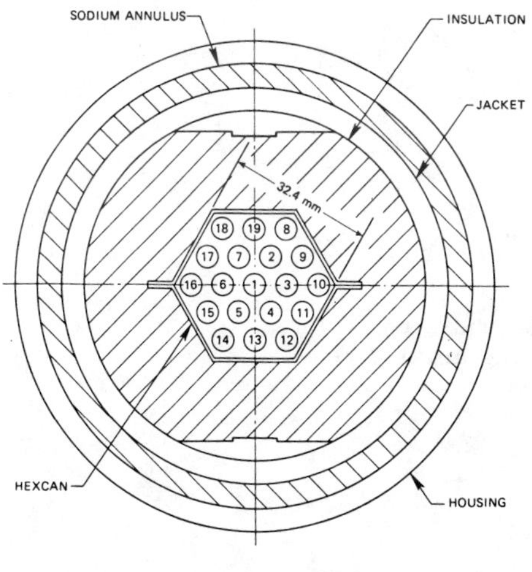

Fig. 2.2. Cross section of Bundle 6A.

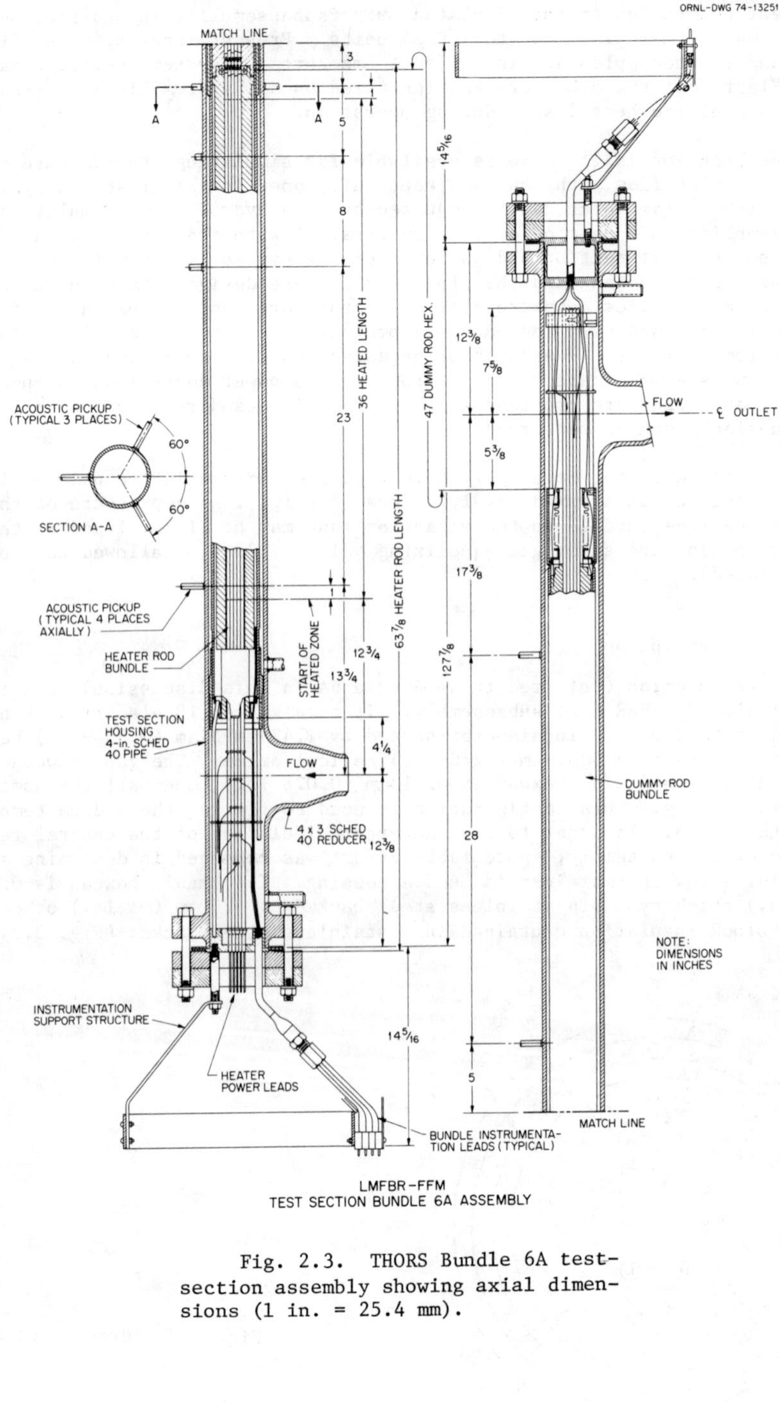

Fig. 2.3. THORS Bundle 6A test-section assembly showing axial dimensions (1 in. = 25.4 mm).

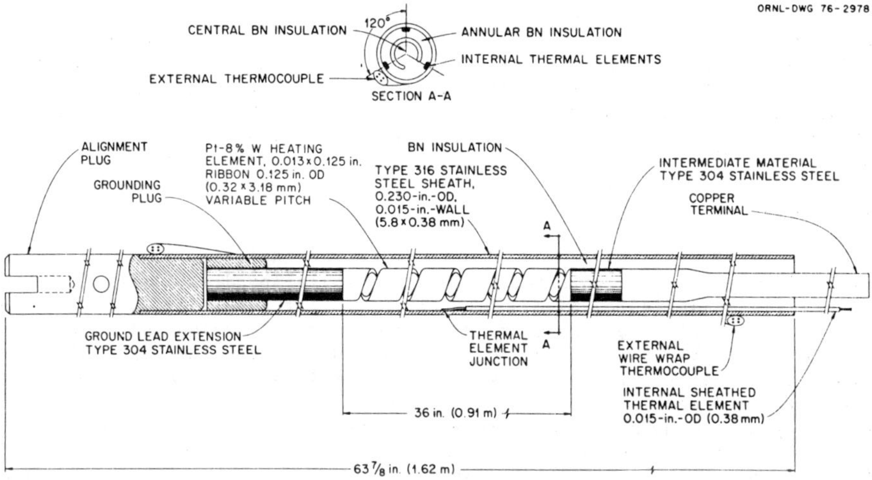

Fig. 2.4. Bundle 6A fuel pin simulator.

ly tests, however, indicated higher values for hexcan thermal inertia than expected and a post-test examination revealed that sodium had indeed permeated the insulation region. The impact of this sodium leakage on the tests will be discussed later.

The heated length of Bundle 6A is 0.9 m (36.0 in.) with variable-pitch heater windings to produce a 1.3 axial peak-to-mean chopped-cosine power distribution. Downstream of the heated length, a nickel reflector and a simulated fission gas plenum have the same length and thermal inertia as an FFTF fuel assembly. Axial dimensions of the bundle are shown in Fig. 2.3.

Both the heated length and the simulated fission gas plenum are instrumented with wire-wrap thermocouples and duct-wall thermocouples. In addition, the heated length is instrumented with heater-internal thermocouples. The installation of heater-internal and wire-wrap thermocouples is indicated in Fig. 2.4. Duct-wall thermocouples are inserted into holes bored to within 0.12-0.25 mm (0.005-0.010 in.) of the inside surface of the hexcan. At the time of the tests discussed here, there were approximately 45 duct-wall, 33 wire-wrap, and 43 heater-internal thermocouples operating reliably in the bundle. Approximately 27 additional thermocouples were strapped to loop piping. Data from these thermocouples and other instrumentation were recorded using a fast response Data Acquisition System (DAS) controlled by a PDP-8E computer. Other details of the THORS DAS may be found in Ref. 6.

2.3 Test Procedure

In addition to the test discussed in this paper (referred to hereafter as Test 39), there were one preliminary test and six other natural convection tests run in this phase of the Bundle 6A test series. Earlier single-phase steady-state forced flow tests run as part of the Bundle 6A test program are reported in Ref. 7; later transient boiling tests are reported in Ref. 8. In the preliminary test, loop trace heater settings were determined so as to minimize heat losses from the sodium in the temperature range of interest. The other six natural convection tests involved changes in the initial conditions (whether a low forced flow or a zero flow), trace heater settings, the state of the cover gas equalization valve, and power skew in the bundle. Details of the other tests may be found in Ref. 9.

Test 39, the one to be analyzed here in some detail, was the only test in which the bypass line was open to simulate a parallel bundle. Under isothermal conditions the valve in the bypass line was placed at a setting which was thought to give approximately equal pressure drops in the test section and bypass for equal flows, and then the pump speed was adjusted to give about 0.13 ℓ/s (2 gpm) through the test section. The pressure drop coefficient for this valve setting had been measured earlier at higher flows but existing instrumentation was not sensitive enough to make this measurement in the flow range of interest. As in all runs the test section inlet valve was fully open. The bundle power was set to 7.3 kW total, while the trace heaters in the bypass line were set at their maximum output (6.6 kW) to simulate a parallel heated assembly. Once temperatures had stabilized, the pump was stopped and the cover gas equalization valve (HV-256) was opened to begin the test. Approximately two hours later the test was concluded.

3. EXPERIMENTAL RESULTS

Plots have been made of the temperature trace in the test section (Fig. 3.1) during the early part of the transient and of the temperature trace around the loop at various times during the whole transient (Fig. 3.2). The abscissa in

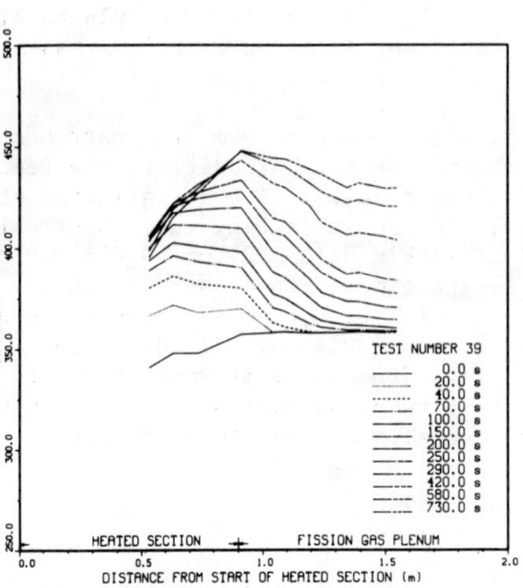

Fig. 3.1. Temperatures recorded along the bundle axis plotted at selected times early in the transient, (THORS Bundle 6A, Test 39, initial low forced flow, bypass open).

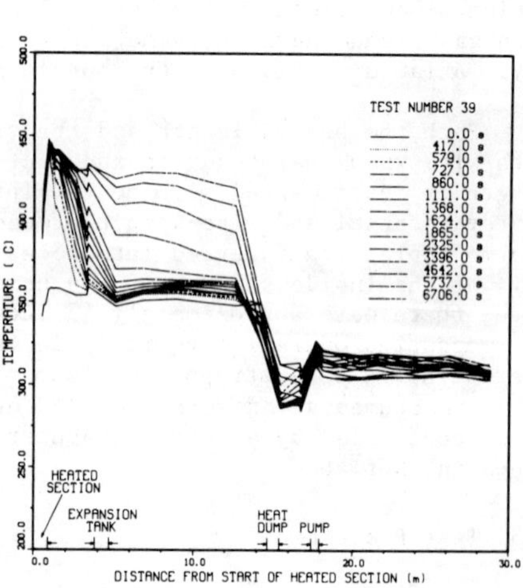

Fig. 3.2. Temperatures recorded around the THORS loop plotted at selected times during the transient, (THORS Bundle 6A, Test 39, initial low forced flow, bypass open).

both plots is distance along the loop in meters measured from the start of the heated section. Data in the test section are from wire-wrap (WW) or heater-internal (HI) thermocouples near the center of the bundle (i.e., thermocouples on pins 1—7). No WW or HI thermocouples were installed upstream of 0.53 m. Data from the rest of the loop are from thermocouples strapped to the outside of loop piping except for two thermocouples that are in wells inserted into the pipes at the test section inlet and outlet. Plots showing the response of selected individual thermocouples as a function of time have been made, but will be discussed later when they are compared to numerical predictions. Data from flowmeters will not be presented because of strong reservations about their validity at such low flows. Experimental heat balances using the indicated flow, temperature increase across the test section, and the bundle power were made throughout the transient. Those made toward the end of the run, when transient heat storage effects in the bundle should be at a minimum, suggest flow meters generally reading some 50—80% high. The quoted accuracy of the flowmeters is about ±70% of the flow predicted by the LONAC code for this case.

Since Test 39 began with a pump coast down from approximately 0.13 ℓ/s (2 gpm, ~5% nominal flow) peak temperatures initially are recorded toward the end of the heated section (Fig. 3.1). During a transitional period extending beyond $\tau = 150$ s, the peak temperature is found nearer the bundle midpoint where the heat generation is at its maximum. Finally after a substantial natural convection flow has been set up, the peak again is recorded toward the end of the heated section. Another phenomenon apparent in the plot of loop thermocouples vs time (Fig. 3.2) is the very large thermal inertia of the expansion tank as evidenced by the delay in the response of the thermocouple just downstream of it. It is also readily apparent that the very large heat loss in the area of the heat dump affects the inlet thermocouple as well as the one at the outlet.

4. ANALYSIS

4.1 Procedure

Analysis of the Bundle 6A natural convection tests has been based on use of LONAC,[5] a one dimensional, single phase, finite-difference model for predicting THORS transients. The early version of this code had no provision for thermal inertia of piping and loop components such as the pump and included no heat losses. Inclusion of structure thermal inertia and piping losses and a fairly detailed model for the insulation and containment around the test bundle itself was essential to the simulation of the Bundle 6A tests. As noted earlier, sodium had permeated the insulation surrounding the test bundle and the resulting heat sink had to be fairly carefully modeled since it was in such close proximity to the pin bundle. Other structural components such as piping, the pump, expansion tank, and heat exchanger were thermally modeled simply as single nodes in intimate contact with the adjacent sodium node.[10]

Since a fairly detailed discussion of the LONAC solution algorithm has been given previously,[5] only a few points will be made here. The loop and bundle are divided into 36 energy nodes, 34 of which are of fixed volume. The other two represent the sodium in the pump bowl and expansion tank. The sodium levels in these two nodes are predicted by the code. Accurate prediction of these levels is particularly important because, for example, in the case of the pipe run between the expansion tank and pump bowl, the temperature field in this line normally would tend to drive the flow in the negative direction were it not for the difference in free surface heights and in some tests argon pressurization in the expansion tank.

LONAC solves in sequence the one-dimensional conservation equations:[11]

$$A \frac{\partial \rho}{\partial \tau} + \frac{\partial (GA)}{\partial Z} = 0 \ , \tag{1}$$

$$A \frac{\partial (\rho h)}{\partial \tau} + \frac{\partial M_s h_s}{\partial \tau} + \frac{\partial (GhA)}{\partial Z} = \phi \ , \tag{2}$$

$$A \frac{\partial G}{\partial \tau} + \frac{\partial}{\partial Z} \frac{G^2 A}{\rho} = -A \frac{\partial p}{\partial Z} - \left(K + \frac{fL}{d} \right) \frac{|G| GA}{2\rho} - \rho g A \ , \tag{3}$$

where

A = cross-sectional area,
d = hydraulic diameter,
f = friction factor,
G = mass flow/area,
g = gravitational acceleration,
h = enthalpy,
h_s = enthalpy of structure,
K = pressure drop coefficient,
L = length,
M_s = mass of structure,
p = pressure,
Z = coordinate in flow direction,
ρ = density,
τ = time,
ϕ = heat source term.

A five node fuel pin radial heat conduction model is coupled to each sodium node in the heated zone. Similarly a five node radial conduction model for the insulation surrounding the test bundle and its containment (Fig. 2.2) is coupled to all test section axial nodes (both in the heated section and in the simulated fission gas plenum section). Runs were made both using manufacturer's data for the insulation thermal properties and using estimated properties for the sodium-saturated insulation.[8] Other than in the test section, piping and other components were assumed to be at the same temperature as the contiguous sodium. Whereas pipe heat capacity is less than half that of the contained sodium for the large pipes used in a full scale reactor,[12] pipe heat capacity ranges from 80 to 120% of the sodium in the small diameter pipes in the THORS loop. The estimated or calculated masses of other loop components which were included in the thermal analysis were: heat dump (but not its support structure), 279 kg; expansion tank, 261 kg; top and bottom flanges of test section, 45 kg; and pump (impeller and bowl only), 241 kg.

Losses from loop piping and other components were estimated using the results of the preliminary test mentioned earlier. An analytical determination of the heat losses would have been difficult if not impossible because of the large number of protuberances such as valves and flowmeters as well as the irregular geometry of elbows, tees and the pump. A scheme was developed for allocating the trace heat between losses to ambient and gains to the flowing sodium based on a comparison between the computed sodium temperature and the temperature for which the trace heater setting had been made. The resulting additional term in the

energy equation for the flowing sodium is

$$q_{el} \frac{T_{TRS} - T_{Na}}{T_{TRS} - T_{\infty}}$$

For Test 39 T_{TRS}, the piping temperature corresponding to the trace heater setting was 480°C (900°F) in the hot leg and 320°C (600°F) in the cold leg. Ambient temperature was about 46°C (115°F). Total trace heat other than on the bypass line was about 12.1 kW. This is nearly equal to the combined electrical input to the test section and the heated bypass line. Assuming about a 350°C (660°F) hot leg temperature early in the transient, about 1/3 of the trace heat in the hot leg would flow into the sodium; later the fraction would be about 1/10. The cold leg does not deviate much from the 320°C (600°F) temperature; therefore, the additional heat input due to the trace heat would be small. Other details on the scheme for allocating the trace heat may be found in Ref. 9.

The momentum equations are set up using the momentum integral method discussed by Meyer.[13] Momentum nodes are set up in a staggered arrangement so that mass flows are computed at the edges of the mass/energy nodes rather than at their centers where thermodynamic properties (density and energy) are defined. Frictional losses, thermal heads, etc., are computed locally at each momentum node. Since the flow through a line at any point may be expressed in terms of the flow at the inlet to that line plus a correction because of thermal expansion or contraction up to that point, it is possible to combine the momentum equations for a series of nodes into one equation for the whole string. When the bypass line is open, four separate momentum equations result: one each for the main line from the pump to the tee, the bypass line from the tee to the expansion tank, the test section line from the tee to the expansion tank and the main line from the expansion tank back through the heat dump to the pump bowl. These equations are algebraically manipulated to eliminate the pressure at the tee. The technique provides for transient flow redistribution because of different thermal heads in the bypass and the test section. For other tests with the bypass line closed, there are basically only two momentum equations to be solved. Data of Engel[14] were used for laminar and transition friction factors in the pin bundle. The stopped-rotor pressure drop of the pump was assumed to be neglible.

4.2 Results

A LONAC simulation of Test 39 was made using the estimated degraded test section insulation thermal properties. In addition, simulations were made using the Test 39 procedure with both degraded and "as built" insulation properties and at several bundle powers in addition to the actual one. Simulations of several other tests were made and are discussed in Ref. 9. Weighted averages of the nearest experimentally measured temperatures were used to determine initial temperatures for each of the finite-difference volumes. Since the argon cover gas in the expansion tank was pressured at $\tau \leq 0$, it was assumed that when the two free surfaces were at the equilibrium no-flow level and the loop was at 320°C (600°F), the cover gas gauge pressure was 34 kPa (5 psig). Pressure changes due to heating or to level changes could then be computed. This calculational procedure has some basis in the experimental procedure used and was necessitated by the lack of pressure and temperature measuring devices in the cover gas space at the time.

As noted earlier, Test 39 was the only run in which the bypass line was open to simulate a parallel assembly. The pressure drop coefficient, determined as noted in Sect. 2.3, was used for the bypass line valve in the simulation; and,

under isothermal conditions, the pump head was iterated until the desired 0.13 ℓ/s (2 gpm) flow was obtained in the test section. The corresponding computed flow in the bypass was about 2.4 times this rather being equal as desired. Trace heat in the bypass was set at its maximum value (6.6 kW).

Computed temperatures in the test section and around the loop and flows are shown in Figs. 4.1, 4.2, and 4.3. Figures 4.1 and 4.2 may be compared to Figs. 3.1 and 3.2 which show the corresponding experimental data. The circled numbers on Fig. 4.2 show the time it would take a parcel of fluid leaving the beginning of the test section at $\tau = 0$ to reach the station indicated. From the expansion tank downstream the flow is, of course, augmented by the bypass flow. The computed flows (Fig. 4.3) for this run are particularly interesting because the experimental procedure involved a simultaneous pump coastdown and expansion tank cover gas depressurization. The effects are particularly apparent in the lightly damped flow oscillation through the heat dump.

Individual thermocouple traces and corresponding LONAC results have been plotted as a function of time for a number of positions around the loop. These are depicted in Figs. 4.4 through 4.7. In most cases the center of the LONAC node did not fall at a thermocouple position. Hence the trace indicated as "LONAC Prediction" is for the node nearest the thermocouple and does not reflect any weighting. Thermocouple identification numbers are keyed to the loop diagram (Fig. 2.1). It may be seen that LONAC has overpredicted temperatures within the heated section (location 1) and at the end of the heated section (location 2). The peak temperature predicted by LONAC was 462°C at 690 s vs the 448°C observed in the loop at about 600 s. At a representative point in the simulated fission gas plenum (location 3) the comparison is somewhat better. At location 4, which

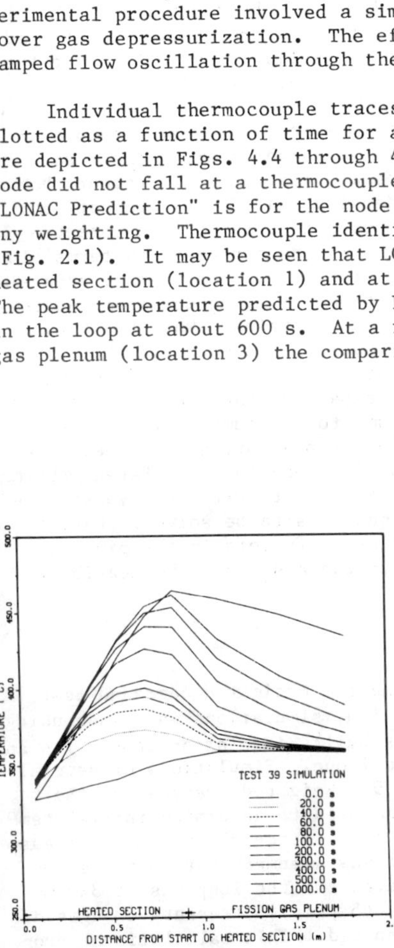

Fig. 4.1. Computed test section temperatures for various times early in the transient (LONAC simulation of THORS Bundle 6A, Test 39).

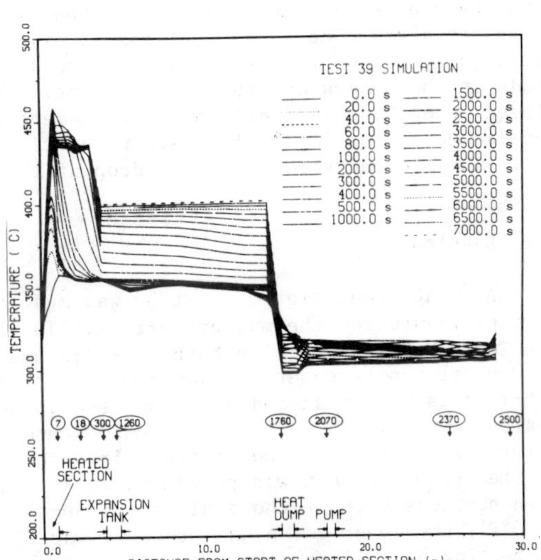

Fig. 4.2. Computed loop temperatures for various times during the transient, (LONAC simulation of THORS Bundle 6A, Test 39). Circled numbers indicate the time (s) it would take a parcel of fluid leaving the beginning of the heated section at $\tau = 0$ to reach the station indicated.

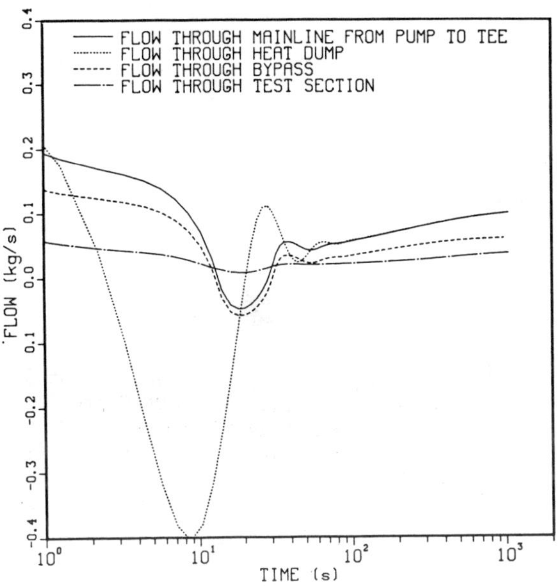

Fig. 4.3. Computed flows vs time (LONAC simulation of THORS Bundle 6A, Test 39).

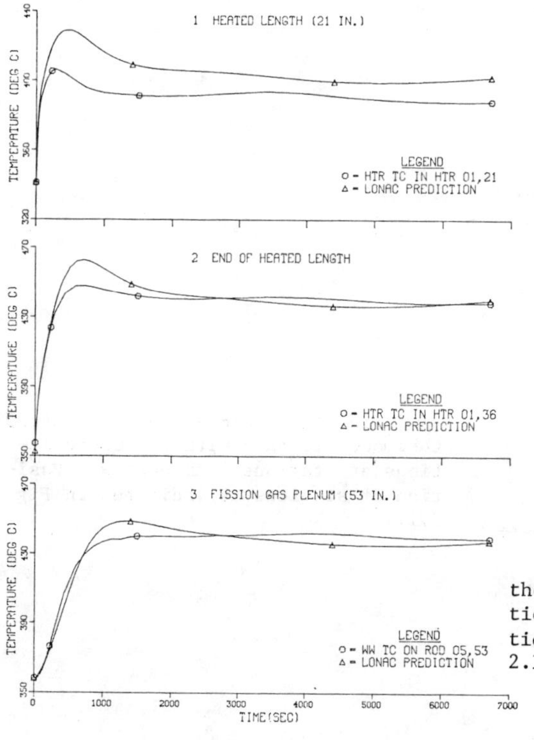

Fig. 4.4. Comparison of Test 39 thermocouple data with LONAC predictions at stations 1 through 3. Position of stations is indicated in Fig. 2.1.

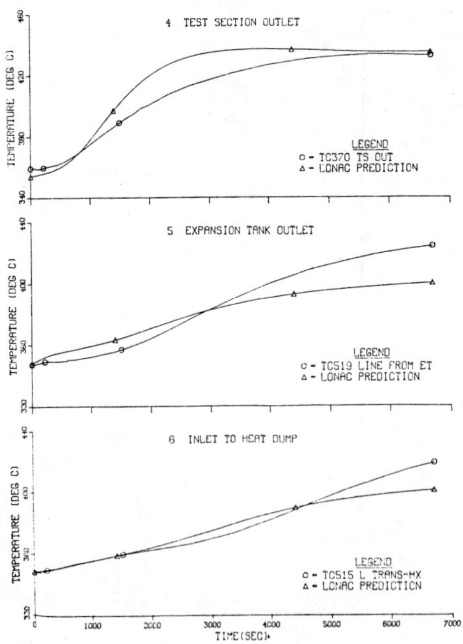

Fig. 4.5. Comparison of Test 39 thermocouple data with LONAC predictions at stations 4 through 6. Position of stations is indicated in Fig. 2.1.

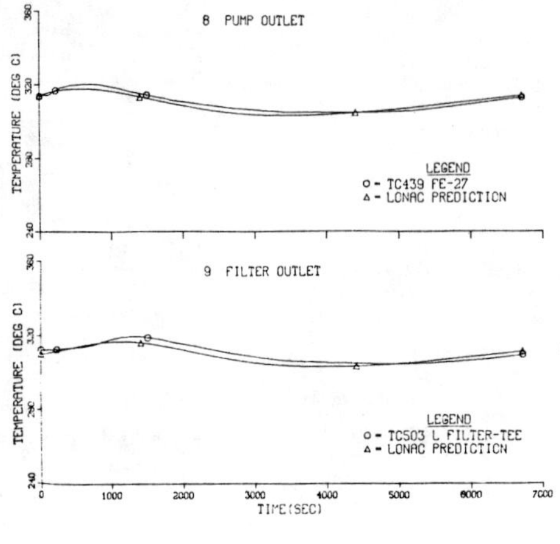

Fig. 4.6. Comparison of Test 39 thermocouple data with LONAC predictions at stations 7 through 9. Position of stations is indicated in Fig. 2.1.

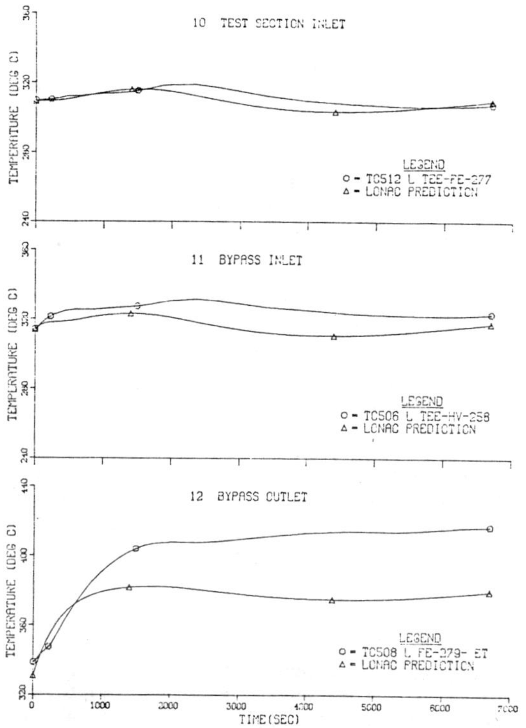

Fig. 4.7. Comparison of Test 39 thermocouple data with LONAC predictions at stations 10 through 12. Position of stations is indicated in Fig. 2.1.

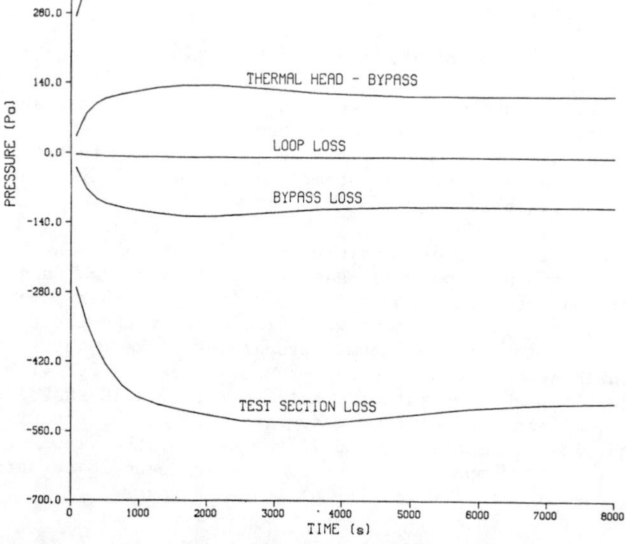

Fig. 4.8. Computed thermal heads and losses vs time for Test 39.

is commonly referred to a "test section outlet" but is actually much closer to the expansion tank inlet, the prototype is seen to respond more sluggishly than the simulation. It must be noted that the heat capacity of some hardware in this area, most notably the flow meter, was not included in this simulation and could account for the difference.

At the next location (4, expansion tank outlet) LONAC first overpredicts and after about 3000 s underpredicts. It is highly likely that the discrepancy is caused by the fact that the fluid in the expansion tank is thermally stratified, while in the one node LONAC representation, it is assumed to be well-mixed. For this reason the LONAC outlet value is representative of the mean rather than of the temperature at the bottom of the tank. The characteristic thermal diffusion time computed for a layer of sodium of thickness equal to the free surface height in the expansion tank is about 900 s. This is virtually the same as the time it would take to flush the expansion tank sodium inventory given the computed test section and bypass line flows. It suggests that insufficient time for thermal equilibration is allowed and that hot fluid is added to the top while cold fluid is drawn off the bottom.

Location 6 has been chosen some distance upstream from the heat dump because the thermocouple just at the inlet reflects the huge losses in the heat dump and its vicinity rather than being representative of the incoming fluid. In other words the losses at the heat dump are really not confined to the dump per se while in the LONAC simulation they are. With this proviso the LONAC results are reasonable although the effect of the stratification in the expansion tank may still be seen.

Locations 7, 8, 9, and 10 show fairly good agreement although it must be pointed out that trace heat in the cold leg of the loop was set to attempt to maintain 600°F (320°C) temperatures throughout the transient. Recorded values did not deviate much from this. In contrast both measured and computed temperatures in the hot leg did vary significantly from the 900°F (480°C) trace heater set point.

The location in the bypass (11 and 12) seems to indicate that the flow in the bypass has been overpredicted somewhat. This is presumably the result of poor characterization of the valve pressure drop coefficient under low flow conditions. The slower transient response seen in both the experimental data and the simulation is directly attributable to the greater sodium inventory in the bypass line as opposed to the test section.

Friction and form losses as well as thermal head for this case are plotted as a function of time in Fig. 4.8. The thermal heads have been computed as $\int \rho g dz$ around the loop with the path taken through the test section in the one case and through the bypass in the other. Hydraulic losses have been computed separately for the bypass, the test section and the remainder of the loop. Losses in the remainder of the loop are seen to be minimal and, other than at very early times, temperal acceleration is also negligible.

In order to assess the effects of higher bundle power and of insulation integrity on the results, five additional runs were made. These include two runs made using the estimated degraded insulation thermal properties but higher powers (to the bundle only, not the heated bypass). Runs were also made at the three bundle powers but using "as built" insulation thermal properties. The results are summarized in Fig. 4.9 which shows the peak bundle temperature and Fig. 4.10 which shows the time to reach the peak temperature. Note that Fig. 4.10 shows an order of magnitude difference between "as built" and degraded cases. In the former case, temperatures and flows rise quickly and in the case of the flow there is even an overshoot before it damps out. Test 39 data are seen to compare quite favorably with the degraded insulation results at the same power.

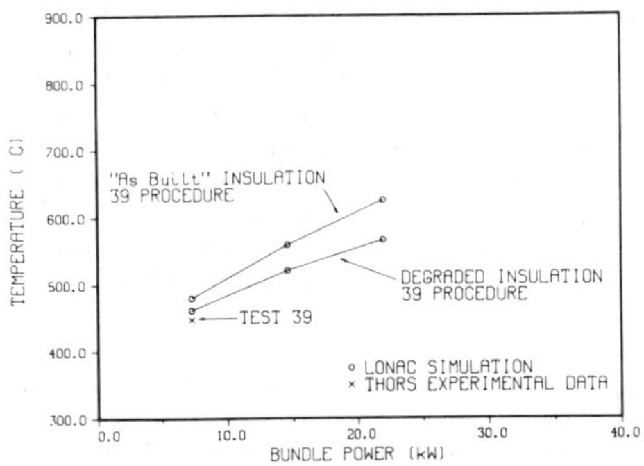

Fig. 4.9. Peak temperature measured in Test 39 compared with LONAC predictions for this case and hypothetical cases using the Test 39 procedure.

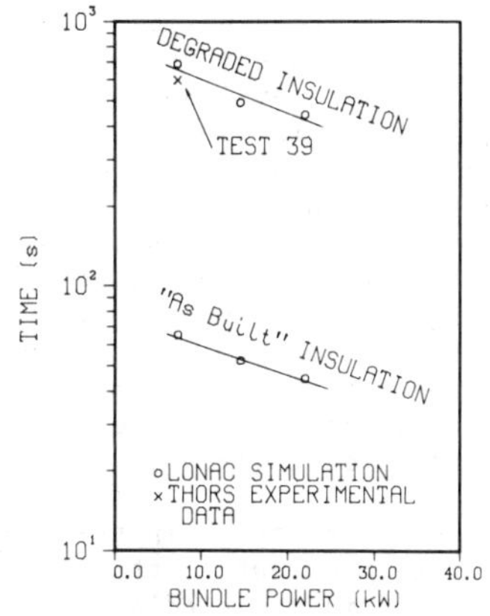

Fig. 4.10. Time calculated to reach the peak temperatures plotted in Fig. 4.9, (LONAC simulation using Test 39 procedure, both with "as built" and degraded insulation and actual and higher bundle powers). Test 39 value is also plotted.

5. CONCLUDING REMARKS

As has been seen, fairly good agreement was obtained between experimentally measured and LONAC-predicted peak temperatures. The temperature profiles early in the transient were also fairly well-matched by the code. Some reasons for this success and other observations will be discussed in this section.

In a large-scale reactor the sodium flow on the primary side of the intermediate heat exchanger (IHX) is in the vertical direction and involves a substantial elevation change over the length.[12] In contrast the sodium flow through the THORS Na-air 0.5MW heat dump is in nearly the horizontal direction. Thus the crudeness of the heat dump thermal model is not felt until the thermal pulse reaches a vertical leg. Further refinement of the heat dump model would be difficult because with the blower off and the damper closed, the heat dump was operating at only a few percent of its design rating. As noted earlier trace heaters on the heat dump tubes and supplementary radiator heaters were off during these tests. A more sophisticated model would also have to include at least part of the massive heat dump support structure. Sodium flow in the new 2.0 MW heat dump to be used for future THORS tests with larger bundles will be upward rather than downward as in the primary side of a reactor IHX. The 0.5 MW heat dump used in these tests will continue to be available.

For similar reasons the crude pump thermal model does not affect the early part of the transient. Future THORS natural convection tests would involve a new electromagnetic (EM) pump which will be located in a vertical piping leg. When not operating, the EM pump may be characterized hydraulically as simply another length of straight pipe. Its thermal behavior should be somewhat easier to characterize than that of the present centrifugal pump.

As illustrated in Figs. 4.9 and 4.10, the thermal properties of the insulation surrounding the test bundle play a dominant role in the transient. Manufacturer's data were available for the "as built" insulation, but the properties of the degraded insulation could only be estimated. The manufacturer's data indicated that the porosity of the material was 53%. Under the assumption that all pore space was full of sodium, a thermal conductivity which was 53% that of pure sodium (and over 200 times that of the "as built" insulation) was used for the degraded insulation. Similarly the heat capacity was computed on the basis of 53% sodium and 47% insulation. Clearly the agreement (or disagreement) between experimental and LONAC values seen in Figs. 4.9 and 4.10 is a strong function of this assumption. In a full-scale reactor the hexcan wall and adjacent subassemblies, if cooler, would have important effects as heat sinks during a slow natural convection transient.

The pressure drops in the bundle and bypass were orders of magnitude larger than that in the rest of the loop. Under the decay heat conditions simulated here the flow in the bundle is in the laminar and transition regimes. This is also the case for the full-scale reactor under decay heat conditions; but in the reactor many more, larger assemblies feed the also larger loop piping. The result is that in THORS the flow in the loop piping is in the laminar and potentially troublesome-to-evaluate transition regimes, while that in reactor piping would be well into the turbulent regime. Again because the bulk of the pressure drop is in the bundle where apparently it has been well-modeled using the Engel et al. results,[14] the LONAC simulation appears not to suffer. Similarly the neglect of the pump stopped-rotor pressure drop seems not to be a problem in THORS.

The THORS tests discussed in this report have generated a wealth of loop and intra-assembly data under simulated decay heat conditions. They have demonstrated the ability of a simplified computer code to model the THORS loop, the older segments of which were designed nearly a decade ago with no intent to determine detailed loop temperature measurements required for testing under natural convection conditions. The tests and analysis discussed here have given insights which will be helpful in planning future tests under natural convection conditions and insights which will help in code improvements for simulating long transients.

REFERENCES

1. S. L. Additon and E. A. Parziale, "Natural Circulation in FFTF, A Loop Type LMFBR," *Symposium on the Thermal and Hydraulic Aspects of Nuclear Reactor Safety*, Vol. 2, Liquid Metal Fast Breeder Reactors, ASME, 1977, pp. 265–283.

2. R. M. Singer, D. Grand, and R. Martin, "Natural Circulation Heat Transfer in Pool-Type LMFBR's," *Symposium on the Thermal and Hydraulic Aspects of Nuclear Reactor Safety*, Vol. 2, Liquid Metal Fast Breeder Reactors, ASME, 1977, pp. 239–264.

3. "LMFBR Demonstration Plant Simulation Model (DEMO)," WARD-D-0005, Rev. 4, Jan. 1976.

4. R. D. Coffield et al., *LMFBR Natural Circulation Verification Program (NCVP) Review of Experimental Facilities and Testing Recommendations*, WARD-NC-3045-1, July 1977.

5. R. J. Ribando, *LONAC: A Computer Program to Investigate Systems Dynamics Under Conditions of Low Forced flow and Natural Convection,* ORNL/TM-6228, April 1978.

6. N. E. Clapp, *The Data Management System for the Thermal-Hydraulic Out-of-Reactor Safety Facility Computer Controlled Data Acquisition System Configuration for Bundle 3C Operation,* ORNL/TM-6745, February 1979.

7. J. L. Wantland et al., *Steady-State Sodium Tests in a 19-Pin Full-Length Simulated LMFBR Fuel Assembly — Record of Steady State Experimental Data for THORS Bundle 6A,* ORNL/TM-6106, March 1978.

8. R. J. Ribando et al., *Sodium Boiling in a Full-Length 19-Pin Simulated Fuel Assembly (THORS Bundle 6A),* ORNL/TM-6553, January 1979.

9. R. J. Ribando et al., *Comparison of Numerical Results with Experimental Data for Single-Phase Natural Convection in an Experimental Sodium Loop (THORS Bundle 6A),* ORNL/TM-6778, September 1979.

10. H. P. Planchon and W. R. Laster, *Loop Heat Capacity Models and Their Effects for DEMO Natural Circulation Transient Analysis,* WARD-NC-3045-2, September 1978.

11. J. E. Meyer, "Conservation Laws in One-Dimensional Hydrodynamics, *Bettis Technical Review,* WAPD-BT-20, September 1960, pp. 61–72.

12. *Clinch River Breeder Reactor Plant Project Design Description,* CRBRP-PMC-77-06.

13. J. E. Meyer and R. P. Rose, "Application of a Momentum Integral Model to the Study of Parallel Channel Boiling Oscillations," *Journal of Heat Transfer, Trans. ASME,* Series C, Vol. 85, 1963, pp. 1–9.

14. F. C. Engel, R. A. Markley, and A. A. Bishop, "Laminar, Transition and Turbulent Parallel Flow Pressure Drop Across Wire Wrap Spaced Rod Bundles," *Nuclear Science and Engineering,* Vol. 69, No. 2, February 1979, p. 290.

Out-of-Pile Experiments for Natural Circulation Decay Heat Removal in LMFBR Fuel Assembly—Power Skew Effects upon Intra-Assembly Hydraulic Characteristics

F. NAMEKAWA, K. MAWATARI, T. TAMAOKI, and R. MAKIURA
Toshiba Nuclear Engineering Laboratory
Toshiba Corporation 4-1, Ukishima-cho, Kawasaki-ku
Kawasaki, 210, Japan

ABSTRACT

Experimental studies to determine hydraulic characteristics in a fuel assembly during natural circulation and low forced circulation conditions were conducted, as a series of experimental programs for an evaluation of decay heat removal capability in LMFBRs. 37-pin bundle test assembly with grid spacers was used in these experiments. As coolant, water was used instead of sodium. In this paper, hydraulic phenomena with radial power skew in pin bundles, under single phase flow conditions, and under boiling flow conditions that were simulated by air injection into pin bundle, are described. Through the examination on pressure loss measurements under single phase flow conditions, there were no significant radial power skew effects upon pressure loss. According to flow pattern observation during single phase natural circulation conditions with radial power skew, recirculation flow between the grid spacers was observed, but the recirculation flow disappeared under combined forced and natural circulation conditions. On the other hand, during two phase flow conditions, the recirculation flow was also observed when air was injected in the pin bundle with local distribution.

1. INTRODUCTION

The natural circulation decay heat removal after reactor shut down or scram is one of the important problems which should be solved to improve the inherent safety systems for LMFBRs. Natural circulation along the primary circuit is induced by the difference in both thermal center elevation and coolant specific weight between the core (hot region) and the intermediate heat exchangers (cold region). Thus, there is a possibility that the deacy heat would be removed by adequate designs for the thermal center difference between two region, without the external forced circulation.

It is, however, estimated that the radial flow redistribution occurs between the core and the radial blanket region under natural circulation conditions, due to the radial power skew.[1] A similar flow redistribution would also occur in the fuel pin bundles. Accordingly, the coolant flow rate at higher heat flux region in pin bundles is greater than that in the lower heat flux region, due to larger buoyancy at the higher heat flux region. Based on the above consideration, the flow reversal may occur inter- and intra-assembly under some magnitude of a radial power skew ratio. These buoyancy effects upon the hydraulic characteristics appear not only under natural circulation conditions (i.e., without external driving force), but also under combined forced and natural circulation conditions (i.e., where flow is driven by pony motors or main pump inertia after the pump trip).

Since most of experimental studies(2),(3),(4) were obtained with regard to the whole-core-wide and/or the primary circuit-wide system, they may not be adequate for an analysis of the intra-assembly natural circulation.

This paper describes preliminary results for hydraulic characteristics in the fuel assemblies during both natural circulation and low forced flow conditions, as a series of experimental programs for natural circulation decay heat removal capability. The purpose of the present experiments is to understand the hydraulic phenomena qualitatively, and then, to obtain information applicable to the next advanced experiments. Accordingly, water was used as a simulated coolant, instead of sodium. Flow patterns were also observed in these experiments.

2. EXPERIMENTAL EQUIPMENT AND PROCEDURE

2.1 Test Loop and Test Section

A series of experiments were carried out in the Heat Transfer Test Loop (HTTL) at Toshiba Nuclear Engineering Laboratory (NEL). Figure 1 shows HTTL flow diagram.

Figure 2 shows the test section that was used in these experiments. A 37-pin bundle with honey-comb type grid spacers was installed in a hexagonal wrapper tube, 50.4 mm flat to flat distance inside. A part of the hexagonal wrapper tube was made of transparent glass. The diameter of each pin is 6.5 mm and the distance between pin centers (i.e., pin pitch) is 7.9 mm. There are nine electrically heated pins in the pin bundle. These electrically heated pins are located along one hexagonal wrapper wall (i.e., as shown in Fig. 2, two rows in parallel with a wrapper wall) to simulate radial power skew in fuel assemblies. Heated length and maximum heat flux for these pins are 1,000 mm and 10 W/cm^2, respectively.

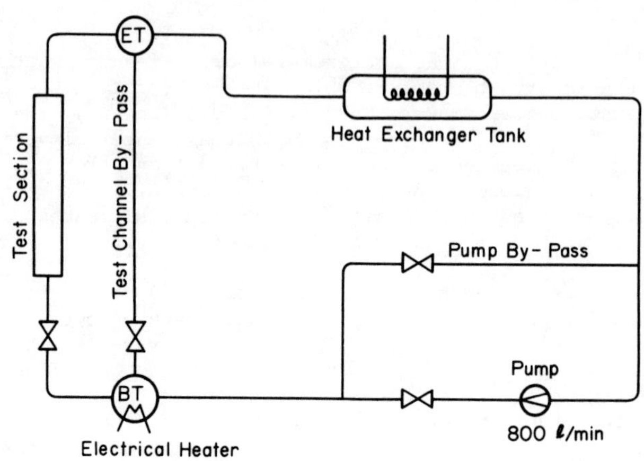

Fig. 1 HTTL Flow Diagram

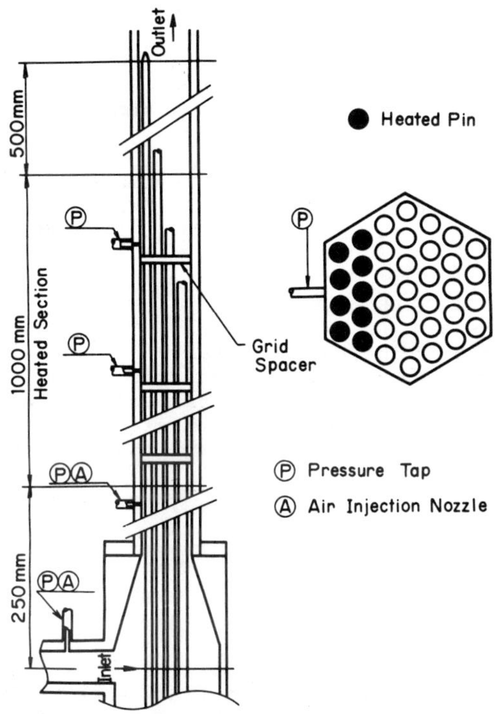

Fig. 2 37-Pin Bundle Test Section

For coolant static pressure measurements, there are three pressure taps along the hexagonal wrapper wall and one pressure tap along the inlet nozzle to the test section. The lowest pressure tap along the wrapper wall and the pressure tap along the inlet nozzle are also used as air injection nozzles during two phase flow experiments (i.e., boiling flow simulation).

2.2 Test Procedures and Measurements

Pressure loss measurements and flow pattern observations were performed in the single phase flow experiments. As basic data, pressure loss along the pin bundle was measured under the isothermal flow condition. Then, natural circulation and low forced flow experiments were performed, changing both heat flux in heated pins and the coolant flow rate, respectively. In the natural circulation experiments, the coolant did not flow through the pump, but flowed through the pump by-pass line, as shown in Fig. 1.

In two phase flow experiments, boiling natural circulation flow phenomena were simulated by air injection into flow water, in order to evaluate the buoyancy effects of vapor bubbles upon the hydraulic characteristics in pin bundles. In these experiments, air was injected through the lowest pressure tap along the wrapper wall or the pressure tap along the inlet nozzle. When air was injected through the inlet nozzle pressure tap, injected air bubbles uniformly spread in the pin bundle. So, this injection nozzle (= pressure tap) is convenient for use to simulate coherent boiling flow. However, when air was injected through the lowest wrapper wall pressure tap, most of injected air flowed along the wrapper wall that had the pressure taps, and injected air did not uniformly spread in the pin bundle. Accordingly, the lowest wrapper wall pressure tap was used as an air injection nozzle to simulate incoherent boiling flow in the pin bundle.

The tracer that was used for flow pattern visualizations, was produced by the solution of rhodamin B and methylene blue in water. This tracer has excellent properties with regard to mass diffusivity and specific weight[5].

3. WATER EXPERIMENT CONDITION DETERMINATION

The purpose of this preliminary experimental study is to evaluate the hydraulic phenomena qualitatively and to obtain information applicable to the next advanced experiments, during both natural flow condition and low forced flow condition in fuel assemblies. However, some attention was paid to the hydraulic simulation and using water as a simulated coolant, instead of sodium, was discussed.

3.1 Single Phase Flow Condition

In the selection of simulation criteria, it is useful to introduce non-dimensional parameters from descriptive equations under single phase flow conditions.

For fully developed laminar flow in the vertical upward direction, the conservation of the momentum and the energy equations can be written, as follow:

$$\mu\left(\frac{\partial^2}{\partial r^2} + \frac{1}{r}\frac{\partial}{\partial r} + \frac{1}{r^2}\frac{\partial^2}{\partial \theta^2}\right)U = \rho g + \frac{dP}{dz} - \rho\beta g(t - t_s) \quad (1)$$

$$K\left(\frac{\partial^2}{\partial r^2} + \frac{1}{r}\frac{\partial}{\partial r} + \frac{1}{r^2}\frac{\partial^2}{\partial \theta^2}\right)t = \rho C_p U\frac{\partial t}{\partial z} \quad (2)$$

where
- r = radial coordinate
- θ = angular coordinate
- z = axial coordinate
- μ = coolant viscosity
- ρ = coolant density
- β = coolant volume expansion coefficient
- C_p = coolant specific heat at constant pressure
- K = coolant thermal conductivity
- g = gravitational acceleration
- P = coolant static pressure
- t = coolant temperature
- t_s = rod surface temperature
- U = coolant velocity

From these equations, Yang[6] introduced the following non-dimensional parameters. That is

Modified Grashof number $\quad Gr^* = \dfrac{g\beta D_h^4 q}{K\nu^2}$

Table 1 Simulation Criteria under Single Phase Flow

Discriptive Parameters	Water	Sodium
Reynolds Number Re	4.8×10^{3} 1)	
Modified Grashof Number Gr*	1.6×10^{4} 1)	
Reference Length (Hydraulic Diameter) Dh m	4.09×10^{-3}	4.09×10^{-3}
Reference Temperature T °C	20	450
Volume Expansion Coefficient β 1/°C	0.2×10^{-3}	0.284×10^{-3}
Kinematic Viscocity ν m^2/s	1.0×10^{-6}	0.312×10^{-6}
Thermal Conductivity K w/m °C	0.518	60.0
Coolant Velocity V m/s	1.17	0.366
Heat Flux q w/m^2	1.51×10^{4}	1.2×10^{5}

1) 6 % of normal operating condition

Reynolds number $$Re = \frac{U_m D_h}{\nu}$$

Pressure parameter $$E = \frac{D^2}{4\mu U_m} (\rho g + \frac{dP}{dz})$$

where D = rod diameter
D_h = hydraulic diameter
ν = coolant kinematic viscosity
q = heat flux from rod surface
U_m = mean coolant velocity

Then, the following correlation was derived.

$$E = f\{(P/D), Gr^*/Re\}$$

where, P = distance between pin centers

Therefore, similarity conditions were determined based on both modified Grashof number Gr* and Reynolds number Re under single phase flow condition. Table 1 shows a comparison of individual operating condition, based on both modified Grashof number Gr* and Reynolds number Re, between water and sodium.

3.2 Two Phase Flow Conditions

Thermal-hydraulic phenomena in two phase flow is more complicated and sophisticated compared with that in single phase flow. It is, therefore, difficult to describe the phenomena under two phase flow conditions clearly. However, some useful information relating to boiling flow phenomena have been recently obtained: correlations between vapor quality and void fraction; pressure loss correlations in pin bundles and critical heat flux correlations. However, those phenomena under boiling flow conditions have not yet been clarified completely. It is, therefore, difficult to postulate, from the water/air experiments, to the actual sodium conditions quantitatively. However, under steady boiling flow conditions, it would be possible to obtain the approximate hydraulic simulation between water/air and sodium, if the specific weight ratio ϕ was made up between water/air and sodium liquid/vapor.

Table 2 Simulation Criteria under Two Phase Flow

Discriptive Parameters	Water / Air	Sodium
Specific Weight of Liquid γ_l kg / m^3	998	720
Specific Weight of Vapor or Air γ_g kg/m^3	1.16	0.513
Specific Weight Ratio γ_g/γ_l	1.16×10^{-3}	0.713×10^{-3}
Reference Temperature °C	20	964
Reference Pressure Pa	10^5	2×10^5

In general, the specific weight ratio ϕ is defined as follow.

$$\phi_{a-w} = \gamma_a/\gamma_w$$

$$\phi_{sv-sl} = \gamma_{sv}/\gamma_{sl}$$

where, ϕ = specific weight ratio
γ = specific weight

subscript

a = air
w = water
sl = sodium liquid
sv = sodium vapor

Results of many experimental investigations(7),(8) show that the hydraulic phenomena under two phase flow is mainly governed by specific weight ratio ϕ. Moreover, the void fraction — that strongly dominates the pressure loss characteristics — is correlated by only the specific weight ratio ϕ within the measurement accuracy limits.

Accordingly, for steady boiling flow, simulation conditions were determined based on the specific weight ratio ϕ. Table 2 shows a comparison of the specific weight ratio between water/air and sodium liquid/vapor.

4. RESULTS AND DISCUSSION

4.1 Single Phase Flow Conditions

The pressure loss in a pin bundle was measured under both isothermal and radial power skew conditions. Figure 3 shows the pressure loss characteristics in the pin bundle. There is no significant difference in the pressure loss coefficient between isothermal and radial power skew conditions, in the range of $350 < Re < 20{,}000$ and $10^4 < Gr^* < 5 \times 10^5$.

Generally, for heating upward flow in a large bore tube, when natural convection strongly governs the flow, pressure loss increases compared to isothermal upward flow(9). The reason for the pressure loss increase under heating upward flow in a large bore tube is explained wherein only fluid near the wall is heated and the fluid velocity near the wall is increased by local natural convection. However, any increase in the flow rate is limited to the

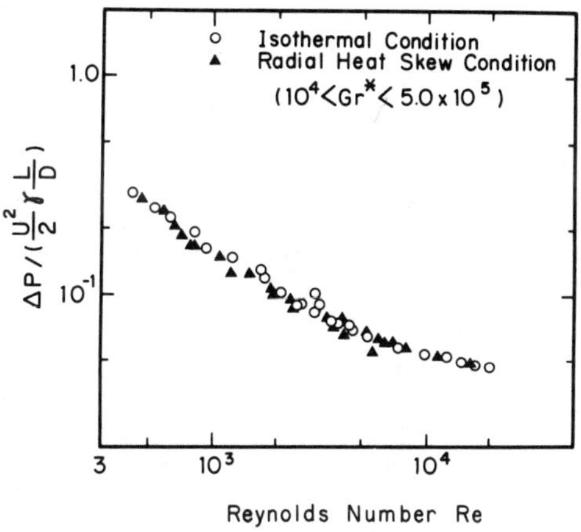

Fig. 3 Pressure Loss Characteristics under Combined Forced and Natural Circulation.

neighborhood of the wall. Thus, total flow rate remains low with increasing wall friction loss. For narrow channels, such as in a fuel pin bundle, the coolant temperature difference between the wall region and the central region in channels would not be large. Thus, local natural convection effects would not work strongly, i.e., total flow rate would be increased by global natural convection. Thus, the increase in pressure loss always matches that in total flow rate. Therefore, there is no significant difference in pressure loss between the isothermal upward flow and the heating upward flow in a pin bundle.

For the above reasons, when sodium is used as a coolant, the local natural convection effects upon the pressure loss might be much less than when water is used, because of its high thermal conductivity. Although water experiment results might provide a conservative estimation, compared to the sodium conditions, they could apply to the sodium condition, as a preliminary estimation.

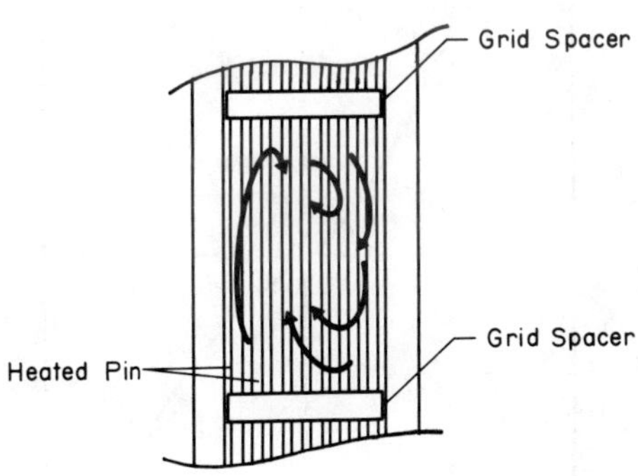

Fig. 4 Flow Pattern under Single Phase Natural Circulation with Radial Power Skew

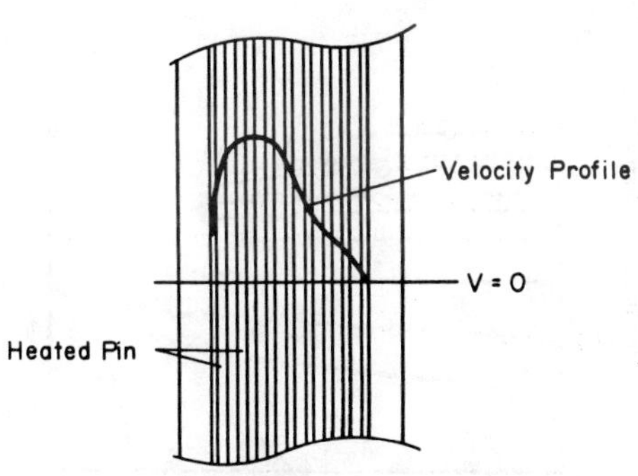

Fig. 5 Flow Pattern under Single Phase Combined Forced and Natural Circulation with Radial Power Skew

According to flow visualization tests, during natural circulation at $Gr^* < 5 \times 10^5$ with radial power skew in the pin bundle, the flow pattern showed that the coolant in the heated region flowed upward due to a local natural convection, but coolant in the unheated region flowed downward. Consequently, recirculation flow occured between the grid spacers in the pin bundle (as shown in Fig. 4). During combined forced and natural circulation conditions at Re < 2,000 and $Gr^* < 5 \times 10^5$, the flow patterns showed that the recirculation flow did not occur in the pin bundle, but the coolant whose velocity profile was not symmetrical (i.e., the coolant velocity in a heated region was higher than that in an unheated region, as shown in Fig. 5), flowed only upward. The flow was redistributed by the local natural convection.

It was, therefore, expected that during natural circulation conditions without any external driving force at all, heat transport would be temporarily disturbed by the flow reversal. On the other hand, during combined forced and natural circulation conditions, it was expected that intra-assembly temperature distribution would remain rather flat, due to the coolant redistribution effect, even though there was radial power skew in the pin bundle.

4.2 Two Phase Flow Conditions

If it was assumed in the following that coolant completely or almost stopped after reactor shut down or scram, coolant boiling might occur in pin bundles. Natural circulation head is, however, promoted by coolant boiling because coolant specific weight in the boiling region decreases due to vapor bubbles. However, by increasing coolant flow rate, two phase friction loss becomes more effective than the natural circulation head. Therefore, coolant flow rate decreases, compared to single flow conditions (i.e., before boiling occurs).

In these experiments, air was injected into the flow water to simulate boiling flow phenomena under natural and low forced flow conditions. Fig. 6 shows the correlation between the air quality and the coolant Reynolds number Re. Fig.7 shows the coolant Reynolds number change ratio ζ (Re $x = 10^{-3}$/Re $x = 0$,

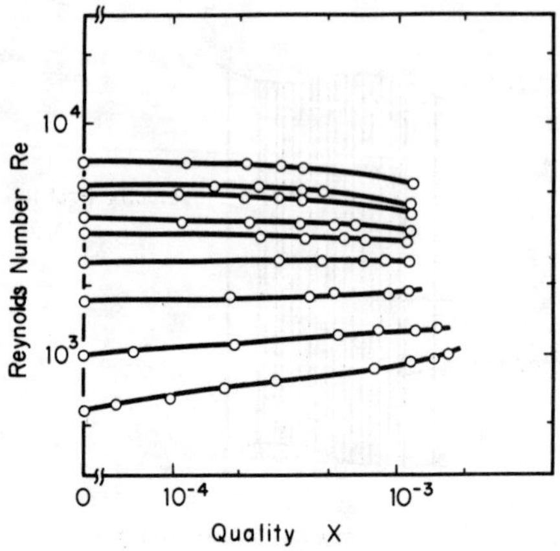

Fig. 6 Air Quality Effect upon Coolant Flow Rate

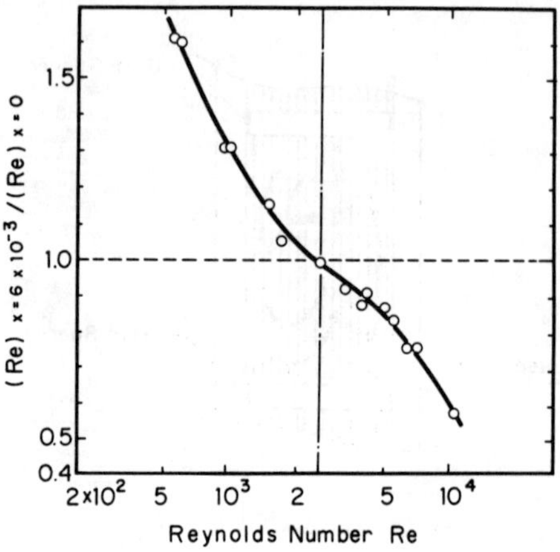

Fig. 7 Coolant Flow Rate Change Ratio as a Function of Initial Reynolds Number

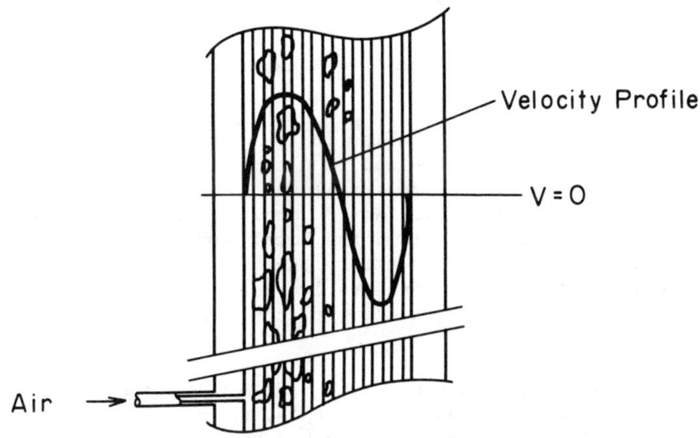

Fig. 8 Flow Pattern under Incoherent Two Phase Flow Natural Circulation

$Re_{x=10^{-3}}$ refers to the coolant Reynolds number at quality $x = 10^{-3}$, $Re_{x=0}$ refers to initial coolant Reynolds number). In the above two cases, air was injected through the inlet nozzle pressure tap. From Figs. 6 and 7, it was estimated that the coolant flow rate after boiling would increase at $Re < 2{,}500$, and then, the coolant boiling would be suppressed, and that the coolant flow rate would decrease at $Re > 2{,}500$. Then, the coolant boiling would further extend.

In general, reactor power decreases less than 6 % of normal operating conditions within a few minutes after reactor shut down. Therefore, less than 2 % of normal condition decrease in the coolant flow rate is required to occur coolant boiling under decay heat removal conditions. Thus, the Reynolds number at boiling inception under decay heat removal conditions is 1,200 to 1,600 (assume the Reynolds number in fuel assemblies at normal operating conditions, $60{,}000 < Re < 80{,}000$). Accordingly, under decay heat removal conditions, it was estimated that the coolant boiling in fuel assemblies would not extend, because coolant flow rate would increase due to buoyancy effects under coherent boiling.

According to the visualization tests, under two phase flow conditions, as shown in Fig. 8, the flow reversal was observed in the lower quality region (i.e., opposite to air injection wall), when air was injected into the coolant from the hexagonal wrapper wall (i.e., incoherent boiling simulation). There was recirculation flow between grid spacers as well as single phase natural circulation conditions. The flow reversal existed not only during natural circulation conditions (without external driving force), but also during combined forced and natural circulation conditions. However, flow reversal disappeared at $Re > 1{,}800$.

Consequently, it was expected that the coherent coolant boiling would be suppressed by an increase in the coolant flow rate due to buoyancy effects. However, it should be emphasized that heat transport would be disturbed by the flow reversal under incoherent boiling flow conditions.

5. CONCLUSIONS

Based on the results of single phase flow experiments, the following conclusions were reached.

(1) For narrow channels such as pin bundles in fuel assemblies, the coolant temperature difference between the wall and central region in channels would so small that significant pressure loss difference between the isothermal and the heating upward flow would hardly appear under low flow rate. For these reasons, when using sodium, the local natural convection effects upon the pressure loss might be much less than when using water. Though water experiments would provide conservative estimation compared to the sodium flow conditions, they could apply to the sodium condition, as a preliminary estimation.

(2) Under natural circulation conditions with radial power skew, recirculation flow would occur between grid spacers. It is, therefore, expected that heat transport would be temporarily disturbed by the flow reversal due to the recirculation flow.

(3) Under combined forced and natural circulation conditions, the recirculation flow does not occur, even though there is radial power skew in the pin bundle. However, the flow in the pin bundle is redistributed by the local natural convection.

If the hydraulic phenomena under steady boiling flow in pin bundles are mainly governed by specific weight ratio ϕ, the followings factors are concluded in regard to the boiling flow phenomena:

(1) The flow rate after boiling inception would increase at $Re < 2{,}500$ and the coolant boiling would be suppressed. It is, therefore, estimated that the coolant boiling — such as coherent boiling in a pin bundle — would not extend during decay heat removal conditions.

(2) The recirculation flow would occur during incoherent boiling under both natural circulation conditions and combined forced and natural circulation conditions. It is, therefore, expected that heat transport would be disturbed due to the flow reversal.

Though the present experimental studies were preliminary and qualitative, it was clarified that the intra-assembly hydraulic characteristics would be affected by the intra-assembly radial power skew. It is, therefore, suggested that radial power skew effects would become more important with regard to an actual analyses of natural circulation decay heat removal capability in LMFBR fuel assembly.

REFERENCES

(1) Cofield, R. D. et al., "Natural Convection Analyses and Verification for LMFBR Cores", IAEA Specialists' Meeting on "Thermodynamics of FBR Fuel Assembly Under Normal and Non-Normal Conditions", Feb. 5-7, 1978, Karlsruhe.

(2) Additon, S. L. et al., "Natural Circulation in FFTF a Loop Type LMFBR", ANS Symposium on the Thermal and Hydraulic Aspects of Nuclear Reactor Safety vol.2, Nov.27∼Dec.2, 1977, Atlanta.

(3) Singer, R. M. et al., "Steady State Natural Circulation Performance of the Experimental Breeder Reactor II Primary Heat Transport Circuit", Nucl. Sci. & Eng., 63, 75-82 (1977).

(4) Lorenz, J. J. et al., "An Investigation of LMFBR Outlet Plenum Thermal Hydraulic Behavior During Reactor Scram Transients", ANL-CT-76-18.

(5) Hain, H. et al., KFK-1895, Jun., 1974.

(6) Yang, J. W., "Buoyancy Effects on Pressure Loss in Rod Arrays Under Laminar Flow", Trans. Am. Nucl. Soc., Vol.32, Jun. 1979.

(7) Colliers, J. G., "Convective Boiling and Condensation" McGraw-Hill, 1972.

(8) Hashizume, K., Toshiba R & D Center, RM-13871, May 1975.

(9) Tanaka, H. et al., Int. J. Heat Mass Transf., 16 (1973), 1267.

Analytical and Experimental Simulation of LMFBR Decay Heat Removal by Natural Circulation Heat Transfer

K. S. CHUNG, W. A. BEZELLA, D. H. THOMPSON, and J. H. TESSIER
Reactor Analysis and Safety Division
Argonne National Laboratory
Argonne, Illinois, USA

ABSTRACT

The capability to remove decay heat by natural circulation only is an essential part of the reactor safety requirements for a properly designed reactor. This paper addresses the development of a calculational methodology for treating natural circulation and its application to a sodium cooled reactor. In the first part of the paper, the basic heat transfer and fluid mechanics related to the natural convection phenomena are studied using the finite element method. The latter part of this paper outlines the computer code development procedure for the study of natural circulation in a loop; the calculated results are compared with SLSF (Sodium Loop Safety Facility) in-pile experimental data.

1. INTRODUCTION

The loss-of-shutdown-heat-removal (LOSHR) accident has been postulated as one of several major types of accidents which could conceivably lead to a core disruptive accident (CDA). Although a LOSHR accident is considered to be extremely remote, some sequence of events can be postulated that will cause a loss of forced cooling of the core following reactor shutdown. Under this condition, the reactor system is required to remove the decay heat by natural circulation alone. Because of its major impact on reactor decay heat removal, the natural circulation phenomenon in reactors has received considerable attention for many years. A substantial amount of literature has reported advancements in the study of decay heat removal by natural circulation at the reactor post-scram condition [1,2,3]. The majority of the earlier research is limited to scoping-nature calculations using a large scale computer code. However, if the reactor channel flowrate becomes relatively small, flow motion becomes more complicated and may not be properly represented by a one-dimensional flow approximation because of the increased buoyancy effect.

The objectives of the present study are two-fold: first, it develops a calculational method which can be used to perform a detailed thermal-hydraulic analysis of an LMFBR reactor channel flow under the natural circulation conditions. Thus, the first part of this paper addresses a numerical procedure for solving the coupled energy-momentum equations. The latter part of this paper outlines the computer code development procedure for the study of natural circulation in a loop and compares calculated results with the Sodium Loop Safety Facility (SLSF) in-pile experimental data. The Sodium Loop Safety Facility [4] is a doubly contained sodium loop designed for installation and operation in the ETR (Engineering Test Reactor), a 175-MW(t) water cooled and moderated high flux test reactor. Figure 1 shows the major SLSF components and their arrangement. As the only facility in the United States capable of conducting safety tests on relatively large, full length liquid metal fast breeder reactor (LMFBR) fuel assemblies at steady-state power levels, the SLSF has been used for the in-pile experiments including the P2 experiment reported in this paper.

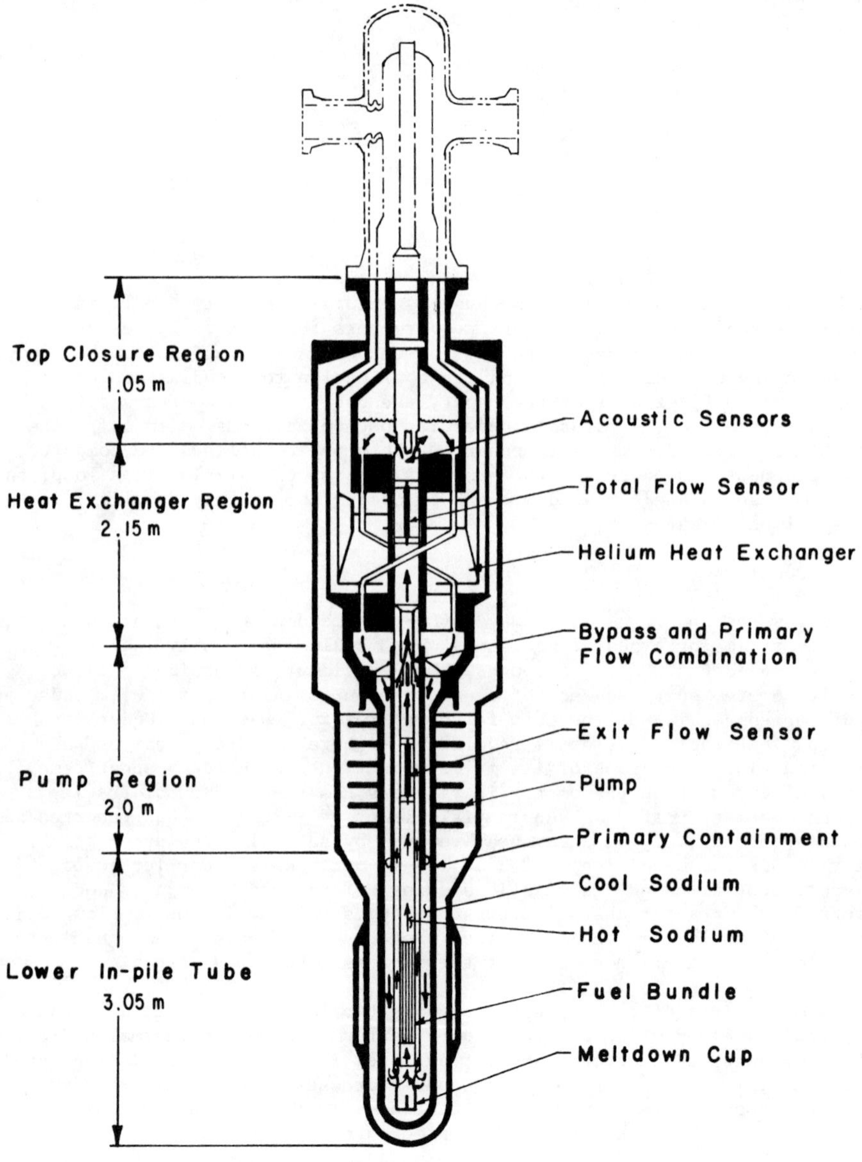

Fig. 1. Sodium Loop Safety Facility

2. TWO-DIMENSIONAL ANALYSIS OF NATURAL CIRCULATION FLOW IN A REACTOR CHANNEL

The combined mode of natural circulation and forced convection phenomenon is expected to exist at a low reactor flowrate condition with the considerable amount of heat generation. Typically, flow behavior after reactor shutdown can be described by this mixed convection phenomenon. Even though no external pumping force is provided under this condition, a substantial coolant flowrate is expected at the inlet of the heating region because of the continuous circulation of the reactor coolant by the buoyancy force.

The natural circulation phenomenon in a reactor channel is generally complicated by nonuniform boundary conditions, heat generation, irregularly shaped boundaries, and the presence of instruments and associated structure. Though the terminology "developing" or "fully developed" are used frequently in the present study, entry conditions of temperature and velocity profiles in a reactor channel are not well defined to deserve such descriptions. This paper idealizes the complex reactor channel geometry as a channel with a uniform cross-sectional area. The channel flow behavior is also simplified so as to be described by the conventional channel flow boundary and inlet conditions.

There exist several calculational techniques [5-8] which are relevant to the present calculational method. However, the physical problems considered in refs. [5-8] are not closely related to the present study. Some of these techniques were more effective for flow with a strong acceleration in one direction. The finite difference method was the numerical technique used in earlier works. Similar problems have been solved by using the finite element method. The reported solutions [9-11] are, however, limited to the lower Reynolds number cases and some of the developed techniques seem inappropriate for the physical problem considered in this study. In this section, a calculational method for the study of combined natural circulation and forced convection flow in an LMFBR has been developed using the Galerkin principle. Similar to one of the previous studies [7], pressure and velocity corrections are made to satisfy the continuity principle. The physical problem considered here is simplified in Figure 2. This is a two-dimensional flow in a channel with a finite channel length.

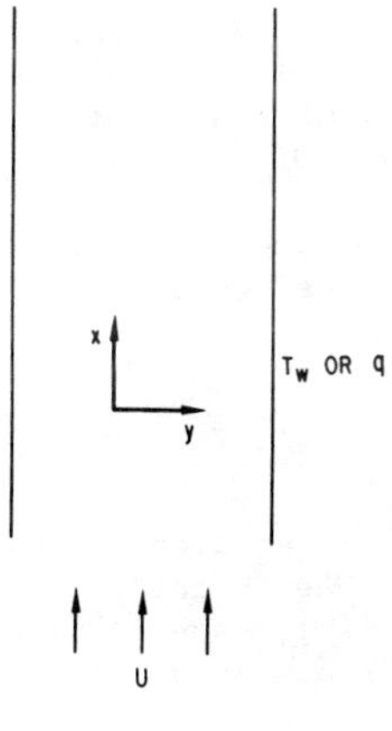

Fig. 2. Two-Dimensional Channel Flow

q, T, and U are heat flux, temperature, and velocity, and subscripts i and w represent the inlet and the wall condition. Following are the governing equations for the flow system shown in Figure 2, written with reference to the Cartesian coordinate system x, y.

Energy equation:

$$\rho[u \frac{\partial I}{\partial x} + v \frac{\partial I}{\partial y}] = \frac{\partial}{\partial x} (k_e \frac{\partial T}{\partial x}) + \frac{\partial}{\partial y} (k_e \frac{\partial T}{\partial y}) \tag{1}$$

where $I = C_pT$, and ρ, C_p, k_e, T, u, and v are density, specific heat, effective heat conductivity, temperature, x-direction velocity and y-direction velocity, respectively.

Continuity equation:

$$\frac{\partial}{\partial x} (\rho u) + \frac{\partial}{\partial y} (\rho v) = 0. \tag{2}$$

Momentum equations:

$$\rho u \frac{\partial u}{\partial x} + \rho v \frac{\partial u}{\partial y} = f_x + \frac{\partial}{\partial x} (\mu_e \frac{\partial u}{\partial x}) + \frac{\partial}{\partial y} (\mu_e \frac{\partial u}{\partial y}) \tag{3}$$

and

$$\rho u \frac{\partial v}{\partial x} + \rho v \frac{\partial v}{\partial y} = f_y + \frac{\partial}{\partial x} (\mu_e \frac{\partial v}{\partial x}) + \frac{\partial}{\partial y} (\mu_e \frac{\partial v}{\partial y}), \tag{4}$$

where f_x and f_y are the driving force terms and given in ref. [12],

$$f_x = - \frac{\partial p}{\partial x} + \rho g_x \beta (T - T_i) \tag{5}$$

and

$$f_y = - \frac{\partial p}{\partial y} + \rho g_y \beta (T - T_i) \tag{6}$$

where μ_e, β are the effective viscosity and the thermal expansion coefficient, and g_x and g_y are gravity force terms in the x and y-directions, respectively. If the flow motion is caused by an external force such as pump torque, the driving force terms are represented by the pressure differential term only and the body force terms are generally neglected. However, if the buoyancy force is the main driving force or has the same order of magnitude as the external force, both the pressure differential term and the buoyancy force term should be kept in equations (5) and (6). For a vertical channel flow considered in the present study, $g_x = g$ and $g_y = 0$.

The viscosity term μ_e in the momentum equations and the thermal conductivity term k_e in the energy equation are given by

$$\mu_e = \mu + \mu_t,$$

and $k_e = k + k_t$, respectively,

where μ_t and k_t are the so-called equivalent turbulent viscosity and the turbulent thermal conductivity. The calculational method for the evaluation k_t and μ_t are described in ref. [13], especially under the natural circulation condition. In this paper, no discussion will be made regarding the turbulent transport properties. Thus, $\mu_t = k_t = 0.0$.

Because all equations are coupled, an iteration procedure must be used to obtain the solutions. The present study calculates the pressure distribution first by assuming the velocity profiles. The calculated pressure distribution is then substituted into the momentum equations to evaluate the velocity profiles. The equation for pressure distribution is obtained by taking the derivatives of equations (3,4) with respect to x and y, respectively, and by summing up the resultant equations. After rearrangement, the pressure equation is given by

$$\frac{\partial^2 P}{\partial x^2} + \frac{\partial^2 P}{\partial y^2} = 2\left[\left(\frac{\partial u}{\partial x}\right)\left(\frac{\partial v}{\partial y}\right) - \left(\frac{\partial u}{\partial y}\right)\left(\frac{\partial v}{\partial x}\right)\right] + f\,\frac{\partial T}{\partial x}, \tag{7}$$

where $f = \frac{Gr}{Re^2}$.

Note that the continuity principle has been utilized to derive equation (7). Consequently, the pressure distribution information obtained from equation (7) should satisfy the continuity principle. Nevertheless, there still exists a possibility of violating the continuity principle due to the truncated error during iteration or due to the numerical instability originated in the finite element technique itself. Thus, the following methods are proposed to minimize the effects of the continuity principle violation. From equation (2),

$$\frac{1}{V}\frac{D}{Dt}(M) = \mathring{m} , \tag{8}$$

where M is the mass of a control volume and $\mathring{m}$ is the mass source term due to the calculation error. D/Dt indicates a total derivative. It is expedient to use the vector notation to see how much effect the calculational error $\mathring{m}$ has on momentum equations (3) and (4). Equations (3) and (4) can be expressed as

$$\frac{1}{V}\frac{D}{Dt}(Mu) = -\rho\,\frac{\partial P}{\partial x} + \frac{Gr}{Re^2}\,T + \frac{1}{UD}\,\nabla\cdot\mu\nabla u \tag{9}$$

and

$$\frac{1}{V}\frac{D}{Dt}(Mv) = -\rho\,\frac{\partial P}{\partial y} + \frac{1}{UD}\,\nabla\cdot\mu\nabla v, \tag{10}$$

where V is the control volume. From equations (9) and (10), one can get

$$\frac{D}{Dt}(Mu) = M\,\frac{Du}{Mt} + u\,\frac{DM}{Dt} = M\,\frac{Du}{Dt} + u\mathring{m} \tag{11}$$

Similarly, the error for the y-momentum equation by violating the continuity principle can be calculated as

$$\frac{D}{Dt}(Mv) = M\,\frac{Dv}{Dt} + v\mathring{m} \tag{12}$$

In the above, $u\mathring{m}$ and $v\mathring{m}$ are the error terms for the momentum equations which require correction. The pressure equation (7) also should include the calculational error given by (8). Thus, equation (7) becomes

$$\frac{\partial^2 p}{\partial x^2} + \frac{\partial^2 p}{\partial y^2} = 2\left[\mathring{m}^2 + \left(\frac{\partial u}{\partial x}\right)\left(\frac{\partial v}{\partial y}\right) - \left(\frac{\partial u}{\partial y}\right)\left(\frac{\partial v}{\partial x}\right)\right] + F\,\frac{\partial T}{\partial x} . \tag{13}$$

Note that variables in the governing equations are normalized by D for length, U for velocity, ρU^2 for pressure, and T by $T - T_i/T_w - T_i$, respectively.

Approximate values for u, v, and p are calculated first by assuming $\mathring{m}$ = 0. In the next iteration, $\mathring{m}$ is calculated from the continuity equation and substituted into the pressure and the momentum equations. Because all governing equations are nonlinear, variable functions are not known. The Galerkin principle is applied to the governing equations with the assumption of constant flow properties in each small element. Now, the governing equations are represented by

$$\int_V N_i \left[u\frac{\partial T}{\partial x} + v\frac{\partial T}{\partial y} - \frac{1}{Pe}\left(\frac{\partial^2 T}{\partial x^2} + \frac{\partial^2 T}{\partial y^2}\right)\right] dV = 0, \tag{14}$$

$$\int_V N_i \left[u\frac{\partial u}{\partial x} + v\frac{\partial u}{\partial y} + \frac{\partial p}{\partial x} - f.T - \frac{1}{Re}\left(\frac{\partial^2 u}{\partial y^2} + \frac{\partial^2 u}{\partial y^2}\right)\right] dV = 0, \tag{15}$$

$$\int_V N_i \left[u\frac{\partial v}{\partial x} + v\frac{\partial v}{\partial y} + \frac{\partial p}{\partial y} - \frac{1}{Re}\left(\frac{\partial^2 v}{\partial x^2} + \frac{\partial^2 v}{\partial y^2}\right)\right] dV = 0, \tag{16}$$

and

$$\int_V N_i \left[\frac{\partial^2 p}{\partial x^2} + \frac{\partial^2 p}{\partial y^2} - 2\left[\left(\frac{\partial u}{\partial x}\right)\left(\frac{\partial v}{\partial y}\right) - \left(\frac{\partial u}{\partial x}\right)\left(\frac{\partial v}{\partial x}\right)\right] - f\frac{\partial T}{\partial x}\right] dV = 0 \tag{17}$$

where N_i represents an interpolation function to approximate the velocities, pressure, and temperature. There exist numerous forms of the interpolation functions which provide a different degree of accuracy and complexity. The present study selects a linear and a quadratic form of interpolation functions with triangular and isoparametric shape elements. In general, a typical interpolation function is given by

$$N_i = a_i + b_i x + c_i y + d_i xy, \tag{18}$$

where a_i, b_i, c_i, and d_i are constants determined by the shape (i.e., triangular or isoparametric) and the size of elements. Calculational procedures for these values can be found in the literature [9-11]. After an appropriate interpolation function is chosen, profiles for temperature, velocities, and pressure are represented by

$$T = \sum_{n=1}^{NE} N_i\, T_i,$$

$$U = \sum_{n=1}^{NE} N_i\, U_i, \tag{19}$$

$$V = \sum_{n=1}^{NE} N_i\, V_i,$$

and

$$P = \sum_{n=1}^{NE} N_i\, P_i$$

where NE is the number of nodal points in one element, and the subscript i represents each nodal point. Interpolation functions for variables do not need to be the same order or the same shape. For some cases, different-order interpolation functions in the same calculation (i.e., a linear interpolation function for pressure and a quadratic interpolation function for velocities) give a better result in terms of the numerical stability and accuracy [10]. This statement is valid especially for the case where the governing equations contain different orders of derivatives for dependent variables. For example, if one considered the second order velocity derivatives and the first order pressure derivative as the dependent parameters, it is recommended to employ different-order interpolation functions for velocities and pressure. Because the governing equations (14-17) of the present study contain the second-order derivative for both velocities and pressure, the same order of interpolation for velocities and pressure can be used without violating the error consistency. In this paper, most of the calculations were performed by using the quadratic form of triangular elements for velocities and the linear form of triangular elements for pressure as recommended in ref. [10]. For the purpose of checking, calculations were repeated using the linear-isoparametric element for both velocities and pressure. After choosing an appropriate interpolation function for variables, we substitute equation (19) into equations (14-17). Integration of the resultant equation gives a set of matrix equations. Evaluation of the matrix equations is straightforward and the calculational procedure is well described in the literature [9-11]. The Newton-Raphson technique is incorporated into the calculational procedure to speed up the numerical convergence.

3. NUMERICAL RESULTS

The buoyancy force in a channel can accelerate flow motion against the gravity force and the friction pressure drop. Further, because of the unilateral heat flux or temperature boundary condition, the buoyancy force redistributes flow in a channel and changes the shape of the velocity profile from that of parabolic or slug flow motion. Both effects are significant from the viewpoint of reactor safety analysis. After a complete pump coastdown, acceleration of flow by the buoyancy force is the only mechanism available for the continuous supply of coolant to the fueled region. The asymmetric velocity profile of channel flow affects the heat transfer in the radial direction. Depending on the physical boundary condition, it can increase or decrease the heat transfer rate.

The boundary conditions and the inlet conditions used in the calculation are chosen to simulate the thermal-hydraulic behavior in the core area of the SLSF. Because the lower plenum at the inlet of the core area functions as a reservoir, a uniform inlet velocity and temperature condition is assumed. To provide a uniform driving force to the channel flow, temperatures at the boundaries are maintained at fixed values. This allows the Grashof number to be controlled easily. Using the equivalent channel width concept, the complicated geometry of the core section is approximated by a channel with a uniform cross-sectional area.

Figures 3 and 4 depict the velocity profile for the developing channel flow. The Reynolds number and Rayleigh number used in the calculations are 2036, 6.58×10^5 for Figure 3 and 2036, 3.24×10^6 for Figure 4 respectively.

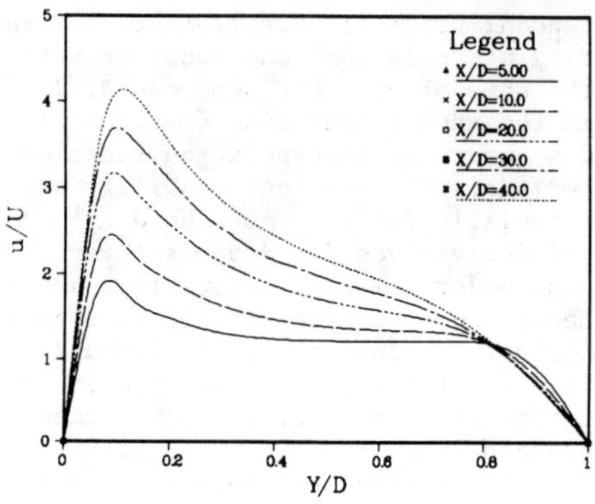

Fig. 3. Velocity Profiles of Developing Flow with Ra = 3.24×10^5

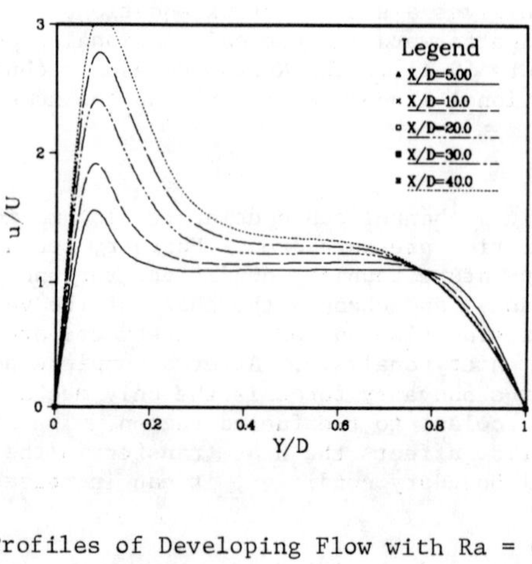

Fig. 4. Velocity Profiles of Developing Flow with Ra = 6.58×10^6

Figure 5 shows the effects of the Grashof number on exit velocity profiles under the same Prandtl and Reynolds number conditions. Note that the radial location of the maximum velocity moves close to the hot wall as the Grashof number increases. The relationship between the velocity profile and the Grashof number could not be easily expressed in a functional form. In Figure 6, the effect of the Prandtl number on the velocity profile is shown. Expectedly, an earlier development of temperature profile induces an earlier development of the velocity profile. From the viewpoint of natural circulation phenomena, employment of a lower Prandtl number liquid is definitely more desirable than a higher Prandtl number fluid.

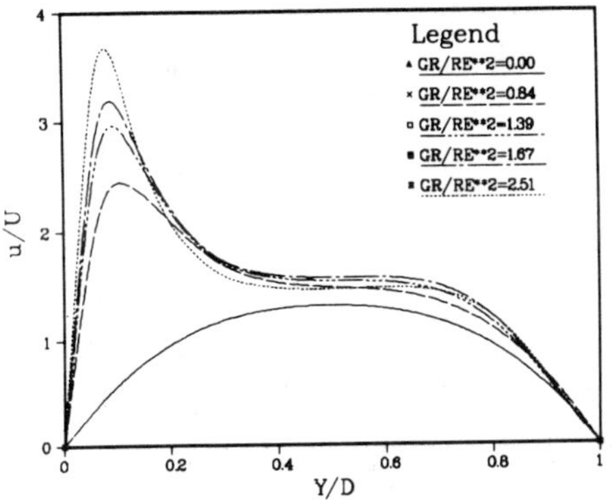

Fig. 5. Effects of the Grashof Number on Velocity Profiles at the Same Prandtl Number

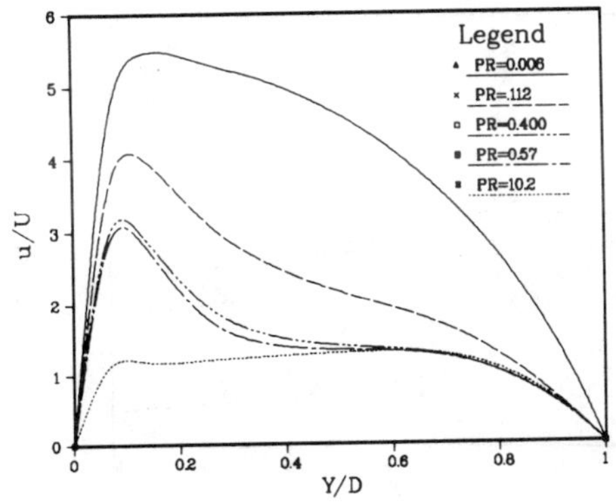

Fig. 6. Effects of the Prandtl Number on Velocity Profiles at the Same Grashof Number

Figures 7 and 8 show the effects of the Prandtl number and the Grashof number on the Nusselt number. Though it is a general practice to represent the Nusselt number in terms of the Peclet number, the present results are given as a function of the normalized axial distance to magnify these two effects. Under the same Grashof number condition, a higher Prandtl number flow results in a higher heat transfer coefficient. However, this does not imply that a higher Prandtl number flow is more effective than a lower Prandtl number flow in transferring heat by natural circulation because the flow rate induced by natural circulation is much larger at a low Prandtl number.

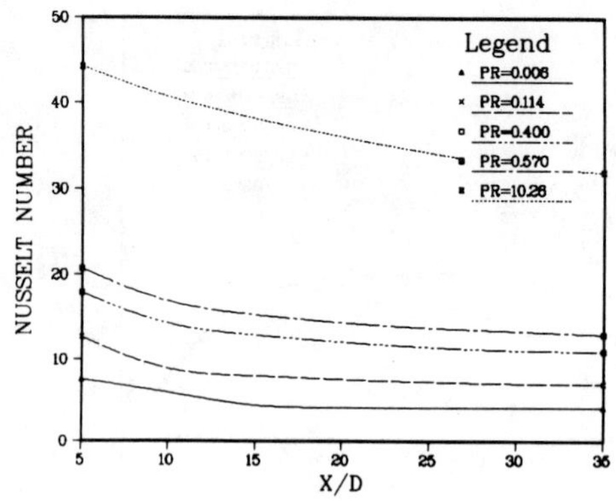

Fig. 7. Effects of the Prandtl Number on Nusselt Numbers at the Same Grashof Number

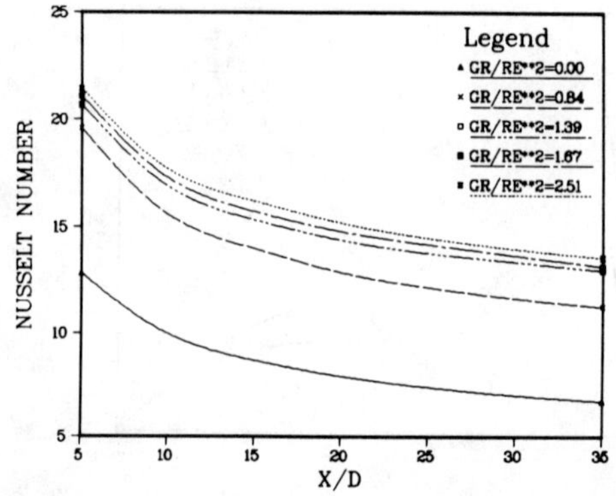

Fig. 8. Effects of the Grashof Number on Nusselt Numbers at the Same Prandtl Number

Refs. 14 and 15 reported experimental data which are closely related to the present study. Comparison between the results of the present study and the results of refs. 14 and 15 show a qualitative agreement. However, a quantitative comparison does not seem possible because of the unspecified experimental condition. There exist several analytical efforts [15,16] for the combined forced and natural convection phenomenon in a vertical channel. Here again, because of insufficient experimental data any direct comparison with these results seem to be impossible or not warranted. Mixed-order interpolation functions (i.e., linear interpolation function for pressure, and quadratic interpolation for velocity and temperature) are used for the calculation results discussed above. For the purpose of comparison, calculations were repeated using the linear-isoparametric elements for all dependent variables. The calculated results using the isoparametric element are in a fairly good agreement with the results obtained using the mixed interpolation functions quantitatively as well as qualitatively for the relatively smaller Grashof number cases. At $Gr/Re^2 = 1.39$, the exit velocity (at $x/D = 40$) by the mixed interpolation is about 2% larger than that by the uniform linear interpolation function. The numerical stability of the developed scheme seems to be favorable. The calculation method can be used for flows with the large Reynolds number without any significant numerical instability problems. The speed of convergence is also reasonably rapid. In most cases, a stable solution is obtained within nine or ten iterations. However, as the Grashof number increases, i.e., the Gr/Re^2 ratio becomes large, the solution becomes unstable. The upper bound on the Gr/Re^2 ratio above which a stable solution cannot be obtained is approximately three. Increasing the number of elements might be needed to correct the instability problem associated with a high Grashof number.

4. ANALYTICAL AND EXPERIMENTAL INVESTIGATION OF NATURAL CIRCULATION PHENOMENA IN THE SLSF

The thermal-hydraulic behavior of the entire SLSF has been studied by using the modified version of the THYME (Thermal-Hydraulic Model for Experiments) code [17]. The THYME code was originally written for obtaining essential information on SLSF experimental design and for SLSF pre- and post-test analysis, and has been successfully used for interpreting the results for several in-pile experiments. In essence, the THYME code describes flow motion in the SLSF loop as a one-dimensional and incompressible. The SLSF loop is divided into ∿75 distinct regions, each having a specific length, flow area, equivalent diameter, and expansion/contraction area. The flow motion in the loop is described by the following form of the governing equation:

$$\ell' \frac{dW}{dt} = H - g\int \rho dz - \Delta P_f \tag{20}$$

where

W = total loop flow

$\ell' = \sum_i \frac{(\Delta L)_i}{Ai}$; summation of length/area ratio over entire loop,

g = gravitational acceleration,

H = pump head,

ΔP_f = total loop friction pressure drop.

The pump head represents the pump torque which provides the driving force to the main coolant; under the natural circulation condition, this term becomes zero. Because the SLSF pump is an electro-magnetic pump, it can stop pumping instantaneously and not offer any impeller resistance.

Under the normal operating condition, the buoyancy head term, $-g\int \rho dz$ is usually small compared with the pump torque and the friction pressure drop term in equation (20). However, when the loop flow rate is relatively small, the contribution of the buoyancy term to the overall loop flow behavior becomes significant. The THYME code calculates the buoyancy at each node and sums up the result over the entire loop. Simpson's rule or Trapezoidal rule are used to integrate the buoyancy terms. For example, at the heat exchanger section, one hydraulic node has seven temperature node spacings. Hence, according to Simpson's rule,

$$g\int \rho dz = \frac{g}{3} * \frac{\Delta\ell}{6} * [\rho(T_1) + 4\rho(T_2) + 2\rho(T_3) + 4\rho(T_4) + 2\rho(T_5) + 4\rho(T_6) + \rho(T_7)] \tag{21}$$

where $\Delta\ell$ = temperature node spacing

$\rho(T_i)$ = density of temperature node i.

When the hydraulic node has irregularly spaced temperature nodes, the Trapezoidal rule is used.

The most important and complicated procedure in evaluating the loop flow rate is the friction pressure drop calculation. The difficulties arise from the fact that the cross-sectional area of channels is not uniform, thus the pressure drop due to the channel expansion/contraction should be considered at each change of loop cross-sectional area. Furthermore, information on the friction coefficient, which is a key parameter in obtaining the friction pressure drop, is limited especially for flow near the transition flow regime and flow under the natural circulation condition. The Reynolds number effect on the channel expansion/contraction loss coefficient is generally recognized as significant. However, the results of existing studies are neither conclusive nor presented in the format that can be used easily for a code development.

Because there exist a relatively large number of friction coefficient correlations and heat transfer coefficient correlations, the THYME code is programmed to select various heat transfer coefficients and friction coefficients optionally. The main reason for including these options is to determine which correlation describes the reactor system more closely and realistically for any particular transient. In most cases, one correlation cannot cover a wide range of flow rate transients. The validity of these correlations are usually restricted by the limited quantity of their experimental data or numerical calculations which these correlations are based on. Therefore, some correlations are recommended only for the certain range of Reynolds number even in the same flow regime.

The buoyancy force effect on the heat transfer coefficient and the friction coefficient is significant only at the heating region. In other regions, heat transfer coefficients and friction coefficients based on the average Reynolds number and the Prandtl number can be used for the natural circulation calculation. Heat transfer coefficients at the test section area increase slightly because of the buoyancy force as shown in Figures 7-8. However, for the conservative reason, these effects were neglected in the present study. Note that heat transfer is considered only in the radial direction including the natural circulation effect and radiation heat transfer effects in the gas gaps.

Using the revised THYME code, calculations were performed to compare with the P2 natural circulation test data. The P2 experiment used a 19-pin test train and had as its principle objective the simulation of the loss-of-flow accident. Natural circulation tests were included as a part of the P2 experiment. These transients were initiated by a pump trip following steady state lower power operation at 3 MW ETR power. The primary heat exchanger was shut off simultaneously by cutting off the helium flow supply to the heat exchanger. Figure 9 shows a comparison between the THYME predictions and the P2 experimental data. A similar calculation was made for the 4.5 MW case and for the 6 MW case. The calculated maximum coolant temperatures are 1305°F for

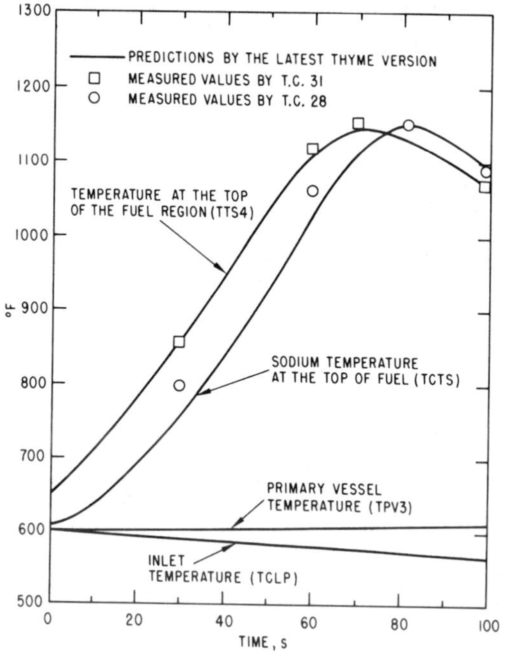

Fig. 9. Comparison Between THYME Predictions and P2 Test Data (Conversion Factor: °C = (°F - 32)/1.8)

the 4.5 MW case and 1446°F for the 6 MW case. The time at which these maximum temperatures occurred is between 60-65 seconds after pump trip. In general, all calculated temperature predictions, including the bypass flow temperatures, are in good agreement with the measured P2 test data. After reaching the steady state condition, flow rates become stable. The test section flowrates by the THYME code calculation are 3.27 Kg/min for the 3 MW case, 3.97 Kg/min for the 4.5 MW case, and 4.35 Kg/min for the 6 MW case, while the P2 test data show 4.08, 4.49, and 5.03 Kg/min respectively. The test section flowrates predicted by the THYME code seem to be reasonable considering the characteristics of flowmeters at low flowrates. However, the calculated loop flowrates are significantly different from those of the P2 data. The loop total flow, eddy-current flowmeter indicated 2.72 Kg/min for all three cases, but the THYME results are 7.62 Kg/min for 3 MW case, 10.1 Kg/min for 4.5 MW case, and 11.7 Kg/min for 6 MW case. A possible explanation for this discrepancy is that the sodium temperatures were well above the temperature range in which the loop total flowmeter was calibrated.

5. CONCLUSIONS

The present study developed a calculational method which can be utilized to study the combined mode of natural circulation and forced convection phenomena in a channel. Calculations were performed for several different combinations of the Prandtl number and the Grashof number. Calculated results seem to be in a qualitative agreement with existing experimental data. The calculated results using the THYME code introduced in this paper are in a good agreement with the P2 experimental data. However, further extensive analytical and experimental efforts are needed to check and improve the present calculational procedures.

REFERENCES

1. Additon, S. L., and Parziale, "Natural Circulation in FFTF, A Loop Type LMFBR", presented at ASME Annual Meeting, 1977.

2. Singer, R. M., et al., "Natural Circulation Heat Transfer in Pool-Type LMFBRs," Presented at ASME Annual Meeting, 1977.

3. Proceedings of the International Meeting on Fast Reactor Safety Technology, Seattle, Washington, August, 1979.

4. Gartside, C. H., et al., "The Sodium Loop Safety Facility," Nuclear Safety, Vol. 19 (3), pp. 334-355, 1978.

5. Harlow, F. H., and Welch, J. E., "Numerical Calculation of Time-Dependent Viscous Incompressible Flow of Fluid with Free Surface," Physics of Fluids, 8, pp. 2182-2189, 1965.

6. Amsden, A. A., and Harlow, F. H., "The SMAC Method: A Numerical Technique for Calculating Incompressible Fluid Flows," Los-Alamos Scientific Laboratory, LA-4370, 1970.

7. Patankar, S. V., and Spalding, D. B., Heat and Mass Transfer in Boundary Layers, 2nd Edition, Intertext Books, 1970.

8. Roache, P. J., Computational Fluid Dynamics, Hermosa Publishers, 1972.

9. Zienkiewicz, O. C., The Finite Element Method in Engineering Science, 2nd Edition, McGraw-Hill Book Company, London, 1971.

10. Huebner, K. H., The Finite Element Method for Engineers, John Wiley & Sons Inc., 1974.

11. Chung, T. J., Finite Element Analysis in Fluid Dynamics, McGraw-Hill Inc., 1978.

12. Schlichting, H., Boundary Layer Theory, 6th Edition, McGraw-Hill Inc., 1968.

13. Chung, K. S., and Thompson, D. H., "Study of Turbulent Natural Circulation Flow and Low Prandtl Number Flow in an LMFBR," to be presented at the 19th National Heat Transfer Conference, 1980, Florida.

14. Ogunba, V., and Barrow, H., "A Visual Flow Study of Mixed Convection in the Entry Region of a Vertical Internally Heated Annulus," Int. J. Heat and Fluid Flow 1(3), pp. 115-121, 1979.

15. Proceedings of Inst. Mech. Engineers, "Heat and Mass Transfer by Combined Forced and Natural Convection," Manchester, 1971.

16. Szewczyk, A. A., "Combined Forced and Free Convection Laminary Flow," ASME J. Heat Transfer 86, pp. 501-507, 1964.

17. Marr, W. W., et al., "Thermal-Hydraulic Dynamic Simulation Model for the FEFPL," ANL/RAS 71-21, 1971.

Numerical Simulation of the 6MW P2 Transient Free Convection Test Using the COMMIX—1A Computer Code

H. M. DOMANUS, M. J. CHEN, T. M. KUZAY, and W. T. SHA
Components Technology Division
Argonne National Laboratory
9700 South Cass Avenue
Argonne, Illinois 60439 USA

ABSTRACT

The single-phase COMMIX (COMponent MIXing) computer code performs three-dimensional, transient, thermal-hydraulic analyses of liquid sodium LMFBR components. It solves the conservation equations of mass, momentum, and energy as a boundary value problem in space and as an initial value problem in time. The concepts of volume porosity, surface permeability and distributed resistance, and heat source have been employed in the quasi-continuum (rod bundle) application. Volume porosity is defined as the ratio of the volume occupied by the fluid to the total volume of the computational cell. Surface permeability is defined as the ratio of the flow area in a particular direction available for fluid to the total area in the same direction across a surface of the computational cell. Distributed resistance is the drag force due to the presence of rods and is distributed continuously throughout the computational cell. Distributed heat source is similarly defined as distributed resistance.

The numerical results obtained from the latest version, COMMIX-1A, for a transient free convection test are presented. This test was conducted in the P2 bundle of the Sodium Loop Safety Facility (SLSF) at a 6 MW steady power level in the Engineering Test Reactor (ETR). The modeling includes: (1) the accurate representation of the volume occupied by the coolant, fuel rods, wire wrap spacers, filler wires and duct wall as well as their respective masses; (2) the transient, three-dimensional temperature distributions of the fuel rods, wire wrap spacers, filler wires and duct walls, where axial and azimuthal conduction have been neglected; (3) the detailed volume porosity and surface permeabilities due to the presence of fuel rods, filler wires, helical wire wraps and duct wall; (4) an additional sweeping force in the momentum equations to simulate swirl flow; and, (5) the gamma heating of the coolant, clad, filler wires, wire wrap spacers and duct wall.

1. INTRODUCTION

The COMMIX-1 (COMponent MIXing) computer code[1] performs three-dimensional, transient, single-phase, thermal-hydraulic analyses of liquid sodium LMFBR components. It solves the conservation equations of mass, momentum, and energy as a boundary value problem in space and as an initial value problem in time. The development of the COMMIX computer code is being carried out in a stage-wise manner, where new and/or improved models are progressively implemented into the code. The code is structured in modular fashion so that new models can readily be implemented.

The most recent version of the single-phase COMMIX code is called COMMIX-1A. This version contains many improvements to the original COMMIX-1 in both physical modeling and computing efficiency. The single-phase COMMIX code has been used to analyze both continuum and quasi-continuum applications[2-6].

2. POROUS MEDIUM APPROACH

The porous medium approach with volume porosity, surface permeability, and distributed resistance and distributed heat source (or sink) is employed in the COMMIX code for rod-bundle thermal-hydraulic analysis. It provides a greater range of applicability and an improved accuracy over subchannel analysis[7]. The volume porosity (γ_V) is defined as the ratio of the volume occupied by the fluid to the total volume of the computational cell. Surface permeabilities (γ_j) are similarly defined as the ratio of the surface area in the direction available to the fluid to the corresponding total surface area in the same direction. The distributed resistance force (R) in the momentum equation represents the sum of the drag force due to the presence of rods and the additional viscous force due to the difference between the fine structure and the cell-averaged velocity distribution in the computational cells. The distributed heat source (or sink) is normally taken as a local volumetric heat source. The concept of surface permeability is an additional feature added in the porous medium formulation. This permits easy modeling of the anisotropic characteristics of a flow domain.

The porous medium can readily be reduced to a continuum by setting volume porosity and surface permeabilities to unity, and distributed resistance to zero. Thus, a single formulation is applicable to both quasi-continuum and continuum analyses.

2.1 Governing Equations

Conservation of Mass.

$$\frac{D}{\gamma_V} = \frac{\partial\rho}{\partial t} + \frac{1}{\gamma_V}\nabla\cdot(\gamma_j\,\rho u)$$

Conservation of Momentum.

$$\gamma_V\rho\frac{\partial u_i}{\partial t} + \gamma_V u_i\left(\frac{\partial\rho}{\partial t} - \frac{D}{\gamma_V}\right) + \frac{\partial}{\partial x_j}(\gamma_j\rho u_i u_j) =$$

$$-\gamma_i\frac{\partial P}{\partial x_i} + \frac{\partial}{\partial x_j}(\gamma_j\tau_{ji}) + \gamma_V\rho g_i + \gamma_V R_i$$

Conservation of Energy

$$\gamma_V\rho\frac{\partial h}{\partial t} + \gamma_V h\left(\frac{\partial\rho}{\partial t} - \frac{D}{\gamma_V}\right) + \frac{\partial}{\partial x_j}(\gamma_j\rho u_j h) =$$

$$\gamma_V\frac{dP}{dt} + \frac{\partial}{\partial x_j}\left(\gamma_j\,\kappa_{eff}\frac{\partial T}{\partial x_j}\right) + \gamma_V Q + \gamma_j\tau_{ji}\frac{\partial u_i}{\partial x_j}$$

Here ρ is the density, $\underline{u}$ is a velocity vector, g is the gravity vector, h is the enthalpy, P is the pressure, x_i is the coordinate direction, and t is the time. The terms γ_V, γ_j, R_i, and Q are the volume porosity, surface permeability, distributed resistance and distributed heat source, respectively. The quantity D is the local mass residue of the continuity equation. In the above equations, a repeated index implies the sum of three terms.

2.2 Fuel Assembly Modeling

The fuel assembly modeling includes: (1) the transient three-dimensional flow and temperature distribution of the sodium coolant; (2) the accurate representation of the volume porosity and surface permeability due to the presence of fuel rods, wire wrap spacers, filler wires and duct walls, and also their respective masses; (3) the transient three-dimensional temperature distributions of the fuel rods, wire wrap spacers, filler wires and duct walls, where axial and azimuthal conduction have been neglected in the fuel rods and the filler wires. Wire wrap spacers and duct walls interact only with the immediately adjacent coolant; (4) an additional sweeping force has been used in the momentum equations to account for the swirl flow; and, (5) the gamma heating of the coolant, clad, filler wires, wire wrap spacers and duct wall has also been accounted for.

3. PROBLEM STATEMENT

The subject of this paper is the application of the COMMIX-1A computer code to simulate the thermal-hydraulic behavior of a 19-rod test bundle P2, which was used in the SLSF (Sodium Loop Safety Facility) in the ETR (Engineering Test Reactor) free convection test conducted by Argonne National Laboratory during 1976[8,9]. The experimental facility and the geometry of the simulated fuel assembly will be described as well as the steady and transient operating conditions.

3.1 Experimental Facility

The primary objective of the Sodium Loop Safety Facility (SLSF) P2 test was an unprotected loss of flow (LOF) safety experiment as a part of the planned LMFBR safety research program by Argonne National Laboratory.

The test loop contained a 19-fuel pin subassembly having a 36-inch long mixed oxide fuel (UO_2, PuO_2). The fuel pins had wire wraps around them and the subassembly was housed in a hexagonal duct which was insulated on the outside to reduce the heat loss from the test section to the flow bypass annulus.

The flow of the sodium coolant was split at the bottom of the test section, approximately 36% flowing through the test section, and the remainder through the bypass. The test section inlet flow was measured by an electromagnetic flowmeter at the entrance to the test section.

Prior to the final LOF experiment, many different types of preparatory and/or simulatory experiments were conducted. Two batteries of natural circulation experiments were among these. In the first battery, three natural circulation experiments were run in succession. During these tests, the power level was raised from 3 to 4.5 to 6 MW successively. These power levels refer to the steady power level of the driver reactor, ETR. The same day, but later, two additonal tests at 4.5 and 6 MW nominal reactor power were conducted. The analysis covered in this paper involves the last 6 MW test. To do the test, the reactor was maintained for a long period at the 6 MW nominal power level to establish the desired loop initial conditions. After the planned flow rate, inlet temperature, and other loop parameters were steadied, the flow was rapidly turned down by an EM pump trip. However, power was maintained at the pretest level of 6 MW.

3.2 Geometry

The 19 rods are contained within a hexagonal duct. Each of the 19-rods has an outside diameter of 0.00584 meter and is arranged in a triangular pitch of 0.00726 meter. Spacing of the rods is accomplished by a wire wrapped helically around each rod. The wire spacers have a diameter of 0.00142 meter when between rods, and .00071 meter when between a rod and the duct wall. The wire is wraped in a counterclockwise fashion (viewed looking upsream) with increasing axial level as shown in Fig. 1. The wire wrap pitch is 0.3048 meters.

The model fuel rods are composed of two materials: the fuel (UO_2, PuO_2) and clad (316SS). The clad occupies the annular region between (a) and (b), while the cylindrical fuel region is of radius (c), as shown in Fig. 2. The axial partitioning of the fuel rods is as shown in Fig. 3. Radially, the clad is partitioned into two regions, while the fuel is partitioned into three regions. Materials used in the model fuel assembly are listed in Table 1.

Table 1

Materials of Model Fuel Assembly

Coolant	Sodium
Duct Walls	316 SS
Wire Wrap	316 SS
Filler Wires	316 SS
Clad	316 SS
Fuel	UO_2, PuO_2

The rod bundle axial length is vertically oriented. The sodium coolant enters from below and flows up through a 0.18593 meter entrance region, followed by a 0.91440 meter heated region, and a 0.11887 meter exit region. The 1.2192 meter overall length of the model fuel assembly is partitioned into 24 equal lengths, each of 0.0508 meters as shown in Fig. 3. These and other geometric parameters are listed in Table 2. The transverse quarter pin partitioning used in the simulation is shown in Fig. 4.

The thickness of the duct walls which form the hexagonal configuration is 0.00305 meters, and is made of 316 stainless steel.

Thermocouples are located within some of the wire wrap spacers and filler wires and also along the inner duct wall. The thermocouple positions within the model fuel assembly are shown in Figs. 3 and 4.

It is to be noted that uncertainties exist in the mesurements: one is the exact location of thermocouples in the wire wrap under operating conditions; another is that the flow rate during the transient is below the calibrated range of the inlet EM flowmeter.

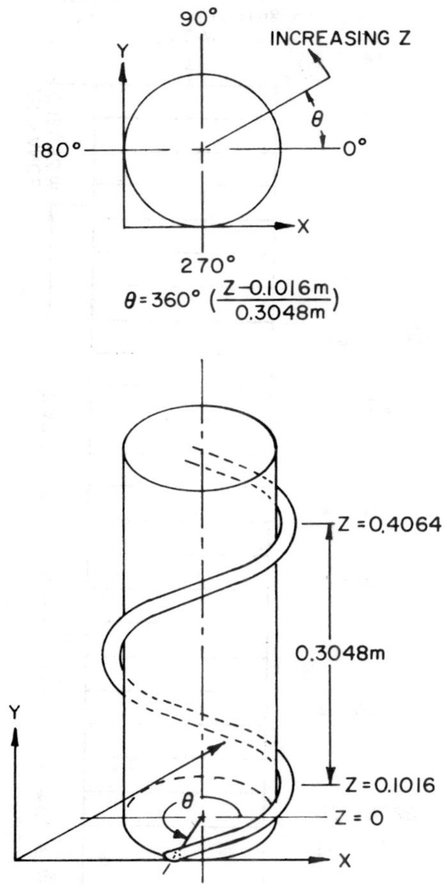

Fig. 1 Orientation and Location of Helical Wire Wrap Spacers

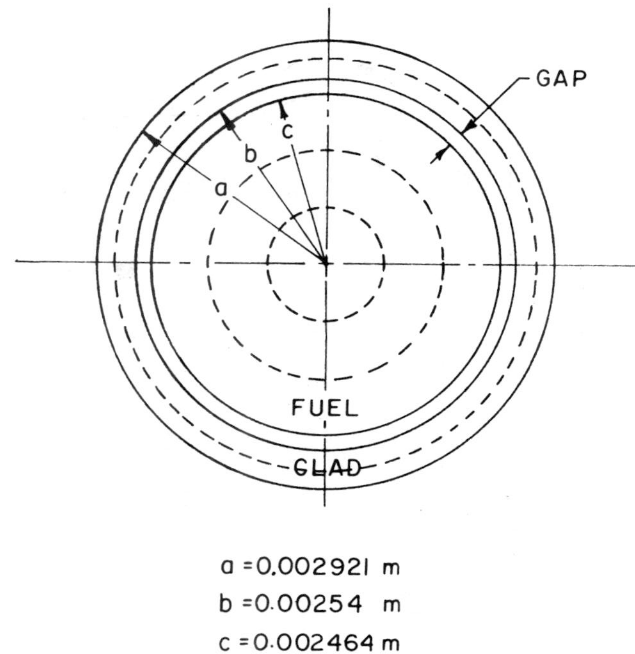

Fig. 2 Partitioning of Model Fuel Rods

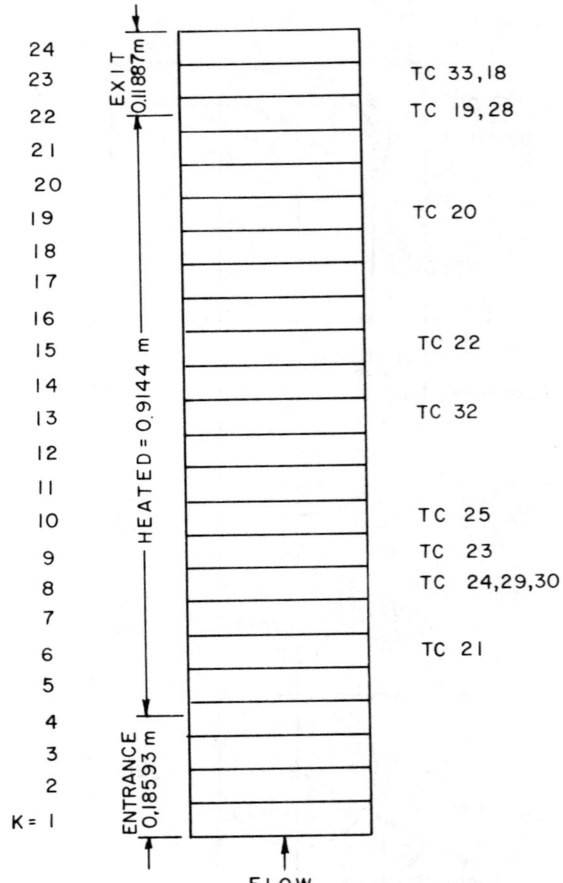

Fig. 3 Axial Partitioning of the Model P2 Fuel Assembly

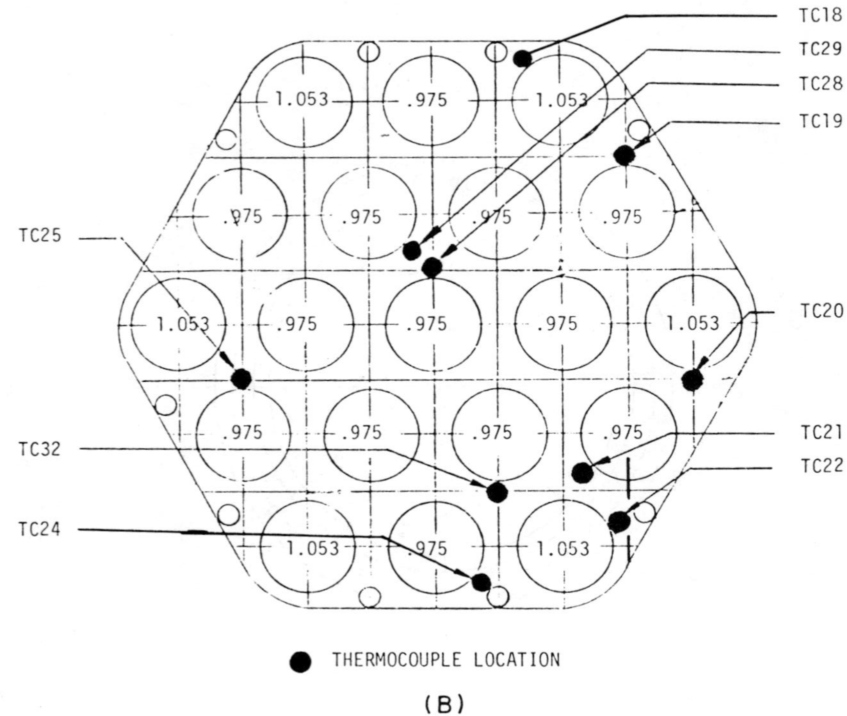

Fig. 4 Transverse Partitioning of the Model P2 Fuel Assembly

Table 2

Geometry of the Model P2 Bundle

Number of fuel rods	19
Triangular pitch	0.00726 m
Outside clad diameter	0.00584 m
Helical wire diameter	0.00142 m (0.00071 m next to wall)
Wire wrap axial pitch	0.30480 m
Inlet flow area	3.58×10^{-4} m^2
Hydraulic diameter	0.00230 m (next to wall)
	0.00412 m (central)
	0.00331 m (overall)
Total axial length	1.21920 m
Entrance length	0.18593 m
Heated length	0.91440 m
Exit length	0.11887 m
Axial partitions	24 equal, each of 0.0508 m
Duct wall thickness	0.00305 m

3.3 Steady State Operating Conditions

Liquid sodium enters the rod bundle uniformly with a temperature of 315°C and a velocity of 3.4 m/sec. The inlet pressure is approximately 634 KPa. A total of 32574 watts of heat is added to the coolant over the heated section. This heat represents both fission and gamma heating sources. The heat source in the fuel is 31271 watts, while the gamma heating of the clad, wire wraps, duct walls, and coolant is about 4%, or 1303 watts. These heat sources are summarized in Table 3. The normalized axial power distribution is shown in Fig. 5, where the normalization is performed over the 0.9144 meter heated length of the model fuel assembly. The normalized pin power distribution is shown in Fig. 4, where the corner rods have a slightly higher power. The volumetric heat sources in the clad, wire wrap, duct walls and coolant, while different from each other, are assumed uniform over an axial level.

Table 3

Heat Sources of Model P2 Fuel Assembly

Fuel (fission + gamma)	31271 watt
Clad (gamma)	275 watt
Wirewraps (gamma)	67 watt
Filler wires (gamma)	34 watt
Duct walls (gamma)	814 watt
Coolant (gamma)	113 watt
TOTAL	32574 watt

3.4 Transient Operating Conditions

From the steady state operating conditions, a flow and temperature transient occurs. The inlet velocity decreases to approximately 2.5% of its initial value in the first 2.5 sec of the transient, as shown in Fig. 6. After leveling off, the inlet velocity slowly increases to 7.5% of the steady state value by 70 sec, then remains constant throughout the rest of the transient.

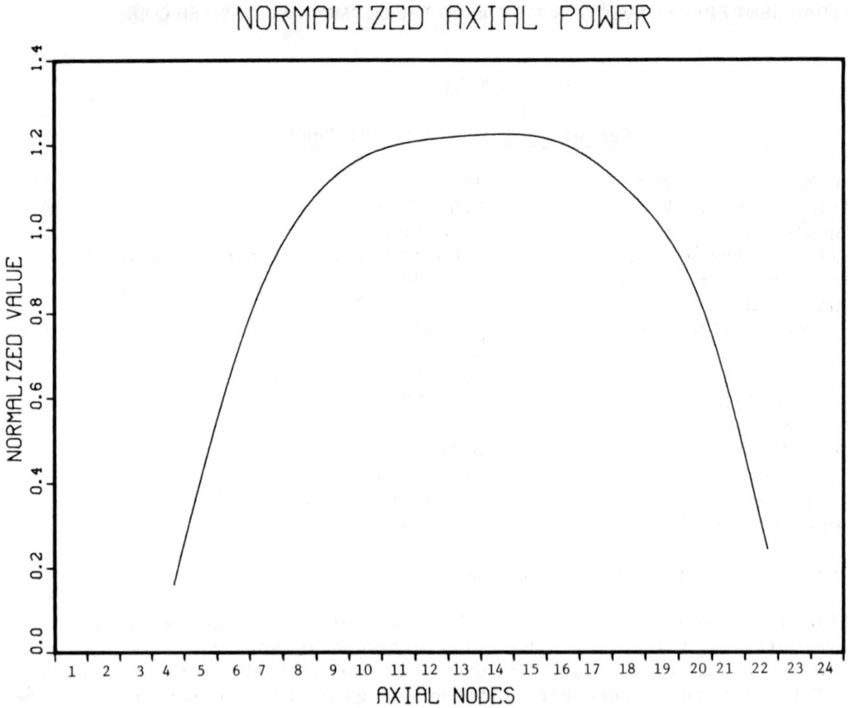

Fig. 5 Normalized Axial Power Distribution

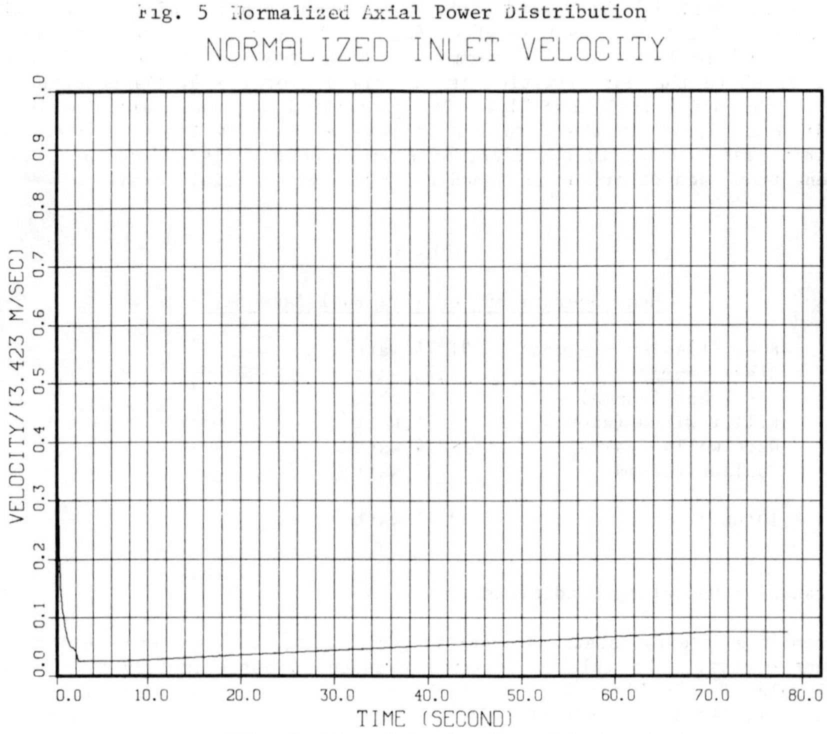

Fig. 6 Inlet Velocity Transient

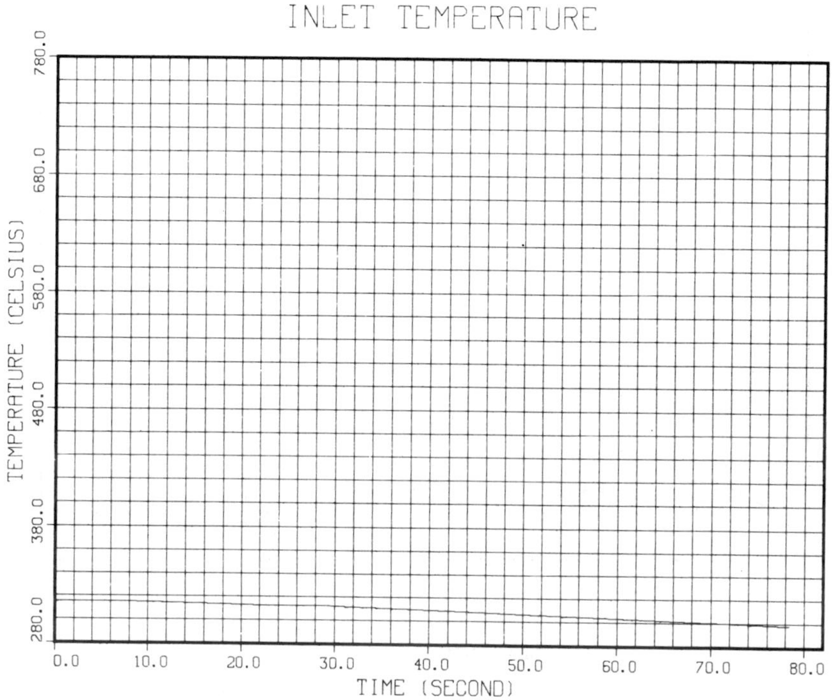

Fig. 7 Inlet Temperature Transient

This flow transient is the result of shutting the pumps off. The flow velocity never goes to zero because the heated fuel assembly acts as a pump due to the bundle heating and creates a density difference, hence a pressure differential across the bundle. Throughout the entire transient the power level of the bundle remains constant.

The temperature transient is shown in Fig. 7. After being a constant 315°C for the first 7.5 sec, it then slowly decreases to 295°C by 88 sec, then remains at that temperature for the rest of the transient.

4. NUMERICAL RESULTS

The steady-state and transient results of the simulation are presented in the following sections.

Since the COMMIX-1A simulation gives the complete transient thermal-hydraulic field of the coolant, radial and axial temperature distributions of each fuel rod, wire wrap, filler wire and the duct wall temperature distributions, only a selected portion of the results are presented.

4.1 Steady State Results

The COMMIX-1A calculated steady-state wire wrap temperatures at the locations measured in the experiment are shown in Figs. 8-14, at time = 0.0, along with the corresponding experimental measurements. As can be seen, the temperature increases as the sodium flows up the bundle. The overall inlet to

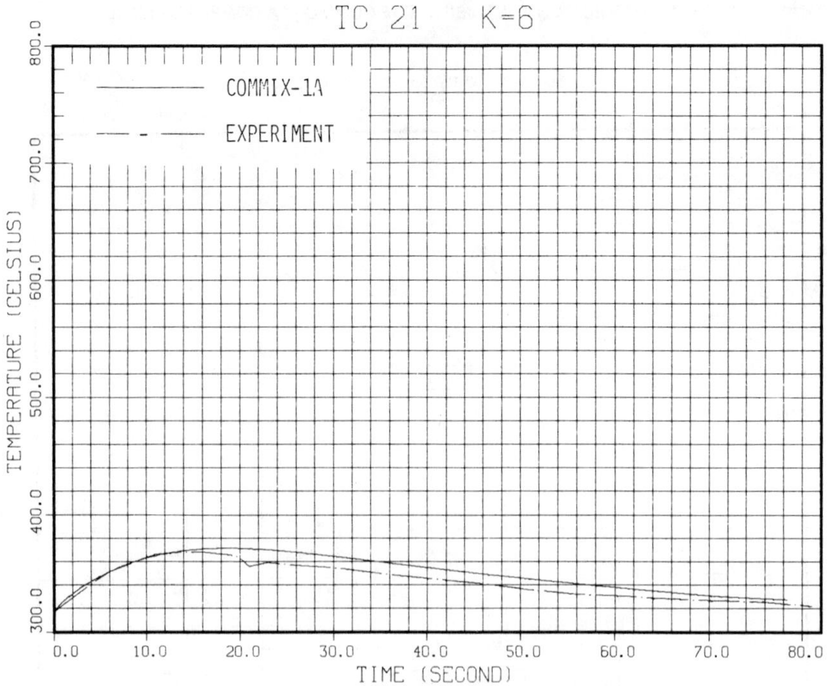

Fig. 8 Calculated Temperature Transient
vs. Measurement of TC 21

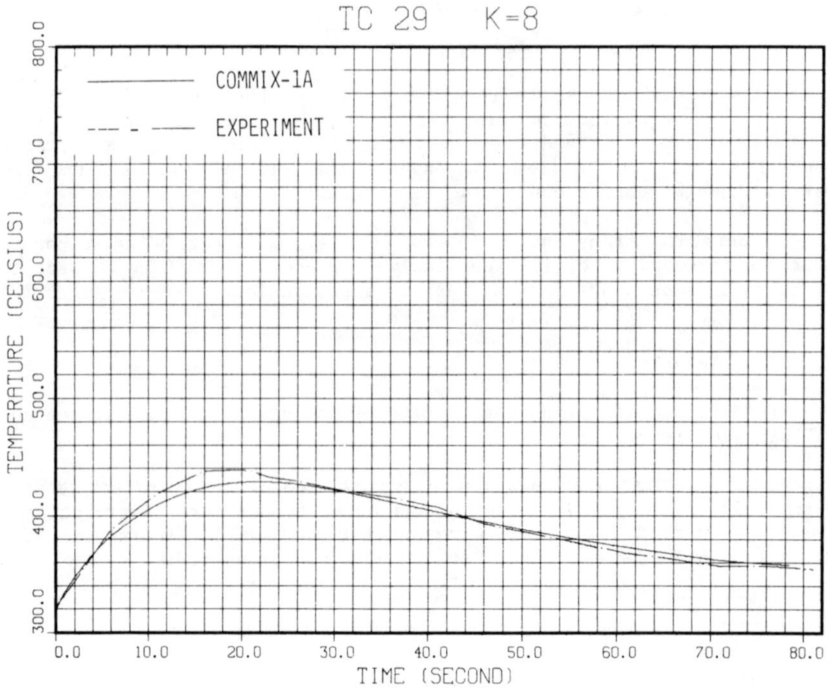

Fig. 9 Calculated Temperature Transient
vs. Measurement of TC 29

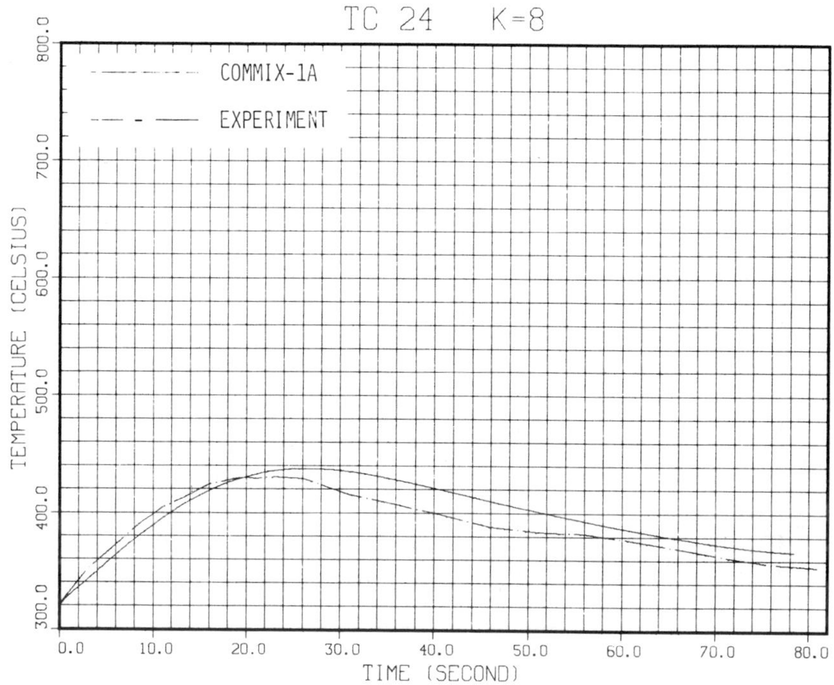

Fig. 10 Calculated Temperature Transient vs. Measurement of TC 24

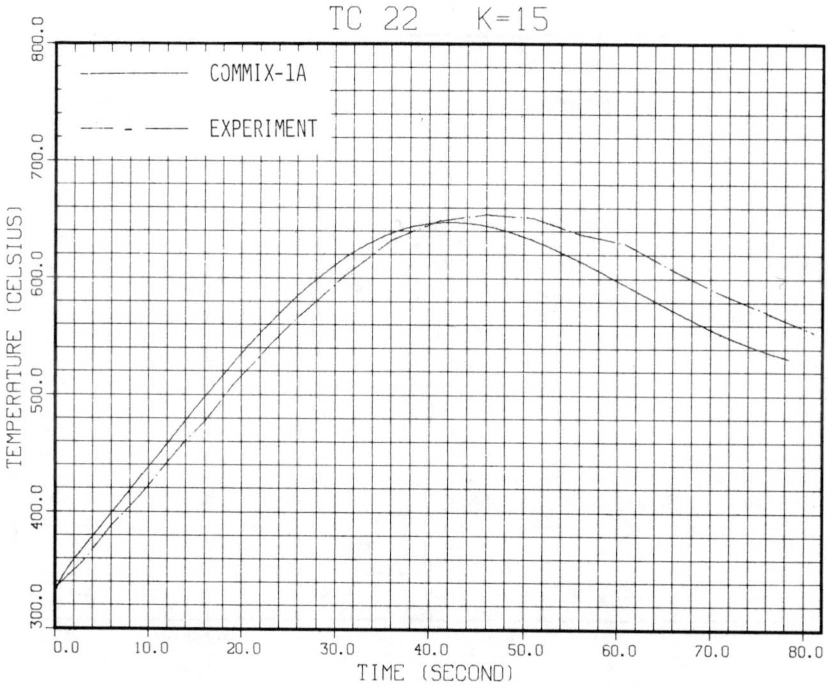

Fig. 11 Calculated Temperature Transient vs. Measurement of TC 22

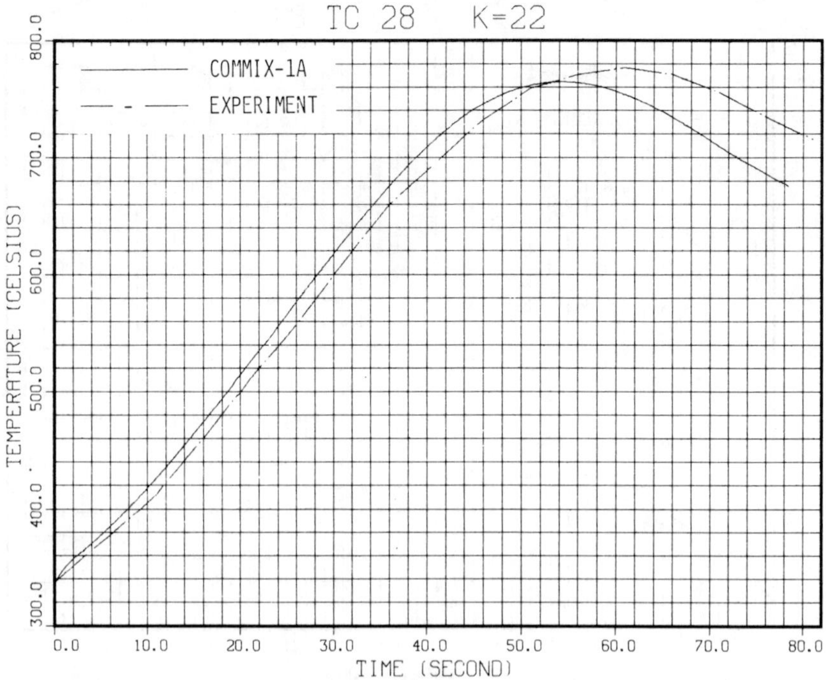

Fig. 12 Calculated Temperature Transient vs. Measurement of TC 28

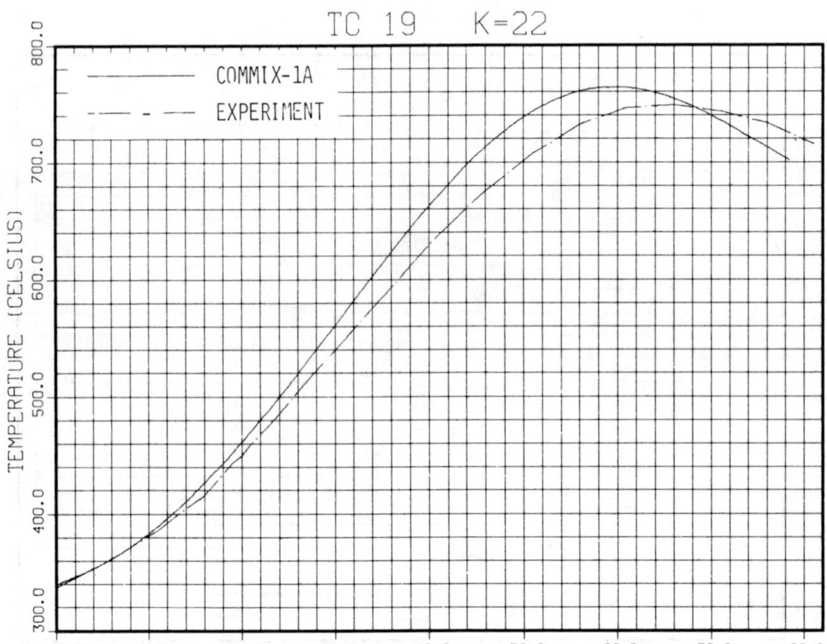

Fig. 13 Calculated Temperature Transient vs Measurement of TC 19

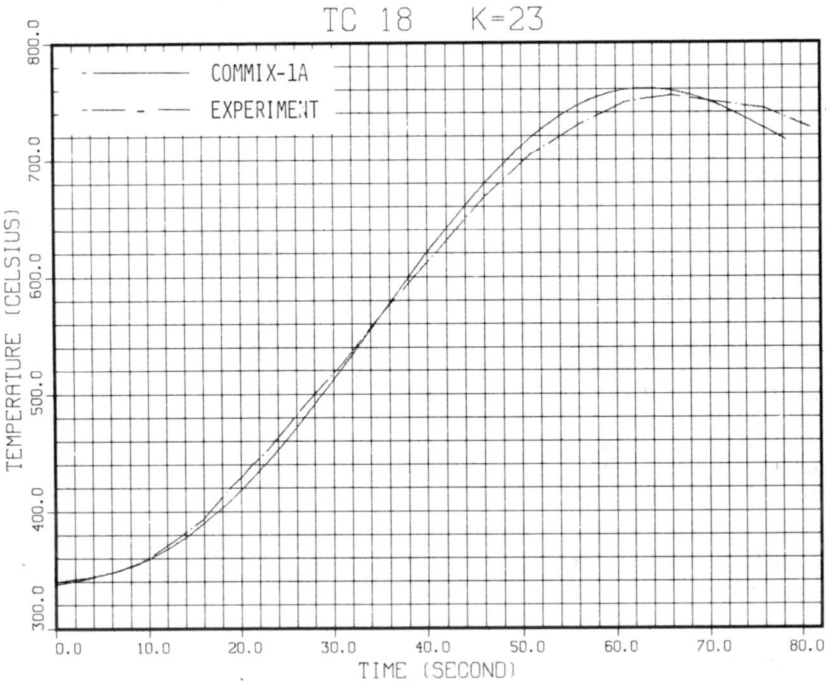

Fig. 14 Calculated Temperature Transient vs Measurement of TC 18

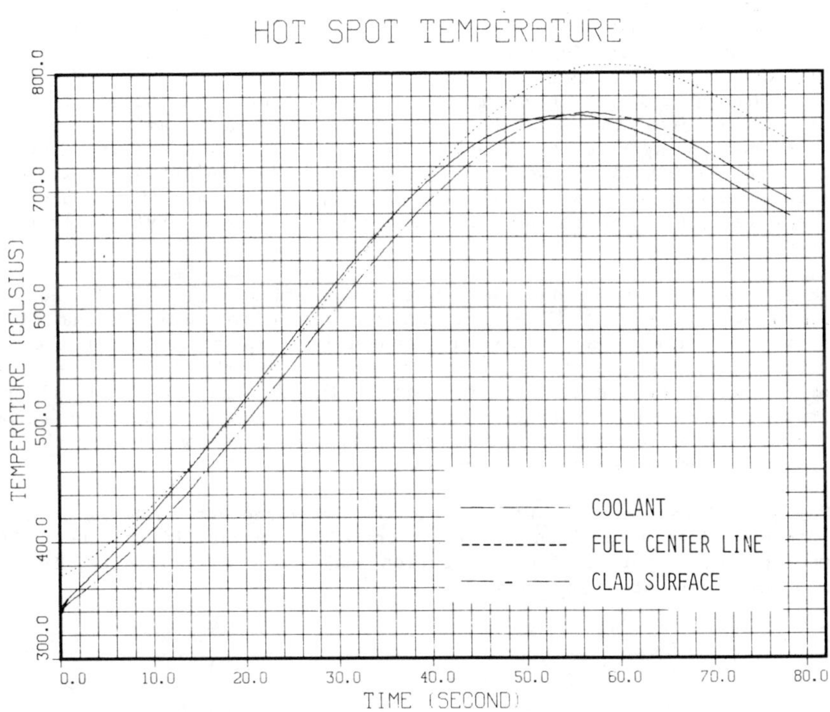

Fig. 15 Fuel, Clad and Coolant Temperatures of Hot Fuel Rod at K = 21

outlet temperature change is 23°C at these steady low-power operating conditions.

4.2 Transient Results

At the start of the transient, the flow velocity decreases. Due to the lower velocity, the coolant is heated for a longer time as it flows through the fuel assembly, thus the coolant temperature begins to increase. This increased coolant temperature and decreased flow velocity reduces the amount of heat transferred from the fuel rods to the coolant; hence, the fuel rods heat up. This is particularly important toward the end of the fuel zone. As the transient proceeds, the inlet flow starts to increase and the inlet temperature starts to decrese. This action causes the sodium temperatures to peak out and decrease until they approach quasi-steady values, which is determined by the buoyancy driven flow and decay heat. The temperature at the start of the heated section peaks first. The temperatures toward the end of the heated section peak at later times.

The COMMIX-1A calculation of the transient temperatures and the temperatures measured by the thermocouples located inside the wire wraps are shown in Figs. 8-14.

5. CONCLUSION

In performing this simulation of a thermal-hydraulic transient leading to a free convection condition, the history of the inlet flow transient is apparently the most important aspect, and has the largest impact on the results, in terms of the maximum cladding temperature. The experimental data clearly show that the sodium coolant temperatures do not asymptotically approach steady-state by heating up. On the contrary, they indicate an overshoot of the coolant temperatures to well above steady-state values, and a subsequent asymptotic cooling approach to steady state. These observations are correctly predicted by the COMMIX-1A code. In this transient simulation, peak coolant temperatures are below the saturation temperature and no boiling occurred. Also, flow recirculation was not observed in the bundle at anytime during the simulation.

An interesting result observed in the simulation occurs at the hot spot location shown in Fig. 15 (K = 21). At the corner rod near the end of the heated section, the adjacent coolant temperature is hotter than the clad temperature indicating heat is being transferred to the fuel rod from the coolant during a portion of the transient. This is due to the axial power distribution where the sodium is heated a greater amount as it flows through the central portion of the heated section and transfers some heat back as it flows past the end of the heated section.

The calculated results give a reasonable agreement with experimental measurements. This simulation has demonstrated the capability of the COMMIX-1A computer code to perform free convection analyses and contributes, in part, to the overall validation of the code.

ACKNOWLEDGEMENTS

The authors want to thank Drs. R. T. Curtis and C. N. Kelber, and Mr. P. M. Wood of the United States Nuclear Regulatory Commission for their support and encouragement, and also Mrs. S. A. Moll for typing the manuscript.

The submitted manuscript has been authored by a constractor of the U.S. Government under contract No. W-31-109-ENG-38. Accordingly, the U.S. Government retains a nonexclusive, royalty-free license to publish or reproduce the published form of this contribution, or allow others to do so, for U.S. Government purposes.

REFERENCES

1. W. T. Sha, H. M. Domanus, R. C. Schmitt, J. J. Oras, and E. I. H. Lin, "COMMIX-1: A Three-Dimensional, Transient, Single-Phase Component Computer Program for Thermal-Hydraulic Analysis", NUREG/CR-9785, ANL-77-96 (Sept 1978).

2. H. M. Domanus, R. C. Schmitt and W. T. Sha, "Numerical Results Obtained from the Three-Dimensional Transient, Single-Phase Version of the COMMIX Computer Code", NUREG/CR-0355, ANL-CT-78-3, (Oct 1977).

3. H. M. Domanus and W. T. Sha, "Numerical Results for a Hexagonal Fuel Assembly with a Planar Blockage using the COMMIX-1A Computer Code", NUREG/CR-0483, ANL-CT-79-8 (Sept 1978).

4. H. M. Domanus, M. J. Chen and W. T. Sha, "Analysis of a Loss of Piping Integrity Transient in an LMFBR Fuel Assembly", Trans. Am. Nucl. Soc., 32, pp. 827-829 (June 1979).

5. H. M. Domanus, M. J. Chen and W. T. Sha, "Computational Results for a 7-Pin Hexagonal Fuel Assembly during a Flow Rundown Transient using the COMMIX-1A Computer Code", NUREG/CR-1285, ANL-CT-80-10 (Jan 1980).

6. M. J. Chen, H. M. Domanus and W. T. Sha, "Simulation of a Thermohydraulic Transient in a Pipe using the COMMIX-1A Computer Code", NUREG/CR-1323, ANL-CT-80-15 (Feb 1980).

7. W. T. Sha, H. M. Domanus, R. C. Schmitt, J. J. Oras, E. I. H. Lin and V. L. Shah, "A New Approach for Rod-Bundle Thermal-Hydraulic Analysis", Proceedings of the International Meeting on Nuclear Power Reactor Safety, Brussels, Belgium (October 16-19, 1978).

8. G. L. Bordner, C. H. Gartside, I. T. Hwang, T. E. Kraft, D. H. Lennox, W. W. Marr, K. J. Miles, J. H. Tessier, D. H. Thompson and D. . Warinner, "Pretest Analysis SLSF in Reactor Experiment P2", ANL/RAS 76-31 (Oct 1976).

9. G. L. Bordner, T. T. Anderson, C. H. Gartside, I. T. Hwang, T. W. Kraft, T. M. Kuzay, B. D. LaMar, W. W. Marr, T. A. McDonald, M. Oats, J. H. Tessier, and D. H. Thompson, "Interim Report on SLSF in Pile Experiment P2: Part I, Experimental Results and Analyses", ANL/RAS 77-8 (Mar 1977).

Impact of Thermal Buoyancy in LMFBR Piping Systems

M. KHATIB-RAHBAR, I. K. MADNI, and A. K. AGRAWAL
Engineering & Advanced Reactor Safety Division
Department of Nuclear Energy
Brookhaven National Laboratory
Upton, New York 11973 USA

ABSTRACT

Mathematical models for transient, one dimensional, incompressible flow inside a pipe are presented; various model assumptions are examined and compared. It is shown that the turbulent mixing and wall heat transfer effects can significantly mitigate the transient. The impact of buoyancy effects on the thermal behavior are discussed by comparing the one-dimensional calculations to parallel three-dimensional predictions. Comparisons show surprisingly excellent agreements at low flow natural convection conditions suggesting that local thermal buoyancy effects although much greater than 10% (based on water data) can be neglected in large system dynamic studies. Further experimentation and analysis of these phenomena are essential.

1. INTRODUCTION

The coolant transport in the primary and intermediate piping network is one of the most important processes that must be accounted for since the longest time the coolant spends in its passage through the heat transport circuit is in the piping system. The long term transient behavior of an LMFBR system is strongly influenced by the fluid behavior in the sodium-carrying piping, especially at low flow natural convection conditions.

Most of the existing system simulation codes,[1,2,3] represent the thermal transport processes by a one-dimensional model. Inherent in this model is the assumption that the influence of radial temperature variations is insignificant. However, under extreme circumstances, distortions in the radial temperature profile may develop leading to local, density driven recirculation and redistribution. For example, cold sodium entering a warm pipe may tend to slide under the warm sodium rather than advance in the plug-like, perfectly-mixed manner as assumed by the one-dimensional model [4].

The influence of local thermal buoyancy and flow redistribution in a horizontal pipe section is examined by first comparing the one-dimensional results for a very severe flow and temperture transient with parallel three-dimensional calculations [5-8]. Finally, a semi-quantitative approach for estimating local free convection effects using the one-dimensional calculations is suggested, which seems to show a remarkable agreement with these effects as evidenced from the three-dimensional predictions.

2. MATHEMATICAL FORMULATION

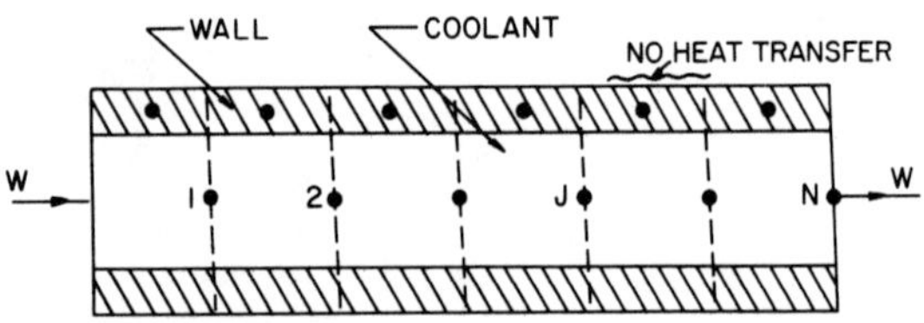

Figure 1. Pipe Section Staggered Mesh Heat Transfer Model

Consider the pipe section of Fig. 1; the coolant transport time, τ , along a subsection J is defined by:

$$\int_{t}^{t+\tau_J} \frac{W(t')}{\rho_J(t')}\, dt' = V_J \qquad (1)$$

where W is the coolant mass flow rate (kg/s), ρ the coolant density (kg/m^3), and V_J the coolant volume (m^3).

For an idealized, adiabatic plug-like flow, the transient temperature (enthalpy) at the outlet of the pipe section is equivalent to the transient temperature (enthalpy) at its inlet delayed by the enthalpy transport time τ. In other words,

$$T_J(t) = T_{J-1}\,(t + \tau_J) \qquad (2)$$

It should be noted, however, that the delay time τ_J can, and indeed does, vary during transients. In the limiting case of a steady state, τ_J is identically equal to $\rho_J V_J/W$. This model for propagation of disturbance is known as a time-delay model. System codes, such as DEMO [9] and BRENDA [10], have used this simplified method to calculate transient temperatures in the piping network.

The time-delay model is not a realistic one because it does not account for turbulent mixing of the coolant as well as heat storage in the pipe walls and it neglects possible energy losses to the environment.

In the more detailed model, the one-dimensional conservation equations based on the Boussinesq approximation [11] are written for individual subsections and solved to obtain the axial temperature and velocity distribution. That is:

Coolant Energy Equation:

$$\rho \frac{\partial H}{\partial t} + \rho u \frac{\partial H}{\partial x} = \frac{\partial q''}{\partial y} \tag{3}$$

Coolant Momentum Equation:

$$\rho \frac{\partial u}{\partial t} + \rho u \frac{\partial u}{\partial x} + \frac{\partial p}{\partial x} + g\rho \frac{\partial z}{\partial x} + \rho \frac{f}{di} \frac{u|u|}{2} = 0 \tag{4}$$

Pipe Wall Energy Equation:

$$\rho_w C_w \frac{\partial T_w}{\partial t} = \frac{\partial q''}{\partial y} \tag{5}$$

where H is the coolant enthalpy (J/kg), u is the coolant velocity (m/s), p is the pressure (N/m^2), z is the vertical elevation (m), x is the axial position (m), f is the friction factor, d_i is the pipe inner diameter (m), C is the heat capacity (J/K-kg), q" is the heat flux (W/m^2), and subscript w refers to the wall.

Defining the coolant mass flow rate W by:

$$W = \rho u A \tag{6}$$

where A is the coolant (pipe) cross sectional area (m^2), and volume averaging equations (3) through (5) for a subsection J, one obtains [11]:

$$\rho_J V_J \frac{dH_J}{dt} = W\,(H_{J-1} - H_J) + Q_J \tag{7}$$

$$\left(\frac{\Delta x}{A}\right)\frac{dW}{dt} = (P_{J-1} - P_J) + \left(\frac{1}{\rho_{J-1}} - \frac{1}{\rho_J}\right)\frac{W^2}{A^2} - \left[f\,\frac{\Delta x}{d_i} + K\right]\frac{W|W|}{2\bar{\rho}A^2} + g\bar{\rho}\,(Z_{J-1} - Z_J)$$

and,

$$\rho_W V_W C_W \frac{dT_w}{dt} = Q_J \tag{8}$$

The heat transfer rate Q_J is defined by:

$$Q_J = U_J A'_J \left[\left(\frac{T_{J-1} + T_J}{2}\right) - T_w\right] \tag{9}$$

where A'_J is heat transfer area (m^2), and U_J is the overall heat transfer coefficient (W/m^2-K) given as:

$$U_J = \left[\frac{1}{h} + \frac{d_i}{2K} \ln\,(1 + \delta/2d_i)\right]^{-1} \tag{11}$$

where h is the surface heat transfer coefficient (W/m^2-K), δ is the pipe wall thickness (m) and k is the thermal conductivity of the pipe wall material (W/m-K).

Equations (6) through (10) are solved numerically to obtain the thermodynamic state of the fluid during the transient [12].

In order to accurately predict the dynamic characteristic of the fluid, a detailed knowledge of axial temperature and density distributions is essential. At steady state, the pipe wall and the coolant are at thermal equilibrium (for a well insulated pipe); thus, the sodium temperature along the pipe is equal to the temperature at the pipe inlet. During a transient, the piping heat capacity influences the sodium temperatures significantly, as will be discussed in the next section.

3. ANALYSIS OF A SEVERE LOSS-OF-FORCED FLOW TRANSIENT

The expected mode of decay-heat removal in most current LMFBR designs is via forced circulation of liquid sodium through the main coolant pumps driven by auxiliary pony motors. However, in the event of a complete loss-of-forced cooling, the plants are designed to utilize buoyancy forces to provide free convection as redundant and diverse means of decay-heat removal.

Since the body forces are often negligible compared to typical frictional losses at full flow, the thermo-physical properties of liquid metals and design parameters of LMFBR systems cause coolant buoyancy and natural convection to be of some limited interest at full power conditions. However, at post-scram or under shutdown decay heat removal conditions thermal buoyancy effects become more significant and often dominant. The dynamic behavior of sodium coolant under these conditions is of utmost importance. The role of thermal buoyancy in maintaining natural circulation flow have been studied extensively [13, 14, 15]. Inherent in these studies is the assumption that the thermal fronts moving through the various components are isothermal in a plane normal to the mean velocity direction. More recently, thermal stratification and buoyancy effects have been studied on component and localized levels using extensive three dimensional methods [5-8].

In this section we will address the problem of a severe loss-of-forced flow transient in a horizontal pipe. The conditions at the outlet of the evaporators during a protected loss-of-flow event are the results of calculations made with the CRBRP system simulation code, DEMO [9,13]; flow and temperature decay are shown in Fig. 2.

Figure 3 compares the transient outlet temperature response, using the 1-D models discussed in the previous section. Excellent agreement is observed between the selected model [12] and the more detailed model

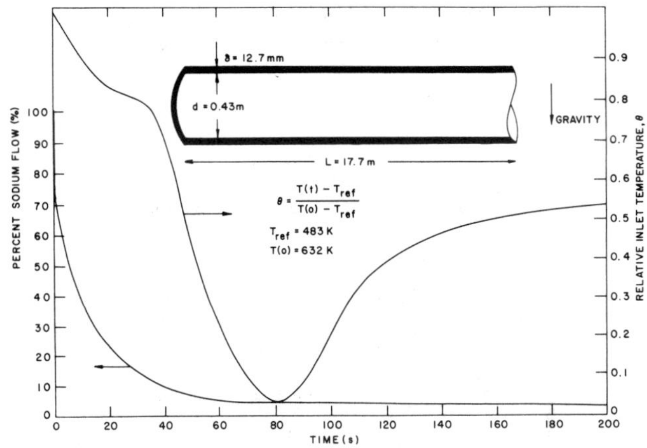

Figure 2. Transient Inlet Flow and Temperature Forcing Functions

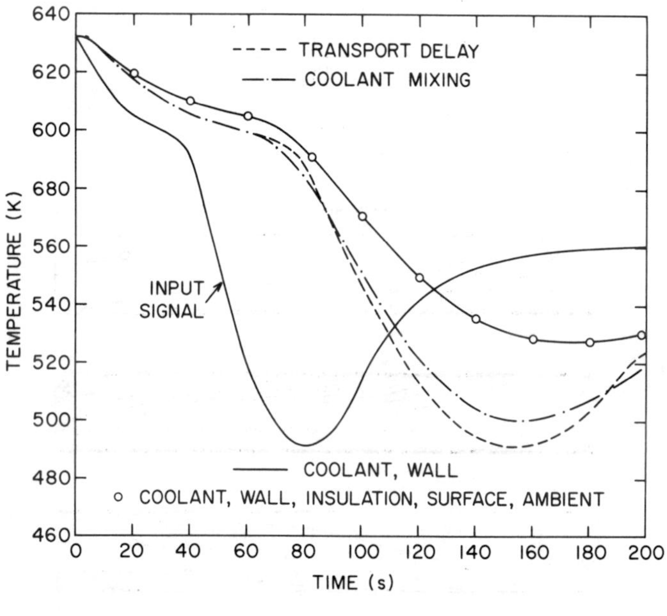

Figure 3. Pipe Section Outlet Temperature Response Using Various Thermal Transport Models

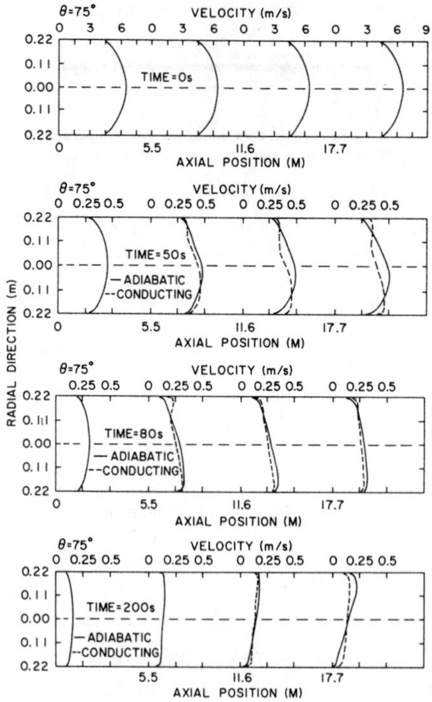

(a)

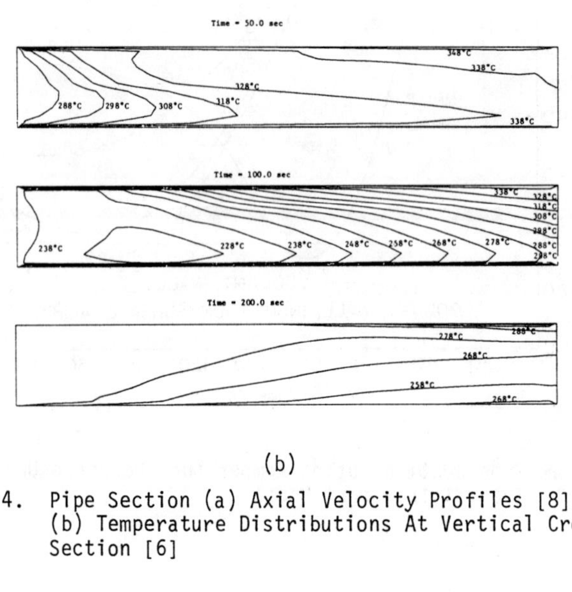

(b)

Figure 4. Pipe Section (a) Axial Velocity Profiles [8], and (b) Temperature Distributions At Vertical Cross Section [6] including heat losses through insulation; the maximum deviation is approximately 0.1K. In contrast, the results using the transport delay and mixing models are seen to deviate considerably, the maximum deviation is greater than 40K with the former and 30K with the latter.

The three-dimensional calculations [7,8] shown in Fig. 4 illustrate the axial velocity profile distortions resulting from a density gradient. The comparison between the adiabatic and non-adiabatic case indicates that the extent by which the velocity profile distortion can be reduced by wall conduction. Also shown in Fig. 4 are the temperature distribution at vertical cross section through center line as calculated by COMMIX-1A computer code [5,6].

During the ramp-down portion of the thermal transient (40 to 80 seconds of Fig. 2) there is a tendency for localized flow stratification to occur near the top of the pipe (with wall conduction). This localized velocity profile distortion is caused by the hot pipe wall warming the fluid to a higher temperature thereby lowering its density relative to the fluid flowing adjacently radially inward. This localized effect is seen to disappear as the temperature of the incoming sodium is increased (80 to 200 seconds of Fig 2). The impact of flow redistribution due to local buoyancy effects is seen to be more pronounced and prolonged near to the pipe outlet. This is due to the extended temperature down-ramp experienced near the outlet.

Figure 5 shows the temperature time history of the fluid at the top and bottom of the pipe. Generally, the heat capacity of the pipe wall slows the rate of temperature decrease at both the top and bottom of the pipe, and decreases the top to bottom temperature differences particularly at locations farther down the pipe, away from the inlet [8]. Comparisons between the 3-D calculations and the present 1-D predictions are seen to be quite favorable. Note that the velocity profile distortions reported by Trent [8] and the temperature distribution reported by Chen, et al [6] (Fig. 4) show a greater sodium flow near the bottom of the pipe. This would bring the mass-averaged fluid temperture closer to the bottom temperature. The agreements begin to improve as local stratification effects become less pronounced (at about 120 to 140 seconds).

4. IMPACT OF THERMAL BUOYANCY AND MULTI-DIMENSIONAL EFFECTS

In the previous section, we examined the dynamic behavior of the fluid by comparing the one- and three-dimensional calculations for a severe transient. It was shown that radial temperature profile distortions can produce local flow redistribution and hence alter the thermal response. On the other hand, it was shown that simple one-dimensional treatments are in generally good agreement with more extensive three-dimensional calculations.

Since computer simulation of the systems behavior should be quite reliable, it is important to be able to estimate the degree of multi-dimensionality, and impact of local free convection phenomena on the overall system.

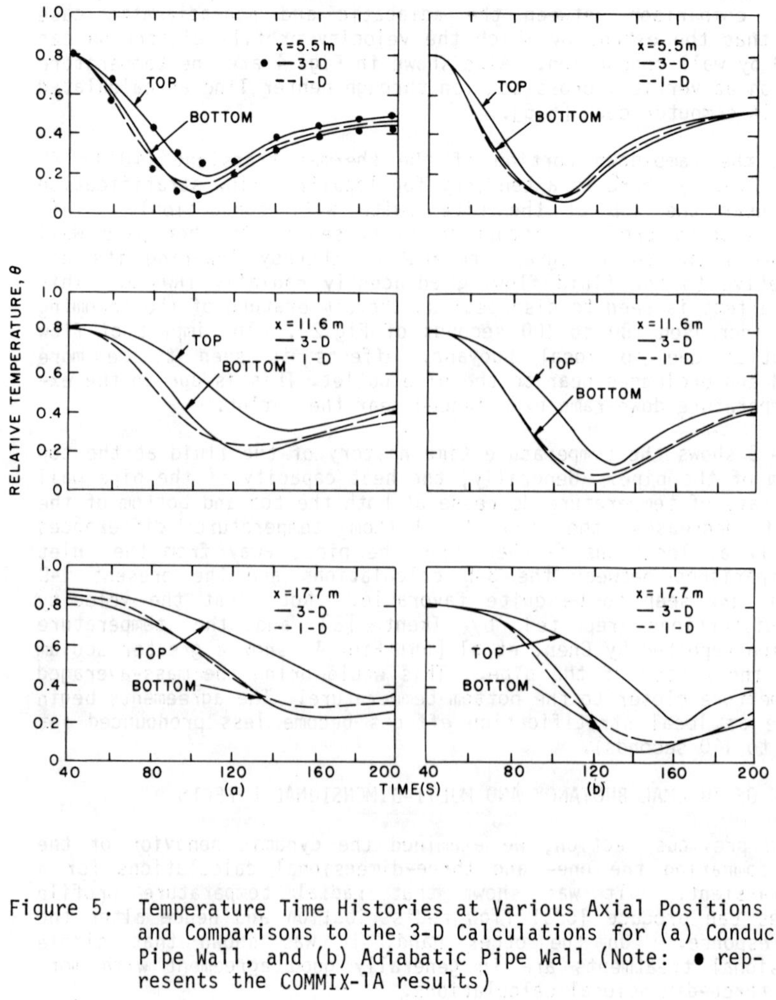

Figure 5. Temperature Time Histories at Various Axial Positions and Comparisons to the 3-D Calculations for (a) Conducting Pipe Wall, and (b) Adiabatic Pipe Wall (Note: ● represents the COMMIX-1A results)

An indication of the relative magnitude of the free-and forced-convection effects may be obtained from the magintudes of the flow transport numbers. This problem has been very extensively studied as reviewed by Gebhart [16], Kasza, et. al. [17] and others for a variety of flow and geometric conditions.

The transport numbers influencing the heat and momentum transport are the familiar Reynolds number, Re; Prandtl number, Pr; and Grashof number, Gr; or other combinations of these primary dimensionless parameters. Depending on the form of the dimensionless equations, the three independant parameters may appear in other forms. In each case the new dimensionless parameter may yield more physical insight than the other. For example the ratio of buoyancy forces to momentum flux may be represented by the Richardson number, Ri; that is:

$$Ri \equiv Gr/Re^2 \tag{12}$$

The impact of buoyancy effects on forced-convection dominated flows for laminar and turbulent regions, under the influence of gravity, has received considerable attention. For turbulent flow inside long tubes, Jackson and Fewster [18] derived an expression for the mixed convection region which is based on equating the total buoyancy force acting on a thin thermal sublayer to the net shear force. They obtained the following dimensionless parameter:

$$\gamma = Gr/Re^{2.625} \cdot Pr^{0.5} = Ri/Re^{0.625} \cdot Pr^{0.5} \tag{13}$$

The available experimental data, obtained from water experiments, for both horizontal and vertical pipes with specified wall temperature or heat flux boundary conditions, shows that the effect of buoyancy can be as high as 10% or more [17] if:

$$\gamma > 1.0 \times 10^{-4} \tag{14}$$

Using the one-dimensional results of the previous section, the value of γ can be estimated as a function of time after the start of the transient for various axial location along the pipe as shown in Fig. 6.

Figure 5 shows that the 1-D calculations start to deviate from the 3-D predictions at about the same time that γ (see Figure 6) exceeds 10^{-4}. The disagreement between the two models increases as parameter γ increases. On the other hand, as γ starts to decrease, substantial improvement in the agreement in observed.

This remarkable trend suggests that the dimensionless parameter, γ , along with the one-dimensional calculations, can be used to provide an indication of the effect of local thermal buoyancy phenomena on the fluid thermal behavior. Furthermore, the impact of free convection effects on the surface heat transfer coefficient needs to be considered as a mean of improving the accuracy of the 1-D predictions.

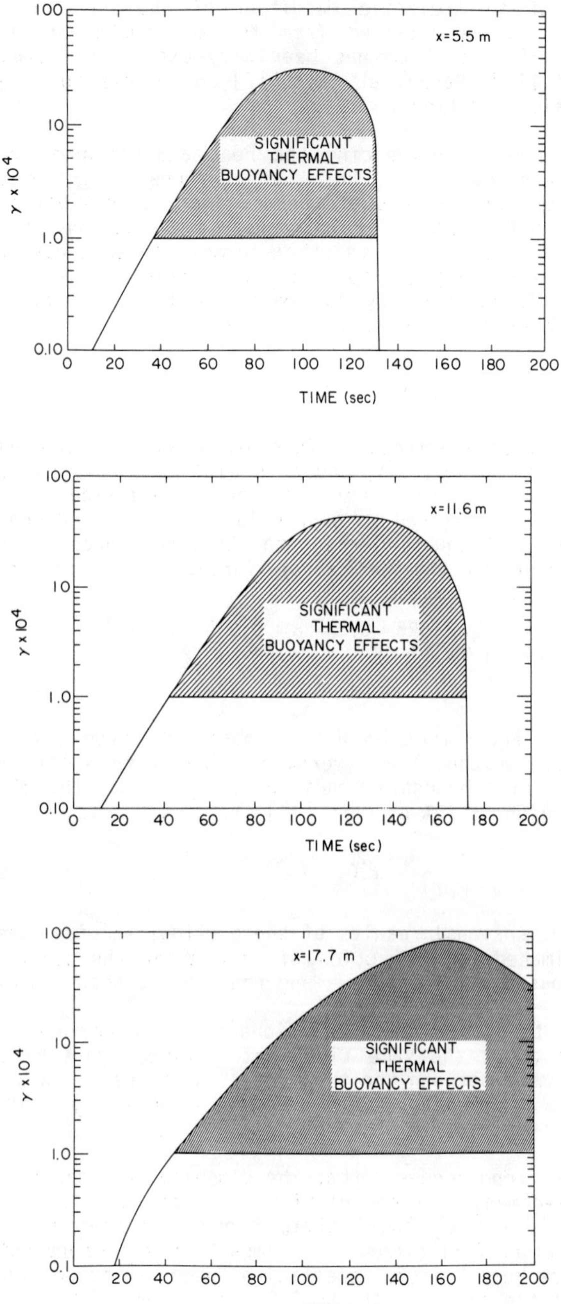

Figure 6. Period of Significant Thermal Buoyancy At Various Axial Positions

5. CONCLUSION

Mathematical models for transient one-dimensional, incompressible flow inside round pipes have been presented and various model assumptions discussed and examined. It is seen that the turbulent mixing and wall conduction have a considerable impact on the transient behavior. Furthermore, the selected results using the one-dimensional model were compared to parallel calculation using detailed three-dimensional models [5,7]. It was also demonstrated that the one-dimensional model along with an appropriate dimensionless parameter can be used to predict the region of significant thermal buoyancy, as predicted by the three-dimensional calculations. Finally it is seen that significant thermal buoyancy and multi-dimensional effects can be neglected in overall system simulation studies as long as local flow maldistribution phenomena are not present for an extended period.

ACKNOWLEDGEMENT

The authors would like to thank Drs. W. T. Sha of ANL and D. S. Trent of BNWL, for providing us with their unpublished results. Thanks are also expressed to Ms. P. Siemen for the excellent job done in preparing the manuscript.

This work was performed under the auspices of the United States Nuclear Regulatory Commission.

REFERENCES

1. Agrawal, A. K., "An Advanced Thermohydraulic Simulation Code For Transients in LMFBRs (SSC-L Code)", Brookhaven National Laboratory, BNL-NUREG-50773 (1978).

2. Cady, K. B., et al, "EPRI-CURL: Dynamic Analysis of Loop-Type LMFBRs", Electric Power Research Institute, Inc., EPRI-NP-1001 (1979).

3. Additon, S. L., McCall, T. B., and Wolfe, C. F., "Simulation of the Overall FFTF Plant Performance," Hanford Engineering Development Laboratory, HEDL-TC-556 (1976).

4. Meyer, J. E., "Some Physical and Numerical Considerations for SSC-S Code", Brookhaven National Laboratory, BNL-NUREG-50913 (1978).

5. Sha, W. T., et al, "COMMIX-1: A Three-Dimensional Transient Single-Phase Component Computer Program For Thermal-Hydraulic Analysis", Argonne National Laboratory, NUREG/CR-0415, ANL-77-96 (1977).

6. Chen, M. J., Domanus, H. M., and Sha, W. T, "Simulation of Thermal-Hydraulic Transient in a Pipe using COMMIX-1A Computer Code", NUREG/CR-1323, ANL-CT-80-15 (1980)

7. Trent, D. S., "TEMPEST: A Three-Dimensional Time-Dependent Computer Program For Hydrothermal Analysis," Battelle Pacific Northwest Laboratory, FATE-79-105, Informal Report (1979).

8. Trent, D. S., Battelle Pacific Northwest Laboratory, Private Communication, (1979).

9. "LMFBR Demonstration Plant Simulation Model, DEMO," Westinghouse Advanced Reactors Division, WARD-D-0005 (Rev. 4), (1976).

10. Hetrick, D. L., and Sowers, G. W., "BRENDA: A Dynamic Simulation for Sodium-Cooled Fast Reactor Power Plant", U.S. Nuclear Regulatory Commission, NUREG/CR-0244 (1978).

11. Agrawal, A. K., and Khatib-Rahbar, M., "Dynamic Simulation of LMFBR Systems", J. of Atomic Energy Review, (In Press).

12. Madni, I. K., "Transient Analysis of Coolant flow and Heat Transfer in LMFBR Piping Systems", Brookhaven National Laboratory Report (In Print).

13. Lowrie, R. R. and Severson, W. J. "A Preliminary Evaluation of the CRBRP Natural Circulation Decay Heat Removal Capabilities", Westinghouse Advanced Reactors Division, WARD-D-0132 (1976).

14. Khatib-Rahbar, M., Cady, K. B., "Establishment of Buoyancy-Induced Natural Circulation in Loop-Type LMFBRs", Trans. Am. Nucl. Soc., 27, 432 (1978).

15. Singer, R. M., et al, "Experimental and Analytical Studies of Natural Convection in EBR-II," in Spalding and Afgan, editors, Heat Transfer and Turbulent Buoyant Convection, Vol, II, 545, Hemisphere Publishing Co. (1977).

16. Gebhart, B., Heat Transfer, 2nd Edition, McGraw Hill Book Co.(1971).

17. Kasza, K. E., et al "Initial Considerations on the Influence of Thermal Buoyancy on Heat Exchanger Performance," Argonne National Laboratory, ANL-CT-78-47 (1978).

18. Jackson, J. D., and Fewster, J. "Enhancement of Turbulent Heat Transfer Due to Buoyancy for Downward Flow of Water in Vertical Tubes", in Spalding and Afgan, editors, Heat Transfer and Turbulent Buoyant Convection. Vol. II, 759, Hemisphere Publishing Co. (1977).

Theoretical Study and Experimental Investigation of Mixed and Natural Circulation in LMFBR Core Subassemblies

D. LETEINTURIER
DRNR/SEDC/CEN
Cadarache (13) France

D. BLANC
NOVATOME
Plessis Robinson (92) France

B. MENANT and G. BASQUE
DTCE/STT/CEN
Grenoble (38) France

ABSTRACT

A presentation is made of theoretical and experimental studies carried out in France on mixed and natural convection in LMFBR wire wrapped bundles. Two codes are described, one for mixed convection THERNAT and the other for natural convection BACCHUS. The related experimental program FETUNA, with electrically heated bundles in sodium loops, is also presented.

NOMENCLATURE

D, d	pin and wire diameters
H	helical wire pitch
A, A_f, A_i	total area, fluid area, subchannel area
V, V_f	total volume and fluid volume
R, Z	radial and axial coordinates
ΔR, ΔZ	radial and axial extents of the control volume
t	time
u, v, w	velocity components
p	pressure
$\vec{F}$	external force per unit volume
τ	component of viscous stress tensor
$\dot{m}_i$	mass flow rate in subchannel i
T	temperature
h	enthalpy per unit mass
Q_i	linear power generated by adjacent pins in subchannel i
q	heat flux
α_R, α_Z	radial and axial power form factors
ρ, a, β_{Th}	density, thermal diffusity and volume expansion coefficient of sodium
$\vec{i}$, $\vec{k}$	unit vector in radial and axial directions
$\vec{n}$	normal unit vector

1. INTRODUCTION

LMFBR core subassemblies may be, during their life, subject to cooling by mixed or natural convection. These situations occur either for nominal conditions (cooling of some blanket subassemblies, transport) or for abnormal ones (break-down of the cooling pumps, blockage of the subassembly inlet).

For the designer it is important to know if subassemblies in these cooling conditions, and particulary fuel and blanket bundles, fulfil the criteria imposed to the cladding and sodium temperatures.

To answer these questions, theoretical and experimental studies are underway in C.E.N Grenoble and C.E.N. Cadarache on wire wrapped bundle geometry.

In the theoretical studies, mixed and natural convection problems are tested in different ways. In mixed convection, a main axial flow and the preponderance of transverse conduction are assumed. With these hypotheses a simplified axial momentum equation is sufficient to calculate the mass flow rate distribution in a subchannel model and to deduce the temperature by thermal balance. This is the purpose of the code THERNAT. In natural convection, axial and transverse momentum equation are necessary, with terms including second derivative in Z coordinate to treat reverse flow. The simplification is introduced in the meshing (annular layers), which limits the number of unknow values, and by the use of a porous model approach : this is the purpose of the code BACCHUS. To support this effort on the calculational tools, experiments are underway with electrically heating bundles in sodium loops (FETUNA program). A preliminary experiment with a seven pins bundle is now achieved and a second experiment of a ninety-one pin bundle is underway, with the purpose of studying both mixed and natural convection.

2. BACCHUS CODE : DESCRIPTION AND RESULTS

2.1 Physical Model

The basis of the analysis upon which the BACCHUS code has been developped is the porous body model . This is an approach which is different from the well known subchannel analysis because the control volumes in which the basic conservation laws are written and averaged can be in principle arbitrary geometrical volumes containing one or more coolant subchannels and including pins and helical wires.

The choice of the control volume results from a compromise between the ability of the code to describe flows in large size bundles and the actual computing possibilities. The basic control volume is an annular layer bounded by two parallel horizontal planes as shown in Figure 1. It results from this choice that the problem is two dimensionnal.

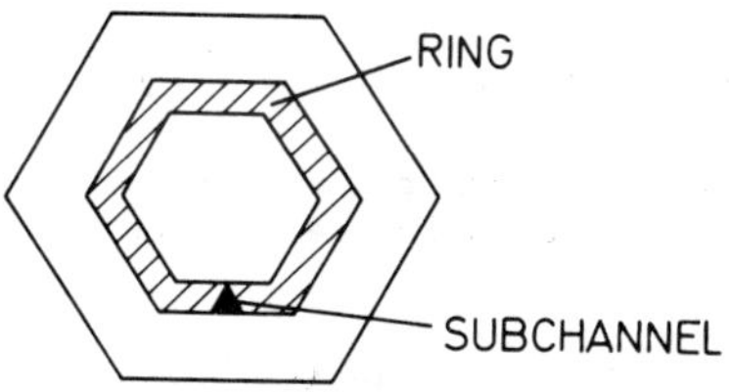

Fig. 1 - Control volume for the description of a subassembly

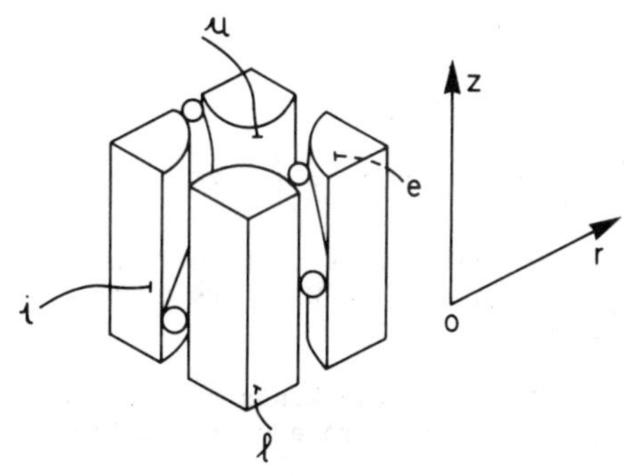

Fig. 2 - Notations related to the control volume

2.2 Mathematical Formulation

The equations that are numerically solved are obtained from the local balance equations for mass, momentum and enthalpy on the control volume. It can be noticed that three different control volumes are used, related to the concept of staggered meshs, classicaly used in fluid flow problems. So the axial and radial momentum equations are averaged over control volumes deduced from the main control volumes V by axial and radial translations of $\Delta Z/2$ and $\Delta R/2$ respectively.

The final form of the equations are :

Mass balance :

$$\varepsilon \frac{\partial}{\partial t} \langle \rho \rangle + \frac{\varepsilon}{\Delta Z} \left(\langle \rho w \rangle_u - \langle \rho w \rangle_l \right) + \frac{1}{\Delta R} \psi_e \left(\langle \rho u \rangle_e - \psi_i \langle \rho u \rangle_i \right) = 0 \qquad (1)$$

Axial momentum balance :

$$\varepsilon \frac{\partial}{\partial t} \langle \rho w \rangle + \frac{\varepsilon}{\Delta Z} \left(\langle \rho w^2 \rangle_u \quad \langle \rho w^2 \rangle_l \right) + \frac{1}{\Delta R} \left(\psi_e \langle \rho u w \rangle_e - \psi_i \langle \rho u w \rangle_i \right)$$

$$= - \frac{\varepsilon}{\Delta Z} \left(\langle p \rangle_u - \langle p \rangle_l \right) + \frac{A W}{V} \langle \vec{F} \cdot \vec{h} \rangle_W$$

$$+ \frac{\varepsilon}{\Delta Z} \left(\langle \tau_{zz} \rangle_u - \langle \tau_{zz} \rangle_l \right) + \frac{1}{\Delta R} \left(\psi_e \langle \tau_{RZ} \rangle_e - \psi_i \langle \tau_{RZ} \rangle_i \right) \qquad (2)$$

Transverse nomentum balance :

$$\varepsilon \frac{\partial}{\partial t} \langle \rho u \rangle + \frac{\varepsilon}{\Delta Z} \left(\langle \rho u w \rangle_u - \langle \rho u w \rangle_l \right) + \frac{1}{\Delta R} \left(\psi_e \langle \rho u^2 \rangle_e - \psi_i \langle \rho u^2 \rangle_i \right)$$

$$= - \frac{1}{\Delta R} \left(\psi_e \langle p \rangle_e - \psi_i \langle p \rangle_i \right) - \frac{AW}{V} \langle p \, \vec{n} . \vec{i} \rangle_W + \frac{AW}{V} \langle \vec{F} \cdot \vec{i} \rangle_W$$

$$+ \frac{\varepsilon}{\Delta Z} \left(\langle \tau_{RZ} \rangle_u - \langle \tau_{RZ} \rangle_l \right) + \frac{1}{\Delta R} \left(\psi_e \langle \tau_{RR} \rangle_e - \psi_i \langle \tau_{RR} \rangle_i \right) \qquad (3)$$

Enthalpy balance :

$$\varepsilon \frac{\partial}{\partial t} < \rho h > + \frac{\varepsilon}{\Delta Z} \left(< \rho hw >_u - < \rho hw >_1 \right) + \frac{1}{\Delta R} \left(\psi_e < \rho uh >_1 - \psi_i < \rho uh >_i \right)$$

$$= \frac{dp}{dt} + \frac{AW}{V} < \vec{q}.\vec{n} >_W - \frac{\varepsilon}{\Delta Z} \left(< qz >_u - < qz >_1 \right) - \frac{1}{\Delta R} \left(\psi_e < qR >_e - \psi_i < qR >_i \right) \quad (4)$$

The subscripts i, e, u, l refer to the sides of the control volumes as shown in Figure 2 and W refers to the fluid /solid interfaces inside the control volume.

There are two kinds of averages. Given a function g :$< g >$ (without subscript) is an average over the fluid volume V_f :

$$< g > \equiv \frac{1}{V_f} \int_{V_f} g \, d V$$

$< g >_u$ is an average over the fluid area at the side u of the control volume :

$$< g >_u = \frac{1}{A_{fu}} \int_{A_{fu}} g \, d A_u$$

The volume porosity ε is defined by $\varepsilon \equiv \frac{Vf}{V}$ and the surface porosities by $\psi_i = \frac{Afi}{A_i}$ and $\psi_e = \frac{Afe}{A_e}$.

The way of obtaining the resulting equations has the great advantage to take into account the anisotropic character of the porosity in a subassembly by including in a single step averaging and space discretization. This is different from the method used by MIAO and THEOFANOUS (Ref. [1]) who have two steps instead of one to derive the final equations,

For closure of the system one needs :

. An assumption on the distribution of variables inside the control volume :

$$< fg > = k < f > < g >$$

Generally a uniform distribution is assumed (k=1) :

. An assumption on the form of terms appearing at the right hand side of the equations, namely :

- Exchanges with surrounding fluid expressed by an eddy diffusivity.
- Exchange at fluid-solide interfaces : the wall shear stress and the heat flux.

The system of equations (1) to (4) is solved as an initial value problem in time and boundary value problem in space. The finite difference scheme is based upon a variant of the ICE method (Réf. [2]) with a high degree of implicitness.

The transient character of the code can be used to calculate a steady-state flow, which is reached asymptotically after a certain numbers of time steps.

This feature has been used to compute situations of natural convection that will occur in the FETUNA-GR91 experiment, i-e a 91-pin bundle with total

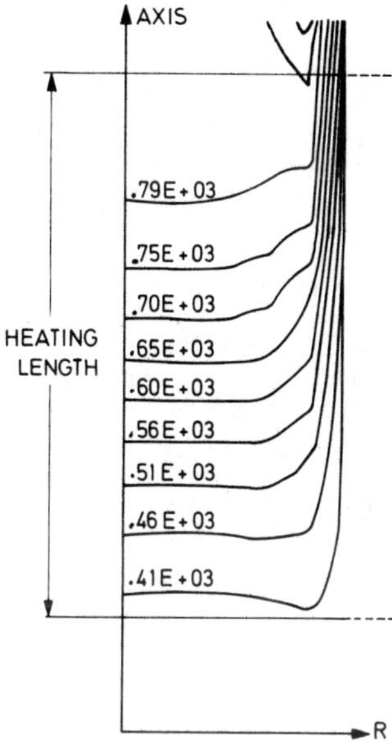

Fig.3 - isothermal lines

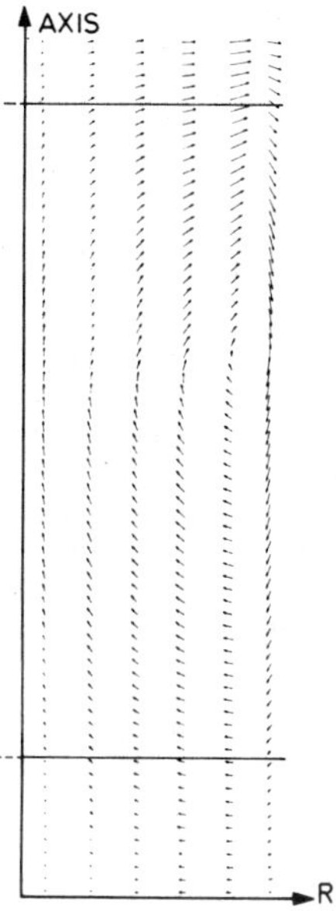

Fig.4 - velocity plot

inlet blockage.

The conditions of computation are summarized below :

pitch of the pin array	=	1.685	cm
pin diameter	=	1.58	cm
wire diameter	=	0.095	cm
heating length	=	160	cm
downstream unheated length	=	34.4	cm
uptream unheated length	=	17	cm
total input power	=	250	kw
number of radial meshes	=	6	
mean axial mesh size	=	3	cm

The inlet and outlet boundary conditions are those of a rigid free slip wall with an inlet prescribed temperature and zero outlet heat flux.

The hexagonal duct is considered as a rigid free slip wall with a heat flux specified as $\Phi (z) = h (T - Te (z))$ where $Te(z)$ is an imposed external temperature.

The results are shown on figures 3 and 4 as a plot of the velocity vector field and isothermal lines.

Predictions of this type will be compared with the FETUNA-GR91 experimental results.

3. THERNAT CODE

3.1 Description of the Model

THERNAT solves a very simplified model of the Navier-Stokes equations in steady-state situations, with a small number of axial levels. At each axial level, the radial mesh consists of three kinds of subchannels (triangular, rectangular, trapezoïdal), the cross sections of which are independent of z coordinate.

The basic hypothesis of the model is to suppose the different control volumes of thickness Δz as independent of one another both hydraulically and thermally. The transverse velocities are neglected with regard to axial velocities inside the control volumes. This hypothesis is justified by the fact that, in the problems we are interested in, heat transfers between subchannels by conduction are preponderant with regard to radial convection heat transfers.

In these conditions, the equations in a control volume of the bundle can be written.

Mass balance :

$$\frac{\partial w}{\partial z} = 0 \qquad \text{w is independent of z coordinate.}$$

$$\rho \sum_i \int_{A_i} w_i \, dA = \rho \sum A_i \, w_i = \dot{M}_b \qquad (5)$$

Momentum balance :

$$u = v = 0 \qquad \frac{\partial w}{\partial z} = 0$$

$$\frac{1}{\rho} \frac{\partial p}{\partial x} = 0$$

$$\frac{1}{\rho} \frac{\partial p}{\partial y} = 0$$

$$\frac{1}{\rho} \frac{\partial p}{\partial z} = g + \nu \left(\frac{\partial^2 w}{\partial x^2} + \frac{\partial^2 w}{\partial y^2} \right) \qquad (6)$$

These momentum equations in x and y lead to a uniform pressure at any axial level of the bundle :

$$p\,(x, y, z) = p(z) \qquad (7)$$

The integration of equation (6) on the control volume (subchannel between axial levels z and z + Δz) gives :

$$p\,(z + \Delta z) - p(z) = \rho g \ \Delta z + \tau_W \cdot \Delta z \qquad (8)$$

τ_W is the axial gradient of wall shear stresses. τ_W can be written as

$$\tau_W = f \, \frac{1}{2} \, \rho \, w^2 \, \frac{1}{D_H}$$

With $\rho = \rho\,(T_o)\,[\,1 - \beta_{Th}\,(T - T_o)\,]$, equation (8) allows to calculate at each level the velocity distribution in the bundle.

Enthalpy balance :

$$\dot{m}_i^j \left(h_i^{j+1} - h_i^j\right) = Q_i^j \, \Delta z - \sum_{a=1}^{3} (\alpha\, a + K)\rho \, \Delta z \left(\frac{h_i^{j+1} + h_i^j}{2} - \frac{h_a^{j+1} + h_a^j}{2} \right) \quad (9)$$

The two coefficients characterizing the heat transfer are : α adimensionned coefficient taking into account the contact widths and heat transfer lengths by conduction between subchannels.

K : eddy diffusivity for the heat transfert due to the wire effect

$$K = 0{,}604 \; . \; d \; . \; (2D - d) \, \frac{W}{H}$$

The linear system corresponding to equation (9) is solved by the GAUSS method.

3.2 Comparison with Experimental Results

THERNAT calculations have been compared with three experimental tests made on 7,19 and 61-pin bundles.

The 7-pin test is part of the FETUNA-GR7 test series.

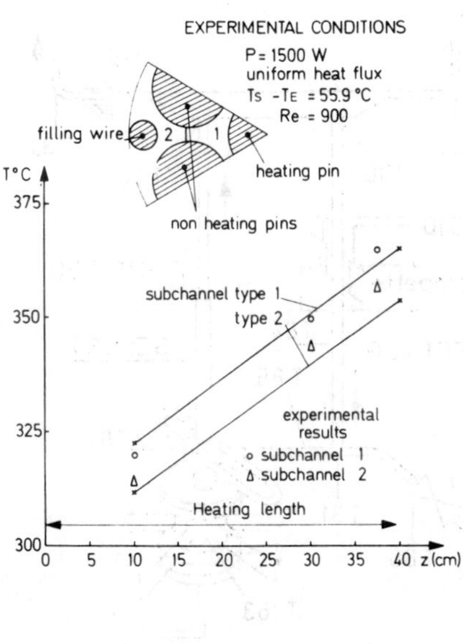

Fig.5 - 7 pin test

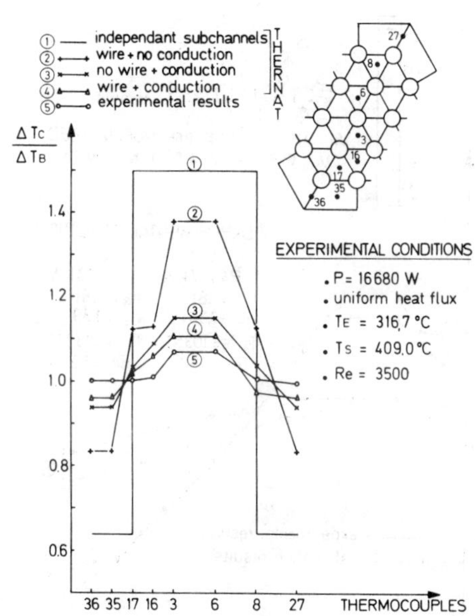

Fig.6 - 19 pin test

Figure 5 shows the axial temperature profile in inner and outer subchannels with only the central heating rod. That permits to emphasize heat exchange between hot inner subchannels and cold outer subchannels. There is a small exchange of mass flow between hot and cold subchannels. On the other hand conduction has an important role and the difference between hot and cold subchannels is only 8°C for an average bundle temperature rise fo 60°C.

For the 19 pin test, experimental results are taken from reference [3] . A comparison with the measured radial temperature profile without radial power skew has been carried out. Calculations show the relative importance of the effects of wires, of conduction, of mixed convection with a Reynolds number of 3500 (Figure 6).

For the 61-pin test, experimental results are extracted from reference [4] . Two tests, with high and low Reynolds numbers (Re = 490 and Re = 9200) and nearly the same temperature rise are considered (Figure 7).

It can be seen a good agreement between calculation and experiments, and a high temperature flattening brought by both conduction and mixed convection when Reynolds number decreases.

4. FETUNA-GR7 EXPERIMENT

The FETUNA-GR7 experiment was designed in order to get (from the experimenter's point of view) a first idea of the basic phenomena of natural and mixed convection in heated bundles simulating LMFBR core elements.

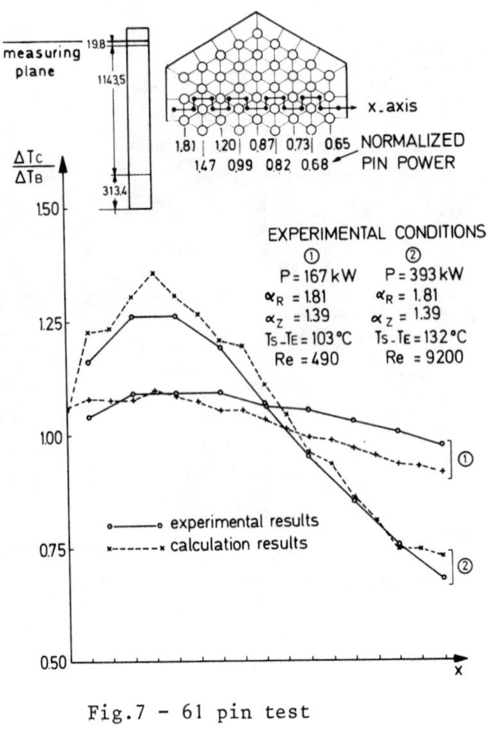

Fig.7 - 61 pin test

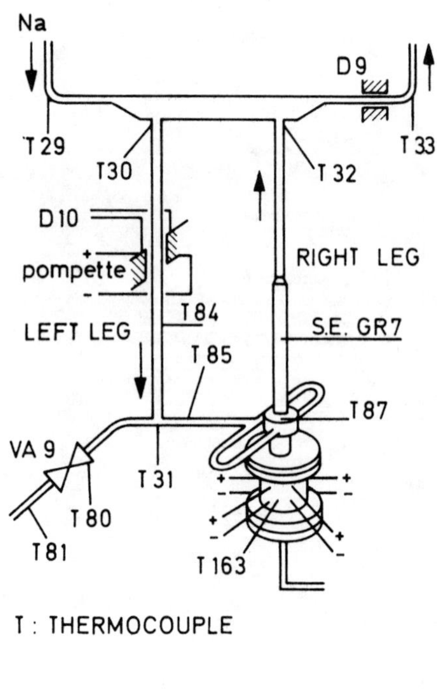

Fig.8 - The test thermosyphon

Due to the effect of radial heat conduction in a seven pin bundle it was clear that this experiment would not be fully representative of the mixed convection which should appear inside a larger subassembly.

Therefore the main aims of this experiment were to :

- Study the establishment and stability of sodium natural convection in a thermosyphon including a heating pin-bundle at the lower part of its hot-leg.

- Evaluate the difficulties (such as flow measurement and heat losses-control) encountered when performing experiments at very low flow and power conditions.

4.1 The FETUNA-GR7 Test Rig

The FETUNA-GR7 test rig is shown in figure 8. It consists of a thermosyphon fed by an upper plenum.

The boundary conditions in the plenum are obtained by the forced flow D9. This flow, and the inlet temperature T29 are stabilised by the CFNa loop. The mean flow velocity at the inlet and outlet of the thermosyphon is generally of about 0.1 m s^{-1} so that the effects of dynamic pressure on the flow through the thermosyphon are negligible.

The thermosyphon itself is mainly constituted of a stainless steel tube of 0.016 m inner diameter. The total height of both cold and hot leg is 1 m. A micropump is included in the left leg in order to improve or contrary the natural convection when it is thought necessary. The heating 7-pin bundle is located at the bottom of the right leg ; its main characteristics are :

pin diameter	D = 0.66 cm
helical spacer wire	d = 0.112 cm
helical pitch	H = 15 cm
heated length	40 cm
downstream unheated length	30 cm
circular duck	2.43 cm
six filling wires	.255 cm

each pin is heated by electrical power from 0. to 2000 W. Heat losses are controlled by accute setting of the electrical heating all along the thermosyphon. This setting is controled on temperature measurements for sodium flows D10 as low as possible.

Besides the mixed-convection experiments performed with a net mass flow through the thermosyphon, some natural-convection experiments were performed after having frozen the sodium at the inlet of the rod bundle.

4.2 Establishment and Stability of Natural Convection in Thermosyphon

Figure 9 is a good illustration of the experimental findings on the establishment and stability of natural convection in the thermosyphon.

The initial conditions are near isothermal natural convection. At t $\simeq$ 2 minutes the power dissipated in the central pin of the bundle is set from 0. to 500 W. A few damped oscillations (see T 32 Figure 9) are observed before establishment of stable final conditions after 10 minutes. Fluctuations of the flow D10 are observed during all the test. It is thought that they represent the hydraulic response of the thermosyphon to eddies flowing accross the upper plenum. It can be seen that these fluctuations have no influence on the mean flow value which is mainly governed by the buoyancy effects due to the temperature differences between hot and cold legs.

It is concluded in accordance with theoretical predictions, that the natural convection in thermosyphons of this kind is fully stable when established.

4.3 Recommandations for Further Experiments on Thermohydraulics in Rod Bundles at very low Flow and Power.

Mixed convection :

The FETUNA-GR7 test rig was designed in view of an evaluation of very small modifications of pressure drops due to mixed convection through the rod-bundle. It was hoped to do this by analyzing the variation of the intensity through the calibrated direct current micro electromagnetic pump "pompette".

The variations of frictional pressure drop due to various power distributions were however so low that this method did not give significant results. We now think that it is hopeless to try to evaluate so low pressure drop variations (some Pascal) : consequently further experiments on low power removal by mixed convection from heating pin bundles will be performed in forced convection loops with setting an accurate measurement of the sodium low flow.

Natural convection :

Heat removal from a subassembly completely blocked at the inlet is only possible radialy through the wrapper duct, or axially to the upper plenum. Some experiments performed with the FETUNA-GR7 were very difficult to analyse : indeed the boundary conditions (i-e repartition of radial and axial heat removal) were not clearly known

It is thought now that is necessary to impose clearly the boundary conditions around the bundle cooled by natural convection : for further experiments, an external sodium forced flow will impose the temperature on the outer side of the hexagonal duct and the distribution of heat flux through the duct will be measured at many locations.

These recomandations were taken into account for the FETUNA-GR91 experiment now in preparation.

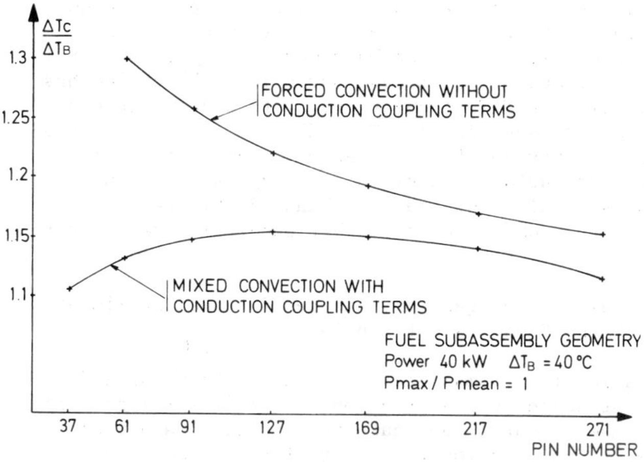

Fig.10 - Temperature form factor and pin number

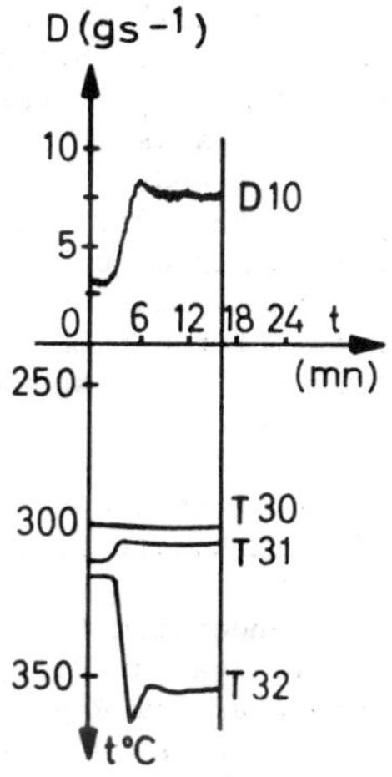

Fig.9 - Response of thermosyphon to a power step.

5. FETUNA-GR91 EXPERIMENT

During the definition of the experiment two purposes, not necessarly consistent, had to be satisfied :

- to have an experiment as pure as possible in order to permit precise comparison with calculations and code qualification ;
- to approach the actual conditions in a reactor.

Therefore, before detailing the instrumentation and the test program the definition choices will be explicited with respect to these two purposes.

5.1 The Definition Choices

The main choices to make are geometrical. Indeed, thermal hydraulic choices are imposed : the heat flux produced by the pins must be close to those existing in pile and the mass flow rate in the bundle must permit to cover the three studied flow regimes, i.e natural, mixed and forced convection. However (as seen in section 4.2) particular care has to be taken to the boundary conditions imposed on the outside wall of the wrapper tube.

Concerning the geometry of the bundle, three main parameters must be defined : the bundle size, its H/D and P/D ratios.

In natural convection, the bundle size is fixed directly by the need to know, for those blanket subassemblies which are not instrumented by an outlet thermocouple, the maximum power which can be removed, in case of total blockage, towards adjacent subassemblies. In mixed convection, the choice has been established, from parametric THERNAT calculations, with a comparison of the temperature form factor (ΔT hot subchannel/ $\overline{\Delta T}$ bundle) in mixed and forced convection. In a fuel bundle geometry and for a given mean temperature increase, the variation of the form factor remains large enough to be seen experimentally for a number of pins equal to or smaller than 127 (Figure 10). In addition table (1) shows that the geometry of a blanket bundle is even better in this sense.

For these different reasons, the geometry choosen for the experiment is a 91-pin bundle close to SUPERPHENIX blanket bundle.

Concerning the H/D ratio, it is smaller for a blanket than for a fuel bundle (12 against 21). Its influence on the friction factor is still not very well known at low Reynolds numbers, but in any case the choice of a blanket bundle is a conservative one, since friction forces are larger than in a fuel bundle and thus limit natural convection.

GEOMETRY OF PIN	HEATING LENGTH (m)	POWER (KW)	ΔT_B (°C)	$\Delta T_c/\Delta T_B$ MIXED CONVECTION	$\Delta T_c/\Delta T_B$ FORCED CONVECTION
Fuel	1.	13.4	70.	1.068	1.26
Blanket	1.6	13.4	70.	1.049	1.296
Fuel	1.	43.6	160.	1.07	1.26
Blanket	1.6	43.6	160.	1.041	1.296

Table 1 : Temperature form factor in mixed and forced convection for fuel and blanket pins. Bundle size = 91 pins.

5.2 Main Characteristics of the Experiment

A scheme of the test section is shown in figure 11.

Geometrical characteristics :

They are close to those of a SUPERPHENIX blanket subassembly, i.e =

H/D = 12
P/D = 1.066
Heating length = 150 cm

Thermal hydraulic characteristics :

Power range : 0. to 240. KW
Axial power distribution = cosine profile
Radial power skew = $1. < P_{max}/P_{mean} < 2$.
Mass flow rate range = 110 gs^{-1} to 2000 gs^{-1}
Inlet blockage possible.

Instrumentation :

Ninety-one thermocouples are distributed inside the bundle in the pin wires. They are located preferentially at the top of the heating length and at mid-lenght. Around the central pins, thermocouples are uniformly distributed along the z axis.

On both sides of the wrapper tube and on the external tube of the test section, one hundred and forty thermocouples are located in relation with the instrumented levels inside the bundle. These thermocouples serve to measure temperatures in edge subchannels and to control the boundary conditions.

5.3 Range of the Tests

Two series of tests are planed, one in mixed convection and the other in natural convection, with the objective of qualifying computational models.

In mixed convection, the main purpose is the determination of the variation of the maximal temperature form factor from forced to mixed convection. This evolution will be measured for several mean temperature increases in the bundle and several power skews.

In natural convection the main purpose is the determination of temperature distributions and the maximum admisible power without sodium boiling. Several power levels up to 240 KW will be used with an imposed external boundary mass-flow rate. The influence of the boundary conditions will be also studied by varying this mass-flow. Tests will be repeated with several power skews.

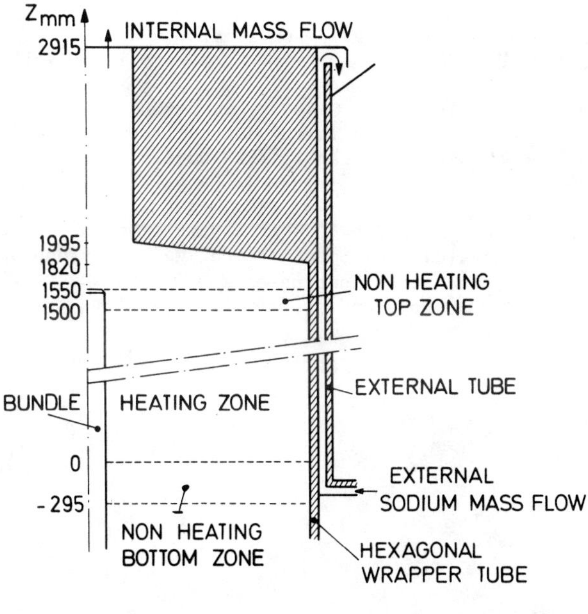

Fig. 11 - Scheme of FETUNA - 91 test section

5. CONCLUSION

The work described here will provide a better understanding of the temperature distribution inside a wire-wrapped bundle subassembly for mixed and natural convection conditions. However the situation is different in the two cases.

In mixed convection the present program seems sufficient to resolve all the situations occuring in a reactor core. The first comparison on 7,19 and 61-pin bundles of the simplified THERNAT model with experimental results shows already a good agreement. A further confirmation will be obtained with the FETUNA-GR91 experiment where core conditions are simulated. In natural convection the situation is more complex. With FETUNA-GR91 and BACCHUS calculations, a direct answer can be obtained on temperatures and admissible power in a blocked blanket subassembly. However this comparison may not be sufficient to permit confidently an extension to other geometries. In particular, an important parameter, the axial and transverse friction factor, and its sensitivity to the geometrical parameters of the bundle are not very well known at low Reynolds numbers. Therefore a good agreement between BACCHUS calculations and FETUNA results for a given friction factor may not allow to predict correctly temperature distribution in a fuel bundle. In this sense, it seems presently that experiments with various bundle geometries would be necessary.

REFERENCES

1. MIAO C., and THEOFANOUS T.G., 1978, Intrasubassembly coolant voiding incoherencies - 8th LMBWG meeting - MOL Belgium.

2. HARLOW F.H. and AMSDEN A.A. 1972. Numerical fluid dynamics calculations for all flow speeds. Journal of computational physics - Vol. 8 p 157.

3. FONTANA M.H., MAC PHERSON R.E, GNADT P.A., PARSLY L.F., WANTLAND J.L. 1974. Temperature distribution in the duct wall and at the exit of a 19-rod simulated LMFBR fuel assembly (FFM bundle 2A). Nuclear Technology Vol. 24.

4. MARKLEY R.A., ENGEL F.C. 1979. Blanket assembly heat transfer and hydraulic test data evaluation. International Working group on fast reactor. I.A.E.A Specialists'Meeting on Thermodynamics of FBR fuel subassemblies under nominal and non nominal operating conditions. KARSLRUHE.

Subchannel Friction Factors for Bare Rod Arrays under Mixed Convection Conditions

SONG-FENG WANG and NEIL E. TODREAS
Nuclear Engineering Department
Massachusetts Institute of Technology
Cambridge, Massachusetts 02139

WARREN M. ROHSENOW
Mechanical Engineering Department
Massachusetts Institute of Technology
Cambridge, Massachusetts 02139

ABSTRACT

The thermal/hydraulic analysis of wire-wrapped rod assemblies under mixed convection conditions is essential to the safety design of liquid metal fast breeder reactors (LMFBRs). In this paper, analytical solutions of the local velocity distribution in subchannels are given for steady, laminar, fully-developed mixed convection flows in finite bare rod arrays. The resultant subchannel friction factors, evaluated from local velocity distributions, are correlated explicitly in simple polynomials. The applicable ranges of these correlations are carefully defined to preclude their use after the transition from laminar to turbulent flow at low Reynolds numbers. The inherent limitation of applying the lumped parameter approach to model a rod assembly is discussed. A conservative criterion is proposed by inspecting the local velocity distributions. The application of the results to typical LMFBR core designs is included.

NOMENCLATURE

A_f	Flow area
D	Rod diameter
D_h	Hydraulic diameter
f	Friction factor
Fr	Froude number $[\equiv U^2/gD_h]$
Gr	Grashof number $[\equiv g\beta\, \rho_o{}^2(T-T_o)D_h^3/\mu_o^2]$
P	Rod pitch
p	Pressure
Pr	Prandtl number $[\equiv \mu_o\, c_{p_o}/k_{w_o}]$
q'	Linear power of fuel rod
R	Dimensionless radial distance $[\equiv r/D_h]$
Ra	Rayleigh number based on D_h $[\equiv g\,\beta\,\rho_o{}^2\, c_{p_o}\, D_h{}^4 \frac{dT}{dz}/k_{w_o}\mu_o]$
Ra_D	Rayleigh number based on D $[\equiv Ra(D/D_h)^4]$
Re	Reynolds number $[\equiv \rho_o\, U\, D_h/\mu_o]$
T	Coolant temperature field

T_c Clad temperature field
U Coolant average velocity
u Coolant local velocity
β Thermal expansion coefficient

Subscript

B Buoyancy force
b Bare rod
w Wire-wrap
o Reference value

Superscript

c Critical
o Isothermal
* Dimensionless quantity

1. INTRODUCTION

Conventional thermal hydraulic analysis of liquid metal fast breeder reactor (LMFBR) fuel assemblies has been mostly concentrated on turbulent forced convection for the past two decades. However, there are occasions where the buoyancy force due to heat addition is appreciable and its effect is superimposed on the forced convection flow. Hence the heat transfer mechanism becomes a combination of natural and forced convection, or mixed convection, as the buoyancy force becomes important. Natural circulation transients and decay heat removal transients are the typical examples in LMFBRs.

During natural circulation transients (also true for any other low flow situation), significant flow redistribution may occur within subchannel regions or between subchannels in an assembly, due to the density changes caused by thermal gradients. In addition, the intra-assembly heat conduction, mainly caused by radial power gradients, becomes important as the flow rate decreases. Both these two effects are found to significantly reduce the maximum clad temperature and transverse temperature gradient in the assembly as found in the WARD [1] 61-pin and HEDL [2] 217-pin tests.

To model the buoyancy induced flow redistribution and intra-assembly heat conduction in an assembly, two analytical approaches are generally being considered: (1) field solution programs using the distributed parameter approach and (2) subchannel programs using the lumped parameter approach.

The large core storage and long computational time required for field solution programs make them less attractive for large assembly calculations. Therefore, subchannel programs become the only viable tool for most applications. The subchannel programs contain several input parameters whose values must be provided analytically or experimentally. They are: subchannel friction factors, heat transfer coefficients and shape factors for conduction heat transfer between subchannels. The shape factor is used to correct the assumed temperature gradient (the difference between subchannel bulk temperatures divided by the distance between subchannel centroids) so that it is equal to the actual temperature gradient at the common boundary between the two subchannels.[3] To date, due to the geometric complexity of the rod assembly and the wide variety of radial power distributions, none of these parameters appear to exist for general cases. A few analytical studies have been done for a much simpler case: an assembly with uniform radial power distribution. In this case, only a characteristic symmetry segment of the subchannel, known as a cell, need be considered. Figure 1 shows the internal, edge and corner cells for both triangular and square rod arrays. Notice that part of an internal subchannel is included in both edge and corner

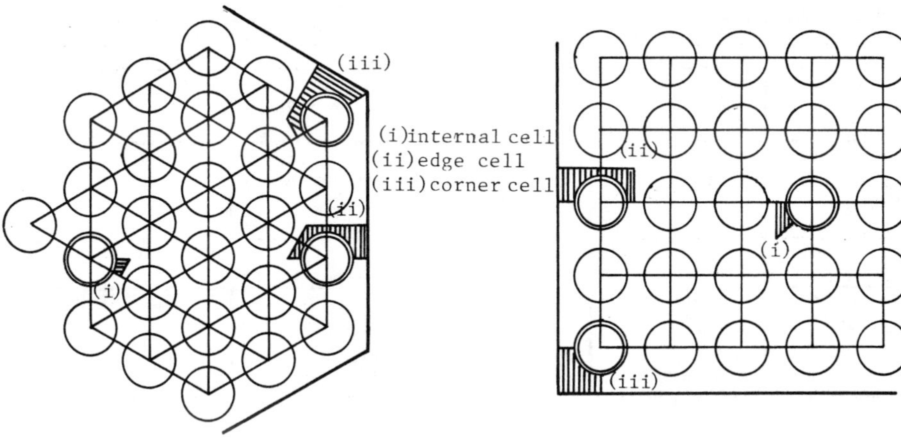

Fig. 1. Characteristic symmetry segment for rod arrays

cells. Iqbal et al. [4] first studied the buoyancy effects on fully-developed laminar heat transfer for upflow in the infinite triangular and square arrays with pitch-to-diameter ratio between 1.3 and 2.5. The buoyancy effects on pressure drop, and therefore the subchannel friction factors, were not investigated. Yang [5] and Kim and Wolf [6] extended Iqbal's analysis to arrays with smaller pitch-to-diameter ratio (P/D=1.05 to 1.50). Both heat transfer and pressure drop parameters were calculated for triangular arrays. Based on these pressure drop parameters, Yang [7] further defined an effective friction coefficient, which combines the effects of the surface friction and gravity. Both Iqbal and Yang considered the coolant region only for internal subchannels. Kim considered both internal and edge subchannels as well as both the clad and coolant regions. This two-region analysis assessed the effects of heat redistribution within the clad. However, because of programming errors in computing the velocity and temperature fields in internal subchannels and the wall shear stress in edge subchannels, the specific numerical values reported by Kim are in error. Yang [8], in his recent study, investigated the buoyancy effects on fully-developed laminar downward flow in infinite triangular and square arrays. Both heat transfer and pressure drop parameters, along with the criteria for flow separation, were given for triangular arrays. However, there is an inconsistency between the pressure drop parameters and the criteria of flow separation. The pressure gradient factors become positive before any flow separation (reversal) occurs.

In order to systematically investigate the buoyancy effects on subchannel friction factors and heat transfer coefficients for different arrays and flow directions, and therefore build a firm foundation for understanding the general behavior of these parameters under nonuniform radial power distributions, the subchannel velocity and temperature distributions were obtained for laminar upward and downward flows in finite triangular and square arrays with uniform radial power distributions. Although the present analysis concentrates on fully-developed flow in bare rod assemblies, the effects of developing flow and wire wrap will be discussed. The resulting friction factors and heat transfer coefficients are correlated in the form of simple polynomials which can easily be applied to subchannel programs. The criteria for local flow recirculation,

Table 1. The relation between present study and previous work

	Aiding Flow		Opposing Flow	
	int. cell	edge cell	int. cell	edge cell
subchannel friction factor	Yang [5] Kim [6] Present Study	Kim [6] Present Study	Yang [8] Present Study	Present Study
subchannel heat transfer coefficient	Iqbal [4] Yang [5] Kim [6] Present Study	Kim [6] Present Study	Yang [8] Present Study	Present Study
local flow recirculation	Present Study	Present Study	Yang [8] Present Study	Present Study
transition from laminar to turbulent flow	Present Study	Present Study	Yang [8] Present Study	Present Study

derived from the detailed velocity distributions, are also given in a simple correlation which can be used to determine whether the use of lumped parameter approach to model the subchannels is appropriate. Table 1 relates the present study to the previous work and highlights the significance of the present study.

2. MATHEMATICAL ANALYSIS

To study the fluid flow and heat transfer characteristics in a rod assembly, two different types of behavior can be found depending on the relative direction of the pressure gradient and buoyancy forces. The heating of a cold fluid flowing upwards is basically the same as the cooling of a hot fluid flowing downwards --the buoyancy force reinforces the pressure force near the rod surface. This pattern of flow is called aiding flow. The other pattern of flow, the cooling of a hot fluid flowing upwards and the heating of a cold fluid flowing downwards, is known as opposing flow because now the buoyancy force opposes the pressure force near the rod surface. The mathematical form of the dimensionless momentum and energy equations appropriate for fully-developed, steady, laminar flow in the vertical direction may be written as follows [9]:

$$- \frac{dP^*}{dz^*} + \frac{1}{Re} \nabla^{*2} u^* \pm \left(\frac{Gr}{Re^2} - \frac{1}{Fr}\right) = 0 \qquad (1)$$

and

$$\frac{\partial T^*}{\partial z^*} = \frac{1}{RePr} \nabla^{*2} T^* \qquad (2)$$

where in the last term of the momentum equation, the minus sign applies for the case of aiding flow and the plus sign applies for the case of opposing flow.

By introducing the new dimensionless temperature ϕ and the pressure gradient factor L, equations (1) and (2) can be further simplied as

$$\nabla^{*2} \overline{V} \pm Ra \overline{\phi} = -1 \qquad (3)$$

and

$$\nabla^{*2} \overline{\phi} - \overline{V} = 0 \qquad (4)$$

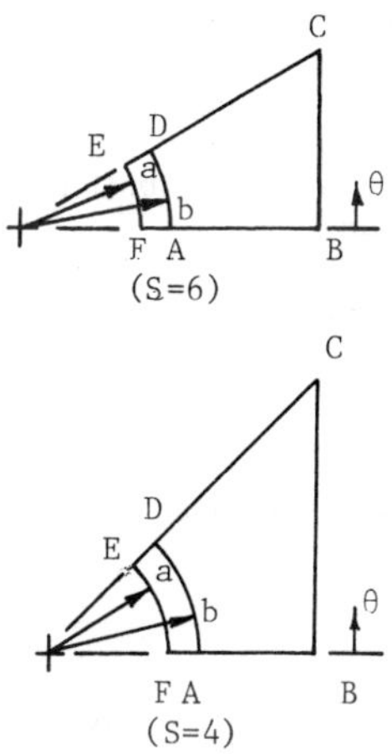

$$\frac{\partial \overline{V}(R,0)}{\partial \theta} = \frac{\partial \overline{V}(R,\pi/S)}{\partial \theta} = 0 \quad (a)$$

$$\frac{\partial \overline{\phi}(R,0)}{\partial \theta} \quad \frac{\partial \overline{\phi}(R,\pi/S)}{\partial \theta} = 0 \quad (b)$$

$$\overline{V}(b,\theta) = 0 \quad (c)$$

$$\frac{\partial \overline{V}}{\partial N} = \frac{\partial \phi}{\partial N} = 0 \quad (d)$$

$$\phi(A) = 0 \quad (e)$$

$$\frac{\partial T_c(r,o)}{\partial \theta} = \frac{\partial T_c(r,\pi/s)}{\partial \theta} = 0 \quad (f)$$

$$\overline{q}'' = -k_c \frac{\partial T_c(a,\theta)}{\partial r} = \overline{q}''_i \quad (g)$$

$$T_c(b,\theta) = T(b,\theta) \quad (h)$$

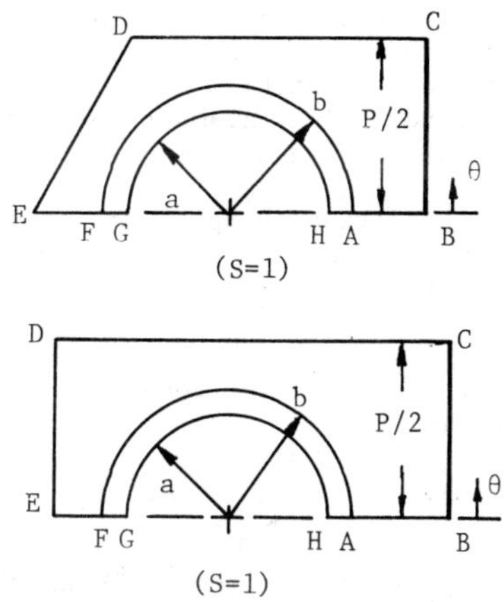

$$V=0, \quad \frac{\partial \overline{\phi}}{\partial N} = -\frac{D\,\overline{q}''_D}{L' k_w}(1-\Omega \sin\theta) \quad (d), \quad -k_c \frac{\partial T_c(b,\theta)}{\partial r} = -k_w \frac{\partial T(b,\theta)}{\theta r} \quad (i)$$

Fig. 2. Boundary conditions

where $\overline{V} \equiv \frac{u^*}{L}$, $\overline{\Phi} \equiv \frac{\phi}{L}$

$$\phi \equiv \frac{T^*}{RePr\ \partial T^*/\partial z^*}, \quad Ra \equiv \frac{PrGr\ \partial T^*/\partial z^*}{T^*} \quad (5)$$

and

$$L \equiv -Re\left[\frac{dP^*}{dz^*} \pm \frac{1}{Fr}\right] \quad (6)$$

The pressure gradient factor L has an important physical interpretation which will be discussed later.

The energy equation in the clad region is simply

$$\nabla^2 T_c = 0 \quad (7)$$

To facilitate applying equations (3), (4) and (7) to the system under study, refer to Figure 2. The sketches are an enlarged view of the characteristic elements (cells) shown in Figure 1. The boundary conditions and the dimensional nomenclature are given in the Figure.

The condition that the normal derivative of velocity and temperature be zero is an expression of symmetry. On the inner surface of the clad, r=a, the heat flux is assigned a value $\overline{q}''_i$ which is independent of angle. Physically, such a boundary condition could correspond to a situation of a fuel rod having a low thermal conductivity clad and high thermal conductivity fuel pellets. On the inner surface of the duct wall, a non-zero heat flux $\overline{q}''_D$ is imposed to simulate the heat transfer between assemblies. In this way, the effects of inter-assembly heat transfer on the velocity and temperature distributions can be studied. Since all quantities are dimensionless, the dimensionless temperature at the reference point, A, must be fixed. Only the scale of the temperature distribution will be changed if a reference point other than A is chosen.

Differentiating equation (3) and combining with equation (4), one obtains

$$\nabla^{*4}\,\overline{V} \pm \eta^4\,\overline{V} = 0 \tag{8}$$

where $\eta^4 \equiv Ra$. The general solution of equation (8) containing even functions of θ is:

For aiding flow:

$$\overline{V} = \sum_{n=0}^{\infty} [a_{ns}ber_{ns}(\eta R)+b_{ns}bei_{ns}(\eta R)+c_{ns}ker_{ns}(\eta R)+d_{ns}kei_{ns}(\eta R)][\cos(ns\theta)] \tag{9}$$

For opposing flow:

$$\overline{V} = \sum_{n=0}^{\infty} [a_{ns}J_{ns}(\eta R)+b_{ns}Y_{ns}(\eta R)+c_{ns}I_{ns}(\eta R)+d_{ns}K_{ns}(\eta R)][\cos(ns\theta)] \tag{10}$$

The corresponding solution for temperature distribution can be obtained from equation (3) as:

For aiding flow:

$$\overline{\phi} = -\frac{1}{\eta^4} + \frac{1}{\eta^2}\sum_{n=0}^{\infty} [a_{ns}bei_{ns}(\eta R)-b_{ns}ber_{ns}(\eta R)+c_{ns}kei_{ns}(\eta R)-d_{ns}ker_{ns}(\eta R)][\cos(ns\theta)] \tag{11}$$

For opposing flow:

$$\overline{\phi} = \frac{1}{\eta^4} + \frac{1}{\eta^2}\sum_{n=0}^{\infty} [-a_{ns}J_{ns}(\eta R)-b_{ns}Y_{ns}(\eta R)+C_{ns}I_{ns}(\eta R)+d_{ns}K_{ns}(\eta R)]\,[\cos(ns\theta)] \tag{12}$$

The general solution of equation (7) containing even function of θ can be written as:

$$T_c(r,\theta)=A_o+B_o \ln r + \sum_{n=1}^{\infty} [(A_{ns}\, r^{ns}+B_{ns}\, r^{-ns})\cos(ns\theta)] \tag{13}$$

There are a total of six sets of coefficients in the general velocity and temperature solutions, which have to be determined from boundary conditions. The coefficients B_{ns} in equation (13) can be eliminated in a straight-forward manner by applying the heat flux boundary condition (g) at the inner surface of the clad. The other three sets of coefficients a_{ns}, b_{ns} and A_{ns} can be expressed in terms of c_{ns} and d_{ns} by applying the no-slip velocity boundary condition (c) and the continuity of temperature and heat flux boundary conditions, (h) and (i), at the rod surface. However, because the solution for clad temperature contains one set of coefficients A_{ns}, the coefficient a_o can only be expressed in terms of b_o, c_o and d_o. The solutions for velocity and temperature are now known in terms of two sets of coefficients, c_{ns} and d_{ns} plus b_o.

The coefficients, c_{ns} and d_{ns}, are determined from the boundary conditions at the symmetry lines of the subchannels. Since these symmetry lines do not match the rod-centered polar coordinates, the boundary conditions can only be satisfied in an approximate manner. The simple point-match method is used, which yields 2m linear algebraic equations for the 2m+1 unknown coefficients, bo, c_{ns} and d_{ns}. We need one more boundary condition to complete the problem. Boundary condition (e) serves this purpose. The sytem of linear algebraic equations are then solved for bo, c_{ns} and d_{ns} by Gaussian elimination. Having determined these coefficients, the velocity and temperature distributions can be obtained from equations (9) to (13). Based on these velocity and temperature distributions, the following parameters of engineering interest can be evaluated: subchannel friction factor, heat transfer coefficient, local heat flux and wall shear stress distribution. The derivation of these parameters can be found in Reference 10.

3. RESULTS

Since there are a total of eight cases to be considered, it is convenient to assign a five-letter code for each individual case. The first three letters indicate the type of cell under consideration: internal (INT) or edge (EDG). The fourth letter represents the relative direction between the forced convective flow and the buoyancy force: adding flow (A) if they are in the same direction, opposing flow (O) if they are not. The last letter indicates the array lattice: triangular (T) or square (S).

3.1 Velocity Distribution

The effect of buoyancy force on the radial velocity distribution is shown in Figure 3. For simplicity, only the results for a typical LMFBR fuel and blanket assemblies are shown. Results for other geometries are given in Reference 10. Notice that the Rayleigh number based on rod diameter, Ra_D, is used to separate the buoyancy effect from the geometric effect [5].

The parabolic-type velocity profiles at low Ra_D are distorted as the buoyancy force increases. In internal cells, since the flow area is simply-connected and is symmetric with respect to the rod, the distortion of the velocity profiles is basically due to radial flow redistribution. No transverse flow crosses the isotach boundaries. However, in edge cells, where the flow region is a combination of internal and edge subchannels, the buoyancy force differs significantly from one side of the rod to the other. Therefore, a significant flow redistribution between edge and internal subchannels due to this asymmetry in buoyancy force is observed.

The effect of inter-assembly heat transfer on velocity distribution is shown in Figure 4. In general, the coolant velocity decreases due to heat loss through duct wall.

3.2 Friction Factor

The pressure gradient factor L, defined in equation (6), can be re-written as:

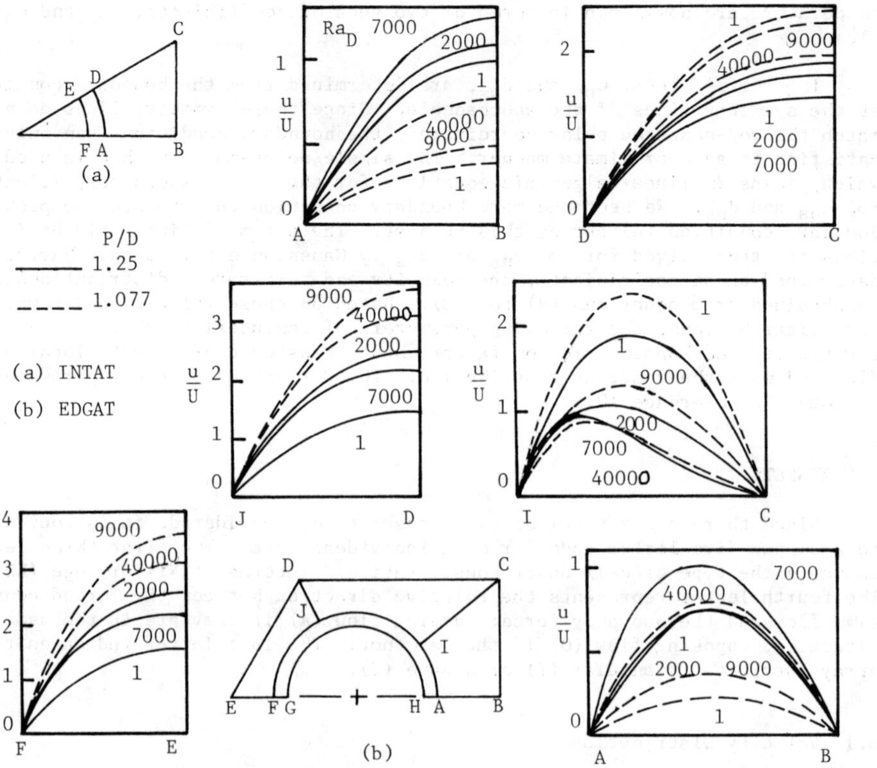

Fig. 3. Buoyancy effect on axial velocity distribution

Fig. 4. Inter-assembly heat transfer effect on velocity distribution

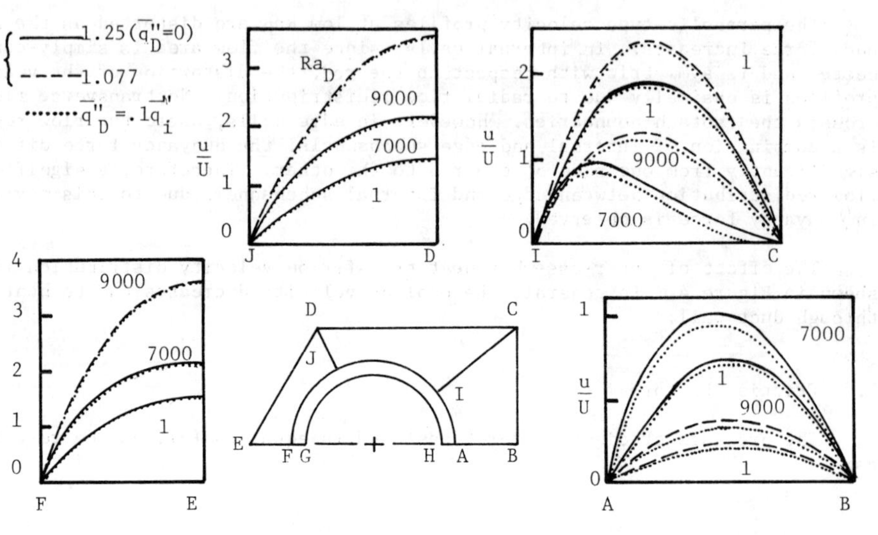

$$L \equiv - Re \left[\frac{dP^*}{dz^*} \pm \frac{1}{Fr} \right] = - Re \frac{D_h}{\rho_o U^2} \left[\frac{dP}{dz} \pm \rho_o g \right] \tag{14}$$

For steady, fully-developed flow, the bracket on the RHS of equation (14) is the pressure gradient due to surface friction. By substituting the friction factor, f, into equation (14), one can obtain

$$L = \frac{fRe}{2} \tag{15}$$

Figure 5 shows the effect of buoyancy force on the pressure gradient factors for different P/D ratios. As buoyancy force approaches zero, these factors should approach their isothermal values. Table 2 shows a comparison of pressure gradient factor L between present calculations and previous work for isothermal cases. The agreement is excellent and therefore assures the validity of present analysis.

In internal subchannels, the velocity gradient at the rod surface monotonically increases (for aiding flow) or decreases (for opposing flow) with increasing buoyancy force. Therefore, the resultant friction factors monotonically

Fig. 5. Pressure gradient factors for rod arrays

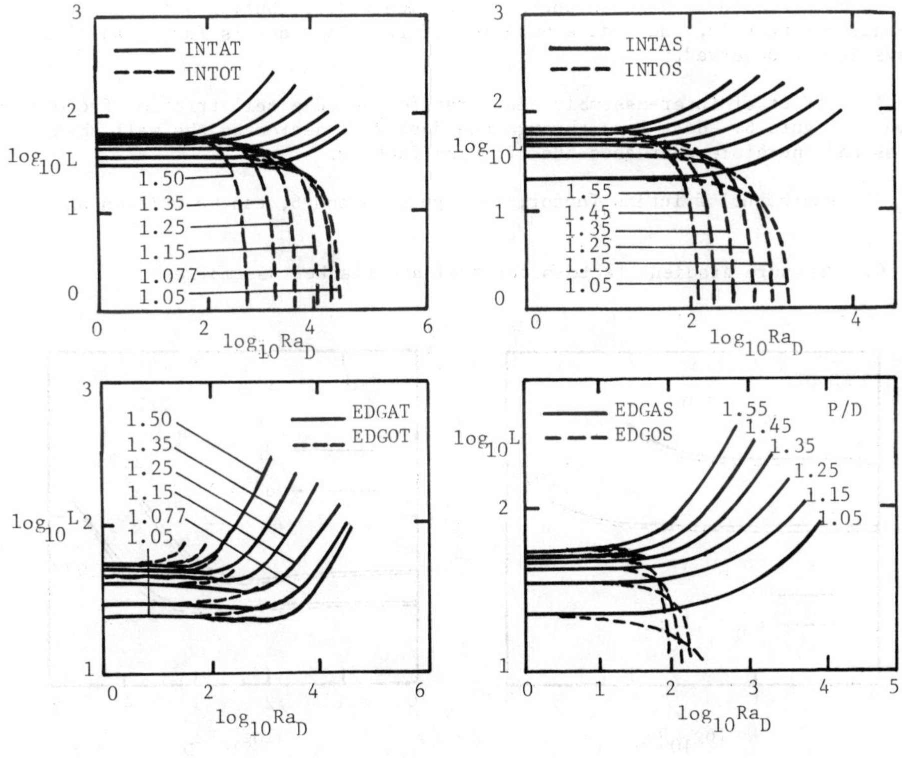

Table 2. Pressure gradient factor L for internal subchannels

Triangular Lattice					Square Lattice			
P/D	Rehme*	Sparrow*	INTAT	INTOT	P/D	Rehme*	INTAS	INTOS
1.050	31.160	30.956	30.8889	31.1400	1.050	21.940	21.3414	22.6938
1.077	--	--	36.9745	37.0736	1.150	--	35.4697	35.5362
1.150	46.440	46.282	46.2757	46.2809	1.250	--	44.7355	44.6719
1.250	--	52.602	52.6004	52.5903	1.350	--	51.5792	51.3769
1.350	--	--	56.8075	56.7780	1.450	--	57.0012	56.7301
1.500	--	62.070	62.0721	61.9609	1.550	--	61.7894	61.2899

* from Reference 13

deviates from their isothermal values accordingly. In edge cells, since there exists transverse flow redistribution between edge and internal subchannels, the behavior of the friction factor is more complicated than that for internal cells.

As buoyancy forces increase, the coolant redistributes from edge to internal subchannels for EDGAT cases, which causes a decrease in friction factor. Further increases in buoyancy force cause the velocity gradient to increase over most of the cell surface and therefore causes an increase in the friction factor. For EDGOT cases, the reverse is true except that a local recirculation zone forms before the velocity gradient decreases over most of the cell surface.

In square arrays, because there is no flow redistribution between edge and internal subchannels, the saddle behavior of friction factors for triangular arrays is not observed.

The effect of inter-assembly heat transfer on edge cell friction factors is shown in Figure 6. Heat loss through the duct wall decreases the wall shear stress and therefore decreases the friction factors.

The subchannel friction factor, in Figures 5 and 6, can be written as

Fig. 6. Pressure gradient factors for fuel and blanket assemblies

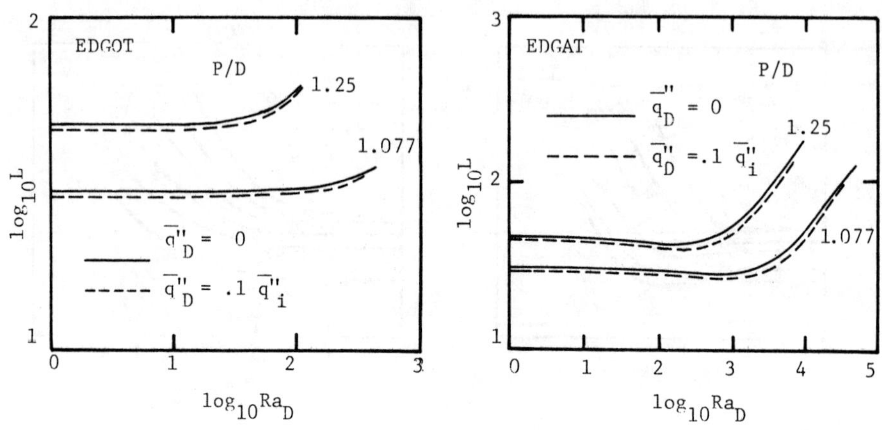

Table 3. Correlated constants for ϕ_B

	P/D	a_o	a_1	a_2	a_3	a_4	a_5	a_6
INTAT	1.077	.9999	$3.6685E^{-5}$	$-2.2474E^{-10}$	$1.5450E^{-15}$	$-9.7145E^{-21}$	$1.8929E^{-26}$	$6.5012E^{-32}$
	1.250	1.0002	$1.9043E^{-4}$	$-2.8664E^{-9}$	$-9.9884E^{-12}$	$1.7682E^{-17}$	$7.3638E^{-22}$	$9.7380E^{-27}$
INTOT	1.077	.9981	$-3.7708E^{-5}$	$2.0834E^{-10}$	$-7.2256E^{-14}$	$5.5651E^{-18}$	$-2.0911E^{-22}$	$2.9917E^{-27}$
	1.250	1.0001	$-1.9445E^{-4}$	$-4.5065E^{-9}$	$-2.3056E^{-13}$	$-5.5107E^{-18}$	$-1.3499E^{-21}$	$2.2012E^{-25}$
EDGAT	1.077	.9759	$-2.9233E^{-6}$	$1.8854E^{-8}$	$-1.3713E^{-12}$	$4.5369E^{-17}$	$-7.1390E^{-22}$	$4.3104E^{-27}$
	1.250	.9747	$-2.7330E^{-4}$	$7.4121E^{-7}$	$-3.2437E^{-10}$	$6.6894E^{-14}$	$-6.5477E^{-18}$	$2.4426E^{-22}$
EDGOT	1.077	.9988	$1.3004E^{-3}$	$-3.2091E^{-5}$	$4.6855E^{-7}$	$-3.4664E^{-9}$	$1.2641E^{-11}$	$-9.4786E^{-13}$
	1.250	.9972	$2.5190E^{-3}$	$-8.7890E^{-5}$	$3.0747E^{-6}$	$-4.6256E^{-8}$	$3.4020E^{-10}$	$-1.7923E^{-14}$

$$f_b = \phi_B \, f_b^o \qquad (16)$$

where f_b^o is the isothermal value and ϕ_B is the buoyancy multiplier. The value of ϕ_B for each of the eight cases can be correlated by a simple polynomial as

$$\phi_B = a_0+a_1\lambda+a_2\lambda^2+a_3\lambda^3+a_4\lambda^4+a_5\lambda^5+a_6\lambda^6 \, , \; 1.050 \le P/D \le 1.500 \qquad (17)$$

where $\lambda \equiv Ra_D$. The calibrated constants a_0 to a_6 for fuel and blanket assemblies are listed in Table 3.

Table 4 shows the typical normal operation and shutdown transient conditions for an LMFBR power plant. According to the present analysis, the effect of buoyancy force on friction factors is very important in blanket assemblies but negligible in fuel assemblies for both normal and shutdown conditions. This is because the blanket assembly has a higher power to flow ratio and a larger flow area. Both of these conditions enhance the buoyancy effect in blanket assembly.

3.3 Critical Ra_D

Having the detailed velocity distributions available, the criterion for the onset of local flow recirculation can easily be determined. The occurrence of this flow recirculation sets a limit on using the lumped parameter approach (LPA) to model the subchannel, because in LPA only one velocity value is assigned in each subchannel. Any attempt to using LPA beyond this criterion will completely miss the velocity distribution.

The limit on using the LPA to model the subchannel can be expressed in terms of a critical Rayleigh number, Ra_D^c. Figure 7 shows Ra_D^c for triangular and square

Table 4. Normal and shutdown conditions for LMFBR

Type	Fuel Assembly		Blanket Assembly	
Operation cond.	Normal	Shutdown	Normal	Shutdown
Peak Power	16 kw/ft	--	7∿16 KW/ft	--
Flow Velocity	25 ft/sec	--	1.6∿8 ft/sec	--
Power/Flow	1.0	peak 1.52*	1.0	Peak 1.97*
Ra_D	135 (@ 1000°F)	205 (@ 1000°F)	5500∿12000(@1000°F)	11000∿24000(")
fb/fb^o(INTAT)	1.03	1.06	1.20∿1.42	1.40∿1.75
fb/fb^o(EDGAT)	.94	.95	1.50∿2.00	1.92∿2.83

* Personnel communication, P. Dickson to N. Todreas, Fall 1977.

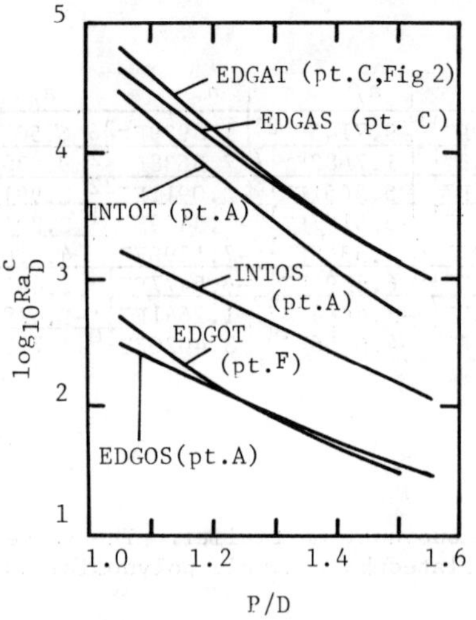

Fig. 7 Critical Rayleigh number

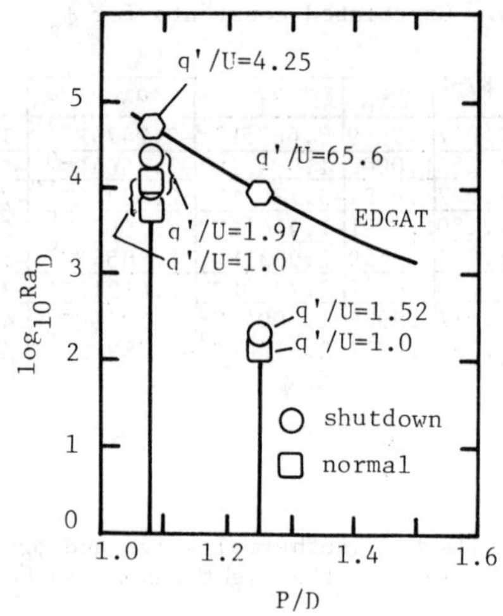

Fig. 8. Rayleigh number for normal and shutdown conditions

arrays along with the location where recirculation first occurs. The Ra_D^c's for INTAT and INTAS cases are not included because their values are extremely high and therefore are unlikely to occur in cases of interest. For a given P/D ratio, flow recirculation will occur when operating at Ra_D greater than Ra_D^c. The numerical value of Ra_D^c can be correlated as

$$\log_{10}Ra_D^c = b_0 + b_1 (P/D) + b_2 (P/D)^2 \quad , \quad 1.050 \leq P/D \leq 1.500 \tag{18}$$

The correlated constants are listed in Table 5.

For the current LMFBR core design, the power to flow ratios corresponding to normal operation, shutdown transient and onset of flow recirculation in an edge cell are shown in Figure 8. According to the present calculation, no flow recirculation will occur in fuel and blanket assemblies. This is because we assume the radial power distribution is flat and the flow is mainly in the axial direction. However, when there is a strong transverse flow caused by a radial power skew across the assembly, the Ra_D^c will be drastically lowered and local flow recirculation can occur during shutdown transients even at normal operation conditions.

Table 5. Correlated constants for $\log_{10}Ra_D^c$

	b_o	b_1	b_2
INTOT	10.545068	-7.08347	1.251772
INTOS	5.927130	-2.718829	.145415
EDGAT	12.556783	-9.866886	2.389166
EDGOT	9.098571	-8.456378	2.245500
EDGAS	10.800161	-7.526179	1.610638
EDGOS	5.756120	-3.827714	.671051

3.4 Transition to Turbulence

For isothermal laminar flow in a rod assembly, the flow distribution approaches its fully-developed value along the flow path and no transition to turbulence occurs. However, when there is heat addition, transition to turbulence can happen at lower average Reynolds number due to the distortion of the velocity profile. For opposing flow, this transition should occur when the local flow recirulcation occurs. For adding flow, this transition may occur either when the maximum local wall shear stress, τ_w^{max}, equals the critical wall shear stress, τ_w^c, or when the local flow recirulcation occurs. The critical wall shear stress, τ_w^c, is the wall shear stress at which the transition from laminar to turbulent flow occurs for isothermal flow. In Reference 10, a maximum pressure gradient factor, L_{max}, corresponding to τ_w^{max} was obtained:

$$L_{max} = -\frac{L^o}{2}\frac{\Psi}{Re^c}\frac{q'_{max}}{Ra_D} \tag{19}$$

where L^o = pressure gradient factor for isothermal flow

$$\Psi = \frac{g\beta\,\rho_o^2\,D^4\,D_h}{k_{wo}\mu_o^2\,A_f}$$

Re^c = transition (critical) Reynolds number for isothermal flow

The intersection between L_{max} and L determines the maximum allowable Rayleigh number, Ra_D^{max}, to avoid the transition to turbulence. The maximum allowable Reynolds number for laminar flow is then determined by the smaller value of Ra_D^c and Ra_D^{max} as

$$Re^{max} = \Psi\, q'_{max}/\min\,(Ra_D^c,\ Ra_D^{max}) \tag{20}$$

For a given q'_{max}, the flow will become turbulent when

$Re > Re^{max}$.

Figure 9 shows the maximum allowable Rayleigh number, Ra_D^{max}, for typical fuel and blanket assembly conditions. In internal cells, the flow always becomes turbulent before any flow recirculation occurs. However, in edge cells, local flow recirculation may occur at low P/D before flow becomes turbulent.

Results for coolant temperature distributions, Nusselt numbers, clad temperature distributions, wall shear stress and heat flux distributions are given in Reference 10.

4. DISCUSSIONS

Since we considered only fully-developed laminar mixed convective flow in bare rod assemblies with isolated cells, the effect of wire-wrap, developing flow and interaction between cells on the results, such as friction factors and criteria for flow recirculation must be considered before these parameters can be applied to real situations.

Hawley et al. [11] have derived subchannel friction factor f_w^o for isothermal flow in wire wrap rod assemblies. By introducing a wire wrap multiplier, ϕ_w, these friction factors can be represented as

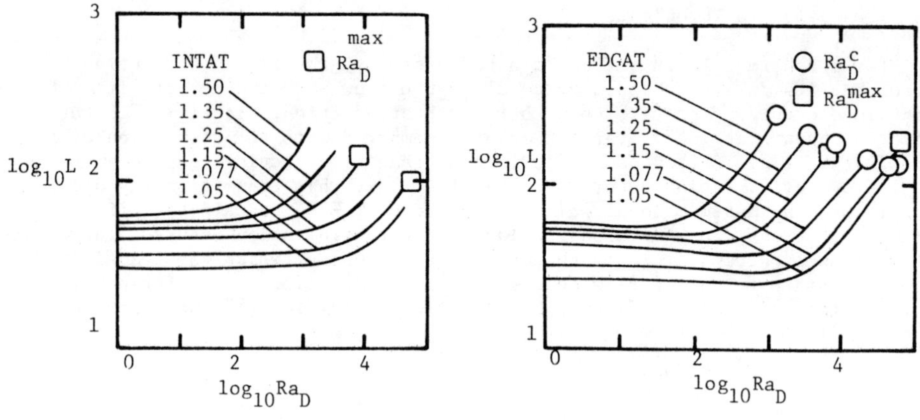

Fig. 9. Relation between Ra_D^c and Ra_D^{max}

$$f_w^o = \phi_w f_b^o \tag{21}$$

By assuming this wire wrap multiplier, obtained from isothermal flow, is applicable under mixed convection condition, the wire wrap effect on friction factor can be included and the friction factor under mixed convection condition is now given as

$$f_w = \phi_w f_b = \phi_w \phi_B f_b^o \tag{22}$$

The effect of wire wrap on the criteria for flow recirculation is not yet clear. For adding flow, the recirculation zone is formed next to the duct wall. Therefore, the wire wrap effect is expected to be small. For opposing flow, however, the flow recirculation zone is generally formed next to the rod surface. The wire wrap effect for this case should be investigated.

The effect of developing flow on friction factor and recirculation criteria has important implications in LMFBR core design, because the flow throughout the core probably never becomes fully-developed under mixed convection conditions. Sherwin [12] investigated the flow recirculation behavior in concentric annulus under mixed convection conditions. He found Ra_D^c for developing flow was much greater than that for fully-developed values. Similar behavior should be expected in rod assemblies and would cause the flow recirculation to be more unlikely to occur. The effect of developing flow in rod assemblies needs to be assessed and verified by experiment.

5. CONCLUSION

A systematic study of adding and opposing flow in finite triangular and square rod arrays has been performed to investigate the fluid flow and heat transfer behavior under mixed convection conditions. The radial velocity distribution, subchannel friction factor and criterion for flow recirculation and for transition to turbulence have been determined under various conditions. For typical LMFBR core designs, the effect of buoyancy force on the subchannel friction factor is important in the blanket assembly but negligible in the fuel assembly. This is true for both normal operation and shutdown transients. In the fuel assembly, flow tends to become turbulent before recirculation occurs, while in the blanket assembly the reverse is true. However, present calculations, based on the uniform radial power distribution assumption, show that flow recirculation in both fuel and blanket assemblies is unlikely to happen under most of the steady-state and transient conditions.

REFERENCES

1. F.C. Engel, R.A. Markely, B. Minushkin, "Buoyancy Effects on Sodium Coolant Temperature Profiles Measured in an Electrically Heated Mock-up of a 61-Rod Breeder Reactor Blanket Assembly," ASME paper 78-WA/HT-25, September 1979.

2. M.L. Millburg, J.A. Hassberg, C.J. Boasso, "Natural Circulation Heat Transfer Testing with a Simulated Full-Scale LMFBR 217-Pin Electrically Heated Fuel Assembly," HEDL-TME 77-3, June 1977.

3. T.E. Greene, "Coolant Mixing in LMFBR Rod Bundles and Outlet Plenum Mixing Transients," Progress Report, September- November 1979, COO-2245-72, MIT.

4. M. Iqbal, S.A. Ansari, B.D. Aggarwala, "Buoyancy Effects on Longitudinal Laminar Flow Between Vertical Cylinders Arranged in Regular Arrays," Proc. of 4th Int. Heat Transfer Conf., Vol. 4, paper NC3.6, 1970.

5. J.W. Yang, "Analysis of Combined Convection Heat Transfer in Infinite Rod Arrays," Proc. of 6th Int. Heat Transfer Conf. Vol. 1, August 1978.

6. J. Kim, L. Wolf, "Fully-Developed Mixed Convection Heat Transfer in the Finite Hexagonal Bundles," MS Thesis, Nucl. Eng. Dept., MIT, February 1977. See also ANS Trans., Vol. 27, November 1977.

7. J.W. Yang, "Buoyancy Effects on Pressure Loss in Rod Arrays Under Laminar Flow," ANS Trans., Vol. 32, June 1979.

8. J.W. Yang, "Heat Transfer and Fluid Flow in Regular Rod Arrays with Opposing Flow." Fluid Flow and Heat Transfer Over Rod or Tube Bundles, ASME, December 1979.

9. S.F. Wang, "Coolant Mixing in LMFBR Rod Bundles and Outlet Plenum Mixing Transients," Progress Report, September - November 1979, COO-2245-72, MIT.

10. S.F. Wang, W.M. Rohsenow, N.E. Todreas, "Steady, Laminar, Fully-Developed Mixed Convection in Finite Rod Arrays," to be issued as COO-2245-73, MIT.

11. J. Hawley, C. Chiu, W.M. Rohsenow, N.E. Todreas, "Subchannel and Bundle Friction Factors and Flowsplit Parameters for Laminar, Transition, and Turbulent Longitudinal Flows through Wire-wrap Spaced Hexagonal Rod Arrays," to be presented at ANS thermal hydraulic topical meeting, Saratoga, October 1980.

12. K. Sherwin, J.D. Wallis, "Combined Natural and Forced Laminar Convection for Upflow Through Heated Vertical Annuli," Heat and Mass Transfer by Combined Forced and Natural Convection, September 1971.

13. R.K. Shah, A.L. London, Laminar Flow Forced Convection in Ducts, Academic Press (1978).

ANALYTICAL METHODS AND MODELING

Facility Requirements for Natural Convection Shutdown Heat Removal System Testing

M. P. HEISLER
Energy Systems Group
Rockwell International

R. M. SINGER
Argonne National Laboratory

ABSTRACT

This paper summarizes similitude requirements for natural convection shutdown heat removal system test facilities. The desired dimensionless similarity parameters are derived from the basic conservation equations for mass, momentum, and energy. Design and operating restrictions imposed on the test facility, and the validity of extrapolating test facility data to the full-scale system are discussed.

1. INTRODUCTION

The development of test facilities for dynamic simulation of shutdown heat removal systems in large Liquid Metal Fast Breeder Reactors (LMFBR's) and verification of codes for describing their performance is a vital part of the fast reactor safety base technology program. Such test facilities are required to guide the design of prototype systems and to support licensing arguments which rely upon detailed knowledge of their performance. In order for codes to be confidently used, an experimental verification program must be conducted in parallel with the mathematical analysis and numerical evaluation efforts. This verification program should include both model and integral testing which provide information on individual components and subassembly effects and on total system phenomenological behavior. Facility testing of individual subassemblies, and whole cores coupled to plenums and to complete heat transport systems, is needed under conditions appropriate to shutdown conditions. Some data are already available, but there are serious shortcomings in several specific areas. The largest uncertainties and the most sparse of all the available data resources occur in systems with buoyancy-affected flows in large plenums coupled to blanket-driver cores and shutdown heat removal systems.

In order to provide data needed to investigate the performance of certain specific shutdown heat removal systems, such as proposed Direct Reactor Auxiliary Cooling System (DRACS) or Intermediate Reactor Auxiliary Cooling System (IRACS), special scale model test facilities are necessary. The prototype systems involve unique and, at times, complex geometries, and there is no apparent way of verifying their phenomenological behavior other than through the testing of carefully designed scale models. The question here is just what is the optimum scale size which will accurately simulate the performance of the full-size system and minimize the total facility cost of fabrication and operation?

2. PROPOSED SHUTDOWN HEAT REMOVAL TEST FACILITY

In order to establish a specific geometrical frame of reference for determining the facility requirements for shutdown heat removal system testing, an assessment of the modeling requirements for a proposed DRACS test facility is presented in this paper. In the proposed natural convection test facility, a controlled sodium temperature and flow conditioning unit will be installed in place of the IRAC system, as shown in Figure 1. This unit will provide prescribed flow and temperature decay ramps to the inlet nozzle of the reactor vessel. Fuel element heat generation will be simulated by electrical resistance heaters.

In the analysis which follows, the scaling requirements for the core, the plenum, and the DRAC system are considered on a component as well as on a system basis. The parameters governing the various flow regimes are identified and the projection of model test data to predictions of full-scale system behavior is discussed. Since complete compatibility of all the similarity parameters is not possible, the sensitivity of the system to the various parameters is considered in some detail.

The prime candidate fluids for the experimental program are the prototypic fluids (i.e., Na, NaK) and water, with its potential for increased flexibility at lower cost. Therefore, an evaluation was conducted of the adequacy of water for natural convection experiments involving the DRACS. A key conclusion emanating from this study is that water cannot adequately model the thermodynamic heat transfer expected under the buoyancy-dominated transient flow conditions typical of shutdown heat removal scenarios.

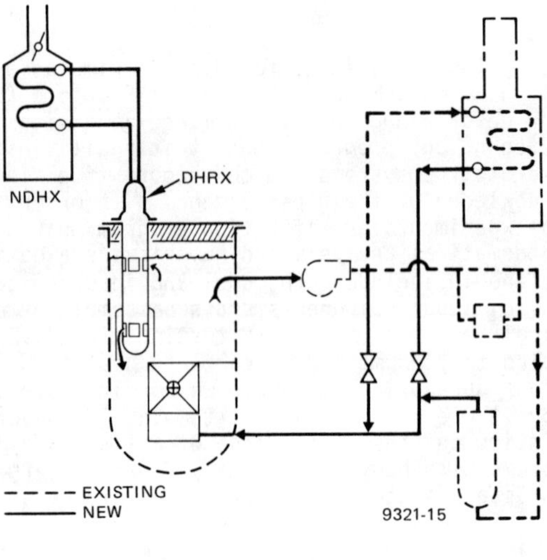

Fig. 1 Preconceptual Natural Convection Program Test Configuration

3. LOOP SIMILARITY STUDIES

The task of selecting an appropriate scale for the test facility is a complex one. The process utilized was to begin with a detailed derivation of the dimensionless parameters from the basic conservation (mass, momentum, and energy) equations for each of the loops to be simulated, with due regard to the effect of the individual components. These include the plenums, the core, the NaK piping, the Na-to-NaK heat exchangers, and the NaK-to-air natural draft heat exchangers.

3.1 The Differential Equations

Conservation and auxiliary equations which are valid for the DRAC and the reactor-plenum loops are given below. In these equations n is the number of components in either of the two loops.

Coolant Momentum Equation (incompressible flow)

$$\frac{dU_r}{d\tau} \sum_{i=1}^{n} \frac{A_r}{A_i} L_i = \Delta P + \frac{gL_o \Delta T_o}{u_o^2} (\theta_h - \theta_c) L_{TC} - \frac{U_r^2}{2} \sum_{i=1}^{n} \left(\frac{fl}{d_h} + \Sigma c \right)_i \frac{A_r^2}{A_i^2} \tag{1}$$

Coolant Energy Equation (for any loop component i)

$$\frac{\partial \theta_i}{\partial \tau} + U_i \frac{\partial \theta_i}{\partial Z} + \frac{4L_o h_i}{\rho u_o c_p d_i} (\theta_i - \theta_{si}) + Q_i = 0 \tag{2}$$

Structure Heat Transfer (component i)

$$\frac{\partial \theta_{si}}{\partial \tau_t} + \nabla^2 \theta_{si} + Q_{si} \tag{3}$$

$$\frac{\partial \theta_{si}}{\partial X_i} = \frac{h_i d_{si}}{k_{si}} (\theta_s - \theta_i) \tag{4}$$

All symbols are defined in the nomenclature. In equation (1) U_r (dimensionless) is the average velocity in the reactor core. Velocities in other segments of the loop scale as the ratio: core flow area to segment flow area. The significant dimensionless parameters of the system can be expressed in terms of the Richardson, Stanton, and Biot numbers, and a dimensionless friction loss number as follows:

$$R_i = \frac{gL_o \beta \Delta T_o}{u_o^2} \tag{5}$$

$$\frac{St_i}{Di} = \frac{L_o h_i}{\rho c_p u_o di} \tag{6}$$

$$\left(\frac{fl}{d_h} + \Sigma c\right)_i \tag{7}$$

$$Bi = \frac{h_i d_{si}}{k_{si}} \tag{8}$$

u_o, ΔT_o, and L_o are a prescribed reference velocity, temperature difference, and length for the system. Convenient values for u_o and ΔT_o are those prevailing in the core at the start of the transient, and for L_o the length of the heat generating region in the fuel assemblies.

3.2 General Loop Scaling Considerations

One of the main objectives of similitude analysis is to develop analytical relationships for the scaling of velocity, temperature, length, and time from the model to the prototype. Since electrical heaters are used to simulate the fuel assemblies, it is convenient for modeling purposes to relate the reference ΔT_o to corresponding heat generation rates in the electrical heaters and the fuel. By equating the total pressure drop around the loop under initial steady flow conditions to the buoyancy head generated by the heaters, a relationship between velocity, power, and length can be obtained.

Equating core average ΔT to steady average power

$$T_h - T_c = \frac{\pi}{4a} \frac{q''' \ell_o}{u_o \rho c} \tag{9}$$

a = function of pitch-diameter

ℓ_o = active fuel length

q''' = average fuel pin power density

Equating loop buoyancy force to Σ (flow losses)

$$\left(T_h - T_c\right) \beta \ell_{TC} = \sum_i \left(\frac{fl}{d_h} + \Sigma c\right)_i \frac{\rho(\varepsilon_i u_o)}{2g} \tag{10}$$

ε_i = flow area ratio

ℓ_{TC} = distance between thermal centers

From (1) and (2)

$$u_o^3 = \frac{2g\beta\ell_o\ell_{TC}q'''}{\sum_i \left(\frac{f\ell}{d_h} + \Sigma c\right) 4\epsilon_i^2\rho^2 a} \tag{11}$$

From (3) it follows, for approximately equal $\left(\frac{f\ell}{d_h} + \Sigma c\right)$ in model and prototype

$$u_R \approx \left(q_R L_R^2\right)^{1/3} \tag{12}$$

For example, if $q_R = 1$ and a scale ratio $L_R = 1/5$ is adopted, then (velocity in the model)/(velocity in the prototype) becomes

$$u_R = 0.342$$

3.3 Structure-to-Fluid Heat Transfer Considerations

The dimensionless time $\tau_t = \alpha t/L_o^2$, which characterizes heat conduction in the structure, differs from the dimensionless time $\tau = u_o t/L_o$, which characterizes convective and hydrodynamic effects in the fluid. However, both equations must lead to the same value of time when extrapolated to the prototype system. That is

$$t_p = t_m \alpha_R / L_R^2 = t_m u_R / L_R \tag{13}$$

It is evident from equation (13) that if the same fluid is used in the model as in the prototype, and since, according to equation (12), u_R cannot scale as $1/L_R$, time in the model will scale as $1/L_R^2$ with respect to heat conduction and, approximately, as $1/L_R$ with respect to convection. Hence, conduction heat transfer in the structure is out of phase with convection heat transfer in the fluid. In fact, if the model dimensions are 1/2 those of the prototype, the thermal relaxation time is proportional to $(1/2)^2 = 1/4$ in the structural components, but it is approximately proportional to 1/2 with respect to fluid flow.

In order to bring flow and conduction back into phase, the conduction heat transfer process must be slowed down. This can be accomplished by increasing the thermal boundary layer resistance by use of a less effective heat transfer fluid, such as water, or by increasing the thermal inertia of the structure, for example, by using a larger scale factor for the pipe. Both methods are examined below.

The structure heat conduction equation, equation (3), when written for a lumped rather than a distributed mass, takes the following form.

$$\frac{d\theta_s}{d\tau} + \frac{HL_o}{u_o(\rho c_p d)_w} (\theta - \theta_s) = 0 \tag{14}$$

In equation (14), H is the effective thermal conductance between the pipe and the fluid. In thin-walled pipes, such as are being considered here, curvature can be neglected, and H becomes simply

$$H = \frac{h}{1 + \frac{Bi}{2}} \tag{15}$$

where h is the surface heat transfer coefficient, and the Biot number Bi is based on the pipe wall thickness d_w. The scaling parameters for the pipe loop then take the following form, depending upon whether a lumped or a distributed mass model is used:

Lumped Mass Model

$$\frac{hL_o}{(1 + \frac{Bi}{2})(\rho cd)_f \, u_o} \quad \text{(fluid)} \tag{16}$$

$$\frac{hL_o}{(1 + \frac{Bi}{2})(\rho cd)_w u_o} \quad \text{(wall)} \tag{17}$$

Distributed Mass Model

$$\frac{hL_o}{(\rho cd)_f \, u_o} \quad \text{(fluid)} \tag{18}$$

$$\frac{hd_w}{k_w} \quad \text{(wall)} \tag{19}$$

If the same fluid properties are used in the model and prototype, equations (16) and (17) for the lumped mass model require that pipe diameter and pipe wall thickness scale in the same ratio. If this is assumed to be valid for the distributed mass model also, then by straightforward algebraic manipulation, it can be demonstrated from equations (18) and (19), and the Lubarsky-Kaufman equation

$$Nu = 0.625 \left(\frac{u_o d_f}{\alpha_f}\right)^{0.4} \tag{20}$$

that

$$d_{hR} = L_R^{0.883} / u_R^{0.1667} \tag{21}$$

where the subscript R denotes the ratio of model to prototype. Values of u_R are obtained from equation (12)

$$u_R = \left(q_R L_R^2\right)^{1/3}$$

For equal power densities in model and prototype, $u_R = 0.342$ for a 1/5-scale model. Substituting this into equation (21) gives $d_{hR} = 0.312$. It is clear, in line with the above discussion of dimensionless time ratios for fluid and structure, Section 3.2, that the pipe diameter does not scale in the same ratio as the length but as 0.312 as compared to 0.2 for the length scaling ratio. Analysis similar to the above can be carried out for other structural components using the appropriate Nusselt relationship.

3.4 Water vs Na-NaK

It may be appropriate here to examine the feasibility of using water in the shutdown heat removal test facility. Water has obvious advantages from the standpoint of cost and operating ease and has been used with some success in a number of forced-convection flow hydraulic studies of plenums. Table 1 lists physical and thermal property data for sodium, NaK, and water.

TABLE 1

SODIUM, NaK, WATER PROPERTY DATA

		500°C 932°F Na	NaK	140°F Water
k	$\frac{Btu}{ft, h, {}^oF}$	38.6	15.1	0.376
ρ	lb/ft^3	51.96	47.6	61.5
c	$\frac{Btu}{lb, {}^oF}$	0.302	0.209	1.0
$\rho \cdot c$	$\frac{Btu}{ft^3, {}^oF}$	15.7	9.94	61.5
$\alpha = \frac{k}{\rho c}$	ft^2/h	2.46	1.52	0.00602
μ	lb/ft, h	0.595	0.475	1.14
ν	ft^2/h	0.0114	0.010	0.0185
Pr	-	0.0046	0.0064	3.02

Thermal conductivity, density, specific heat, thermal diffusivity, viscosity, and Prandtl number of sodium and NaK differ significantly. Therefore, while it may be possible to separately simulate sodium or NaK with water, it does not appear possible to use water to simulate both fluids at the same time, as would be required for the proposed Na-NaK test facility.

As was pointed out previously, if equal power densities are used in model and prototype and a 1/5-scale model is adopted for the test facility, then according to equation (12), $u_R = 0.342$. In Table 2, values of u_R calculated from equation (16) are presented for $d_R = 1/5$ and water. Steady flow conditions were assumed and 1% reactor power. A power ratio $q_r = 1.0$ was assumed.

It appears from Table 2 that using a 1/5-linear-scale model requires the velocity ratio u_R to be greater than 0.342. From equation (12) it then follows that the power density ratio q_R for water must be much greater than 1.0. This

TABLE 2

LOOP SCALING PARAMETERS FOR WATER

	u_R (Equation 12)	L_R	d_R	u_R (Equation 14)
Water	0.342	1/5	1/5	0.784

is an unattractive design condition for the test facility and weighs heavily against adopting water for the facility.

3.5 Friction Scaling Considerations

It follows from equation (7) that the friction loss term $(f\ell/d_h)_i + \Sigma c_i$ must be the same in the model and the prototype. Σc_i is the sum of entrance, exit, expansion, contraction, and turning losses for the ith component of the loop. For equal linear scaling in both length and width, values of L_o/d, A_r/A_i, L_i, and L_{TC} for the model will be equal to those for the prototype. However, the friction factor f varies not only with diameter but also with flow. In addition, heat transfer dictates that when the length ratio L_R is 0.2, the diameter ratio d_{hR} must be 0.312. Thus, $1/d_h$ in the model scales as 0.2/0.312 = 0.640. The scaling theory therefore states that a distorted length/diameter ratio is required in order to correct the phase error due to the different conduction/convection time ratios. This may run counter to other scaling requirements and needs further evaluation. Such evaluation has not been addressed in the preliminary evaluations.

Representative design flow velocities and pressure drops for the DRAC loop, at its design capability of 1% of full reactor power, are given in Table 3.

TABLE 3

DRAC SYSTEM FLOW PARAMETERS

Component	Velocity (ft/s)	Re	ΔP (psi) due to ($f\ell/d$)	ΔP (psi) due to (Σc)
Pipe Spools	2.8	1.8×10^6	0.603	0.447
Air Blast HX Tubes	2.76	2×10^5	0.272	0.268
NaK-Na HX Tubes	1.26	5×10^4	0.155	0.178
			1.03	0.893

It can be seen that the pressure drop around the DRAC loop is divided almost equally between friction and contraction and turning losses. Thus, only about one-half of the total pressure drop around the loop depends on velocity and dimensional scaling factors. Using the initial flow velocities for the proposed 1/5-scale model, values of friction factors for the pipe spools become, roughly, 0.013 in the prototype and 0.018 in the model. Thus, the friction factor scales as 0.018/0.013 = 1.38. Hence, $f\ell/d$ in the model scales as 0.64 x 1.38 = 0.885. Since the pipe spools are the major contributors to the friction pressure drop (see Table 1), the maximum relative error in flow in the model as compared to the flow in the prototype is, approximately

$$1 - \sqrt{0.885 \times 1.03 + 0.893} / \sqrt{1.03 + 0.893} = 3\%$$

As would be expected from Reynold's analogy between friction and heat transfer, the error is small. If required, most of this error can be eliminated by reducing the value of the contraction and turning loss term, Σc, to compensate for the increase in the friction term, for example by eliminating one or more bends.

4. THE REACTOR CORE (To be reported in a later phase of this study)

5. PLENUM SCALING CONSIDERATIONS

5.1 The Differential Equations

Assuming incompressible flow, the mass, momentum, and energy conservation equations for the plenum can be written as follows using vector notation:

Mass

$$\nabla \cdot \rho U = 0 \tag{22}$$

Momentum

$$\frac{\partial U}{\partial \tau} + \nabla \cdot U^2 = - \nabla P + Ri\theta + \frac{1}{Re} \nabla \cdot \sigma \tag{23}$$

Energy

$$\frac{\partial \theta}{\partial \tau} + \nabla \cdot U = \frac{1}{Pe} \nabla^2 \theta \tag{24}$$

The molecular and turbulent eddy terms are included in σ. The set of six stress equations for these terms is contracted to a single one in Cartesian notation:

$$\sigma_{i,j/\mu} = \frac{\xi i,i}{\mu} \left(\frac{\partial u_i}{\partial x_j} + \frac{\partial v_j}{\partial x_i} \right) \tag{25}$$

i,j,k = 1, 2, 3, with Einstein's summation convention

Coolant-Structure Interface

$$\frac{\partial \theta}{\partial n} = Bi \; (\theta_s - \theta) \tag{26}$$

Structure Heat Conduction

$$\frac{\partial \theta}{\partial \tau_t} = \frac{\partial^2 \theta}{\partial X^2} \tag{27}$$

Hydrodynamic similarity requires matching the Richardson and Reynolds numbers. This requires matching ratios of the length to some power of the velocity (i.e., L/u^2 for Ri and Lu for Re).

Thermodynamic similarity requires matching of the Peclet number, which, since it is derived from the Reynolds number, also requires matching the product of a length time velocity (i.e., Lu).

Clearly both of these conditions cannot be satisfied within a scaled model (<1.0). An important part of this study, therefore, is to determine what errors are incurred by the conflicting demands imposed by thermal and hydrodynamic scaling requirements.

5.2 Momentum Scaling Considerations

In most hydraulic applications, the shear terms $(1/Re)\nabla\cdot\sigma$ are small in comparison to the inertial terms $\nabla\cdot U^2$ and are, therefore, neglected. Additionally, in the case of free convection, there is no impressed pressure, and the ∇P terms can be removed from the equations. Flow within the plenum is therefore characterized primarily by the Richardson number $gL_0\beta\Delta T_0/u_0^2$. Hydrodynamic scaling therefore requires satisfying Richardson similitude for both free and forced convection flow conditions.

5.3 Heat Transfer Scaling Considerations

Peclet number considerations. The coolant energy equation contains the Peclet number. In analogy to the Reynold's number, the Peclet number, RePr, is generally large and can be neglected except for liquid metals at low flow conditions. It is therefore, necessary to examine the Peclet number in some detail since its effect cannot arbitrarily be dismissed for sodium at low flow conditions.

During the Atomics International - Electric Power Research Institute (AI-EPRI) Phase A and B pool reactor design studies, Argonne National Laboratory (ANL) tests on a 1/5-scale of the pool plenum and supporting VARR-II plenum analysis showed strong flow stratification effects with the fluid spreading quite uniformly over the top and bottom of the plenum for up and down transients, respectively. Furthermore, no significant shadow effect was observed in the sheltered region of the plenum behind the pumps and intermediate heat exchangers (IHX's). Indications were that the hot fluid floated on top of the cold underlayers and an essentially one-dimensional flow existed.[1]

Under these flow conditions, the velocity in the pool is dominantly vertical. In typical pool designs, the average vertical velocity for stratified flow is about 0.002 ft/s at the DRACS design flow of 1%. A corresponding loop design would have an average vertical velocity of about 0.007 ft/s. It is difficult to believe that appreciable turbulence can exist at these creeping flow conditions. One can assume (subject to later confirmation) that turbulent transport of heat is small in comparison to molecular transport through the stratified sodium layers. A simple analysis is then possible which sheds additional light on the scaling requirements.

For stratified flow, the heat transfer process between the hot and cold layers in the pool can be expressed approximately by the following equation:

$$\rho cu\frac{dT}{dz} - k\frac{d^2T}{dz^2} = 0 \qquad (28)$$

The ratio of conduction (plus small eddy diffusive) to convective heat transport is given by Q_r, where:

$$Q_r = k\frac{\partial T}{\partial \xi}/\rho cuT \qquad (29)$$

A solution of equation (1) for a unit step change in temperature over a small length L is:[2]

$$T = (e^{-Pe\xi} - 1)/(e^{-Pe} - 1) \qquad (30)$$

where: $Pe = \rho cuL/k$

$\xi = z/L$

Inserting equation (3) into (2) leads to the following expression for the ratio of diffusive to convective heat transfer in the boundary strata which separate the hot and cold fluid layers.

$$Q_r = e^{-Pe\xi}/(e^{-Pe} - 1) \qquad (31)$$

The heat transfer ratio Q_r for $\xi = 1/2$ is plotted in Figure 2. The characteristic length L used for the calculations is the distance traveled by the fluid in 1 s. Peclet numbers for loop and pool reactors are given on the figure for 1% flow. Sodium velocities are also shown. It is evident from these velocities that at 100% flow, Pe would be 2.0 and 7.0 for pool and loop, respectively, and convection heat transport would completely overshadow molecular conduction and eddy diffusivity heat transfer. The relative heat transport effects for 1% flow are the reverse of this. The Peclet numbers for sodium are now 0.07 and 0.02, approximately. These put the heat transfer in the steeply rising region of the graph where conduction-eddy heat transfer can be many time larger than convection transport.

Peclet numbers for water are also shown on the figure. Even at flows well below 1%, convection transport by water dominates the heat transfer process.

Figure 3 shows temperature distributions in sodium and water pools at the time that a step temperature change would have been convected from the pool surface to the inlet of the DHRX. There is considerable heat transfer between the hot and cold sodium layers but virtually none in the water layers. Evidently, the buoyancy effects in the external loop will be significantly different for sodium and water during flow transients.

Figures 4 and 5 suggest that the diffusion contribution may be quite important at low flows and that the Peclet number effect should be investigated

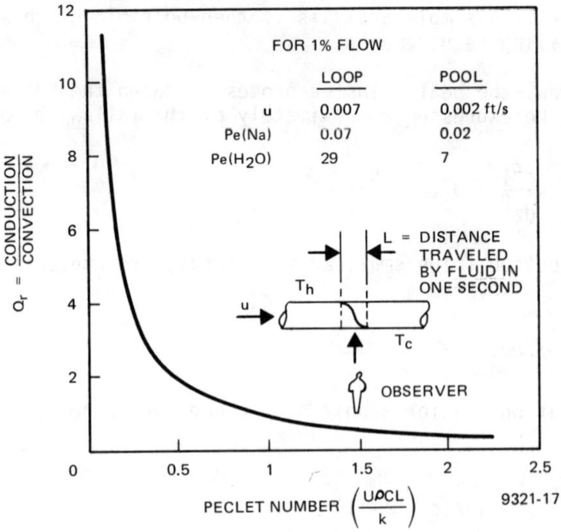

Fig. 2 Effect of Peclet Number on Plenum Heat Transfer

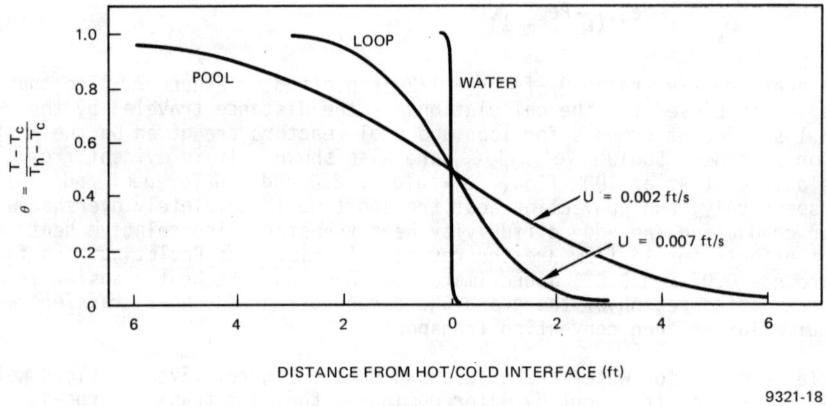

Fig. 3 Temperatures in Water and Sodium Pools Under Stratified Flow Conditions (time = distance to H inlet/fluid velocity)

for the design of the natural test facility. If the Peclet number must be considered, then thermodynamic similarity requires that:

$$(u\rho cL/k)_{model} = (u\rho cL/k)_{prototype}$$

For water/NaK simulation this leads to

$$u_R L_R = \alpha_R = \frac{0.006}{1.5} = 0.004$$

This is certainly not a practical scaling ratio and is a further demonstration against water as the simulant fluid. It also does not appear feasible to match the Biot number in a sodium system with water because of the poorer heat transfer between structure and coolant in certain flow regimes. This is discussed in more detail below.

Structure and Flow Passage Simulation

Figure 4 compares the heat transfer coefficients of water to both NaK and Na for a range of velocities and effective flow diameters. The figure highlights key design points which reflect scaled versions of the current Conceptual Design Study (CDS)-DRACS concept. As the velocity and pipe diameter are reduced (i.e., reflective of either low flow natural convection conditions and/or reduced scale modeling) conduction through the laminar boundary layer becomes a larger portion of the overall heat transfer coefficient, thus accounting for the increase in Na/NaK-to-water ratios. Even at full-scale simulations, utilization of water to model the behavior of core channels at low flow conditions would introduce significant errors in the heat transfer coefficient. It is also apparent that water cannot simultaneously simulate both Na and NaK.

While models can be constructed to compensate for these differences during steady-state simulations, extrapolation of water test results for transient behavior could lead to erroneous conclusions. Although the effect in the NaK 12-in. diameter piping is not nearly as pronounced, consideration of the smaller diameter NaK tubing in the heat exchangers would lead to similar arguments regarding the adequacy of a water simulation. Analagous arguments for the thermodynamic behavior of the plenum (based on matching Peclet numbers) and for structure heat transfer (based on matching Biot and Stanton numbers) constructed above further demonstrate the inadequacy of water under low flow conductance influenced conditions.

CONCLUSIONS

Natural convection systems require different hydrodynamic and structure scaling laws. Hydrodynamic scaling requires the equivalent of Richardson's number (i.e., L/u_o^2), thermodynamic requires matching of the Peclet number (i.e., Lu_o). Therefore scaling requires some compromise in hydrodynamic and thermodynamic similarity, and test scale should be as large as possible.

Water cannot adequately simulate thermodynamic heat transfer during buoyancy dominated transient flow conditions. Key phenomenon is the increasing importance of conduction effects at low flows — laminar boundary layer effects in small flow passages, Peclet number similitude in large plenums, and Biot and Stanton number similitude for structural heat transfer.

A single-fluid water test facility cannot simulate the two-fluid prototype system.

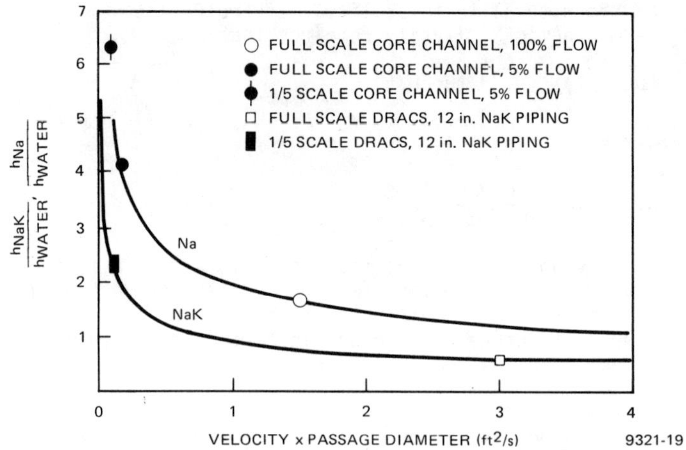

Fig. 4 Comparison of Na/water and NaK/water Heat Transfer Coefficient Ratio in Flow Passages

NOMENCLATURE

A	coolant flow area
C_p	specific heat
d	pipe diameter, pipe wall thickness
h	heat transfer coefficient between coolant and structure
k	thermal conductivity
L	depth (or height) of fluid region, length of component
P	dimensionless pressure
Q	dimensionless volumetric heat generation rate
q'''	power density
R	dimensionless radial coordinate
ΔT_o	reference temperature difference
T	temperature
U	dimensionless velocity
u	velocity vector
W	dimensionless axial velocity
X	dimensionless axial coordinate
Z	dimensionless vertical coordinate

Greek symbols

α	thermal diffusivity
β	isobaric thermal expansion coefficient of coolant
ν	kinematic viscosity

μ	viscosity
ρ	coolant density
θ	dimensionless temperature difference
τ	dimensionless time

Subscripts

c	cold
h	hot, hydraulic diameter
i	loop component i
o	initial value at steady-state prior to transient
R	ratio (test facility/full-scale facility)
TC	thermal centers

REFERENCES

1. A. F. Lillie, "Pool Reactor Model Hydraulic Test — Final Report," N322TI310-002, Rockwell International, Energy Systems Group (May 21, 1979)

2. H. S. Carslaw, J. C. Joeger, "Conduction of Heat in Solids," 1947, pp 270-271

[illegible] Gravity [illegible]

[illegible] ... (ΔT) [illegible]

Development of Computer Codes for Decay Heat Removal Analysis in Reactor Vessel and Primary Loop System

HIDEHITO KINJO, HOZUMI WADA, YOSHIYUKI OKUBO, and TAKASHI SAWADA
Advanced Reactors Engineering Department
Nuclear Development Center
Mitsubishi Atomic Power Industries, Inc.
1--297, Kitabukuro, Omiya, Saitama 330, Japan

ABSTRACT

Transient thermal-hydraulic behavior in a primary coolant loop system and a reactor vessel under forced and natural circulation conditions have been investigated using two newly developed computer codes, PANAC and PTAR. PANAC is a one-dimensional plant dynamics simulation code for overall thermal-hydraulic analysis in a whole primary loop system, while PTAR is a detailed two-dimensional code to be used in a reactor vessel. The calculated results of the codes have been compared with experiments; thermal transient tests in simulated reactor upper plenum, natural circulation tests in 50 MW steam generator test facility and natural circulation tests in plant dynamics simulation test facility. It has been found that both codes are capable of long-term simulation of thermal-hydraulic coolant behavior in a primary loop system and a reactor vessel with sufficient accuracy and reasonable computing time, and more accurate and reliable results are obtained from the coupled version of these codes to be used in actual designing and research.

1. INTRODUCTION

Decay heat removal from the core during various transients and accident conditions can be accomplished if a sufficient flow through the reactor vessel continues to exist within the primary coolant system. In evaluating the decay heat removal capability through the primary system of a Liquid Metal Cooled Fast Breeder Reactor (LMFBR) plant, considerations should be given to detailed prediction of thermal-hydraulic behavior of the coolant for a long time from the initiation of the transients [1].

The principal purpose of this study is to develop and verify experimentally an analytical procedure to predict the transient thermal-hydraulic behavior in a primary coolant system with sufficient accuracy. The transients to be covered extend from forced convection to natural circulation regimes. Both detailed analyses and long-term simulations are required for predicting various phenomena in key components in the whole primary system, e.g. especially the transient heat transfer mode between complicated boundaries and the buoyancy effect on the coolant flow at low flow rate in a reactor vessel. Two computer codes, PANAC and PTAR, have been developed balancing accuracy with cost of computation under the above requirements. PANAC is a one-dimensional plant dynamics simulation code for overall thermal-hydraulic analyses in a whole primary loop system, while PTAR is for two-dimensional detailed thermal-hydraulic analyses in a reactor vessel. Furthermore, it is expected that the coupling of these codes would yield more improved and more reliable results, because PTAR can use the PANAC results as input conditions, and PANAC can refine its solution by using the PTAR solution.

This paper describes the models in these codes, and the results of code validation with the experimental data on natural circulation tests and thermal transient tests conducted at Takasago Technical Institute of Mitsubishi Heavy Industries, Ltd. (MHI) and Oarai Engineering Center of Power Reactor and Nuclear fuel Development Corporation (PNC).

2. DESCRIPTION OF ANALYTICAL TOOLS

2.1 Analytical Model of PANAC

PANAC has been developed to calculate the temperature and flow transients in primary loops taking into account natural convection and pump driving forces. This code contains models for main components in the primary coolant system, such as the reactor, intermediate heat exchangers (IHXs), auxiliary core cooling system intermediate heat exchangers (ACCS-IHXs), pumps, and hot-leg and cold-leg pipings. The governing equations in the model are based upon a one-dimensional form of mass, momentum and energy conservations [2]. Components within a reactor vessel are treated as a bulk model. For example, the reactor core is modeled by a single average channel, and perfect mixing of coolant temperature in both lower and upper plena is assumed. The IHX model represents a vertical shell-and-tube type with straight heat transfer tubes of counterflow design. The IHX tube bundle is modeled by the single equivalent tube approximation. The coolant temperature in the loop in calculated by a one-dimensional energy equation taking into account the time-delay in energy transport, the decay heat generation in the coolant and the heat exchange with the piping wall. The piping wall is assumed to be perfectly insulated from the atmosphere. Although incompressible single phase liquid is assumed for the coolant, the density change of the coolant with its temperature is considered.

The mass flow rate is calculated from the following equation in the loop taking into account the pump head, natural convection head and pressure drops throughout the loop, assuming a single mass flow rate at any instant;

$$\frac{1}{g_c} \frac{\partial}{\partial t} \frac{w}{A}\, dX = P_{pump} - \Delta P + P_{natural} \qquad (1)$$

where

$$P_{pump} = - \oint \frac{\partial P}{\partial X} dX$$

$$\Delta P = \oint \frac{\zeta}{D} \frac{w^2}{2 g_c \rho A} dX$$

$$P_{natural} = \oint \rho \frac{g}{g_c} C_X dX$$

The natural convection head $P_{natural}$ is calculated by integrating the coolant density along its height throughout the entire loop.

PANAC is capable of simulating coastdown of the primary loops and switching over to pony motor operation. An asymmetric operating mode among the multi-loops, however, is not simulated in the current version. Transients in the secondary loop are specified as a user's input.

2.2 Analytical Model of PTAR

The reactor vessel to be modeled is divided into liquid and non-liquid regions, where the liquid region represents coolant, and the non-liquid and/or solid regions include core assemblies, core debris bed if any, upper core structures, neutron shield, core barel, reactor vessel wall, etc. The sodium coolant is assumed to be of single phase and incompressible, but spatial density changes due to temperature difference are taken into account, which are the source of buoyancy effect in the reactor vessel. The fluid is assumed to be Newtonian, and to follow the Stokes' hypothesis regarding its relation between the stress and rate-of-deformation tensors. The turbulence effects on momentum and energy transfer are effectively included in the molecular diffusion term with semi-empirical mixing parameters. With these assumptions, the governing equations are:

Conservation of Mass

$$\frac{\partial \rho}{\partial t} + (\nabla \cdot \rho \vec{v}) = 0 \quad (2)$$

Conservation of Momentum

$$\frac{\partial \rho \vec{v}}{\partial t} + [\nabla \cdot \rho \vec{v}\vec{v}] = - [\nabla P] - [\nabla \cdot \tau] + \rho \vec{g} \quad (3)$$

$$\text{where } \tau = - (\mu + \rho \varepsilon_M)(\nabla \vec{v} + t\nabla \vec{v})$$

Conservation of Energy

$$\frac{\partial \rho CpT}{\partial t} + (\nabla \cdot \rho \vec{v} C_p T) = S(\vec{r},t) + \{\nabla \cdot (k + \rho C_p \varepsilon_H)\nabla T\} \quad (4)$$

The code calculates the two-dimensional transient temperature and velocity distributions in the liquid region by solving the above equations as a boundary value problem in space and an initial value problem in time. For the non-liquid region, temperature distributions are calculated as the solution of heat conduction problems. Heat transfer between the solid and the liquid region can be consistently determined as a function of temperature difference with flow and geometry dependent heat transfer coefficients.

In the code, the thermal coupling constants between finite meshes can be altered from the original ones obtained by direct integration of the above equation. This is achieved by introduction of the effective heat transfer area and volume, heat flux and equivalent diameter. The numerical techniques employed in this code are based upon a finite difference scheme obtained by integrating the conservation equations under specified thermal-hydraulic boundary conditions. In this finite difference scheme, a technique of fractional mesh division and inclined boundary mesh is employed to treat the complicated geometry and the resulting coolant behavior. This technique allows a much smaller number of meshes than ordinary mesh divisions. Automatic time step control is also incorporated to guarantee its accuracy. Furthermore, for reducing computational time, a semi-automated choice of initial guess, upwind differencing scheme and overrelaxation method are introduced to accelerate convergence [3-5].

3. VERIFICATION OF PANAC BY EXPERIMENTS

3.1 Comparison with Natural Circulation Tests in 50MW SG Test Facility

In order to check the component models in PANAC, such as pipings and IHX models, under natural circulation, comparisons were made with the data of the natural circulation tests carried out in the secondary loop system of the 50MW Steam Generator Test Facility. As shown schematically in Fig. 1, the loop used for the tests consists of: a vertical shell-and-tube type IHX with straight heat transfer tubes of counter-flow design; a helical coil type evaporator; a vertical single-stage centrifugal pump; a 12 inch flow control valve; and 12 inch pipings. In this configuration, the IHX functions as a heat source and the evaporator, a heat sink. The experiments included: steady-state natural circulation tests with temperature difference between the heating and cooling sides and pressure drop through the loop taken as parameters; and transient-state natural circulation tests by tripping the sodium pump from forced convection flow and changing the inlet temperature of the IHX primary side in cyclic transients. Good agreements were observed between the analyses and the measurements for the steady-state test results.

As the next step, the component models in PANAC, such as IHXs and pipings, were also compared with extremely low flow transient data. As the result, IHX models were converted so that the heat transfer rate of the tube bundle would be evaluated low in the natural circulation flow. In the piping models, it is reasonable to consider an appropriate heat capacity for the piping wall in the low flow regime. The resultant thermal responses at two points in the hot-leg pipe are shown in Fig. 2.

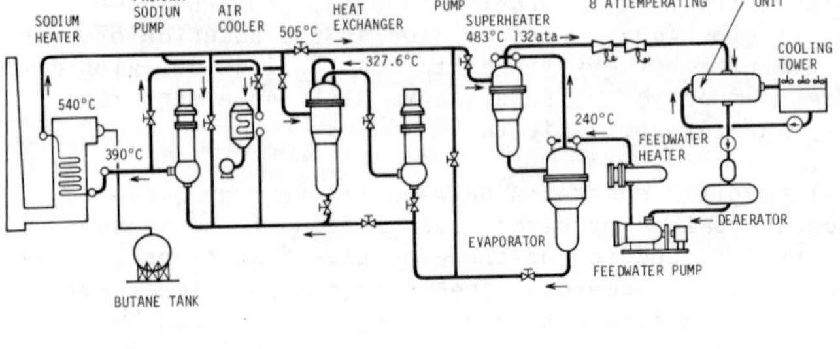

Fig. 1 Schematics of 50 MW SG Test Facility

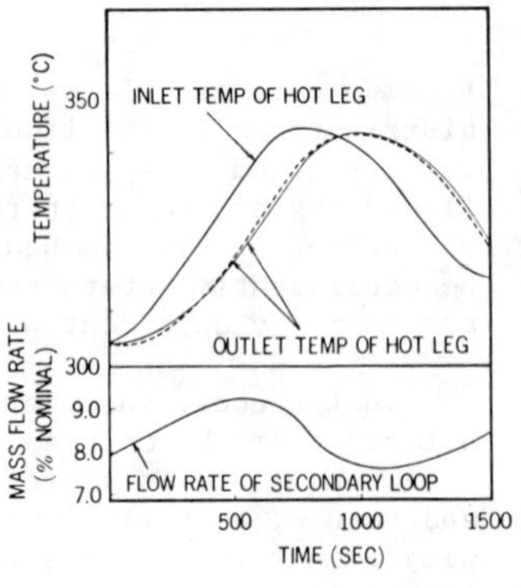

Fig. 2 Comparison of Thermal Response in Hot-Leg Pipe

3.2 Natural Circulation Tests in 1/5-Scale Plant Dynamics Simulation Facility

Description of experiment. Natural circulation tests were performed with a small scale model (approximately 1/5) of the primary coolant loop system of the prototype LMFBR to obtain data for code verification using water as a test fluid, at Takasago Technical Institute, MHI, under the auspices of PNC. The test loop used in this investigation, shown in Fig. 3, has a simulated reactor vessel, an intermediate heat exchanger, an air cooler, a heat exchanger simulating a steam generator, a primary and a secondary flow pump. The ratios of the test model design parameters to those of the prototype LMFBR are about 1/2 for the elevation difference of the key components, 1/2 for the entire loop pressure loss, 1/40 for decay heat capacity, and 8/10 for fluid transport time.

The typical experimental procedure is to simulate the phase of transition to the natural circulation regime from the nominal operating conditions after a pump trip. Trasients to natural circulation were obtained by first establishing the initial hot-leg and cold-leg temperatures at 60°C and 110°C,respectively, and $7m^3$/hr of primary flow rate, and then switching off the pumping power. The model-reactor power is decreased simulating the actual reactor decay heat curve. The feature of the typical test cases compared with the calculations of PANAC and the coupled code are described below:

(1) Natural circulation test with an abrupt increase of core flow immediately before pump trip (Run 1). The procedure of the abrupt flow increase is adopted in order to simulate the initial core outlet temperature drop in the actual reactor.

(2) Natural circulation test with no abrupt increase of core flow (Run 2).

(3) Forced convection flow test with the lowest pumping power (Run 3).

The test parameters and flow conditions are shown in Table 1.

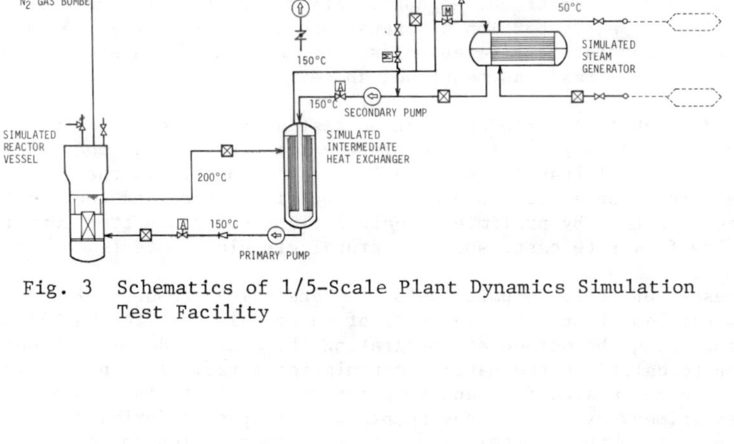

Fig. 3 Schematics of 1/5-Scale Plant Dynamics Simulation Test Facility

Table 1 Test Parameters and Flow Conditions

Run	Primary System					Secondary System					Remarks
	Hot-leg Temp (°C)	Cold-leg Temp (°C)	Flowrate [Initial] (m^3/hr)	Flowrate [Increased] (m^3/hr)	Flow Coastdown Mode	Hot-leg Temp (°C)	Cold-leg Temp (°C)	Flowrate [Initial] (m^3/hr)	Flowrate [Increased] (m^3/hr)	Flow Coastdown Mode	
1	110	60	7	44	Controlled (45 sec)	90	40	7	27	Controlled (45 sec)	*
2	110	60	7	7	Controlled (45 sec)	90	40	7	7	Controlled (45 sec)	**
3	110	60	7	44	Free	90	40	7	27	Free	***

*Reference case, ** Hot-shock, *** Forced convection flow at the lowest pumping power

Results and discussions. The calculated flow transients on the natural and forced circulation tests are compared with the experimental results in Fig. 4. The natural circulation flow rates of the primary loop are calculated to be about 1 to 2% of nominal which agree well with the test results of Runs 1 and 2. The calculated flow rates increase quickly after the flow coast down, and at about 300 sec the flow rates settle nearly at a constant value, while the experimental values increase very slowly. The calculated flow rates agree well with those of the tests although the calculated results are slightly larger than those of experiments for the primary loop. The simplified heat transfer model for plenum mixing employed in PANAC has its limitations in treating complex flow transients such as buoyancy-induced flow stratification at low flow regimes. Therefore, the calculated temperature transients at, for example, the core outlet and the reactor outlet nozzle are somewhat different from the test results during the early portion of transients as shown in Figs. 5 and 14. In spite of these temperature discrepancies, the calculated transient flow rates agreed well with the tests as mentioned above.

In Run 3 which is a forced convection test at the lowest pumping power, both the calculated flow and the temperature transients show good agreement with the experimental results as shown in Figs. 4 and 5. In the well mixed case where forced convection is dominant, the mixing mode and the input data of PANAC which is set up by preliminary calculations do much better than in the extremely low flow rate cases such as natural circulation tests.

As a result of these comparisons of the analytical results with the sodium loop and water loop tests, the validity of the models used in the PANAC code is affirmed including the method of integrating the coolant density throughout the entire loop to calculate the natural circulation force. It can be said that the PANAC code calculated flow and temperature transients in good agreement with the experiment except for the temperature responses during the early stage of the transient to the natural circulation regimes, which do not make much effect on the general flow behaviors.

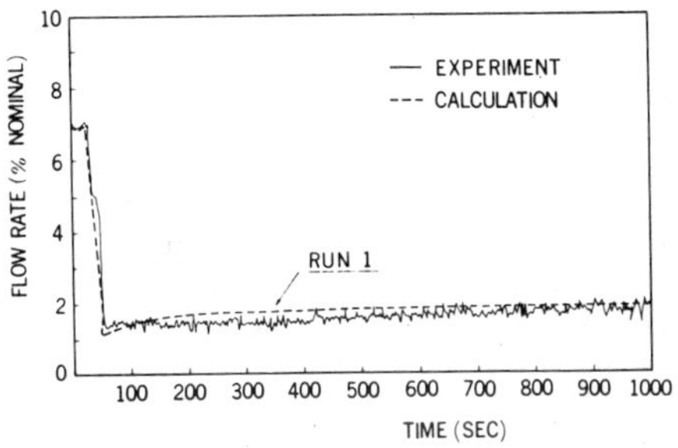

(a) Run 1

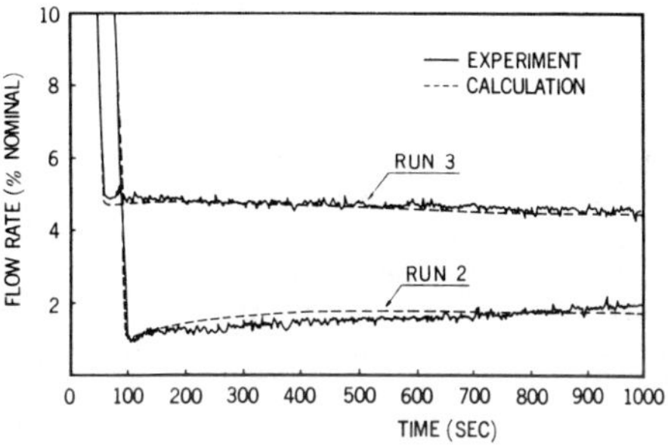

(b) Runs 2 and 3

Fig. 4 Comparisons of Flow Transients in 1/5-Scale Tests

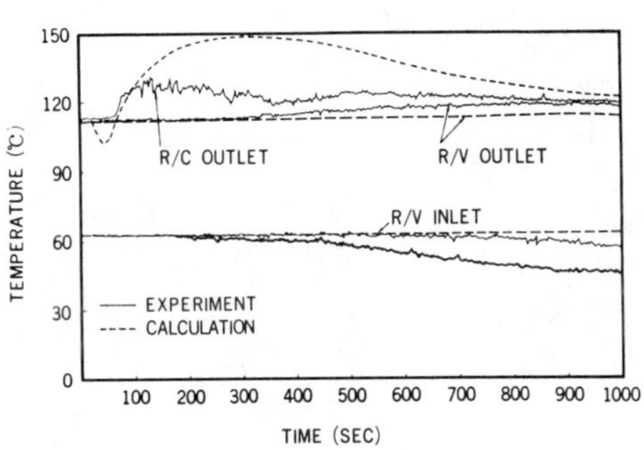

(a) Run 2

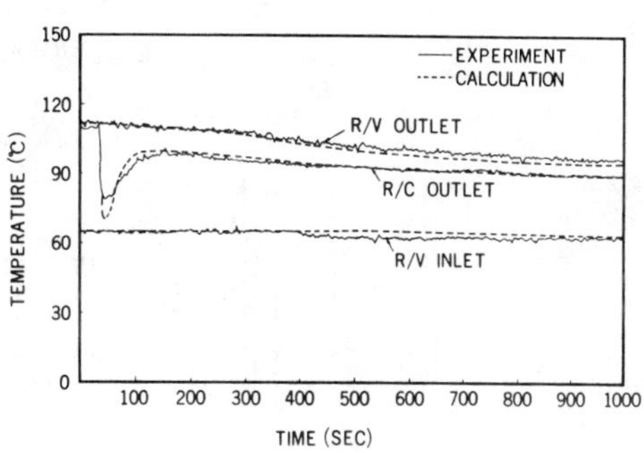

(b) Run 3

Fig. 5 Comparisons of Temperature Transients in 1/5-Scale Tests

4. VERIFICATION OF PTAR AND THE COUPLED CODE BY EXPERIMENTS

4.1 Thermal Transient Tests on 1/6 Scale Model of Prototype Reactor Vessel Upper Plenum

Description of experiment. An accurate prediction of the thermal response of the fluid in the upper plenum under various transient and accidental conditions is important for structural and safety evaluations. Thermal transient tests were conducted on a 1/6 scale model of the prototype reactor vessel upper plenum with and without an inner barrel, at PNC Oarai Engineering Center, to study the efficiency of the barrel on sodium mixing and to obtain various data for structual designs and plant dynamics analyses. Typical upper plenum configuration of the test section along with thermocouple locations is shown in Fig. 6. The thermocouples were located in 4 grid planes passing through the plenum at 0°, 100°, 180° and 280° circumferential direction along the axis for measurement of sodium temperatures. Inlet flow distribution from simulated core assemblies was controlled by the flow collector. A schematic representation of the sodium test loop is shown in Fig. 7. The loop is basically composed of a main pump, a high and a low temperature tank, a storage tank, a mixer, a plenum test section and a purification system. The inlet conditions

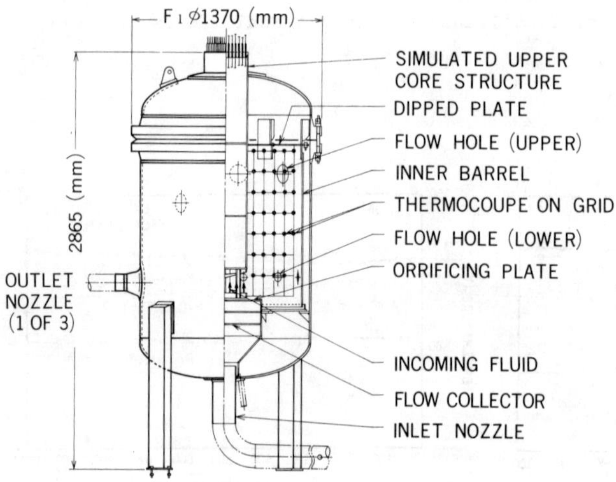

Fig. 6 Configuration of Test Section

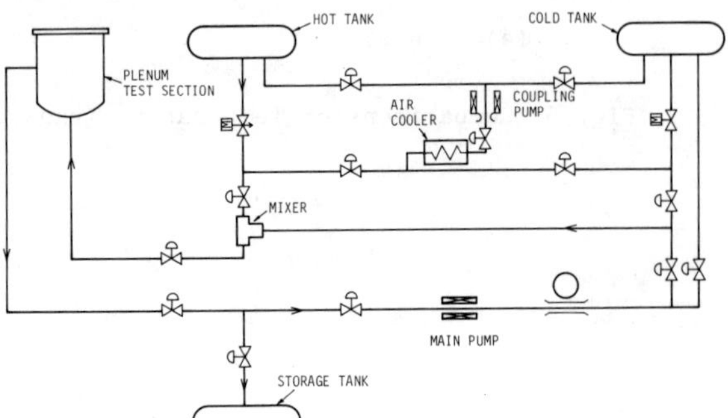

Fig. 7 Schematics of Sodium Flow Loop

such as the rate of the temperature changes are arbitrarily controlled by the flow valves V-1 and V-2 in the constant flow scram transient tests. Typical test conditions compared with the calculations of PTAR are given in Table 2. The test parameters were selected to study the effects of the inlet flow rate, the inner barrel and inlet temperature variation modes on the thermal transients.

Results and discussions. Preliminary investigations have shown that the measured temperature data have sufficient reproducibility and axial symmetry in the plenum under various flow conditions and geometrical configurations. Sensitivity analyses on the turbulent mixing parameters have also been performed, showing that the turbulent diffusivity of heat ε_H is the relatively stronger parameter to the temperature transients in the plenum than that of momentum ε_M, and that the flow patterns are not influenced by ε_M at this range of inlet flow rates.

Fig. 8 shows the calculated flow patterns in the plenum with and without the inner barrel. The calculations are in good agreement with the previously conducted test results with water loops which indicated the presence of two torroidal recirculation zones above and below the principal jet flow. The calculated axial flow rate distributions at the flow holes on the inner barrel have been adequately close to the measured data. For no inner barrel plenum case, the similar tendencies of the flow pattern are observed except that the entering fluid flows straightly toward the outlet nozzles.

Transient temperatures for Runs 1-4 are compared with the experiments in Fig. 9. At the outlet nozzles and its neighbnoring positions, the numerical calculations slightly overpredict the transients both with and without inner barrel. This may be explained since the main flow quantity toward the flow holes and outlet nozzles tends to be overestimated in the present two-dimensional slit modeling of flow holes and three discrete outlet nozzles, while the residual upward recirculation flows are underestimated. However, the overall trends of agreements with the measured data are satisfactory at almost all locations in the plenum, and as stated above, it has been found that both the flow and the temperature fields within the plenum are nearly symmetric despite the presence of the flow holes and the outlet nozzles. It has also been found that a constant mixing parameter used in the code could be widely applied to the various conditions such as inlet flow rate, temperature change and geometrical configurations.

Table 2 Test Parameters and Conditions

Run No.	Inlet Temp. Ramp Mode [(Hot/Cold)-Shock]	Maximum Temp. Variation (°C)	Inlet Flowrate (10^{-3} m^3/sec)	Inner Barrel
1	Cold	-160	85	With
2	Hot	+160	106	With
3	Cold	-90	130	Without
4	Hot	+160	89	Without

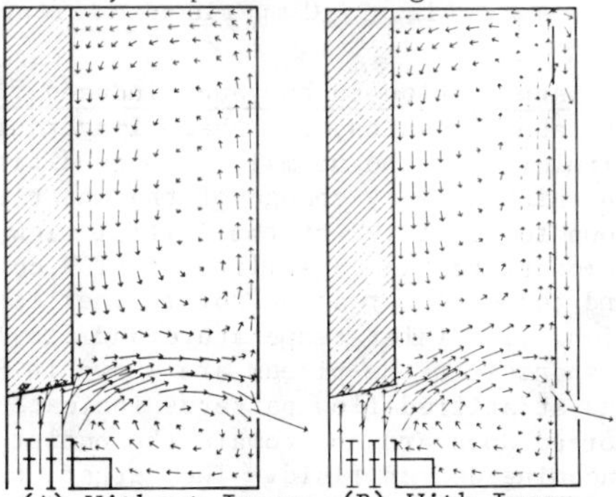

(A) Without Inner Barrel (B) With Inner Barrel

Fig. 8 Calculated Flow Patterns in the Plenum

(A) With Inner Barrel

(B) Without Inner Barrel

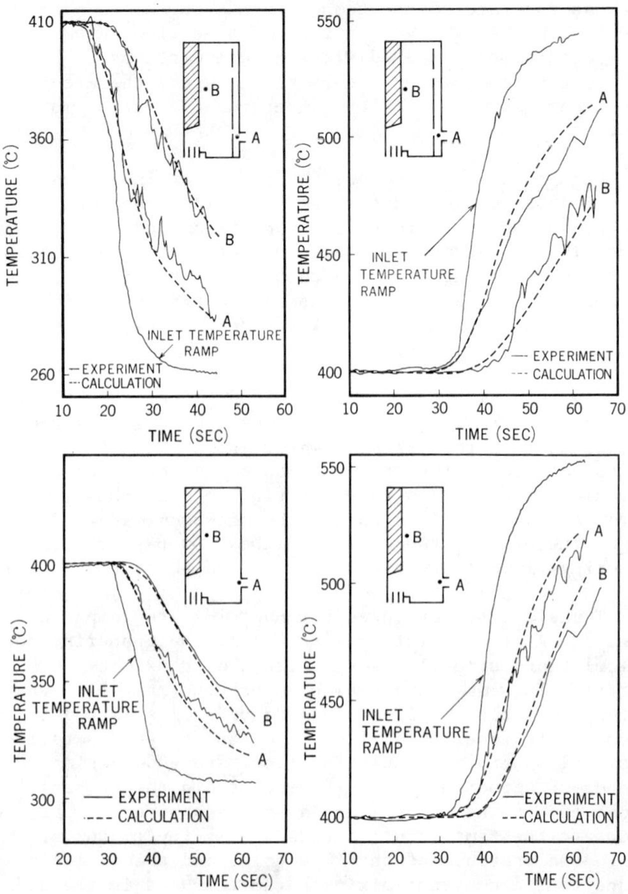

Fig. 9 Comparisons of Temperature Transients in the Plenum

<u>Application to buoyancy-induced flow stratification analysis</u>. Based upon the above results, the thermal-hydraulic transients have been numerically simulated on the normal scram condition of the prototype reactor upper plenum in which density change of the entering fluid is accompanied by a flow coast down to about 10% of the initial flow rate during the first fifty seconds, to demonstrate the capability of PTAR code for long-term and detailed analyses, and to get information for further experimental tests on the flow stratification. The inlet temperature and flow conditions with resultant outlet thermal responses in the plenum are shown in Fig. 10. As shown in Fig. 11, the initial and stratified flow patterns indicate that the main flow at about 50 sec are forced downward and toward the outlet due to negative buoyancy force caused by incoming denser fluid. The larger, but very weak, recirculation zone continues to exist for several hundred seconds, and gradually decays with the disappearance of negative buoyancy forces due to the isothermal-state recovery. At 2000 sec, the stratified flow patterns are diminished, and flow patterns similar to the initial ones are recovered. These processes of the buoyancy-induced flow stratification phenomenon show similar tendencies as those investigated by Lorenz, et. al. [6-7]. It may be said that these results would be

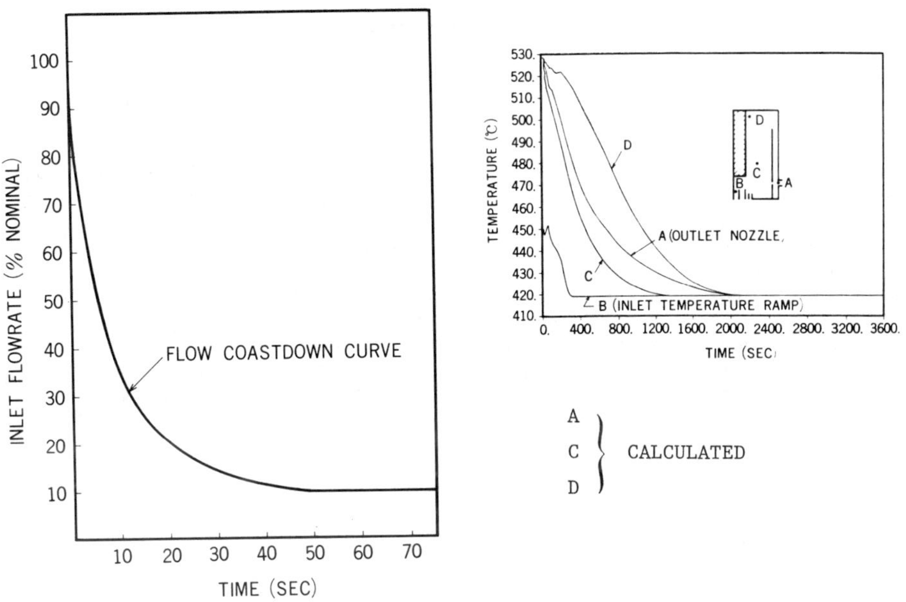

Fig. 10 Temperature and Flow Conditions and Calculated Results

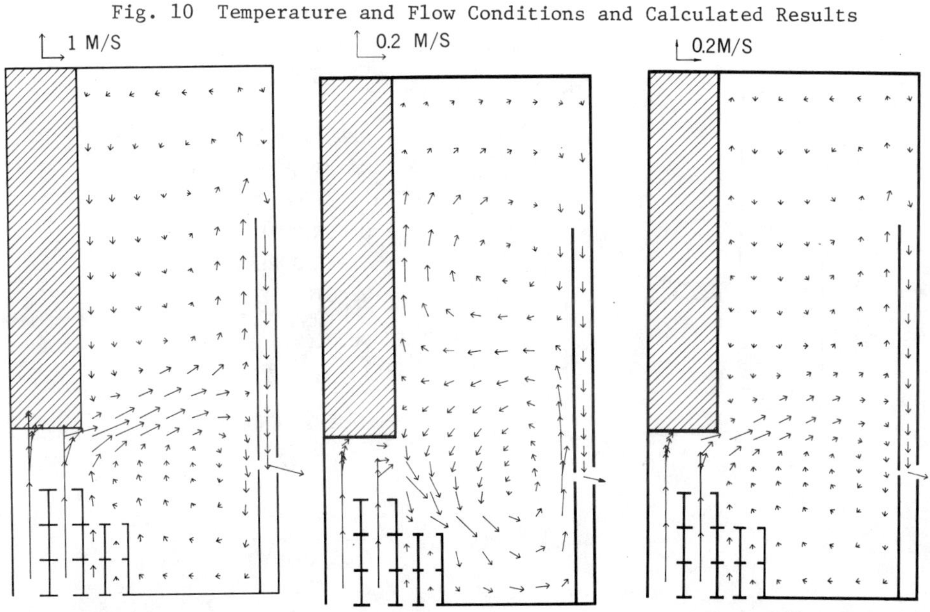

(A) Initial (B) Stratified (C) Recovered

Fig. 11 Calculated Flow Patterns in the Plenum

acceptable, and further validations by temperature and flow transient tests are planned in the near future.

4.2 Natural Circulation Tests in 1/5 Scale Plant Dynamics Simulation Test Facility

PTAR verification in the reactor vessel. To verify the model of PTAR concerning thermal-hydraulic behavior in heated core bundle geometry, the calculated results have been compared with the data of a heated-up test in the model reactor vessel conducted as one of the preliminary investigations on the static and dynamic characteristics of each key component. The schematic representation of the reactor vessel and the calculated flow pattern are shown in Fig. 12. The calculated temperature responses at the core outlet show, as given in Fig. 13, very good agreement with the experiments compared with the PANAC results. At the reactor outlet the superiority of the PTAR results is again very clear as in Fig. 13. This proves the importance of the corrected treatment of the heating section such as core bundle and the upper plenum mixing. These facts combined with the results mentioned in 4.1 show the validity of the PTAR code.

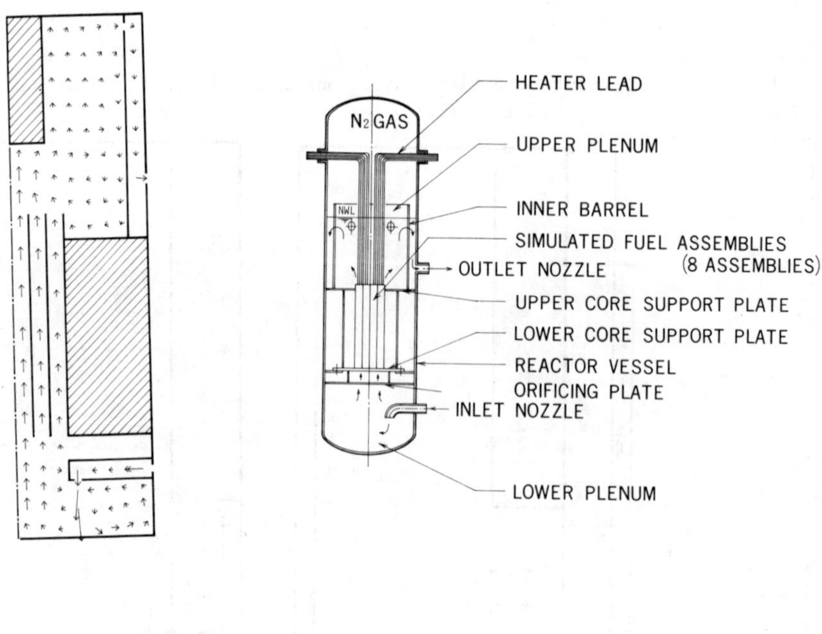

Fig. 12 Configuration of Model Reactor Vessel and Calculated Flow Pattern

Coupled code verification. It is expected that incorporation of detailed modeling of thermal mixing in the upper and the lower plena in the reactor vessel would improve the PANAC results of the natural circulation tests with the 1/5 water loop system in 3.2. The analyses of Run 1 in natural circulation tests in 3.2 have been repeated by the coupled code to check to what extent the calculated results could be improved compared with those of PANAC. The improvements are quite obvious as shown in Fig. 14. These may be due to the accurate thermal mixing model taking into consideration the upward and the downward transient convection flows in the reactor vessel as was pointed out in 4.1, and due to proper transmission of the thermal-hydraulic quantities between these codes at the inlet/outlet nozzles of the reactor vessel. Especially, the total pressure loss, the density head and the otulet temperature from the detailed analyses should be consistently connected into the governing equations of momentum and energy transfer in PANAC through the entire loop, and the boundary conditions for PTAR such as the inlet flow rate and temperature are derived from the PANAC part. It is to be noted however, that the analytical results obtained above are one of the sample testing cases using a simplified reactor vessel model and that the consistency of these thermal-hydraulic quatitites transmitted in the coupled code should be further examined under various conditions in complicated geometries of an actual reactor.

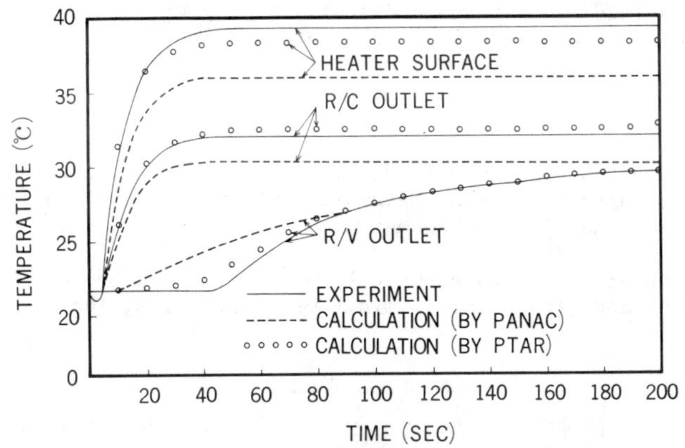

Fig. 13 Comparisons of Temperature Responses

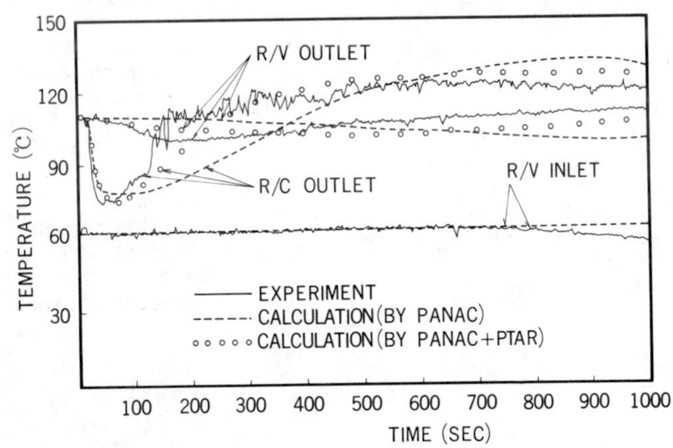

Fig. 14 Comparisons of Temperature Responses (Run 1)

5. CONCLUSIONS

Two computer codes, PANAC and PTAR, have been developed to evaluate decay heat removal capabilities. Numerical results of each code have been compared with the experimental data both from the natural circulation tests in plant dynamics test facility and the thermal transient tests in the upper plenum of the model reactor vessel under a scram condition. The results obtained are summarized as follows. 1) PANAC code is capable of long-term simulations of thermal-hydraulic behavior of coolant in a primary loop system with sufficient accuracy and reasonable computing time. Incorporation of more accurate pressure loss data, however, is to be done. A more elaborate model for the reactor vesel is necessary in order to consider, for example, the plenum mixing more properly. 2) On the other hand, PTAR does solve the detailed thermal-hydraulic behavior in the reactor vessel, therefore, more accurate and reliable results have been obtained from the coupled version of these codes. 3) It is clear that this type of numerical simulation yields detailed flow and temperature distribution information, which are helpful in safety analysis and nuclear reactor design.

It can be concluded that these codes will be a powerful tool for analyzing the decay heat removal under various transient and accident conditions.

ACKNOWLEDGEMENTS

Authors wish to express their thanks to Dr. N. Tanaka and Dr. A. Watanabe of PNC and Dr. A. Sugawara of FBR Engineering Office for their helpful suggestions and continuing encouragements.

*Work performed under contracts between Power Reactor and Nuclear Fuel Development Corp. and Mitsubishi Atomic Power Industries, Inc.

NOMENCLATURE

A	Flow area, Eq.(1)
C_p	Specific heat capacity, Eq.(4)
C_χ	Component along the gravitational force, Eq.(1)
D	Hydraulic diameter, Eq.(1)
$\bar{g}$	Gravitational acceleration, Eq.(3)
g_c	Gravitational constant, Eq.(1)
k	Thermal conductivity, Eq.(4)
P	Pressure, Eq.(2)
$\bar{r}$	Position vector, Eq.(4)
$S(\bar{r},t)$	Volumetric heat generation rate, Eq.(4)
T	Temperature, Eq.(4)
t	Time
$\bar{v}$	Velocity vector, Eqs.(2), (3) and (4)
W	Mass flow rate, Eq.(1)
χ	Flow path direction, Eq.(1)
$\varepsilon_M, \varepsilon_H$	Turbulent diffusivities of momentum and heat, respectively, Eqs.(3) and (4)
ζ	Pressure loss coefficient, Eq.(1)
μ	Viscosity, Eq.(3)
ρ	Coolant density, Eqs.(2) and (4)
τ	Viscous stress, Eq.(3)

REFERENCES

1. Agrawal, A. K. et al., "An Advanced Thermohydraulic Simulation Code for Transients in LMFBRs (SSC-L Code)," BNL-NUREG-50773, 1978.

2. Alstad, C. D., Isbin, H. S., and Amundson, N. R. "Transient Behavior of Single-phase Natural Circulation Loop System," AIChE Journal, Vol 1, No. 4, 1955.

3. Hirt, C. W. and Nicholes, B. D., "SOLA-A Numerical Solution Algorithm for Transient Fluid Flows," LA-5852, 1975.

4. Harlow, F. H. and Amsden, A. A., "A Numerical Fluid Dynamics Calculation Method for All Flow Speed, " J. Comp. Phys., 8, 197, 1971.

5. Sha, W. T. et al., "COMMIX-1: A Three-Dimensional Transient Single-Phase Component Computer Program for Thermal-Hydraulic Analysis," ANL-77-96, 1978.

6. Lorenz, J. J., Carlson, R. D. and Howard, P. A., "The Investigation of Buoyancy-Induced Flow Stratification in a Cylindrical Plenum," Proceedings of the 1976 Heat Transfer and Fluid Mechanics Institute, Stanford University Press, 1976.

7. Yang, J. W., "An Analysis of Transient Thermal Response in the Outlet Plenum of an LMFBR," BNL-NUREG-50521, 1976.

Effects of Buoyancy on LMFBR Plant Design

G. S. DRUCKER
Energy Systems Group
Rockwell International

ABSTRACT

Liquid Metal Fast Breeder Reactor (LMFBR) plant dynamic analyses have been performed in support of the design effort. Both pool and loop plants have been simulated to provide system and component requirements and inputs to safety studies. The scope and features of the Plant Pool Model (PPM) and the Plant Loop Model (PLM) codes are briefly described. Finally, the following examples of the impact of buoyancy and natural convection on dynamic system performance and component requirements resulting from these studies are presented:

1) Effect of buoyancy on low-flow pump performance requirements

2) Natural circulation and transient mitigation

3) Impact of buoyancy on loop hydraulic analysis.

1. INTRODUCTION

Dynamic performance analysis is an integral part of the Energy Systems Group, Atomics International Division (ESG-AI) LMFBR plant design effort. Numerous dynamic studies have been performed to determine operational and plant control requirements, [1,2,3] to provide inputs to safety-related studies, [4,5,6] and to determine the impact of transients on subsystem and component requirements [2,3,7,8]. These studies have been done in support of both loop and pool plant designs with carbide and oxide cores and once-through and saturated steam cycle combinations.

The PPM and the PLM digital simulation codes have been developed for this purpose. The scope of these models includes the primary, intermediate, and steam-feedwater heat transfer circuits. All major components are incorporated in the model. Emphasis was placed on providing sufficient detail of the dominant plant characteristics to provide realistic transient results. For example, natural circulation, fluid inertia, pump dynamics and characteristics, valve characteristics, control and protection system logic and response times, etc., are simulated.

Resulting from these studies have been a number of examples of the impact of buoyancy and natural convection on a dynamic system performance and on system and component requirements. Before these examples are discussed, a brief description of typical pool and loop dynamic models is given.

2. MODEL

The pool and loop LMFBR 1000 MWe electrical power plants simulated for the Conceptual Design Study (CDS) is depicted schematically in Figures 1 and 2 respectively. Only one of the four primary, intermediate, and feedwater circuits are shown in these figures. The primary source of energy for both the pool and loop Nuclear Steam Supply System (NSSS) is a carbide-fueled heterogeneous reactor core. A complete description of the core and its performance characteristics is given in Reference 9.

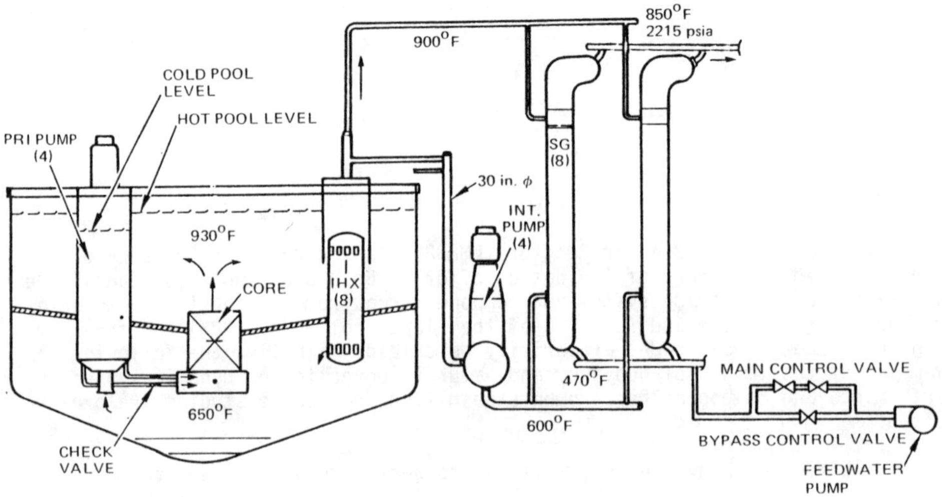

Figure 1. CDS Pool Plant Schematic

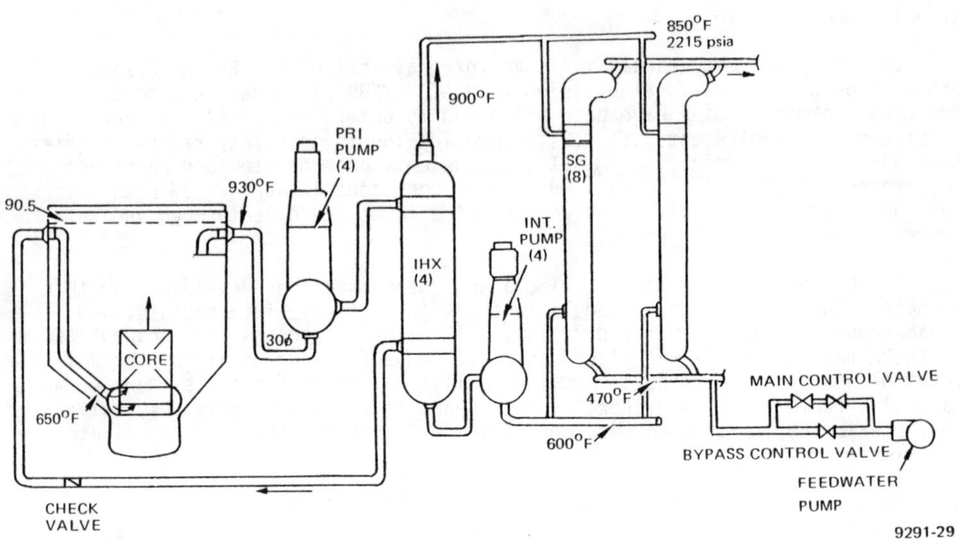

Figure 2. CDS Loop Plant Schematic

The pool-type NSSS primary system is contained within a single large vessel. Four primary pumps and eight intermediate heat exchangers (IHX)* circulate primary sodium within the reactor vessel and transport reactor thermal power to four independent intermediate heat transfer system (IHTS) loops. Superheated steam at 2215 psia and 850°F is generated by the eight once-through-type steam generators (two steam generators per IHTS loop). The heat transferred to the IHTS via the IHX is in turn transferred to the steam turbine-generator system via the steam generators.

The loop-type NSSS primary heat transport system (PHTS) consists of four loops each containing a pump and IHX. Each PHTS loop is linked with a IHTS loop to form four independent heat transfer circuits. The steam generator system is the same as described for the pool plant. The principal plant performance characteristics and operating parameters simulated are given in Table 1 for the pool and loop plant.

The PPM and the PLM digital dynamic codes were developed utilizing the Continuous System Modeling Program (CSMP)[10]. CSMP is an IBM program for dynamic simulation applications. The codes are capable of simulating multiple, primary, and intermediate heat transfer circuits to allow investigation of failures in a single heat transfer circuit. Major model features and scope are listed in Table 2. Even for this level of modeling detail, the ratio of real time to computing time on a 3033 IBM machine varies between 6 to 12.

3. EFFECT OF BUOYANCY ON LOW FLOW PUMP PERFORMANCE REQUIREMENTS

Sodium pump hydrostatic bearing flow is a function of the head developed by the pump. The developed pump head is in turn a function of its speed and flow. For steady-state isothermal operation, flow is proportional to speed and, therefore, developed pump head is proportional to pump speed squared (affinity laws). For the usual heat transport operating mode, the pump head is also a function of the loop or circuit buoyant head. For pump speeds corresponding to the plant operating range, the buoyant head has an insignificant effect on loop flow and, thus, pump head. However, for low speed (pony motor operation), the buoyant head is dominant and acts as a natural convection "pump". This is illustrated by the data shown in Figure 3. The pump and buoyant heads given are prior to and 600 s after a reactor trip. Figures 4 and 5 show the effect that natural convection flow has in reducing primary and intermediate pump developed head at about 600 s following plant trip. Thus, it is possible to have practically no, even negative, pump head available for inducing bearing flow at low pony motor speed and plant configurations having relatively large buoyant drives. It is apparent from Figures 4 and 5 that three factors determine the sodium-bearing flow for low-speed operation: (1) pony motor speed, (2) plant hydraulic profile (buoyant head), and (3) pump characteristics. Increasing pony motor speed, decreasing buoyant drive, and flat pump characteristics all will increase the pump head for low-speed operation. Even at relatively higher pony motor speeds, the magnitude of the bearing flow will be less than what would be predicted by applying the affinity laws (speed-flow ratio units). The reduction in pump head is greater for the intermediate pump. This reduction is even more severe in a plant with a saturated steam cycle. Since the steam generator is filled with saturated water, the sodium temperature approaches saturation temperature, except for a short sodium inlet region. This causes the thermal center of the saturated cycle steam generator to be near the top (see Figure 6). Thus the buoyant head is large which causes pump "runout" and reduced pump head. It is noted that the

* The most current CDS design employs four IHX's for the pool plant.

TABLE 1

PLANT PERFORMANCE AND OPERATING PARAMETERS

	Loop	Pool
Gross System Power (MWe):	1000	1000
Number IHTS Loops	4	4
Number IHX's	4	8
Number Primary Pumps	4	4
Reactor:		
Core Type	Carbide	Carbide
Power (MWt)	2600	2600
Outlet Temperature (^oF)	930	930
Reactor ΔT (^oE)	291	291
Flow Rate 10°	104.25	104.14
ΔP (psi)	109.4	113
IHX:		
Primary Flow/IHX lb/h (10^6)	26.04	13.02
Power (MWt)	650	325
Secondary Outlet Temperature (^oF)	899	899
Secondary ΔT (^oF)	311	311
Secondary Flow/IHX (10^6 lb/h)	24.205	12.104
Tube OD (in.)	1.0	1.125
Number Tubes	4244	2124
Active Tube Length (ft)	50.5	28
Shroud ID (in.)	104.5	85
Pumps:		
Primary Speed (rpm)	680.5	882
Primary Flow Rate (gpm)	62,451	59,877
Primary Head (ft)	500	342
Secondary Flow Rate (gpm)	54,600	54,600
Secondary Head (ft)	253	253
Secondary Speed (rpm)	846.0	828
Steam Generator:		
Sodium Flow (x 10^6 lb/h)	12.103	12.103
Number Modules/Loop	2	2
Power Number/Module (MWt)	325	325
Steam Pressure (psia)	2215	2215
Steam Outlet Temperature (^oF)	850	850
Feedwater Inlet Temperature (^oF)	470	470
Steam Flow Rate (10^6 lb/h)	1.217	1.217
Tube OD (in.)	0.625	0.625
Number Tubes	2435	2435
Active Tube Length (ft)	77	77
Shroud ID (in.)	57.25	57.25

TABLE 2

MODEL FEATURES AND SCOPE

Hydraulic:

- Polynominal Pump Characteristics (flow, speed, head, torque)
- Low-Speed Pump Characteristic
- Fluid Inertia
- Natural Convection Head Along the Flow Path
- Two Zone Upper Reactor Plenum Mixing
- Separate IHX and Reactor Flows (pool model only)
- Flow Reversal
- Reactor and Pump Vessel Levels (pool model only)

Thermal:

- Discrete representation of Variable Transport Delays (heat capacitance of walls included)
- Average and Peak Power Reactor Driver and Radial Blanket Fuel Assemblies Channels
- Hot Channel Temperatures
- Fixed 10 Node IHX (including shell)
- Variable 10 Node Steam Generator
- Pump Thermal Capacitance
- IHX Downcomer

Neutron Kinetics:

- Six Neutron Delay Groups
- Doppler and Sodium Worth Effects
- Prompt Jump Approximation
- Decay Heat

Mechanical:

- Pump and Pump Drive Coastdown

Plant Protection System (PPS):

- Sensor and PPS Dynamics
- PPS Logic

Plant Control System (PCS):

- Plant Load Demand
- Sensor Dynamic
- Controller Compensation
- Power Range Controls
- Post-Shutdown Controls

	PUMP HEAD (psi)	BUOYANT HEAD (psi)
PRIMARY PUMP		
100% SPEED AND FLOW	114.9†	0.264
4% SPEED AND FLOW*	0.184	0
4% SPEED AND 5.5% FLOW	0.142	0.205
INTERMEDIATE PUMP		
100% SPEED AND FLOW	85.2†	0.947
4% SPEED AND FLOW*	0.136	0
4% SPEED AND 8.5% FLOW	0.007	0.611

*ISOTHERMAL PUMP OPERATION
†PHASE A NOMINAL VALUES

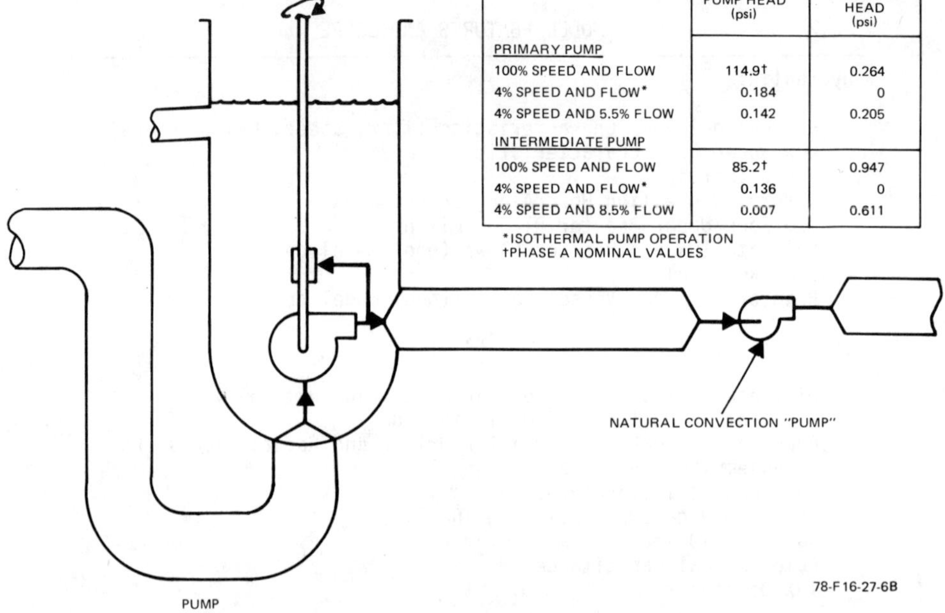

Fig. 3. Natural Convection "Pump" at Pony Motor Operating Conditions

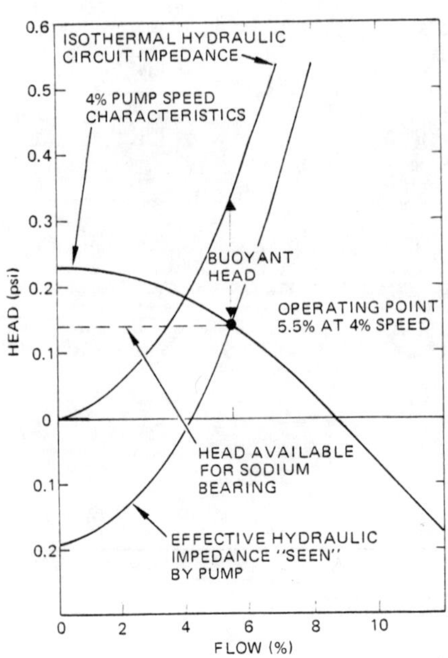

Fig. 4. Effect of Buoyancy on Low Speed Primary Pump Operation

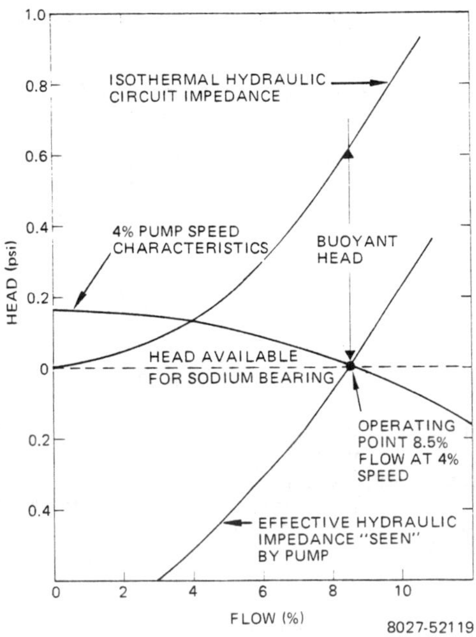

Fig. 5. Effect of Buoyancy on Low Speed Intermediate Pump Operation

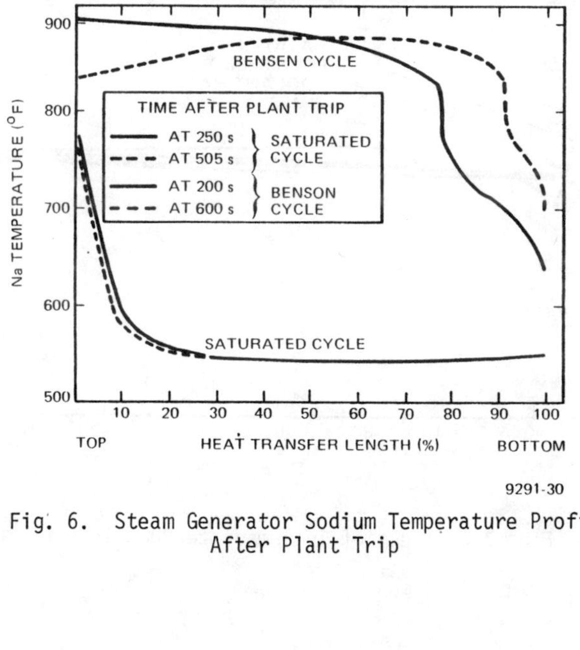

Fig. 6. Steam Generator Sodium Temperature Profile After Plant Trip

reduction in pump head decreases with time following plant trip as the loop thermal gradients, and thus, buoyant head collapses.

It is also possible to have reduced or even negative developed pump head during the transition to pony motor speed following a reactor trip. If the fluid inertia of the loop is not negligible, the coolant flow will decay more slowly than the pump speed, and the pump will provide a drag which reduces the developed head. The pump speed and flow decay half times for the primary and intermediate loops used in a typical loop plant are as follows:

	Half Times(s)	
	Pump Speed	Coolant
Primary Pump Circuit	3.10	0.192
Intermediate Loop	1.92	3.23

It is noted that the fluid half time is relatively small for the primary hydraulic circuit but is the dominant factor in the flow decay half time for the intermediate circuit. This is typical for most LMFBR plants, since the primary circuit is designed with short piping runs to minimize containment. This leads to a relatively small fluid inertia. Conversely, the intermediate circuit piping runs are long because the steam generators are located at the NSSS periphery.

The intermediate pump head for during the transition to pony motor speed is shown in Figure 7 for various ratios of pump speed to coolant decay

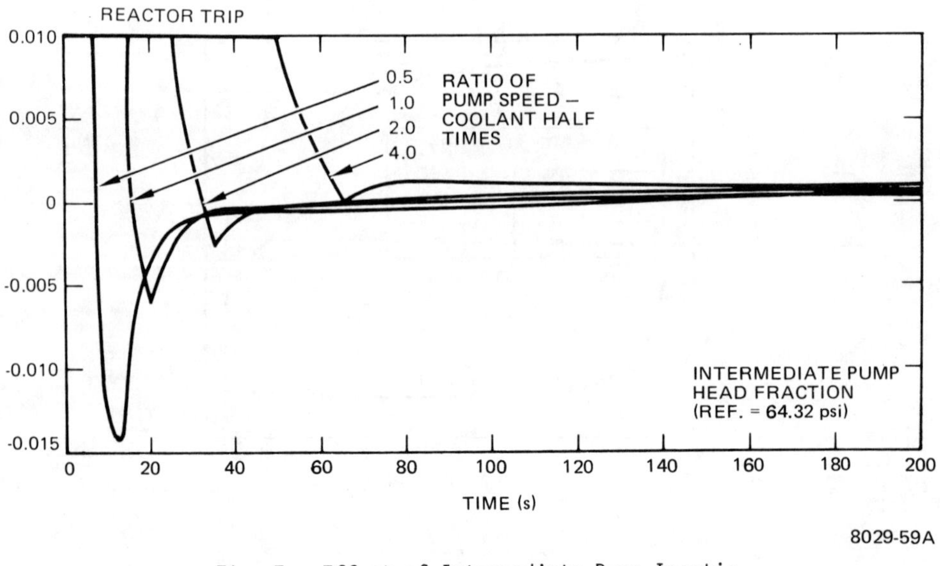

Fig. 7. Effect of Intermediate Pump Inertia

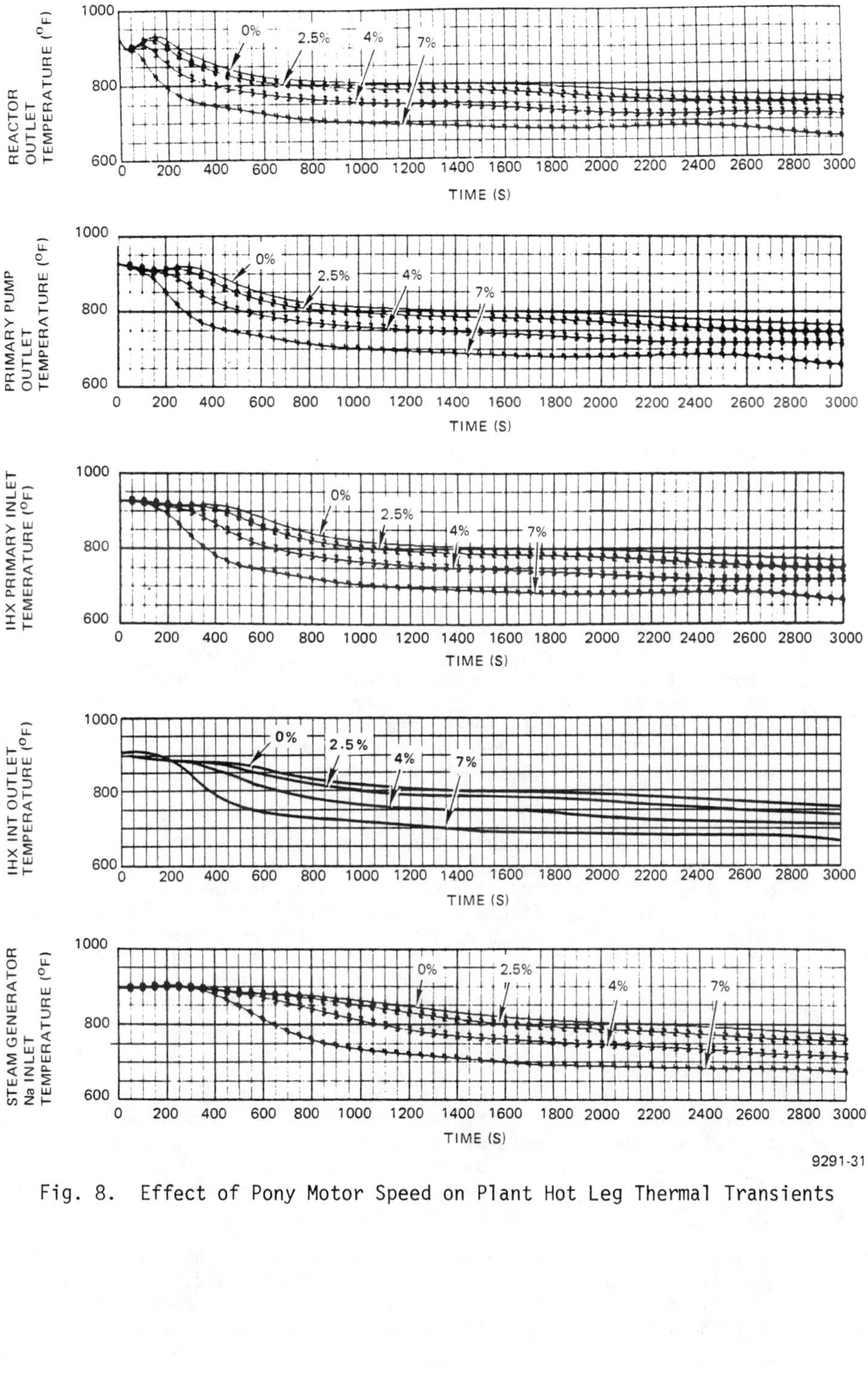

Fig. 8. Effect of Pony Motor Speed on Plant Hot Leg Thermal Transients

halftimes (T_R/T_F). A halftime ratio of four or more is required to prevent a flow reduction below the steady-state pony motor speed level. The pump rotating kenetic energy, thus half time, can be varied by selection of the motor drive L/D ratio, the addition of a fly wheel, or the type of pump drive used.

It is essential that the system effects discussed be considered in the pump design to assure its reliable operation under all expected conditions. This information was transmitted to Argonne National Laboratory (ANL) and has resulted in a revision to the LMFBR pump specification which now includes system buoyancy effects.

4. NATURAL CIRCULATION AND TRANSIENT MITIGATION

Because of the large number of reactor trips and its significance for most plant components, schemes which mitigate this transient were concentrated on for investigation.

During the first few minutes following reactor trip, the transient is governed solely by the match between the flow and fuel thermal energy decay characteristics. Thereafter, the plant shutdown control system determines the transient. Several post-shutdown control schemes to mitigate the effects of this transient were analyzed.

Early attempts to mitigate the hot leg by trimming feedwater flow with a steam temperature signal proved not feasible because of the excessive process delay. The next transient mitigation approach evaluated was to trim feedwater flow on steam generator outlet temperature control. For this concept, single-speed pony motors drive the sodium pumps. The cold leg temperature is held constant by controlling feedwater flow and maintaining steam pressure and feedwater temperature at essentially their pretrip values. For this shutdown control approach, the severity of the hot leg transient is dependent on the pony motor speed (see Figure 8). Reducing the pony motor speed "softens" the hot leg transients but impacts pump low speed design requirements, as discussed previously.

There was also concern about the potential for excessive peak clad temperatures for low pony motor speeds, natural convection being the limiting case. To investigate this, the parametric effects of pony motor speed, primary pump coastdown, and loop elevation effects on natural circulation were analyzed for peak power pins in both the driver and blanket core regions (oxide bullseye core)[11] for a reactor trip transient. The results (Table 3) underscore the fact that the decay of thermal energy stored within the pin must match the heat removal rate of the decaying sodium flow after a reactor trip to avoid high cladding transient temperatures in the peak power blanket pin. Consequently, matching the pump half time to the stored energy in the core can significantly reduce blanket pin cladding temperature transients (see Figure 9). However, increasing the pump half time also causes the average driver assembly exit temperature to decrease more rapidly. This may cause thermal downshock problems for the reactor upper internal structures. Therefore a balance must be struck in selection of the pump half time. The analytical results also show that elevation changes and pony motor speed produce minimal and no effect, respectively, on the peak cladding temperature.

Another method of softening the hot leg transients and increasing the IHX and steam generator design margins without a substantial reduction of pony motor speed involves only a minor modification to the above scheme. In this mitigation approach, the control scheme causes a programmed initial rise in

TABLE 3

PEAK 2-SIGMA MID-WALL CLADDING TEMPERATURES FOLLOWING REACTOR TRIP

Run Conditions	Parameter	Central Blanket		Outer Drive	
		Temperature (°F)	Time (s)	Temperature (°F)	Time (s)
7% Pony Motor Speed 0.5 x Nominal Primary Pump Inertia		1638	22	1009	36
Natural Convection (0% pony motor speed)	Primary Pump Inertia				
Primary Pump	0.25 x nominal*	1897	18	1231	5
Inertia Varied	0.5 x nominal	1638	22	1087	54
	0.75 x nominal	1469	17	1066	70
	1.0 x nominal	1370	13	1054	86
	1.5 x nominal	1271	9	1045	120
Natural Convection	Change in Core to IHX Center Distance				
0.5 x Nominal Primary Pump Inertia	-15.0 ft	1648	25	1124	55
	0 ft	1638	22	1087	54
	+15.0 ft	1629	20	1057	50

*Nominal Primary Pump Inertia = 225,000 lb-ft^2

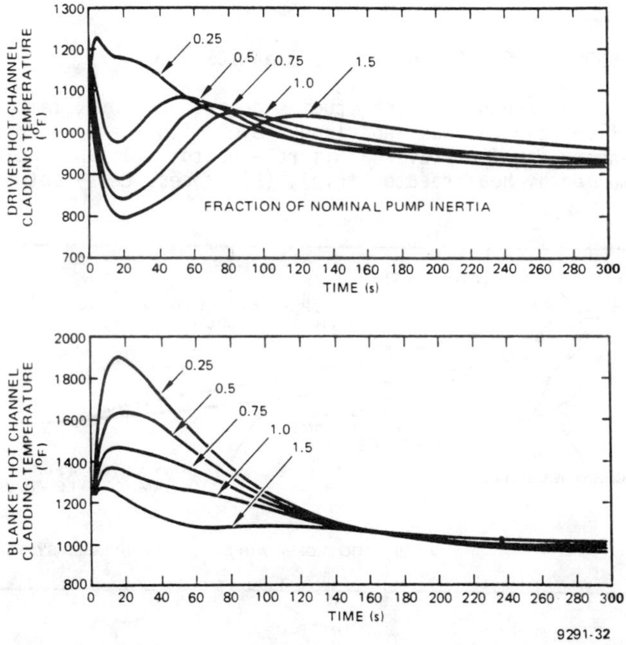

Fig. 9. Driver and Blanket Hot Channel Temperature Transient Following a Reactor Trip with Natural Convection

steam generator sodium outlet temperature and then a slow decrease to saturation temperature. The resulting rise in a steam generator sodium outlet temperature reduces the negative temperature rate of change at the steam generator sodium inlet. This, coupled with feedwater flow control to maintain the feedwater-to-sodium-flow fraction ratio less than unity, assures that the steam temperature will be tied to the sodium inlet temperature (superheat pinch point operation) and that a sudden increase in steam temperature to saturation temperature does not occur as the sodium inlet temperature slowly approaches saturation temperature. The results for 7% and 4% pony motor speed using the programmed cold leg temperature set point are shown in Figure 10. Although the steam generator upper tubesheet (hot leg) transient is milder for the 4% case, low-speed hydrostatic pump bearing design problems make the 7% pony motor speed a prudent selection as the reference approach. For the purpose of comparison, the steam generator upper tubesheet transients for the constant cold leg temperature control scheme and the Clinch River Breeder Reactor Program (CRBRP) steam generator with and without Protected Air-Cooled Condenser System (PACCS) are also presented in Figure 10.

In the event that additional transient mitigation schemes must be employed for higher outlet temperature plant designs, sodium flow control can be added to reduce transient rates even further. Two variable sodium flow control methods were investigated. One uses constant speed pony motors plus control valves, and the other employs variable speed pony motor pump drives to modulate the flow. Both of the variable sodium flow control methods are used in addition with the reference feedwater (cold leg temperature) control scheme. This permits independent control of the hot and cold leg temperatures. The control valve approach can be used with natural circulation and pony motor operation as a backup. This was the method used successfully for the Hallam Nuclear Facility.

5. IMPACT OF BUOYANCY ON LOOP HYDRAULIC DYNAMICS

In the course of conducting the studies described above, temporary intermediate sodium flow reversal was experienced even with the pony motors operating. The propensity for flow reversal is related to: (1) the rapid system cooldown (minimum decay heat reactor trip), (2) process delay between the heat

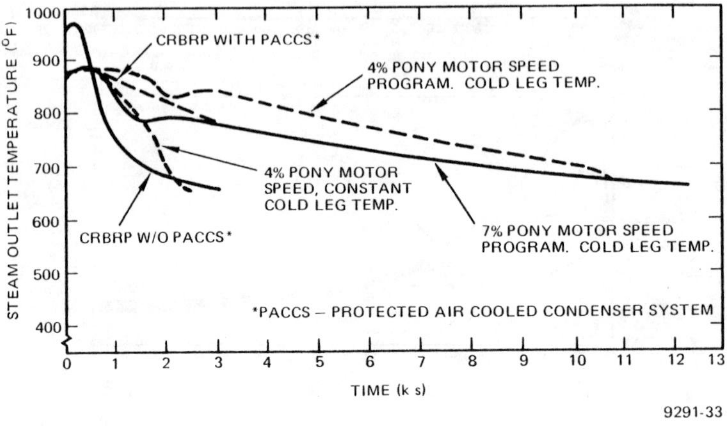

Fig. 10. Effect of Control Schemes in Mitigating Steam Generator Upper Tubesheet Trip Transient

source and sink, (3) temperature distributions (particularly in the steam generator), and (4) pony motor speed. Figure 11 shows the intermediate loop hot and cold leg temperatures following a plant trip with nominal and minimum decay heat. The transient histories are for a loop plant with a once-through steam generator. The drop in the IHX temperatures are not immediately transmitted to the steam generator because of the process delay. Therefore there is a temporary condition in which temperature profiles in the IHX and steam generator result in a decreasing and finally negative loop buoyant forces (see Figure 12). As the loop ΔT decreases, the buoyant head starts to increase. As indicated by Figure 11, the decrease in the loop buoyant forces is more severe for the minimum decay heat case where the peak temperature difference between the IHX outlet and steam generator inlet sodium temperatures is greater. The downward shift in the steam generator thermal center (see Figure 6) is also a strong factor in the reduction of the buoyant drive. The maximum developed pump head (at zero flow) for 7% pony motor speed is ∿0.49 psi. Therefore the intermediate flow reverses at ∿650 s after plant trip for the minimum decay heat case. For this case of a minimum decay heat reacter trip and a once-through steam generator, a pony motor speed of greater than 8% is required to prevent temporary flow reversal.

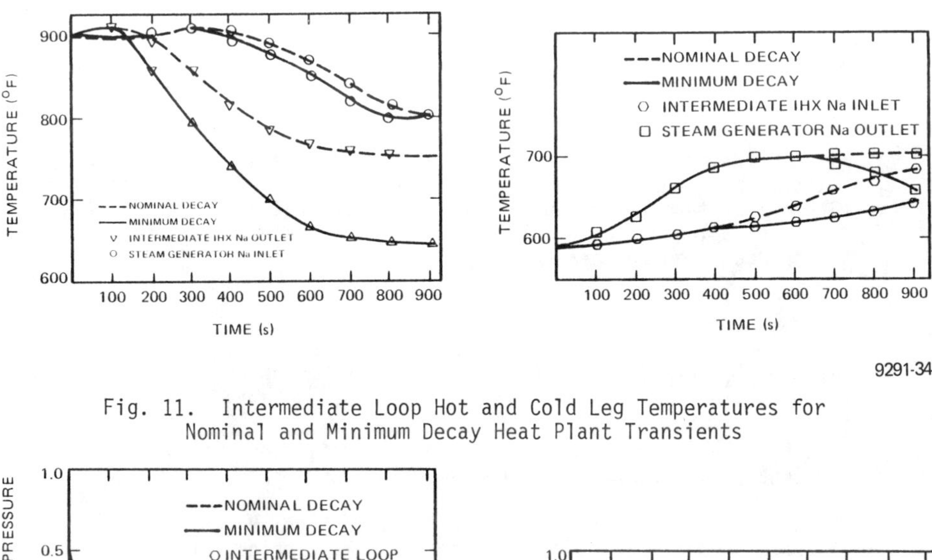

9291-34

Fig. 11. Intermediate Loop Hot and Cold Leg Temperatures for Nominal and Minimum Decay Heat Plant Transients

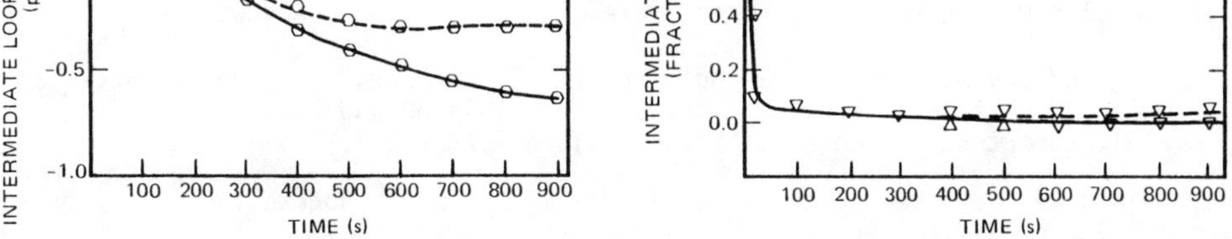

9291-35

Fig. 12. Intermediate Loop Flow and Buoyant Head

6. CONCLUSIONS

Continued efforts in simulation development and verification of natural convection effects are essential to LMFBR design. ESG-AI dynamic performance analysis points to the fact that buoyancy and natural convection have a significant impact on component requirements, selection of alternate design features and component designs, and control and plant operation approach. The following conclusions are drawn from the material presented:

1) Milder plant component transients are experienced with reduced pony motor speeds.

2) No transient or operational problems encounter with NSSS natural circulation heat removal.

3) Although plant component generally experience milder transients with reduced pony motor speed, the controlling consideration is adequate low-speed hydrostatic bearing flow. Buoyancy forces tend to decrease bearing flow. The problem is more acute in the intermediate loop and for plants employing a saturated steam cycle. Therefore to supply adequate hydrostatic bearing flow, pony motor speeds of >7% are recommended.

4) Ratios of pump rotating to fluid kinectic energies < 4 cause a reduction of pump hydrostatic bearing flow during the transition to pony motor speed. Ratios of < 2 can cause complete bearing flow starvation.

5) With inclusion of buoyancy forces in reactor core simulation, pony motor speed and loop elevation changes have small effect on peak fuel clad temperature following reactor trip. The match between primary flow and fuel pin stored energy decay is the dominate factor in determining peak clad temperatures.

REFERENCES

1. Presson, A. J., "Feedwater Shutdown Control Study," Rockwell International, Energy Systems Group, N323TI000046, November 11, 1979.

2. O'Neill, R. J. and Jam, M. W., "Transient Mitigation Analysis," Rockwell International, Energy Systems Group, N323TI000035, September 28, 1979.

3. "LMFBR Pool Plant — 1000 MWe," Volumes 2 and 3, "Pool Reactor Engineering — NSSS," prepared by Rockwell International, Atomics International Division for EPRI, Report ref EPRI NP-1015, February 1979.

4. "Further Examination of Pool Type 1000 MWe LMFBR, Volume 3: Examination of NSSS Concepts," prepared by Rockwell International, Atomics International Division for EPRI, Report reference EPRI NP 881, August, 1978.

5. "Engineering Aspects of the Pool Type LMFBR, 1000 MWe," prepared by Rockwell International, Atomics International Division and Bechet Corp. for EPRI, Report reference EPRI NP-645, Volume 2, April, 1978.

6. Auleta, J. J., "Decay Heat Removal System Analysis," Rockwell International, Energy Systems Group, N254TI410001, February 3, 1978.

7. Presson, A. J., "Comparative Analysis of Carbide vs Oxide Core for Plant Transients," Rockwell International, Energy Systems Group, N323TI000040,

November 11, 1979.

8. Drucker, G. S. and Bost, D. S., "Primary System Temperature Limits and Transient Mitigation," Rockwell International, Energy Systems Group, N254TI122008, October 1978.

9. Vitti, J. A., Felten, L. D., Brettingham, J. C., Galluzzo, N. G., Ostermier, B. J., and Von Ark, A. V., "Summary Report Preconceptual study of 1000 MWe Carbide-Fueled LMFBR Designs," Rockwell International, Atomic International Division, ESG-DOE 13244, September 30, 1978.

10. Continuous System Modeling Program III (CSMP III) Program Reference Manual, Program number 5734-XS, International Business Machines Corporation.

11. "Liquid Metal Fast Breeder Reactor Conceptual Plant Design — 1000 MWe," prepared for ERDA and EPRI by Rockwell — Atomics International Division, FBR-77-3, Volume II Plant Description, May 1977.

A Computer Simulation of a Pool-Type LMFBR Passive Decay Heat Removal System

K. KESAVAN, J. D. BRADLEY, and D. E. NOE
Westinghouse Electric Corporation
Advanced Reactors Division
Madison, Pennsylvania

ABSTRACT

The cooling of a large pool-type LMFBR by a totally passive decay heat removal system, following a complete loss of electric power, is analyzed with the aid of a computer program designed to simulate the total plant system. The results of the analysis indicate that adequate core cooling can be obtained by natural circulation means.

1. INTRODUCTION

A diverse and redundant means of decay heat removal in a liquid metal-cooled fast breeder reactor (LMFBR) plant enhances the safety margin of the reactor in extreme emergency situations. In addition to the normal means of heat removal through the steam generating system, an LMFBR may employ separate, dedicated "active" and "passive" decay heat removal systems for shutdown cooling of the reactor. The Westinghouse conceptual design of a commercial-size, pool-type LMFBR employs such shutdown heat removal systems (SHRS) which are diverse and redundant. The passive shutdown heat removal (SHR-P) system, unlike the active (SHR-A) system which utilizes pumps and blowers, relies solely on natural circulation of coolants in the reactor and heat transport loops. The SHR-P system consists of multiple sets of an independent liquid metal coolant heat transport loop, a shutdown heat exchanger (SHX) located directly within the reactor vessel above the intermediate heat exchanger (IHX) that transfers the decay heat to the loop, and a natural draft heat exchanger (NDHX) that rejects this heat to atmospheric air. The positioning of the SHX above the core and of the NDHX above the SHX, and the configuration of the shutdown heat transport loop are such that the reactor can be adequately cooled without any active means. Such a passive shutdown heat removal system that is driven by coolant buoyancy forces brought about by the temperature difference between the heat source and the heat sink can offer the ultimate defense against loss of all forced cooling.

The purpose of this paper is to describe a digital computer simulation of the shutdown cooling dynamics of the Westinghouse pool-type reactor following a loss of all electric power event.

A previously reported work [1] in this continuing effort was based on a general purpose computer program that modeled the thermal-hydraulic system of the reactor as a network of flow channels and plena. This computer program relied on externally supplied forcing functions for the early part of the transient, viz., the initial forced circulation caused by the primary pump coastdown, and the heat removal through the IHXs. Moreover, the reactor upper plenum was distributed over a number of vertically connected control volumes and the mixing of the core effluent in the upper plenum was simulated by a sequential, complete mixing process in these control volumes.

The present work has been performed with the aid of a modified version of the DEMO computer code [2] . The modified version, called COMTRAN (Westinghouse Proprietary), is used extensively to analyze the Westinghouse commercial plant design transients. Both loop- and pool-type reactor design options are available in the COMTRAN code. The reactor components represented in the pool-type model are the cold and hot pools, the reactor fuel and blanket assemblies, intermediate heat exchangers and the primary pumps. In addition, the intermediate sodium system, steam generator and the steam power conversion system are modeled to give a full plant simulation. Thus the COMTRAN code offers an integrated model of the entire plant system. Furthermore, with the explicit representation of the reactor upper plenum, a model of the mixing process based on test results is used. In addition, the various components of the shutdown heat removal systems are now a part of the COMTRAN code.

2. SYSTEM DESCRIPTION

2.1 Large Pool Reactor

The system analyzed is a large pool reactor (LPR) in which the reactor and various primary heat transport system components are integrated into a single structure. This contrasts with a loop-type reactor which requires piping to interconnect the reactor and the individual elements of the primary heat transport system.

A cutaway view of the reactor is shown in Figure 1. The cylindrical reactor vessel is 75 ft in diameter with a torospherical bottom head and is supported from a cold structural deck. The deck also supports six intermediate heat exchangers (IHX), four primary sodium pumps (PSP) and other reactor components, including the upper internals structure (UIS) and the control rod drive mechanism (CRDM). The reactor core is supported by the double cone, lower support structure (LSS) which is attached to the reactor vessel. The interior of the reactor vessel is baffled to separate the primary sodium into three main regions; the hot pool which contains the core effluent, the cold pool which is fed by the primary flow exiting the IHXs, and the stagnant intermediate plenum which provides a thermal buffer zone between the hot and cold pools.

The UIS is positioned above the core in the hot pool and serves to align the control rod drivelines. It also provides secondary holddown for all core assemblies and supports thermocouple instrumentation.

The six IHXs are equally spaced around the periphery of the reactor. In addition to the IHX tube bundles, the IHX support cylinders include coils for the shutdown heat removal (SHR) system. Each IHX is nominally 10 ft in diameter with a 25 ft active tube bundle length. They are positioned axially to locate their thermal centers 11 ft above the thermal center of the core. This promotes natural circulation of the sodium during complete loss of pumped circulation. A shutdown heat exchanger (SHX) is positioned at the top of each of the six IHXs. Its function is to remove reactor decay heat when heat

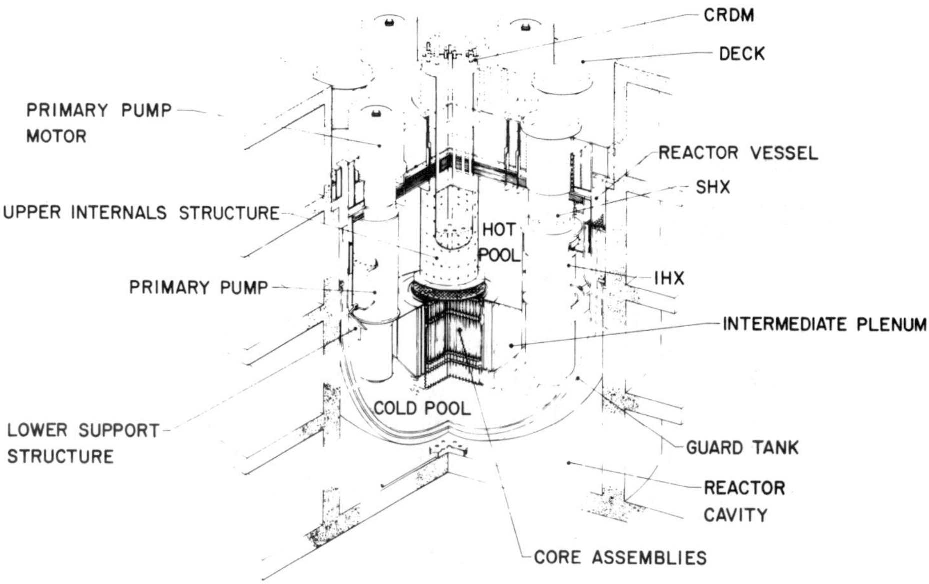

Fig. 1 Cutaway View of the Large Pool Reactor

removal through the IHXs is disrupted. Its position above the IHX tube bundle enhances natural circulation through the reactor circuit.

The four primary pumps are symmetrically located within the reactor vessel. The two-stage pumps are of modular design with concentric inlet and outlet chambers. They plug into the LSS allowing sodium to be drawn from the cold plenum and discharged into the integral flow distribution passages in the LSS.

2.2 Primary Coolant Flow Path

The coolant flow path in the LPR is illustrated in Figure 2. As previously mentioned, the pumps draw from the cold pool and discharge into the core inlet chamber. The coolant is distributed to the various core assemblies (fuel, blanket, control and shielding) and flows upward into the core exit chamber of the UIS. The core effluent mixes within the UIS and exits into the hot pool. Fluid from the hot pool enters the heat exchanger casings near the top of the pool and flows downward through the SHXs and the IHXs, discharging into the cold pool.

2.3 Shutdown Heat Removal System

In the normal mode of operation, heat generated in the reactor is transferred across the IHXs to steam generators, through the intermediate heat

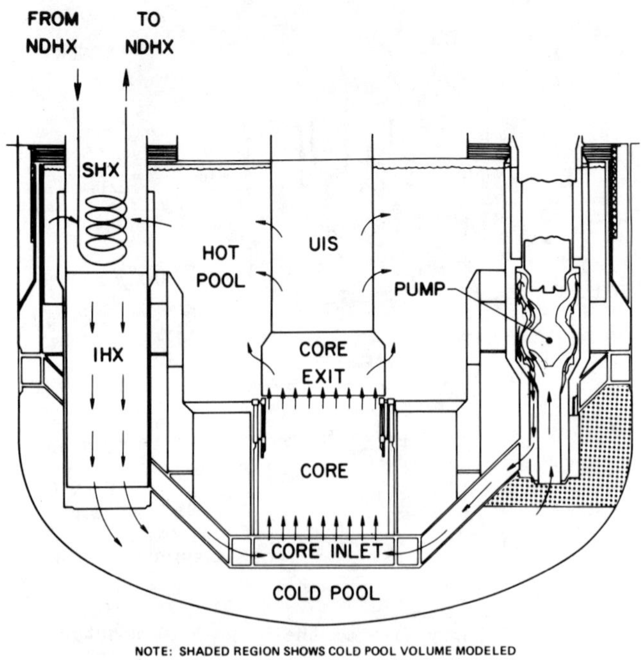

Fig. 2 Flow Schematic of the Large Pool Reactor

transport system (IHTS) consisting of six independent intermediate sodium loops. High pressure, superheated steam produced in the steam generators is utilized in a turbine-generator to generate electricity. The shutdown heat removal system (SHRS) provides emergency cooling to transfer fission product decay heat and other residual heat from the reactor when the normal system of heat removal is inoperative. Both an active (SHR-A) system and a passive (SHR-P) system are provided, giving separate, diverse, and redundant capability.

During the SHR-A function, sodium from the IHX is pumped from the IHTS hot leg piping, passed through an air blast heat exchanger (ABHX) and into the IHTS cold leg piping to return to the IHX. When all pumping capability is lost in the plant, the SHR-P can be activated by opening the damper on the natural draft heat exchanger (NDHX). In this mode of operation, heat is removed from the primary reactor sodium through the SHXs, transported through the SHR-P loops to the NDHX and discharged to the ambient air. The SHR-P, which is the subject of this paper, is a totally passive system on the primary, secondary, and air sides. A performance criterion specified for the SHR-P is that the maximum values of sodium temperature in the average fuel and blanket assemblies not exceed 1350°F under all anticipated operating or failure conditions.

2.4 LPR Operating Parameters

The plant is designed to produce 1000 MW of electric power with 2550 MW of heat generated in the reactor. Salient operating parameters of the LPR at rated power are listed in the table below:

LARGE POOL REACTOR
OPERATING PARAMETERS AT RATED POWER

Total Reactor Power	2550 MW
Total Reactor Flowrate	102.2×10^6 lb/hr
Power Distribution: Fuel/Inner Blanket/Outer Blanket	0.794/0.157/0.049
Reactor Flow Fraction: Fuel/Inner Blanket/Outer Blanket/Bypass	0.7272/0.1618/0.051/0.060
Core Inlet Temperature	670°F
Core Outlet Temperature (mixed)	950°F
Average Assembly Outlet Temperatures:	
Fuel	975°F
Inner Blanket	942°F
Outer Blanket	939°F

3. ANALYTICAL MODEL

The simulation of the overall plant dynamics under the natural circulation heat removal conditions was performed with the aid of the COMTRAN code which is a Westinghouse Proprietary version of the DEMO code [2]. For the present discussion, description of the models in COMTRAN will be limited to those of the pool reactor and the passive shutdown heat removal system.

3.1 Modeling of the Large Pool Reactor (LPR)

The thermal-hydraulic modeling of the Large Pool Reactor (LPR) is briefly outlined here. Beginning with the primary pump, the model description of the various components of the LPR follows the coolant flow path sequence as illustrated in Figure 2. The calculation of the coolant flow in the circuit is described at the end of this section.

Primary pumps. The head, flow, speed and hydraulic torque characteristics of the primary pumps are provided in the COMTRAN code in the form of curve-fit polynomials. During a transient, the hydraulic torque of the pump is determined from a knowledge of existing impeller speed and flowrate. Then, from an angular momentum balance on the rotating assembly of motor, hydraulic and frictional torques, the new resulting impeller speed is calculated. From the updated pump speed and flowrate, the resulting pump head is determined. The newly calculated pump head is used in the flow momentum equation around the primary circuit to update the flowrate. In analyzing events involving total loss of power, the motor torque is set to zero.

Reactor core. The power generation in the core is modeled by neutron kinetics and decay heat equations. The decay heat data is based on nominal conditions for a heterogeneous core. The reactor internals are represented by models of a lower plenum, lower axial blanket, active core, upper axial blanket, inner radial blanket, outer radial blanket, and a bypass channel.

The average channel of the active core is represented by 7 axial nodes and 5 radial nodes. Two of the radial nodes represent the fuel; the clad, the coolant and the duct are represented by one radial node, each. The average channel of both the inner and outer radial blankets is represented by 7 axial nodes and 6 radial nodes. Four of the radial nodes represent the fuel; the clad and the coolant are represented by one radial node, each. The metal heat capacity of the various reactor components is included; however, the stagnant intermediate plenum is not modeled. The core inlet flow is distributed based on a fixed flow split composed of fuel, inner radial blanket, outer radial blanket and bypass fractions. The temperature of the coolant at the core outlet prior to entering the outlet plenum is derived by assuming perfect mixing of the flow fractions from the various channels.

Hot pool. Since the natural circulation driving heads are thermal dependent, it is important to characterize the dynamics of the hot pool sodium flow and temperatures as realistically as possible. The COMTRAN hot pool model handles thermal down-transients and up-transients. The UIS is not modeled since its exact configuration was not determined at the time of the analysis. For the case of up-transients, the hot core effluent is completely mixed with the upper plenum (hot pool) fluid using a mixing volume determined at full flow condition. The mixing volume is based on a two-dimensional axisymmetric model with the aid of the VARR-II code [3]. For an up-transient following a plenum fluid stratification (result of down-transient), the hot core effluent is mixed with the volume below the hot/cold interface. Further refinement of this COMTRAN up-transient mixing model is pending experimental tests at Westinghouse.

For the down-transient, the upper plenum model is based on ANL tests [4] . The temperature and flow rate of the core effluent decrease and primary sodium stratification occurs with its attendent hot/cold sodium interface that gradually moves upward. This upward movement is due primarily to fluid displacement rather than entrainment because the flow outlet is located at the top of the plenum (IHX inlet) for this pool reactor design (See Figure 2) rather than near the plenum bottom as it is in a loop-type reactor. Subsequent to a reactor trip, the core effluent jet height is determined using the Yang correlation [5]. When the jet height is less than that of the suppressor plate (24 feet above the core fuel assembly exit plane) jet collapse is assumed and complete mixing ceases; sodium stratification then ensues. The upper plenum fluid behavior is assumed to go through 6 phases as follows:

- Phase 1: Complete mixing at full flow and prior to core effluent jet collapse (single effective mixing volume).
- Phase 2: Formation of a hot/cold sodium interface subsequent to jet collapse (stratification: single node mixing volume below interface and no mixing above it). The interface lies between the hot pool floor and the core exit plane. (For a schematic of the hot pool mixing model, see Figure 3.)

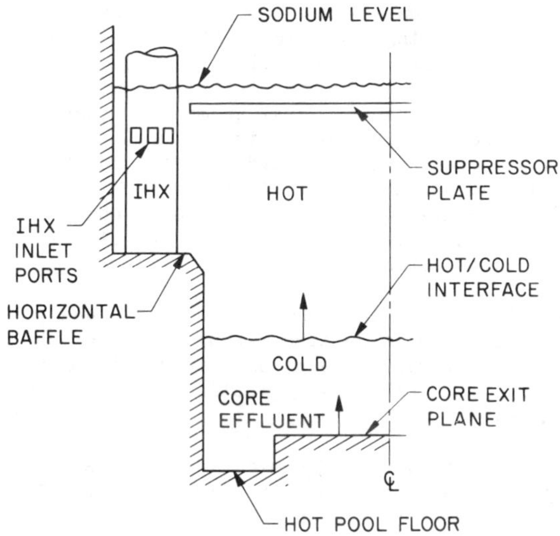

Fig. 3 Stratified Hot Pool Schematic

- o Phase 3: Hot/cold interface rises from the core exit plane to the top of the horizontal baffle (rise rate based upon fluid displaced by core effluent).
- o Phase 4: Hot/cold interface rises from the top of horizontal baffle to the bottom of IHX inlet ports (similar to Phase 3).
- o Phase 5: Hot/cold interface rises from the bottom to the top of IHX inlet ports (hot sodium above interface and cold sodium below it mix and enter the IHX inlet ports).
- o Phase 6: Hot/cold interface rises past the top of the IHX inlet ports and disappears as fluid is mixed in a single effective mixing volume as in Phase 1.

The temperature and flowrate of the hot pool effluent provide the primary side inlet conditions for the SHX. During stratified flow Phases 2 through 5, there is a volume of cold sodium below the interface and a volume of hot sodium above the interface. The effective hot pool sodium temperature used for sodium level calculations and natural circulation buoyancy head determination is the volume weighted average temperature of these two volumes.

Intermediate heat exchanger (IHX). The primary coolant after leaving the shutdown heat exchanger (SHX) enters an inactive volume before entering the primary tubes of the IHX. This inactive volume is divided into a number of segments to simulate temperature transport delay. The temperature of the coolant leaving the SHX is then transported at a speed corresponding to the existing coolant flowrate. The primary coolant then enters the IHX tubes. The exit conditions of the IHX serve as the input boundary conditions for the cold pool mixing model.

The IHX is modeled as a counter-flow, shell and tube heat exchanger with the primary reactor sodium flowing downward on the tube-side and the intermediate loop coolant flowing upward inside the shell. The primary and intermediate coolant volumes along with the intervening tube wall and shell metal are discretized into a number of axial segments. Heat exchange between the intermediate sodium and the shell metal is included. With flowrates and inlet temperatures on the shell- and tube-sides specified, the outlet temperatures are calculated using finite difference equations of thermal exchange. Additionally, shell- and tube-side pressure changes due to friction and static head are computed to be used in the reactor circuit and intermediate loop flow calculations.

Cold pool. The cold pool is characterized for this analysis as a single mixing node using a conservative effective mixing volume of 50% of the actual. No cold pool stratification is assumed to exist for hot IHX effluent. This model will be refined when Westinghouse experimental test results become available.

Flow calculation. The flowrate of the coolant in the primary system is calculated from a pressure balance around the flow circuit. The one-dimensional momentum equation used in the calculation includes head developed by the pump during coastdown, the net buoyancy head around the circuit, and pressure losses in the various parts of the flow path due to form, friction and acceleration. The resistance offered by the stopped pumps is considered in the equation following completion of coastdown.

3.2 Modeling of the SHR-P System

A schematic arrangement of the SHR-P system is illustrated in Figure 4. The system consists of the SHX, the SHR-P loop, and the NDHX.

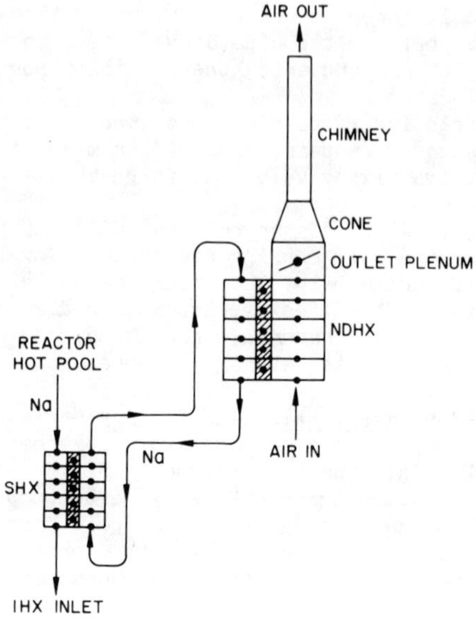

Fig. 4 SHR-P System Model

Shutdown heat exchanger (SHX). The SHX is represented as a counter-flow, shell and tube heat exchanger with the primary reactor sodium from the hot pool exit flowing downward on the shell-side and the secondary SHR loop coolant flowing upward inside the helically wound tubes. The thermal-hydraulic modeling of the SHX is similar to that of the IHX as described in Section 3.1. With flow rates and inlet temperatures on the shell- and tube-sides specified, the temperature and pressure distributions of the coolants are computed for use in the reactor circuit and SHR-P loop flow calculations.

SHR-P loop hydraulics. The SHR-P loop circulates the sodium between the SHX and the NDHX. The coolant piping is divided into a convenient number of major lengths, each of which is further divided into small segments to simulate temperature transport delay. The temperatures of the coolant leaving the SHX and the NDHX are transported around the loop at a speed corresponding to the existing coolant flowrate. The transport delay model functions much like a "shift register", neglecting any coolant mixing effects in the piping. Also neglected in the model are thermal interactions between the coolant and the piping walls, and axial conduction between coolant segments.

Using the SHR-P loop coolant temperature distribution obtained from the transport model, the static heads in the loop are determined. Taking into account head losses such as those due to friction and pipe elbows, the net driving head in the loop is obtained, and the flowrate in the pipe is calculated from a one-dimensional momentum model around the loop.

Natural draft heat exchanger (NDHX). The modeling of the NDHX is similar to that of the SHX. The shell-side of the heat exchanger containing air and the tube-side fluid and the tube wall are discretized axially into a number of equal axial segments. With flowrate and inlet temperature on the tube-side determined from the SHR-P loop model, and the shell-side inlet temperature specified, the temperatures at the heat exchanger outlets are calculated from finite difference equations. Additionally, shell- and tube-side pressure changes due to friction and static head are computed to be used in the SHR-P loop and air-side flow calculations.

The flow of air through the shell of the NDHX is caused, as in the case of the SHR-P liquid metal loop, by natural convection. Ambient air entering the shell-side is moved by the buoyancy force resulting from the static head difference between the hot column of chimney air and an equivalent cold column of atmospheric air. In calculating the net driving force, pressure losses across the NDHX tube bundles, in the outlet plenum and cone regions and through the chimney are taken into account. Air flow in the outlet plenum, cone and chimney regions is treated as adiabatic.

4. ANALYSIS

Using the COMTRAN program, the cooling of the LPR by the SHR-P system following a loss of all electric power event was analyzed.

4.1 Scenario of the Loss of Power Event

Prior to the loss of electric power, the plant is assumed to have been operating at rated conditions for an extended period of time. The event is initiated by a loss of off-site electrical power followed immediately by a

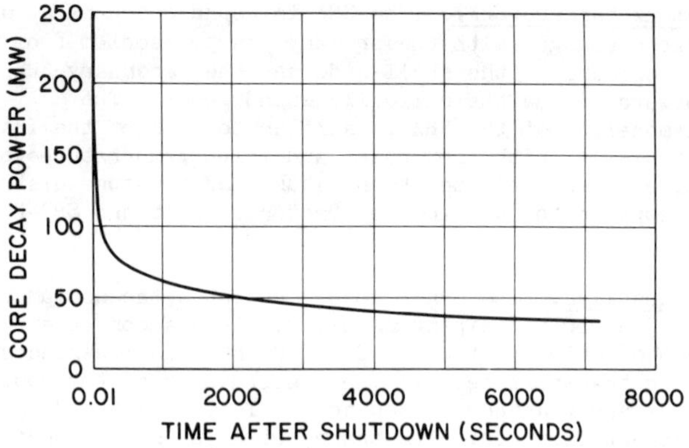

Fig. 5 LPR Decay Power

reactor scram. With an assumed unavailability of all on-site power, all primary and intermediate pumps coastdown and stop. Heat rejection to the steam generators is assumed to be instantaneously terminated; however, some heat removal due to the thermal capacitance of the intermediate loops occurs during the coastdown period. Simultaneous with the reactor scram, the air-side damper of the NDHX opens activating the SHR-P system which was in thermal equilibrium with the reactor hot pool during the steady state operation. Ambient air enters the NDHX and begins to cool the SHR-P loop coolant which, in turn, driven by buoyancy forces, begins to circulate. After a transport time delay, the cooled SHR-P coolant reaches the SHX and begins to cool the reactor primary sodium. Following the coastdown of the primary pumps, flow through the reactor is maintained by natural circulation and the residual decay heat is removed by the SHR-P system.

The time-dependent decay heat load of the reactor following shutdown is shown in Figure 5.

5. RESULTS

Results obtained from the simulation of natural circulation cooling of the LPR following an assumed loss of all electric power event are presented here. The simulation covers the first two hours of the transient during which time the core decay heat has decreased to the level of the shutdown heat removal system capability (30 MW), and the significant temperature increases have occurred.

The behavior of the reactor system during the transient is illustrated in Figure 6 through Figure 9 and that of the SHR-P system in Figure 10 through Figure 13. The history of the coolant flowrate in the reactor is shown in Figure 6. Following coastdown of the primary pumps which occurs during the first 60 seconds, the reactor flowrate is seen to establish itself between 3% and 4% of its initial value under natural circulation. As explained in Section 3.1, the distribution of this total flowrate among the fuel and blanket assemblies and bypass is held constant throughout the transient.

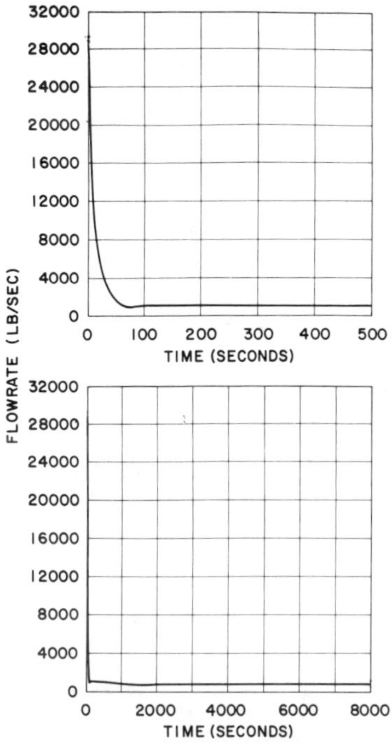

Fig. 6 Reactor Flowrate Following Loss of Power

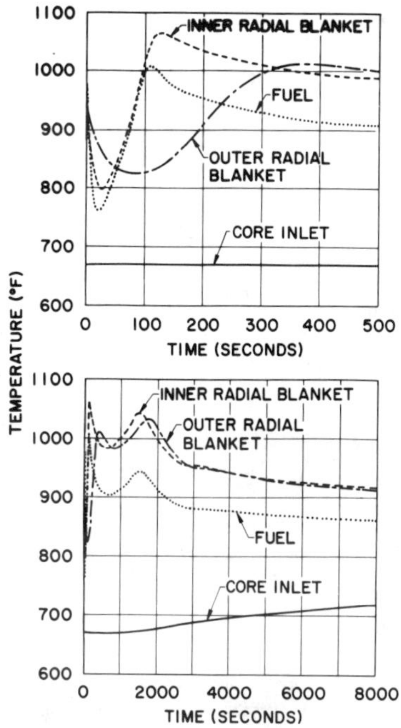

Fig. 7 Core Inlet, and Fuel and Blanket Assembly Outlet Temperatures

Figure 7 shows the coolant temperatures at the inlet of the core and at the outlets of the fuel and blanket assemblies. Due to the differences in the decay power histories, thermal capacities and heat transfer areas among the fuel, inner and outer blanket assemblies, the dynamic responses of the outlet temperatures are seen to vary. The increasing flow-to-power ratio in the various assemblies during the early part of the pump coastdown causes an overall reduction in the assembly exit temperatures. However, following the cessation of forced flow, the coolant outlet temperatures are seen to rise. Maximum values of 1010°F, 1045°F and 1036°F are reached during the transient for the fuel, inner blanket and outer blanket, respectively. The second set of peaks in these temperatures at around 1700 sec. is brought about by the slight, transient reduction in the natural circulation flowrate through the core. This transient reduction in the flowrate is caused by the increasing temperature and decreasing density of the primary coolant leaving the IHX due to loss of heat removal through the IHX (See Figure 9).

The mixed core outlet temperature and the temperatures of the lower and upper parts of the hot pool are shown in Figure 8. As a consequence of the rapidly decreasing power-to-flow ratio following scram, the mixed core effluent experiences a cooldown initially. During this period, with jet collapse, the hot pool coolant inventory stratifies into upper hot and lower cold zones. At about 20 sec. into the transient, the mixed mean temperature of the core effluent begins to rise. With the current mixing model of the hot pool, the effect of this rising hot core effluent is to increase the temperature of the lower volume. Following the establishment of natural

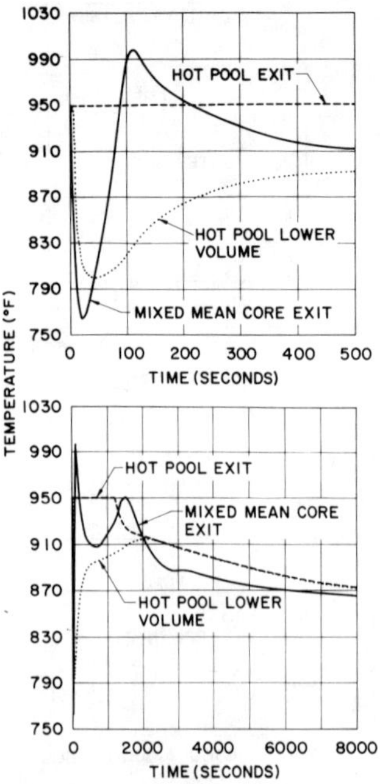

Fig. 8 Mixed Mean Core Outlet and Hot Pool Temperatures

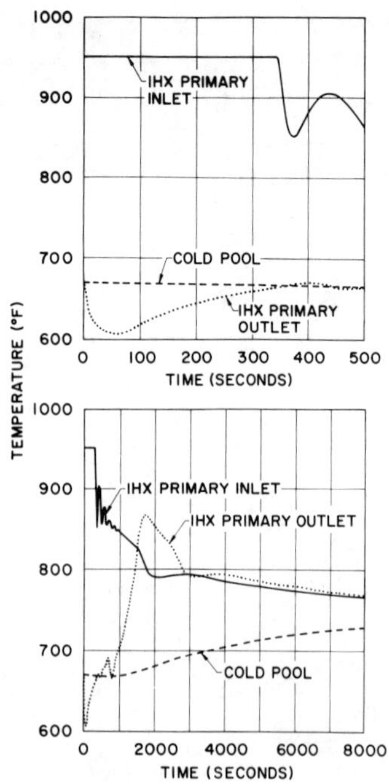

Fig. 9 IHX Inlet and Outlet and Cold Pool Temperatures

circulation in the reactor, the core effluent temperature begins to fall after about 110 sec. as seen in Figure 8. A second, lower peak in the core outlet temperature is seen to occur at approximately 1500 sec. due to the transient reduction in the reactor flowrate. Prior to this, at about 1200 sec., the hot/cold interface has risen above the bottom of the IHX inlet ports as signified by the sudden drop in the hot pool exit temperature from the initial value. The effect of the second peak in the core outlet temperature is seen in the hot pool exit temperature between 2000 and 2500 sec. as the interface passes the inlet ports. After this the hot pool lower volume temperature and exit temperature merge to identical values as stratification disappears.

The temperature of the primary coolant at the inlet and outlet of the IHX are shown in Figure 9. Following the initial plateau at 950°F due to the delay in heat removal through the SHX and transport delay between the SHX and the IHX, the temperature at the IHX inlet is seen to fall. (The oscillations in this temperature are explained later in connection with the performance of the SHX and the NDHX.) Following the loss of heat sink effect of the intermediate loops at about 1500 sec, the IHXs begin to release their stored heat. Although cooler primary flow continues to enter the IHX from the SHX, the IHX outlet temperature is seen to rise. This is largely due to transport delay through the IHX. The attenuating effect of the IHX thermal capacitance on the outlet temperature is noticeable. Figure 9 also shows the moderating effect of the cold pool on the IHX outlet temperature.

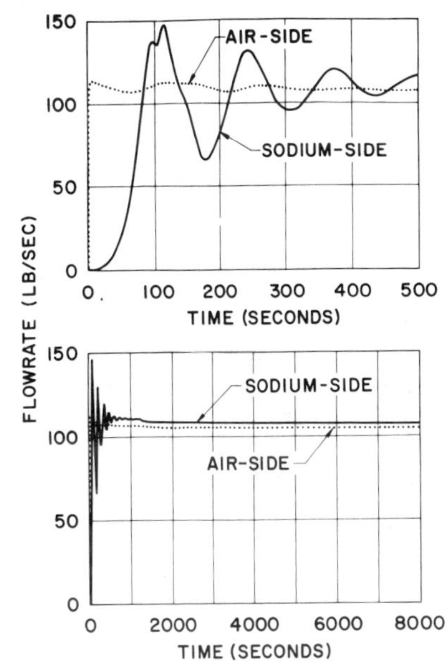

Fig. 10 Sodium and Air Flowrates in the NDHX Following Loss of Power

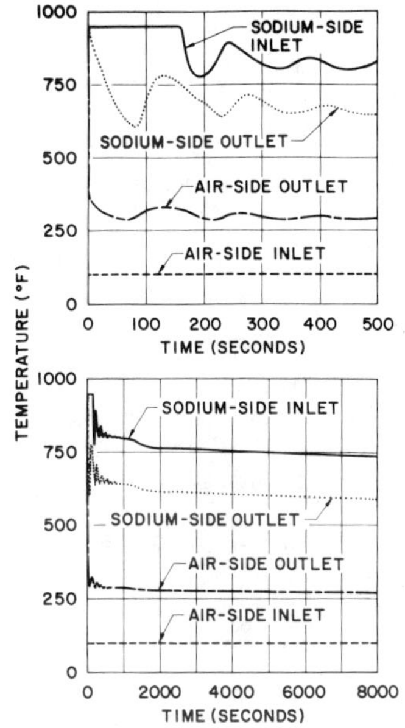

Fig. 11 Sodium and Air Temperatures at the NDHX Following Loss of Power

The flowrates in the SHR-P system on a per loop basis are shown in Figure 10. On the air-side, the flow builds up rapidly as soon as the damper is opened. The loop coolant slowly begins to move, and after a few short oscillations prior to 1000 sec, the flow levels out at about 110 lb/sec.

Figure 11 shows coolant temperatures in the NDHX during the transient. The oscillations in the flow (Figure 10) and temperature (Figure 11) histories are caused by natural circulation effects. As ambient air rises through the NDHX, the initially hot column of air in the chimney experiences a cooldown. The sudden surge of cold air on the shell-side begins to cool the stagnant sodium in the tubes creating an imbalance of static head distribution in the SHR-P loop. As the sodium begins to move, it brings an increasing amount of energy into the NDHX. Due to the long hot leg piping connecting the SHX and the NDHX, the sodium temperature entering the NDHX remains constant at its initial value for a period of time as evidenced by the long plateau in Figure 11. This plateau of about 160 sec demonstrates the effect of transport delay in the loop and indicates the time required to sweep the hot leg piping of its initial fluid. The increased energy input to the NDHX on the tube-side and the simultaneous fall in the air flow causes a gradual rise in the exit temperatures of sodium and air. The alternating rise and fall of the fluid temperatures leads to changes in their driving heads. A similar explanation applies to the coolant temperatures in the SHX which are shown in Figure 12. It may be noted that although the SHR-P system exhibits an oscillatory behavior, the oscillations are well-damped as shown by their decreasing amplitudes and increasing periods.

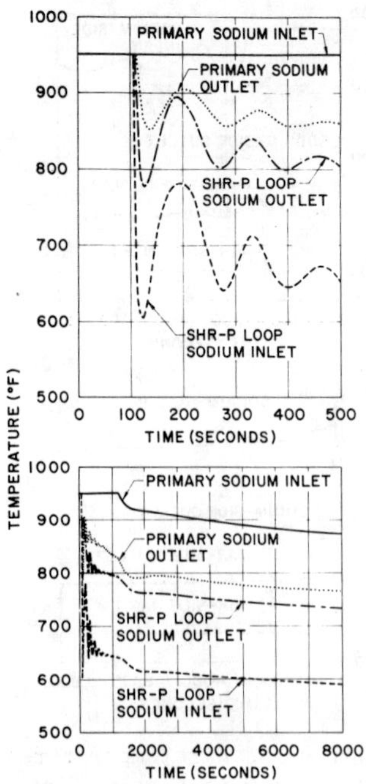

Fig. 12 Reactor Primary and SHR-P Loop Sodium Temperatures at the SHX

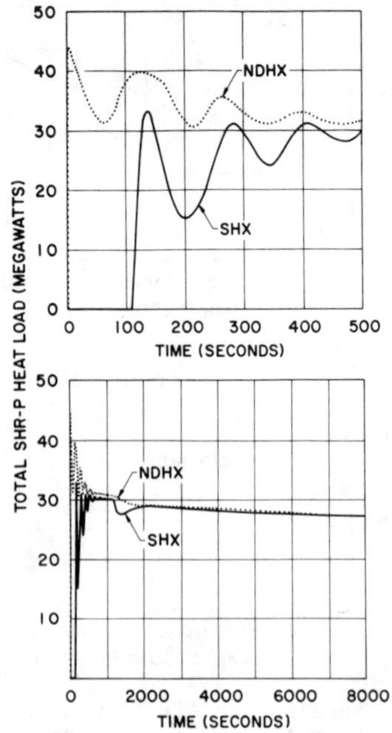

Fig. 13 SHX and NDHX Heat Loads Following Loss of Power

The total heat loads of the SHX and NDHX units in the plant are shown as functions of time in Figure 13. After initial oscillations, the heat-removal rates level out at about 29 MW and begin to fall steadily.

6. CONCLUSIONS

This analysis indicates that adequate decay heat removal by natural circulation can be obtained in the Large Pool Reactor with a passive shutdown heat removal system. The results show that the average core outlet temperatures following a simulated loss of power event are kept well below the specified maximum of 1350°F.

Further work in this area involving improvements in the hot pool mixing model, inclusion of variable flow distribution at core inlet and core hot channel factors, and consideration of pipe heat capacitance of the SHR-P loop, among others, are in progress.

7. REFERENCES

1. K. Kesavan, D. A. Wiseman, M. D. Woods and J. R. Barrett, "Dynamic Simulation of Natural Circulation Cooling of a Large Pool LMFBR,"ASME Paper 79-WA/HT-60.

2. W. H. Alliston, A. B. Batenburg, B. J. Dvorznak, F. C. Luffey and H. P. Planchon, "Clinch River Breeder Reactor Plant: LMFBR Demo Plant Simulation (DEMO), Topical Report", CRBRP-ARD-0005, February 1978. (Availability: US DOE Technical Information Center).

3. J.L. Cook, and P. I. Nakayama, "Clinch River Breeder Reactor Plant: VARR II, A Computer Program for Calculating Time-Dependent Turbulent Fluid Flows with Slight Density Variation," CRBRP-ARD-0106, November 1976. (Availability: US DOE Technical Information Center).

4. P. A. Howard, and J. J. Carbajo, "Plenum-2: A Two-Region Outlet Plenum Model," ANL-CT-78-16, March 1978. (Availability: US DOE Technical Information Center).

5. J. W. Yang, "Penetration of Core Flow in Upper Plenum of an LMFBR," Trans. Am. Nucl. Soc., 23, pp. 414-415, (1976).

Buoyancy Induced Flow and Heat Redistribution during LMFBR Core Decay Heat Removal

R. D. COFFIELD, JR., Y. S. TANG, J. S. KILLIMAYER, and R. A. MARKLEY
Westinghouse Electric Corporation
Madison, Pennsylvania 15663

ABSTRACT

Natural convection cooling of the core exhibits one of the few core design transients where low power/high temperature conditions exist. Core inter-assembly and intra-assembly flow/heat redistribution: a) become significant with regard to accurately predicting temperatures; and b) decrease the maximum hot element temperatures. A technique is described herein for linking dynamic core-wide transient effects as predicted by CØBRA-WC to the localized hot element temperatures calculated with FØRE-2M. These latter temperatures are used in evaluations with regard to core design adequacy.

1. INTRODUCTION

The normal mode of LMFBR core heat removal upon reactor shutdown is by forced circulation from AC powered pony motors (which have emergency backup power from batteries and/or diesel generators) driving the primary pumps. However, plants are being designed to have the added capability of adequate cooling by means of natural circulation. This inherent emergency coolant flow is provided by the thermal driving head developed by the thermal center of the IHX being elevated above that of the core (plus the respective elevation differences in the intermediate loops and steam generator system).

At the ~10% pony motor flow level after shutdown insignificant flow redistribution occurs between the parallel flow core assemblies. However, for the core natural convection cooling mode, the effect of rapid flow reduction with slower changing decay heat loads results in a power-to-flow ratio greater than one. Consequently, core temperatures increase and natural convection phenomena such as inter- and intra-assembly flow redistribution become significant once low flow conditions are reached. In general, the core thermal head becomes significant relative to the form and friction losses across the core below 5% of full flow. Coupled with the flow redistribution, significant heat redistribution on an inter- and intra-assembly basis occurs throughout the core due to radial temperature differentials and an increased heat transport time. Both of these effects (i.e., natural convection flow and heat redistribution) have been found to significantly reduce maximum core temperatures as demonstrated in the EBR-II natural circulation experiments[1].

Several analytical studies on the CRBR have been published which show a significant decrease in predicted maximum core temperatures due to reactor flow redistribution during natural circulation conditions. For example,

Agrawal, et al.[2], using the SSC-L code, predicted localized flow increases as large as 20% in the hot fuel assembly and 40% in the hot blanket assembly for the CRBR during natural convection cooling. Corresponding reductions in the predicted maximum transient coolant temperature on the order of 16 and 22% ($\sim 130^{o}F$ and $210^{o}F$) were shown for the hot fuel and blanket assemblies, respectively, relative to the maximum temperatures predicted without flow redistribution. Similar results were found in Reference 3 using the CURL-L code. For these studies, inter-assembly heat transfer as well as intra-assembly flow redistribution and heat conduction effects were neglected. Inclusion of these effects would further reduce the maximum core temperatures.

Relationships governing core intra- and inter-assembly flow and heat redistribution were discussed in Reference 4. Also given are various aspects and data sources for verifying computer codes that are used for core analyses. This paper addresses some of the experience gained to-date from two of the computer codes, i.e., CØBRA-WC[5] and FØRE-2M[6].

2. ANALYTICAL APPROACH

To perform core analyses under the natural circulation mode, a two-stage calculational approach can be adopted. A detailed whole-core flow and heat redistribution analyses of all the parallel core assemblies and bypass regions is first performed by the CØBRA-WC code. This is done for a sector of symmetry of the reactor core. By necessity, the calculational matrix or nodal representation is coarse. Clusters of elements rather than individual subchannels are modeled. The results of this analysis provide the inter- and intra-assembly flow and heat redistribution data. The data are then used for a detailed analysis on a single, "hot element" using the FØRE-2M code. These latter analyses include effects of localized phenomena and uncertainties in nuclear/thermal-hydraulic/mechanical data.

A linkage between the CØBRA-WC and FØRE-2M codes has been developed to incorporate the inter- and intra-assembly phenomena into the localized hot element transient analyses[6]. For each axial node of the hot element modeled in FØRE-2M, a heat balance is performed using the expression for the heat transferred to the coolant at that section, $Q_c(\chi,\tau)$ as

$$Q_c(\chi,\tau) = Q_r(\chi,\tau) + Q_{ex}(\chi,\tau) \qquad (1)$$

where

$Q_r(\chi,\tau)$ = heat transferred from the element surface at axial location χ and time τ;

$Q_{ex}(\chi,\tau)$ = coolant heat input or loss due to radial conduction heat transfer and flow redistribution to adjacent coolant channels; directly input from CØBRA-WC.

Coupled with this, the axial mass flow rate for each axial node $G(\chi,\tau)$ is also input from CØBRA-WC analyses. Both $Q_{ex}(\chi,\tau)$ and $G(\chi,\tau)$ are based on nominal conditions in the CØBRA-WC code. It is conservative to use $Q_{ex}(\chi,\tau)$ and $G(\chi,\tau)$ data in this fashion because these values are lower than those calculated for the hot channel temperature conditions, and thus, result in a conservatively higher predicted hot channel temperature.

Boundary conditions of CØBRA-WC analysis depend on the characteristics of the entire heat transport system. These boundary conditions (e.g., rate of total reactor flow coastdown and coolant inlet temperature) are to be supplied to the CØBRA-WC code from a plant-wide code, e.g., DEMØ(7). Worst case system parameters are used in the plant-wide code calculations for the boundary conditions of the CØBRA-WC analysis. Likewise, a worst case modeling of the core parallel flow network, with regard to pressure drop uncertainties and decay heat uncertainties, is included in the CØBRA-WC analyses for input to the FØRE-2M hot element temperature predictions.

Figure 1 shows the level of detail which can be used in the CØBRA-WC to maintain a reasonable computer running time(8). In this case a 217 element FFTF fuel assembly has been divided into 37 channels. Each channel shown is represented by one element with power and flow conditions corresponding to the average for all elements contained in it. For the interior, hexagonal shaped regions, this averaged element would represent a total of 12 elements.

CØBRA-WC uses a smaller axial node size than FØRE-2M to obtain a better modeling of the rod bundle dynamic effects between the 37 channels. For example, in the analyses to be described later, 2 inch axial increments were used in the CØBRA-WC calculation and 7 equal length nodes (over the 36 inch active fuel length) were used in the subsequent FØRE-2M analyses. The $Q_{ex}(\chi,\tau)$ and $G(\chi,\tau)$ data are appropriately integrated axially for input to the axially coarser FØRE-2M model.

To evaluate the effects of localized core uncertainties, hot channel factors such as those shown in Table 1 are used in FØRE-2M. Currently, these direct and statistical type factors are conservatively applied to what is termed the "hot element" by the semi-statistical method. More information on this approach and the bases of the uncertainties can be found in References 9, 10 and 11. In addition to the factors in the Table 1, hot element neutronic and decay powers are also evaluated using 3σ uncertainties. The hot element temperature predictions resulting from the application of all these hot channel factors are used in evaluations with regard to core design adequacy.

Currently, the uncertainty factors of Table 1 are formulated on a steady state basis. Consideration of the net effect of each uncertainty indicates that a corresponding transient factor would not be larger for an undercooling event such as natural circulation. Any change in the localized flow uncertainty for the hot channel will be derived from comparisons of code predictions to experimental data by the verification approach described in Reference 4.

3. TYPICAL RESULTS

For conditions typical of the FFTF descending into and operating in the natural circulation cooling mode, a series of studies were performed to test the practicality of the modeling method. Localized hot channel predictions were made for the instrumented fuel assemblies FOTA's, in Row 2 (near core center) and Row 6 (outermost fuel region) of the core. These 217 element assemblies are geometrically the same as the driver fuel assemblies. The Row 2 FOTA had a larger power generation (6.4 MW vs. 5.0 MW), higher coolant flow rate (2.1×10^5 lb/hr vs. 1.7×10^5 lb/hr) and smaller radial power variation (1.03 vs. 1.23) than the Row 6 FOTA.

Figure 2 shows a typical match between the hottest channel temperatures of CØBRA-WC to those of FØRE-2M using $Q_{ex}(\chi,\tau)$ and $G(\chi,\tau)$ data from the

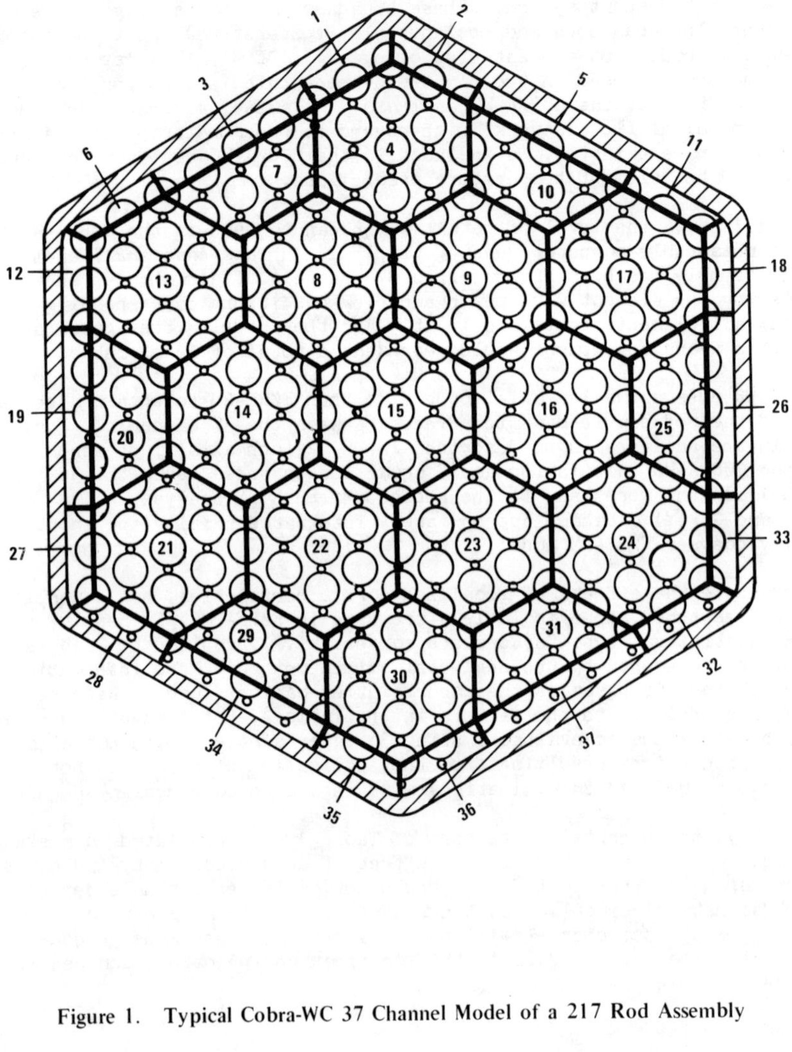

Figure 1. Typical Cobra-WC 37 Channel Model of a 217 Rod Assembly

TABLE 1

TYPICAL LMFBR UNCERTAINTY FACTORS APPLIED TO HOT ELEMENT IN FØRE-2M

	Coolant 1	Film 2	Cladding 3	Gap 4	Fuel 5
A. DIRECT					
1. Inlet Flow Maldistribution	1.050				
2. Flow Distribution Calculational Uncertainty (Simulation Bias)	1.030	1.022			
3. Power Control Band	1.020	1.020	1.020	1.020	1.020
4. Wire Wrap Peaking		2.000(1.214)			
B. STATISTICAL					
1. Fissile Fuel Maldistribution	1.035	1.035	1.035	1.035	1.035
2. Power Level Measurements	1.079	1.079	1.079	1.079	1.079
3. Nuclear Power Distribution	1.060	1.065	1.065	1.065	1.065
4. Subchannel Flow Area	1.011				
5. Film Coefficient		1.266			
6. Gap Coefficient				1.480	
7. Fuel Conductivity					1.100
8. Pellet - Cladding Eccentricity		1.15	1.15		
9. Cladding Conductivity and Thickness			1.120		
10. Coolant Properties	1.01				
11. Flow Distribution Calculational Uncertainty (Calibration)	1.054	1.015			
12. Wire-Wrap Orientation	1.01				

a. The number in parentheses should be used only for calculating fuel temperatures; the other number (which represents under the wire-wrap temperature peaking) is used for cladding temperatures.

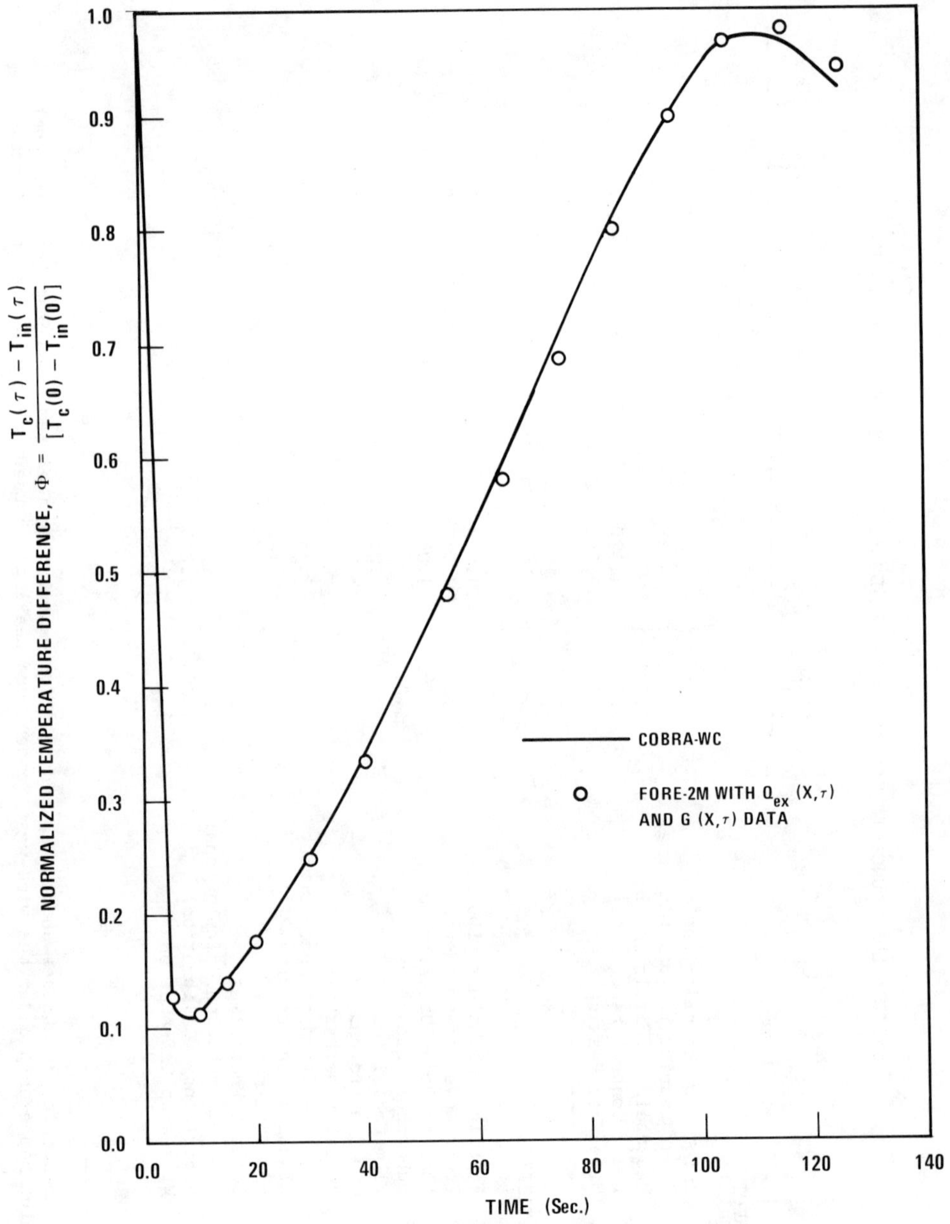

Figure 2. Comparison of Cobra-WC and Fore-2M Transient Coolant Outlet Temperatures (Channel 15 – Row 2 FOTA – Nominal Conditions)

CØBRA-WC. This particular case is for channel 15 in the Row 2 FOTA (see Figure 1). Similar good agreement was found in the Row 6 FOTA (channel 13 modeled).

Using the CØBRA-WC data four cases were analyzed with FØRE-2M for the Row 2 and Row 6 FOTA's to illustrate typical results predicted by this technique for the peak element in each assembly:

- Case 1 - Fuel assembly peak coolant channel transient flow and temperature calculations including both inter- and intra-assembly flow and heat redistribution; nominal conditions (no uncertainty factors applied);
- Case 2 - Fuel assembly peak coolant channel transient flow and temperature calculations including both inter- and intra-assembly flow and heat redistribution; 3σ uncertainty factor conditions;
- Case 3 - Fuel assembly peak coolant channel transient flow and temperature calculations only including inter-assembly flow redistribution; 3σ uncertainty factor conditions; and
- Case 4 - Fuel assembly peak coolant channel transient flow and temperature calculations without inter-assembly and intra-assembly flow and heat redistribution; 3σ uncertainty factor conditions.

Results for these four cases are presented in Figures 3 and 4 for Rows 2 and 6, respectively, as normalized temperature differences relative to the steady state temperature difference in Case 1, i.e.,

$$\phi = \frac{\left[T_c(\tau) - T_{in}(\tau)\right]_i}{\left[T_c(o) - T_{in}(o)\right]_1} \qquad i = \text{Case } 1,\ 2,\ 3 \text{ or } 4$$

where

$T_c(\tau)$ = maximum hot channel coolant temperature;

$T_{in}(\tau)$ = inlet temperature.

4. DISCUSSION OF RESULTS

The differences in the hot channel coolant temperature rise for the different cases in the Row 2 FOTA (Figure 3) and in the Row 6 FOTA (Figure 4) clearly indicate the inter- and intra-assembly flow and heat redistribution effects. These effects reduce the predicted transient coolant temperatures in the "hot channel". In Figure 3, the reduction in temperature rise is 11% when inter-assembly flow redistribution alone is considered (Cases 3 and 4), while the reduction becomes 14% when both inter- and intra-assembly flow and heat redistribution are included in the computation (Cases 2 and 4). At the same time, for the nominal case without the uncertainty factors (3σ factors used in this example), the maximum coolant temperature rise would be only 65% of the value predicted with the uncertainty factors (Cases 1 and 2).

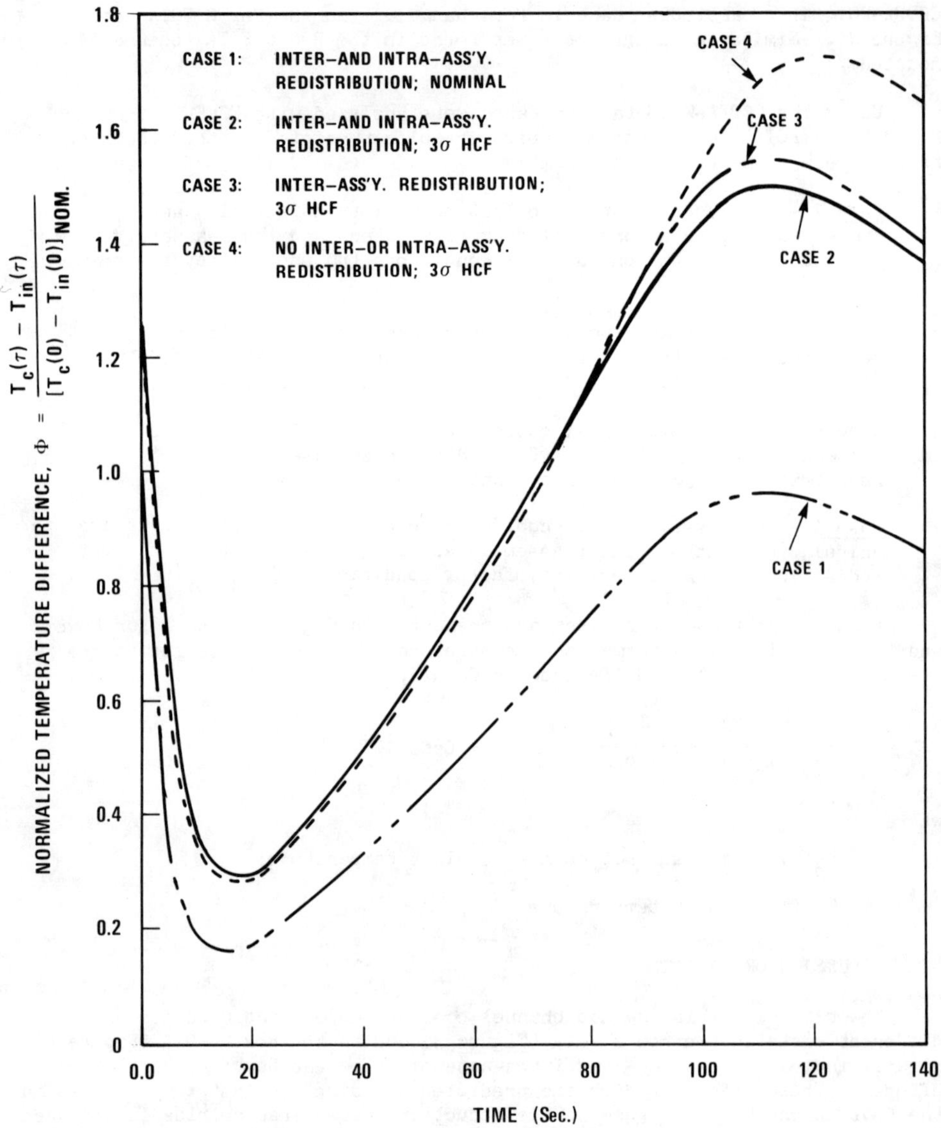

Figure 3. Typical Effect of Uncertainties and Inter-/Intra-Assembly Flow/Heat Redistribution on Coolant Temperatures During Natural Convection Cooling (Row 2 FOTA)

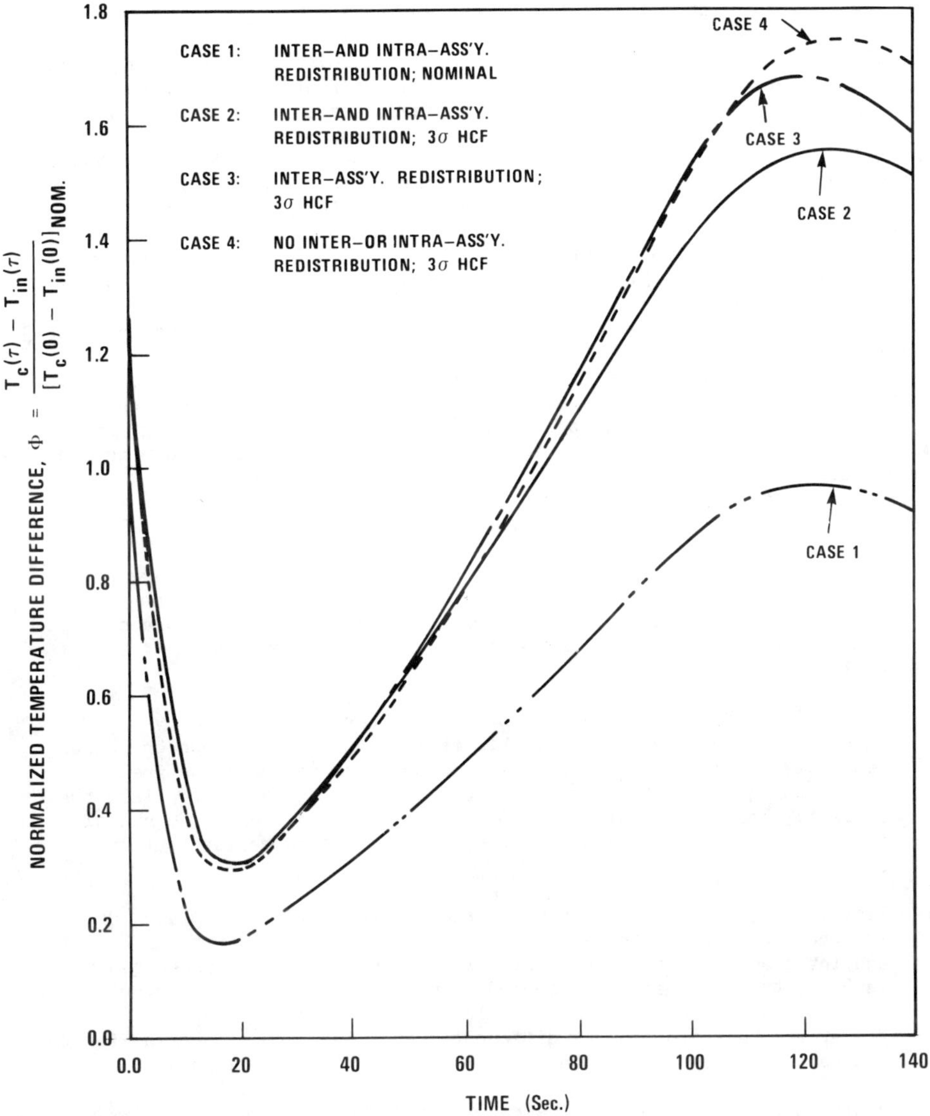

Figure 4. Typical Effect of Uncertainties and Inter-/Intra-Assembly Flow/Heat Redistribution on Coolant Temperatures During Natural Convection Cooling (Row 6 FOTA)

The effect of inter-assembly flow redistribution can be seen from the variation of inlet axial mass velocity $G(o,\tau)$, during the transient for the nominal conditions. This variation for the Row 2 FOTA is shown in Figure 5, as expressed in inlet mass velocity fraction $G(o,\tau)/G_R(\tau)$ at the time τ relative to the mass velocity fraction at steady state, $G(o,o)/G_R(o)$, where $G_R(o)$ and $G_R(\tau)$ are total reactor flow rates at time o and τ, respectively. The inter-assembly redistribution becomes appreciable after 75 seconds, which corresponds to the point of deviation shown in Figure 3 for Cases 3 and 4. The intra-assembly flow redistribution can be seen by comparing the axial mass velocity $G(\chi,\tau)$ for the top (χ=7) and bottom (χ=0) axial position of the fuel element (also shown in Figure 3). The difference between the solid curve (χ=7) and dotted curve (χ=0) indicates the flow change due to intra-assembly cross flow. The largest difference occurs at t=90 sec. which corresponds to the maximum deviation shown in Figure 3 for Cases 2 and 3.

The heat redistribution effect cannot be shown explicitly. This effect consists of that due to radial temperature gradient (conduction effect) and the cross flow mixing (transport effect) as well as the stored energy in the fuel element. Figure 6 shows the two components in Equation (1) during the nominal case in the top axial node of the hot element of Row 2 FOTA. There appears to be three periods during the transient with respect to the mode of énergy transfer from the element to the "hot channel" at the top axial node:

1) Period (0 to 7 sec) - The heat transfer from the element is larger than the power generated due to stored heat effects from the element internals. The absolute value of $Q_{ex}(\chi,\tau)$ decreases due to the overall assembly cooling during this period, i.e., radial temperature gradient collapses.

2) Period (7 to 85 sec) - The coolant temperature is montonically increasing. The heat deposition to the coolant is less than the power generated since part of this power has to first increase the fuel temperature to a sufficient level to drive the heat generated into the coolant. The absolute value of $Q_{ex}(\chi,\tau)$ increases due to the radial temperature gradient buildup in the bundle and flow redistribution effects.

3) Period (85 to 125 sec) - After 85 seconds the $Q_{ex}(\chi,\tau)$ decreases due to a decrease in radial temperature gradient in the bundle with high inter- and intra-assembly flow redistribution. Eventually, the power generated is lower than the heat transferred from the element to the coolant due to the relative increase in coolant flow (which increases the fuel-to-coolant temperature difference) in this channel coupled with the decrease in power generation.

Additionally, by comparing the results for the Row 2 and 6 FOTA's, one can see that the larger radial power skew of the Row 6 FOTA and its longer flow transport time enhances the $Q_{ex}(\chi,\tau)$ attenuating effect on maximum coolant temperature (i.e., larger difference between Cases 2 and 3 for Row 6 in Figure 4 than that in Figure 3). During the transient $Q_{ex}(\chi,\tau)$ reached a maximum value of 37% of the heat from the element surface in Row 6 as opposed to $\sim$18% for Row 2. However, part of this $Q_{ex}(\chi,\tau)$ effect is off-set by a smaller increase in flow from inter-assembly flow redistribution (Row 6 is cooler than Row 2 above the heated length due to inter-assembly heat loss).

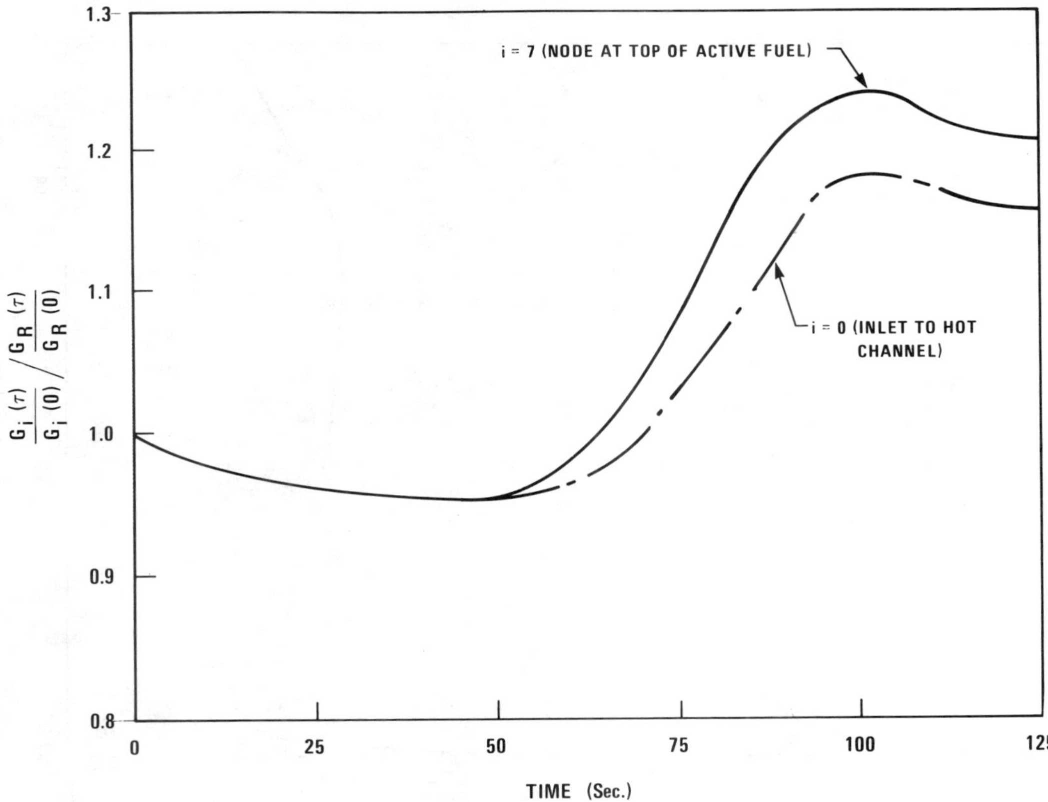

Figure 5. Hot Element Flow Relative to Reactor Flow Coastdown (Row 2 FOTA – Nominal Conditions)

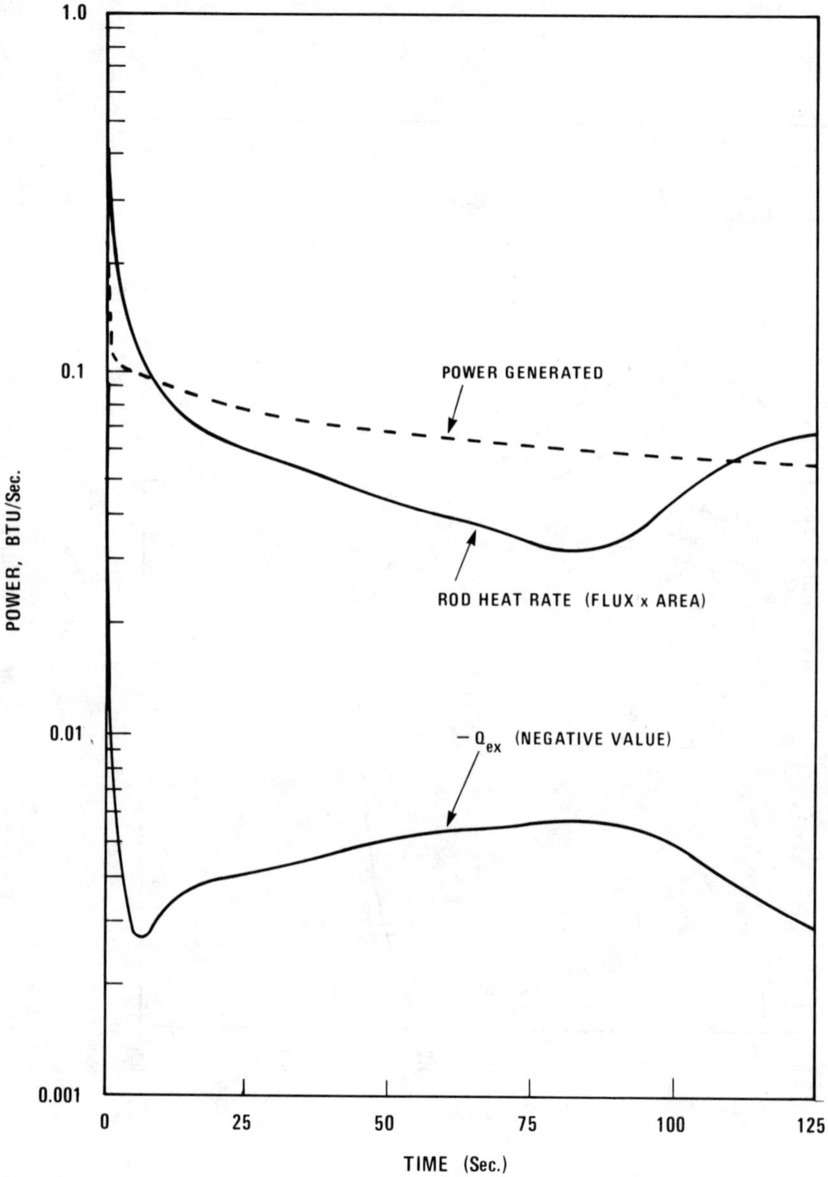

Figure 6. Variation in Power Data for Hot Element at Top Node of Active Fuel Length (Row 2 FOTA–Nominal Conditions)

The heat and flow redistribution could be sufficiently large in some cases as to necessitate analyzing several channels within the assembly to assure that the hottest channel during the transient is being evaluated.

5. CONCLUSIONS

In summary, natural convection cooling of the core exhibits one of the few core design transients where low power/high temperature conditions exist. Due to the long coolant transport time and low pressure drop for the core while descending into and operating in this mode, core inter-assembly and intra-assembly flow/heat redistribution effects become important for accurately predicting temperatures; and, these effects significantly decrease the maximum hot element temperatures.

The technique described herein for linking the CØBRA-WC core system analyses to the FØRE-2M localized hot element analyses is shown to be a viable means of accounting for all the coupled inter- and intra-assembly flow and heat redistribution effects which influence the hot channel temperatures during natural convection cooling. Through the $Q_{ex}(\chi,\tau)$ and $G(\chi,\tau)$ terms design calculations can be performed for the hot elements with proper credit taken for the temperature attenuating phenomena.

6. ACKNOWLEDGEMENTS

This work was performed under US/DOE Contract E(11-1)-2395. The authors are indebted to Thomas L. George of PNL for providing the CØBRA-WC code calculations and Figure 1 (used in the text) and to J. V. Miller of Science Applications, Inc., who was largely responsible for the modified version of FØRE-2M.

7. REFERENCES

1. R. M. Singer and J. L. Gillette, "Measurements of Subassembly and Core Temperature Distributions in an LMFBR," AIChE Symposium Series, 73, p. 97-104, 1977.

2. A. K. Agrawal, et al., "Dynamic Simulation of LMFBR Plant Under Natural Circulation," ASME Paper 79-HT-6, 1979.

3. M. Khatib-Rahbar and K. B. Cady, "Establishment of Buoyancy-Induced Natural Circulation in Loop-Type LMFBRs," Trans. Amer. Nucl. Soc. 28, pp. 432-433, June 1978.

4. R. D. Coffield, R. A. Markley and E. U. Khan, "Natural Convection Analyses and Verification for LMFBR Cores," to be published in Nuclear Enginering and Design.

5. T. L. George, R. A. Masterson and K. L. Basehore, "A Modified Version of CØBRA for Whole-Core LMFBR Transient Analysis," Trans. Amer. Nucl. Soc. Vol. 322, pp. 531-532, June 1979.

6. J. V. Miller and R. D. Coffield, "FORE - 2M: A Modified Version of the FORE-II Computer Program for the Analysis of LMFBR Transients," CRBRP-ARD-0142, November 1976 (Availability: US/DOE Technical Information Center).

7. W. H. Alliston, et al., "Clinch River Breeder Reactor Plant; LMFBR Demo Plant Simulation Model (DEMØ)," CRBRP-ARD-0005, February 1978 (Availability: US/DOE Technical Information Center).

8. Personal communication, T. L. George (PNL), March 1979.

9. Y. S. Tang, R. D. Coffield, Jr. and R. A. Markley, Thermal Analysis of Liquid-Metal Fast Breeder Reactors, pp. 82-104, American Nuclear Society, LaGrange, Il. 1978.

10. A. J. Friedland, "CRBRP Core Assemblies Hot Channel Factors Preliminary Analysis," CRBRP-ARD-0050, February 1980 (Availability: US/DOE Technical Information Center).

11. "Users Guide for Irradiation of Experiments in the FTR," HEDL-MG-22, Rev. 2, May 1978 (Availability: Hanford Engineering Development Laboratory, Richland, Washington).

EXPERIMENTS AND ANALYSES

Response of EBR-II to a Complete Loss of Primary Forced Flow During Power Operation*

RALPH M. SINGER, JERRY L. GILLETTE, DALE MOHR, JOHN E. SULLIVAN, JOHN V. TOKAR,† and ERIC M. DEAN
EBR-II Project
Argonne National Laboratory
Argonne, Illinois and Idaho Falls, Idaho, USA

ABSTRACT

An important aspect of the ongoing program at the Experimental Breeder Reactor No. II (EBR-II) is the study of the thermal, hydraulic, and neutronic response of the core and plant to off-normal conditions. Included among the goals of this study is an investigation of the inherent characteristics of an LMFBR which permit it to benignly respond to various upset and emergency events. Some of the more important of such postulated events are those in which the removal of decay heat is a dominant factor. The reliability and licensing value of decay heat removal systems can be markedly improved if natural circulation of the coolant can be shown to provide adequate core cooling, especially during the transition period immediately following the loss of forced circulation. In order to provide this demonstration, it is necessary to develop a phenomenological understanding of the dynamics of buoyancy-driven flows in LMFBR's and to generate reference or benchmark data for use in system code development and verification. This paper will present detailed experimental data and the associated analytical model predictions for a whole plant natural circulation transient initiated from power operation by a complete loss of electrical power to the primary pumps. These results provide documented evidence of the benign and predictable response of a small but prototypical LMFBR power plant to a loss of primary pumping power.

1. INTRODUCTION

A basic element of the evolving fast reactor safety philosophy in the United States is the development of technology to aid the design of highly reliable reactor systems. The objective of this effort is the demonstration that an LMFBR can be designed, constructed, and operated to have an extremely low probability of accidents which adversely affect either public or operator safety. This particular objective is embedded in the Lines-of-Assurance (LOA) structure as the goal of LOA-1. In order to assure the attainment of this goal within the prescribed limits of uncertainty, it is necessary to demonstrate the reliability of the basic reactor and heat transport systems as well as the reactor shutdown and shutdown heat removal systems. An improvement of several

---

*Work supported by the U.S. Department of Energy.

†Presently at Bell Telephone Laboratories, Naperville, Illinois, U.S.A.

orders of magnitude in the reliability of the later system can be achieved if natural circulation can be demonstrated to be an effective mode of core cooling during shutdown conditions.

It appears to be quite clear that it will be necessary to develop a well documented and experimentally verified system simulation code which includes representation of buoyancy-driven flows in order to aid reactor designers to take advantage of the inherent attributes of natural circulation as well as to convince licensing authorities to accept systems relying upon this mechanism to improve reactor safety. Such codes have been and are being developed, for example, DEMO-IV for the CRBRP [1], IANUS for FFTF [2], NATDEMO for EBR-II [3], SSC for general LMFBR plants [4], MELANI for pool-type LMFBR's [5], and several others. However, the data base upon which these codes are based, especially in the case of natural circulation, is quite limited. Data are required on the dynamics of reactor and plant response to a wide variety of initiating events and initial operating modes which lead to buoyancy-driven flows. This includes information on the performance of individual driver and blanket subassemblies as well as the integrated whole core in order to understand the dynamics of heat transport from the reactor to the primary coolant. The subsequent transport of heat to an ultimate heat sink, whether via a dedicated decay heat removal system or the main heat transport system, must also be experimentally studied in order to verify those aspects of the system simulation code.

It is the purpose of this paper to present the result of one step in this verification process by the description of a whole plant natural circulation transient conducted in EBR-II and the associated analysis of the event by the NATDEMO code. This particular test was the first attempt in EBR-II to examine the transition to natural circulation from an initial at-power condition and therefore was specifically designed to involve only the primary system. Earlier whole plant experiments conducted in EBR-II were initiated from a previously shutdown reactor [6] and therefore at low decay power.

2. DESCRIPTION OF REACTOR AND INSTRUMENTATION

The Experimental Breeder Reactor No. II (EBR-II) is a sodium-cooled fast breeder reactor with a complete steam-electric system. The plant is located in the Idaho National Engineering Laboratory (INEL) and is operated by Argonne National Laboratory for the U. S. Department of Energy. The reactor consists of 16 rows of subassemblies, the inner 7 constituting the fueled core and the outer nine containing reflector and blanket subassemblies. The active fuel length in the core is 0.343 m, with 91 elements in a metal fuel driver subassembly. These driver fuel elements are 4.42 mm in diameter and are contained within a hexagonal can which has a flat-to-flat dimension of 56.1 mm. The spacing of the individual fuel elements is maintained by wire wraps with a diameter of 1.24 mm and an axial pitch of 152 mm. The normal operating power of the reactor is 60.0 MWt at a total coolant flow rate of 0.516 m^3/s. Additional descriptive information on EBR-II is available in [7] and a sketch of the primary heat transport circuit is shown in Fig. 1.

The instrumentation utilized in the EBR-II tests includes both the normal plant sensors and special in-core sensors. The in-core instrumentation is located in a modified driver subassembly placed in a converted control-rod position in the fifth row of the core. This experimental subassembly, designated XX08, consists of 61 elements, 58 of which are fueled, and are contained within a hexagonal can which measures 46.4 mm across its inside flats resulting in a pin bundle hydraulic diameter of 2.76 mm. Within this subassembly, there are two inlet, permanent magnet flowmeters, six fuel centerline thermocouples,

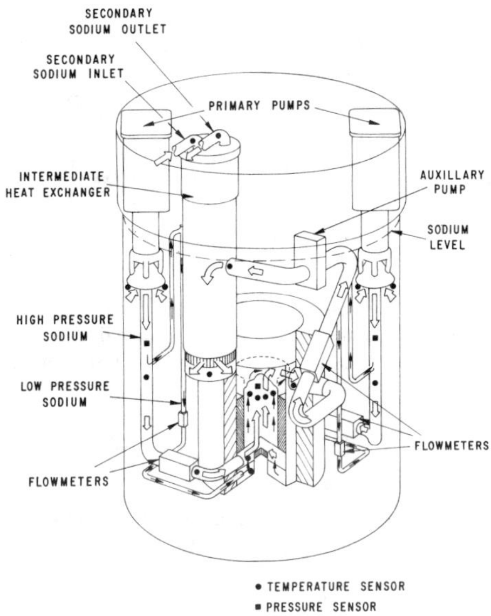

Fig. 1. Schematic of the EBR-II Primary Heat Transport System

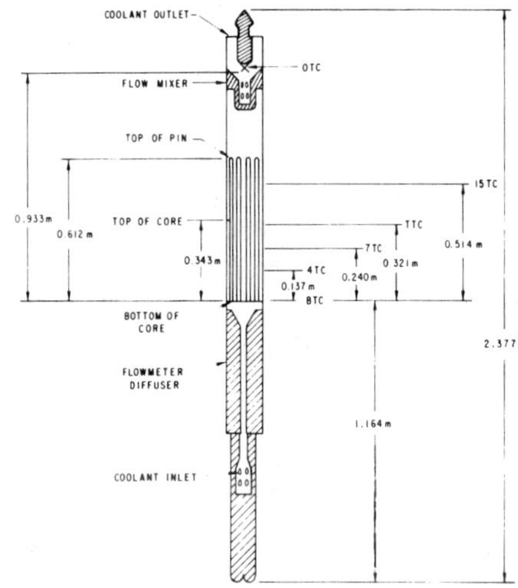

Fig. 2. Schematic of the XX08 Instrumented Subassembly

referred to as "FTC," located 21.6 mm below the top of the fuel (at beginning of fuel life), and 16 coolant thermocouples mounted as wire wrap spacers. Two of these coolant thermocouples, called "BTC," are located at the core bottom, two, called "4TC," are at 0.4 of the core height (0.137 m above core bottom), two, called "7TC," are at 0.7 of the core height (0.240 m above core bottom), nine, called "TTC," are near the core top (same height as the fuel thermocouples), and one, called "15TC," is at 1.5 of the core height (0.514 m above core bottom or 0.171 m above core top). In addition, two coolant thermocouples, called "OTC," are located above a flow mixer near the subassembly exit, but within the hex can and 0.933 m above core bottom, and measure the mixed-mean coolant outlet temperature. A schematic diagram indicating the axial location of the XX08 instrumentation is shown in Fig. 2. Additional details on the instrumented subassembly XX08 are available in [8]. Several selected plant sensors which were utilized in the test reported in this paper are also shown in Fig. 1.

The general plant instrumentation is calibrated on a periodic basis, however special calibration efforts are undertaken for whole plant testing. The temperature sensors (including both thermocouples and resistance thermometers) and their associated signal conditioning equipment were checked for off-sets by operating the plant at full (100%) flow and extremely low decay power. This resulted in an isothermal primary system (except for the slight temperature rise across the core which was accounted for) and all of the signals from the temperature sensors could thereby be corrected. With only a few exceptions, the corrections obtained from this calibration were less than 1 C, and therefore, it is felt that all of the temperature measurements herein reported have uncertainties of about this magnitude. Although the flowmeters were not calibrated in-place, extensive out-of-pile calibration tests were performed for those units located within the XX08 subassembly. Thus, the sensitivity of these flowmeters is known explicity over the range of about 0.3×10^{-4} to 25×10^{-4} m^3/s (about 1 to 100% of full flow), and was obtained for lower flow rates by extrapolation between 0 and 0.3×10^{-4} m^3/s. The calibration uncertainty was estimated to be about ±0.2% for flow rates greater than about 6×10^{-4} m^3/s, about ±1% for flow rates between 0.3×10^{-4} and 6×10^{-4} m^3/s, and about ±5% for lower flow rates. Because of the additional uncertainties involved with the signal conditioning equipment and the actual reactor environment, it is estimated that the final reported flow rates have inherent uncertainties of about twice these calibration values.

Signals from all operating sensors were continuously monitored during the plant test and were digitized, converted to engineering units, and recorded on magnetic tape at a rate of 2 samples/sec/signal by the EBR-II Data Acquisition System (DAS).

3. EXPERIMENTAL DESCRIPTION

This experiment, which was designated XX08 Test 7A in recognition of the availability and importance of the in-core instrumented subassembly XX08, was designed to measure the response of the EBR-II reactor and primary heat transport system to a complete loss of electrical power to the primary pumps during power operation with the auxiliary pump inoperative. Since this transient was the first of its type ever conducted in EBR-II, it was decided to establish appropriate initial conditions which would assure that a relatively large margin to normal full power operating temperatures would exist throughout the test. Accordingly, the primary tank temperature (reactor inlet) was reduced to 351 C from its normal value of 371 C, the initial reactor power and flow levels were set at 28.5% and 32.1% of their respective full values, and the secondary pump trip circuit was by-passed in order to continue secondary IHX cooling following the loss of primary forced flow.

The establishment of the desired initial conditions for the experiment commenced following fuel handling for the EBR-II Reactor Run Number 97A [9]. The core loading for this run included a wide assortment of special experimental fuels and materials irradiation subassemblies as well as the standard EBR-II metal driver fuel subassemblies, reflector subassemblies, and depleted uranium blanket subassemblies. The ascent to power occurred over a 150 minute period at which time the following plant parameters were established: reactor power, 17.1 MWt (28.5% of normal), primary coolant flow rate, 0.166 m^3/s (32.1% of normal), overall reactor temperature rise, 93 C, reactor inlet temperature, 351 C, secondary system coolant flow rate, 0.113 m^3/s (29.9% of full flow), and IHX secondary inlet and outlet coolant temperatures of 302 C and 437 C, respectively. These operating conditions were maintained constant for 192 minutes in order to permit the entire plant to reach a fully stable initial condition as well as to permit a sufficient build-up of an inventory of short-lived isotopic fission products so that the resulting decay heat generation transient during the first several minutes following reactor scram would reasonably represent that which would be expected after long term reactor operation.

The transient was initiated by manually opening the 2400 volt power supply and clutch circuit breakers to the primary pumps. This action resulted in the following automatic sequence of events: 0 seconds, both primary pumps start to run down; 1.8 seconds, the reactor is scrammed by the plant protection system (PPS) on a low flow signal; 36.5 seconds, pump number 1 reaches zero speed; and, 42.5 seconds, pump number 2 reaches zero speed. Electrical power to the secondary pump and the steam drum feedwater pump was maintained throughout this test. The thermal, hydraulic, and neutronic response of the plant to this event is described in the following section.

4. MEASURED AND CALCULATED PLANT RESPONSE

The NATDEMO code [3] was used to predict the response of EBR-II to this event. Since this code provides a complete thermal, hydraulic, and neutronic representation of the plant, the only required input, aside from the geometric, hydraulic, and reactivity feedback characterization parameters, are the following functions or parameters: (1) transient impeller speeds for each pump; (2) transient negative reactivity from control rod motion; (3) initial reactor power, coolant flow rate, and inlet temperature; (4) secondary system coolant flow rate and IHX inlet temperature; and (5) pre-transient reactor power operating history. Based upon this information, along with the time sequence of primary pump trip and reactor scram, the resulting system temperatures, coolant flow rates, and fission-product decay heat plus neutron power was calculated. The following discussion and figures will describe both the measured and predicted reactor and primary system behavior.

4.1 Neutronic and Primary System Hydraulic Response

The input and measured primary pump impeller speeds are shown in Fig. 3. An interesting feature shown here is the significantly different rundown times for each pump. Although each pump is of identical design and were simultaneously tripped, they have sufficiently different internal frictional resistances to cause somewhat different coastdowns. The implication of this result is that the potential of reverse flow through the inlet piping leg associated with the stopped pump exists during the time the other pump is still running. Thus, it is possible that during the period when one pump is running and the other is stopped, some of the coolant flow supplied by the operating pump can bypass the core through the inlet plenum towards the other pump. Thus, it was necessary to include proper modeling in NATDEMO to account for the potential of reverse flow in the inlet piping.

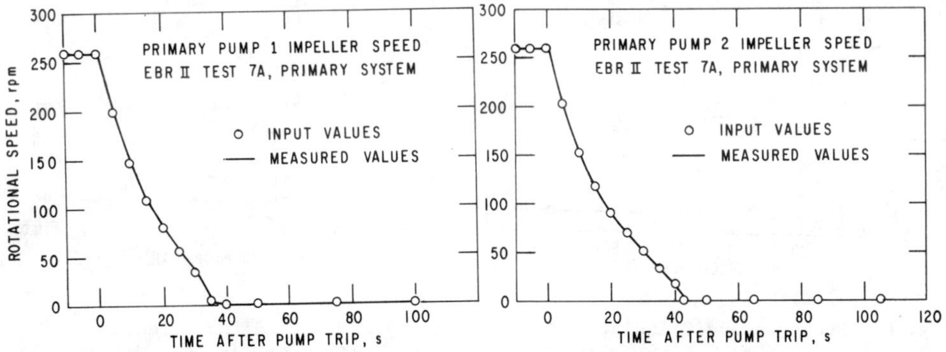

Fig. 3. Input and Measured Rundowns of Primary Pumps

A comparison of the measured and predicted coolant flow rates within the core (Fig. 4) and in the inlet to the high pressure plenum (Fig. 5) indicates that more flow is being supplied to the cores inlet plenum by pump No. 2 than is flowing through the core during a period of time consistent with the pump speed mismatch. This effect is not noticed in the inlet to the low pressure plenum (Fig. 6) which feeds the blanket and reflector subassemblies due to the large pressure drop across a valve between the outlet of the pumps to the low pressure plenum piping. An examination of Figs. 4, 5, 6, and 7 also reveals an excellent agreement between the measured and predicted coolant flow rates within the core, inlet to the high pressure plenum, inlet to the low pressure plenum, and reactor outlet.

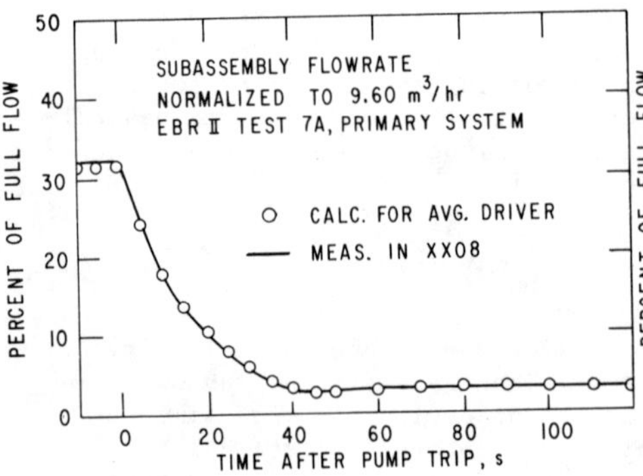

Fig. 4. Measured and Calculated Coolant Flow Rate in an Instrumented Driver Subassembly

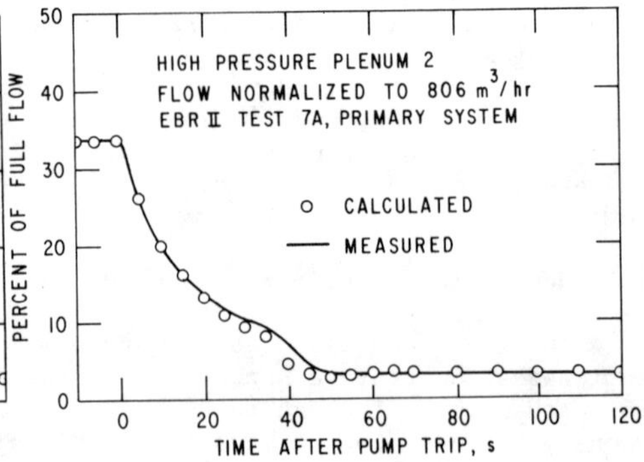

Fig. 5. Measured and Calculated Coolant Flow Rate at the Inlet to the High Pressure Plenum

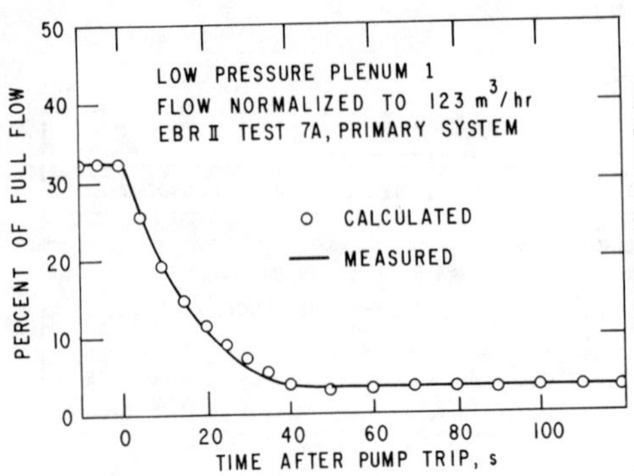

Fig. 6. Measured and Calculated Coolant Flow Rate at the Inlet to the Low Pressure Plenum

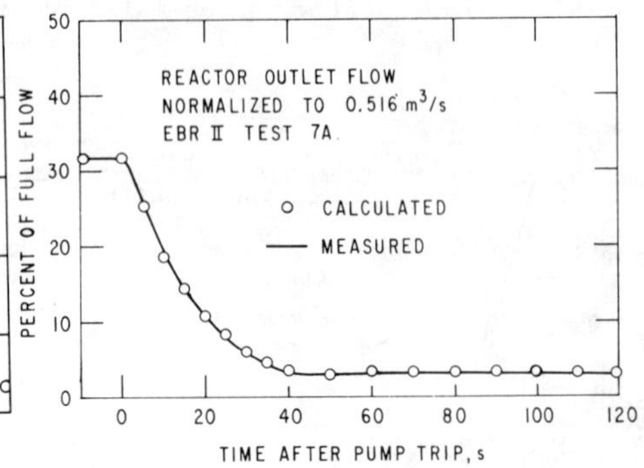

Fig. 7. Measured and Calculated Coolant Flow Rate at the Reactor Outlet (Total Primary Flow Rate)

Referring to Fig. 4, which shows the transient coolant flow rate in a driver subassembly, it is noted that the flow reaches a minimum value of 2.8% at 47 seconds after the pump trip. Both pumps were completely stopped at 42.5 seconds, and only buoyancy forces were available to drive the flow thereafter. From this data, as well as the predicted behavior, it is clear that the transition from forced to natural circulation flow was quite smooth under these conditions. After reaching this minimum value, the flow gradually increased over the next 20 to 30 seconds to about 3.3% and then slowly decreased, following the declining fission-product decay heat in a quasi-steady manner.

The reactivity transient which occurred due to the reactor scram on a low flow signal is shown in Fig. 8, where both the measured and code input values are indicated. The initial loss of reactivity was at 1.8 seconds following the primary pump trip and was due to a PPS-activated control rod scram. As an added conservatism, the safety rods were manually scrammed at 11 seconds which removed about an additional $1 of reactivity from the core. The calculated and measured power transient resulting from this reactivity loss is shown in Fig. 9. Since only the neutron power can be directly measured in EBR-II (the fission-product decay power must be obtained from caloriometric measurements), only this portion of the total transient power is plotted. As can be seen, the calculated values of the prompt jump and the delayed neutron power agree quite well with the corresponding measured values. There is a slight overprediction of about 0.02 to 0.04% for times greater than about 15 to 20 seconds which is accentuated by the semi-log presentation.

4.2 Thermal Response

As was described earlier, temperature measurements were made within an instrumented, fueled subassembly, at the outlet of a number of subassemblies, and at various locations along the primary heat transport circuit. The transient coolant temperature at the top-of-core (top of fueled zone) location within a fueled subassembly is shown in Fig. 10, where the measured values were obtained from XX08 and the calculated values are based upon an average driver. Since XX08 operates at a somewhat larger power-to-flow ratio than an average driver, some differences between the measured and calculated values are to be expected.

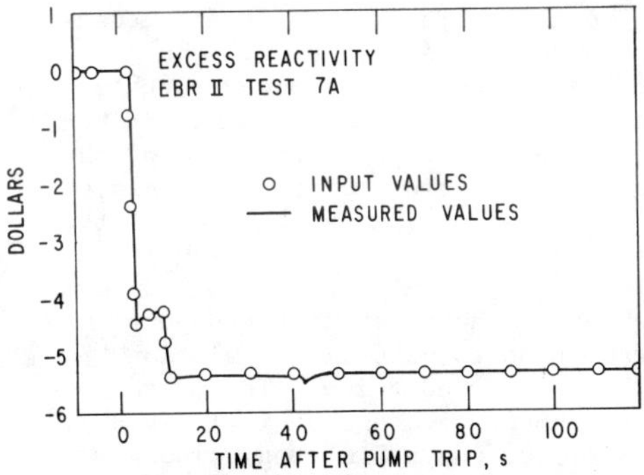

Fig. 8. Input and Measured Values of the Reactivity Transient

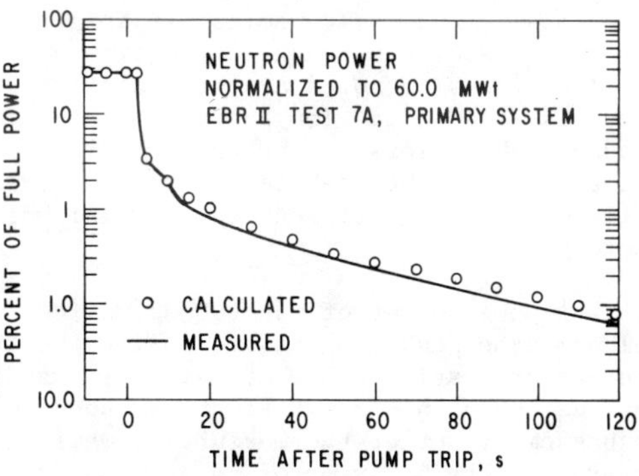

Fig. 9. Measured and Calculated Neutron Power Transient

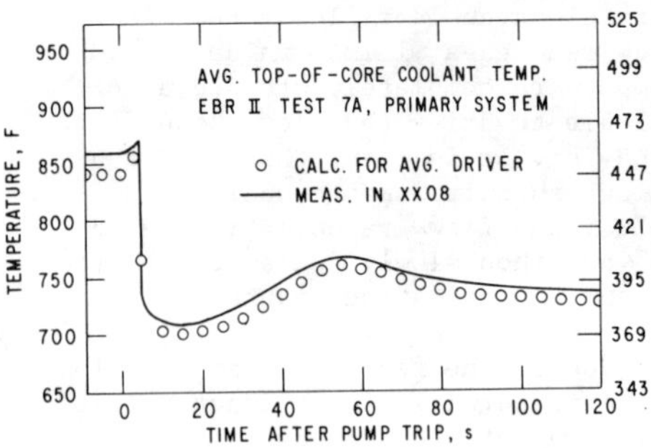

Fig. 10. Measured and Calculated Coolant Temperature at Core-top in Driver Subassemblies

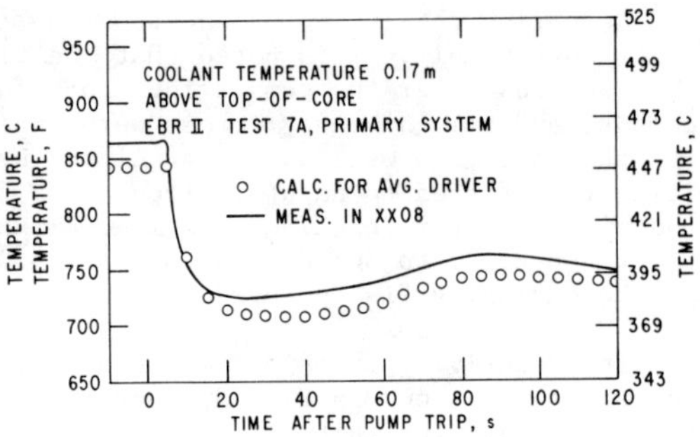

Fig. 11. Measured and Calculated Coolant Temperature 0.17 m Above Core-top in Driver Subassemblies

Immediately following the start of the primary pump rundown at zero time, the top-of-core coolant temperature is seen to increase. At 1.8 seconds, the temperature drops rapidly as the control rods are scrammed. Due to the large and rapid drop in power and the relatively slower decrease in primary flow, the core is overcooled for a short period. However, this condition shortly reverses due to the continuing decrease in the coolant flow rate and a relatively slower decrease in the neutron plus decay power, resulting in a gradual increase in coolant temperature. This increase in temperature is reversed as the convective flow develops, and a peak (in XX08) of about 410 C occurs at 55 seconds, and a slow monotonic decrease continues thereafter. As is apparent from Fig. 10, the overall agreement between the measured and calculated dynamic thermal response of the coolant at this location is excellent. The small differences in temperature levels can be accounted for by the different initial power-to-flow ratios and heat capacities of the measured and modeled subassemblies.

A similar comparison is made in Fig. 11 for a location within the subassembly and 0.17 m above the top of the fuel zone. Again, there is a difference between the measured and calculated temperature levels as discussed previously, but the dynamics are in good agreement. The initial temperature peak caused by the reduction in flow prior to reactor scram as was seen at the core top is also noted at this location well above core top. However, this peak is delayed in time due to coolant transport and reduced in magnitude because of heat loss to the cooler structure in the above core region. For the same reasons, the peak temperature at this axial location occurs at a delayed time and is somewhat reduced in magnitude (406 C) from that at the top of the core.

At the outlet of the subassemblies, which is the inlet to the upper plenum, the peak temperature during the transition to natural circulation occurs at a still somewhat later time and has a reduced value. The measured values shown in Fig. 12 are based upon an arithmetic average of 17 individual subassembly outlet thermocouples, while as before, the calculated values represent the behavior of an average driver subassembly. As was the case for the internal subassembly measurements, the actual temperature level calculated for an average subassembly outlet is not expected to precisely agree with that measured at the outlet of a collection of various subassemblies. However, reasonable agreement between the calculated and measured dynamic response is expected, and in fact is observed in this figure. Furthermore, when the measured temperature at the outlet of a particular fueled subassembly which happens to be operating close to the average driver power-to-flow ratio is compared to the same prediction as was shown in Fig. 12, excellent agreement results. This comparison is shown in Fig. 13.

The dynamic response of blanket and reflector subassemblies to a protected loss of flow transient such as that discussed here is expected to markedly differ from that of driver. This is due to the significantly different decay heat levels in these non-fueled subassemblies, as well as to the much larger thermal heat capacities and different hydraulic characteristics. The measured temperatures at the outlets of two particular blanket and reflector subassemblies are shown along with the corresponding predictions of average blanket and reflector conditions in Figs. 14 and 15. The quantitative agreement is within what would be expected considering the differences between these subassemblies and the corresponding hypothetical average units. In both case, the qualitative agreement is also quite good, wherein only very modest and slow thermal responses to the rather wide swings in power-to-flow changes are noted.

Somewhat further down the primary heat transport circuit, a temperature measurement is available just beyond the upper plenum outlet nozzle which leads to the IHX. The response at this location is dependent not only upon the overall reactor thermal-hydraulic behavior, but also the dynamic mixing and heat transfer to the cover and other structures which occurs in the reactor outlet plenum. In addition, the particular thermocouple which is located at this position is mounted within a well so that a significant thermal lag exists between the coolant and the thermocouple junction. Analytical studies indicated that this structure had a thermal response time constant of 7.5 seconds; both the calculated fluid temperature and thermocouple indicated temperature with several assumed time constants are compared with the measured signal in Fig. 16. A comparison between these calculated results and the measurements indicate a slight discrepancy which appears to be only partially explained by the thermal lag of the thermocouple assembly. Phenomena in addition to this lag which could affect the measured temperature at this location include non-uniform coolant mixing in the upper plenum and thermal stratification in the

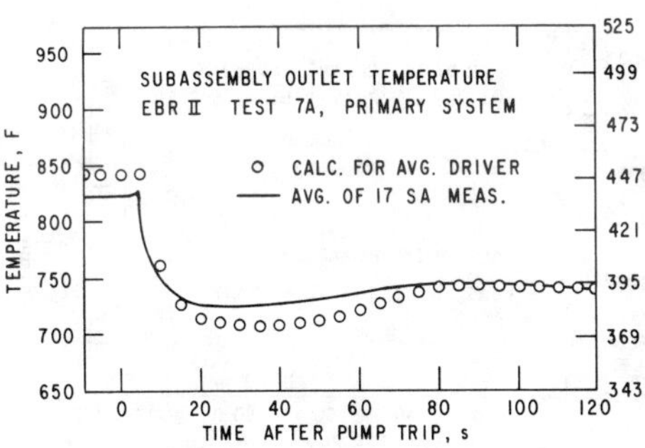

Fig. 12. Calculated Average Driver Subassembly Outlet Temperature Compared to an Unweighted Measured Average of Core Subassemblies

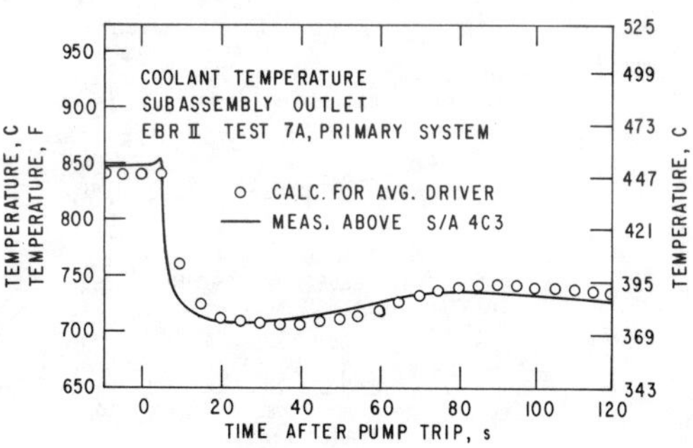

Fig. 13. Calculated Average Driver Subassembly Outlet Temperature Compared to a Particular Fueled Subassembly

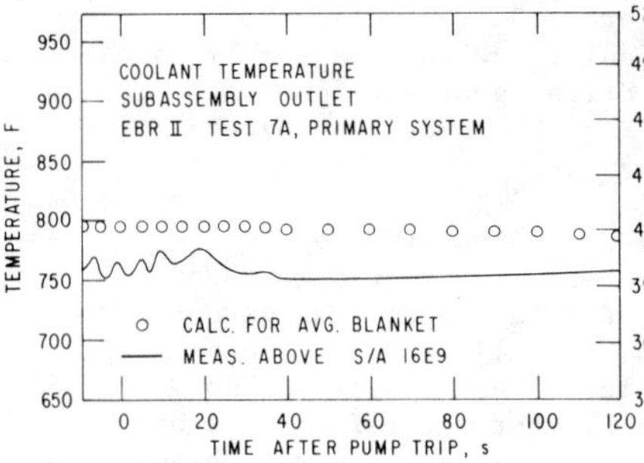

Fig. 14. Measured and Calculated Coolant Outlet Temperature from Radial Blanket Subassemblies

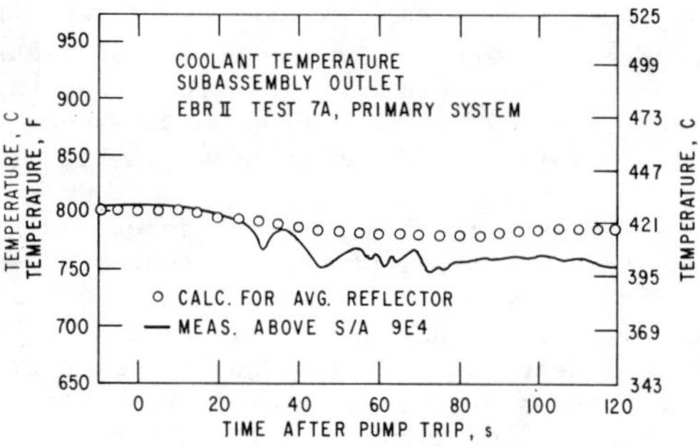

Fig. 15. Measured and Calculated Coolant Outlet Temperature from Radial Reflector Subassemblies

355 mm (14 inch) outlet pipe. With a single temperature measurement at one circumferential position (90° from top vertical) on this pipe, it is not possible to substantiate the specific cause for the difference between the measured and calculated temperature at this location.

The next measurement location along the primary heat transport circuit is at the IHX inlet. The thermocouple is mounted on a dummy tube within the IHX and is located directly in the incoming stream of sodium from the piping. Although this dummy tube does not have secondary sodium flowing through it, the results of the comparison between the temperature measurement and prediction at this location as shown in Fig. 17 indicate that there is some influence of the secondary sodium. The measured temperature agrees quite well with the predictions for about the first 80 seconds of the transient, but then drops off quite rapidly to a level which is below anything existing in the primary system, but roughly corresponding to the secondary system sodium temperatures. Only speculative explanations can be offered for this anomalous behavior.

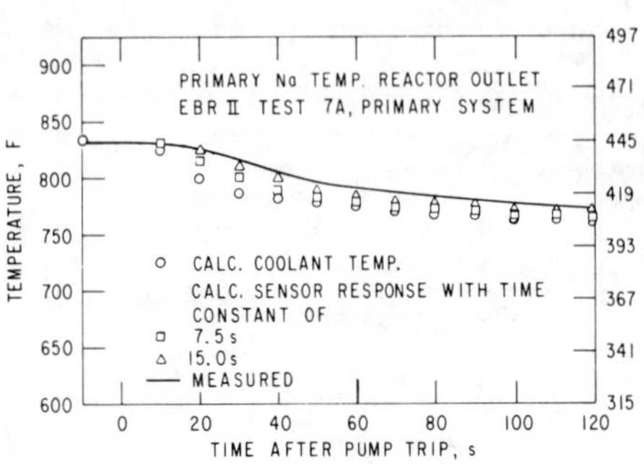

Fig. 16. Measured and Calculated Coolant Temperature at Reactor Outlet Nozzle

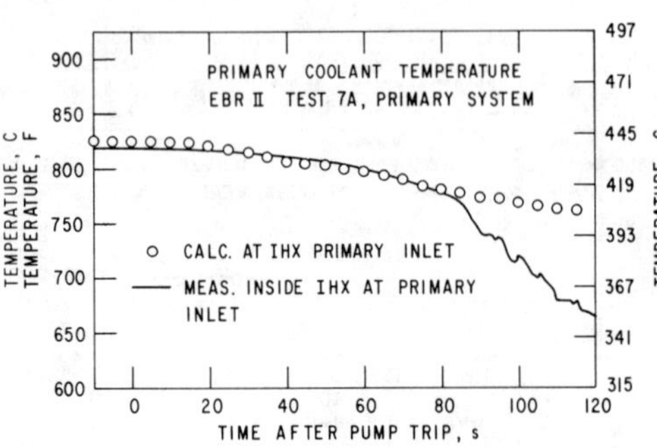

Fig. 17. Measured and Calculated Coolant Temperature at the IHX Primary Inlet

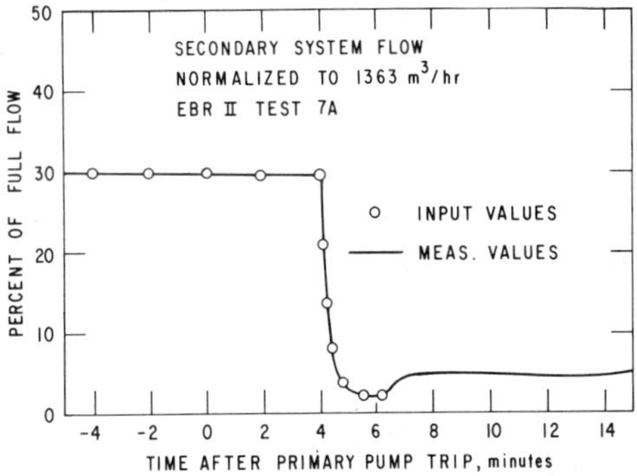

Fig. 18. Input and Measured Values of the Secondary Coolant Flow Rate

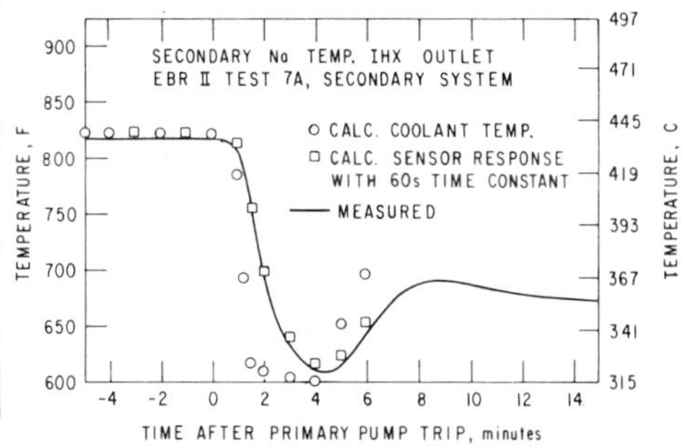

Fig. 19. Measured and Calculated IHX Secondary Coolant Outlet Temperature

4.3 Secondary System Response

As described previously, the secondary system coolant flow rate was maintained at its initial forced flow value well into the transient. A manual trip of the secondary pump occurred at 4 minutes as is seen in Fig. 18; this measured flow transient was used as input to NATDEMO. In addition, the temperature of the secondary sodium entering the IHX was constant at 302 C and was an input parameter. However, the secondary sodium IHX outlet temperature changed substantially throughout the transient in response initially to the changing IHX inlet primary sodium temperature and flow rate, and after 4 minutes, to the changing secondary flow rate. The resulting secondary side IHX calculated and measured (or input) temperatures are shown in Fig. 19. Due to the slow response of the secondary outlet resistance thermometer, both the calculated fluid and sensor responses are shown.

4.4 Measured Temperature Profile Within XX08

Since nine wire-wrap coolant thermocouples were present at the top-of-core location within XX08, it was possible to examine the effect of this loss-of-forced-flow transient upon the transverse temperature profile within the subassembly. The results of these measurements are shown in Fig. 20. While the plant was operating at power with forced flow, a normal spread in coolant temperatures is observed in which there is about a 70 C difference between the coolest and hottest subchannel. However, almost immediately following the loss of forced flow, the transverse temperature gradient across the subassembly collapses and the hottest subchannel is only about 5 to 10 C above the coldest. An explanation of this response is the greatly increased importance of transverse thermal conduction and buoyancy-induced flow redistribution at the low coolant flow rates present. An important implication of this result is that the appropriate hot channel peaking factor for loss-of-flow events is substantially less than that at normal full power conditions. Recognition and implementation of these phenomena in thermal-hydraulic codes can provide significantly larger margins to specified design and/or safety limits.

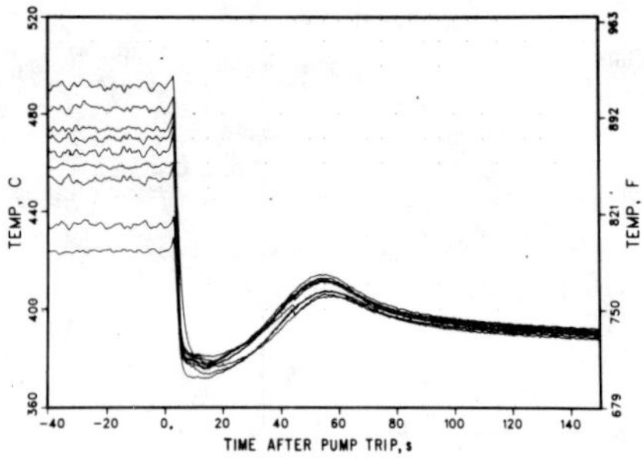

Fig. 20. Effect of a Loss-of-Forced-Flow Transient Upon the Transverse Coolant Temperature Profile Within a Fueled Subassembly

5. SUMMARY AND CONCLUSIONS

In this paper, detailed measurements of the thermal, hydraulic, and neutronic response of EBR-II to a complete loss of primary forced flow followed by a PPS-activated scram are presented. The experimental results clearly indicate a smooth transition to natural convective flow with a quite modest in-core temperature transient. The accompanying calculations using the NATDEMO code agree quite well with the measured temperatures and flow rates throughout the primary system. The only region of the plant where a significant discrepancy between the measurements and calculations occurred was in the IHX. The reasons for this result could not be definitively determined, but it is spectulated that the one-dimensional assumptions used in the modeling may not be valid in the IHX during buoyancy driver flows.

It is concluded from these tests and the supporting analysis that there is an extremely high probability that natural circulation can be demonstrated to be a reliable method of decay heat removal under emergency conditions in EBR-II, and by implication, other LMFBR's. Additionally, the detailed response of a plant to this type of event appears to be predictable, contingent upon an a priori knowledge of the hydraulic, thermal, and neutronic characteristics of the system. The prediction of the detailed temperature distribution within the core, especially the local hot channel temperature rise, appears to require the use of a multi-channel core model which accounts for flow redistribution and inter-subassembly heat transfer. These two phenomena will be of even greater importance in correctly modeling the response of a heterogeneously loaded core.

ACKNOWLEDGMENTS

The authors would like to express their appreciation to the EBR-II engineering team (John Poloncsik, Orville Seim, and Ernest Hutter) who designed and fabricated the instrumented subassembly, XX08, without which this test would not have been conducted. Acknowledgments are also due to Gary Lentz and Nadine Allen for their efforts in developing the detailed plant operational procedures required to conduct this test as well as to the reactor operating crews for their flawless performance of the experiment. The data collection program used to collect and display on-line the many plant parameters was developed by William Booty.

REFERENCES

1. "LMFBR Demonstration Plant Simulation Model 'DEMO'," Westinghouse Advanced Reactor Division Report, WARD-D-0005, Revision 4 (1976).

2. Additon, S. L., Strait, A. E., and Wolfe, C. F., "FFTF Natural Circulation Evaluation: Transition to Natural Circulation," Hanford Engineering Development Laboratory Report HEDL TC-557 (April 5, 1976).

3. Mohr, D. and Feldman, E. E., "A Dynamic Simulation of the EBR-II Plant During Natural Convection with the NATDEMO Code," presented at this conference.

4. Agrawal, A. K., Madni, I. K., Guppy, J. G., and Weaver, W. L., III, "Dynamic Simulation of LMFBR Plant under Natural Circulation," ASME Paper 79-HT-6, presented at the 18th National Heat Transfer Conference, San Diego, CA (August 6-8, 1979).

5. Durham, M. E., "Optimization of Reactor Design for Natural Circulation Decay Heat Removal in a Pool-Type LMFBR," Optimization of Sodium-Cooled Fast Reactors, British Nuclear Energy Society, London (1977) p. 45.

6. Gillette, J. L., Singer, R. M., Tokar, J. V., and Sullivan, J. E., "Experimental Study of the Transition from Forced to Natural Circulation in EBR-II at Low Power and Flow," ASME Paper No. 79-HT-10, presented at the 18th National Heat Transfer Conference, San Diego, CA (August 6-8, 1979).

7. Koch, L. J., et al., "Hazard Summary Report: Experimental Breeder Reactor II (EBR-II), Argonne National Laboratory Report No. ANL-5719 (May 1957) and ANL-5719 (Addendum) (June 1962).

8. Smaardyk, A., et al., "EBR-II Environmental Instrumented Subassembly XX08: Engineering and Assembly," Argonne National Laboratory Report ANL-78-9 (May 1978).

9. Gillette, J. L., et al., "Compilation of Data from EBR-II Natural-Circulation-Test Procedure EX-140, Section IV, Addendum 7A," Argonne National Laboratory Report ANL/EBR-104 (April 1979).

A Dynamic Simulation of the EBR-II Plant during Natural Convection with the NATDEMO Code*

D. MOHR and E. E. FELDMAN
EBR-II Project
Argonne National Laboratory
Argonne, Illinois, USA

ABSTRACT

A prime requisite in the ongoing program of the Experimental Breeder Reactor No. II (EBR-II) is a thorough understanding of the thermal, hydraulic and neutronic behavior of the reactor and balance of plant over a wide variety of operating conditions. One of the key elements in this behavior study has been the overall simulation code, NATDEMO, which describes the primary, secondary, and steam systems along with the essential control systems. The models described in this paper form the basis of the two NATDEMO modules, NATCON and DEMO, which are linked together to form NATDEMO. The NATCON model describes the primary system while the DEMO model describes the secondary and steam systems and is an adaptation of the original DEMO-IV code which simulates the Clinch River LMFBR. The NATCON module drives the program with the aid of a special subroutine that synchronizes the two modules with a master clock/counter.

Although one spatial dimension was considered in the basic solution, multidimensioned effects of flow redistribution and power generation were derived for the reactor core. Two-dimensional temperature distributions were effectively modeled for the outlet plenum structure and the primary sodium pool. Some additions to the overall model are planned in the near future, including the feedwater-drum level control system. Also, modeling improvements are planned for the steam drum and associated recirculation piping, and possibly the plant power demand control system involving primary and/or secondary flow. Additional test results from dynamic tests as currently planned could indicate that still other improvements to the code may be desirable.

The NATCON code itself has been checked against numerous natural convection tests in EBR-II, including low power tests with the specially instrumented XX07 subassembly installed and higher power tests with XX08. In this paper, the NATDEMO predictions are compared with measured results for Test 8A, which was a natural convection test conducted from about 36% rated power and 39% rated primary flow. The close comparison of these results gives substantial confidence

---

*Work supported by the U. S. Department of Energy

in the overall model. Moreover, the NATDEMO modeling methodology, which has been verified with experimental data taken from the EBR-II plant should be of considerable aid in the dynamic simulation of future LMFBR plants.

1. INTRODUCTION

This paper describes a new code, NATDEMO, developed to simulate the dynamic response of the EBR-II plant to a variety of transients ranging from normal operational maneuvers to upset conditions caused by disturbances such as loss of flow, control rod insertion, or loss of electrical load. While numerous analyses of EBR-II individual components and systems have been performed in the past, a whole-plant analytical model is required to predict the dynamic performance of the coupled systems. The salient features of the physical models are given along with a comparison of measured and calculated results for a recent natural convection test on EBR-II, XX08 Test 8A, which was performed with the specially instrumented XX08 subassembly installed.

A key element in establishing the basis of design reliability and safety of an LMFBR plant is the dynamic simulation of the overall plant. Such a system can have a wide range of behavior, depending upon the type of fuel, core geometry, relative sizes of the core and blanket, layout of the plant, and types of transients imposed. The overall NATDEMO code is the result of successfully linking two separate codes, NATCON [1] and DEMO-IV, [2] where the NATCON module simulates the primary system, and the DEMO portion simulates the secondary sodium and steam systems.

Important features of the NATCON module are its detailed descriptions of decay heating, reactivity feedback, the intermediate heat exchanger (IHX), the primary and auxiliary pumps, and the primary sodium pool. A two-region reactor tank model has been derived which includes a heat balance on the reactor tank and the mixing effects of tank sodium on the primary coolant stream as it flows from the IHX to the two primary pump inlets. NATCON has dynamic flow-reversing capability in its thermal-hydraulic models of the core blanket and reactor inlet piping. Another important feature is its initial condition subroutine (a modified CONVECT [3]) that readily calculates the initial values of all required state variables by an iterative scheme. DEMO-IV was developed by the Advanced Reactor Division of Westinghouse Electric Corporation (WARD) for analysis of the Clinch River Breeder Reactor Plant (CRBRP). Although the CRBRP is a loop-type and EBR-II is a pool-type design, their secondary and steam systems are fundamentally similar, as shown in Fig. 1, and basic features of the DEMO-IV models for the secondary and steam system have been retained in the EBR-II application. In addition, a subroutine was developed to initialize the DEMO module at any desired operating condition. The NATCON module is simulated via the IBM CSMP package [4] with a special integration scheme for stiff equations, while DEMO is coded in standard FORTRAN and utilizes the Euler integration technique. The overall NATDEMO integration amounts to a multi-step process wherein no a priori criterion exists for regulating the ratio of NATCON or DEMO calculational time steps.

2. DESCRIPTION OF THE NATCON MODULE (PRIMARY SYSTEM)

2.1 Power Generation

The 16 rows of reactor subassemblies are divided into three basic regions: driver fuel (rows 1-7), radial stainless steel reflector (rows 8-10), and radial

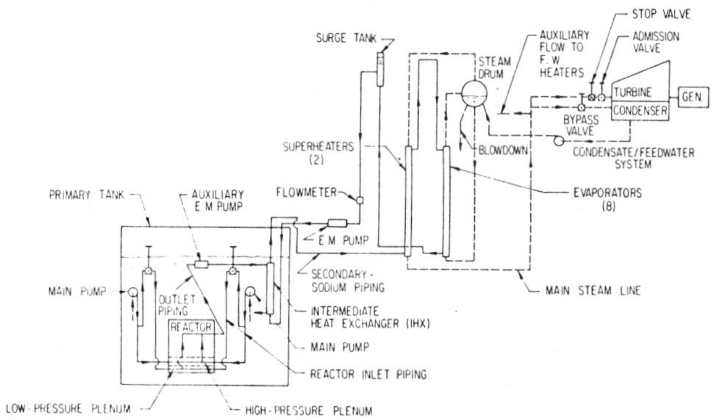

Fig. 1. Schematic of EBR-II Plant for NATDEMO Code

depleted ^{238}U blanket (rows 11-16). Each of these regions has separate power generation functions containing both prompt and delayed components. Separate power functions were found to be important in NATCON because of the varying fraction of delayed power decay (following a reactor shutdown) depending on whether the power generation medium is driver fuel, blanket fuel, or stainless steel. In addition to the residual heating of gamma and beta decay of fission products, the decay of fission neutrons is also significant in most transients for at least the first minute because the shutdown reactivity in EBR-II is usually in the range of -\$4.00 to -\$5.00. The Shure correlation [5,6] for decay heat is utilized, but adapted so that any desired reactor operating history prior to final shutdown can be input as the prerequisite to the decay heat evaluation. In addition, the effect of current power level (with a critical core) upon the level of decay power present is included in the simulation.

2.2 Reactivity Feedback

NATDEMO includes a reactivity model which comprises the following effects: driver fuel expansion, driver sodium expansion, upper reflector sodium expansion, radial reflector sodium expansion, control rod bank extension, and nonlinear effects (primarily subassembly thermal bowing). These feedback terms will now be further described individually.

Driver Fuel Expansion. This effect is assumed to be directly related to the expansion of the ^{235}U fuel and therefore is confined to the active fuel height of the core and also to the first seven rows of subassemblies. Another assumption is that the effect is related to the average fuel temperature.

Driver Sodium Expansion. This effect is directly related to the driver sodium coolant temperature and again is confined to the fueled driver region in the first seven rows of subassemblies. This effect is strongly dependent on density changes near the core boundary.

Axial Reflector Sodium Expansion. This reactivity effect is related to the sodium coolant temperatures above the core including the upper stainless steel shield region in each driver subassembly within rows 1-7. The nominal reactivity contributions for each of the four uppermost axial nodes are included.

Radial Reflector Sodium Expansion. This effect is evaluated for the sodium region within the radial stainless steel reflector subassemblies of rows 8-10 for the total axial height of each subassembly; i.e., from the upper grid to the subassembly outlet (six axial nodes).

Control Rod Bank Extension. The control rods (movable subassemblies) in EBR-II are suspended from the control rod drive supports mounted above the vessel on the reactor tank cover. Thus, the actual control portion (fueled zone) is suspended by long rods extending from the core upward through the outlet plenum, reactor vessel cover, and tank sodium. The thermal response of the rod extension effect is determined by dividing the subassembly and extension rod into sections, each of which responds to changes in the surrounding sodium temperature during a reactor transient. It is also assumed that the control rods operate approximately in a banked configuration. This reactivity effect must incorporate the thermal expansion of the extension rod and the hex can. In EBR-II the fueled portion of the control subassembly is below the poison portion; thus, control rod insertion--downward motion--actually drives fuel below the active core region and replaces it with neutron absorber. The total rod bank effect is obtained by summing the individual contributions of axial displacement, weighted by node length, and multiplying by the rod bank differential worth appropriate for a specific insertion of the bank. Hence, this overall reactivity effect is negative with respect to temperature increases. The dynamic response of the rod bank to imposed temperature changes is described in detail elsewhere [7].

Non-linear Effects (Primarily Subassembly Bowing). The greatest contribution to this effect is assumed to be subassembly bowing which can be a positive or negative effect, depending on the specific core loading. It is related to the radial displacement of subassemblies, driven by the temperature differences across hex cans principally in the radial reflector region. The tops of subassemblies normally "flower" outward with increasing coolant temperature, usually causing the center portions of subassemblies to move radially inward after the tops meet a restraint. Extensive steady-state analyses of this effect were performed with the BOW-V code [8], verifying that the reactivity effect can be reasonably well estimated and often exhibits a threshold phenomenon with respect to power/flow ratio. In the model described here, the effect is treated on a quasi-steady basis and is assumed to be related directly to coolant temperatures in the radial reflector.

The total reactivity feedback thus comprises six terms and all are evaluated relative to a base temperature of 371°C, i.e., the nominal sodium pool temperature. For many of the NATDEMO cases involving a reactor scram, the option is exercised to bypass the feedback reactivity calculation. A significant reduction in running time can thus be made while good accuracy is retained because the reactivity feedback term is normally small compared with negative reactivity inserted via control rods.

2.3 Core, Reflector, and Blanket Thermal-hydraulics

These three regions of the core are described by similar thermal-hydraulic models which treat these channels in parallel (with equal pressure drops from inlet to outlet plena). Each region is described by an equivalent (average) channel which allows heat generation over the entire length of hex can and axial heat conduction in the sodium coolant. However, radial heat transfer between channels is not included. Each channel is described in terms of an equivalent (average) element which has three radial nodes in the fuel element (or reflector block) plus one node in the coolant. The driver region has 12 axial nodes and the radial reflector and blanket regions each have six axial nodes.

The transition from turbulent to laminar flow is allowed in each channel by appropriate input, and the pressure drop vs. flow characteristics can differ in each channel. The effects of the lower and upper reflector shields are fully included in the driver model by describing them as a number of equivalent pins with appropriate diameters, power generation, heat capacity and coolant velocity. The massive amount of heat capacity associated with the upper shield blocks has a significant damping effect on the temperature transient entering the outlet plenum.

2.4 Outlet Plenum

The coolant streams leaving the three regions described above are assumed to be mixed in an upper plenum model which includes the effects of delayed mixing, heat losses and storage with respect to the reactor cover and vessel wall, and leakage of coolant through escape paths in the cover. The nodalization in the plenum includes one node in the coolant, four in the reactor cover, and three in the vessel wall. The effects of coolant holdup in the upper plenum, heat storage in the metal structure and heat losses through the reactor vessel cause a significant difference between the coolant temperature entering the outlet piping compared with that entering the plenum. NATDEMO predicts the magnitude and direction of "leakage" from the outlet plenum which can change from outward to inward at low flow conditions.

2.5 Reactor Outlet Piping (Z-pipe from Reactor to IHX)

This model includes the effects of coolant transport, heat storage, and losses to the reactor tank. The number of axial nodes is 12 and the three radial nodes include one each in the coolant, the pipe and liner combined, and the surrounding tank sodium. The 12 axial nodes adequately describe the transport of a temperature wave as it leaves the outlet plenum on its passage to the IHX.

2.6 Intermediate Heat Exchanger

The IHX model has ten axial nodes and includes axial heat conduction in the sodium. Increasing the number of axial nodes (from 4 to 10) increased the heat balance accuracy of the model, especially at flowrates frequently experienced in the range of 0.5% to 5% of rated. The model contains three radial nodes (primary coolant, tube wall, and secondary coolant) and simple one-point back differencing for the formulation of the axial convective derivative. In addition, a special technique developed by Additon [9] has been adopted to obtain flow-dependent weighting factors used to obtain nodal average temperatures describing the radial heat transfer. Without this specialized treatment, previous studies had shown that a heat exchanger model suffers several deleterious effects if the number of axial nodes is less than about 40 and a large primary/secondary flow imbalance exists, or if the flows become very small. These effects include: oscillatory temperatures with respect to distance along the tube, inaccurate heat balance between primary and secondary sides, and improper response to rapid changes in inlet coolant temperature. These effects are essentially absent with the current model.

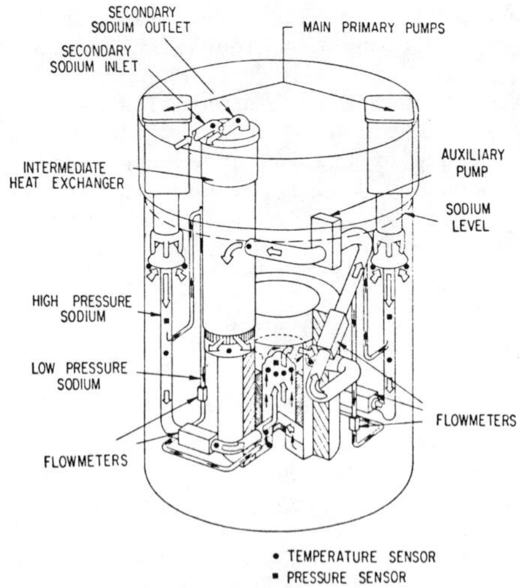

Fig. 2. EBR-II Primary Coolant System

2.7 Reactor Tank

The primary system components are located in the primary tank, as shown in Fig. 2. A two region reactor tank model has been derived which includes a heat balance on the reactor tank and the thermal effects of the reactor tank sodium on the primary coolant stream as it flows from the IHX to the two primary pump inlets. These latter effects are described by a "leaking pipe" mixing model wherein the main coolant stream is imagined to flow within a pipe with a porous wall and thus mix with the tank sodium. The volume taken up by the porous pipe region is assumed to be 10% of the reactor tank sodium. The local stream "leaking" into (and out of) the porous pipe is adjusted as a fixed fraction the primary flowrate. A heat balance on the metal primary tank and sodium is modeled which allows a dynamic description of the bulk sodium temperature in the primary tank.

2.8 Reactor Inlet Piping Thermal Model (Main Pumps to Inlet Plena)

This model is basically identical to the reactor outlet piping model, except that four piping paths are described instead of just one, as shown in Fig. 1. Thus, two piping runs are described for each inlet plenum, i.e., the high pressure plenum (HPP) and low pressure plenum (LPP). The piping for both the HPP and LPP in each run are described by three axial and three radial nodes.

2.9 Main Primary Pumps, Auxiliary Pump, and Fluid Dynamics

The two main pumps are modeled separately and require pump speed as input. Pump speed can be generated either from measured data or by the FLOWDOWN code [10]. Both main pumps are described such that the pump head dynamics are involved in determining the flow coastdown transient following the loss-of-flow trip signal. The code can describe two separate pumps which may have different characteristics and may be tripped at different times. The NATCON model allows the auxiliary pump to operate at any desired head or to be turned off completely at any desired time.

Extensive coastdown testing at EBR-II has shown that identical coastdown characteristics of the primary pumps are the exception rather than the rule. Thus, it has become important to use the option of inputting the two individual pump speeds in LOF transients since the "average" of two different speeds does not yield the same total flow rate as the sum of the individual flows from two pumps with different characteristics. Furthermore, negative flow in one loop will result even for small speed differences during the coastdown transient.

Some of the studies with NATDEMO involving LOF/Scram incidents have indicated a tendency for reverse flows to develop in the depleted blanket region (in some cases where the total reactor flow becomes very low). The model has been modified so that the thermal-hydraulic response will be calculated properly if flow reversals do occur. For example, in the case of the blanket region, the fuel and coolant temperature distributions at the instant of flow reversal are momentarily retained while the entering flow to the subassembly switches from the inlet to the outlet plenum. In this down flow condition, the coolant temperature entering the blanket is the mixed-mean outlet plenum temperature and a "doubly heated" coolant emerges from the lower end of the blanket, if down flow persists, to be mixed in the inlet plenum.

The fluid dynamics are expressed in terms of the net head produced by each pump; the frictional characteristics of the various components, piping sections and reactor flow channels; the buoyancy terms in each region; and the fluid inertance terms. The model allows a transition in flow regime from turbulent to laminar in several components.

2.10 Simulation of Temperature Sensors

Two coolant temperatures of principal interest are those measured at the reactor outlet and at the IHX secondary outlet. In each case, however, the measurements must be corrected for coolant transport lag and thermowell heat capacitance effects. Although the transport effect is small for the reactor outlet sensor, it is much more significant for the IHX outlet sensor. In both cases, the effective thermowell time constants are significant. For convenience in comparing NATDEMO predictions with measurements, the above dynamic effects have been added to the appropriate variables generated in the code. Thus, the "actual" and the "delayed" versions of these variables are part of the output.

2.11 Automatic Power and Flow Demand Control Systems

A control system package has been added to NATDEMO that requires feedback and reference signals as input and provides a control signal as output. This simulation includes the actual control system planned for EBR-II to the extent that it has been defined at this time. The model basically consists of a three-element controller operating in series with an on-off relay which includes the effects of an adjustable dead band and hysteresis. As an option, however, the discontinuous mode of the control system can be replaced by a continuous (proportional) version while retaining the three-element controller. The control system also has adjustable limits for the range of the output control signal and its rate of change. This power demand control system has two principal purposes; one is to cause power changes that can follow a prescribed (demanded) function of time, and the other is to cause the reactor to operate in a load-following mode where the demand signal may be desired plant power level, turbine inlet valve position, IHX secondary outlet temperature, etc.

The automatic flow control system also uses the control algorithm described above but has its own separate inputs, output, and controller action. This simulates the most probable mode of automatic primary flow control to be adopted for operational safety testing in EBR-II; i.e., to supply a flow demand function of time and allow the control system to generate a current control signal for the eddy-current clutches of the motor-generator sets driving the primary pumps.

2.12 Evaluation of Initial Conditions

The NATCON module includes a subroutine that calculates the initial values of variables as a function of the desired reactor power, total flowrate, tank temperature, etc. This capability has been found to be especially convenient in studies where the initial power and/or flowrate are frequently changed, or where these quantities themselves are treated parametrically.

Initial values are internally calculated by solving the set of equations obtained after setting the time derivatives of all the first order differential equations equal to zero. The following basic input quantities must be

1. Reactor power level
2. Reactor vessel forced flow rate (if any)
3. Secondary flow rate
4. Auxiliary pump condition (on or off) (desired capacity must be specified if it is to be energized)
5. Flow split between HPP and LPP
6. Radial reflector flow rate
7. Leakage flow rate from outlet plenum to pool
8. Reactor tank bulk sodium temperature
9. IHX secondary sodium inlet temperature
10. Reactor operating time (or power history) prior to shutdown
11. Reactor state (critical or subcritical)
12. Heat gains and losses for tank heaters, instrument thimble cooling system, shutdown coolers, shield cooling system, purification cell and conduction loss from primary tank.

The initialization routine has an option whereby the code automatically adjusts the secondary flow rate to cause the derivative of tank temperature to equal zero i.e., corresponding to a stable tank temperature condition. In this case, an initial value of secondary flow need not be specified by input since the rated value of secondary flow (part of a separate input segment) is automatically iterated to the appropriate initial value corresponding to the steady-state criterion specified.

Another feature of the initializer is that it can accommodate initial conditions corresponding to:

- Main pumps and auxiliary pump operating
- Main pumps on, auxiliary pump off
- Main pumps off, auxiliary pump on
- All pumps off (natural convection only)
- Initial flow regimes either turbulent or laminar (in those components where a transition can occur)

If the main pumps are operating, their initial pump speeds are internally evaluated, and if the auxiliary pump is initially operating alone, its shut-off head is determined for the desired flow rate it is to supply.

3. DESCRIPTION OF THE DEMO MODULE (SECONDARY SODIUM AND STEAM SYSTEMS)

3.1 Introduction

While the IHX temperatures are calculated by the NATCON module, the IHX hydraulics and the balance of plant (shown to the right of the primary tank in Fig. 1) are modeled in the DEMO module. The overall DEMO-IV methodology, degree of sophistication, and basic assumptions, to a large extent, have been retained in the DEMO module, and the original steam tables and empirical correlations are essentially unchanged. In spite of these similarities, a number of code changes were necessary to accommodate differences between the EBR-II and CRBR plants. A steady state initializer and a thermal model of the secondary sodium piping were developed along with other DEMO module improvements.

3.2 Secondary Loop

The fluid dynamic behavior of the entire secondary loop is represented by a single momentum equation containing terms for the fluid acceleration, the net pressure rise developed by the secondary electromagnetic (EM) pump, the gravitational head due to sodium density variations around the loop, and the frictional pressure drop of the entire loop (except for the pump itself). The net pressure rise developed by the secondary pump is a function of applied voltage, flowrate, and sodium temperature and is assumed to be independent of electrical dynamic effects. The DEMO module simulates three sets of sodium piping in the secondary loop as shown in Fig. 1. The first connects the IHX to the superheaters, the second connects the superheaters to the evaporators, and the third completes the circuit by connecting the evaporators to the IHX.

In the EBR-II plant, the net buoyancy head in the secondary loop represents only about 1.3% of the total losses. When the secondary flow rate is 10% of its nominal value, however, the frictional and buoyancy pressure drops are of the same order of magnitude. Typically, under purely natural-convective conditions, the resulting flowrates are usually 5-10% of nominal. Hence, an accurate simulation of natural-convective secondary flow rates requires accurate predictions of secondary sodium temperatures which, in turn, necessitate a detailed sodium piping thermal model. The new model developed for the DEMO module includes separate heat capacitance effects for the flowing sodium and the piping walls, and is similar to the primary system model in NATCON. Piping lengths, elevation lengths, elevation changes, cross-sectional geometries and material properties are explicitly represented throughout the system. The pipe cross section comprises the flowing sodium, pipe wall, air gap, and thermal insulation. Some of the piping contains a loosely-fitted carbon steel jacket which together with its associated annular air gap is explicitly modeled. The insulation of the secondary sodium piping is so effective that the total heat loss through the pipe walls of the entire secondary loop causes the sodium to lose only about 0.15°C at rated conditions. Since this temperature loss is so small, the outer boundary was assumed to have perfect insulation. In the finite-difference model, each axial segment has one sodium node and one pipe wall node; a radial node is added where the pipe jacket is present. The finite-difference representation is implicit in both space and time which results in a model which is unconditionably stable numerically. At each axial node within each time step only two or three simultaneous linear equations must be solved, depending on whether or not the pipe jacket is present.

3.3 Steam Generator

The EBR-II steam generator comprises eight evaporators, two superheaters, a steam drum and the recirculation piping. The DEMO-IV code contains two steam generator models. Because the detailed (optional) model in DEMO-IV is not needed for the current EBR-II application, only the simpler model was incorporated in our DEMO module. In NATDEMO, a single evaporator and a single superheater represent the eight evaporators and two superheaters, respectively.

The model introduces the feedwater flow into the recirculation loop at a mixing node located in the downcomer immediately below the steam drum. The steam and water in the drum are assumed to be in thermodynamic equilibrium throughout its volume. The mass and energy transported into and out of the steam drum volume are used to determine the specific volume and energy within the volume. The steam table subroutines are used to evaluate the temperature, pressure, and enthalpy of the steam and water volumes which are required to calculate the water level in the drum. The recirculation loop momentum equation is similar to that of the secondary loop, except that the EBR-II recirculation loop does not have a pump. Transport delay times and pipe wall heat capacitance are ignored in the present recirculation model. Hence, the enthalpy of the water entering the evaporators is the same as the enthalpy at the feedwater mixing point described above, and the enthalpy of the two-phase mixture entering the drum is identical to that leaving the evaporators.

All evaporator tubes are assumed to behave identically as are all superheater tubes, thus a single tube is represented in each model. In both the evaporator and the superheater models the active heat transfer length of the single tube is divided into seven equal-length axial nodes. In the radial direction, each level has a water/steam node, a tube wall metal node, and a sodium node. The EBR-II superheater units are essentially identical to the evaporators except that the superheater tube has a central core tube which causes the steam to flow in the resultant annular channel. In the original DEMO-IV code, the calculation of heat transferred from the sodium to the water/steam side of an axial node gave equal weighting to the sodium-to-water/steam temperature difference at each end of the node. This is a conventional approach which works reasonably well for the simulation of normal and some off-normal reactor plant operating conditions. The difficulty with this numerical scheme, however, is that the simulation of many off-normal conditions can result in a solution which is spatially unstable. As described above for the IHX, the sodium-to-water/steam temperature differences at each end of the node are unequally weighted which greatly improves numerical stability without requiring a large number of nodes. These weighting factors, which depend on flows on both sides of the tube, are calculated for each axial node. After these factors were added to the evaporator and superheater models, all cases tested were found to have numerically stable solutions, for both steady-state and transient conditions.

The steam exiting the two EBR-II superheaters flows into a common steam header which is part of the main steam line. This pipe carries the steam to the turbine and the turbine bypass line. The fluid in the main steam line is assumed to be in thermal equilibrium throughout its volume. The thermal model for this volume, which contains superheated steam, is analogous to the steam drum model except that no provision is made for the presence of liquid.

3.4 Plant Pressure Control and Feedwater Flow Rate Models

The EBR-II plant has two paths through which the superheated steam can flow from the steam header to the condenser. One path goes through the turbine and the other goes through the turbine bypass which is capable of carrying all of the steam flow produced by the steam generator at rated power conditions. In the model each path has a valve whose stroke can be controlled manually or automatically. Normally the bypass valve is closed and the turbine admission valve is automatically regulated to maintain the turbine inlet pressure at a reference value. Since the DEMO-IV model of this part of the steam system was not representative of EBR-II, a new subroutine was developed to model these two steam paths. In the EBR-II simulation, both valve models have the same control system features; each has proportional and reset action as well as an input signal range limit, a first order actuator lag, a stroke velocity limit, and an output valve position range limit. When the manual control option is used in NATDEMO, the relative valve position is input as a specified function of time.

In the initial EBR-II model, it was assumed that all steam flow had to pass through either the turbine or the bypass valve. With this assumption, automatic control of either valve assured that the turbine inlet pressure could always be restored to its setpoint value. Measured plant data, however, obtained during scrams to natural convective conditions showed the importance of additional steam flow paths taken from the main steam header for heating feedwater. These flows are modeled as an auxiliary steam flow originating just ahead of the turbine inlet valve. The feedwater train is not explicitly represented at this time in NATDEMO. In the EBR-II adaptation of DEMO-IV either the feedwater enthalpy is specified via a feedwater flow rate function, or the feedwater temperature is specified as a function of time. Furthermore, depending on the option used, the feedwater flow rate is either set equal to the total steam header exit flow rate or the feedwater flow rate is a specified function of time.

3.5 Steady-state Initialization

A special routine was developed for the DEMO module which initializes and fine tunes the DEMO portion of the model. The core of the initializer is a simultaneous nonlinear set of 52 functional relationships obtained from the appropriate model equations by setting all time-derivatives to zero. These relationships contain 52 unknown variables in addition to numerous model parameters, such as heat transfer areas and hydraulic resistance coefficients. Seven relationships of the form $x - a = o$, where x is an unknown and a is a constant of specified value, were added to the above 52 to form a simultaneous set of 59 equations and unknowns. Since the seven unknown x's can be virtually any seven of the total set of unknowns, the initializer allows the remaining unknowns to be solved in terms of the specified seven.

If the option is chosen to fine tune the model, it is convenient to be able to specify, for example, three experimentally measured flow rates and to obtain values for three model parameters from the solution of the above 59 simultaneous equations. This set of relationships, many of which are transcendental, is solved by the Newton-Raphson techniques. Some of the variables in the DEMO model not among the 52 are iteratively calculated in conjunction with the Newton-Raphson process, while others are evaluated afterwards.

The DEMO module can be initialized in less than a minute of computing time and the values obtained are mathematically accurate to about one part in a million. In the operation of the NATDEMO code, NATCON may access the DEMO initializer several times as part of an optional overall interation process, to make certain that the steady-state conditions calculated for the IHX satisfy both the NATCON and the DEMO modules.

4. PROGRAM EXECUTION

4.1 Program and Operation Size

The NATCON model of the primary system is simulated with the IBM CSMP-III simulation program package for the IBM-370. The model comprises about 240 integrators, 12 subroutines and several optional integrating algorithms. The DEMO program module is comparable in size and complexity to NATCON and contains some 60 subprograms. The CSMP program package, however, treats all of the subroutines in the same manner in relationship to the main program since all are compiled after the translation phase of CSMP is completed.

Because DEMO-IV was written for the CDC-7600 computer, the DEMO module was converted to the IBM-370 in double precision to approximate its computational accuracy with the CDC machine. In the overall NATDEMO code, NATCON contains the main and controlling portion of the program, while the DEMO module is treated as a subprogram. A special subroutine was developed to insure proper time sequencing of the two schemes and to allow both NATCON and DEMO to vary their computational step sizes independently during the solution.

4.2 Input and Output

The present version of the NATCON module requires a combination of about 600 input parameters and functions. These, in general, are input according to standard CSMP rules, while DEMO input data are read in via standard FORTRAN READ statements. Up to 220 output variables can be specified for printing in the NATCON module. The DEMO module has its own optional printout document which normally accompanies the NATCON output. The DEMO output contains about 110 variables and has a printout interval independent from that specified in the NATCON module.

4.3 Input Disturbance Functions

The optional input disturbance variables that can drive the NATDEMO simulation are as follows:

1. Primary pump speeds
2. Auxiliary pump current
3. Control rod reactivity, or neutron fission decay power (optional)
4. Reactor power demand
5. Primary flow demand
6. Secondary pump voltage
7. Secondary flow rate through IHX } (When the NATCON module operates
8. Secondary inlet temperature to IHX } separately from DEMO, by option)
9. Turbine inlet pressure set point
10. Turbine inlet valve position

11. Turbine bypass valve position
12. Auxiliary steam flow
13. Feedwater flow rate
14. Feedwater temperature

Most of the above variables are input as normalized functions and can be used in any combination.

5. DESCRIPTION AND ANALYSIS OF TEST TRANSIENT

The transient reported here for analysis is the final natural convection test in the XX08 test series, Test 8A. This test was conducted on EBR-II on March 29, 1979, and plant signals were recorded via the Data Acquisition System (DAS). The basic input for the NATDEMO analysis included the following initial conditions:

Reactor power level:	21.8 MWt
Reactor power operating history:	(See Fig. 4)
Control rod reactivity inserted during scram:	-$5.30
Primary tank temperature:	350.6°C (663.9°F)
Primary flow rate:	0.199 m^3/s (3157 gpm)
Delay time between loss of primary pump power and scram:	2.0 s
Secondary flow rate:	0.137 m^3/s (2167 gpm)
Delay time between loss of primary and loss of secondary pump power:	17 s
Primary auxiliary pump condition:	De-energized
Auxiliary steam flow to feedwater heaters:	2.50 kg/s (5.5 lb/s)
Blowdown flow rate leaving steam drum:	2.27 kg/s (5.0 lb/s)
Steam drum level:	0.43 m (17.0 in.)
Turbine/bypass valve condition:	Turbine shutdown, bypass operating

The following variables were required as input functions of time:

Control rod reactivity insertion

Primary pump speeds (2)

Auxiliary steam flow rate

Feedwater flow rate to drum

Feedwater temperature at drum

Because this was a planned test at reduced power level, the primary auxiliary pump had been de-energized several minutes before the transient was initiated. In addition, the turbo-generator was not operating and steam flow was bypassed to the condenser. As is the case with turbine operation the main header steam pressure (near the turbine inlet) was automatically controlled--in this case with the bypass valve instead of the turbine admission valve. The initiating action was a manual trip of the primary pumps, followed by an automatic scram due to low flow. Finally, the secondary pump was manually tripped at 17 s into the transient, because its automatic trip with scram had been bypassed.

The sequence of events is depicted in Figs. 3 through 10, in which measured data is superimposed on the NATDEMO predictions in most of the plots. It should be pointed out that the individually measured pump speeds are input to NATDEMO because of the complexity of the modeling required for the motor-generator sets that power the primary pumps. The M-G sets have been simulated separately by the FLOWDOWN code, but the pump speed transients resulting from a loss-of-power are somewhat different than those observed due primarily to a lack of detailed information for the generators. The HP plenum flow, compared in Fig. 3 with that measured by XX08, is found to respond rapidly to pump speed because of very low loop inertance. It should be mentioned that the primary pumps were tripped somewhat differently in Test 8A than for the other tests in the latest XX08 test series. The net effect was that the pumps coasted down approximately 37% faster, although peak core coolant temperatures were predicted to be less than 1.5°C higher than with the slower coastdown. The principal effects of this faster coastdown were slightly higher natural convection rates (both primary and secondary) and higher peak temperatures in the balance of plant of 5°C, or less.

The reactor operating history prior to Test 8A is shown in Fig. 4. Predicted and measured values of driver region power are presented in Fig. 5, which includes the predicted prompt component and the measured ion chamber response. The core thermal-hydraulics were of course driven by the sum of the prompt plus delayed components. The predicted and measured subassembly outlet temperatures are compared in Fig. 6. The calculated IHX secondary outlet temperature, superheater sodium inlet temperature and the sodium temperature at an intermediate pipe location are shown together in Fig. 7; note that the pipe transport effect is evident by comparing these three responses. In Fig. 8, a comparison of the predicted and measured IHX secondary outlet temperature is made; in this case, however, the temperature sensor has a time constant of about 60 s which is accounted for in the NATDEMO calculation. The predicted secondary flow transient is compared with the measured response in Fig. 9, revealing a very close agreement between the two. Coincident with reactor scram, the steam drum level control system was switched to manual with the result, as seen in Fig. 10, that feedwater flow rate exceeds the steam flow rate causing the drum level to rise monotonically with time.

In conclusion, the predicted results compare favorably with the measurements giving confidence that NATDEMO does simulate the system sufficiently well to predict transients of this type. It is especially noteworthy that the predicted secondary flow is very close to the measured response since this transient is very important for a good prediction of natural convection in the primary system. When better instrumentation is available and the planned drum level tests have been run, the steam generator model will be thoroughly investigated. Effects that we contemplate adding to the code include phase nonequilibrium within the steam drum, transport delay and heat capacitance for the recirculation piping, and heat capacitance of structure within the drum. In addition, a three-element control system for regulating drum level is being developed for NATDEMO.

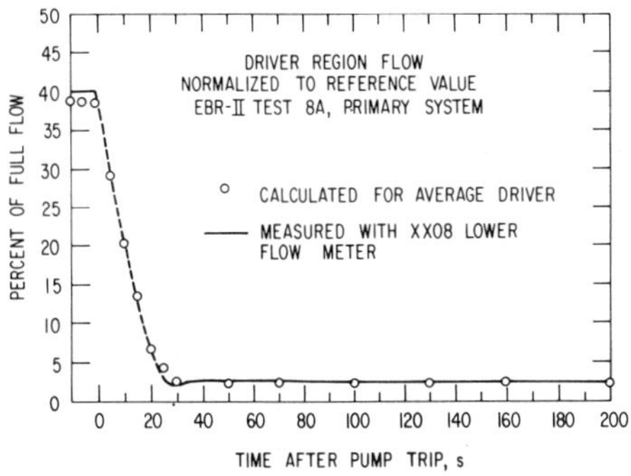

Fig. 3. Measured and Calculated Driver Region Flow Rates

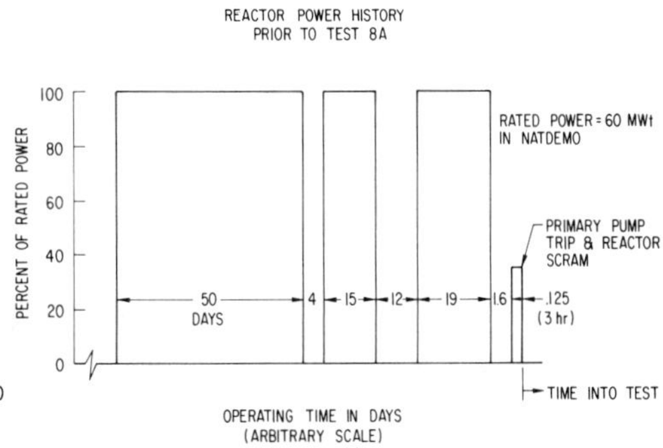

Fig. 4. Reactor Power Operating History Prior to Test 8A

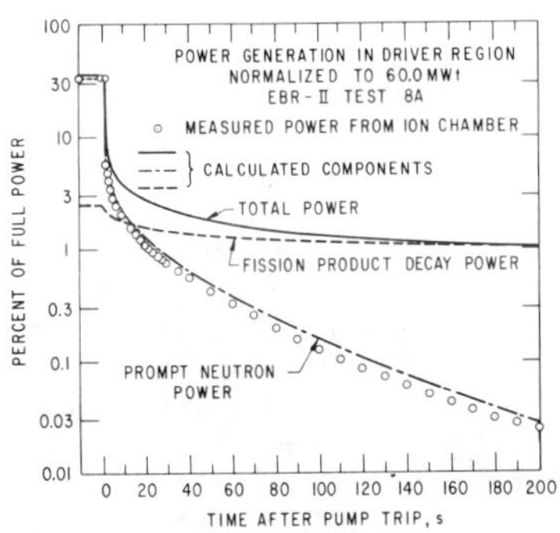

Fig. 5. Measured and Calculated Power Generation in Driver Region

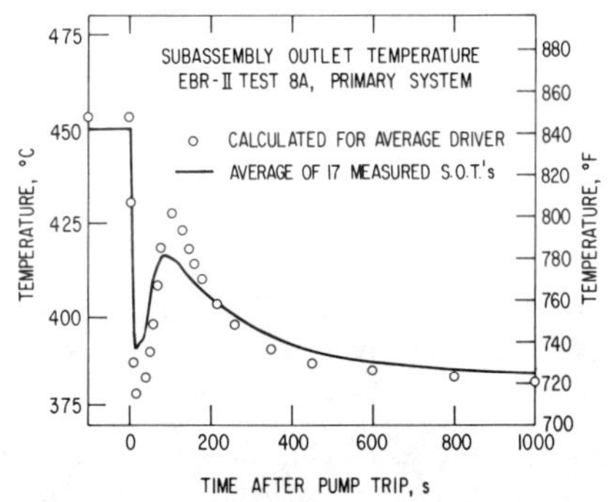

Fig. 6. Calculated Average Driver Outlet Temperature Compared with Measured Average of 17 Outlet Temperatures

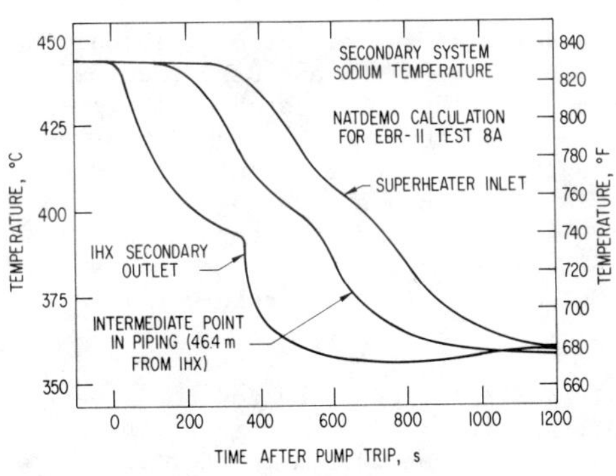

Fig. 7. Calculated Response of IHX Secondary Outlet and Downstream Piping Temperatures

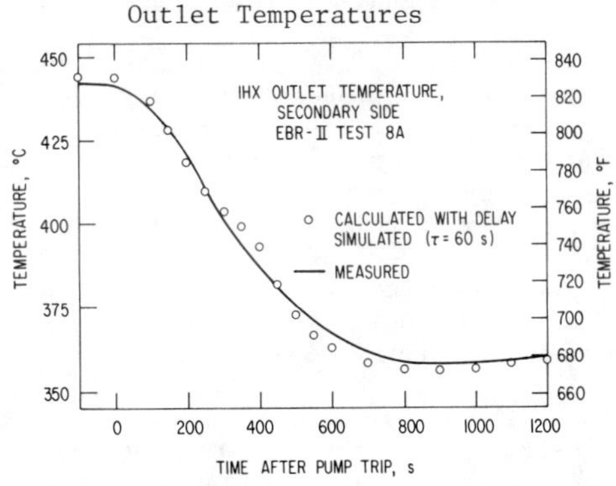

Fig. 8. Measured and Calculated IHX Outlet Sodium Temperatures on Secondary Side

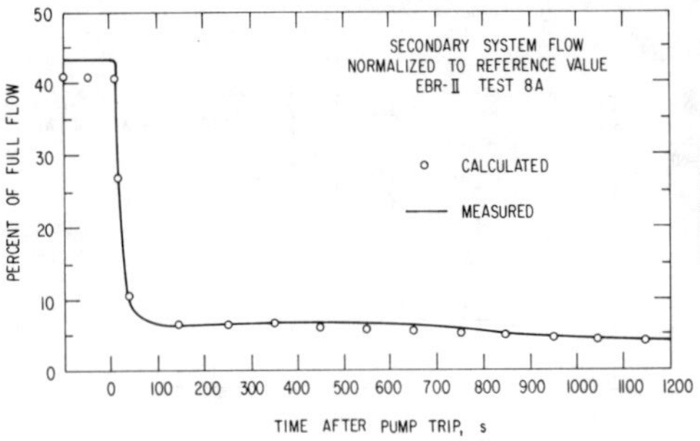

Fig. 9. Measured and Calculated Response of Secondary Sodium Flow Rate

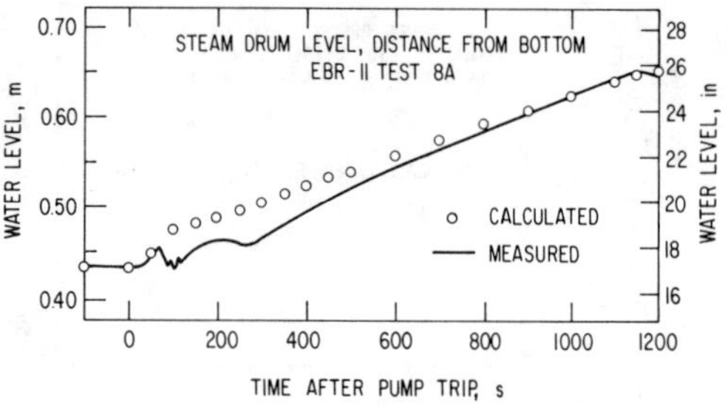

Fig. 10. Measured and Calculated Response of Steam Drum Water Level

6. CONCLUSIONS

The NATDEMO code is a complete and usable tool for the study of EBR-II plant dynamic behavior. The NATCON module itself has been checked against numerous natural convection test measurements in EBR-II, during both steady-state and transient conditions. Although it appears that all important processes in the plant have been modeled, we plan to enhance certain submodels, including the steam generator, the drum level control system and possibly the power level/primary flow control system. Some of the important effects that were added or improved during prior model verification studies are the effect of power history on decay power, heat capacity of structure in the outlet plenum, flow-reversing capability in the primary system, stabilization of nodal heat transfer in all heat exchangers, and flow transport/heat capacitance effects in primary and secondary piping. The process of verification is continuing with the analysis of XX08 natural convection tests conducted at reduced power as well as reactor scram and loss-of-flow/scram incidents at full power. In addition, tests planned for the operational safety testing program at EBR-II are being designed and analyzed with the aid of NATDEMO.

The linking of two separate, dynamic codes with the aid of a synchronizing subroutine has proven to be very satisfactory. The NATCON and DEMO modules have separate integration routines that have individual time-step size criteria. The time step can by option be made variable in NATCON and be made process dependent in the DEMO module. Substantial savings in computer running costs are realized with this procedure compared with a simultaneous updating process for the entire set of difference equations with a single time step throughout. The NATDEMO modeling methodology, which has been verified with EBR-II plant data, should aid considerably in the dynamic simulation of future liquid-metal-fast-breeder reactors.

All natural convection tests and analytical studies conducted to date on EBR-II indicate that the reactor can safely survive an incident involving a total loss of primary and secondary pumping coincident with a reactor scram. Although the latest series of four tests did not exceed 37% of rated power, the natural convection performance in both the primary and secondary systems was sufficient to remove core decay heat. NATDEMO predictions have compared closely with test results, and studies do indicate that natural convection will limit peak core temperatures to less than 650°C following a total loss of pumping incident at full power coincident with a reactor scram.

ACKNOWLEDGMENTS

The use of the DEMO-IV code in developing the DEMO module of NATDEMO is a good example of technology developed at one laboratory being effectively transferred to another. This could have proved to be a rather difficult, if not impossible, task if it were not for the gracious and willing cooperation of the staff at the Advanced Reactor Division of the Westinghouse Electric Corporation. The authors are particularly grateful to Dr. A. Batenburg who has provided invaluable assistance. We are also indebted to Gary Lentz and Nadine Allen of Argonne for their work in developing the detailed test procedures and to the reactor operators for their cooperation during the experiment. A special commendation must also be made to William Booty and John E. Sullivan whose continuing efforts on the data collection and reduction programs are invaluable.

REFERENCES

1. D. Mohr, "NATCON, A Dynamic Code for Evaluating the Thermal-Hydraulic Behavior of the EBR-II Primary System, Including the Effect of Natural Convection," Reactor Development Progress Report, ANL-RDP-74, p. 1.11 (August 1978).

2. "LMFBR Demonstration Plant Simulation Model 'DEMO'," Westinghouse Advanced Reactor Division Report, WARD-D-0005, Revision 4 (1976).

3. D. Mohr and J. L. Gillette, "A Code for Analyzing Parallel-Channel Thermal-Hydraulics in EBR-II During Natural Convection Conditions (the CONVECT Code)," Trans. Am. Nucl. Soc., 21, p. 299 (1975).

4. IBM Application Program, "System/360 Continuous System Modeling Program," Program Reference Manual, SH19-7001-3 (1975).

5. K. Shure, "Fission Product Decay Energy," WAPD-BT-24, Bettis Technical Review 1-17, 1961.

6. K. Shure, "^{235}U Fission Product Decay Energy--1972 Reevaluation," USAEC Report WAPD-TM-1119, Westinghouse Electric Corp., NTIS 1972.

7. J. A. DeShong, Jr., "Dynamic Analysis of Liquid-Metal-Cooled Fast Power Reactors," ANL-7529, Argonne National Laboratory (January 1969).

8. D. A. Kucera and D. Mohr, "BOW-V: A CDC-3600 Program to Calculate the Equilibrium Configurations of a Thermally Bowed Reactor Core," Argonne National Laboratory Report ANL/EBR-014 (January 1970).

9. S. L. Additon and W. L. Frisch, "Transient Heat Exchanger Simulation Employing Analytic Nodal Average Temperatures in the Finite-Difference Equations," Trans. Am. Nucl. Soc., 15, p. 259 (1972).

10. D. Mohr and A. Gopalakrishnan, "Coastdown Characteristics of the EBR-II Primary Cooling System," Trans. Am. Nucl. Soc., 17, p. 442 (1973).

Post Test Evaluation of Natural Circulation in FFTF Secondary Loops

T. R. BEAVER, D. M. TURNER, and S. L. ADDITON
Hanford Engineering Development Laboratory
Richland, Washington, USA

ABSTRACT

Two transient tests in one of the FFTF secondary heat transport loops were performed (March-May 1979) in order to verify that the transition to natural convective flow could be effected from near isothermal refueling conditions without excessive cooling at the air dump heat exchangers. Following the tests, the best estimate computer model was calibrated against the data, yielding in the process insights about the loop parameters and the probable suitability of the model structure. Key empirical parameters of pressure drop and heat loss were found to be at 71% and 81% of the pretest safety evaluation model parameter values, respectively. Pretest piping thermal transport and flow calculational models required no further revision to produce good agreement with the test data. Additional detail in the air dump heat exchanger heat loss model, accounting for long structural thermal time constants, was found to be necessary to obtain model agreement with transient outlet temperature data. The pretest model had conservatively employed steady-state heat losses for transient calculations.

1. TEST RESULTS

The demonstration of the capability of sodium in the secondary loops of the Fast Flux Test Facility (FFTF) to undergo a transition to a natural circulation condition from a shutdown (near isothermal) operating condition following a loss of electrical power to the plant is one objective of the Plant Acceptance Test Program. This transition to natural circulation must be accomplished without freezing the sodium in the Heat Transport System (HTS) so that the Dump Heat Exchanger (DHX) modules will provide an ultimate heat sink for the reactor decay heat.

The original plans for the secondary loop testing are described in Ref. 1. The first test performed was a transient initiated at a loop temperature of approximately 370°C (700°F). The high initial loop temperature afforded a large margin to sodium freezing and to abort limits established to limit loop hot-cold leg temperature differences. It also allowed substantial margin beyond nominal test predictions, thus assuring that the test could be run to completion.

Upon the successful completion of the initial elevated temperature test, a test initiated at prototypic refueling conditions (∿205°C (400°F) hot leg temperature) was designed and performed. The minimum loop temperature obtained during this experiment was approximately 54°C (130°F) above the sodium freezing temperature, proving the ability of the secondary loops to undergo transition to natural circulation and provide heat sinks for reactor decay heat even from the lowest initial loop temperatures (refueling conditions) permitted during plant operation.

The secondary loop Heat Transport System consists of a series of pipes interconnecting the various components (pumps, mixing tees, IHX, DHX), as shown on Fig. 1. During the tests, temperatures of the sodium in the hot and cold legs were measured with resistance temperature detectors (RTDs). Thermocouples were used to measure module tube bundle outlet temperatures. Magnetic flowmeters provided data for determination of the loop and three of the four individual module flowrates. An on-line computer system, data logger and strip charts were used to collect the data.

Figs. 2 and 3 present the DHX outlet temperatures following the pump trip for the 370°C and 205°C tests respectively. The delay in temperature response for three of the four modules during the 205°C test was a result of a period of reverse flow through the three modules caused by DHX preheater operation prior to the start of the transient. Three of the module preheaters were initially adding heat to the system in excess of the module heat losses, leading to a thermal head reversal. However, as the modules cooled the flow became positive again and the colder sodium from the modules finally reached the downstream thermocouples, resulting in the sharp temperature drops shown in Fig. 3. Comparison with the computer simulation shown on the figures is described below.

2. APPROACH TO DATA EVALUATION

A thermal/hydraulic computer simulation of an FFTF secondary Heat Transport System (HTS) loop was generated as part of the overall program to demonstrate that natural circulation is both a satisfactory and sufficient means of removing decay heat from the FFTF core following a complete loss of electrical power. The initial "safety" version of the code was used for design analysis and acceptance test predictions (Ref. 1).

In the post test data evaluation our objective was to search for evidence which possibly indicates that this safety evaluation includes non-conservative features. Our initial post-test data evaluation focused on defining those revisions necessary to bring the safety model in conformance with the measured test results.

Employing all available test data, best estimates of the test plant conditions and of the model parameters were generated and potential model structural defects were investigated until the best estimate model results were consistent with all test data within the measurement uncertainty bounds. This process alone does not constitute validation of the best estimate model. Test data will always suffice to prove that models which predict outside the measurement uncertainty bands are invalid, but it is not necessarily true that a model which is consistent with the data has been uniquely determined. Thus the process of generating a feasible best estimate model is a necessary but insufficient condition for validity. Validation requires further investigation to determine the degree to which the test design and instrumentation may have been sufficient to uniquely determine all relevant model parameters and structure. The evaluation of the data collected in the FFTF secondary loop indicates that the only significant model structural defect was in the DHX thermal model, where the simplified heat loss representation was unduly conservative. The remainder of the paper discusses the evaluation process which led us to this conclusion.

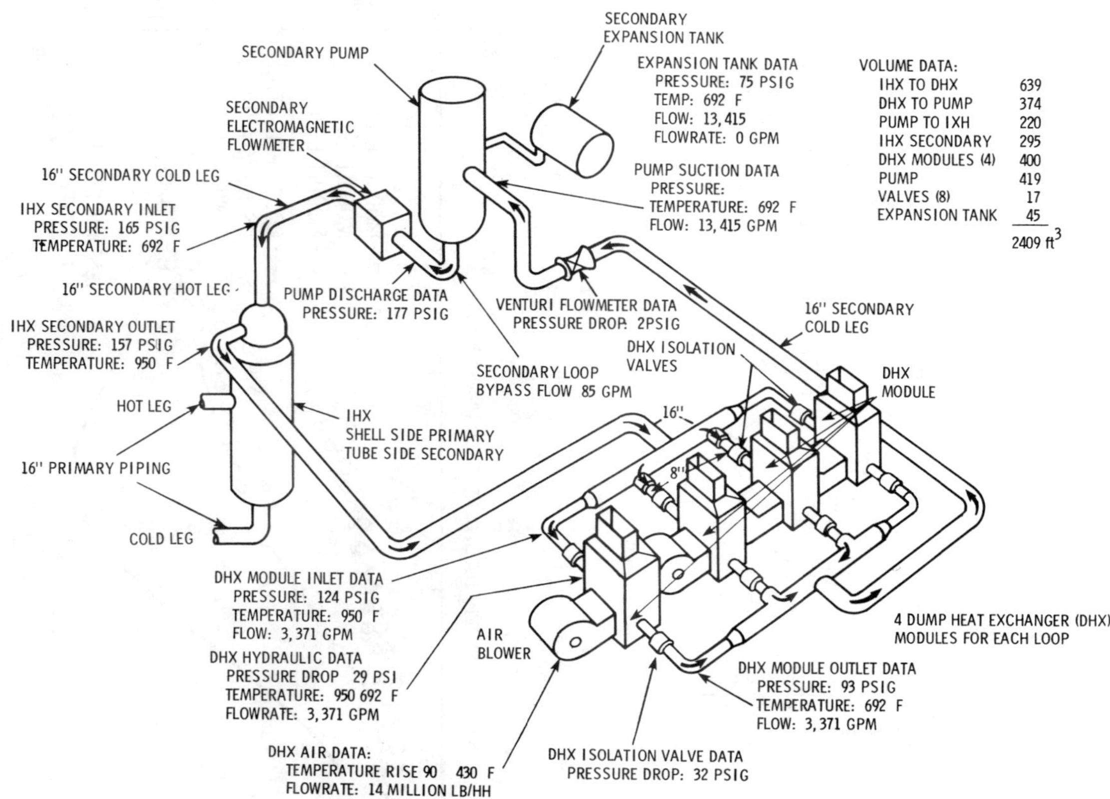

HTS SECONDARY LOOP SCHEMATIC DIAGRAM

FIGURE 1

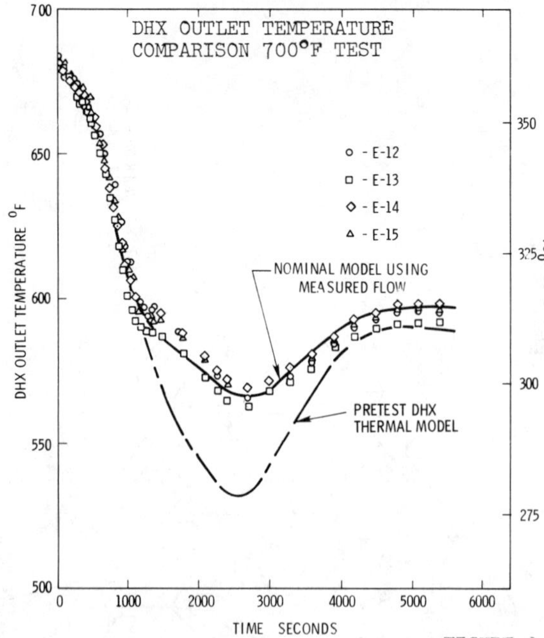

FIGURE 2

DHX OUTLET TEMPERATURE COMPARISON 400°F TEST

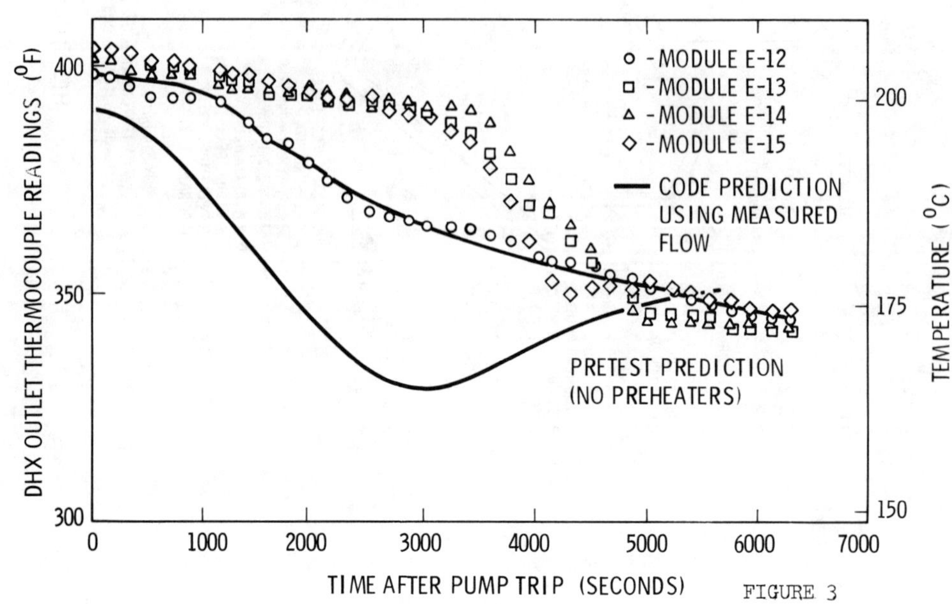

FIGURE 3

3. MODEL CALIBRATION

All major components of the secondary loop (including the pump with stopped rotor (have been represented in the secondary loop code by a single circuit of equivalent piping which is divided into a series of short segments (nodes) across which the governing flow equations (momentum balance) are integrated to provide loop flows as a function of time. Empirical data on component flow resistance in the high turbulent region near design operating conditions has been used in representations for the DHX and pump locked rotor pressure drop. For the laminar and transitional flow regions all component flow resistances rely on calculation and extrapolation. Thermodynamically, the model makes use of the same series of axial (piping) nodes plus a series of radial nodes placed along the piping across which heat transfer to ambient takes place. The number and arrangement of these radial nodes depends upon the particular component being modeled (piping heat loss, IHX, DHX, etc.) and the detail to which it is defined. Temperature variation along the axial nodes is governed by the one-dimensional (slug flow) energy equation utilizing a "mixed" differencing scheme (Refs. 2 and 3) to approximate the axial temperature gradient. Axial conduction and viscous heating are assumed to have negligible effect and are ignored. Heat loss from process sodium in the DHX to ambient is assumed quasi-steady and is based on empirical data for the "average" module. Only one of the four parallel circuits for the DHX modules has been simulated in detail. The other circuits, when "on-line", are assumed identical.

Comparison of the safety code predictions with measured results from the steady-state portion of these tests showed good agreement with two exceptions: the actual DHX heat loss was only 81 percent of the value estimated from earlier tests at the Liquid Metal Engineering Center (Ref. 4) and total system pressure drop was overpredicted by 40 percent. Comparison of the code with secondary loop transient results showed good agreement with the code hydraulic and component thermal models with the exception of the DHX thermal model.

3.1 Pressure Loads

A pressure loss calculation requirement for the nominal "best estimate" version of the code was that it match the total system resistance with the accuracy of the measurements and that it correctly approtion this pressure drop between the main loop and the parallel flow paths connecting the DHXs. As can be seen from Fig. 4 both the Westinghouse Proprietary IANUS model (Ref. 5) and the initial version of the secondary loop code conservatively over-predict the system resistance by approximately 33 to 40 percent*. Overall accuracy of the data shown on Fig. 4 ranges from 6 percent at pony motor flow to an average of 10 percent at higher flow. System resistance for flow rates less than pony motor (approximately 90 liters/sec (1430 gpm) were derived by calculating the loop thermal head from steady-state temperature measurements obtained during natural circulation operation. These calculations are estimated to be accurate within approximately 10 percent.

The major source of the pressure loss conservatism noted in the figure was attributed to the piping and DHX pressure loss expressions. Minor modifications were also made to loss expressions for the other components. Values for the "equivalent length" of various pipe fittings, taken from Crane (Ref. 6), were

---

* The break in the system resistance curves at approximately 1000 gpm results from the addition of the pump stopped rotor pressure drop to the rest of the system resistance for loop flows less than 1000 gpm.

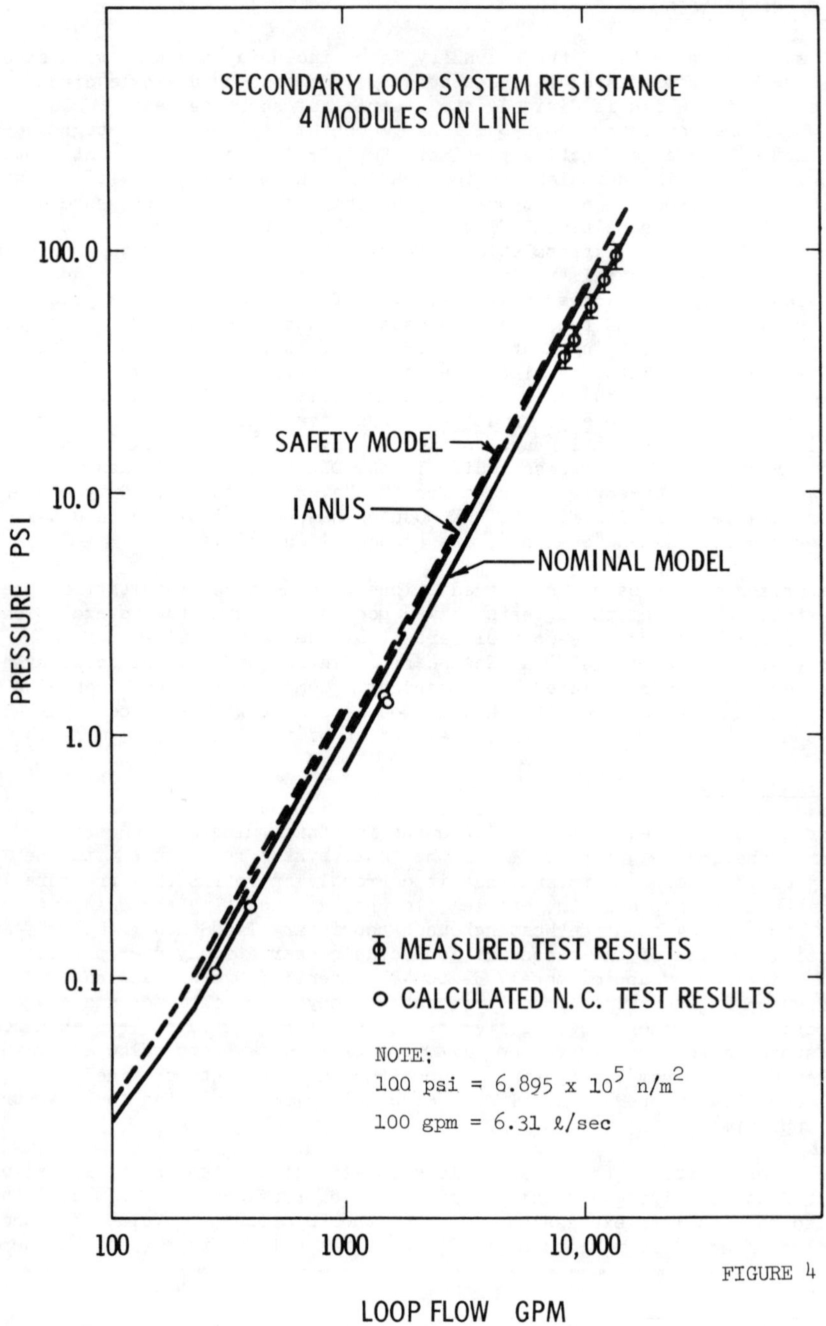
SECONDARY LOOP SYSTEM RESISTANCE
4 MODULES ON LINE
100.0
10.0
1.0
0.1
PRESSURE PSI
SAFETY MODEL
IANUS
NOMINAL MODEL
MEASURED TEST RESULTS
CALCULATED N. C. TEST RESULTS
NOTE:
100 psi = 6.895 x 10^5 n/m^2
100 gpm = 6.31 ℓ/sec
100
1000
10,000
LOOP FLOW GPM

FIGURE 4

considered to be too conservative and were revised downward by approximately a factor of 2. The present values more nearly correspond to the recommended elbow losses of Piggott (Ref. 7).

As previously pointed out, Fig. 4 indicates good agreement between the total loop pressure drop predicted by the revised code and the measured data. To evaluate how well the modified code apportions this pressure drop between the main loop and the parallel flow paths connecting the DHXs, the calculated main loop pressure loss was subtracted from the measured total pressure loss and the results plotted vs. the per module flow. Fig. 5 presents the results of this calculation as well as the code prediction for the per module pressure drop. As can be seen, all data points fall nearly on a straight line coincident with the code predictions.

3.2 Sub-Model Verification

Accuracy of the individual code sub-models, during transient conditions, was determined by decoupling the models and using actual test data as the input forcing function for different model segments. The model output was then compared with measured test results. The combined pump and piping thermal models were first checked by decoupling the code hydraulic and DHX thermal model and forcing flow and DHX outlet temperature histories. Predicted temperatures at the location of the cold leg RTD (approximately 20m (67 ft) downstream of the pump and 95m (310 ft) downstream of the DHX) is compared with the measured value on Fig. 6. These predictions agree very well with the measured results.

A similar test of the code for the transition to natural circulation from 205°C (400°F) was not made because of the initially inverted loop temperatures and resulting reverse flow. However, a test of the loop thermal transport was made for one module operation following startup of its preheater. This event results in an initital cooldown caused by startup of the preheater fan followed by a gradual temperature rise as the preheater adds heat to the sodium. Again the code was uncoupled at the DHX outlet by overriding the predicted temperature with the measured value. Flow for the thermal portion of the code was also forced with the measured value. Temperatures at the cold leg RTD are compared with the measured value on Fig. 7. Thermal transport and initial temperature response appear to agree very well with the measured data, however, after approximately 3000 seconds the prediction possibly deviates from the data. Potential causes have been postulated for this effect. Variation in heat transfer between the flowing sodium and the surrounding pump tank or by transient-induced natural convection between pump and surge tank causing mixing with colder sodium from the surge tank are two promising candidates. Further investigation will be required to formulate a recommendation, however, the results do not appear to have any significant programmatic consequences since the 5°C (8°F) maximum error is within the nominal error bounds of the data.

The code hydraulic model was checked, after first revising the system resistance calculations to achieve a pressure loss consistant with the steady-state results described above. To do this the test transient was rerun while overriding the predicted DHX outlet temperature with the measured test results. The predicted sodium flow rate resulting from use of the real thermal forcing function was then compared with the measured flow. Results shown on Fig. 8 again indicate close agreement between measurement and calculation.

The above two consistency checks imply that the code hydraulic calculations are substantially correct and that thermal transport is adequately accounted for; implying that no significant three dimensional piping flow problems

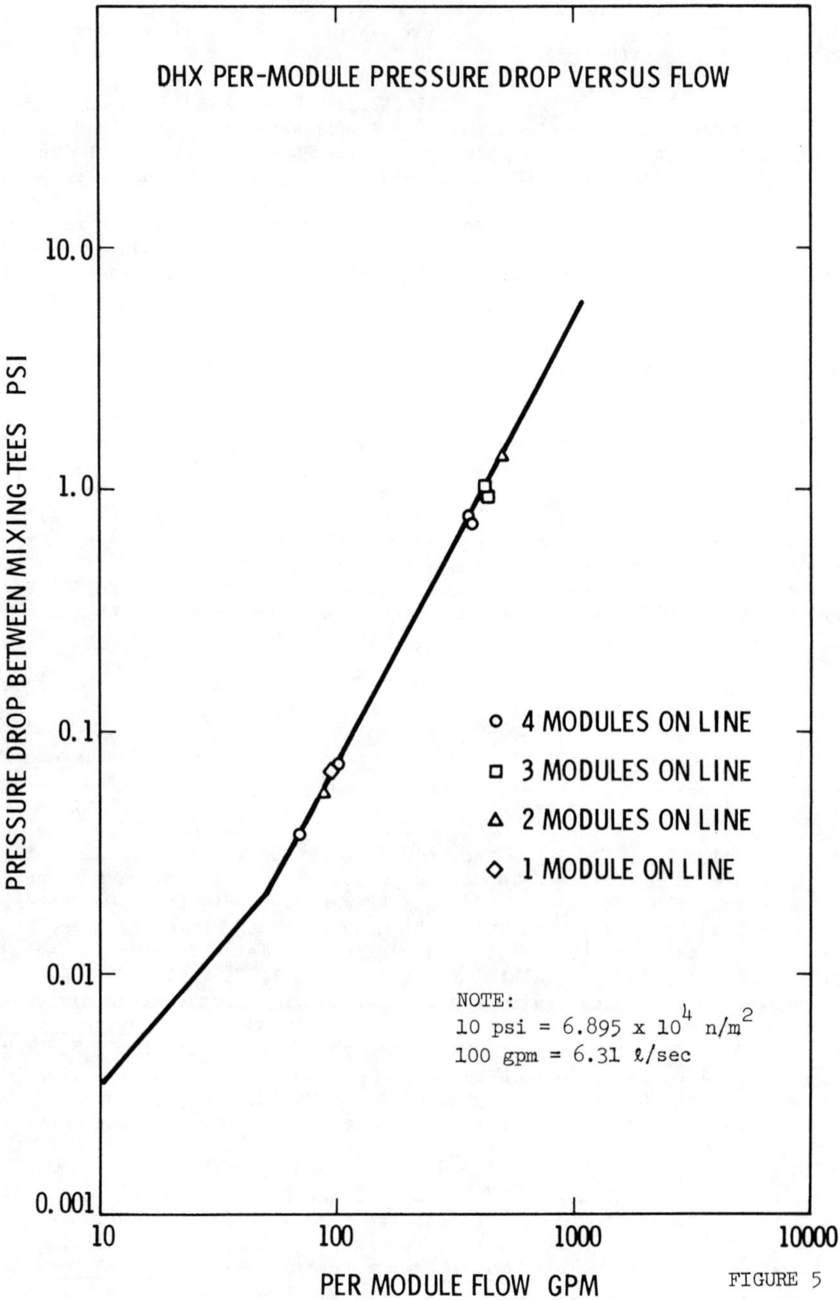
DHX PER-MODULE PRESSURE DROP VERSUS FLOW
PRESSURE DROP BETWEEN MIXING TEES PSI
10.0
1.0
0.1
0.01
0.001
10
100
1000
10000
PER MODULE FLOW GPM
○ 4 MODULES ON LINE
□ 3 MODULES ON LINE
△ 2 MODULES ON LINE
◇ 1 MODULE ON LINE
NOTE:
10 psi = 6.895 x 10^4 n/m^2
100 gpm = 6.31 ℓ/sec

FIGURE 5

SECONDARY LOOP TEMPERATURES 700°F TEST

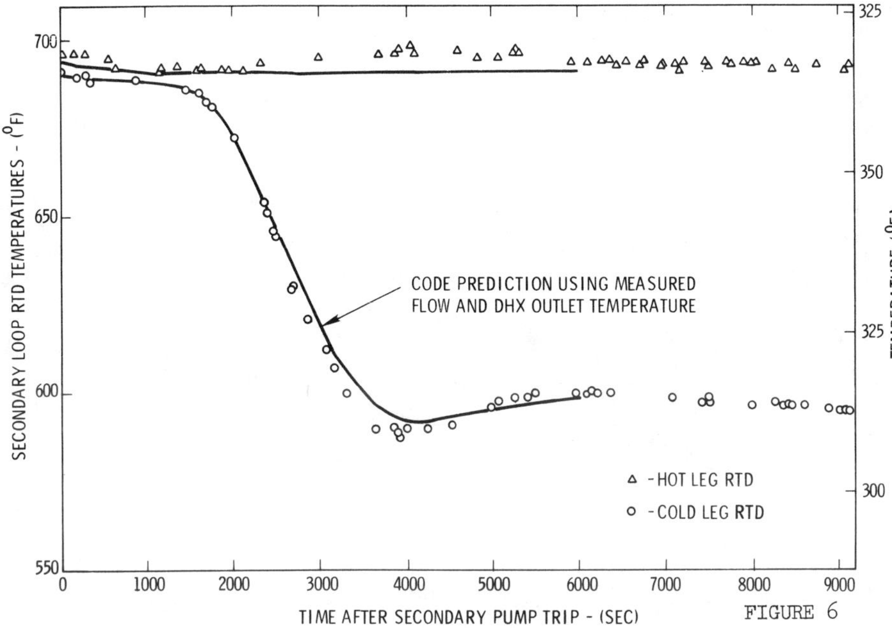

FIGURE 6

SECONDARY LOOP TEMPERATURES 400°F TEST

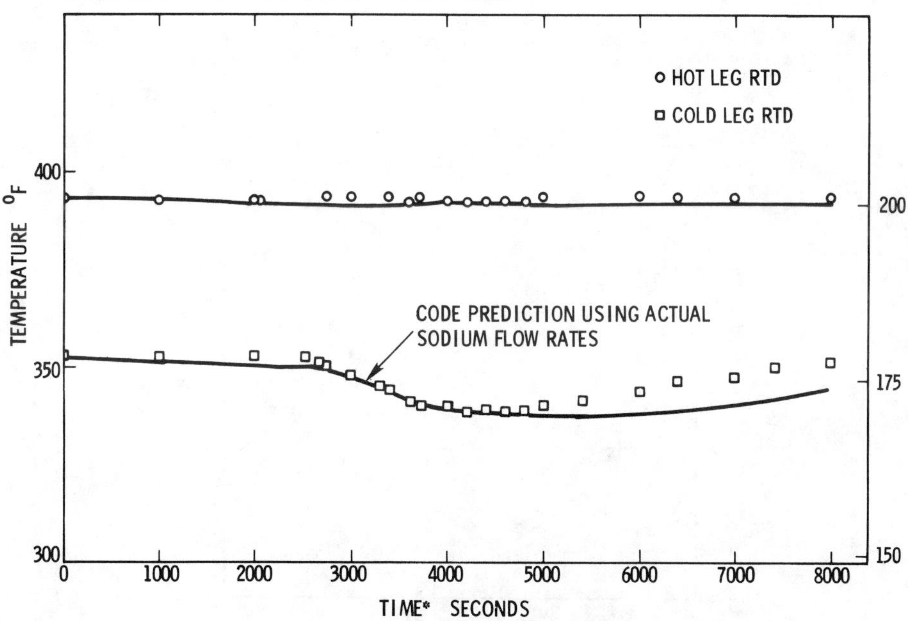

*TIME ZERO CORRESPONDS TO 13:40 ON MAY 30, 1979

FIGURE 7

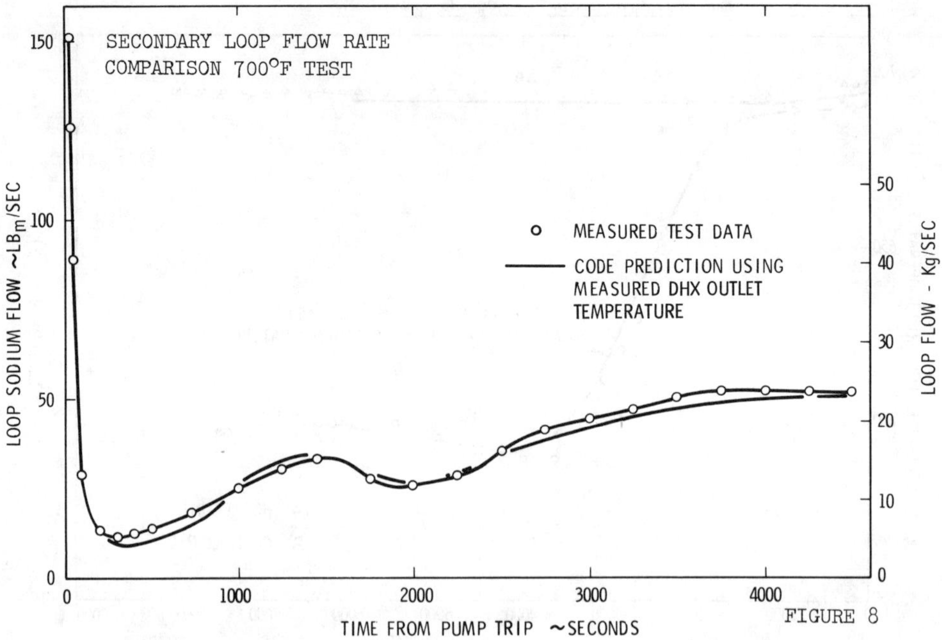

FIGURE 8

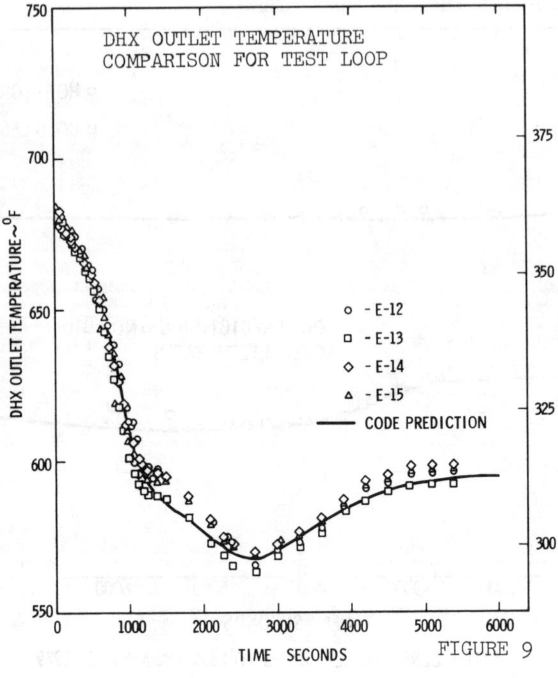

FIGURE 9

existed over the range of sodium flows and transient rates of temperature change involved in the tests (1.4 to 26 liters/sec (22 to 417 gpm)). Futhermore, the pump thermal model is shown to be acceptable.

3.3 DHX Thermal Model

A comparison of the four DHX module outlet temperatures measured during the initial transient phase of the 370°C natural circulation test is shown on Fig. 2. As can be seen from the figure, module to module temperature variation was not large with 5°C being the largest deviation. Total accuracy of the measurements shown is estimated to be ±5.6°C (±10°F).

Also shown on Fig. 2 is the initial post test analysis with the safety evaluation model. Although the prediction matches the initial cooldown rate, a significant, though conservative, deviation from the data occurs in the region of minimum temperature. Driving the model with the measured flow did not significantly alter these results. Cause of the difference between measurement and prediction was postulated to be due to thermal interaction between air inside the DHX and the DHX structure, which the quasi-steady heat loss assumption used in deriving the model could not match.

Using the above hypothesis, the DHX model was revised to include a single air node which exchanges heat with a seven node tube bundle. Conductance between the bundle and air was assumed to be a function of distance along the bundle. This was done to compensate for the actual decrease in sodium to air ΔT as the air rises through the tube bundle. The air node, in turn, was assumed to transfer heat both to a single average liner node and to ambient directly. A single wall node was interposed between the liner node and ambient. Thermal mass of the air node was adjusted to obtain a best fit with the measured data. While the new model structure was derived from physical hypotheses, viewed more abstractly it added a general capability to adjust both the spatial bundle heat loss distribution and the heat loss time dependence. The final balance of the two adjustments was controlled by the measured data. Predictions of the DHX outlet temperature for the 370°C test using this revised model is also shown on Fig. 2. As can be ssen, this prediction now lies within the envelope of measured data.

To further verify the DHX thermal model a prediction was made of the outlet temperature history for Module E-12 during transition to natural circulation from 205°C (400°F) (Fig. 3). For this test preheaters in all four modules were on until the pump was tripped. In three of the modules, temperatures were inverted by the preheaters resulting in zero or reversed flow through the modules for some period of time after test initiation. Actual loop flow was also reversed for approximately 2000 to 3000 seconds. Temperatures in E-12 were not inverted and, consequently, positive flow developed through this module. Measured flow from a temporary EM flow meter placed on the E-12 piping was used as a forcing function for the DHX code resulting in the prediction shown in Fig. 3. Due to the ducting configuration, the oil fired preheater heats not only the tube bundle but also the entire DHX structure above and below the tube bundle. The additional volume (thermal mass) available for thermal interaction with the tube bundle was calculated to be approximately 1.5 times the value used in the code. Consequently, thermal masses of the air, liner and wall nodes along with their heat transfer areas were increased by a factor of 1.5 for this prediction. While agreement with the test data is very good, this check does not constitute complete validation of the model since the above mentioned portion was modified to simulate the different preheat boundary condition.

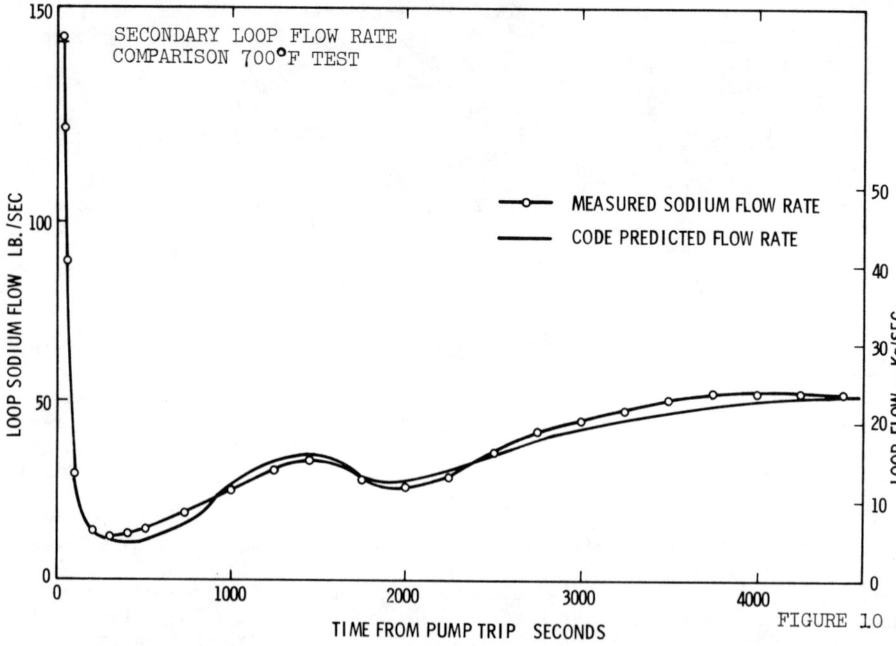

FIGURE 10

SECONDARY LOOP TEMPERATURE COMPARISON 700°F TEST

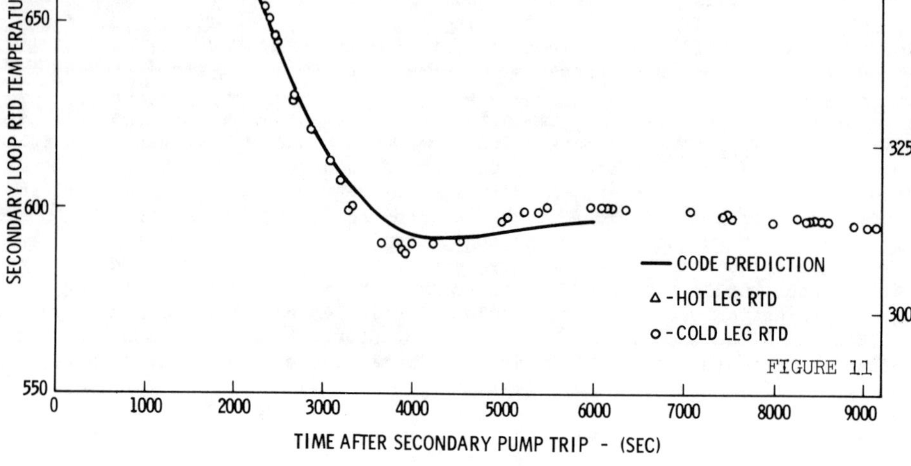

FIGURE 11

Comparison of the measured data with the revised code predictions for transition to natural circulation from 370°C (700°F) is shown on Figs. 9 thru 11. As can be seen, the predictions by the revised version of the code agrees very well with the measured data.

4. CONCLUSIONS

The main conclusions drawn from the FFTF secondary loop natural circulation tests and ensuing data analysis are:

1. The tests conclusively demonstrated the ability of the FFTF secondary loops to effect the transition to natural circulation flow from both near isothermal refueling and higher temperature conditions.

2. The safety evaluation model is confirmed to be conservative both by the direct test result, which shows a greater margin to freezing than predicted, and by the subsequent analysis of the data, which indicates favorable parameters and conservative model structure.

3. Acceptability of the one-dimensional flow assumption used in the model is indicated for the range of natural circulation conditions tested.

ACKNOWLEDGEMENTS

These secondary loop natural circulation tests could not have been successfully conducted without the direct technical support of many HEDL and DOE personnel. We would like to especially acknowledge R. D. Redekopp and all of the Testing and Operations people for their efforts in developing the procedures and setting up the instrumentation for the tests and W. L. Knecht for his early efforts on the Test Specification.

REFERENCES

1. Additon, S. L. and Parziale, E. A., "Natural Circulation in FFTF, a Loop Type LMFBR," Symposium on the Thermal and Hydraulic Aspects of Nuclear Reactor Safety, Volume 2: Liquid Metal Fast Breeder Reactors, Winter Annual Meeting of the ASME, Atlanta, Georgia, November 27-December 2, 1977.

2. Gunby, A. L., "Simulation of Piping Lengths in Sodium Systems," WHAN-FR-22, December 1970.

3. Gunby, A. L., "Intermediate Heat Exchanger Modeling for FFTF Simulation," BNWL-1367, May 1970.

4. LMEC Document, "FFTF HTS/DHX Prototype Module Testing in SCTI Final Report," LMEC-76-4, October 1976.

5. Additon, S. L., McCall, T. B., Wolfe, C. F., "Simulation of the Overall FFTF Plant Performance," HEDL-TC-556, December 1975.

6. Crane Company, "Flow of the Fluids Through Valves, Fittings and Pipe," Technical Paper No. 410, 1957.

7. Pigott, R. J. S., "Losses in Pipe and Fittings," Transactions of the ASME, November 1957.

FFTF Primary System Transition to Natural Circulation from Low Reactor Power

W. T. NUTT and S. L. ADDITON
Hanford Engineering Development Laboratory
Richland, Washington, USA

G. D. BOUCHEY
U.S. Department of Energy
Richland, Washington, USA

ABSTRACT

Plans for reactor and primary loop natural circulation testing in the Fast Flux Test Facility (FFTF) are summarized. Detailed pretest planning with an emphasis on understanding the implications of process noise and model uncertainties for model verification and test acceptance are discussed for a transition to natural circulation in the reactor core and primary heat transport loops from initial conditions of 5% of rated reactor power and 75% of full flow.

1. INTRODUCTION

The FFTF has been designed to provide shutdown heat removal utilizing natural convective cooling in the event that the normal forced flow cooling capability is not available. [1] A series of tests are planned during the FFTF startup program to confirm plant decay heat rejection capabilities by natural circulation. [2] The primary loop and whole plant testing is focused on satisfactory transition to natural circulation assuming loss of all electrical power to the pumps during operation at full power, end-of-life burn-up conditions. This paper briefly summarizes this series of planned tests. It describes the instrumentation to be used in these tests and provides detailed test plans and analyses for the first test in this series; a reactor scram to natural circulation from 5% power and 75% full flow. Tests in an FFTF secondary loop of transition to natural circulation from refueling conditions, a second major objective of the natural circulation plant acceptance test program, have been completed and are described in more detail in Reference [3].

2. NATURAL CIRCULATION TEST PLANS

The principal objective of the FFTF natural circulation tests is to confirm adequate FFTF shutdown heat rejection capability utilizing a series of tests which culminate in prototypic transients that require minimal extrapolation to the limiting design cases. To facilitate required extrapolation and to ensure that the relevant processes are understood, analytical models will be sufficiently validated to establish their acceptability for safety evaluation. The progression of tests has been designed to assure that transient fuel cladding temperatures are maintained within normal steady state operating values to minimize the potential for damage to FFTF fuel; protection of plant components has been given similar attention. A secondary objective of the testing is to collect data for further refinement of analytical models used in the prediction of natural circulation performance.

The planned sequence of tests for confirming an acceptable transition to natural circulation from full power, maximum burn-up conditions consist of (1) a reactor scram to natural circulation from 5% power, 75% flow, (2) low power (<5%) steady-state natural circulation tests, and (3) high power scram transients to natural circulation conditions.

The first test (scram from 5% power) is the principal subject of this paper and will be conducted prior to reactor operation at higher power. This test is designed to provide early confirmation of sufficient shutdown heat removal capability in the reactor vessel and the primary heat transport loops to assure adequate decay heat removal during early phases of plant power testing. The detailed test design and planned evaluation is addressed below.

Following the initial ascent to power, a series of natural circulation tests is planned at low fission power (<5%). These tests will measure loop and core performance during steady-state operation utilizing natural convection in the primary and secondary Heat Transport Systems. These steady-state tests are designed to provide key model verification data including loop pressure drop/thermal head correlations, in-core inter- and intra-subassembly flow redistribution information, and unbalanced loop heat rejection data.

A final series of reactor scrams to natural circulation is planned from the following initial power, flow, and decay heat conditions:

- 35% power, 75% flow, 1 hour at power
- 75% power, 75% flow, 1 hour at power, and
- 100% power, 100% flow, 25 hours at power.

The final transient is highly prototypic of the limiting case total loss of pump electrical power design transient, providing a rather direct confirmation of the adequacy of FFTF decay heat removal capability. Other tests in the series will support extrapolation from this test to the maximum burn-up condition.

Final pretest predictions have not been completed for the low power scram, the steady-state, or the high power transient tests. Preliminary predictions were previously reported in Reference [2]. Detailed analyses for the 5% scram are reported later in this paper. Final pretest predictions will be issued approximately one month prior to actual test performance.

3. NATURAL CIRCULATION TEST INSTRUMENTATION

In-core instrumentation for the primary system natural circulation tests includes two Fueled Open Test Assemblies (FOTAs), a Vibration Open Test Assembly (VOTA) and an Absorber Open Test Assembly (AOTA). The core map of Figure 1 shows the location of these assemblies in the core. The FOTAs, one in row 2 and the other in row 6, are prototypic of the FFTF driver fuel assemblies with thermocouple temperature instruments provided in pin wire wrap, on the duct walls and in the outlet stalk which extends above the sub-assembly outlet. Each FOTA has approximately 44 thermocouples distributed throughout the sub-assembly (e.g. see Figure 2). A specially calibrated eddy current flowmeter is located in each FOTA sub-assembly stalk. Several of the FOTA thermocouples will be used to provide a high temperature scram signal during the steady-state natural circulation tests; thereby providing automatic

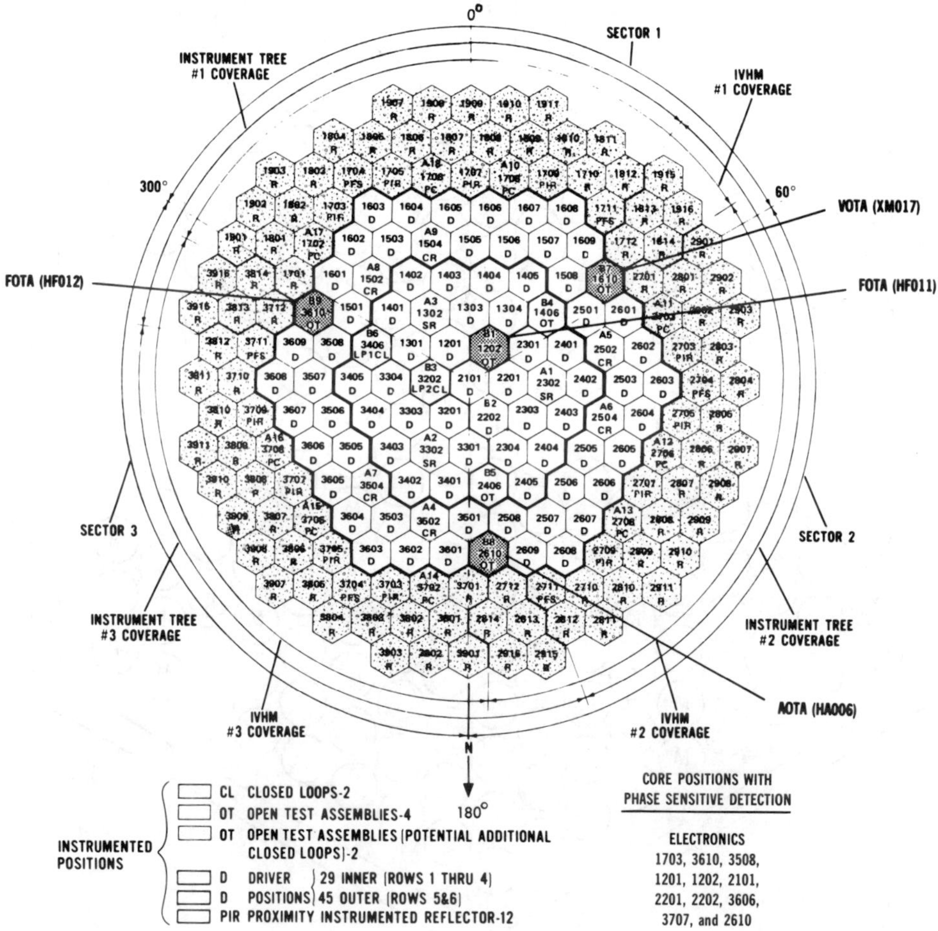

Figure 1

FFTF Core Map for Natural Circulation Tests

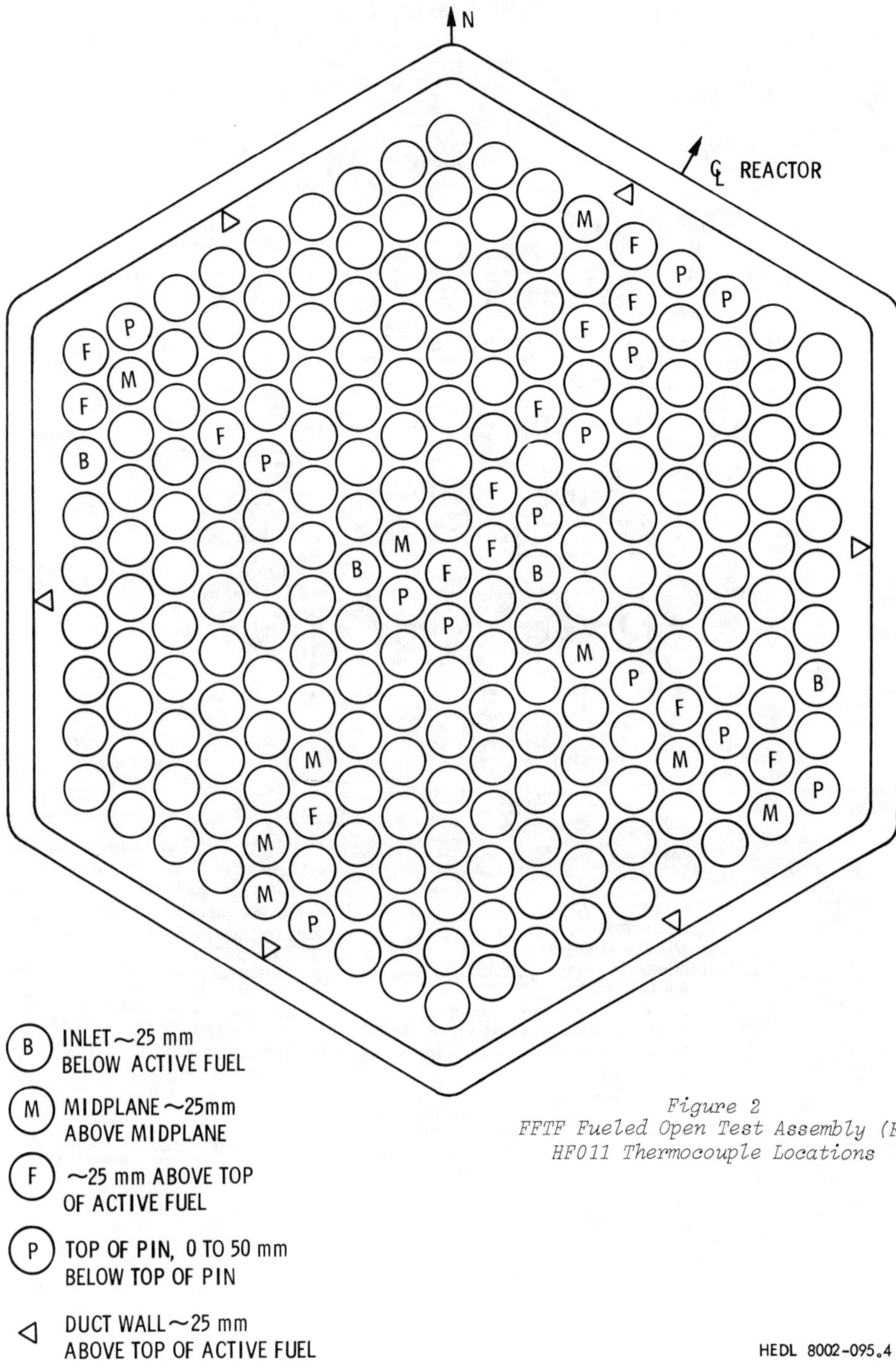

Figure 2
FFTF Fueled Open Test Assembly (FOTA), HF011 Thermocouple Locations

HEDL 8002-095.4

test termination if unexpected cladding temperatures are measured. The AOTA contains absorber pins prototypic of a control rod assembly and is also instrumented with numerous thermocouples and an eddy current outlet flowmeter. The VOTA is a structural assembly which contains thermocouples, an outlet eddy current flowmeter, and self-powered fission chambers for measuring neutron flux levels. The two FOTAs and the VOTA are currently installed in the core. The AOTA will be loaded after the initial ascent to power and will not be available for the 5% scram test.

The in-core thermocouples are expected to have an accuracy of ±1° - 3°C depending on the success of planned averaging and isothermal calibration procedures. The FOTA eddy current flowmeters utilize phase detector electronics and have been calibrated in ex-reactor sodium loops at low flows to an accuracy of ±60 cc/sec. which is less than 26% of the expected fueled sub-assembly flow during the 5% scram test.

In addition to the instrumentation in the OTAs described above, each core position is instrumented with an eddy current flowmeter and thermocouples located in the instrument tree above the sub-assembly outlet. This instrumentation is expected to be of limited value for model calibration due to the slow response time of the instrument tree thermocouples and the possibility of cross flow between sub-assemblies through the gap between the sub-assembly handling sockets and the bottom of the instrument tree. Nonetheless, all of the flowmeters have been calibrated in an ex-reactor sodium test loop at high flows and selected flowmeters (as shown in Fig. 1) have been provided with phase detection electronics and have been calibrated in the natural circulation flow range. The selected assemblies include FOTA "twins" and neighbors to isolate effects attributable to the FOTA stalk or to cross flow.

Reactor power will be measured using installed neutron flux instrumentation (compensated ion chambers) located in the cavity outside of the reactor vessel. These chambers will be calibrated using calorimetric methods with an expected accuracy of (±7.4%) during the initial ascent to power. The linearity of the ex-vessel chambers down to the low powers used in the natural circulation tests will permit transfer of this calibration. The assumption involved can be checked using in-core self-powered fission chambers in the VOTA. Since the reactor thermal power calibration will not be done until after the 5% scram test, the reactor power level uncertainty is expected to be ±20% for this early test (though the subsequent more accurate calibration will be transferable to the test data). The decay heat computation has an uncertainty of ±12% for an overall power uncertainty of ±32%.

The instrumentation in the heat transport loops during the natural circulation tests includes permanently installed Resistance Temperature Detectors (RTDs) in the primary and secondary loops for measurement of loop ΔT. Primary loop flow is measured with the magnetic flowmeter located in the cold leg which has been calibrated using a Pulsed Neutron Activation (PNA) [4] flow meter (±5% of reading) which will be operating in a least one of the secondary loops. In addition, temperature data will be obtained from numerous process thermocouples located in both the reactor vessel and primary and secondary loops.

4. TEST DESCRIPTION, SCRAM FROM LOW POWER (5%)

The first test planned to address natural circulation in the FFTF primary loop and reactor vessel is a plant scram from 5% power (20 Mw), 75% flow. The primary pump pony motors will be de-energized just prior to reactor scram so that the reactor will undergo a transition to natural circulation following automatic trip of the pump main motors upon scram. The test will be conducted prior to power operation above 5% power and after steady operation for approximately 1 hour at 5% power. The calculated decay heat curve following the transient as shown in Fig. 3. The secondary loops will be operated with pump pony motors and with cold leg temperature controlled (via air flow modulation) during the transient.

This particular test was originally incorporated into the test program to provide experience in planning, conducting, and evaluating a transient natural circulation experiment at a sufficiently low reactor power level that risk to the fuel and the plant components would be minimal even if the plant response exhibited unanticipated behavior. Implicitly, a requirement was established that any significant unanticipated phenomenon be identified. The role of the test was extended somewhat when the project elected to condition subsequent higher power plant testing upon demonstration by this test of sufficient plant natural circulation capability to safely remove the decay heat associated with the low total burn-up planned through the subsequent test series. To meet both of these objectives, it was necessary to characterize the feasible range of normal plant response (hence defining an unanticipated result) and to establish a criterion for permitting follow-on tests to proceed as planned. These tasks necessitated studying the relevant uncertainties (in plant data, model parameters, test measurements, and model structure) and then determining whether the test, the instrumentation, and the planned evaluation, given the uncertainties, could be expected to meet the objectives. This process is the detailed focus of this paper; such focus providing direct information for those following the FFTF tests and serving as an exercise in the development of verification and validation technology.

The analyses performed to study this test have employed primarily the Westinghouse proprietary computer program, IANUS [5]. IANUS is a complete plant simulation similar in structure and purpose to the more widely available DEMO program. [6] The principal simplifying assumptions in the model pertinent to the low power test are characterized by the inclusion of only one fully dynamic thermal/hydraulic channel in the reactor and one-dimensional modeling of the fluid flow process. To compensate for the first assumption, which can readily be shown to be invalid with EBR-II data, the model is augmented with a quasi-dynamic treatment of the fuel-bypass flow (18% of total reactor flow at steady-state for FFTF) and the use of flow dependent factors to extrapolate to a particular assembly from the calculated core average assembly. The second assumption is not expected to be important for this test since the fuel assemblies will dominate the loop response and since a flow dependent multiplier is again employed to reflect the effect of intra-channel thermal/hydraulic mixing processes within an assembly. The two flow dependent factors are generated with FLODISC and COBRA as discussed in [1]. The sensitivity of the predictions for this test to the assumptions made in deriving the flow dependent factors and to the uncertainties in actual plant multi-channel behavior, are assessed with FLODISC and CORA [7], which model multiple hydraulically-coupled, adiabatic-boundary channels and fuel assembly clusters with the ENERGY method for interchannel energy transfer, respectively. These programs require an IANUS-generated core driving pressure or total flow boundary condition, with attention required to ensure model compatibility. Sensitivity results are discussed below.

As a baseline for later comparisons, the nominal IANUS predictions for the planned test are shown in Figures 4 and 5, with temperatures and flows respectively. Temperatures shown include average assembly mean coolant at top of active fuel zone, and row 2 FOTA (HF011) peak coolant at top of active

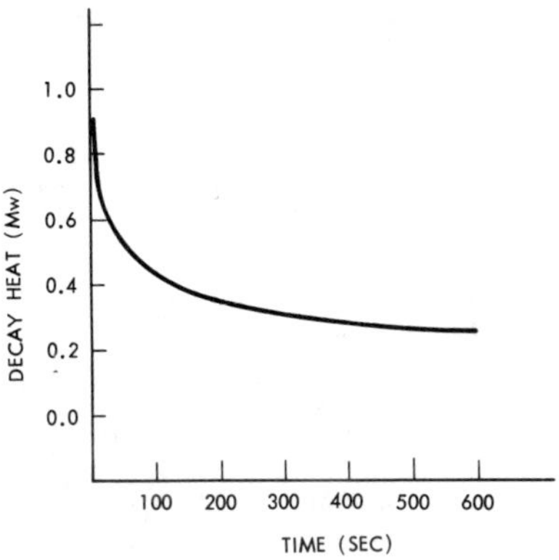

Figure 3. DECAY HEAT IN LOW-POWER TEST

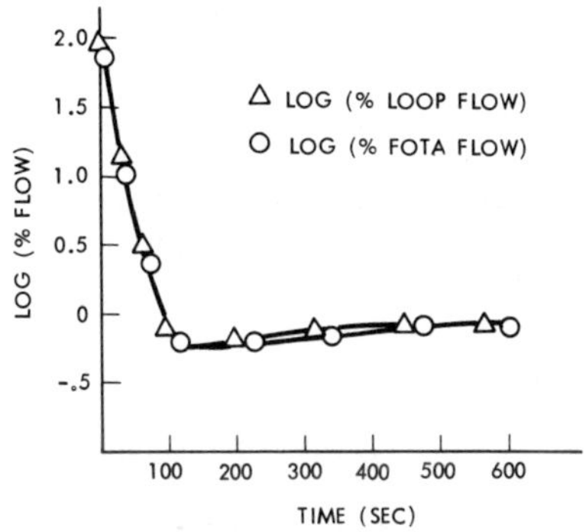

Figure 5. LOOP AND FOTA FLOW FOR LOW-POWER TEST

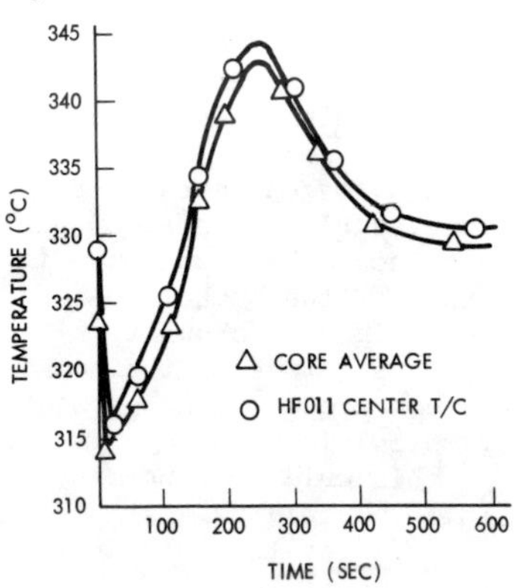

Figure 4. TRANSIENT TEMPERATURE AT THE TOP OF THE ACTIVE ZONE

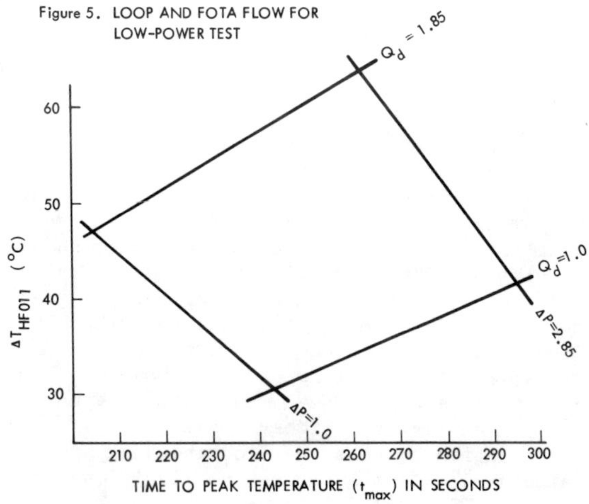

Figure 6. GRID FOR DETERMINING Q_d AND ΔP

HEDL 8002-095.6

fuel zone. Flows include the total loop flow and the estimated HF011 peak temperature rise is predicted to be 31°C occurring 243 seconds into the event. Extrapolation from the core average state variables calculated by IANUS to the HF011 peak measured value was performed using the same flow dependent analysis previously used to provide the hot channel estimate present in the hot channel model. The common model assumptions for the hot channel factor and the HF011 predictions helps assure that the accuracy (or conservatism) of the HF011 T/C prediction can be related the IANUS hot channel model. The nearly linear temperature rise rate from 60-120 seconds has also been selected for evaluation as it represents a significant measurable process characteristic. The heatup rate during this period is effectively a function of decay heat, core mass, and heat capacity with relatively low sensitivity to core hydraulics. A quasi-equilibrium "steady-state" is reached after 600 seconds though upper portions of the outlet plenum, irrelevant to loop thermal head, will still be heating.

5. SENSITIVITY

In order to assess possible alternate outcomes of the test, sensitivity studies were performed with the IANUS model. Two parameters selected for variation in the initial sensitivity study were decay power and pressure drop, the most sensitive design-event parameters. [1] The transient from 5% power is not sensitive to pump coastdown time since peak temperatures occur relatively long after the pump stops and, further, tests already completed have eliminated most coastdown uncertainty. In the period immediately following the pump coastdown a definite relationship is expected between power, flow and ΔT. Decay power, treated as a parameter, directly represents uncertainty by well-known system geometry, ΔT, pressure drop, and some tranient phenomena. Treating the pressure drop as a parameter emulates the flow uncertainty in the relationship. Consequently, it is inferred that results generated over a broad range of these two parameters encompass most foreseeable outcomes of the actual test. Two significant dependent variables which could be measured during a test are the peak temperature rise and the time of the peak temperature rise from the start of the transient in seconds. Figure 6 shows calculated points of peak HF011 temperature rise and elapsed time to the peak for various combinations of decay power and core pressure drop. Higher pressure drop causes a higher peak temperature later in time, while higher decay power leads to an earlier, higher peak. The test result on this grid will permit interpolation of pressure drop and decay power multipliers yielding a simply "calibrated" model, provided no significant deficiencies exist in other areas of the IANUS model.

Additional sensitivity runs were made to analyze the effects of other potential contributors to test uncertainty. The sensitivity to mass in the upper pin structure was obtained. The mass above the core was doubled in one run and was distributed very nonuniformly in a second run. The effects in both cases were small. Other sensitivities examined in this study were the sensitivity of the HF011 T/C reading to the degree of coupling to the mass of the duct wall, the sensitivity to fuel gap conductance, and the sensitivity to initial power. Predictions for the flow and temperature measurement for the test under each of these cases are listed in Table I. These predictions are for: (1) the temperature rise at the top of the active zone in HF011, ΔT_{HF011}, at its peak in the transient, (2) the time of the transient peak, τ_{max}, (3) the change in T_{HF011} between 1 and 2 minutes into the test, T_{1-2}, (4) the loop flow, W_{loop}, at τ_{max}, (5) the flow through HF011, W_{HF011}, at τ_{max}, and 10 minutes after the initiation of the test, (6) ΔT_{HF011}, (7) W_{loop} and (8) W_{HF011}. Table II consists of the results of the Table 1 analyses expressed as a matrix of sensitivities, reflecting predicted changes in each measured

Table I

Sensitivity Analysis Runs for 5% Transient Natural Circulation Test

				TRANSIENT PREDICTIONS					STEADY STATE PREDICTIONS		
No.	Q_d	P	Q_o(%)	ΔT_{HF011}(°C)	T_{max}(sec)	T_{1-2}(°C)	W_{loop}[1] $(t=t_{max})$	W_{HF011} $(t=t_{max})$	ΔT_{HF011}(°C)	W_{Loop}(%)	W_{HF011}(%)
1	1	1	5	31	243	9	0.512	0.645	18	0.625	0.712
2	1	2.85	5	43	295	9	0.410	0.469	24	0.474	0.528
3	1.85	1	5	49	205	17	0.572	0.871	26	0.718	0.887
4	1.85	2.85	5	66	262	17	0.469	0.563	37	0.528	0.614
5[5]	1	1	5	32	251	9	0.492	0.665	17	0.660	0.732
6[2]	1	1	5	32	218	11	0.490	0.600	16	0.612	0.689
7[2]	1	2.85	5	41	257	11	0.387	0.441	22	0.463	0.510
8[3]	1	1	5	30	242	9	0.520	0.646	17	0.629	0.715
9	0.83	1	6	33	235	9	0.572	0.675	17	0.687	0.757
10	1.25	1	4	34	250	9	0.444	0.630	18	0.557	0.670
11	1.25	2.85	4	44	305	9	0.377	0.452	26	0.440	0.511
12[4]	1	1	5	34	243	9	0.512	0.645	18	0.625	0.712

1 Flows are expressed as percent of design flow

2 This run had no coupling of the coolant at the top of the active zone to the duct.

3 Gap conductivity 3/16 nominal

4 Redistributed mass above active zone

5 Mass above core doubled

HEDL 8002-095.5

Table II

Sensitivity Matrix for 5% Transient Natural Circulation Test

MEASUREMENTS	PARAMETERS					
Transient	ΔP	Q_d	Upper Structure Mass	Duct	Q_o(%)	Axial Distribution of Mass
ΔT_{HF011}(°C)	6.3	21	0.83	1.1	3.4	-2.8
t_{Max}(sec)	28	-45	7.5	-25	-16	0
T_{1-2}(°C)	0	8.8	0	1.1	1.4	0
W^2_{Loop}(%)	-0.0541	0.071	-0.02	-0.02	0.079	0
W_{HF011}(%)	-0.0973	0.271	0.02	-0.05	-0.079	0
Steady State						
T_{HF011}(°C)	3.6	9.4	-0.3	-2.2	1.1	0
W_{Loop}(%)	-0.0811	0.106	0.035	-0.01	0.086	0
W_{HF011}(%)	-0.0973	0.212	0.040	-0.02	0.087	0

1 The variation in predicted values for a 100% change in the parameter (except Q_o).

2 All flows are expressed as percent of design flow.

HEDL 8002-095.2

system state assuming a unit change (i.e. 100% change) in each of the uncertain parameters (except for Q_o where the uncertainty was bounded by a 20% change).

A most useful result of these studies is the insensitivity of the temperature rise between one and two minutes to all parameters except the decay power and the duct mass coupling. This insensitivity to flow means that one can estimate the decay power from the HF011 T/C reading which, in turn, allows estimates of the reactor pressure drop from the flows at peak temperature or from the quasi-steady state temperature and flows. Estimates obtained from these temperature and flow measurements should be consistent with the values of Q_d and ΔP obtained from the grid in Figure 6. In the next section the appropriate degree of consistency is discussed.

The impact of the uncertainties in the parameter values was assessed by bounding them and then using the sensitivity matrix to calculate the resulting uncertainties in the interpretation of the measured values summarized in Table III. Column 3 in Table III presents the measured state uncertainties due to a possible 6% change in the shape of the decay heat curve between the time of the peak and the quasi-steady state at 10 minutes. The uncertainties in Columns 4-7 reflect the results of the Table II sensitivity studies. The estimated uncertainties in M_{us} and in M(Z) represent plausible bounds; the uncertainty in the duct mass effect was based on a CORA run which indicated a comparatively flat temperature profile across the assembly for the test. Column 8 in Table III consists of the additional uncertainties in the measured values due to statistical hot channel factors.

Column 9 in Table III is an estimate of the effects of three known model structural approximations whose effects were estimated without a direct comparison of IANUS sensitivity runs. The three model approximations are:

- no radial heat transfer in the reactor core or vessel,
- no flow mixing under the instrument tree,
- a single node upper plenum without a stratification model.

All three effects are expected to be small in the low power test. The reliance of the analysis of HF011 temperatures for verification minimizes the effect of heat transfer since all of the sub-assemblies have power-to-flow ratios which are close to that of HF011. HF011 has a stalk which prevents cross flow for the assembly. Finally low flow and plenum ΔT should minimize stratification. Nevertheless, the effects are relevant for this test and in the design case and some uncertainty allowance was judged to be appropriate.

Radial heat transfer beyond immediate neighbor assemblies will act to effectively deposit more decay heat in the fuel bypass channels, resulting in a lowered flow through the core and a later peak temperature - much like increased ΔP and decreased Q_d. Coolant mixing under the instrument tree will tend to reduce the adverse chimney effect caused by the cold sodium in the instrument tree and reduce instrument tree pressure drop, both effects causing a reduction in core ΔP, but without benefit in the FOTA. Since flow will be low and the bypass sodium temperature will be quite close to the upper plenum temperature, plenum stratification causing bypass streaming into the outlet nozzle will be negligible. The simplified upper plenum model would be more significant in the design transient. All three model approximations affect only the core pressure boundary conditions. Hence, the rate of heatup is insensitive to these uncertainties, and an estimate of the decay power in HF011 based on heatup rate would be unaffected by the uncertainties. Estimates for the effects of these structural approximations on flow sensitive measurements were obtained by treating them as effective ΔP and using the

Table III

Uncertainty Summary for 5% Transient Natural Circulation Test

Measurement	Uncertainty[2]	Anticipated Model Uncertainties[1]								
		$Q(t)$	M_{us}	Duct	Q_o	$M(z)$	Hot Channel[3] Statistical Factor	Other[4]	Subtotal[6] Prediction Uncertainty	Total[7] Prediction Uncertainty
<u>Transient</u>										
ΔT_{HF011} (°C)	±3	0	±0.2	±0.1	±3.4	±1.4	±1.6	±2	±5.4	±6.5
t_{max} (sec)	0	0	±1.5	±2.6	±16.5	0	±4.4	±10	±20	±24
T_{1-2} (°C)	±1	0	0	±0.1	±1.4	0	±0.6	0	±1.8	±2.1
W_{Loop} (%)[5]	±0.05	0	0	0	±0.08	0	±0.01	0	±0.09	±0.09
W_{HF011} (%)	±0.18	0	0	0	±0.08	0	±0.02	0	±0.2	±0.21
<u>Steady-State</u>										
ΔT_{HF011} (°C)	±3	±0.6	±0.06	±0.02	±1.1	0	±0.06	0	±3.3	±3.7
W_{Loop} (%)	±0.05	0	±0.01	0	±0.09	0	±0.01	0	±0.1	±0.11
W_{HF011} (%)	±0.18	±0.01	±0.01	0	±0.09	0	±0.02	0	±0.2	±0.21

1. These estimates are based on the assumption that $\Delta Q(t) = \pm 6\%$, $\Delta M_{us} = \pm 0.2$, $\Delta \text{Duct} = \pm 0.1$, $\Delta Q = \pm 1\%$ and $\Delta M(z) = \pm 0.5$.
2. Represents 3σ values on the data acquisition.
3. Statistical factors due to inlet flow maldistribution, rod bowing, and power distribution.
4. Estimated uncertainty due to radial heat transfer, upper plenum modeling and sub-assembly cross-flows.
5. All flows are expressed as percent of design flow.
6. This is the root-mean-square of all the uncertainties relevant to the calculation of Q_d and ΔP.
7. Uncertainties of 0.4 in ΔP and 0.12 in Q_d are included.

HEDL 8002-095.3

Table IV

Uncertainty in Decay Heat (Qd) and Pressure Drop (ΔP)

Method	Measurement #1	Measurement #2	Uncertainties Qd	Uncertainties ΔP
1	ΔT_{HF011} [t_{max}]	t_{max}	0.29	0.53
2	T_{1-2}	W_{loop} [t_{max}]	0.24	1.17
3	"	W_{HF011} [t_{max}]	"	2.2
4	"	T_{HF011} [SS]	"	1.0
5	"	W_{loop} [SS]	"	1.3
6	"	W_{HF011} [SS]	"	2.1

HEDL 8002-095.1

sensitivity matrix. The estimates are listed in Column 10 of Table III.

The subtotals of uncertainties in Column 11 represent the appropriate uncertainties in the measurements when the test results are to be used to estimate Q_d and ΔP. That is, these uncertainties will indicate the precision with which a determination of Q_d and ΔP can be made, given this test. To establish the best possible estimate, there are six methods of obtaining Q_d and ΔP given in Table IV. The uncertainty for each estimate of Q_d and ΔP is calculated from the sensitivites in Table II and the uncertainties in Column 11 of Table III.

6. EVALUATION OF TEST RESULTS

To satisfy the initial test objective, we need only observe that an unanticipated result is one which is outside the bounds specified by the total a priori prediction uncertainty. Alternately stated, the total a priori prediction uncertainty represents the resolution limit for this test to uncover unanticipated phenomena in the FFTF. Failure to uncover an unanticipated result will provide a technical basis for proceeding with the natural circulation test plan for the steady state tests. In these tests and subsequent transient tests, more precise screening for unanticipated phenomena can be performed. Should an unanticipated result occur in the low power test, a more detailed analysis and a closer examination would precede the next natural circulation test; some changes in the follow-on test plan might be made to resolve any outstanding anomaly.

The second test objective, verification of sufficient plant natural ciruclation capability to permit plant testing at higher power levels, can be met by performing an approximate calibration of the model by estimating the most sensitive parameters, Q_d and ΔP, using the low power test result. Since these are only two of many actual uncertainties (plant data parameters, model structure), the estimates for Q_d and ΔP are expected to be distored from real values. To compensate for the known approximations involved, the error bounds from the intersection of the six methods in Table IV will be used to generate an extreme, conservative estimate of Q_d and ΔP. This constitutes a "worst case" estimate of results of the low power test. Further, to allow for potential unkown error sources, we use this result as we would use a nominal value and we apply the conservatism of the "safety evaluation model". Even with this conservative procedure, the margin to the established limit is sufficiently large to allow an expectation of a successful test result. From this result we conclude that the low power test, while inherently imprecise, is suited to the intended purpose. Should the final estimate be beyond the established limit, additional evaluation or supplemental testing will be performed.

7. SUMMARY AND CONCLUSION

In the FFTF natural circulation acceptance test plan a low power scram test will be performed prior to operation of the plant at a power level above 5%. This test result will be used to determine effective values of decay power and reactor pressure drops; these values will be used as nominal imputs to the "safety evaluation model", providing an approximate, but sufficiently accurate calibration to the model to permit demonstration of safe decay heat removal through the ensuing test series, pending completion of additional, more precise natural circulation tests. The meaning of the low power test measurements will be partially obscured by IANUS model simplifications, parameters and plant data uncertainties, and the measurement inaccuracies. As a result the determination of decay power and pressure drop is less certain than measurement errors alone would indicate. This study establishes a priori bounds for total uncertainty and also prescribes the use of redundant measurements to help to improve the quality of the determination. An uncertainty-broadened criterion for model acceptance is also developed. The conclusion is that the proposed test is designed to satisfy the established objectives. Further the techniques applied here, when uncertainties are large, may also prove useful for designing and evaluating future tests planned for model verification and validation.

ACKNOWLEDGEMENTS

The work summarized in this paper was sponsored by the U. S. Department of Energy under contract No. EY-76-C-14-2170. The contribution of Mr. H. G. Johnson in performing multi-channel core analyses used for obtaining the test predictions is especially acknowledged.

REFERENCES

1. Additon, S. L. and E. A. Parziale, "Natural Circulation in FFTF, a Loop Type LMFBR," Symposium on the Thermal and Hydraulic Aspects of Nuclear Reactor Safety, Volume 2: Liquid Metal Fast Breeder Reactors, Winter Annual Meeting of the ASME, Atlanta, Georgia, November 27-December 2, 1977.

2. Bouchey, G. D., S. L. Additon, and E. A. Parziale, "Planned Natural Circulation Testing in the Fast Flux Test Facility," TRANS, 28 592-594, June 1978 (HEDL-SA-1458).

3. Beaver, T. R., D. M. Turner, and S. L. Additon, "Post Test Evaluation of Natural Circulation in FFTF Secondary Loops," to be presented at Specialists' Meeting on Decay Heat Removal and Natural Convection in FBRs, Brookhaven National Laboratory, February 28-29, 1980.

4. Kehler, P., "Feasibility of Calibration of Liquid Sodium Flow Meters by Neutron Activation Techniques," ANL-CT-76-16, July 1976.

5. Additon, S. L., T. B. McCall, and C. F. Wolfe, "Simulation of the Overall FFTF Plant Performance," HEDL-TC-556, March 1976.

6. Alliston, W. H., A. Batenburg, B. Dvorznak, et. al., "Clinch River Breeder Reactor Plant — LMFBR DEMO Plant Simulation Model (DEMO)," CRBRP-ARD-0005, February 1978.

7. Johnson, H. G., "CORA-Transient Analysis Code for a Cluster of Reactor Core Assemblies," HEDL-SA-1838-FP, presented at ANS meeting Altanta, Georgia, June 1979.

Decay Heat Removal of the SNR-300 with Special Consideration of the Natural Convection Effects

F. ROESGEN, M. DUEWEKE, F. TIMMERMANN, and H. VOSSEBRECKER
International Atomreaktorbau GmbH
Bergisch-Gladbach, Federal Republic of Germany

ABSTRACT

The paper reports on a set of analyses on the behavior of systems of the SNR-300 at several decay heat removal situations with special regard to the natural convection effects. The following cases are described:

- decay heat removal following incidents
- decay heat removal at emergency conditions
- complete loss of all active heat sinks

The response of the SNR-300 demonstrates that natural convection helps to mitigate the temperature loadings at incidents and at emergency cooling and represents a remarkable inherent safety potential of LMFBR's.

1. INTRODUCTION

The safety discussions of the past years show an increasing interest in inherent safety features. Natural convection is an inherent, self-regulating and surveyable phenomena. Its utilization for the decay heat removal of LMFBR's improves reliability and helps to mitigate temperature loadings.
The aspects of natural convection have been taken into account at the design of the sodium loops of the SNR-300. But normally the decay heat is removed by forced convection.
Nevertheless, the analyses of various incidents have proven that the advantages of natural convection can be utilized, which results in a large safety potential.

2. SYSTEM DESCRIPTION

The decay heat removal system of the SNR-300 /1/ consists of five independent chains to fulfill essential general safety and specific reliability criteria. Fig. 1 shows a general view of the SNR-300 decay heat removal system.

The three main heat transfer chains are nearly identical. The coolant is picked up from the hot reactor outlet plenum by the primary pumps and transported to the IHX's. From there the heat is transported by the secondary loops to the steam generators. Special decay heat removal systems are installed within the water/steam-system.
Chains are separated spatially. Normally the primary and secondary pumps are driven by emergency power supplied pony motors at 5% of their nominal speed. The sodium mass flow is controlled by butterfly valves to mitigate thermal shock loadings.

The emergency core cooling system is diverse, spatially separated and independent from the operation of the main decay heat removal system. It meets the requirement of a 2 * 100% decay heat removal system. Each of the 2 redundancies consists of three closed secondary sodium loops with an immersion cooler, situated in the reactor vessel and a sodium-to-air heat exchanger. The three heat exchangers are located in one chimney and served by one blower. The sodium in the closed loops is driven by EM pumps. The decay heat is being transported to the immersion coolers by in-tank natural convection.

3. DECAY HEAT REMOVAL FOLLOWING INCIDENTS

In the first part a general view concerning the significance of the natural convection is given in the case of normal decay heat removal through the 3 main heat transfer chains. After a reactor scram, the rotational speed of the pumps is being turned down. The pressure caused by the pumps with low r.p.m. shows nearly the same value as the pressure of natural convection. Therefore, these buoyancy forces have a great influence on the normal course of the decay heat removal. In order to demonstrate this three analyses are being presented here:

The first case 'blockage of a primary pump' was chosen, because it is very appropriate to study the different type of behavior of two chains under normal post scram conditions with well working pumps and one chain with a blocked one. The incident 'blockage of a pump' means that the rotational speed of the pump suddenly becomes zero. Thereby a reactor scram is caused immediately which shuts down the nuclear power and the pumps. Some essential results are shown in Fig. 2. The mass flow in the disturbed primary circuit is determined only by buoyancy forces and the difference of pressure between the inlet and the outlet of the reactor vessel. In the first phase, the high pressure in the inlet plenum created by the orderly working pumps causes a negative mass flow rate in the disturbed circuit for about 30 seconds. Later, when this pressure has decreased, the mass flow is essentially determined by natural convection and reaches a value of about 3.0%. The essential conclusion is that only the buoyancy forces normalize the heat removal in the disturbed chain.

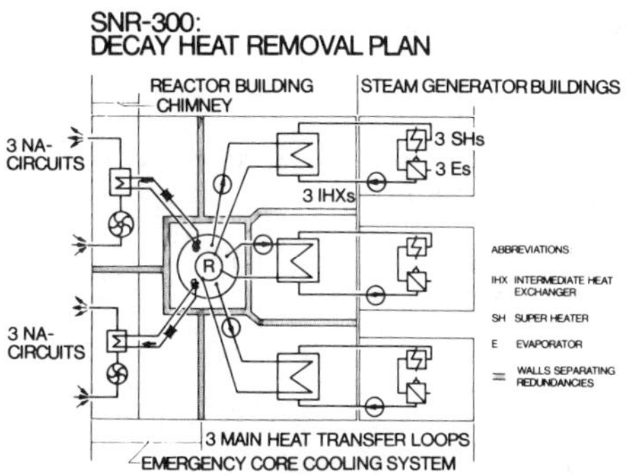

-NA-SIDE-

FIG.1

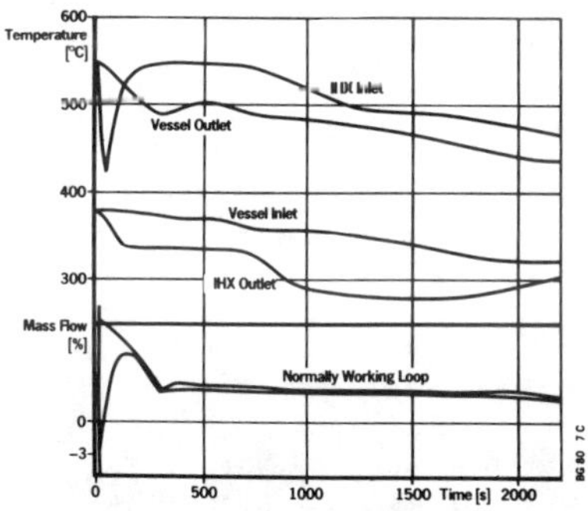

SNR 300
Temperature and Mass Flow in the Disturbed Loop after Blockage of a Primary Pump

FIG.2

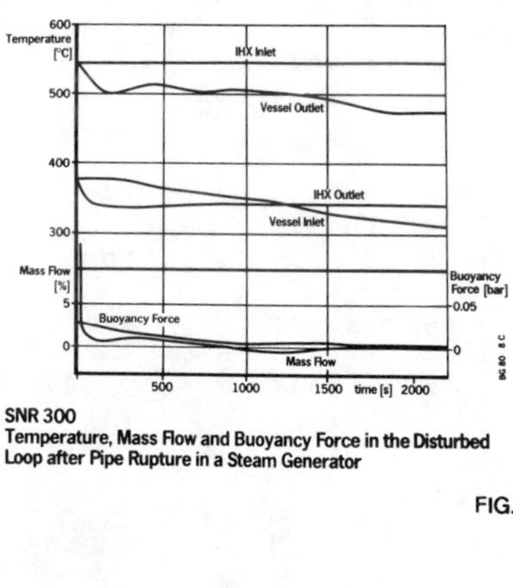

SNR 300
Temperature, Mass Flow and Buoyancy Force in the Disturbed Loop after Pipe Rupture in a Steam Generator

FIG.3

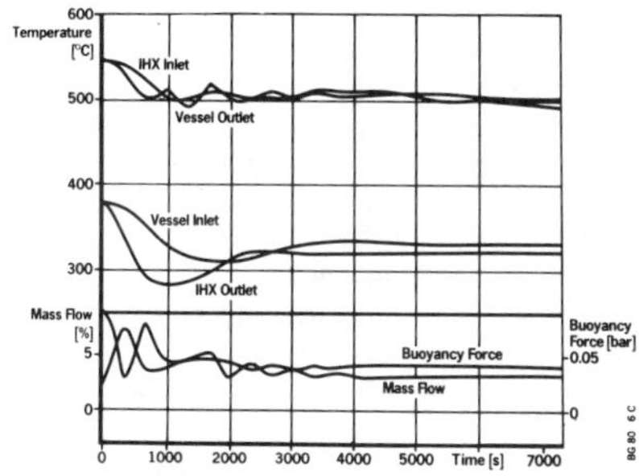

SNR 300
Temperature, Mass Flow and Buoyancy Force in a Chain at Decay Heat Removal by Natural Circulation only

FIG.4

SNR-300:
ECCS OPERATING LEVELS & FLOW PATHS

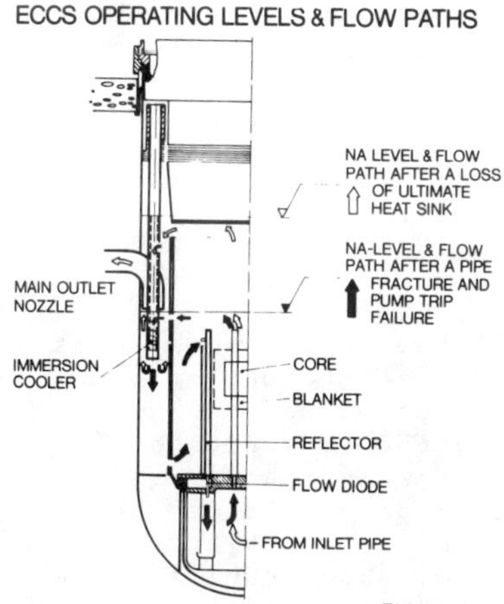

FIG.5

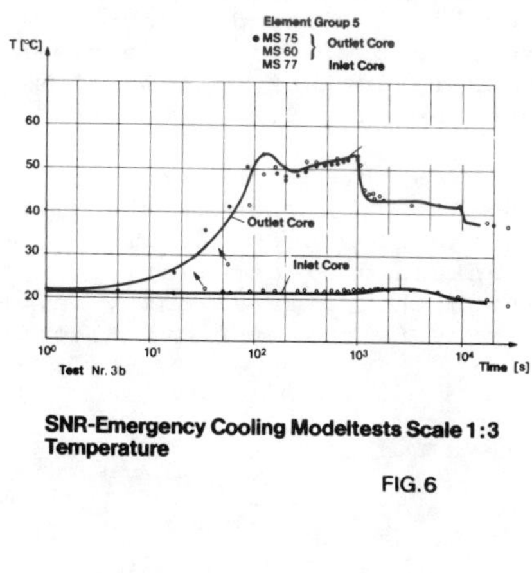

SNR-Emergency Cooling Modeltests Scale 1:3
Temperature

FIG.6

In the second case, a pipe rupture in the steam generator is being investigated and some of the possible protection measures are as follows: The valves are closed and the pumps are turned down in the primary and secondary loops of the disturbed chain. The secondary valve is being totally shut off, whereas the primary valve leaves a defined hole in the shutter. Due to these measures, the IHX is no longer cooled and the sodium temperature in the primary part of the IHX is getting hot gradually. For this reason, the buoyancy force and consequently the mass flow decrease to nearly zero, as shown in Fig. 3. Therefore, the cold leg of the primary loop is protected against the hot sodium.

Thirdly, the hypothetical case 'Loss of all pumps and closing of the valves in two chains' is being discussed and shown in Fig. 4. This case only shows the possibility of removing the decay heat by natural circulation through one chain only. The temperature difference in the heat exchangers (core, IHX, SG) becomes great enough to reach mass flow rates of 10% by buoyancy forces only. The temperature at the reactor outlet vessel never exceeds the normal operation value of 546° C.

4. IN-TANK NATURAL CIRCULATION

The heat transport from the core to the immersion coolers by natural convection is an essential part of the emergency core cooling system. The flow paths of the in-tank natural circulation are shown in Fig. 5.

The sodium in the core, heated by the residual heat, streams upward into the outlet plenum and through the orifices in the shield vessel to the immersion coolers. The colder sodium leaving the immersion coolers, streams downward within the annular space between reactor vessel and shield vessel. It enters the annular space outside the core barrel through the lower orifices of the shield tank and streams up to the perforation zone of the reflector subassemblies. The cold sodium streams then downward through special reflector subassemblies equipped with flow diodes into the middle plenum and through the blanket-diodes into the inlet plenum. The diodes are of the scoop-cascade type promoting natural circulation for emergency decay heat removal. Nearly the whole in-tank sodium takes part in the natural circulation, caused by this flow pattern.

In Fig. 5 the black arrows show the flow path of the in-tank natural circulation at the emergency level and the open arrows show the flow path at normal operating level. In this case, the in-tank natural circulation is overlapped by the main flow through the primary loops, which is driven by primary pumps and buoyancy forces or by the latter only.

For the calculation of the dynamic behavior of the in-tank natural circulation and its instationary temperature distribution within the vessel and the loops, the special computer code NØTUNG has been developed /3/.

5. VERIFICATION OF NØTUNG COMPUTER CODE

This program solves the energy equation and the equation of motion by a forward-difference technique and takes into account:

- The inertia of the sodium mass flow
- The distribution of heat capacities
- The local distribution and time decrease of the residual heat
- The different pressure drop characteristics of the various flow paths
- The heat transport by condution and mass flow and
- the insulation heat losses of the pipes

The interior of the vessel is subdivided into about 600 nodes. Each primary and secondary loop can be subdivided into 500 nodes. In the emergency cases the velocities are low in the larger plenums which are - inlet and outlet plenum and the annular spaces outside the core. Hence, a significant stratification occurs through the temperature difference. This phenomena allows a description by use of simple calculation models, which in turn afford short calculation times.

The licensing authorities have required a verification of the program by suitable model tests. Besides a series of tests, performed in order to study special phenomena, an integral test in a mock-up of the reactor vessel was performed at the Karlsruhe Nuclear Research Center /2/.
The mock-up was a detailed reconstruction of the reactor vessel and its internals on a longitudinal scale of 1:3, but limited to a 120 degree section. Water was used as a coolant. The fuel and breeder elements were simulated by electrically heated rod bundels. Radial power variations in the core and the decrease of power with time were selected on the basis of reactor conditions.
The test parameters, heating power, immersion cooler capacity, heating power transient, blanket power, several cross-section reductions, coolant level and intial temperature were varied in a test series of 25 experiments.

The comparison between calculation and experiment was made in three selected cases. Fig. 6 shows the comparison of the temperatures in the core as the reference case, which simulates the emergency case at the emergency level. The temporal course is similar to that of the reactor. The natural convection starts a few seconds after the heating power is turned on and the core outlet temperature increases rapidly to a first maximum until the mass flow is fully developed. The agreement between the calculated inlet and outlet temperatures and the experimental results were good, even after the heat power was lowered by a step. On the whole, the agreement between calculation and experiment was good and the NØTUNG computer program was found to give a good representation of the dynamic behavior of the system.

6. DECAY HEAT REMOVAL UNDER EMERGENCY CONDITIONS

The temperature loadings of the system of the SNR-300 in the following three emergency cases

1) the hypothetical case of the loss of all primary loops
2) the destruction of all main heat sinks by external events
3) the loss of three main decay heat removal systems, because of random failure of components

were calculated by the NØTUNG computer code under conservative conditions.

The ECCS has been laid out to remove the decay heat after a loss of all three primary loops. This can occur in the hypothetical case, if after a major leak in one of the main primary loops the shut-off of the main pumps fails simultaneously. The sodium level in the vessel decreases to below the suction nozzles and gas enters the primary loops and interrupts the circulation of the sodium. The emergency sodium level in the vessel is being reached within about 30 s. The temperature of the outlet plenum decreases to 400° C, caused by the high mass flow rate. As shown in Fig. 7, there is a first temperature maximum with about 780° C after a further 100 s, while the natural circulation has been fully developed. At about 3000 s a second temperature maximum of about 780° C occurs, because the temperature in the backstream positions - blanket and middle plenum - is higher than the core inlet temperature. Thereafter, the system is cooling down continuously.

Some essential results of the emergency case caused by the destruction of the main heat sinks by external events are shown in Fig. 8. The following has been assumed: All secondary loops are being destroyed simultaneously, all primary valves are closed with the exception of a defined hole in the shutter. One primary pump works at 5% of the normal speed, the others are turned off and the initial temperatures in the entire primary system are 23° C higher than during normal operation. (Increase up to the first scram level).

The core outlet temperature reaches a maximum of about 700° C, which is slightly higher than the maximum clad temperature at normal operation of 685° C. The maximum temperature within the primary system is about 610° C. The mass flow through

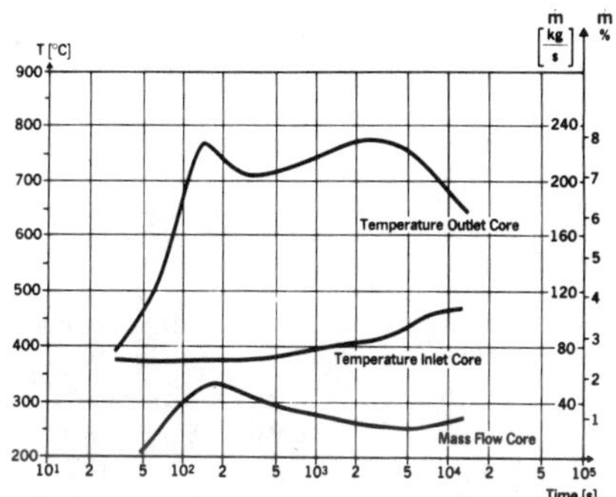

SNR-300
Temperature and Mass Flow
after Loss of all Primary Loops

BG 803 C

FIG.7

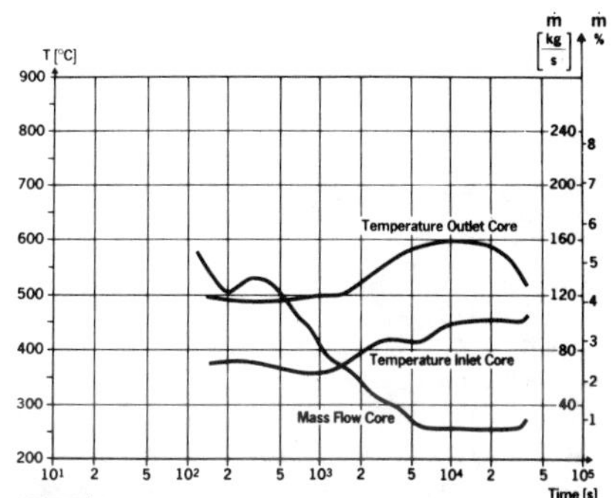

SNR-300
Temperature and Mass Flow
after Loss of the Main Decay Heat Removal Systems

BG 801 C

FIG.9

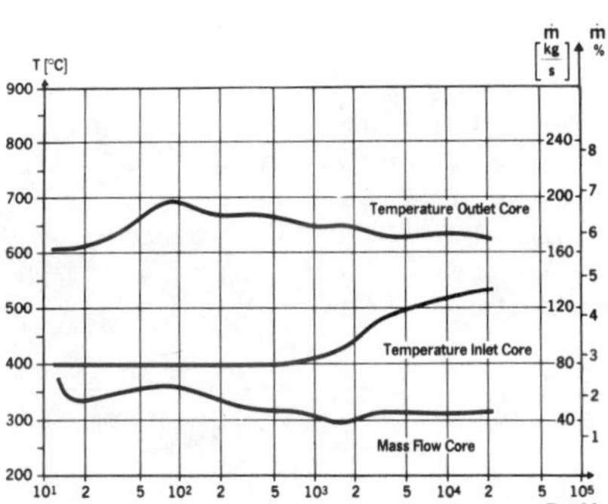

SNR-300
Temperature and Mass Flow
after Distruction of the Main Heat Sinks by External Events

BG 802 C

FIG. 8

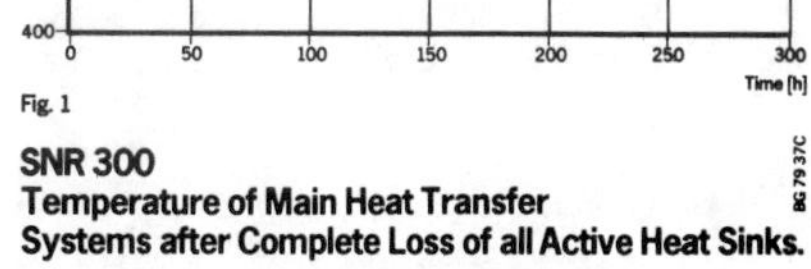

Fig. 1

SNR 300
Temperature of Main Heat Transfer
Systems after Complete Loss of all Active Heat Sinks.

BG 79 37C

FIG.10

the core comes to a minimum when the cold leg of the primary loop is filled with hot sodium and the buoyancy forces drop to nearly zero in the loops. Then the core cooling is being carried out chiefly by in-tank natural circulation.

With regard to a sufficient reliability of the decay heat removal system of the SNR-300, investigations have been made concerning the loss of the three main decay heat removal systems, because of random failure of components. The results are shown in Fig. 9. A leak in the steam generator of one chain is assumed to be the basic incident. Simultaneously the failure of the main decay heat removal systems, as well as one redundancy of the ECCS is assumed. The pumps of the main heat transfer chains coast down.

During the first 170 s the water evaporates within the steam generators. It takes about one hour until the vessel inlet temperature approaches the temperature of the hot leg, because of the high heat capacity of the primary-and secondary loops of the system. The flow rate through the loops, driven by buoyancy forces only, drops to nearly zero. The core outlet temperature increases to 590° C. The normal operating temperatures are practically not surpassed in the hot loops of the system.

7. LOSS OF ALL ACTIVE HEAT SINKS

In order to demonstrate the inherent safety of the SNR-300 with regard to the decay heat removal, the behavior of the heat transfer system in case of complete loss of all active components of all decay heat removal systems has been investigated by using realistic computation data. This situation would be given in case of the complete failure of energy supply combined with the assumption that no action would be taken during the accident, but reactor shut down takes place.

The behavior of the system is analyzed by a transient circuit program in which the temperature-dependent insulation heat losses and the natural convection are treated carefully. During the first 100 seconds the transient temperature fields in the vessel are similar to those after a normal scram. The thermal power decreases faster than the flow rate, which results in a reduction of the core outlet temperature. An immediate transition to natural convection in all sodium circuits takes place. The convection is driven incipiently by the temperature differences between hot and cold legs as still existing from previous normal operation.

It takes approximately 1 h until the core outlet temperature again reaches the level of normal operation. 90% of the energy produced during the first hour is being stored in the vessel and the circuits resulting in an average temperature increase by 20 K. The remainder has left the sodium systems by natural insulation losses.

The more than proportional increase of heat losses with temperatures and the decreasing decay power cause a continuous retardation of the temperature increase, which finally results in a temperature maximum after 70 h. Fig. 10 shows the development of the temperature.

Stress analyses have shown that the fuel claddings and the primary system - which represent the first and second radiological barrier - could accommodate the temperature loadings of this accident.

The inner containment represents the third barrier. The relatively large surfaces and capacities of the concrete walls of loop-type LMFBR's enable the accommodation and removal of the insulation heat losses.

8. CONCLUDING REMARKS

The investigations presented here demonstrate that natural convection helps to lower the temperature loadings after incidents and that it represents a remarkable inherent safety potential of the SNR-300. These advantageous results are based to a great extent on the specific properties of the sodium and the design features of the system and components resulting from these properties.

REFERENCES

/1/ F.H. Morgenstern, W. Gyr, D. Stötzel, H.Vossebrecker: "The Decay Heat Removal Plan of the SNR-300 A Licensed Concept" Conference on Fast Reactor Safety and Related Physics, Chicago 1976

/2/ H. Hain, F. Hofmann, D. Kirsch, K. Kleefeldt, W. Kramer, R. Meyder: "Modellversuche zum tank-internen Notkühlsystem des Prototyp-Reaktors SNR-300" KfK-Bericht 1895, Juni 1974

/3/ F. Timmermann, H. Vossebrecker "SNR Emergency Cooling-Heat Transfer Calculations on In-Tank Natural Circulation with Blanket Back-Flow" Conf. on Engineering of Fast Reactors for Safe and Reliable Operation KfK Karlsruhe, October 1972

Decay Heat Removal in SUPER PHENIX and Related Design Basis Plant Conditions

MICHEL DEBRU and PHILIPPE LAURET
Novatome

JURGENS DECKERT and JEAN-CLAUDE SCHNEIDER
Nersa

ABSTRACT

The various decay heat removal systems equipping SUPER PHENIX are described togother with their performances. The decay heat removal situations considered as design basis data are discussed.

1. INTRODUCTION

SUPER PHENIX is a 3000 MWth/1200 MWe sodium cooled fast breeder reactor. Its construction started in 1977 at Creys Malville and commercial operation is scheduled for the end of 1983.

The various decay heat removal systems, their characteristics and performances together with the corresponding design basis criteria are presented in the present paper.

The table below lists the main characteristics of this reactor, which is of the pool type.

Table 1 : Main characteristics of SUPER PHENIX

Number of primary pumps	4
Primary sodium mass flow	15750 kg/s
Number of secondary loops	4
Number of IHX	8
Secondary sodium mass flow per loop	3280 kg/s
Number of steam generators	4
Number of turbine generators	2 x 600 MWe
Primary sodium mass	3250 t
Primary circuit thermal inertia	5600 MJ/°C
Core inlet temperature	395°C
Core outlet temperature	545°C
(IHX inlet temperature (SG outlet temperature	345°C
(IHX outlet temperature (SG inlet temperature	525°C
Water temperature at SG inlet	237°C
Steam temperature at SG outlet	490°C

The main characteristics of the power supply system are given on figure 1.

The redundant safety motors and actuators are supplied from two separate power supplies (channels A and B).
- two off-site sources (380 kV and 225 kV)
- two on-site sources each comprising two diesel generators.

The decay heat values, including margins used for NSSS (Nuclear Steam Supply System) sizing are listed in Table 2 below. These data were validated by comparison with similar calculations and measurements performed on PHENIX.

Time	Decay heat (MW)
1000 s	74.5
3000 s	56.2
5000 s	48.1
10000 s	39.8
50000 s	25.9
1 day	22
10 days	9

Table 2 : Decay heat values versus time

The objectives assigned to the various decay heat removal systems equipping the nuclear boiler are of two orders :

1 - Ensure core cooling and restrict vessel temperatures to values compatible with containment structure sizing criteria.
2 - Cool quickly and hold the NSSS at a low temperature (180°C to 300°C) for servicing and handling operations.

The decay heat can be removed :
- by the normal decay heat removal circuits : water-steam circuits and condensers
- by the NSSS cooling systems
  - BPR sodium-air exchangers on the secondary circuits
  - emergency systems : emergency cooling circuits (RUR) and vessel cavity cooling circuits (RUS).

2. DECAY HEAT REMOVAL BY STEAM-WATER CIRCUITS (figure 2)

The plant is provided with two 600 MWe turbine generators, each equipped with 2 extraction pumps, a turbo-driven feed pump (680 kg/s), 2 motor-driven feed pumps (170 kg/s each) and a by-pass line designed to discharge the full rated flow to the condenser transiently or 50% of the rated flow permanently. Each condenser is Rhône river cooled.

Further, for the start-up/shutdown phases, each steam generator is equipped with a moisture separator, designed for 12,5% of the rated flow (42,5 kg/s) and 140 bars. The pressure of the water thus removed is reduced in a flash tank common to all four separators.

The condensed water in the flash tank is discharged to the condensers.

The normal heat removal procedure after turbine tripping is as follows :

Figure 1 : Structure of power supply

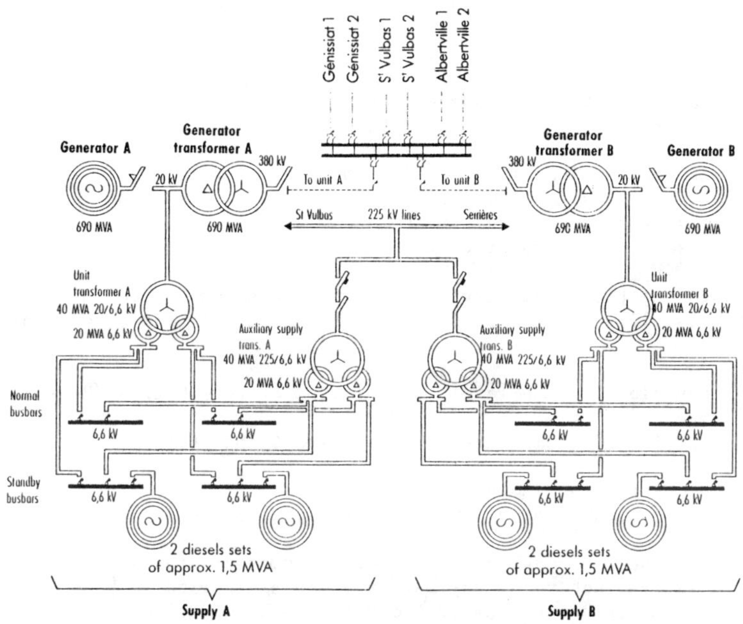

Figure 2 : Decay heat removal by steam-water circuits

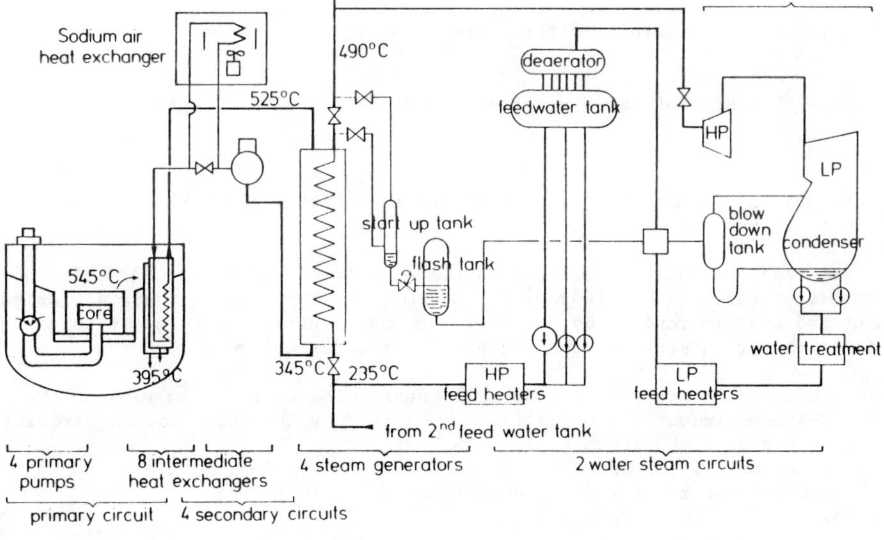

- scram the reactor and insert control rods
- open turbine by-passes
- reduce secondary pump speed to 20% of rated speed in 80 seconds (the steam generator water flowrate is automatically reduced proportionately by the control systems)
- in the initial phase, heat is discharged to the condensers through the turbine by-passes
- as the temperature of the hot sodium drops, the control system gradually reduces the water flow to 10% (34 kg/s per steam generator). When the steam temperature reaches 370°C, the start-up/shutdown circuits are automatically switched in. The steam leaving the steam generators is routed to the moisture separators. As soon as the steam is saturated, the level control systems start operating and the steam is discharged through the turbine by-passes and the water to the flash tank.

After isolating the steam leaving the separators, the shutdown temperature can be stabilized by adapting the water flowrate (a few percent of rated flow).

- separator functioning with level control (temperature between 280°C and 330°C
- or separator flooded (temperature between 180°C and 280°C).

This brief description shows that the water-steam circuits provide for quick cooling of the plant to cold shutdown (drop to 300°C in about 3 hours).

However, this heat removal system involves numerous steam-water circuits components. It is also not very easy, after a certain period to hold the temperature constant, since the water flowrates to be adjusted become too low.

One of the SUPER PHENIX design options was to simplify the water-steam circuits, which have no decay heat removal safety function. The power supplies for the corresponding auxiliary systems have consequently no emergency backing. Specifications for these circuits do not include seismic requirements.

Plant cooling safety criteria are thus satisfied by means of complementary NSSS systems.

3. DECAY HEAT REMOVAL THROUGH THE NSSS COOLING CIRCUITS (figure 3)

3.1. Conditions of use

Provision is made for decay heat removal through the NSSS cooling circuits in the following cases :

- After reactor cooling to a temperature of between 180°C and 250°C. If this condition has to be prolonged, the steam generators are deliberately dried out and heat is removed by means of the sodium-air exchangers. In this way, it is easier to keep the cooling operation under control.

- Following incidental (or accidental) conditions resulting in loss of the water-steam circuits, the cooling is achieved by the NSSS cooling circuits. The origin of such circumstances can be :
  - loss of off-site power
  - steam generator water flow cut-off
  - loss of the two condensers.

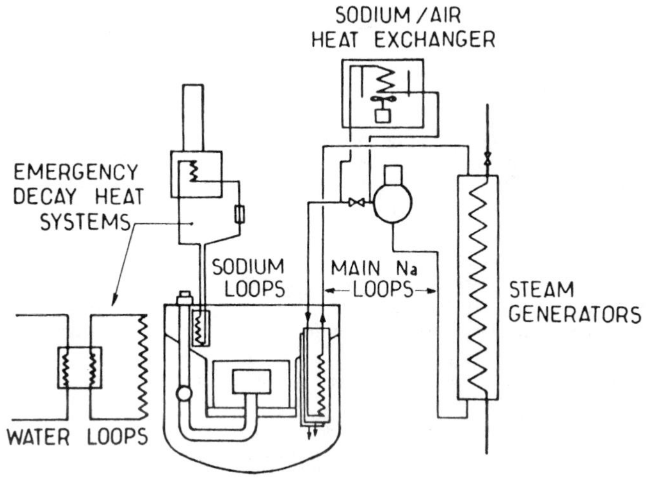

Figure 3 : NSSS Decay heat removal systems

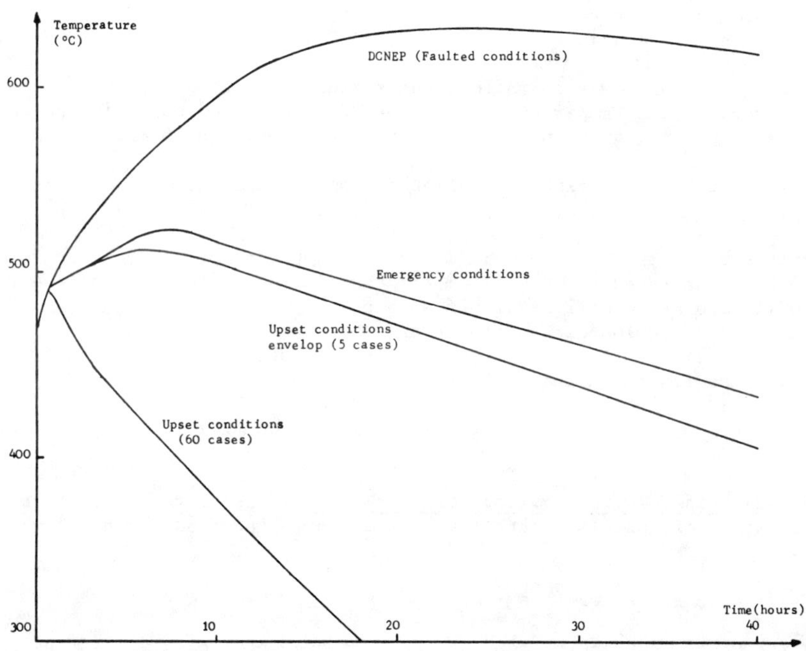

Fig.4 : Decay Heat Removal Transients with NSSS Systems

The plant design basis data comprises 65 such situations, occurring over 30 years, starting under full power conditions.

3.2. Description and performances

3.2.1. BPR sodium-air exchangers. There are 4 BPR sodium-air exchangers (1 on each secondary loop), each installed in parallel with butterfly valves located on the cold legs between the pump and IHX at the +30m level of the SG building (the core mid-plane is at the +10m level and that of the IHX at the +13 m level).

On the air-side, the exchanger is equipped with a motor-driven fan (no back-up power supply) and with a 30m stack. The air-side can be isolated by means of air shutter-panels, which also permit air-flow control.

Each exchanger is designed for the following conditions :

- forced convection exchange on both air and sodium sides :
  - power removed : 11,5 MW
  - sodium flowrate : 115 kg/s
  - sodium inlet temperature : 450°C
  - air inlet temperature : 32°C

- natural convection exchange on the air side :
  - power removed : 5 MW
  - sodium flowrate : 65 kg/s
  - sodium inlet temperature : 450°C
  - air inlet temperature : 32°C
  - maximum pressure drop : 980 Pa.

During reactor power operation, the sodium air exchangers are isolated on the air side. The temperature level (345°C under rated operating conditions) is maintained by means of a sodium flow of between 16 and 80 kg/s.

Operation of a BPR exchanger after drying out the steam generators requires :

- operating the pump from the main motor at a speed comprised between 20% and 50% of the rated speed or from the auxiliary motor at 20 % of the rated speed (in the event of loss of off-site power)
- opening the isolation air shutter panels
- opening the air shutter panels upstream and adjusting the air-flow
- starting the fan (when the off-site source is available).

All these operations are controlled from the control room by the operator.

3.2.2. Vessel cavity cooling circuits (RUS). These circuits consist of cooling water coil panels, installed outside the safety vessel in the vessel cavity.

Their purpose is two-fold :

- keep the vessel connecting courses cool and prevent excessive heating of the vessel cavity under normal operating conditions,
- remove the heat radiating from the vessel in the hypothetical event of loss of all secondary loops (DCNEP).

The circuits comprise 2 independent lines and remain in service under all operating conditions. The actuators on each line are duplicated and provided with an emergency power supply.

The heat is transmitted by radiation between main vessel and safety vessel and between safety vessel and RUS tubes.

The circuits are cooled by raw water circuits, which are themselves redundant, emergency power supplied and cooled by water from the Rhône.

Using very pessimistic postulated calculation data, we can assess the heat radiating from the main vessel as follows :

- 2.1 MW at 400°C and 7.3 MW at 670°C with 2 lines functioning
- 1.8 MW at 400°C and 6.2 MW at 670°C with 1 line functioning.

To this should be added the 1 MW removed by the upper closure cooling circuits.

2.3. Sodium emergency cooling circuits (RUR)

These circuits, together with the vessel cavity cooling circuits, are designed to remove the decay heat when the normal cooling systems are inoperable (loss of the 4 secondary loops).

Four independant sodium loops cool the primary sodium by means of sodium/sodium exchangers installed in the hot collector and remove the heat by means of sodium/air exchangers.

The sodium/sodium exchangers are removable components. The exchange tube bundle immersed in the primary sodium consists of straight tubes housed in a 400 mm ∅ guard shell.

Primary sodium cooling operates on a thermo-syphon principle ; the sodium enters the exchanger through holes in the guard shell and is discharged vertically at the bottom of the component.

The sodium/air exchangers are similar in design to those equipping the secondary loops (finned U-tube bundle placed between 2 collectors installed in a heat insulated casing topped by a 24 m stack and isolable both up and down-stream by means of shutter panels). Heat is removed by natural convection alone on the air-side.

During power operation, the exchangers are isolated on the air side. The sodium is permanently circulated by the electromagnetic pump. The loop is kept at the temperature of the hot plenum (545°C).

The operating procedure for this system is similar to that described for the BPR sodium/air exchangers. The sodium circulates by forced convection ; the electromagnetic pumps have emergency power supplies (a different diesel generator for each pump). However, the circuits are so designed as to favour efficient natural convection operation.

The system is put into operation from the control room (by opening the shutter panels). Each RUR loop is sized for the following conditions :

- for a primary sodium temperature of 650°C, each loop has to remove 5.7 MW with a flowrate of 25 kg/s (forced convection) and 5 MW with a flowrate of 12.4 kg/s (natural convection). Under these conditions, the sodium inlet temperatures for the sodium/air exchangers are respectively 513°C and 553°C.

Tests are presently being performed in Hengelo (Netherlands) to check the sodium/sodium exchanger efficiency. First results show good agreement with expected figures.

4. SITUATIONS TAKEN INTO ACCOUNT AND DESIGN BASIS CRITERIA (figure 4)

4.1. Upset and Emergency conditions use of BPR + RUR + RUS systems

Normal cooling conditions after reactor shutdown imply the use of the systems described in Chapter 2.

Decay heat removal through the NSSS cooling circuits, under normal conditions, is restricted to maintaining cold shutdown. No design basis requirements are specified for this situation.

We take into account 65 cases of unavailability of the water-steam circuits (Upset conditions), where cooling is consequently assured by the systems described in Chapter 3.

For 60 out of these 65 cases, all NSSS cooling circuits are available. In the case of an off-site power breakdown, off-site sources are presumed to be available after a short period (less than 1 hour).

Under these conditions, decay heat is removed by the vessel cavity cooling circuits and the BPR and RUR sodium-air exchangers. The sodium in the primary and secondary circuits and in the emergency loops circulates by forced convection. Further, we assume that the sodium-air exchangers only start operating after half-en-hour, which is pessimistic.

Temperatures do not exceed 490°C and revert to 300°C within 18 hours.

The following Upset condition envelope situation is taken into account 5 times :
- 2 RUS loops operating
- 3 BPR exchangers operating with natural convection air circulation after half-an-hour
- 3 RUR exchangers operating after half-an-hour.

This case could correspond to the prolonged loss of all off-site power sources and the failure of a diesel generator to start operating thus preventing forced convection circulation in a secondary loop and a RUR loop.

Under these conditions, the maximum temperature on the structures does not exceed 515°C, reached after 5 hours.

For these situations, the design basis criteria applicable are the ASME code Upset criteria.

Emergency situations. Plant design data is based on verifications of the ASME code Emergency criteria, for the following envelope situation :

1 RUS loop operating
2 BPR exchangers operating with natural convection air circulation
2 RUR exchangers operating with forced convection sodium circulation
2 RUR exchangers operating with natural convection sodium circulation.

It corresponds to the prolonged loss of off-site power and the total loss of one of the two electric channels.

The maximum temperatures reached, after 7 hours, are 521°C. If 1½ hours instead of ½ hour is assumed as the time required to get the 2 RUR loops functioning with natural convection circulation, the maximum temperature increases by only 5°C.

4.2. Hypothetical DCNEP situation (loss of the 4 secondary circuits - design basis data for the RUR circuits)

This hypothetical situation is taken into account in NSSS sizing in order to guarantee plant safety, even in the event of inoperability of all secondary circuit heat removal systems. This situation is the design basis postulate for the RUR circuits.

To guarantee primary containment integrity, Faulted type criteria are specified for the main vessel.

Furthermore, the safety authorities have demanded that loss of integrity of the main vessel during a thermal transient be a design basis postulate, implying that ASME code Emergency criteria be applied for safety vessel sizing.

Given the technological limits of the vessel construction materials and allowing a suitable safety margin to cover uncertainties, the specified design basis requirements for the emergency systems is as follows :

Applying the single failure criterion to all the decay heat removal systems (RUR + RUS) and assuming simultaneously the loss of off-site power, the mean primary sodium temperature shall not exceed 650°C.

On this basis, the decay heat to be removed by a RUR loop is 5.7 MW for a primary sodium temperature of 650°C.

Checking the validity of the above design basis postulates involved consideration of several situations :

First of all, satisfactory primary pump operation cannot be guaranteed at such high temperatures nor disregarded. We are thus led to consider two hypotheses : forced convection (even temperature distribution) and natural convection in the primary circuit.

The latter problem will be further discussed in paper II.9 presented at this congress (ref. 1).

Application of the single failure criterion revealed that total loss of a RUR loop was the most penalizing situation. In particular, this case is more severe than the loss of a power supply channel, which implies operating 4 RUR circuits, two of which work on a thermo-siphon principle.

It is clear from the theoretical and experimental studies made on natural convection in the primary circuit (the results of which are presented in paper II.9) that forced convection is far more penalizing for the vessels. This is because temperature homogenization results in maximum temperatures (around 630°C) at the bottom of the vessel, where mechanical stresses are also the most severe.

As regards fuel cladding behaviour, the maximum cladding temperature calculated, making allowances for uncertainties, is about 800°C in the case of natural convection primary sodium circulation.

4.3. Consideration of natural convection as a means of decay heat removal in SUPER PHENIX

4.3.1. Natural convection in the primary circuit. We have seen that this case is taken into account in design basis data for a situation involving loss of all secondary loops. It is the subject of theoretical and experimental studies discribed in paper II.9.

4.3.2. Natural convection in the emergency cooling circuits (RUR). The natural convection operation of the RUR circuits is a design basis requirement. Investigations are presently proceeding to define accurately the thermal transients associated with the starting of the thermo-siphon.

4.3.3. Natural convection in the secondary circuits. Natural convection in the secondary circuits is not taken into account in plant sizing.

However, design trends were such as to favor natural convection circulation (sodium-air exchanger located on upper level piping).

Simulation studies are presently being performed. The first results show that natural convection gets under way satisfactorily. However, the size of the secondary piping (up to 1 m in diameter) is such that calculation postulates have to be validated, in particular by experimental work, presently underway, to study stratification phenomena liable to be encountered under these conditions.

CONCLUSIONS

We have attempted in the present paper to describe the solutions adopted in SUPER PHENIX for decay heat removal with particular emphasis on the NSSS cooling circuits.

To conclude, we feel that two aspects should be stressed : the redundancy and operation simplicity of these systems.

Decay heat is removed by the 4 secondary loops, the 4 RUR loops and the vessel cavity cooling circuits. On-site power is provided by 4 diesel generators. The power supplies to the pumps on the different circuits can thus be separated. Furthermore, primary sodium circulation is assured by natural convection.

Given the inherent thermal inertia of a pool type reactor, the sodium-air exchangers are not required to function immediately. Even after several hours without cooling, the specified limit temperatures are not reached.

These characteristics, together with the extremely simple operation procedures devised for these systems (opening of shutters on the air-side) provide the plant with a remarkably high intrinsic safety level.

REFERENCES

1 Theoritical and experimental analysis of SUPER PHENIX 1 thermohydraulic problems in natural convection/

Astegiano, Martin, Gesi

Specialists'Meeting on Decay Heat Removal and Natural Convection in FBRS Brookhaven February 28-29, 1980.

Theoretical and Experimental Analysis of SUPER PHENIX 1 Thermohydraulique Problems in Natural Convection

J. C. ASTEGIANO and R. MARTIN
Commissariat à l'Energie Atomique
C.E.N. Grenoble
38041 Grenoble Cédex (Isére), France

E. GESI
Commissariat à l'Energie Atomique
C.E.N. Cadarache
13115 St. Paul-Lez-Durance, France

ABSTRACT

This paper covers the problems involved in simulating the SUPER PHENIX 1 primary sodium system in natural convection conditions. We discuss the experimental facilities implemented to assess the thermohydraulique phenomena which determine natural convection in the primary system, and describe the analytical model developed on the basis of the test results.

1. INTRODUCTION

In the safety analysis of LMFBR, hypothetic accidental configurations appear where decay heat removal is needed with compatible temperature conditions for structures.

For Creys-Malville, 1200 MWe nuclear plant, with a pool type breeder reactor, the hardest accidental configuration taken into account for decay heat removal is the following :

- loss of any pumping power ; primary pumps speed decreases following their inertia at least during the transient beginning,
- emergency shutdown is effective
- loss of any heat removal by intermediate heat exchangers
- decay heat is removed only with the emergency heat removal system, that is to say :
  - . radiation through the main and safety vessels to the external emergency water loops,
  - . convection towards emergency heat exchangers plunging into the hot plenum.

Bases on an analysis of the thermal consequences of the reference accident, an analytical model was developed to evaluate the major hydraulic and thermal phenomena characterizing the SUPER PHENIX 1 primary circuit.

The object of this paper is to present :

- the problems involved in simulation of SPX 1 primary circuit in natural convection,
- the experimental facilities for testing this natural convection in transient,
- the analytical model developed on the basis of the test results.

2. PREDICTION DIFFICULTIES OF THE PRIMARY CIRCUIT BEHAVIOR IN NATURAL CONVECTION

French design fast breeder reactors are of the pool-type, i.e. the core, the primary pumps and heat exchangers are contained within the primary vessel. An example of this concept is shown schematically in figure 1.

The design of pool-type reactor shows a number of significant advantages resulting from its considerable heat capacity and from the main vessel design free of external connections.

Conversely, the large volume of the primary vessel and the thermal segregation of various regions raises problems with analytical prediction of natural convection in the primary heat transport system. Main difficulties of prediction are :

- natural coolant circulation can result in the accumulation of large cold masses of static sodium in confined regions ;
- the low sodium flow rate at the core and heat exchanger outlets and the large hot and cold plenum volumes may very likely create hot plumes resulting in thermal stratification ;
- by thermosiphon action, the heat exchangers immersed in the hot collector produce density currents which significantly disturb the natural primary flow and temperature distribution in the hot plenum ; it is difficult to set up thermohydraulic models of these recirculation streams.

Since an adequate representation of the temperature gradients within the flowstream requires that the above phenomena be correctly simulated, an experimental small scale test program has been undertaken to fit the analytical model with respect to the physical reality.

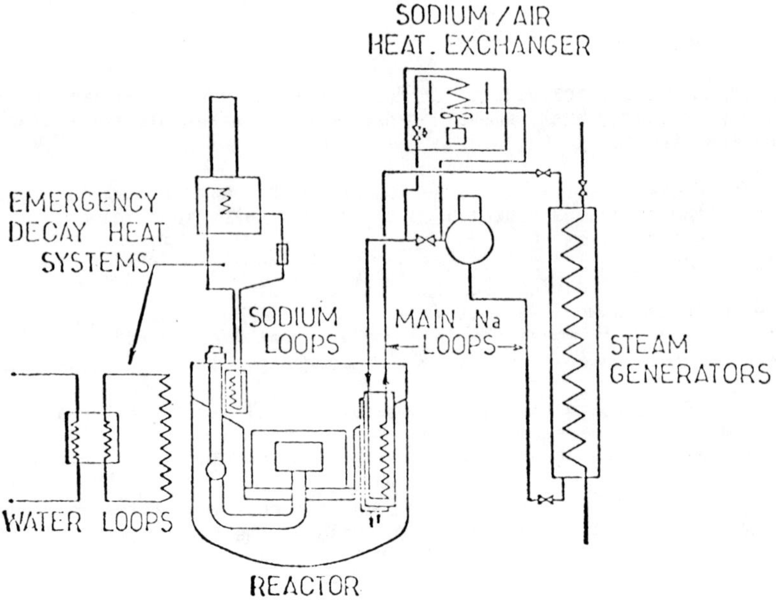

FIG: 1 SUPER PHENIX 1 - POOL-TYPE REACTOR

3. EXPERIMENTAL FACILITIES (figure 3a)

Testing was conducted on a 1:20 scale model in geometrical similarity to the reactor using water as the experiment fluide. The choise of scale was determined by the need for representing some geometrical details and for obtaining measurable temperature differences.

The scale model was designed on the basis of a close dimensional analysis in which distorsion affects the Prandtl number, while the Péclet and Boussinesq numbers are maintained.

The natural convection which occurs in the primary system is simulated by forced circulation. The following conditions are also simulated in the model :
- core heat decay
- power removed via emergency decay heat system-sodium loops
- radiation thermal losses through the primary vessel wall.

These conditions are based on calculations after transposition taking into account the similitude criteria. Under these conditions, the scale model simulation provides information primarily on the following points :
- flow characteristics
- fluid stratification
- reactor division into characteristic zones

used to adjust the analytical model (refer to figure 2).

3.1 Water test results

Numerous tests were conducted to assess the influence of various parameters.

The experimental results indicated in figure 3b illustrate the overall test results. This test shows that :
- efficient power dissipation occurs event though the integrated heat exchangers operate under thermosiphon conditions ;
- the primary system is divided by a thermal stratification into two zones : the boundary between them lies at the pump inlet level. Flow velocity estimates in the boundary region show that the lower portion of the experimental model does not interact with the primary coolant. In that region the fluid is stably stratified and heat is transferred by conduction only.

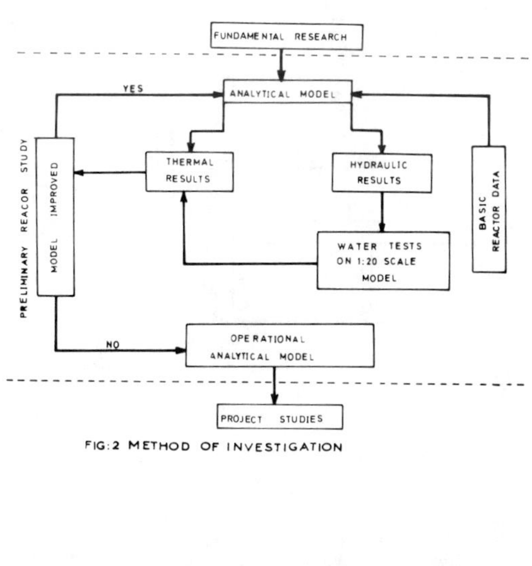

FIG:2 METHOD OF INVESTIGATION

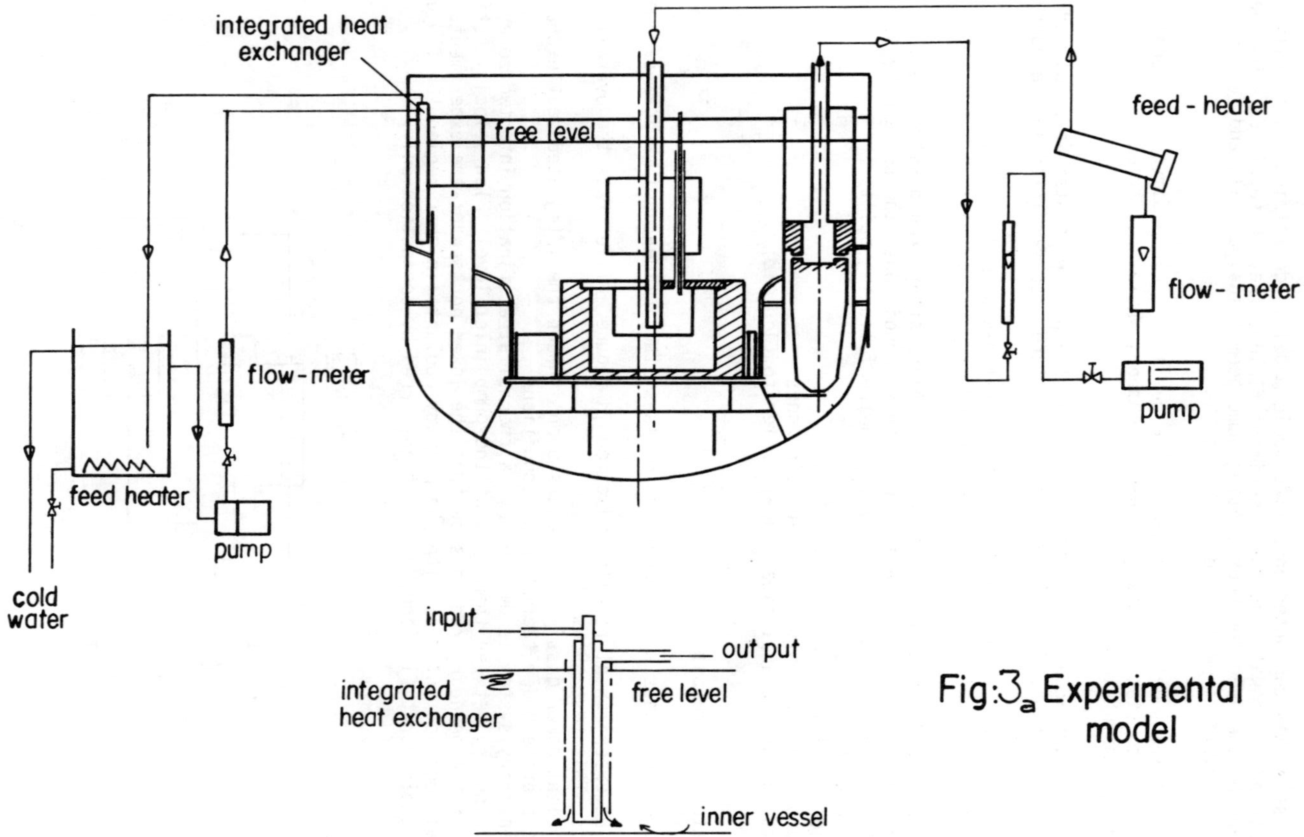

Fig:3a Experimental model

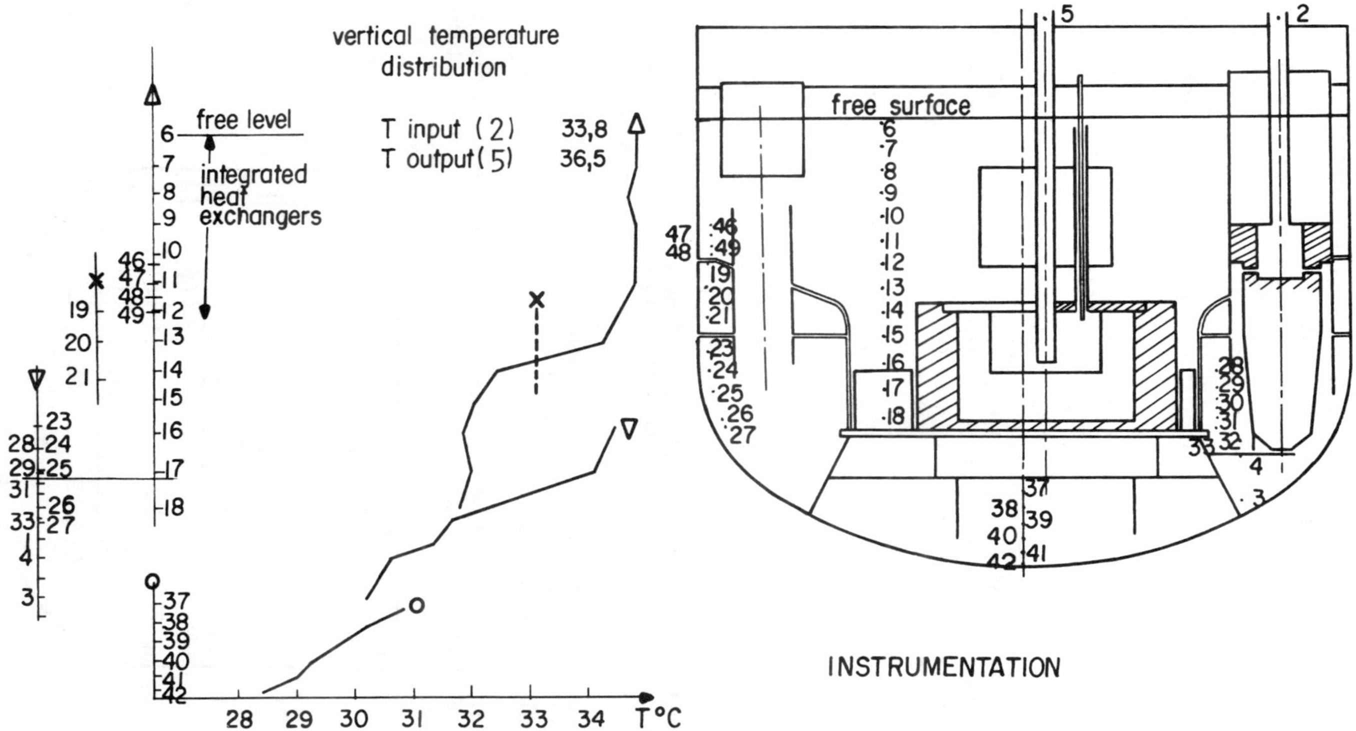

Fig: 3b WATER TEST RESULTS

The temperature distribution reveals four qualitative regions :

- a layer with a very slight decreasing temperature gradient from the free surface to the integrated heat exchangers outlet this "hot layer" can practically be considered as isothermal. The difference between its temperature and the core outlet temperature accounts for about 70 % of the core ΔT, indication that 70 % of the power is removed before reaching the free surface. This heat is transferred to the recirculating flow from the integrated heat exchanger.
- a zone with a temperature gradient extending from the bottom of the integrated heat exchangers to the core exit. This is the region in wich power dissipation occurs in the core out-flow via strong recirculating flow induced by the integrated heat exchangers.
- a zone included between the inner vessel and the lateral neutronic protection with a slight decreasing temperature gradient.
- a contained "inter-redan" zone in which the temperature is uniformly higher than that of the step upper face which is washed by the cold-flow from the integrated heat exchangers. There is heat transfer cold collector ↠"inter-redan"↠ lower part of the hot collector.

Visual representations show that inside the inner vessel the core outlet flow creates a plume effect. However, the cooling flow at the exit of the integrated heat exchangers induces a mixing region above the core exit and, thus, a reduced plume effect.

The following fig. 4 represents the fluid circulation flow in the main vessel.

From the outlet of the integrated heat exchangers, the cold fluid flows down along the conical redan filling the upper volume between the inner vessel and the lateral neutronic protection and mixing with the core outflow.

Two phenomena predominate in the cold plenum :

- a plume effect at the intermediate heat exchanger outlet,
- the horizontal displacement of a fluid layer at the pump inlet level.

These studies are currently being continued by a fundamental sodium test program designed to assess the influence of the Prandtl number on the temperature distribution.

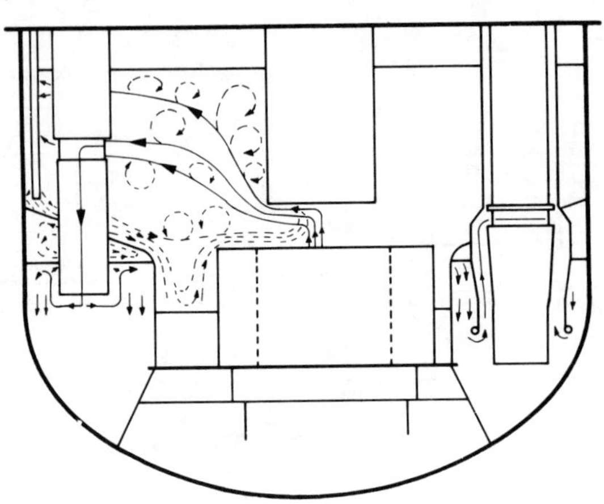

FIG. 4 FLUID CIRCULATION FLOW IN THE MAIN VESSEL

4. ANALITICAL MODEL SUPER PHENIX 1 NATURAL CONVECTION

This computer code calculates the transient response of the primary and emergency cooling systems in the event of the accident described in section 1.

Since natural convection predominates from a thermal viewpoint, the reactor was divided into a fixed number of regions (figure 5) bases on the test results presented in section 3.

Each region may itself be further subdivided into a variable number of layers for analysis of thermal stratification phenomena (refer to figure 6).

From a hydraulic viewpoint, however, an overall "conservation of momentum" equation was defined for the various hydraulic circuits (figure 7).

Allowance is made for the pumping force resulting from natural convection by calculating the integral of the buoyancy forces along each hydraulic loop.

Plume effects in the hot and cold collectors are taken into account along with the thermosiphon flow set up by the emergency system heat exchangers.

The entire core is considered hydraulically as a single assembly, and thermally as a single point

Initially it was considered unnecessary to divide the core into subchannels. This hypothesis was subsequently verified analytically using nine core flow channels.

The existing model thus does not allow for possible spurious flow phenomena :
- a possible reverse flow circulation in the subassemblies with little or no power output (e.g. fertile or reflector subassemblies) ; this flow circuit returns via the grid plate plenum.
- circulation of a downward flow among the lateral neutron protection, then between the assembly bases, and using in the midst of the fuel assemblies.

However, it should be pointed out that in its present state the experimental model is incapable of providing data on these spurious flow paths, and that - since this flow is beneficial to the core - failure to take it into account results in pessimistic evaluations.

The thermal model kept for emergency decay heat removal sodium loops is founded on the principe that such transients can be simulated as sequences of steady state conditions. To this end, the time delay and the thermal capacity of the emergency sodium loops have been neglected.

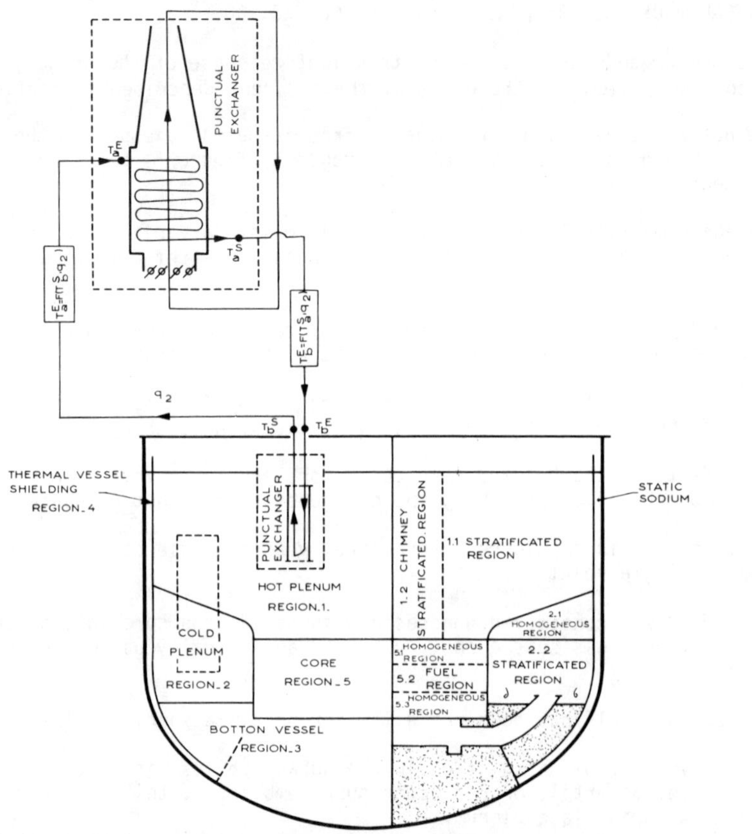

FIG: 5 THERMAL MODEL

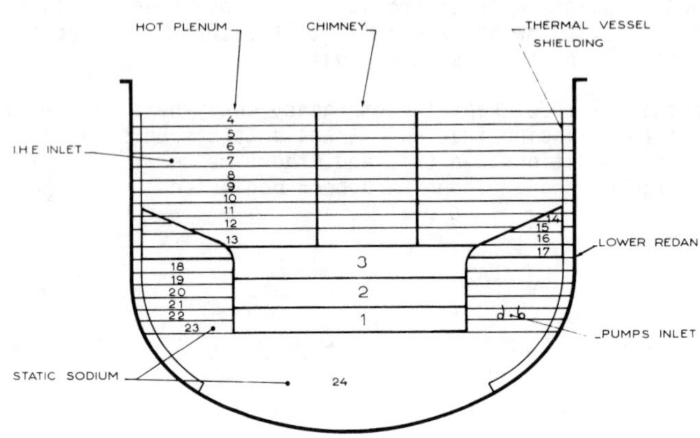

FIG:6 MODEL FOR THERMAL STRATIFICATION ANALYSIS

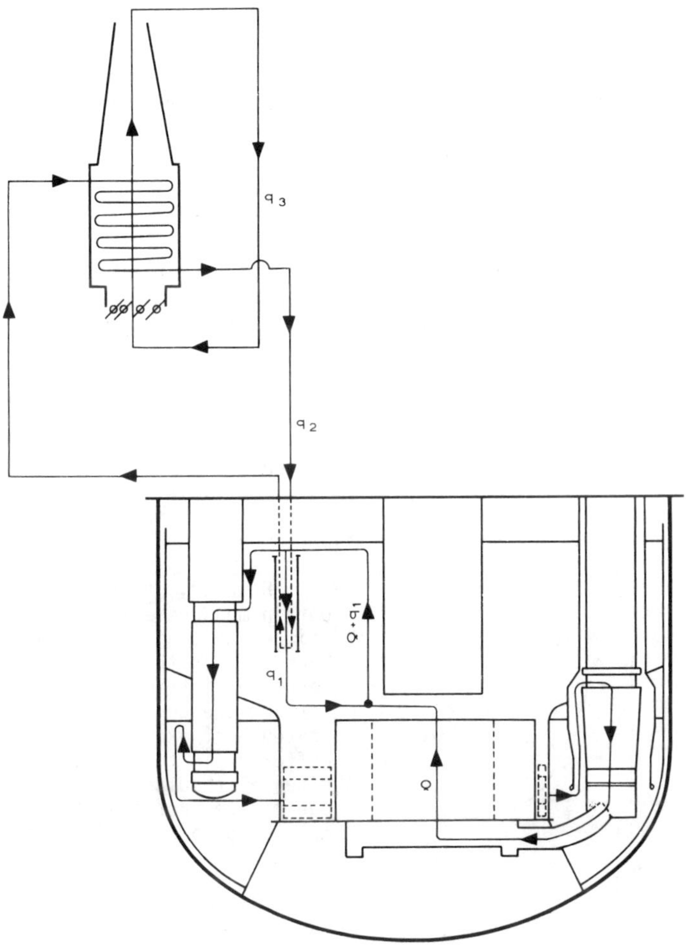

FIG: 7 HYDRAULIC _ MODEL

4.1 Calculated results

Figure 8 shows the uncorrected results at the temperature peak as given by the analytical code. Figure 9 plots the transient behaviour of some characteristic parameters of the primary circuit.

The figure 10 shows the sodium temperature map in the main vessel.

It may be seen that the primary flow-plume effect at the core outlet is strongly attenuated by the presence of the integrated heat exchangers, and that their location in the hot collector is optimal with regard to the temperature at the upper part of the vessel.

5. COMPARISON OF ANALYTICAL AND EXPERIMENTAL RESULTS

The figure below compares the calculated and experimental values for the test with 3 integrated heat exchangers operating. The temperature were transposed according to similitude criteria without Prandtl correction.

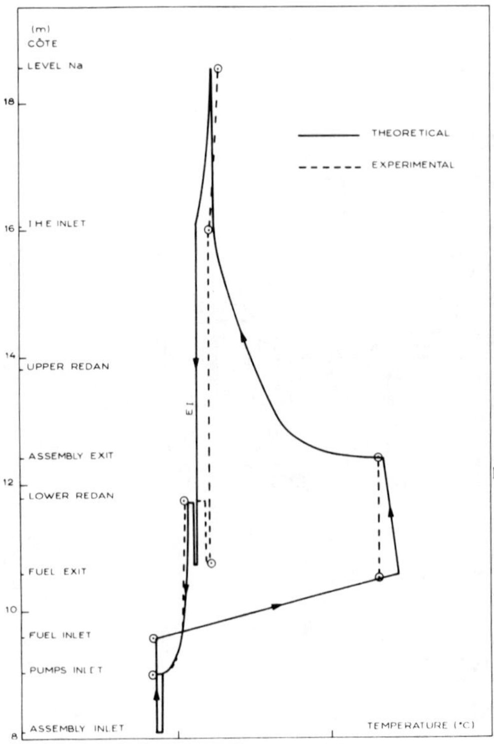

THEORETICAL AND EXPERIMENTAL RESULTS IN PRIMARY HYDRAULIC LOOP

The temperature discrepancy at the intermediate heat exchanger inlet results from the higher heat transfer between the hot and cold collectors in the scale model than in the calculated analysis. When this correction is computed, the results are in good agreement.

However, it is not possible to obtain measurement points in the chimney to confirm the calculation. Moreover, in the hot plenum the temperature distribution in the zone between the core outlet and the integrated heat exchanger's outlet, is different in the calculated and experimental results (figure 11) ; this is due to the fact that this zone is simulated by a zone of uniform temperature.

The temperature discrepancy at the bottom of the vessel (figure 11) can be explained by minor differences in modelling, and have no practical consequences.

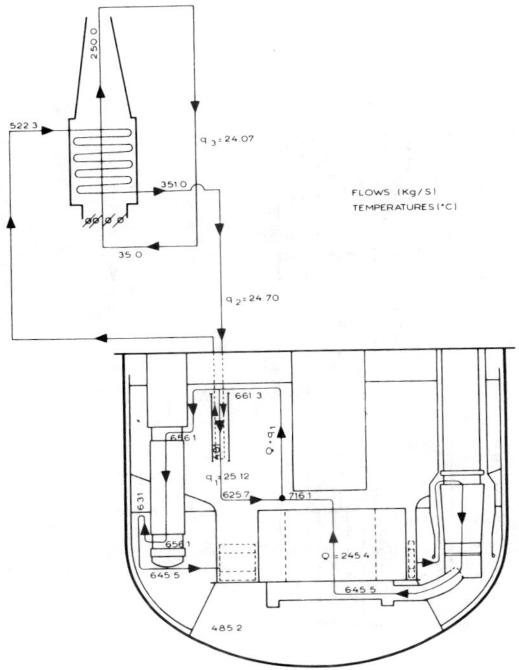

FIG:8 THERMOHYDRAULIC STATE OF PRIMARY AND EMERGENCY LOOP

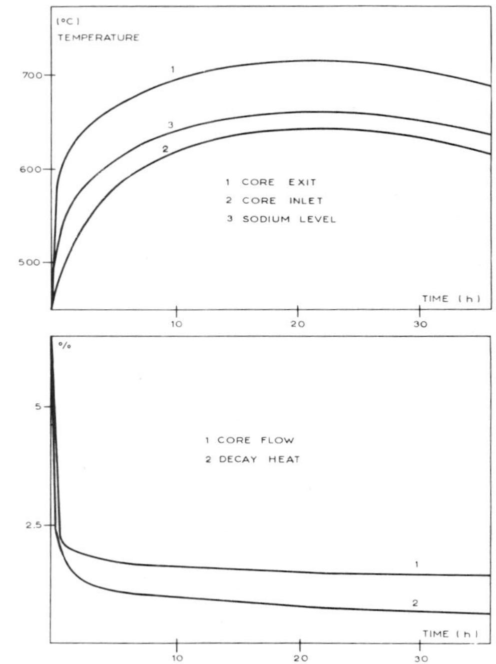

FIG:9 TEMPERATURES AND FLOW TRANSIENT

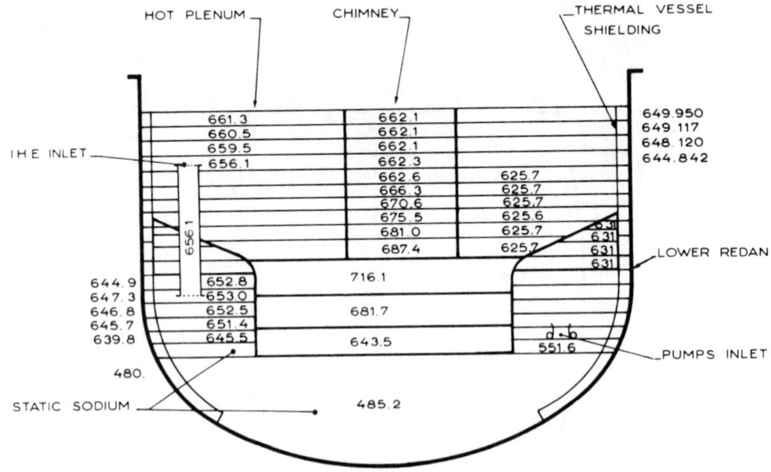

FIG:10 SODIUM TEMPERATURE MAP IN THE MAIN VESSEL

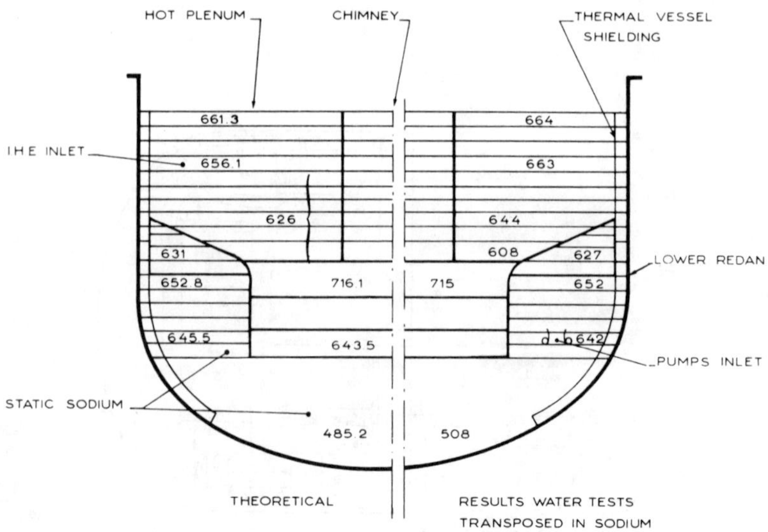

FIG:11 THEORETICAL AND EXPERIMENTAL RESULTS

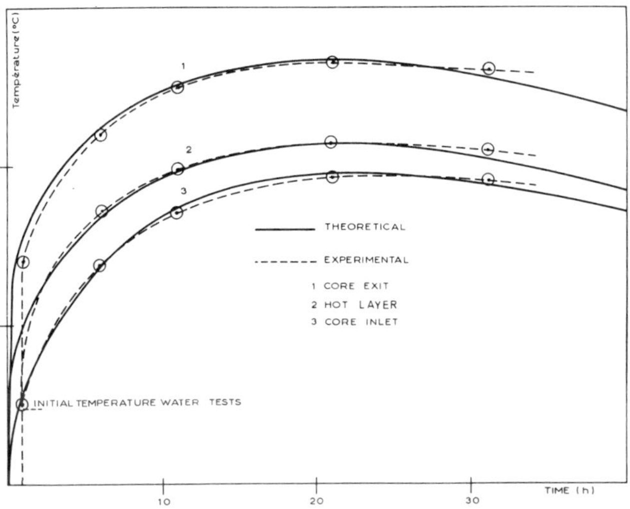

FIG : 12 ANALYTICAL AND EXPERIMENTAL RESULTS OF THE MAIN TEMPERATURES TRANSIENT

The above figure compares the analytical and experimental results for core inlet, core outlet and hot layer temperatures during the transient.

The temperature deviation observed after the peak results from the fact that the boundary conditions are held constant on the scale model after peaking.

Analytical model gives results generally in good agreement with the experimental findings, especially for the most characteristic temperatures, i.e. :
- upper part of the vessel
- internal structure bearing point
- core outlet

The analytical model may be further improved by better approximating the thermal behavior at the bottom of the vessel and the region between the integrated heat exchanger inlets and the core outlet.

6. CONCLUSION

The code model with stratification and plume effects is on the whole satisfactory for calculating the accident conditions described in section 1.

The experimental validation is at present based entirely on water testing, but will be continued using sodium test results as they become available.

The code version discussed in this paper was used to check the thermal dimensioning of the decay power dissipation system for the SUPER PHENIX 1 reactor (refer to paper N° II.8 discussed at this conference).

Observations on Coolant Flow Patterns in the PFR Primary Circuit during Natural Circulation Experiments

R. ANDERSON, C. W. DAWSON, C. V. GREGORY, D. J. LORD, and R. WEBSTER
Fast Reactor Technology Group
Dounreay Nuclear Power Development Establishment
Dounreay, Thurso, Caithness, Scotland

ABSTRACT

A brief review is given of experiments performed on the Prototype Fast Reactor (PFR) at Dounreay to study the transition from forced to free (natural) convective flow and the stability of the reactor during longer term operation with natural circulation cooling. A number of outstanding problems from the tests are discussed and a common explanation sought in terms of a postulated flow field; such an explanation must at this stage be speculative because of the reliance upon instrumentation designed and located for operational rather than experimental purposes.

1. INTRODUCTION

A loss of power to the PFR primary pumps causes the reactor to trip; adequate coolant flow is ensured by continuously running auxiliary motors, driven from batteries, which automatically engage the pump drive shafts as the primary motors run down. Although these motors are highly reliable the consequences of a failure of all three to engage are being studied at Dounreay. A brief review of the experimental programme is given in the first part of the paper.

In experiments with complex operational plant there are always features which, although they are not expected to have a major influence upon the main conclusions, offer potential benefits in increased understanding if they can be resolved. The major part of this paper is concerned with the description and discussion of four such factors. The factors, in order of appearance, are

(i) The sub-assembly coolant outlet temperature spike following a pump restart

(ii) The sub-assembly outlet temperature radial profile

(iii) Coolant stratification

(iv) The effect of heat removal capacity

In each instance an explanation is discussed in terms of the flow and temperature fields believed to exist in the primary circuit.

TABLE 1

SUMMARY OF PFR NATURAL CIRCULATION TESTS

Test	Date	Description
1	1.12.74	DHR loop operation with zero core power
2	6.12.74	1.5 MW core power no control rod movement
3	22.12.75	3.0 MW core power no control rod movement
4	16.11.77	Test 3 repeat with additional instrumentation
5	28. 1.78	4.5 MW test
6	28. 6.78	4.5 MW test from low pump speed
7	25. 8.78	4.5 MW test from full pump speed
8	20. 2.79	5 MW constant power test
9	5. 3.79	10 MW constant power test
10	21. 3.79	15 MW constant power test
11	8. 4.79	5.0 MW constant power test initially with no DHR loop. Final power 18.5 MW
12	27. 8.79	50 MW trip test 1
13	5. 9.79	50 MW trip test 2

2. REVIEW OF NATURAL CIRCULATION EXPERIMENTS AT DOUNREAY

Descriptions of natural circulation experiments carried out on PFR are given in References [1] and [2]; a description of some of the complementary theoretical work is provided in a companion paper to this meeting, Reference [3]. Table 1 lists the major tests carried out to date; a brief history of the programme is given there.

Early calculations (1973) using design data suggested that the onset of natural circulation might be delayed beyond the complete cessation of pumped flow and as a result high coolant temperatures (exceeding 800 deg C) might be attained. These calculations contained a number of simplifications and assumptions which could only be tested by comparison with the reactor itself. Accordingly in 1973 an exploratory series of PFR decay heat removal tests were proposed and carried out during the first low temperature commissioning run; further tests involving the simulation of decay heat by fission power were begun in 1974 but limited to initial power levels less than 1.5MW. The coolant temperatures measured above the core centre behaved much as expected during each test but those above the outer regions of the core were far lower than expected and could be interpreted as indicating coolant stagnation, or even flow reversal.

Alternative explanations of the experimental results which did not depend upon coolant stagnation were proposed; in general these explanations were based upon the existence of a complex flow field in the above core plenum which, in effect perturbed the outlet thermocouple signals under low flow conditions. Between 1974 and 1978 seven reactor experiments were carried out with the objective of disproving the stagnation hypothesis and demonstrating the existence of natural circulation flows throughout the entire core. The use of prompt neutron power together with on line reactivity measurement showed that the mean core temperature throughout each of these tests was very much as expected, thereby conclusively demonstrating that no significant region of stagnant coolant existed in the core. Careful examination of the temperature of coolant 'flushed' from each sub-assembly at the end of each test by the pump restart confirmed this result; that is the temperature spike associated with the coolant flushed from the core was broadly consistent with the expected outlet temperature and there was no evidence of excessively high coolant temperatures. More convincingly in the later loss of electrics tests the magnitude of the spikes did not vary with the different durations of test, (if stagnation had occurred the magnitude of the spike would be expected to vary linearly with time).

In June 1978 a major test was performed in which, under steady natural circulation conditions; the reactor power was raised in a series of steps to an eventual 14 MW and steady conditions maintained for half an hour. A later test increased this level to 18.5 MW. These tests provided a clear demonstration that once established, natural circulation was capable of removing substantial heat from the core with acceptable outlet temperatures. It remained to be demonstrated that high temperatures would not arise during the setting up of natural circulation under realistic conditions; that is with a power level of 20 MW during the important transition from forced (pumped) to free (natural) convection compared to powers up to 2 MW used in the earlier tests and with a temperature distribution, in the Intermediate Heat Exchanger (IHX) and plenum, more representative of the conditions existing following a genuine loss of electrics trip.

To cover these points a new series of experiments was proposed at the end of 1978. The first half of the series was largely completed during February and March 1979. In these tests the reactor power was held constant at a predetermined level and the coolant pumps tripped with the auxiliary motors bypassed. Tests were performed with reactor powers of 5MW, 10 MW and 15 MW. The results of these experiments are discussed later in this paper and in Reference [3]. Suffice it here to say that they have convincingly demonstrated the smooth transition from forced to steady natural circulation flows as the pumps run down and, even with the small extrapolation required to the maximum possible decay heat level, they show that the corresponding core temperatures are comparable with those achieved during normal operation.

The tests in the second half of the series aim to reproduce all facets of a reactor trip and loss of cooling including the adverse temperature distributions in the plenum and Intermediate Heat Exchangers. Two tests have already been completed successfully, both from initial powers of 50 MW with corresponding decay heat levels of 1 MW after 200 s. The results are discussed and compared with theoretical predictions in Reference [3]; the important conclusion is that they appear consistent with those from the constant power tests. A test from a higher initial power planned for the current reactor run is expected to confirm the reliability of natural circulation as a means of cooling a fast reactor core under decay heat

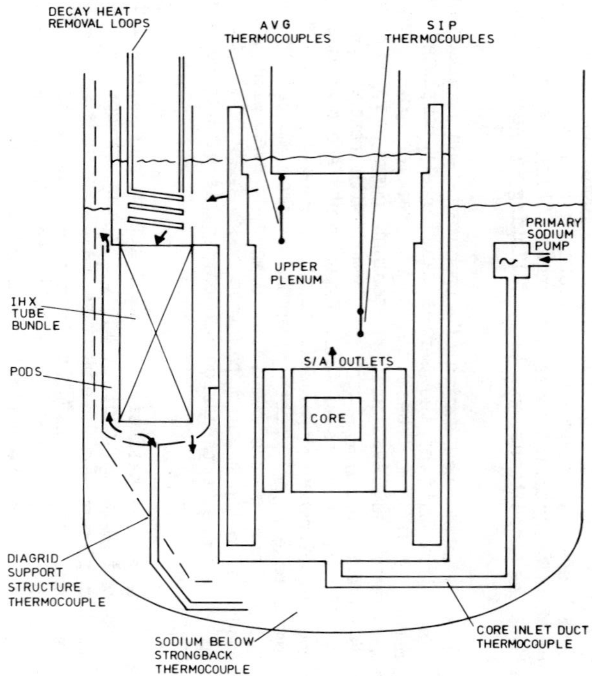

FIG. 1. PFR PRIMARY CIRCUIT.

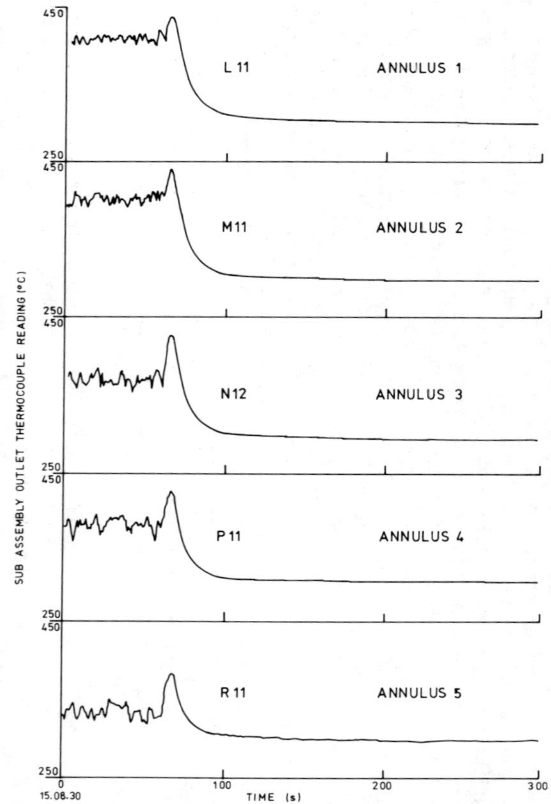

FIG. 2. TEMPERATURE TRANSIENT ON RESTARTING FORCED FLOW AFTER 15 MW TEST.

conditions.

If the results of all the tests carried out to date are taken into account it is possible to extrapolate to the full power trip conditions, initial power 600 MW reducing to 20 MW after 200s, and show that during such an incident even if the auxiliary motors fail to engage the core temperatures will be comparable to those experienced during normal operation.

3. THE PFR PRIMARY CIRCUIT

Before discussing some of the individual features of the experimental results a brief description of the PFR primary circuit is provided.

Figure 1 shows that like PHENIX but unlike EBRII, PFR is a double pool design with two separate free sodium surfaces. The primary pumps force sodium from the outer pool through the core inlet duct, through the core and breeder and into the upper plenum. This plenum is completely surrounded, over its entire height, by two rows of closely spaced neutron shield rods. The sodium flows through the narrow shield rod interspaces over the Decay Heat Rejection (DHR) loops and through the IHX, driven by the level difference between the inner and outer pools.

There are thermocouples at the sub-assembly outlets and several thermocouples, identified by the prefix AVG, sense sodium temperatures near the surface of the inner pool (Fig 1). Apart from a few tests in which a Special Instrument Probe (SIP) was used to monitor positions closer to the sub-assembly tops, these are the only temperature measurements available in the upper plenum. Thermocouples are also positioned between the DHR loops and the IHX inlet. In the outer pool there are thermocouples at the pump inlets, in the core inlet duct and at various other strategic points.

4. SUB-ASSEMBLY OUTLET TEMPERATURE SPIKE

In this section the first of the four factors listed in the Introduction is discussed. The experimental practice has been to conclude each test by restarting a primary pump; observation of the sub-assembly outlet thermocouples during the corresponding flow transient has in every case shown a marked temperature spike, Fig 2, indicating the existence during the tests of coolant within the sub-assemblies with a temperature higher than that measured at the sub-assembly outlet.

A number of properties of the spike can be described. For a given sub-assembly, it is not strongly power dependent; for example at the end of the 5 MW, 10 MW and 15 MW constant power tests the spikes observed above a central sub-assembly were 18°C, 25°C and 31°C respectively. The spike does not depend upon the length of time natural circulation conditions have existed; thus in the two 50 MW loss of electrics tests the spike was identical, for the same sub-assembly, even though the pump restart occurred after 360s in the first test and after 600s in the second test. Both these properties together with the small absolute value of the spike are convincing evidence that coolant stagnation, or even very low flow, is not occurring in the core.

The fact that the outlet thermocouples measure an equilibrium temperature during a test which is lower than the peak when the pumps are restarted seems to be the result of some factor or factors extraneous to the individual sub-assemblies. A possible explanation can be sought in the postulated existence, under low flow conditions, of a cold coolant flow from the periphery to the core centre across the top of the core. The coolant temperature measured at the outlet thermocouple positions, 25mm above the core, would be reduced compared to the genuine core outlet temperature either through mixing of the hot efflux with the cold cross flow or through diversion of the coolant issuing from the sub-assembly away from the thermocouple position or through a mixture of these processes. Restarting a pump re-asserts the dominance of the forced core flow over the cross flow - hence the spike.

5. RADIAL OUTLET TEMPERATURE PROFILE

Section 2 described how, from the earliest tests, the radial distribution of the sub-assembly outlet temperatures has shown a remarkably 'peaked' profile, leading at one time to the suggestion that coolant stagnation may be occurring in the outermost sub-assemblies where the measured outlet temperature only marginally exceeded the inlet temperature. Reactivity monitoring and the observed temperature spike behaviour have been used as evidence to dispose of the stagnation argument; in this section the profile is discussed against the expected behaviour under natural circulation conditions and it is shown how the apparent discrepancy might be resolved in terms of a postulated flow field in the plenum.

Figure 3 shows, for two of the constant power tests, how the average sub-assembly outlet temperature varies with distance from the core axis. A peaked distribution is seen with a smooth transition from the core centre across the boundaries between the inner zone, outer zone and radial breeder. This distribution was unexpected but has proved to be a feature of all the natural circulation tests.

Under pumped flow conditions the flow through a sub-assembly, to a first approximation, is independent of the power generated in the sub-

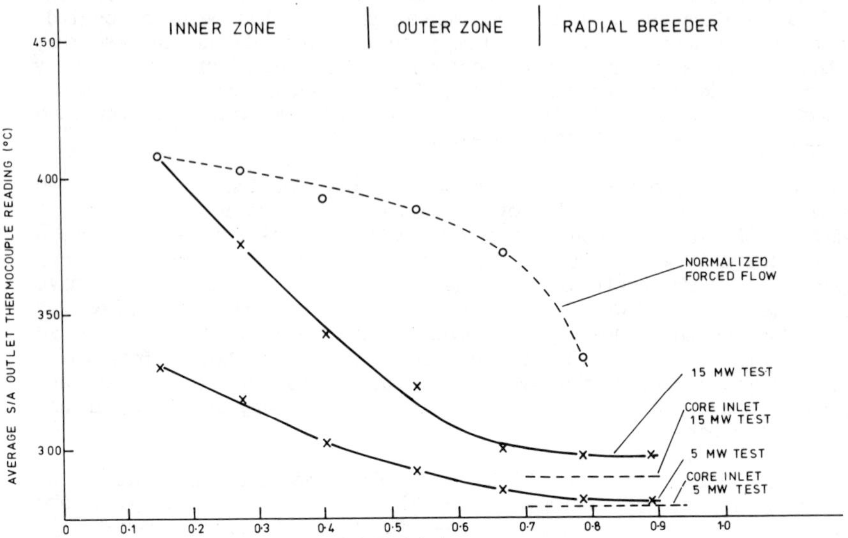

FIG. 3. RADIAL TEMPERATURE PROFILE FROM SUB ASSEMBLY OUTLET THERMOCOUPLES.

TABLE 2

SAMPLE THERMOCOUPLE READINGS FPR ADJACENT HIGH AND LOW POWER SUB-ASSEMBLIES IN ANNULUS 5 AND ANNULUS 3

Annulus	5 MW Test 8		15 MW Test 10		Relative Channel Power	
	High Power	Low Power	High Power	Low Power	High Power	Low Power
5	281°C	283°C	294°C	301°C	5.0	0.35
3	315°C	308°C	346°C	346°C	8.9	0.02

assembly and, consequently, the temperature rise through a given sub-assembly should be proportional to the sub-assembly power. Under natural circulation conditions the coolant flow depends upon the temperature rise through a sub-assembly and hence upon the sub-assembly power. Under such circumstances a flattening of the radial profile is to be expected because high power sub-assemblies induce a correspondingly higher coolant flow and vice versa.

The problem may be summarised thus,

(i) flow exists through the core

(ii) because the pumps have stopped the flow must be caused by natural circulation

(iii) natural circulation should result in a flattening of the outlet temperature profile

(iv) the converse is observed.

Radial conduction can be dismissed as a possible explanation; if it were the dominant mechanism the temperature distribution would be convex (the gradient increasing with radial position) rather than concave.

The explanation proposed for the temperature spike behaviour was that the sub-assembly outlet coolant flows were being influenced by a cross flow of cold plenum coolant. This hypothesis is now developed further to cover the radial temperature profile discrepancy. It has already been postulated that the response of the outlet thermocouple is influenced by the plenum sodium above it. This point is made strongly in Table 2 in which the outlet temperatures for adjacent high and low power sub-assemblies in the same annulus (radial position) are compared with a similar pair from a different annulus in two different tests. The conclusion is that the measured outlet temperature depends primarily upon radial position than upon details of the sub-assembly beneath the thermocouple. It is also significant that the magnitude of the pump restart spike (Fig 2) also varies with radial position, being small near the core axis and increasing with distance from the axis. If the measured radial profile on Figure 3 is corrected to allow for the pump restart spike then the profile approaches more closely the one obtained under forced flow conditions.

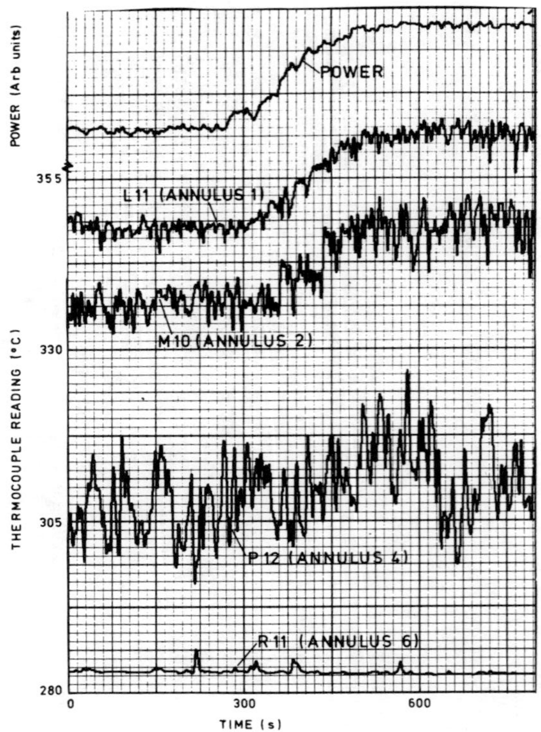

FIG. 4. SUB-ASSEMBLY OUTLET TEMPERATURE RESPONSES TO AN INCREASE IN POWER.

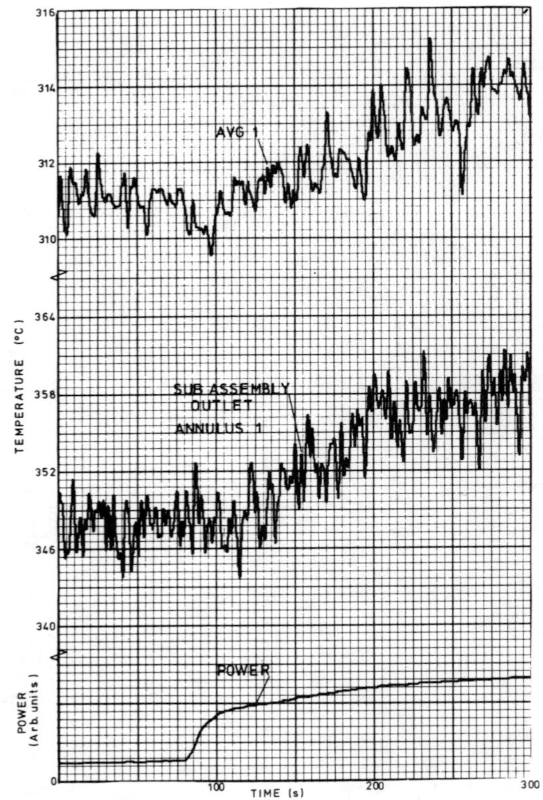

FIG. 5. TEMPERATURE RESPONSES IN THE UPPER PLENUM FOLLOWING AN INCREASE IN POWER.

To explain the variation with radial position requires that the degree of interaction between hot sub-assembly coolant and the cold plenum cross flow also varies with position, being high at the edge of the core and decreasing towards the core axis. Such a picture is supported by the temperature noise and response to power changes observed on the outlet thermocouples. Figure 4 illustrates the change in thermal noise with radial position; in the central regions it is low and is combined with a precise response to power changes; further from the core centre the noise is high (tens of degrees) but the response to changes has become "blurred"; finally at the core edge the thermal noise is very small but there is no direct response to any power change. This might be consistent with very high mixing at the core edge leading to complete domination of the thermocouple measurements by the plenum coolant, with strong mixing further inboard where the influence of the plenum coolant is strong but not dominant, and with relatively weak mixing near the core axis (at least at the outlet thermocouple position), allowing the sub-assembly coolant to dominate the thermocouple response.

The total core sodium flow rate during natural circulation has been estimated to be a few percent of the full flow, of the order of 50 kg/s. Assuming that the transit times scale directly with flow rate it should take approximately 200s for coolant to pass from the sub-assembly tops to the plenum surface. A measure of this transit time can be obtained from the results of Test 11 during which relatively rapid changes in power were made from 5 MW leading to corresponding responses on the sub-assembly outlet and AVG thermocouples. The time lapse between these responses is a fair indication of the sodium transit time through the upper plenum. Figure 5 shows the delay to be of the order of 50s, implying that the effective upper plenum flow area is smaller than the physical area and more importantly less than the equivalent area under forced flow conditions; this is consistent with a plume of hot sodium, fed by most, if not all, of the sub-assemblies, rising vertically along the core axis above the central sub-assemblies. In turn this must imply some form of cross flow across the top of the core, from edge to centre.

This and the preceding section have shown that the temperature spike behaviour and the outlet temperature radial distribution can both be explained by a common hypothesis, namely that there exists a strong coolant flow across the top of the core which either involves or interacts with cold coolant in the plenum. Because PFR is not equipped for detailed experimental measurements of temperature and flow patterns this hypothesis remains speculative.

6. COOLANT STRATIFICATION

Before dealing with the last of the four factors listed in the Introduction, namely the effect of heat removal capacity, which will allow further development of the hypothesis, it is first necessary to elicit one important fact from the observations made of coolant stratification. Therefore at the risk of losing continuity the topic of stratification is dealt with in this section; stratification in the inner pool (plenum) and outer pool are covered in separate subsections.

6.1 Inner Pool Upper Plenum

There are thermocouples at three different vertical positions on the AVG probe down to about 1m below the surface of the inner pool and these indicate a decreasing temperature with depth. This hot stratum is very stable and, following the start of a main primary pump to produce about 8%

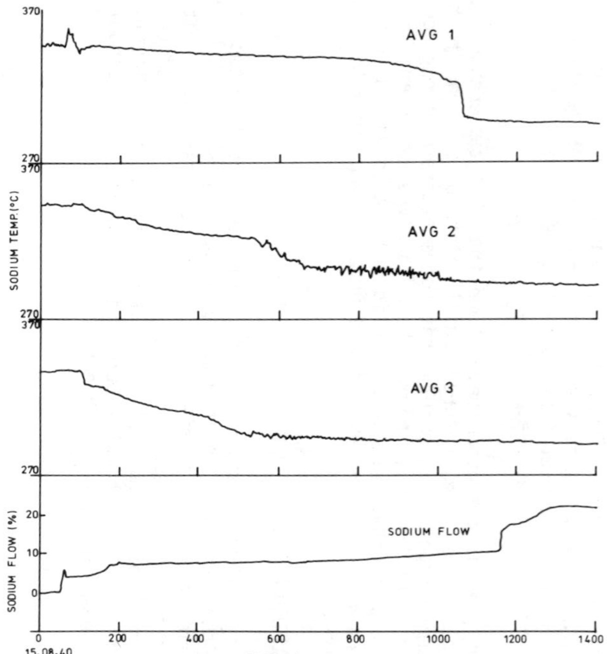

FIG. 6. UPPER PLENUM TEMPERATURE RESPONSES ON RESTARTING FORCED FLOW AFTER 15 MW TEST.

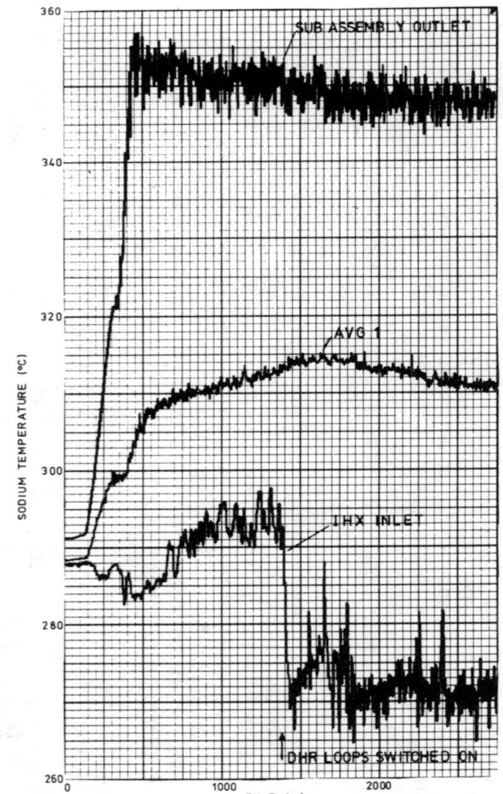

FIG. 7. INNER-POOL RESPONSES FOLLOWING THE LOSS OF ALL FORCED COOLING AND THE SUBSEQUENT START OF DHR LOOPS.

flow it persists much longer than the time required to displace the sodium inventory of the upper plenum. This is illustrated in Figure 6 where temperatures at the AVG are traced as functions of time. On starting a pump, core sodium flows rise to about 8% of full flow in a few seconds. The cold front should then reach the probe in about 30s. The lowest thermocouple showed a change on a timescale of about 30s but it is clearly seen that the cool forced flow did not reach the top thermocouple AVG 1 until several hundred seconds after forced flow was restored. The cold pumped sodium must therefore have been flowing through the neutron shield rods at a lower level and passing directly to the IHX without disturbing the hot layer at the top of the pool.

6.2 Outer Pool

During natural circulation tests with functioning DHR loops, coolant temperature differences between the top and bottom of the outer pool (stratification) of up to 30°C have been observed. The rate of formation of a cold layer at the bottom of the pool may be used to estimate the flow out of the IHX. It is found that when the DHR heat removal rate greatly exceeds the core power the rate of formation of the cold layer implies a flow of cold sodium out of the IHX's considerably greater than the core flow (a factor of five). In the case of DHR power removal not being significantly greater than core power, large flows are not required to account for the stratification.

It can be postulated that under conditions under which DHR power removal exceeds core power an additional natural circulation loop operates, namely from the outer pool, up through the IHX tube bundle, over the DHR loops and back to the outer pool. Thus the observed stratification in the outer pool leads to the postulate of a fresh natural circulation loop between the outer pool and the DHR loops.

7. EFFECT OF HEAT REMOVAL CAPACITY

The main heat removal facility under natural circulation conditions is the DHR loops and operation of these loops should produce measurable temperature changes from which flow and temperature patterns might be deduced. The measured data is not ideal for such an exercise but in this Section an attempt is made to complete the picture sketched in the preceding Sections.

Figure 7 shows results from the early stages of Test 11 (Table 1) in which the reactor power was held at about 5 MW with no heat removal from the DHR loops. Seventeen minutes into the test heat removal commenced through the DHR loops, producing a rapid change on thermocouples in the vicinity. The important point is that although the sub-assembly outlet temperatures remained sensibly unchanged the AVG thermocouple readings started to decrease approximately 300s after switching on the DHR loops. Sodium flow must be returning directly to the plenum from the DHR loops.

Evidence for a downward flow around the periphery of the core, and therefore in support of the above deduction, comes from Test 4 in which a number of extra thermocouples were available in the plenum. Figure 8 shows how one such thermocouple, SIP 9, situated only 55cm above core sub-assembly R12 in annulus 5 at the breeder boundary registered a decreasing temperature consistently lower than the AVG thermocouples (situated far higher in the plenum) and also, significantly, lower than that measured by the outlet thermocouple above R12. It can also be noted from Figure 8 that whereas the outlet thermocouples show a very noisy response consistent with being situated in a region of mixing between hot and cold flows, the SIP thermocouples are nearly noise free, indicative of a well mixed essentially constant temperature region.

There is evidence from the two 5 MW constant power tests (Numbers 8 and 11) that DHR heat removal is not significant in determining core flow. This observation is reinforced by the two 50 MW trip tests (Numbers 12 and 13) which had substantially different IHX heat removal conditions. Figure 7 shows the stage in Test 11 at which the DHR loops started operating; the sub-assembly outlet thermocouples showed no evidence of any change. This is at variance with the predictions from BUNTY [3] which suggests a very strong correlation between the IHX conditions, including the DHR, and the magnitude of the induced natural circulation flow through the core. However on the basis of the natural circulation flow from the outer pool over the DHR loops postulated in the preceding section it is not surprising that changes in the DHR state have little immediate effect upon the induced core flow. This is because the enhanced coolant flow results in a reduced temperature drop over operating DHR loops and the mean IHX temperature will be closer than expected to the inner pool sodium temperature.

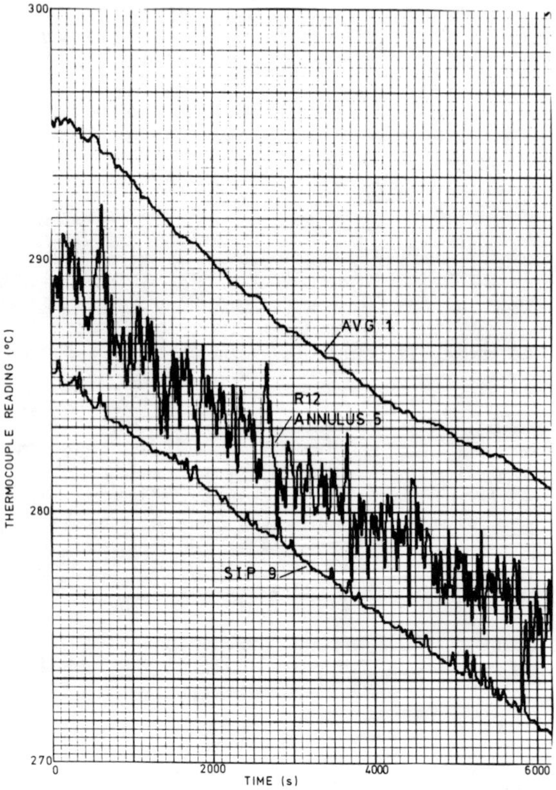

FIG. 8. UPPER PLENUM TEMPERATURES UNDER NATURAL CIRCULATION CONDITIONS

8. CONCLUSIONS

The paper has briefly reviewed the programme of natural circulation tests carried out at Dounreay on PFR. The results from these experiments show that,

(i) once established natural circulation can remove from the reactor, core heat, comparable with that expected after a full power trip, at temperatures which are acceptable.

(ii) the transition from forced (pumped) flow to free (natural) convection can be accomplished smoothly without any recognisable flow minimum.

Experiments designed to simulate all the features of a loss of electrics trip are now being carried out; the first two tests in this series, both from 50 MW, produced results consistent with those from the earlier tests.

If account is taken of the results from all the tests carried out to date then extrapolation to the full power case suggests that following a full power loss of electrics test the peak core temperatures will be comparable with or less than their normal operating values.

The paper has identified several factors from the tests which might be of interest to specialists in the field and has attempted an explanation in terms of the flow patterns existing in the primary circuit under low flow conditions. In particular it has been postulated that,

(i) the behaviour of the temperature spike and the radial distribution of the sub-assembly outlet temperatures can be explained by the assumption of a strong cross flow across the top of the core which either involves or interacts with cold coolant in the plenum.

(ii) two additional natural circulation loops have been identified; the first involves the outer pool, DHR loops and IHX; the second involves the upper plenum and DHR loops. These additional natural circulation loops may be responsible for the apparent insensitivity of core flows to the DHR status.

(iii) the effect of changes in DHR status upon circuit temperatures provides further evidence of cold coolant passing from the IHX trays directly back into the plenum.

The work described in this paper has resulted in a unified view of coolant flow fields in the primary circuit of PFR, which may explain several disparate phenomena observed under natural circulation conditions. This is encouraging and will allow the further development of theoretical models for the reactor against which other observations can be checked.

9. ACKNOWLEDGEMENT

The programme of natural circulation tests has been carried out with the assistance of many colleagues and the authors gratefully acknowledge this help. In particular the assistance provided by Mr J D C Henderson and Mr D S Crowe Dounreay, Mr R Hampshire of the CEGB, Mr G A Brown and Mr R Bell of CTS Risley is especially acknowledged.

10. REFERENCES

1. Natural circulation studies in support of the Dounreay PFR. C V Gregory et al. Paper VD/2 ANS/ENS International Meeting on Fast Reactor Safety Technology. Seattle USA 19-23 August, 1979.

2. Natural circulation boosts safety of pool-type fast reactor. C V Gregory et al. p29-33 Nuclear Engineering International, December 1979.

3. Theoretical investigation of the establishment of natural circulation in the Dounreay Prototype Fast Reactor - C W Dawson. Paper to this meeting.

A Theoretical Analysis of the Establishment of Natural Circulation in the Dounreay Prototype Fast Reactor

C. W. DAWSON
Fast Reactor Technology Group
Dounreay Nuclear Power Development Establishment
Dounreay, Thurso, Caithness, Scotland

ABSTRACT

A digital computer code is described which predicts the events following a loss of primary pumps in the Dounreay PFR. The code is compared with experiment and predictions for a full power trip are presented. Results are also given for two non-typical situations: boiling during natural circulation; and a loss of flow with failure to scram.

1. INTRODUCTION

Loss of electrical supplies to the Dounreay Prototype Fast Reactor (PFR) causes the coolant pumps and the reactor to trip; sufficient coolant flow to remove the core decay heat is provided by three highly reliable continuously running auxiliary motors which automatically engage as the main drive units run down and which together can sustain a flow equal to 10% of normal full flow.

In the unlikely event that all three auxiliary motors fail to engage the primary pumps stop after 200s. With a core decay power of 18 MW, compared to full power of 600 MW, and no coolant flow, coolant saturation temperatures would be reached within the core after a further 50s. By contrast if the decay heat can be convected from the core to the bulk coolant and thence to the decay heat rejection (DHR) loops it can be removed from the reactor vessel entirely. Even if the DHR loops fail to operate transfer of the decay heat to the bulk primary coolant (1000 Te) will delay the onset of boiling for many hours.

To maintain core temperatures below normal operating temperatures if the auxiliary motors fail to engage requires the establishment within 200s of a natural circulation flow through the core equal to approximately 3% of normal full flow. The programme of experiments already carried out at PFR has demonstrated the feasibility of such flows occurring in PFR and the further tests still to be performed are expected to confirm the extrapolation to the high power case (1, 2, 3), thereby illustrating how advantage can be obtained from the large thermal capacity of the primary pool thus providing an intrinsic and powerful safety feature.

In parallel with the experiments described in References 1-3 a digital computer code (BUNTY) is being developed to model the PFR primary circuit and to predict the events following a loss of electrics trip. The intention is to use the results from the reactor experiments to validate

BUNTY, using identified discrepancies to improve the modelling, and thence to use BUNTY as a predictive tool able to describe the initiation of natural circulation for a wide range of reactor conditions.

The paper first briefly describes the PFR primary circuit and the problems it poses for the initiation of natural circulation; it then describes the code and continues with a comparison between results from the code and experiment. The code is applied to a full power PFR trip and results presented. Finally the paper shows how the code has been extended to deal with extreme off-design conditions such as boiling following pump rundown and an unprotected loss of flow incident.

2. THE PFR PRIMARY CIRCUIT

The PFR primary circuit is illustrated in Figure 1. Coolant is pumped from the outer pool via the core inlet duct and diagrid to the core and thence to the above core plenum from where it passes into the intermediate heat exchanger (IHX) trays before falling over the DHR loops and entering the IHX. From the IHX the coolant returns to the outer pool via the IHX pods. The heights of the free surface levels in the plenum and the outer pool differ by 1.2 m under full flow conditions, this difference being the head required to overcome the hydraulic resistance of the IHX.

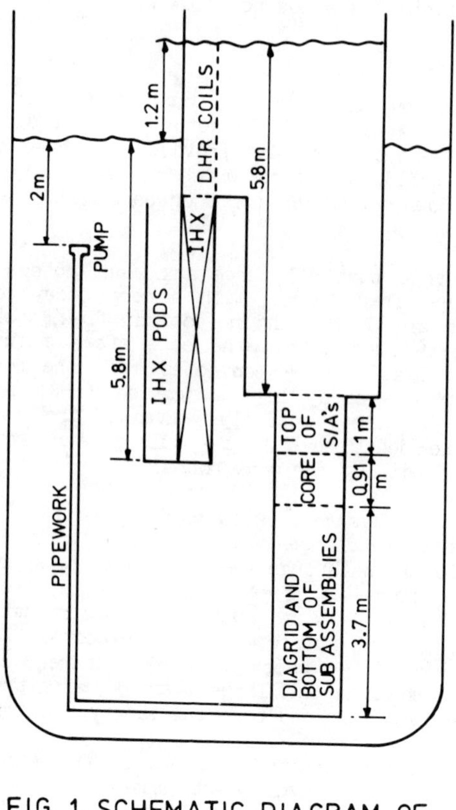

FIG. 1. SCHEMATIC DIAGRAM OF PFR PRIMARY CIRCUIT.

Following a station trip from full power as a result of a loss of electrical supplies the primary coolant pumps begin to run down with a halving time of 12s, finally stopping after 200s if the auxiliary motors fail to engage, the reactor power decreases almost instantaneously to approximately 90 MW and to 18 MW within a further 200s. Thus the core outlet temperature falls until the flow has decreased to a level commensurate with the decay heat power (150s); in the interim the plenum fills with colder coolant and thereby reduces the natural circulation head. Initiation of natural circulation is opposed by the decaying free surface level difference and by the temperature conditions in the IHXs; the rapid rundown of the secondary pumps (4s halving time), the delay before the DHR system becomes fully effective (240s after trip) and the efflux of hot sodium from the plenum all combine to set up a buoyancy head in opposition to that from the core.

The smooth and speedy establishment of natural circulation is therefore a transient problem which, viewed in terms of the three heads - pump head (falling to zero within 200s), net thermal head and the level difference head (which decays at a rate determined by the difference in flow rate through the core and the IHX) - requires that the thermal head becomes sufficient to overcome the decaying level difference before the pump head falls to zero. Confidence that the net thermal head combined with the pump head will always exceed the level difference comes from reactor experiments and from an understanding of the phenomena involved; reactor experiments have already demonstrated that natural circulation, once established, can limit core temperatures to normal operating levels (1, 2, 3) and further experiments are expected to provide a satisfactory demonstration of the early transient. In order to interpret the experiments and to allow extrapolation to a wider range of conditions than that covered by the experiments the BUNTY code has been developed.

3. THE DIGITAL COMPUTER CODE 'BUNTY'

BUNTY is a lumped parameter code - that is it provides a description of the transient flow and temperature behaviour of the PFR primary coolant by assuming that the circuit can be broken down into a series of compartments which behave essentially as pipes. In each of these compartments flow enters at one end and leaves the other at a time and with properties determined by the lags, delays, sources and sinks associated with that particular compartment. As will be seen later this approach works well for some components (eg the core) but less well for some of the larger volumes (eg the above core plenum) where the flow field is far too complex to be represented in this simple fashion.

BUNTY considers the circuit as a 'broken syphon' consisting of two distinct sections:

(i) Core section - flow is driven from the outer pool via the pipework and the diagrid and through the core to the inner pool or plenum (the 'positive' direction). The driving heads are the pump head; thermal head with the diagrid, core and plenum forming the hot let; and the level difference between inner and outer pools.

(ii) IHX section - flow is driven from the inner pool via the IHX to the outer pool. The driving heads are the head due to the level difference and a thermal head with the trays and IHX forming the hot leg and the pods and pool forming cold legs.

The flows in the two sections are considered separately in BUNTY; they are coupled only by the fact that the difference between them controls the volumes of the inner and outer pools and thus the level difference between them.

BUNTY's treatment of the driving heads can be analysed in the following way. In the core section the pump head (h_p); the thermal head (h_{tc}), and the level difference head (h_d) must overcome the hydraulic resistance of the core section (R_c) giving (ignoring accelerational terms)

$$h_p + h_{tc} - h_d = R_c \qquad (1)$$

and if positive flow is to be maintained

$$h_p + h_{tc} > h_d \qquad (2)$$

In the IHX section the driving heads are the level difference head (h_d) and the thermal head (h_{ti}). If R_i is the hydraulic resistance of the IHX section then

$$h_d + h_{ti} = R_i \qquad (3)$$

and if positive flow is to be maintained,

$$h_d > - h_{ti} \qquad (4)$$

Equation (2) expressed the point brought out in the previous section, viz that the positive driving heads through the core (ie the pump and the thermal head) must be greater than the adverse head due to the level difference. Equation (4) expresses the less obvious fact that the level difference must at all times be kept large enough to provide a head sufficient to overcome any adverse (or negative) thermal head in the IHX section.

A fundamental feature of BUNTY's representation of the primary circuit is therefore that the level difference is controlled by the temperature distribution in the IHX and the level difference reacts on the core flow, ie the core flow is sensitive to the IHX temperature distribution and thus to the heat removal capacity of the IHX and DHR loops.

Although natural circulation may also be established in the secondary circuits thus providing additional cooling in the IHX, BUNTY pessimistically assumes that after a few seconds all heat removal capacity in the IHX is lost.

4. COMPARISON WITH EXPERIMENT

References 1-3 describe the wealth of experimental data available from the PFR natural circulation tests; the references also describe the difficulties associated with the interpretation of measurements for instruments positioned for operational rather than experimental reasons. In this section attention is concentrated on two different types of test, constant power tests in which the decay heat is simulated by a constant neutron power during the pump transient, and a loss of electrics test from 50 MW.

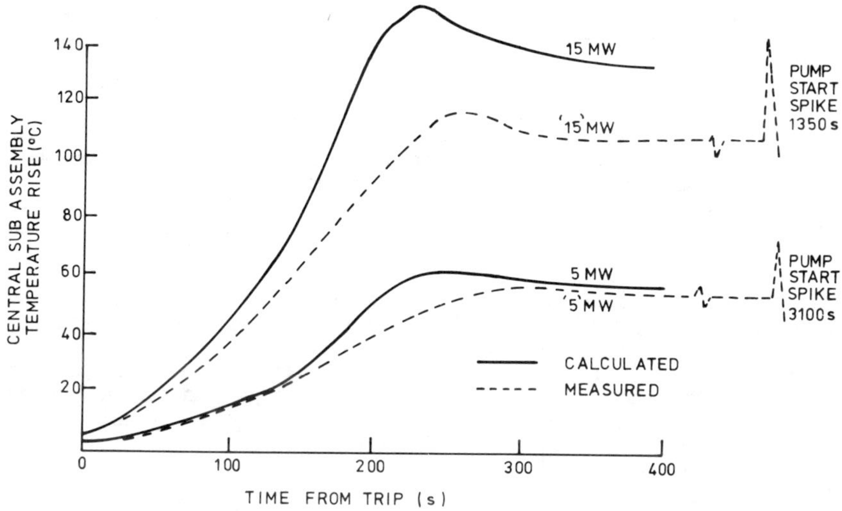

FIG. 2. COMPARISON OF BUNTY WITH PLANT DATA FROM STEADY POWER TESTS.

Figure 2 compares the measured delta T (inlet minus outlet temperature) for a particular central sub-assembly with BUNTY predictions for an average central sub-assembly during two constant power tests. In reference 1 it is argued that the sub-assembly measured outlet temperatures have to be corrected to account for heat loss between the core outlet and the thermocouple position by an amount approximately equal to the temperature spike observed at the end of the test when a pump is restarted. Figure 2 shows reasonable agreement between prediction and the 'corrected' measured values for sub-assembly outlet temperature; a more detailed comparison however, particularly with other circuit temperatures, yields some substantial discrepancies.

One of the most important discrepancies is the effect of the heat removal capacity (ie the DHR loops and the IHX) upon the circuit temperatures. It is a consequence of the assumptions in the present version of BUNTY that the heat removal capacity has a large effect on the induced core flow and therefore the sub-assembly delta Ts. When this hypothesis was tested on PFR by performing two 5 MW steady power loss of flow experiments (one with the DHR loops fully operational from time zero and the other with the DHR loops removing negligible heat during the pump rundown period) little difference was found in the measured core outlet temperatures. The same insensitivity of core outlet temperature to heat removal was also noted during two simultaneous pump and reactor trips from 50 MW. These two tests although nominally identical had different initial secondary circuit flows and therefore different IHX heat removal rates. Once again little difference was found in the measured sub-assembly outlet temperatures.

Although little change was observed in the sub-assembly outlet thermocouples there was a noticeable difference in the behaviour of the plenum thermocouples during the two 5 MW steady power tests. In the first test, with the DHR loops operating, the temperatures at the top of the plenum settled down to a value 20°C higher than the inlet temperature whereas in the second test, without the DHR loops, the plenum temperatures continued to rise slowly past the value reached in the first test and only dropped when the DHR loops were switched on. Thus although the measured core outlet temperatures, and by implication the core flow, seem to be unaffected by the heat removal capacity there are noticeable differences in the plenum temperature distribution.

This discrepancy between theory and experiment may be explained by postulating a secondary recirculating flow in the plenum possibly involving some of the sodium passing over the DHR loops and spilling back into the plenum. Such a flow would also account for many of the other discrepancies between calculated and measured effects eg. theory predicts a longer transit time through the plenum than measured as well as a larger than measured temperature difference across the DHR loops.

The flow pattern in the plenum is discussed in more detail in Reference 1. Suffice it to say that at the moment BUNTY is incapable of dealing with such a complicated flow path and this is one of the first areas in which the modelling will be improved.

Figure 3 shows the measured delta T across a central sub-assembly during a simultaneous pump and reactor trip from 50 MW to approximately 1 MW decay heat. Also shown are two predictions from BUNTY; one with the DHR loops building up from zero to full heat removal capacity between 140 seconds and 240 seconds (as in the experiment); the other with no heat removal through the DHR loops.

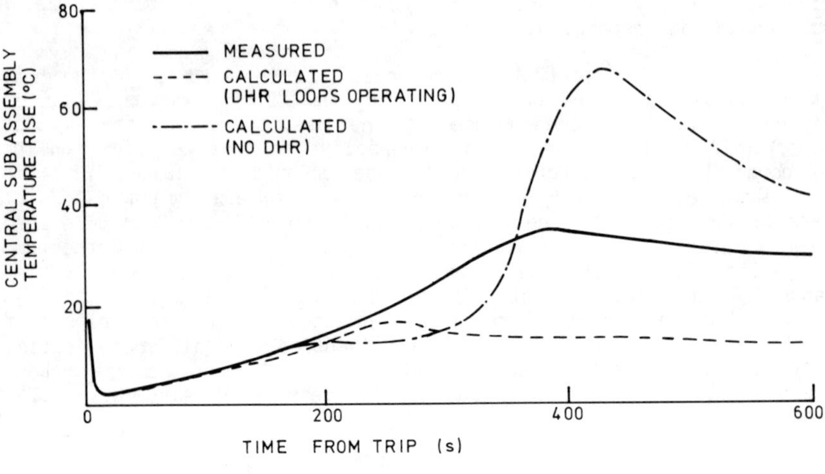

FIG. 3. COMPARISON OF BUNTY WITH PLANT DATA FROM 50MW SIMULTANEOUS PUMP AND REACTOR TRIP.

The predictions when the DHR loops are operating are seen to be significantly optimistic. This can be explained by the observation that the measured delta T across the DHR loops is much lower than predicted. Thus the calculation predicts too low a temperature in the IHX and thus too high a thermal head driving sodium through the IHX. This is reflected in a lower level difference which increases the flow through the core and leads to the lower sub-assembly outlet temperatures shown in Figure 3. The discrepancy in the delta T across the DHR loops can be explained by the presence of an additional recirculating flow over the loops.

For the case in which the DHR loop operation is delayed beyond 600 seconds the prediction is clearly pessimistic but the difference between this and the case in which the DHR loops operate is evidence of the extent to which BUNTY's predictions are unrealistically sensitive to the temperature distribution in the IHX.

It is worth noting that the IHX conditions in this case are not representative of a full power trip. In the 50 MW trip the core power falls rapidly to approximately 1 MW compared to a DHR loop removal capacity of 7 MW, thus once the DHR loops become operational there is substantial cooling of the sodium entering the IHX resulting in a large positive thermal head through the IHX. In a full power trip the decay heat is approximately 18 MW compared to a DHR loop capacity of 15 MW (the decay heat falling below the DHR capacity only after 400 seconds) and thus during the important transition to natural circulation the sodium entering the IHX will be relatively hot leading to a negative thermal head through the IHX. The effect of this is that BUNTYs predictions for a full power trip will not be so optimistic as for the 50 MW trip. Indeed evidence from the 15 MW steady power test suggests that BUNTY may be pessimistic at these higher powers (Figure 2).

5. PREDICTIONS FOR A FULL POWER TRIP

Despite the known inadequacies of the present version of BUNTY demonstrated by the preceding comparisons it is of interest to use BUNTY to predict the outcome of a full power loss of electrics trip in PFR, assuming the auxiliary motors fail to engage, if only to gain a feel for the scale of transient involved. Figure 4 shows the mean core outlet temperature, core flow and power for a best estimate case; the outlet temperatures, it should be noted, barely exceed normal operating temperatures even in the hottest sub-assemblies.

The slight temperature peak shown shortly after the pumps stop arises from a transient minimum in the driving head which largely results from the estimated adverse buoyancy head in the IHX. It is believed, from the results of the precceding comparisons, that the simplified IHX model in BUNTY together with the lack of a model to describe the recirculatory flow in the plenum region cause the adverse buoyancy head (or at least its effect) to be overestimated. An improved prediction might therefore lead to no significant peak and generally lower temperatures than those in Figure 4. The calculated core driving heads are shown in Figure 5.

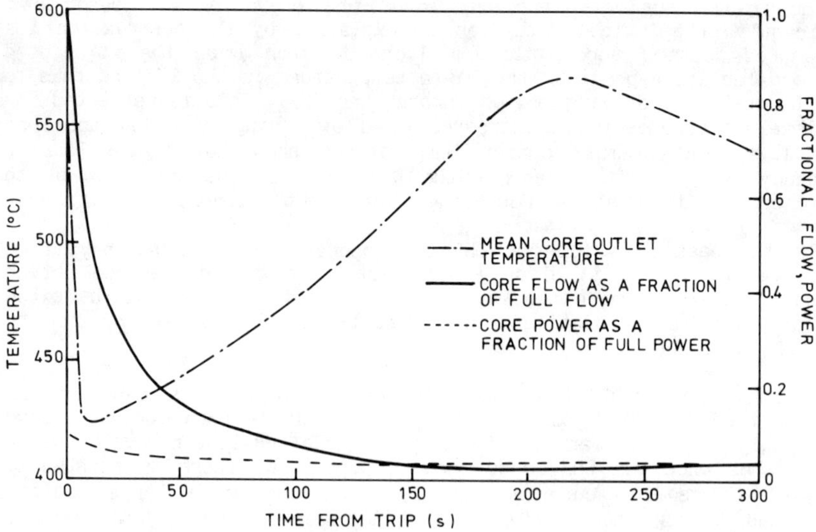

FIG. 4. PREDICTIONS FOR A FULL POWER TRIP.

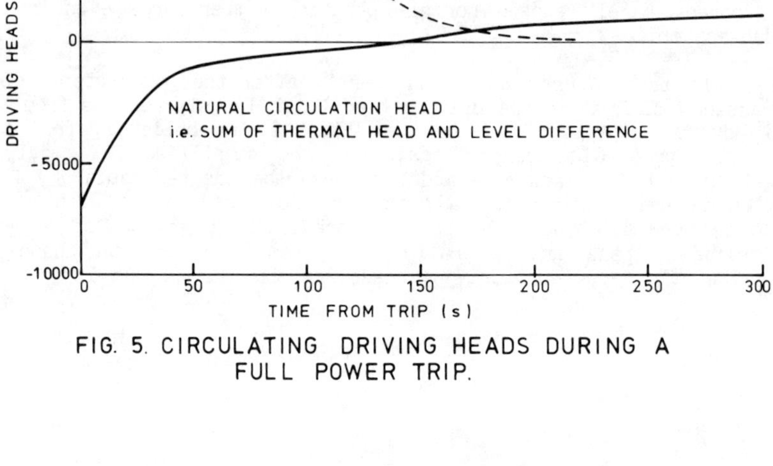

FIG. 5. CIRCULATING DRIVING HEADS DURING A FULL POWER TRIP.

6. BOILING AND NATURAL CIRCULATION

The experimental and theoretical work carried out to date give great confidence in the ability of PFR to ride out a loss of electrics incident (with auxiliary motors failing to engage) without the primary circuit temperatures exceeding their normal operating values. Nevertheless by combining a number of very pessimistic (and highly unlikely) assumptions (short pump run down time, large inital level difference) it is possible to predict a peak outlet temperature comparable with the coolant saturation temperature. BUNTY was extended by including a simple boiling model to allow the consequence of such boiling to be assessed.

In simple terms coolant boiling has two competing effects on the induced core flow: a large increase in the natural circulation driving head (a result of the 1000: 1 density change between liquid and vapour); and an increase in the hydraulic resistance. A simple one dimensional, homogeneous flow boiling model in which the two phases were assumed to be in thermal equilibrium was incorporated into BUNTY together with the Lockhart Martinelli correlation for two phase resistance. Tests with the model against boiling induced natural circulation flows in a water rig showed agreement with the measured flow rates to within 15%.

Figure 6 shows the core outlet temperature and flow for a case in which the reactor parameters were adjusted to give boiling at core outlet after the pumps had stopped. The temperature and flow exhibit an oscillatory behaviour; the increased head due to boiling increasing the flow sufficiently to bring the temperatures below saturation which decreases the flow until boiling cocurs once again. The maximum quality in this case was 0.5% and the maximum extent into the core was 3.5cm (the major boiling region was in the region of the sub-assemblies above core outlet). Eventually events in other parts of the circuit produce conditions in which single phase natural circulation can be maintained and the core outlet temperature falls.

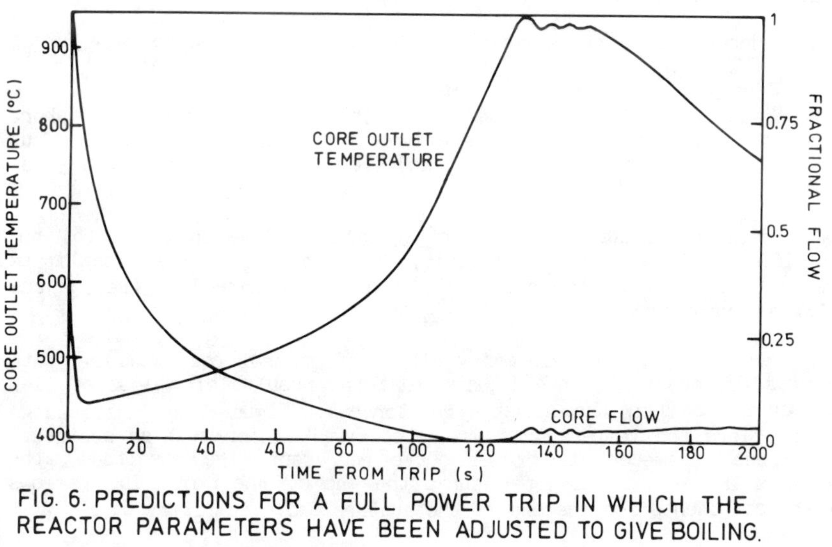

FIG. 6. PREDICTIONS FOR A FULL POWER TRIP IN WHICH THE REACTOR PARAMETERS HAVE BEEN ADJUSTED TO GIVE BOILING.

Thus even in this pessimistic and hypothetical circumstance the combination of boiling and natural circulation leads to a quasi stable condition which the reactor is able to accommodate until heat losses elsewhere are able to decrease the core outlet temperatures. In this context it should be noted that tests carried out in the Dounreay Fast Reactor (DFR) showed the ability of fuel pins to survive boiling for extended periods of reactor operation (4).

7. LOSS OF FLOW WITH FAILURE TO SCRAM

Any decrease in the primary cooling power of PFR, whether by a loss of electrical supplies or other means, is detectable by a number of different routes, each of which will trip the reactor; it is therefore inconceivable that an accident in which all pumping power is lost could occur, without tripping the reactor. However it is possible, because of the reactivity properties of PFR (and similar reactors), that even were the unthinkable to occur the consequence might be confined to coolant temperatures less than saturation.

Reactivity feedback effects ensure that rising sodium temperatures decrease the reactor power leading to the possibility of maintaining temperatures at an acceptable level. To investigate this a reactivity feedback model was included in BUNTY. The model uses the standard neutron kinetics equations in six delayed neutron groups and employs experimentally derived coefficients for the various components of reactivity feedback (5). Four separate reactivity effects have been included:-

(i) Expansion of control rod supports which effectively inserts the control rods as the supports heat up. The size of this effect is -0.5 F_Z ¢/K where F_Z is the differential control rod worth normalised to 1 at the start of a reactor run and falling to 0.2 at the end of the run. Thus the reactivity effect of control rod expansion varies from -0.5 ¢/K to -0.1 ¢/K during a reactor run.

(ii) A core delta T dependent term -0.16 ¢/K to -0.04 ¢/K

(iii) Diagrid temperature dependent term -1 ¢/K

(iv) Power dependent term: varying from -0.1 ¢/MW at 600 MW to -0.4 ¢/K at 100 MW

The feedback model used in BUNTY does not explicitly include the short time constants associated with fuel temperature and Doppler etc. Thus while long term effects are adequately modelled short term transients (less than one second) will not be correctly described.

A similar investigation has been reported by Freslon et al (6) but with a different breakdown of reactivity effects. In order to facilitate comparison Table 1 shows the reactivity effects considered by Freslon and their PFR equivalents.

The variation of the reactivity effect of control rod expansion with control rod differential worth implies that the results of a loss of flow without scram will vary throughout a reactor run. Figure 7 shows the range of powers, flows and temperatures predicted by BUNTY for such an accident. For all three parameters the lower curve represents the results at the beginning of a run and the upper curve the end of the run. The various reactivity components for the end of the run are shown in Figure 8.

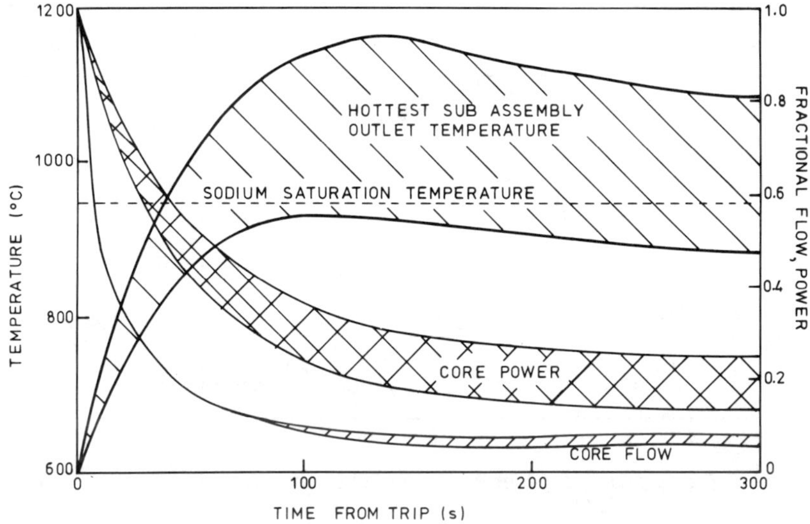

FIG. 7. PREDICTIONS FOR A LOSS OF FLOW WITHOUT REACTOR TRIP.

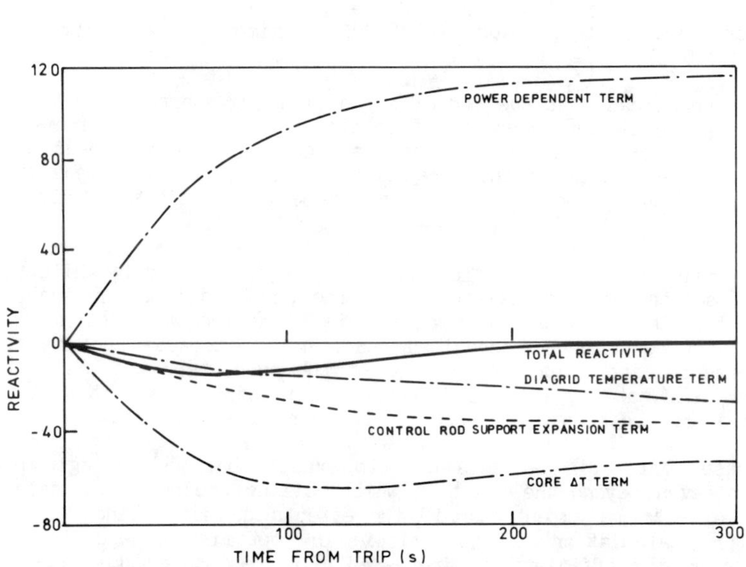

FIG. 8. CALCULATED REACTIVITY COMPONENTS FOR A LOSS OF FLOW WITHOUT REACTOR TRIP AT END OF REACTOR RUN.

	PFR	PHENIX	SUPERPHENIX
Fissile Doppler TdK/dT	-143 ¢	-163 ¢	-241 ¢
Sodium density in core	-0.03 ¢/°C	-0.07 ¢/°C	-0.22 ¢/°C
Diagrid radial expansion	-0.4 ¢/°C	-0.36 ¢/°C	-0.31 ¢/°C
Flowering or pad effect	No such effect	-0.22 ¢/°C	-0.23 ¢/°C
Control rod system reactivity worth	0 to -3.5 ¢/mm	-0.9 ¢/mm	-1.4 ¢/mm
Axial expansion of core	-0.08 ¢/°C	Not mentioned	

TABLE 1: Comparison of reactivity coefficients for PFR, PHENIX and SUPERPHENIX

It can be seen from Figure 7 that at the beginning of the run coolant boiling is avoided altogether and the circuit settles down into a quasi-steady state with a core power of 75 MW being convected from the core with a mean core outlet temperature of 720°C (maximum sub-assembly outlet temperature 850°C).

In the worst case, at the end of a run, boiling starts after 40s, but even so the power at the inception of boiling has been reduced from 600 MW to 350 MW. At present the calculations have only been performed in single phase and any results after the inception of boiling are only indicative. A means of assessing the reliability of these calculations is discussed below.

In an unprotected loss of flow boiling occurs (if at all) while there is still a substantial driving head from the pump and thus any increased effect on the flow. Thus boiling in this situation will not be self-limiting in the manner described earlier. Because under these circumstances boiling occurs at the top of the core where the sodium void coefficient is negative, it will decrease the reactor power; it will also lead to an increased flow resistance and hence reduced core flow.

If these conflicting effects were to exactly balance a single phase calculation, even beyond the point at which boiling occurs, would describe the core flows to be expected. If the effects do not balance then the single phase calculation is pessimistic or optimistic depending upon whether the void coefficient or increased flow resistance dominates. A parameter X can be defined such that if X is less than unity the single phase calculation is pessimistic, ie the effect of boiling on reactivity and buoyancy head outweigh the increased resistance to flow and if X is greater than unity the single phase calculation is optimistic. X is defined as

$$X = \left[Q \left(\frac{R}{h_p + h_{nc}} \right)^{1/2} \right]_{TP} \left[Q \left(\frac{R}{h_p + h_{nc}} \right)^{1/2} \right]_{SP}^{-1}$$

where Q is the reactor power, R the core flow resistance, h_p the pump head and h_{nc} is the natural circulation head in the core, each quantity being evaluated for two phase (TP) and single phase (SP) conditions.

This simple analysis has been applied to the results of a number of BUNTY calculations and shows that for the majority of the cases where boiling is predicted X lies between 1 and 2, ie boiling will continue to develop. Nevertheless the values are not so far from unity that improvements to the recognisably pessimistic model could not indicate stable conditions.

8. CONCLUSION

A digital computer code, BUNTY, which models the establishment of natural circulation of the PFR primary coolant following a loss of electrics trip has been described. The code is being developed in parallel with a comprehensive series of reactor experiments, the results of which are being used to validate and improve the models in the code.

Comparisons of BUNTY with experiment have been presented together with predictions for a full power trip, establishment of natural circulation with boiling, and a loss of flow without scram.

9. REFERENCES

1. ANDERSON, R, DAWSON, C W, LORD, D J and WEBSTER, R. 1980. Observations on coolant flow patterns in the primary circuit of PFR during natural circulation experiments. Specialists meeting on decay heat removal and natural convection in FBRs. 28-29 February 1980. New York.

2. GREGORY, C V et al. 1979. Natural circulation studies in support of the Dounreay PFR. ANS/ENS International Meeting on Fast Reactor Safety Technology. 19-23 August 1979. Seattle, USA.

3. GREGORY, C V et al. 1979. Natural circulation in the Dounreay Prototype Fast Reactor. Nuc Eng Int, December 1979.

4. GREGORY, C V et al. 1978. IAEA-SM-225/49. Vienna.

5. LORD, D J and WEBSTER, E B. 1979. Reactor physics experience on the Prototypoe Fast Reactor. International symposium on Fast Reactor Physics, 24-28 September 1979. Aix-en-provence.

6. FRESLON, H et al. 1979. Analysis of the dynamic behaviour of the Phenix and Super Phenix reactors during certain accident sequences. ANS/ENS International Meeting on Fast Reactor Safety Technology. 19-23 August 1979. Seattle USA.

The Effects of System Configuration and Initiating Conditions on Natural Convection in an LMFBR

JOHN H. GERMER and FRANK W. SCIACCA
Advanced Reactor Systems Department
General Electric Company
Sunnyvale, California

ABSTRACT

This paper discusses the conditions which must be met in order to establish satisfactory natural convection in an LMFBR system after shutdown of the main system pumps. Curves showing the results of transient calculations are presented for both pool and loop systems, under differing conditions of system configuration, pump coastdown rate, and with heat removed from the system at different points. Several important conclusions are reached that influence the design of such a system, as well as recommended operating procedures.

1. INTRODUCTION

The safety of any reactor system is strongly influenced by its natural ability to maintain safe temperatures under all conditions of circulating pump failure. The very large heat capacity in an LMFBR should be available for limiting temperatures, even with a delay in initiating the operation of a shutdown heat removal system. The reactor system designer must thoroughly investigate the natural convection characteristics of his system to understand its performance uhder a range of possible conditions. With such an understanding he can modify the design, if needed, to enhance natural convection, and thus enhance the safety characteristics of the plant.

The computer program GENCON was developed by General Electric to investigate natural convection in large LMFBR systems. Particular emphasis was placed upon flexibility in the program to handle a wide variety of system configurations and starting conditions, rather than precise modeling of the small details. This paper discussed the basic GENCON model, and a few specific results. Several important conclusions are reached. The circulation method can be very useful in understanding and developing overall system design concepts, possibly pointing out unanticipated problem areas in the early stages of design.

Natural convection in an LMFBR involves two distinct realms: the short term convection beginning at the time when the main circulation pumps stop, and the long term when the shutdown heat removal system has a prominant influence on temperatures. These two realms tend to be relatively independent, and will be discussed separately.

The temperature distribution throughout the primary system when the pumps stop will determine the initial convection flow that will occur at that time. Under some conditions, this temperature distribution can be unfavorable, and cause the sodium to reverse direction and flow downward through the reactor core. In that case, the reversed flow will continue until sufficient temperature is reached in the core to return the flow to its normal direction. This problem tends to occur when the pump operation continues too long after reactor scram. Initiating conditions can be either too long a pump coastdown, excessive delay in tripping the pumps, or operation on pony motor power followed by total pump failure. Conditions in the remainder of the system, such as continued operation of the steam generators, can make the consequences less severe, but tend to have little influence upon the convection immediately after an early pump failure.

Long term natural convection is most favorable when heat is removed at a high point in the system, particularly at the top of a high down-flowing component such as the intermediate heat exchanger. One can consider the shutdown heat removal system to serve two functions; to limit the rise in the average temperature of the system, and to enhance natural convection of sodium through the reactor core. The long term transient is influenced by both of these functions. A system that effectively promotes natural convection will tend to minimize the difference between the peak temperature (at the reactor core outlet) and the mean system temperature (which is a function of the heat added and the heat capacity of the system). These relationships will be discussed with the following presentation of temperature and flow transient calculation results.

In this paper, several specific aspects of natural convection flow in LMFBR's have been studied. For the loop system they include:

- effect of pump coastdown rate and delay of pump trip
- effect of outlet plenum size
- effects of heat removal or no heat removal
- raising elevation of the IHX
- effects of the intermediate heat transport system.

For the pool system they include:

- effect of delay in pump trip
- effect of heat removal in the hot plenum or no heat removal
- effect of lowering the upper part of the IHX

2. CALCULATION METHOD WITH GENCON

The calculation model describes a system as consisting of one or more "sections", or parts of the system where flow is the same throughout. A section may be a closed loop, or may join other sections at junction points. By the use of junction points, complicated systems can be analyzed with parallel flow paths. Each section is made up of "components", such as pipes, heat exchangers, reactor cores, plenums, etc. Input data for a section include its flow resistance, rated flow, fluid properties, pump characteristics, etc. Each component is specified by its elevation, sodium volume, stationary heat capacity, initial inlet temperature, etc. Further component information is specified for the thermal admittance of a heat exchanger, and heat generation rate in a reactor core. Large plenums are further assumed to maintain stratified temperature conditions, and a simplified model is assumed.

Flow rates are calculated on a pseudo-steady-state basis by equating the friction pressure drop to the sum of pump and convective driving pressures. Inertia, except for local effects, is not considered to be significant in the low flow rates that are normally associated with natural convection flow. Temperatures are calculated based upon flow mixing between adjacent elements, with allowance for heat capacity of fixed metal, and heat flow from sources.

3. SYSTEM DESCRIPTION

Both loop and pool systems are typical of current designs being proposed for large power plants producing 1000 megawatts of electric power. They operate with normal reactor inlet and outlet temperatures of 670°F (354°C) and 950°F (510°C).

3.1 Loop System

A schematic hydraulic profile of a loop-type LMFBR is shown in Figure 1. This configuration serves as the basis for transient studies that will be discussed. The primary sodium system includes a reactor and reactor vessel, piping to a hot-leg pump, the primary side of an intermediate heat exchanger (IHX), and return piping to the reactor vessel. The intermediate sodium system was simplified in this model to include only the secondary side of the IHX, the sodium side of a single-unit steam generator, and direct interconnecting piping.

There are four parallel loops from the reactor hot outlet plenum to the reactor inlet plenum, each with a separate intermediate system. These four loops are combined into a single composite for calculational purposes.

The hydraulic profile is shown also in Figure 2 in the form of a block diagram as it is described for the GENCON calculation. The system is broken up into various components that represent pipes, a reactor core, a hot plenum, the primary and secondary sides of the heat exchanger, and the primary side of the steam generator. The pumps are described as pipe elements of the appropriate volume. Flow resistances and pump characteristics are specified as characteristics of an entire section of the system.

3.2 Pool System

A simplified model for a typical pool system is described schematically in Figure 3 and as a block diagram in Figure 4. It differs from the loop system in that the entire primary sodium system is contained within a single large reactor vessel. The multiple intermediate heat exchangers are combined as a single unit for calculation. Sodium flows from the bottom of the IHX's into a very large cold plenum before entering the pump inlets and being pumped into the reactor inlet plenum. The intermediate sodium system is not described, since the pool concept results presented in this paper consider only the case when intermediate sodium flow is stopped.

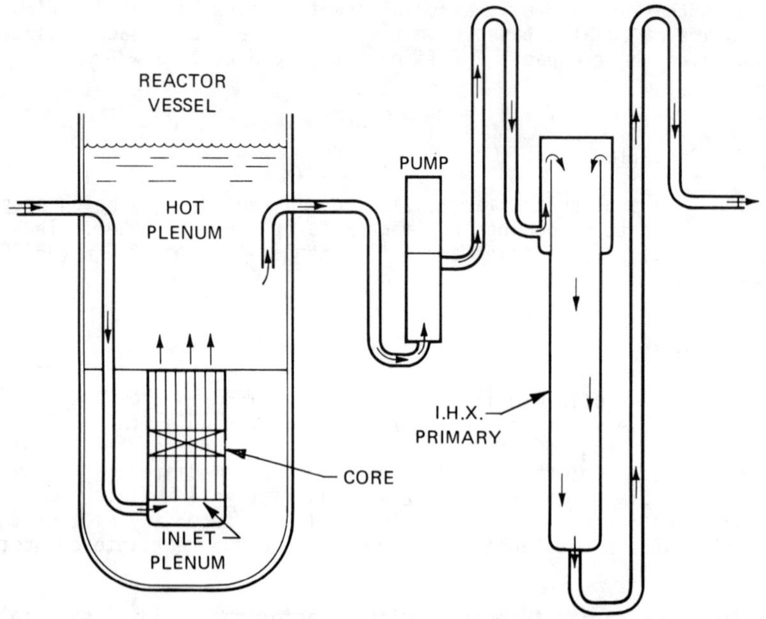

Figure 1. LOOP SYSTEM – HYDRAULIC PROFILE

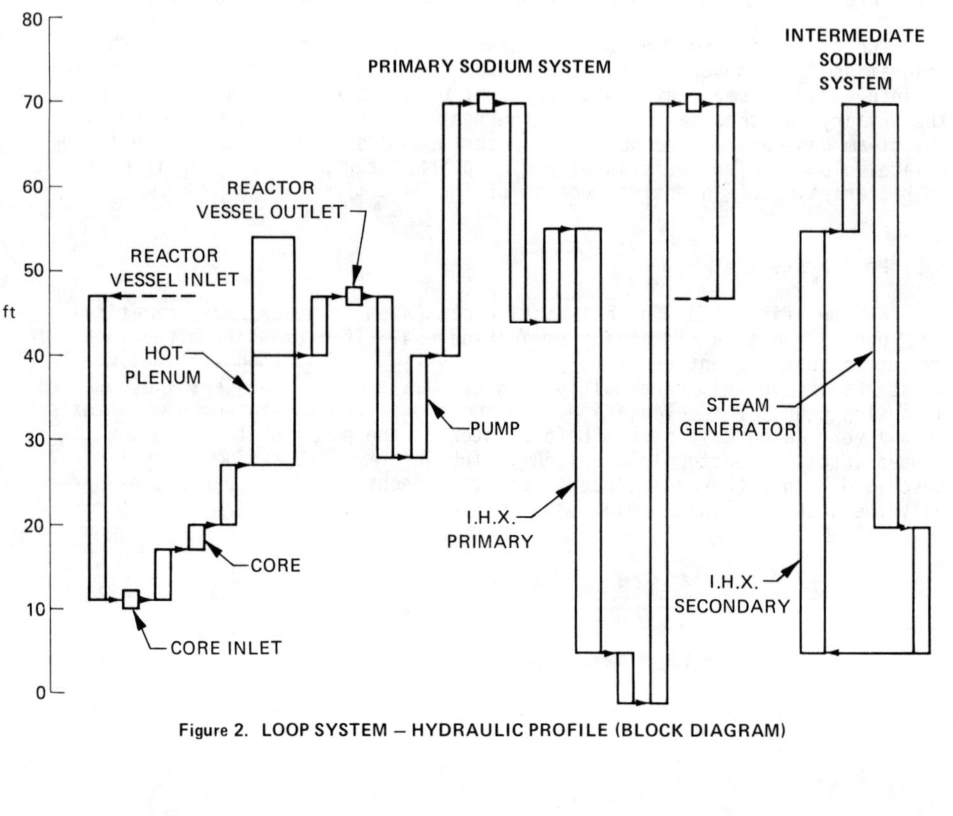

Figure 2. LOOP SYSTEM – HYDRAULIC PROFILE (BLOCK DIAGRAM)

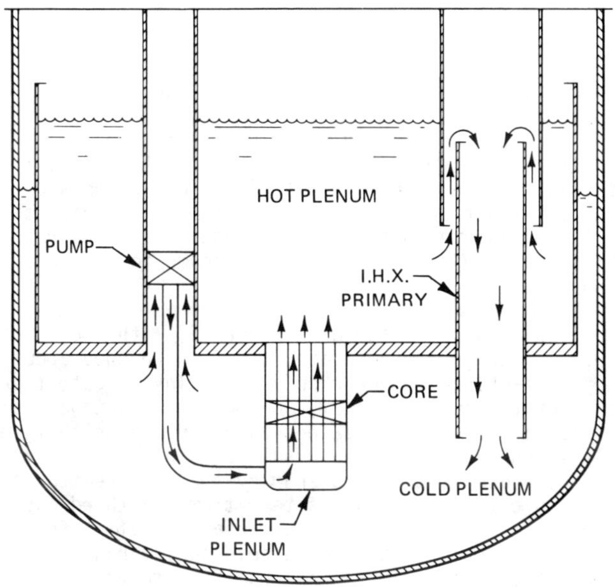

Figure 3. POOL SYSTEM – HYDRAULIC PROFILE

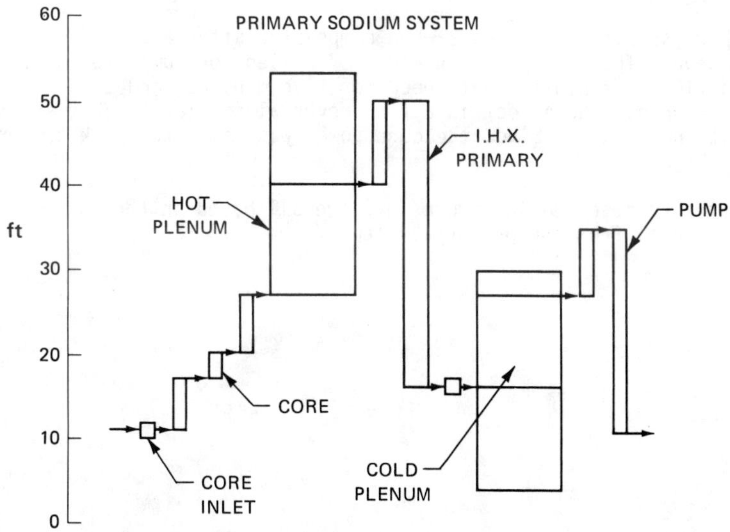

Figure 4. POOL SYSTEM – HYDRAULIC PROFILE (BLOCK DIAGRAM)

4. LOOP SYSTEM TRANSIENTS

4.1 Short-term Transient

Figures 5 - 7 show the short-term flow and power decay in the reference loop system described in Figures 1 and 2. These curves ignore the intermediate system since it has little effect on short-term transients. Three different pump decays were assumed. In each case the flow followed closely to the assumed pump coastdown until the flow reached about 5 percent of full flow. The flow after the pump came to rest was caused by natural convection only.

A rapid pump coastdown was assumed in Figure 5, with the pumps coming to rest at 60 seconds. In the reference case, the flow reached about 1.6 percent before rising and approaching the power curve. To investigate the effect of increasing the size of the outlet plenum, another transient was calculated with 5 times the plenum volume. The dashed curve indicates that the natural convection is only slightly better, the result of a decrease in the thickness of the cold layer introduced into the plenum by the cold reactor outlet after scram. In order to investigate the opposite extreme, a third case was run with the plenum completely eliminated and the reactor outlet feeding directly into the vessel outlet pipes. In this case cold sodium has filled a greater portion of the reactor outlet piping during pump coastdown. The effect is a complete reversal of flow for 120 seconds. Although peak core temperatures were not plotted, they were found to reach about 1450°F (788°C).

The effect of a slower pump coastdown is shown in Figure 6. In this case it was assumed that the time scale of the pump coastdown was doubled, with the pumps coming to rest in 120 seconds. The reference case indicates that the flow almost reaches zero, but for too short a time to reach excessive temperatures. The effect of a larger plenum size is much more pronounced than in the previous example.

A much worse situation occurs when pumps stop after a period of pony motor operation. The assumption in Figure 7 called for pumps to coast to motor speed (10%), remain at that speed for five minutes, and then slow to a stop in 20 seconds. The effect is a flow reversal similar to that shown in Figure 5 with no plenum. Since the core power was less, the peak temperatures were not as high.

The cases represented in Figures 5-7 are all based on the assumption of no heat removal from the primary system.

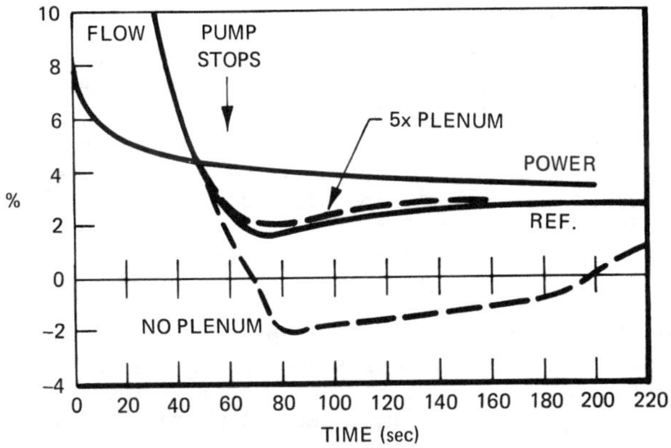

Figure 5. LOOP SYSTEM – FLOW TRANSIENT (PUMPS STOP IN 60 sec)

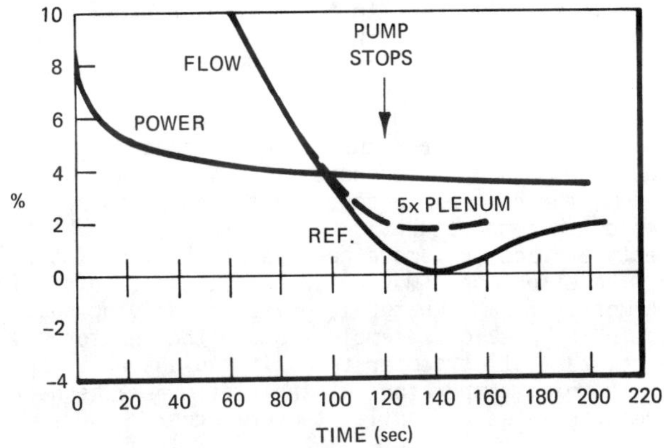

Figure 6. LOOP SYSTEM – FLOW TRANSIENT (PUMPS STOP IN 120 sec)

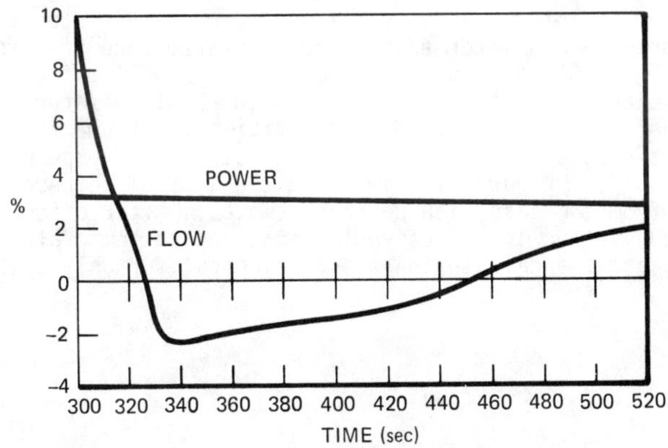

Figure 7. LOOP SYSTEM – FLOW TRANSIENT (10% PONY MOTORS, PUMPS STOP 300-320 sec)

4.2 Long-term Transient

Figures 8 - 11 show the long-term temperature transient for a loop system. Each curve plots the core inlet, the core outlet, and the mean temperature of the primary sodium system. In Figure 8, it was assumed that there was no flow in the intermediate sodium system. Pump coastdown was 60 seconds as in Figure 5. No heat was removed from the system. It can be seen that the core outlet temperature required 4200 seconds to reach 1100°F (593°C). Thus, these calculations predict a substantial grace period for this type of system, with temperatures remaining relatively low for some time even without heat removal from the system.

The effect of raising the elevation of the intermediate heat exchanger by 15 feet (4.6 m) was calculated, and is shown in the dashed line curves of Figure 8. The effect is most pronounced in the first 1000 seconds, when the natural convection was improved. The later improvement is not very great.

The effect of removing a constant 32 megawatts of heat from the top of the hot plenum is shown in Figure 9. A somewhat higher reactor outlet temperature results from the somewhat lower temperature in the lower part of the hot plenum and in the piping up to the heat exchanger inlet, both important components of the hot leg. In the long range, temperatures remained close to the 1100°F (593°C) level. A further continuation of the transient would show that the temperature eventually turns down, since the reactor power drops to 32 megawatts in about 10000 seconds. It will be noted that both the reactor inlet and the reactor outlet temperatures exceed the average temperature of the system at about 5000 to 6000 seconds. This is because of the large volume of the "hot" plenum, the heat exchanger, and the loop piping are all operating at essentially cold leg temperature.

A more effective place to remove heat is at the top of the intermediate heat exchanger. The effect of removing 32 megawatts at this point is shown in Figure 10. A more favorable natural convection driving head results in higher flow than in the previous example. Since the reactor outlet plenum operates at the reactor outlet temperature, it can absorb a considerable amount of heat. The mean temperature of the system remains near to the average of the reactor inlet and outlet temperatures.

A more usual shutdown transient occurs when the steam generators remain active after reactor scram (a condition that would be encouraged to prevent thermal shock damage to the steam generators). In Figure 11, it was assumed that the primary and intermediate system pumps coast to a stop in 60 seconds, and that both primary and intermediate sodium systems are free to flow by natural convection. At low flow, a steam generator will maintain high heat transfer coefficients on both the sodium and the boiling water sides. Therefore it was considered to be a good approximation to assume that the entire sodium side of the steam generator is at the saturation temperature of the steam (650°F, 343°C). Figure 11 shows that the initial temperature transient differs little from the case with no heat removal. After a few minutes, however, natural convection is effective in reducing the temperature of the system rapidly, and the calculation was carried out for only 3000 seconds.

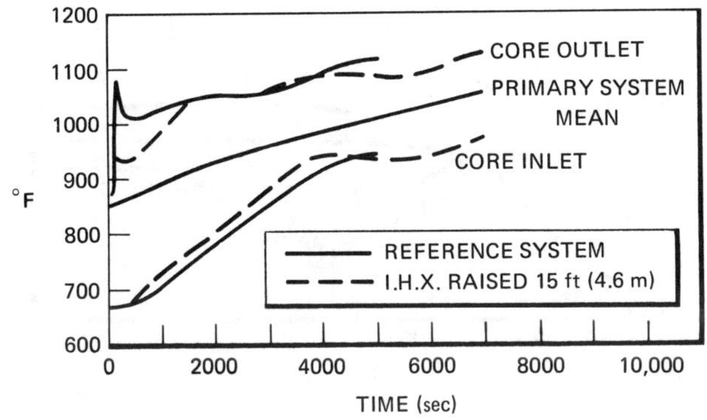

Figure 8. LOOP SYSTEM – LONG TERM TEMPERATURE TRANSIENT (NO HEAT REMOVAL FROM PRIMARY SYSTEM)

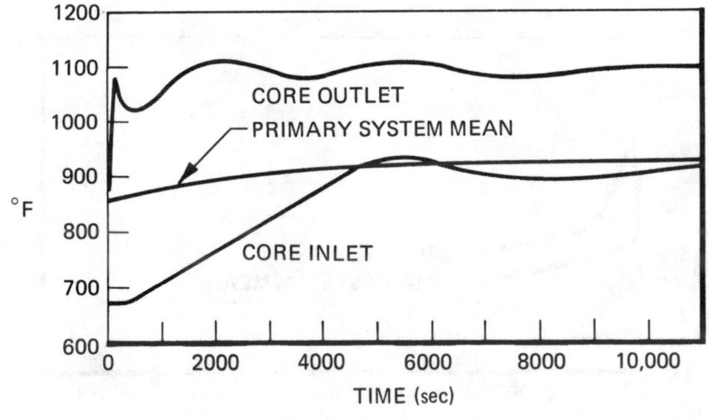

Figure 9. LOOP SYSTEM – LONG TERM TEMPERATURE TRANSIENT (32 MW HEAT REMOVED FROM TOP OF HOT PLENUM)

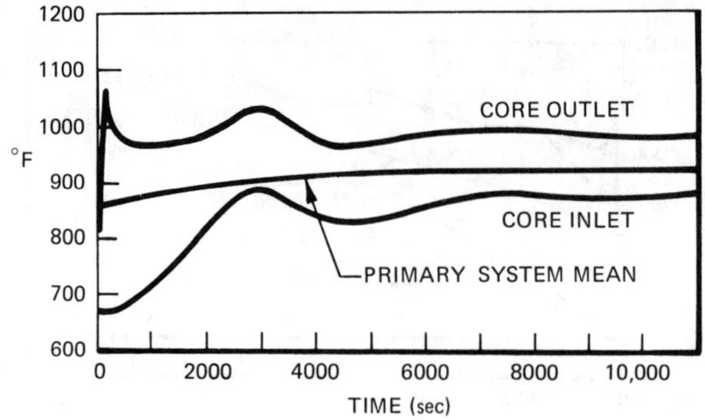

Figure 10. LOOP SYSTEM – LONG TERM TEMPERATURE TRANSIENT (32 MW HEAT REMOVED FROM TOP OF I.H.X.)

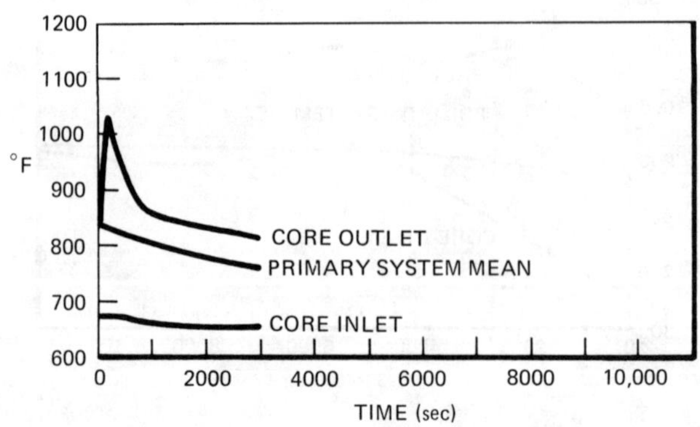

Figure 11. LOOP SYSTEM – LONG TERM TEMPERATURE TRANSIENT (INTERMEDIATE SODIUM SYSTEM AND STEAM GENERATORS)

5. POOL SYSTEM TRANSIENTS

5.1 Short-term Transients

Most of the work in this study was addressed toward the loop system. For comparison a few transient calculations were made on a typical pool system. The flow transients shown in Figure 12 show the effect of 300 seconds of 10% pony motor operation followed by a 20 second coastdown to stop. As in the loop system, the flow reversed for somewhat over one minute, resulting in high reactor temperatures. The dashed curve shows an adverse effect of relocating the top of the IHX. It was assumed that the upper end of the IHX, as well as the entrance level from the hot plenum were both lowered by 6 ft. (1.8 m), but that the bottom of the IHX and its volume remained constant.

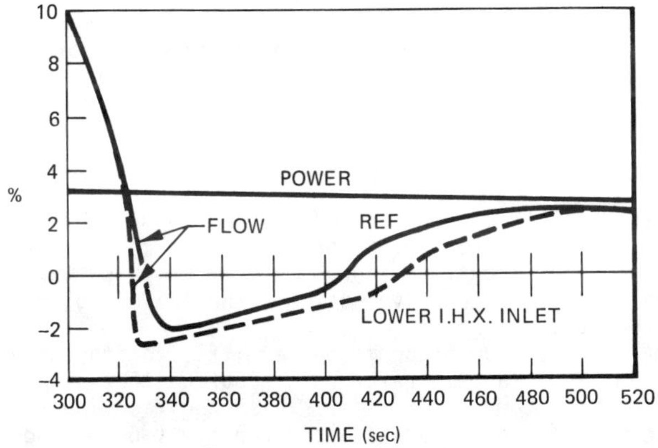

Figure 12. POOL SYSTEM – FLOW TRANSIENT (10% PONY MOTORS, PUMPS STOP 300-320 sec)

5.2 Long-term Transient

Figure 13 shows the temperatures in a pool system with no heat removal and with 32 megawatts of heat removed from the top of the hot plenum. Heat removal at that location has an adverse effect upon core outlet temperature for about the first 9000 seconds, but eventually lowered the temperature. It should be noted that the very large cold plenum at the bottom of the vessel has a pronounced effect upon the rate at which the core inlet temperature rises, compared with a loop system. A more detailed study of reactor core temperatures was made by dividing the core into four parallel sections with different heat generation rates. The results are not presented here, but were as would be expected. Convection causes the flow to be greatest in the hot channel in upward flow, tending to reduce radial temperature peaking. In downward flow, the hot channel flow becomes less than average, and in some cases both upward and downward flow will occur in parallel channels.

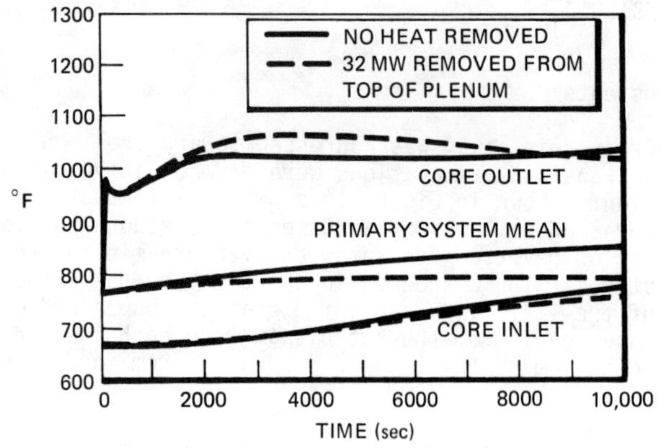

Figure 13. POOL SYSTEM – LONG TERM TEMPERATURE TRANSIENT

6. CONCLUSIONS

It is recommended that a computer code such as GENCON be used early in a design study to determine the most favorable system configuration, and to uncover potential problem areas. Although this paper presents only a few selected transients in representative systems, the following general conclusions can be reached:

- The primary sodium pumps should be permitted to stop only when the primary sodium has a temperature distribution that will result in positive natural convection flow. This condition appears to be most favorable when the pumps are stopped rapidly (∿60 seconds) after reactor shutdown. Delayed pump stopping can cause core flow reversal.
- The initial reactor outlet temperature transient after the pumps stop is only slightly affected by heat removal through a shutdown heat removal system.
- If the initial flow transient is satisfactory, either a loop or a pool system can continue for an extended period of time without pump flow and without heat removal. Under those conditions, core outlet temperatures will remain below about 1100°F (593°C) for about one hour in a loop system and several hours in a pool system.
- Heat removal from a shutdown heat removal system appears to be most favorable at the upper end of the intermediate heat exchanger. Although it tends to decrease core flow, heat removal from the hot reactor outlet plenum appears to be a satisfactory method of limiting long-temperatures in either a loop or a pool system.

Hypothetical Loss-of-Heat-Sink and In-Vessel Natural Convection: Homogeneous and Heterogeneous Core Designs

M. KHATIB-RAHBAR, J. G. GUPPY and A. K. AGRAWAL
Engineering & Advanced Reactor Safety Division
Department of Nuclear Energy
Brookhaven National Laboratory
Upton, New York 11973 USA

ABSTRACT

The preboiling response of homogeneous and heterogeneous LMFBR core designs to hypothetical loss-of-heat-sink (LOHS)accidents leading to in-vessel natural convection cooling with a detailed system representation using the SSC-L code is presented. The results indicate that considerable flow reversal occurs in the core leading to substantial sodium flow recirculation. This recirculation rate is shown to be much lower for the heterogeneous as compared to the homogeneous design. Sodium boiling is predicted to occur at about 90s after LOHS at high decay heat (6% of nomial power). The temperatures remain well below saturation for a case in which LOHS occurs 10 hours after reactor scram (1% of nominal power). Post saturation analyses are required for prediction of clad failure and subsequent relocation.

1. INTRODUCTION

The analysis of hypothesized loss-of-flow and overpower accidents leading to disassembly has received a great deal of attention in LMFBR design and development. To a great extent the concern with these scenarios stems from a number of features that tend to be unique to fast reactor cores. Unlike most thermal reactors, a fast reactor is not in its most reactive configuration. A collapse of the core into a more compact form can lead to a more reactive state and hence recriticality, which could subsequently lead to pressure driven disassembly.

Most investigations of recriticality have focused around unprotected accidents with scenarios leading to disassembly.[1,2] More recently protected accidents with potential for core disruption have been studied by various investigators [3,4,5]. In view of the accident at Three Mile Island-2 nuclear power station [6], the likelihood of protected loss-of-heat-sink accidents (LOHS) is realistically more significant, and should therefore be included in the overall safety evalution of LMFBR systems.

The protected loss-of-heat-sink calculations have used simplistic models to predict the heat transfer and fluid dynamic behavior of the coolant inside the reactor vessel [3,5,7]. Inherent in these models is the assumption that the in-vessel natural circulation flow path is known prior to the analysis; and therefore, neglecting the dynamic interassembly flow redistribution effects which have a significant impact on core thermohydraulic behavior, can seriously alter the transient results.

The purpose of this paper is to present the impact of LOHS and in-vessel natural convection cooling for homogeneous and heterogeneous LMFBR core concepts under two LOHS conditions, namely: (1) loss-of-heat-sink occuring 90 seconds after reactor scram, and (2) loss-of-heat-sink occuring 10 hours after shutdown.

The calculations are performed with a very detailed reactor vessel representation using the Super System Code (SSC) [8]. The Clinch River Breeder Reactor Plant (CRBRP) is used as the reference design [9,10].

2. METHOD OF ANALYSIS

SSC models the entire primary, intermediate and tertiary system of the fast reactor using detailed state-of-the art modeling techniques as described in Reference [8]. In this section, we will set forth the essential features relating to the present study.

The core internals are represented by fuel, axial and radial blanket, control and bypass regions. These regions are made up of closely packed individual rods (except for the bypass region) separated by spiral spacer wire wraps. In SSC, different types of assemblies are represented as a set of one-dimensional, parallel, equivalent channels which are only coupled with each other through (1) hydraulic coupling in lower and upper plena and (2) neutronic coupling. The thermal interactions between individual channels are neglected (this may introduce some inaccuracies in the results during transient where large radial temperature gradients exist).

Under normal conditions, the coolant flow is upward through the vessel. However, under severe accident situations, coolant flow may change direction in some of the channels. The mass flow rate, fluid presure and temperature at the vessel inlet and outlet nozzles interconnect the system dynamics of the vessel and the primary loops.

Calculations of sodium flow redistribution in the reactor vessel are based on the one-dimensional, incompressible flow equation using the Boussinesq approximation, and applied to any of the parallel channels from the bottom of the to the top of the core through various axial regions. That is:

$$\left(\frac{L}{A}\right)_J \frac{dW_J}{dt} = (P_{out} - P_{in}) + g\int_{in}^{out}\rho_J dz + \left(\frac{1}{\rho_{out}} - \frac{1}{\rho_{in}}\right)\left(\frac{W}{A}\right)_J^2 + \left(\frac{fL}{D_e} + K\right)\left(\frac{W|W|}{2\rho A^2}\right)_J \quad (1)$$

where P is the pressure (N/m^2), z is the vertical elevation (m), L is the channel length (m), A is the flow area (m^2), D_e is the flow equivalent diameter (m), ρ is the coolant density (kg/m^3), W is the coolant mass flow rate (kg/s), g is the gravitational acceleration (m/s^2), f is the flow friction factor, and K is loss coefficient due to expansion, contraction and other obstructions in the flow.

The coolant temperature changes as a function of time and position in the channels which influences the density and thus the buoyancy term ($g\int \rho dz$). At steady-state full power, full flow condition, the effect of body forces is insignificant as compared to the frictional forces. However, at post scram, low flow conditions, the buoyancy forces become increasingly significant and often dominant.

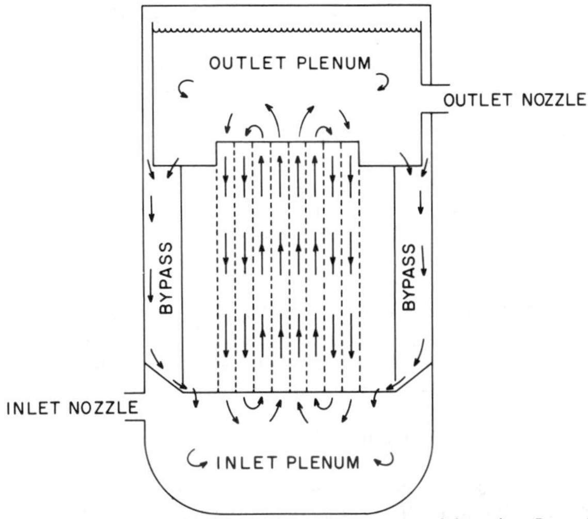

Figure 1. Schematic of the Flow Patterns Inside the Reactor Vessel

Following the LOHS accident the reactor vessel is essentially isolated (see Figure 1) and therefore relying solely on in-vessel circulation for cooling. Earlier studies have assumed that the coolant flow is upward through the fuel zone, and downward through the blanket, control and bypass regions [5] or the bypass region only [7]. Furthermore, the flow was assumed to be quasi-steady and governed by the competition between frictional and buoyancy forces alone; and hence a simplified relationship was obtained which related the coolant circulation rate to the square root of the density difference.

Close inspection of equation (1) reveals that acceleration forces can play a significant role immediately following the loss-of-heat-sink. Furthermore, the flow path is directly linked to the thermal behavior of the parallel channels and can therefore change magnitude and direction during the transient. Another important phenomenon which must be considered is the flow regime as it influences flow redistribution [10]. Fluid flow through wire-wrapped rod bundles does not appear to show the classical transition that is observed for flow in piping between the turbulent and laminar flow regimes. It is important to note that the flow will remain well within the laminar regime following the LOHS event with Reynolds numbers ranging from 0 to 1200. There exists a considerable uncertainty in the type of correlation which must be used. HEDL [12] data shows that the friction factor varies as 84/Re, while WARD [13] measurements seem to indicate 110/Re for Reynolds numbers below 400. The current analysis utilizes the HEDL correlation.

The impact of fuel compostion and irradiation history on core thermal behavior is quite significant. Therefore, channel-dependent decay-heat rate is supplied to the code using the CRBRP data [14]. The fuel and coolant thermal calculations are based on a detailed solution of the transient, one-dimensional energy equation as described in Reference [8].

It is important to note that the present analysis is limited to single phase preboiling range, since the only interest is to determine the time required to reach sodium boiling; however, system cooling can prevail, and internal circulation can continue after boiling inception as studied by Perkins, et al [7] and Dunn [15].

3. SIMULATION OF LOHS FOR HOMOGENEOUS AND HETEROGENEOUS CORE DESIGNS

To improve the safety features of LMFBRs, the use of reactor cores with low positive sodium void reactivity effect has been considered for large power reactor designs. A number of alternative designs have recently been examined as substitutes to the conventional cylindrical core with two zones of fuel enrichment and surrounded by breeder material [10,16]. Among those core designs, the introduction of internal blanket layers into the fissile fuel regions is often foreseen according to various patterns. These designs are termed heterogeneous or parfait as compared with the homogeneous concept.

The parfait designs are generally more advantageous from the point of view of breeding and doubling time. Another intrinsic characteristic of this type of core is the presence of internal blanket assemblies which have relatively slow thermal response characteristic and produce an incoherency in the time and sequence of thermal excursions.

Therefore, from the safety point of view it is important to understand the core dynamics behavior under accident conditions to determine the merits of either concept. For example, the impact of radial temperature distribution coupled with widely different time response characteristic can alter the nature of flow redistribution phenomenon at low flow conditions.

For the purpose of this analysis the recently completed Alterntive Fuel Management Scheme (AFMS-M) of CRBRP will be compared to the Reference Fuel Management Scheme (RFMS) based on homogeneous design. A detailed description of the two designs is given in References [9,10].

The CRBRP-RFMS design consists of 198 fuel and 150 blanket assemblies arranged in a homogeneous configuration. The fuel assemblies are serviced by five orifice zones; the blanket assemblies by four orifice zones; in addition there are 19 control assemblies in the reactor core.

The CRBRP-AFMS-M design consists of 156 (odd cycle) or 162 (even cycle) fuel and 214 (odd cycle) or 208 (even cycle) blanket assemblies aranged in a heterogeneous configuration. The fuel and blanket assemblies are serviced by twelve orifice zones (six each); in addition, there are 15 control assemblies in the reactor core.

For an SSC-L simulation, assemblies can be grouped together. The grouping can proceed according to either similiar hydraulic characteristics, similar power characteristics or similiar power and flow characteristics. In this study, the assemblies were grouped based on similiar power and flow characteristics. This assures preservation of possible core wide incoherencies as they can influence the core transient response characteristics.

Figure 2 shows various orifice zones and the SSC-L channel number for a 30° sector of the RFMS design and a 60° sector of the AFMS-M design. The steady state power and flow fractions assigned to the twenty channel representation are noted in Table I for the homogeneous core and in Table II for the heterogeneous core design. It should be pointed out that only one fuel assembly in the entire core was used to represent the hot fuel channel. Similarly, the hot blanket channel represents only one blanket assembly in the reactor. Note that the power fraction values are expressed in terms of fraction of core-generated heat of 967.4 MW(t). When heat generated by sodium pumps (six of them) are included, the total power becomes 975 MW(t). Heat losses

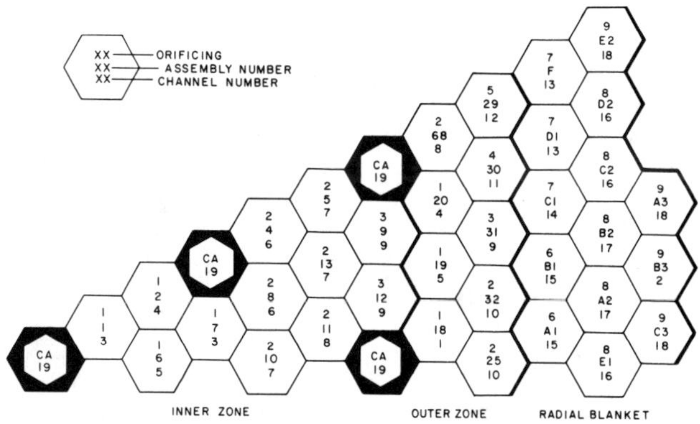

(a)

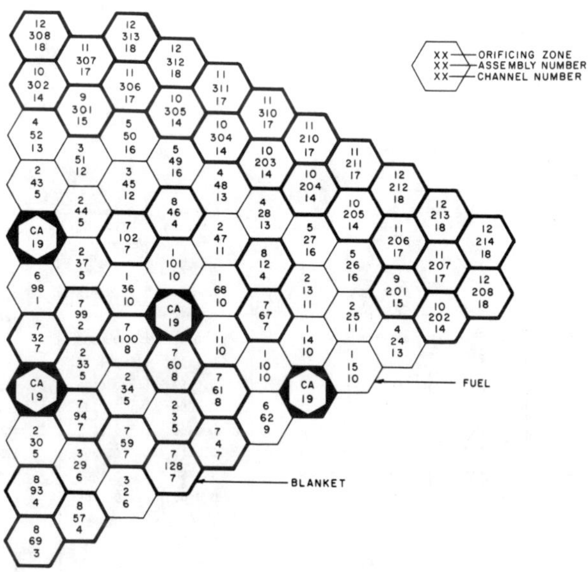

(b)

Fogure 2. Orificing Scheme and SSC-L Channels for CRBRP (a) Homogeneous Core, and (b) Heterogeneous Core.

Table I RFMS Steady State Data (EOEC7)

Channel	Orificing Zone	Type	No. of Assemblies	Flow Fraction	Power Fraction	Temperature Rise ΔT*(k)
1	1	Fuel	1	0.0044	0.0096	315
2	9	Blanket	1	0.0004	0.0008	289
3	1	Fuel	29	0.1290	0.1569	175
4	1	Fuel	18	0.0800	0.0769	138
5	1	Fuel	18	0.0800	0.0890	160
6	2	Fuel	18	0.0718	0.0865	173
7	2	Fuel	24	0.0958	0.1003	150
8	2	Fuel	18	0.0718	0.0890	178
9	3	Fuel	36	0.1352	0.1496	159
10	2	Fuel	18	0.0718	0.0717	143
11	4	Fuel	12	0.0419	0.0402	138
12	5	Fuel	6	0.0183	0.0185	145
13	7	Blanket	18	0.0214	0.0112	75
14	7	Blanket	12	0.0142	0.0100	101
15	6	Blanket	24	0.0308	0.0255	119
16	8	Blanket	30	0.0211	0.0162	110
17	8	Blanket	24	0.0168	0.0171	146
18	9	Blanket	41	0.0154	0.0174	163
19	-	Control	19	0.0383	0.0136	51
20	-	Bypass	--	0.0416	0	0

*Inlet Temperature = 655 K

Table II AFMS-M Steady State Data (EOEC4)

Channel	Orificing Zone	Type	No. of Assemblies	Flow Fraction	Power Fraction	Temperature Rise ΔT*(k)
1	6	Fuel**	1	0.0043	0.0062	208
2	7	Blanket†	1	0.0021	0.0025	171
3	8	Blanket	1	0.0019	0.0019	144
4	8	Blanket	18	0.0333	0.0358	155
5	2	Fuel	36	0.1512	0.1730	165
6	3	Fuel	9	0.0357	0.0428	172
7	7	Blanket	33	0.0696	0.0748	154
8	7	Blanket	23	0.0485	0.0540	160
9	6	Fuel	5	0.0213	0.0264	178
10	1	Fuel	39	0.1760	0.1863	152
11	2	Fuel	18	0.0756	0.0766	146
12	3	Fuel	12	0.0476	0.0566	171
13	4	Fuel	18	0.0656	0.0665	146
14	10	Blanket	36	0.0417	0.0390	134
15	9	Blanket	12	0.0179	0.0164	132
16	5	Fuel	24	0.0852	0.0852	144
17	11	Blanket	48	0.0403	0.0327	116
18	12	Blanket	36	0.0222	0.0150	97
19	--	Control	15	0.0260	0.0083	46
20	--	Bypass	--	0.0340	0	0

*Inlet Temperature ≃ 655 K

**162 Fuel Assemblies

†208 Blanket Assemblies

through insulation and other components are small and, hence, neglected. Total in-vessel flow of sodium is 5224 kg/s. A steady-state plant characterization is performed by the code from the above information and other design data, including power shapes. All conditions here are representative of end-of-life equilibrium cycle (EOEC7 for RFMS and EOEC4 for AFMS-M); except, the fuel hot channel in the RFMS is assumed to be at the beginning-of-life equilibrium core. The AFMS-M hot channel temperatures include the 3σ uncertainty factors only. The decay heat values were taken to be 125% of the nominal values for the fission product heating and 110% of the nominal values for the decay of the transuranic elements. It should be noted that there are 217, 61, and 37 pins in each of the fuel, blanket and control assemblies, respectively.

The LOHS simulations were performed for two different cases; namely (1) LOHS occuring at 90 seconds following reactor shutdown, and (2) LOHS occuring at 10 hours following reactor shutdown.

3.1 LOHS At High Decay Heat Level

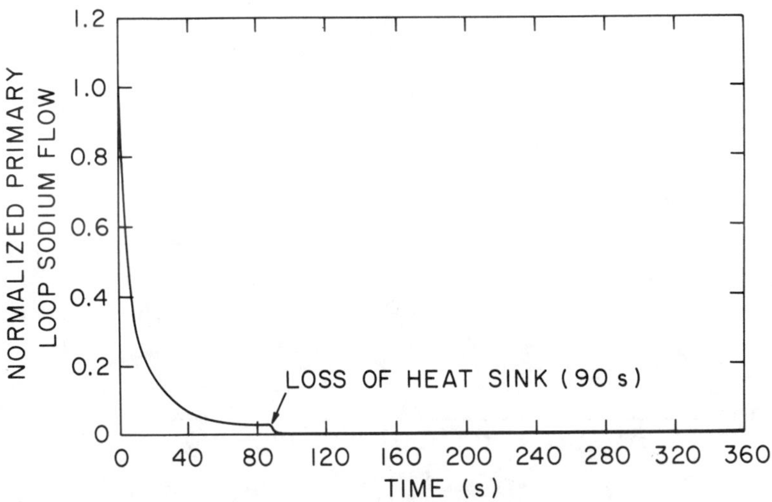

Figure 3. Normalized Primary Loop Sodium Flow Rate

Figure 3 shows the primary loop sodium flow rate as a function of time after scram. It is observed that the flow rate goes to zero at 90 seconds following reactor scram, causing the LOHS. At this time the reactor vessel is totally isolated from the heat transport loops and can rely only on in-vessel natural circulation caused by nonuniform radial temperature gradients. The effect of interchannel flow redistribution is displayed in Fig. 4. Here, the coolant flow fractions, $W_J(t)/W_J(0)$, are shown as a function of time after LOHS for various channels. Immediately following the accident, the flow fractions are seen to drop to nearly zero flow and as interchannel buoyancy starts the circulation, the warmer channels draw more fluid at the expense of the cooler channels. Channels 1 and 2 (hot channels) draw a substantial portion of the coolant, and channel 1 tends to draw coolant as the sodium temperature in this channel is rising (see Fig 5) much faster compared to the hot blanket channel. This incoherency effect is caused by the difference in the characteristic times of the two subassemblies. Also shown in Fig. 4 are the flow fractions in channels 6, 8, 10, 13, 14, and 19 (control assemblies). In general all of the blanket and control and cooler fuel assembly channels undergo flow reversal; and the reversal is more severe in the blanket and control channels as compared to

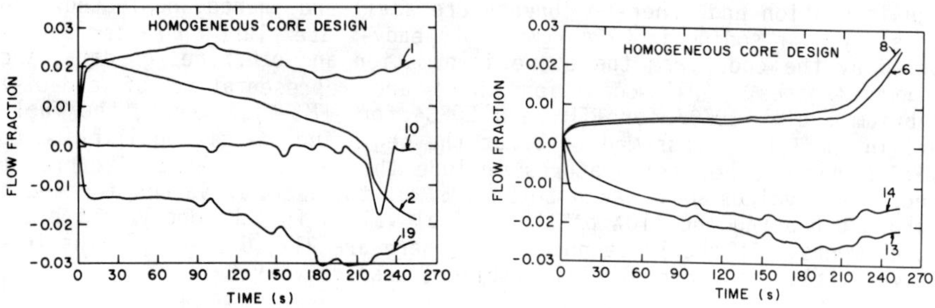

Figure 4. Flow Fractions in Various Channels of the Homogeneous Core Following LOHS (6% Power)

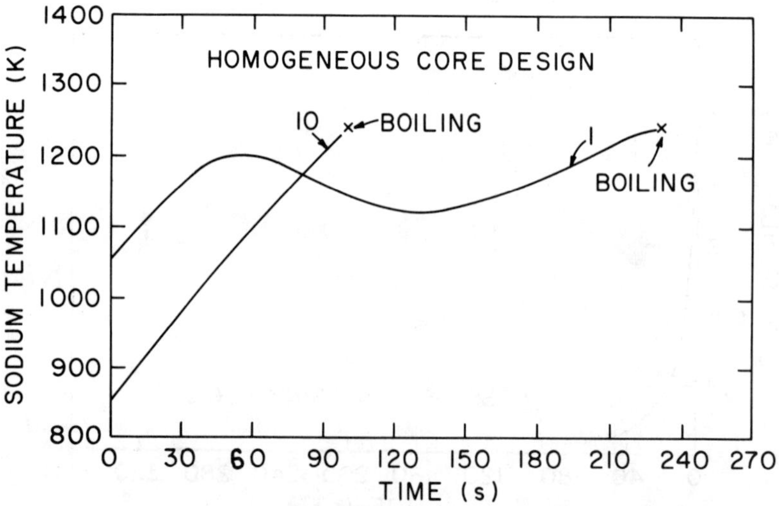

Figure 5. Maximum Sodium Temperature in Channels 1 and 10 of the Homogeneous Core Following LOHS (6% Power)

the cool fuel channels. The most interesting effect is observed in channel 10 (a fuel channel), where the sodium flow rate is essentially zero for an extended period of time leading to an exponential rise in temperature as observed from Fig 5. This figure reveals a very interesting phenomenon which is in contrast to the common belief that the worst conditions are normally reached in the hot channel; as in this case, sodium flow stagnation causes boiling to occur in channel 10 about 2.5 minutes earlier than in channel 1 (the fuel hot channel).

The results also showed no significant temperature rise in the channels with substantial flow reversal, indicating sufficient heat removal capability. It is important to note that the current version of SSC-L is limited to single phase flow only, therefore the oscillation observed in the channel flow rates following boiling initiation are not physical, but only indicative of variations in sodium temperatures in those channels.

Figure 6 shows the typical sodium flow fractions in the heterogeneous core design. It is seen that the behavior of the core is nearly the same as the homogeneous core, except the more radially uniform temperature gradient has reduced the magnitude of the flow fractions therefore an enhancement in the rate at which the coolant temperatures increase as observed from Fig. 7. The sodium temperatures reach boiling in channels 1 and 9 almost simultaneously. In this case some of the radial blanket channels achieved positive flow; and a higher number of fuel channels underwent flow reversal as expected.

Figure 8 and 9 illustrate the transient behavior of the sodium temperature in the reactor inlet and outlet plena for homogeneous and heterogeneous designs, respectively. It is seen that the rate at which the sodium temperature in the plena increase is very small, since the heat capacities of the coolant as well as the plena structural material are quite high, leading to a considerable mitigation of the transient.

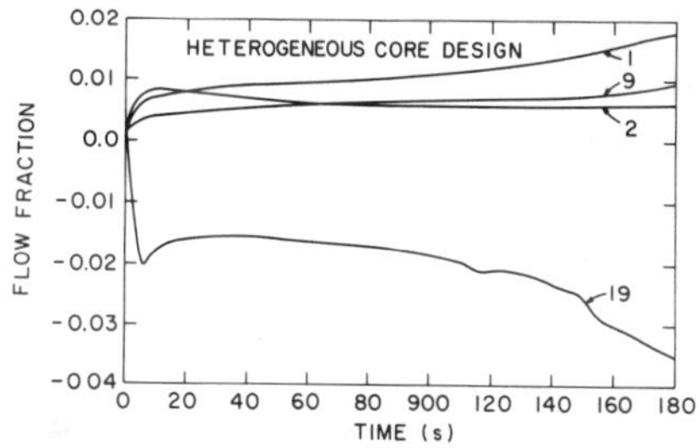

Figure 6. Flow Fractions in Various Channels of the Heterogeneous Core Following LOHS (6% Power)

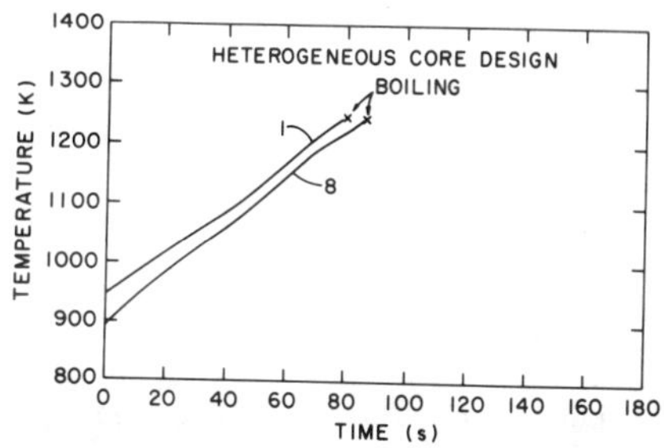

Figure 7. Maximum Sodium Temperature in Channels 1 and 8 of the Heterogeneous Core Following LOSH (6% Power)

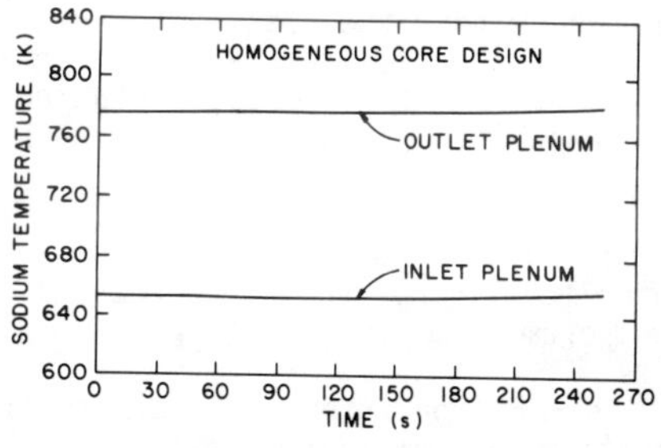

Figure 8. Sodium Temperatures in the Inlet and Outlet Plena of the Reactor Vessel for Homogeneous Core

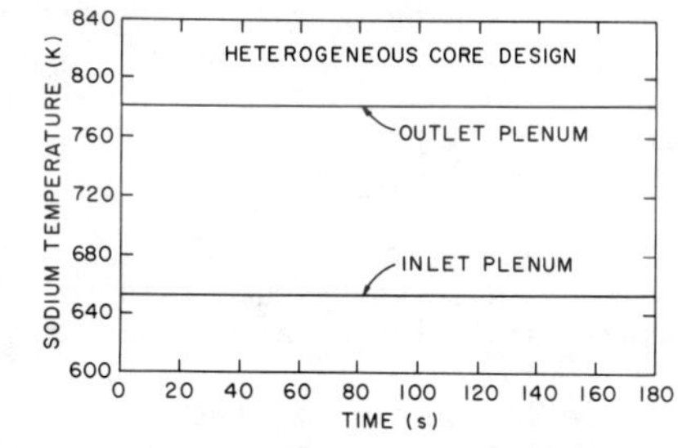

Figure 9. Sodium Temperatures in the Inlet and Outlet Plena of the Reactor Vessel for Heterogeneous Core

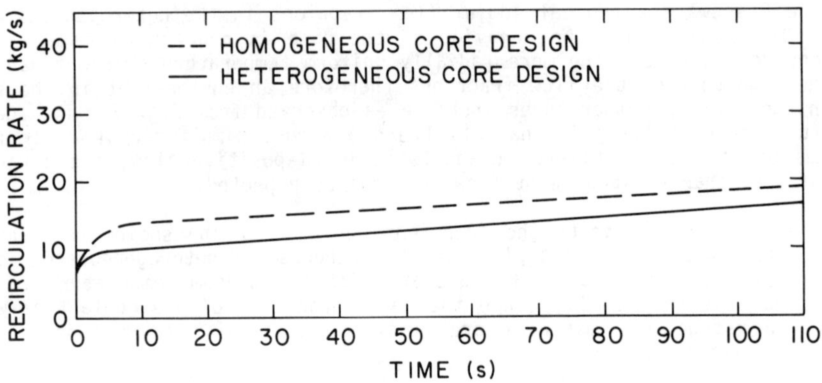

Figure 10. Total Recirculation Rate in the Vessel Following LOHS (6% Power)

The total recirculation rate within the vessel is calculated as the sum of all positive or negative flow rates in the reactor, that is:

$$W_{Ric} = \sum_J \text{positive } (W_J) = \sum_J \text{negative } (W_J) \qquad (2)$$

Figure 10 illustrates the natural circulation rate as a function of time after LOHS for both homogeneous and heterogeneous core design concepts. Immediately following the loss-of-heat-sink the recirculation rate starts from about 7 kg/s which is just the negative flow through the bypass channel (note bypass channel flow reverses prior to the LOHS event, during the initial flow coastdown, since there is no heat produced within the bypass region). As the natural convection flow is established, the recirculation rate starts increasing. It is important to note that the recirculation rate should reach a maximum after which it should start to decrease due to continuous reduction in decay-heat. The total recirculation rate for the homogeneous core is about 20% higher than its heterogeneous counterpart. This is due to smaller radial temperature gradient in the latter case.

3.2 LOHS At Low Decay Heat Level

The decay heat for CRBRP is estimated to decrease to about 1% of the total reactor power (~10 MW (t)) in about 10 hours after reactor scram [14]. It was argued earlier that the natural circulation rate within the vessel is higher if the radial temperature gradient is larger; therefore, a delayed loss-of-heat-sink occuring several hours following the reactor shutdown was also studied to determine the impact of isothermality on the establishment of in-vessel natural convection cooling.

The sequence of events leading to the LOHS consists of (1) reactor scram at time zero, followed by pump trip to pony motor level. (2) decay heat removal for 10 hours, and (3) LOHS at 10 hours following scram.

Table III summarizes the reactor vessel thermal condition prior to the LOHS at 1% power level, 10 hours after reactor shutdown. Comparisons of these values to those at 100% full power condition of Tables I and II show a considerable reduction in the radial temperature gradient for both design concepts.

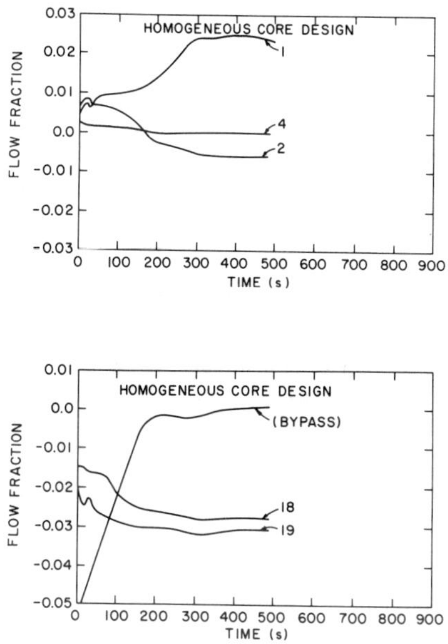

Figure 11. Flow Fractions in Various Channels of the Homogeneous Core Following LOHS (1% Power)

Table III Reactor Temperature Rise at 1% Power

Channel No.	Temperature Rise, ΔT*(K)	
	RFMS	AFMS-M
1	46	31
2	45	23
3	26	20
4	20	21
5	23	24
6	26	25
7	22	21
8	26	22
9	24	26
10	21	22
11	20	21
12	21	25
13	12	25
14	16	23
15	19	22
16	19	25
17	26	22
18	31	21
19	4	3
20	0	0

*Inlet Temperature = 557 K

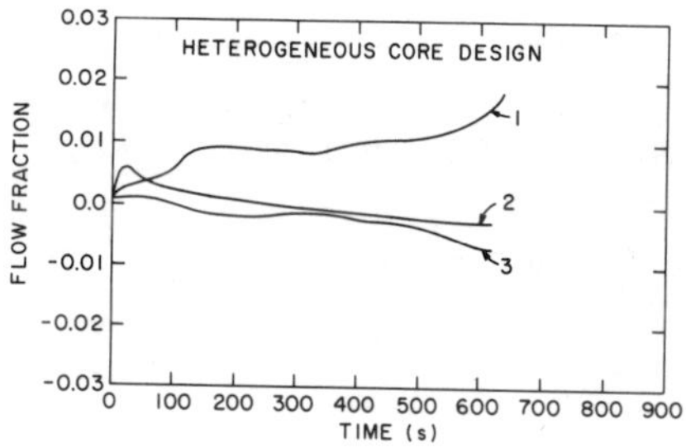

Figure 12. Flow Fractions in Various Channels of the Heterogeneous Core Following LOHS (1% Power)

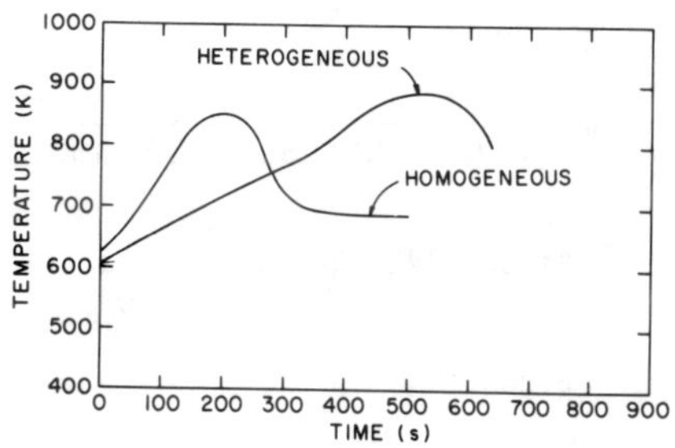

Figure 13. Maximum Sodium Temperatures in the Hot Channel of the Homogeneous and Heterogeneous Core Designs Following LOHS (1% Power)

Figure 11 illustrates the flow fractions in several typical channels for the RFMS design. It is evident that small radial temperature graident produces a period of oscillation before natural convection cooling is established. Also evident is the incoherency effect between the hot fuel and blanket channels caused by the substantial difference in their corresponding characteristic times.

The bypass channel flow fraction shown in Fig. 11 is seen to be the first to reverse and then, as the natural circulation starts to stablize it tends to increase and approach stagnation.

A similar behavior was observed for the heterogeneous core; with the exception of greater oscillation for the first 2 minutes following the accident as seen in Fig 12.

Figure 13 shows the maximum sodium temperature for both homogeneous and heterogeneous core designs. It is seen that the initial excursion causes a sharp increase in sodium temperatures in the hot channel. The temperature and flow behavior is also seen to be similar, and these variations are expected to continue until the initiation of boiling. It was also observed that all blanket and control assemblies and five cooler fuel assemblies experienced flow reversal and/or stagnation. The total recirculation rate within the vessel is shown in Fig 14. It is seen that it takes approximately 2 minutes before the natural circulation flow is established, and as expected the flow rate is much higher for homogeneous cores as compared to its heterogeneous counterpart.

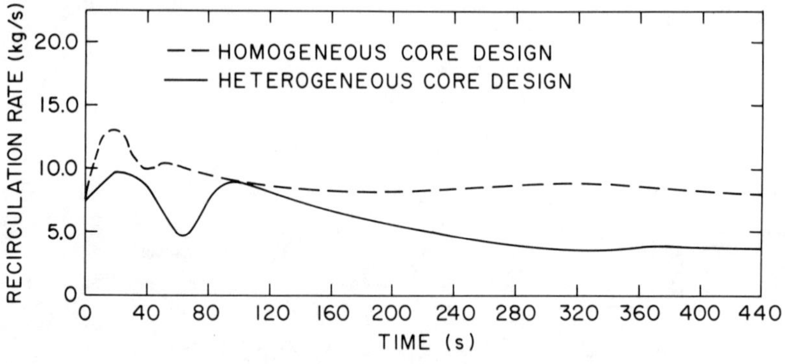

Figure 14. Total Recirculation Rate in the Reactor Vessel Following LOHS (1% Power)

4. DISCUSSIONS

The results for hypothetical loss-of-heat-sink with reactor scram were presented under conditions of high and low decay heating levels for both homogeneous and heterogeneous core designs in the CRBRP. The study shows that, detailed representation of reactor core is essential for accurate prediction of the onset of natural convection cooling; the results also show substantial flow reversal in most of the reactor with possibility for flow stagnation leading to rapid boiling in certain channels.

The study shows the need for further investigations using more detailed subchannel codes, such as COMMIX [17] and TEMPEST [18] for determination of local subassembly flow recirculation; and examination of occurance of thermal instability in the channels with substantial flow reversal (in these channels the higher temperatures tend to occur towards the lower end of the subassembly). The impact of axial conduction at very low Reynolds number conditions is quite significant and requires further treatment [19]. Furthermore, the study should be extended to include post saturation simulation. Limited analytical and experimental [7,15,20,21] evidence exists to justify successful decay heat removal for an extended period.

ACKNOWLEDGEMENT

The authors acknowledge the invaluable help of R. J. Kennett and E. G. Cazzoli in exercising the code. Thanks are also expressed to P. Siemen for the excellent job done in preparing the paper

This work was performed under the auspecies of the United States Nuclear Regulatory Commission.

REFERENCES

1. Speis, T. P., et. al., "Studies of Core Disruptive Accidents and Licensing aspects of Fast Breeder Reactors", First European Nuclear Society Conference, Paris, France (1975).

2. Stevenson, M. G., et. al., "An Overview Assessment of Energetic core Disruptive Accidents", in Proc. of the International Meeting on Fast Reactor Safety Technology, Seattle, Washington, USA (1979).

3. Bari, R. A., et. al., "Preliminary Assessment of the Meltdown Progression of the Loss-of-Heat-Sink Accident with Scram in the LMFBR", Brookhaven National Laboratory Informal Report, BNL-NUREG-23137 (1977).

4. Bari, R. A., et. al., "Phenomena and Scenarios Related to a Loss-of-Heat-Sink Accident (with scram) in an LMFBR", in Proc. of the International Meeting on Fast Reactor Safety Technology, Seattle, Washington, USA, (1979).

5. Chan, C. K., Min, T. K., and Okrent, D., "A Look At Alternative Core Disruption Accidents in LMFBR's", University of California, Los Angeles, UCLA-ENG-7720, (1977).

6. "Report of the President's Commission on the Accident at Three-Mile Island, Washington, D.C., (1979).

7. Perkins, K. R., Bari, R. A., and Pratt, W. T., "In-Vessel Natural Circulation During A Hypothetical Loss-of-Heat-Sink Accident in the Fast Flux Test Facility", ASME Winter Annual Meeting Paper No. 79-WA/ HT-66, (1979).

8. Agrawal, A. K., et. al., "An Advanced Thermohydraulic Simulation Code for Transients in LMFBRs (SSC-L)," Brookhaven National Laboratory, BNL-NUREG-50773 (1978).

9. Clinch River Breeder Reactor Plant Preliminary Safety Analysis Report, with amendments through 51, Project Management Corporation (1975-1979).

10. Clinch River Breeder Reactor Plant, Technical Review, Project Management Corporation, CRBRP-PMC-79-01, (1979).

11. Khatib-Rahbar, M., and Cady, K. B., "Establishment of Buoyancy-Induced Natural Circulation in Loop-Type LMFBRs", Trans. Am. Nucl. Soc. 28, 432 (1978).

12. Tang, Y. S., Coffield, Jr., R. D., and Markley, R. A., Thermal Analysis of Liquid Metal Fast Breeder Reactors, American Nuclear Society Monograph Series, (1978).

13. Engel, F. C., Markley, R. A., and Bishop, A. A., "Laminar Transition and Turbulent Parallel Flow Pressure Drop Across Wire-Wrap-Spaced Rod Bundles", Nucl. Sci. Eng. 69, 290 (1979).

14. McGinnis, C. A., and Tomlin, D. O., "CRBRP Decay Power Analysis", Westinghouse Advanced Reactors Division, WARD-D-0090 (1976).

15. Dunn, F. E., "Severe FFTF Natural Circulation Transients with Boiling", Argonne National Laboratory, ANL/RAS 76-26 (1976).

16. Renard, A., Evrard, G., and Wehmann, U., "Comparative Study of Heterogeneous and Homogeneous Liquid-Metal Fast Breeder Reactor Cores in Some Accident Conditions", Nucl. Tech. 46, 453 (1979).

17. Sha, W. T., et. al., "COMMIX-1: A Three-Dimensional Transient single-Phase component Computer Program For Thermal-Hydraulic Analysis", Argonne National Laboratory, NUREG/CR-0415, ANL-77-96, (1977).

18. Trent, D. S. "TEMPEST: A Three-Dimensional Time-Dependent computer Program For Hydro-thermal Analysis, " Battele Pacific Northwest Laboratory, FATE-79-105, Informal Report (1979).

19. Meyer, J. E., "Some Physical and Numerical Considerations for SSC-S Code", Brookhaven National Laboratory, BNL-NUREG-50913 (1978)

20. Klein, G., and Dunn, F., "SAS3D Analysis of Natural Convection Boiling Behavior in the sodium Test Facility", Trans. Am. Nucl. Soc. 33, 517 (1979).

21. Senglaub, M. and Ishii, M., "Predictive Method for Na Boiling Under Low Heat Flux Condition Typical of Loss-of-Heat-Sink Accidents," Trans. Am. Nucl. Soc. 33, 519, (1979).

A Test Loop Concept for Verification of Natural Convection Decay Heat Removal in LMFBR'S

J. C. MILLS, R. T. LANCET, and M. J. GABLER
Atomics International Division, Energy Systems Group,
Rockwell International

H. R. ZWEIG
Energy Technology Engineering Center, Energy Systems Group,
Rockwell International

ABSTRACT

This paper summarizes the preconceptual design of an experimental facility to demonstrate natural convection decay heat removal in LMFBR's. Experimental program objectives and their translation into specific facility requirements are presented. A description of the proposed test loop and facility configuration is provided to establish the feasibility and economic attractiveness of conducting a meaningful experimental program.

1. INTRODUCTION

1.1 Technical Problem

The two primary safety requirements for LMFBR's are to safely shut the core down and remove the core decay heat for all credible transients. Currently, base technology programs are in place which should provide shutdown systems with sufficient reliability. A corresponding effort is required to assure the required shutdown heat removal system (SHRS) reliability. Current large LMFBR designs have highly reliable SHRS composed of diverse and redundant subsystems. Studies to date, however, show that to achieve the required reliability by the most technically feasible and cost-effective means, at least the diverse system must be capable of performing its functions without external power sources, i.e., under natural convection. Demonstration of this attribute of the SHRS is also a prime mission of the U.S. LMFBR safety research program, Line-of-Assurance 2 (LOA-2) objective to demonstrate safe shutdown via inherent processes. The importance of inherent natural convective shutdown cooling is further enhanced as an outgrowth of the lessons learned from Three Mile Island (TMI).

A supporting data base for natural convection, both analytical and experimental, is required if full credit for enhanced reliability of the SHRS is to be realized. In the analytical area, computer codes, comprised of detailed models are needed to adequately predict the behavior of shutdown heat removal systems under low power, low flow conditions. These codes would be utilized to both guide the design development of the SHRS and perform confirmatory analysis of designed systems to demonstrate safety margins. A companion experimental program is required to provide a data base to verify these analytical codes. In addition, a well-conceived experimental program would provide proof-of-principle data to compensate for uncertainties and limitations associated with even the most sophisticated analytical predictions. Included in this latter class of experiments are carefully designed scale model tests of fully developed SHRS concepts to support licensing acceptance of their performance. The SHRS for the Conceptual Design Study (CDS) would be a prime concept requiring such a program.

Accordingly, the technical problem being addressed in this paper is the development of an experimental concept which can provide the required prototypic scale model testing of natural convection mode operation of the CDS SHRS. The concept should be developed on a schedule compatible with the CDS program and within the funding constraints imposed by cost-effective utilization of U.S. LMFBR safety research resources.

1.2 Background

The SHRS for CDS is to be comprised of the normal heat removal system (i.e., through the main condenser) and the DRACS-1 and IRACS-1 systems shown in Figure 1. At least one of these dedicated systems must perform its safety function:

a) without any sources of power
b) using natural convection to promote coolant flow
c) with no operator action
d) with the minimum of active components.

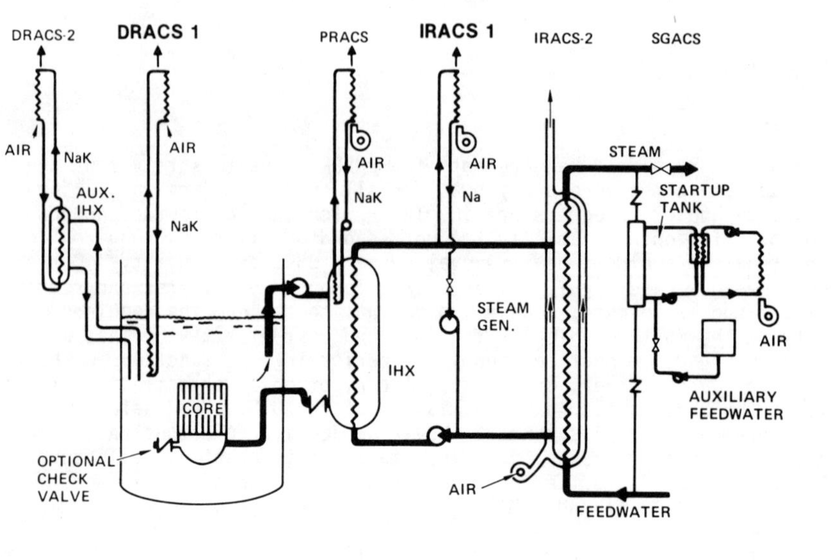

Fig. 1. Alternate Dedicated Heat Removal Systems for CDS LMFBR

The system in Figure 1, which clearly satisfies these requirements, is the Direct Reactor Auxiliary Coolant System (DRACS) concept. As illustrated, the reference DRACS is an in-vessel heat exchanger which transfers heat from the reactor hot pool to natural draft NaK-to-air heat exchangers. The concept may include a check valve (normally held closed by coolant flow from the primary loops) or flow control devices to create a flow circulation path within the reactor. However, the concept should still perform its safety function without these flow control devices using inherent flow recirculation paths within the core (i.e., upflow in the hot fuel assemblies and downflow in the colder blanket or reflector assemblies).

It is anticipated that CDS will require prototypic experimental confirmation of the selected DRACS concept to meet licensing requirements. Key areas to be addressed in the experimental program relate to the resolution of uncertainties in both steady-state and transient natural convection performance. These major uncertainties include flow stratification in the reactor vessel, secondary flow in large piping, response to thermal and hydraulic transients, initial flow reversal, grace period, and inherent performance. The intent of this paper is to develop a preconceptual design of an out-of-pile experimental facility to meet these objectives.

1.3 Program Interfaces

As mentioned earlier, prototypic scale model testing comprises just one facet of the natural convection verification program. Numerous analytical efforts in the natural convection area are underway which potentially relate to the program proposed here. For instance, WARD and GE are developing detail reliability codes which can be used to define reliability requirements for the various DRACS operational modes. These requirements can be directly translated into temperature limits for key natural circulation scenarios against which experimental results can be measured to characterize safety margins. Several natural circulation codes, varying in the degree of detail, have been or are being developed. For example, ANL is developing a large scale computer code SASSYS, which couples complex plant simulation codes (i.e., SSC) with a detailed core treatment (i.e., SAS4A). SASSYS, when completed should be capable of being utilized to conduct confirmatory analysis of the performance of the CDS DRACS and as such would not only assist in defining test conditions for the out-of-pile scale model experimental program, but directly benefit from test results (i.e., model development and verification).

In the experimental area, key programs are in progress at ORNL, FFTF, and EBR-II. ORNL is proposing to modify its THORS facility to conduct experiments which specifically address core coolability under low flow, low power conditions expected during natural convective cooling by the SHRS. The results of this program can be utilized to verify the viability of recirculation flow paths within the core for a case where DRACS flow control devices are inoperative. In addition, the existence of the ORNL program can both simplify and define the core modelling requirements for the out-of-pile system test facility. FFTF natural circulation experiments are in progress and will provide a good experimental data base for certain aspects of computer code verification, particularly, with respect to the performance of the primary and secondary heat transfer loops. EBR-II can significantly add to the FFTF data base, because of its increased experimental flexibility and its unique ability to simulate the electrical generation portion of the plant. As such, EBR-II is highly suited to provide experimental verification of portions of the major computer codes.

The out-of-pile experimental concept proposed here provides two essential characteristics not found with a single facility elsewhere.

a) Prototypical representation of the CDS DRACS system. The unique and, at times, complex geometries of prototype design, dictate a need for carefully designed scale model testing.

b) An out-of-pile facility offers the flexibility to go to higher system temperatures than in-pile facilities (i.e., FFTF and EBR-II) due to both safety and investment cost advantages associated with non-fuel operations.

The remainder of the paper is devoted to a discussion of the proposed out-of-pile facility including program objectives, facility requirements, and facility design.

2. PROGRAM OBJECTIVES

The overall objective of the proposed out-of-pile experimental program is to demonstrate, via testing of a truly prototype system, the natural circulation performance of the CDS DRACS. The top level objective will be satisfied within the constraints of the following subobjectives.

a) The selection of test scale, test fluid, and test configuration should be made on a technical basis whereby the major uncertainties associated with natural convection (e.g., flow stratification, transient behavior) are adequately modeled such that their resolution is achievable.

b) The experimental facility should possess sufficient flexibility to test the DRACS performance under a wide variety of possible operating modes (e.g., with and without flow control devices, interactions with other SHRS systems).

c) The schedule for implementation of the experimental program should be compatible with the CDS program which requires initial results be made available to DOE in the 1984-85 timeframe.

d) The costs of the experimental program should be kept to the minimum level that is consistent with providing technically sound results.

3. FACILITY REQUIREMENTS

3.1 Test Fluid

The prime candidate fluids for the experimental program are prototypic fluids (i.e., Na, NaK) or water with it potential for increased flexibility at at lower cost. Therefore, AI-ESG, in conjunction with ANL, conducted an evaluation[1] of the adequacy of water for natural convection experiments involving the DRACS. This evaluation included a derivation of the dimensionless similarity parameters from the basic conservation equations and an assessment of the abilities of a water model to match these parameters as they exist in the full-scale prototype.

The key conclusion emanating from the study is that water cannot adequately model the thermodynamic behavior expected under buoyancy-dominated transient low flow conditions ($<3\%$), typical of natural convection scenarios. The low flow promotes a heat transfer regime where conduction effects become increasingly important, and thus water, with a thermal conductivity significantly less than sodium, cannot be expected to adequately model thermal behavior in this regime. Figure 2 compares the heat transfer coefficients of water to both NaK and Na for a range of velocities and effective flow diameters. For reference, Figure 2 highlights key design points which reflect scaled versions of the current CDS/DRACS concept. As the velocity and pipe diameter are reduced (i.e., reflective of either low flow natural convection conditions and/or reduced scale modelling) conduction through the laminar component of the boundary layer becomes a larger portion of the overall heat transfer coefficient, thus accounting for the increase in Na/NaK to water ratios. Even at full-scale simulations, utilization of water to model the behavior of core channels at low flow conditions, would introduce significant errors in the heat transfer coefficient. While models can be constructed to compensate for these differences during steady-state simulations, extrapolation of water test results for transient behavior could lead to erroneous conclusions. Although the effect in the NaK 12-in. diameter piping is not nearly as pronounced, consideration of the smaller diameter, NaK tubing in the heat exchangers would lead to similar arguments regarding the adequacy of a water simulation. Analogous arguments for the thermodynamic behavior of the plenum (based on matching Peclet numbers) and for structure heat transfer (based on matching Biot numbers) can be constructed to further demonstrate the inadequacy of water under low flow conductance influenced conditions. The reader is referred to Reference 1 for the details.

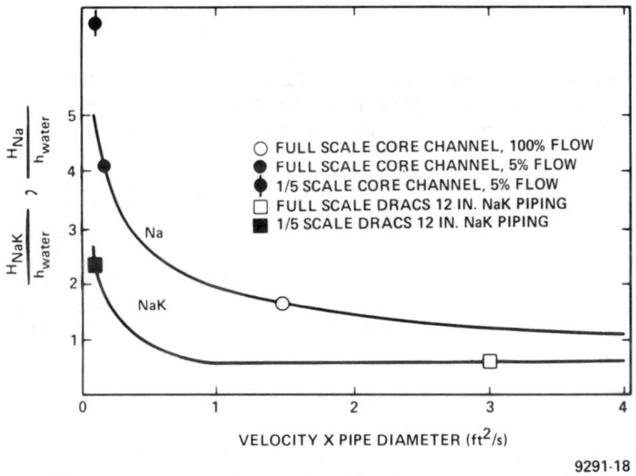

Fig. 2. Comparison of Na/Water and NaK/Water Heat Transfer Coefficient Ratios

It is significant to note that the above conclusions are not in conflict with those reported by ANL[2] in their outlet plenum behavior studies. While their results indicate reasonable agreement between water and sodium even with reduced scale models, those experiements were performed at higher flow conditions (10% of full flow) where turbulent, reasonably well mixed fluid conditions exist. The conduction heat transfer effects mentioned here would not be a significant factor under those circumstances.

Based on the above conclusions, a decision was reached to dismiss water as a viable candidate fluid for the experimental investigation of DRACS natural convection behavior.

3.2 Test Scale

The task of selecting an appropriate test scale is a complex one and at best still involves a compromise. The process utilized was to begin with a detail derivation of the dimensionless parameters from the basic conservation (mass, momentum, and energy) equations for each of the significant components to be simulated. These include the outlet plenum, the core, the NaK piping, the Na-to-NaK heat exchangers, and the NaK-to-air natural draft heat exchangers. Then, to reduce the number of variables to be studied, the following model parameters were fixed:

a) prototype test fluid (i.e., NaK and Na)
b) core average power density identical to prototype
c) identical ΔT as the prototype

Under these conditions, such key dimensionless parameters as Biot number, Prandtl number and $\beta\Delta T$ are inherently matched, and complete similarity requires the following[1].

a) Hydrodynamic similarity requires matching of the Richardson number (or its equivalent in certain geometries). This requires matching ratios of length to some power of the velocity (i.e., L/V^2).

b) Thermodynamic similarity requires matching of the Peclet, or by inference the Reynolds number. This requires matching the product of length time velocity (i.e., LV).

Clearly, both of these conditions cannot be satisfied within a scaled model (<1.0). The situation is further complicated by the fact that structure heat transfer considerations dictate a need to scale diameter different than the length.

These preliminary findings lead to a conclusion that the test scale should be as large as practical for the test simulation. Fortunately, the ANL outlet plenum studies[2], while not totally applicable here because of higher flow rates, tend to indicate that some approximate behavior in the thermodynamic modelling can be accepted. While a water simulation would introduce too large a thermodynamic uncertainty, a reduced scale sodium model should be acceptable with a limited sacrifice in thermodynamic similarity.

Based on the above, the following scaling requirements have been selected for the initial scoping of an experimental facility.

a) A general 1/5 test scale, primarily based on the availability of existing equipment at the Energy Technology Engineering Center (ETEC), the proposed test site. The economic advantages of this selection are presented in Section 4.

b) Implement the three assumptions mentioned above for test fluid, core power, and ΔT.

c) For certain structural components (i.e., NaK piping, NaK heat exchangers), the diametric scaling factor should be enlarged to ~ 0.3 to more accurately model structural heat transfer.

d) The length scaling factor should remain 1/5 with the key criteria being the maintenance of the vertical distance between thermal centers at precisely 1/5 scale.

It is significant to note that the above scaling requirements were for the expressed purpose of conducting a meaningful design feasibility assessment. Therefore, additional refinements will be explored (e.g., pie-shaped scaling) which have the potential of effectively increasing the scale of the test while falling within the economically attractive facility envelope described in Section 4.

3.3 Experimental Conditions

The DRACS must be proven to provide reliable operation for a wide range of circumstances. A list of potential DRACS operational modes is shown in Table 1 along with typical performance limits. The conditions shown in Table 1 are not complete but probably bracket the scenarios of interest and therefore provide a reasonable basis for developing facility requirements which assure that these conditions can be tested.

TABLE 1

DRACS TEST FACILITY SIMULATION FLEXIBILITY REQUIREMENTS

DRACS Configuration Options	
In-vessel or ex-vessel heat exchangers	
Various flow control options	
None (Super Phenix)	
Check valve (CDS)	
Flow diodes (SNR)	
Typical DRACS Operational Modes	Current CDS Recommended Safety Limit (Service Condition)
In parallel with normal DHRS	A
In parallel with IRACS	B
With pony motors on/ check valves open or closed	B
With pony motors off/check valves open	B
With pony motors off/check valves closed	
Primary loops functional	C
Primary loops not functional	Maintain Cladding Integrity

Since the intent of the experimental program is to determine the behavior of the DRACS, prototypic scaling is being confined to the reactor vessel, vessel internals, the core, and everything within the DRACS loop including check valves if they are part of the design. Under these circumstances, the key facility requirements relate to the mass flow rate, temperature ΔT, and temperature rate change capabilities the supporting facility must possess to: (1) simulate the dynamic behavior of any systems shown in Table 1 which are downstream of the reactor vessel, and (2) establish initial test conditions within the vessel. In addition, the supporting facility should be able to provide the electrical power needed for the core simulation.

Rigorous definition of these requirements can be provided by an overlay of the scaling assumptions on a detail analysis of the prototypic CDS system dynamic response for the scenarios shown in Table 1. Unfortunately, there is a limited amount of available analysis of the CDS DRACS transient behavior to draw upon. Therefore, an alternate approach of selecting a conservative set of bracketing conditions was employed to set preliminary requirements. Upon inspection of Table 1 scenarios for prototypic systems, the supporting facility must be able to simulate the following conditions.

a) A maximum of 8% of full flow with a core ΔT of 280°F (i.e., representing transients under pony motor operation)

b) A maximum core ΔT of 750°F at near zero-vessel inlet flow (i.e., represents a worst case where the PHTS is loss)

c) A collapsed ΔT condition (e.g., less than 100°F ΔT)

d) A maximum rate of change in either vessel inlet or vessel outlet sodium temperature of 2°F/s

e) A maximum core power of 8% of full power.

Using the scaling requirements defined in Section 3.2, 8% flow corresponds to $\sim$30 lb/s and 8% power to 1.7 MWt. The ΔT and temperature rates remain the same as in the prototype.

These requirements are sufficient to permit sizing of the facility heat rejection system, sodium storage capacity, pumping system and piping system. In addition, the facility must incorporate sufficient design flexibility (i.e., variable control of electrical power, pump speed, inlet temperature) to be able to provide controlled test conditions. For example, simulation of certain scenarios will require maintenance of prescribed vessel inlet temperature and flow histories in the presence of an unknown vessel outlet temperature (a desired experimental output and therefore not a controlled variable).

4. FACILITY DESIGN

A preconceptual design of a DRACS experimental facility has proceeded on the basis of the requirements defined in the previous section. Where possible, utilization of existing equipment and capability was implemented to develop the most economically attractive design. The intent of the design study was to resolve any feasibility issues.

The Sodium Components Test Loop (SCTL) at ETEC will be the site for testing of a scale model natural circulation decay heat removal system. It is an operational sodium facility which has been used in the past for testing a variety of components such as the FFTF secondary isolation valve, the HEDL Type II cold trap, the AI L-100 EM pump, and the AI subscale inducer pump. The SCTL is currently in use for testing Class 1 valves for CRBR and the CRBR IVTM.

Following completion of the current test efforts, the SCTL will be modified into a two-loop configuration: each loop will have a 3500-gpm centrifugal pump for primary circulation. Also, a new control room will be added and the electrical switchgear capability will be expanded. The facility will resume operation in December 1980.

Figure 3 is a process flow diagram of the STCL, as modified. A principal feature is the core tank, a 8-ft-diam vessel, 1/5 scale of the CDS reactor vessel, which is mounted vertically. It has a flat top lid and a standard semi-elliptical bottom head. The overall length of the tank is 31 ft. The design temperature of the vessel is 1200°F. Assuming a 400°F sodium inlet temperature, the 750°F maximum ΔT requirement should be readily accommodated. This existing tank will be used to simulate the reactor vessel in the DRACS experimental program.

Modifications to the tank will be required to install the electrically heated, simulated reactor core. Also, a new lid will have to be built to accommodate interfaces with the NaK system, and to provide access from above to the tank internals. With regard to the latter, the design of the simulated in-vessel sodium-to-NaK heat exchanger and the core size envelope will be based on exact replicas of the CDS design, scaled according to guidelines defined in Section 3. A key design feature is the prototypic simulation of the CDS heterogeneous core design. Current plans call for a design comprised of five flow zones separated by cylindrical partitions. Within each zone, sodium-immersed

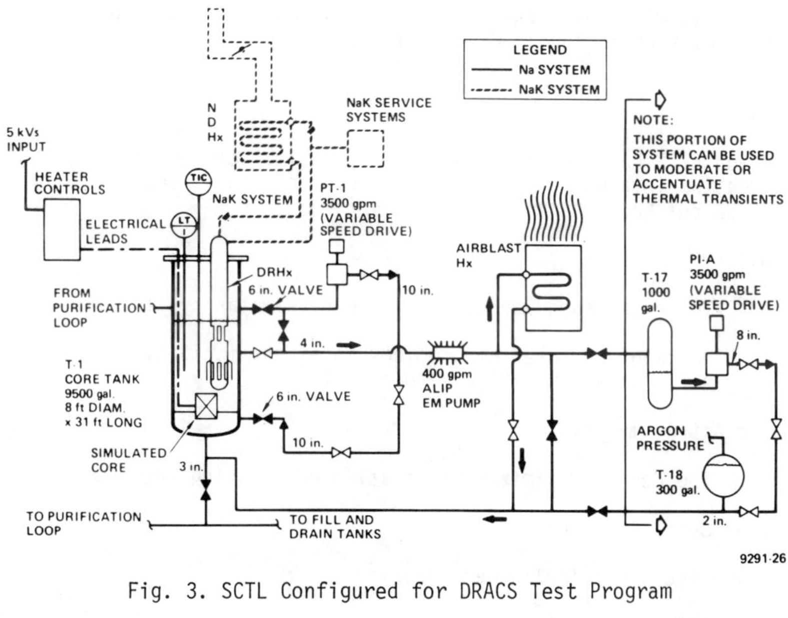

Fig. 3. SCTL Configured for DRACS Test Program

Inconel-sheathed heater elements will be used to simulate core power. On a zonal basis, these elements will be controlled and, in conjunction with flow orifices, can provide adequate simulation of power/flow ratios prototypic of heterogeneous cores.

As part of the further modifications to SCTL to accommodate the DRACS test program, a NaK piping system designed to simulate a large plant installation will be added between the core tank and the natural draft-air-dump heat exchanger. The heat exchanger, with an appropriately scaled heat rejection capability, will be supplied by AI and installed by ETEC at an elevation which will locate its thermal center consistent with the simulation requirements of the test program.

The primary and intermediate loops of the large plant will be simulated using the existing SCTL sodium system with the addition of a forced-draft cooler (utilizing a tube bundle now in storage at ETEC) and a low-flow circulation system. The prime mover for the latter will be the 400-gpm CRBR EM pump soon to be tested in the SCTI at ETEC; the EM pump will be available for facility use after completion of that test.

Distribution and control of the electricity for the approximate 2-MW in-vessel core simulator will be a major consideration in design of the test system. ETEC is presently serviced from a 10-MVA substation; the test facilities are serviced by 4160-V feeders. Within these capabilities, the feeder to the SCTL should be able to handle the peak-load requirement of the DRACS test.

Both AI and ETEC have been authorized by DOE to proceed with conceptual design studies of the DRACS test system in the SCTL. The primary objective is to develop cost and schedule planning data. Also, any areas of major design uncertainty and items for long-lead procurement will be identified. It is anticipated that if Title I design were to start in January 1982, then the modified facility, with test articles installed, could be ready for acceptance test and subsequent operation in January 1985.

5. CONCLUSIONS

AI-ESG and ETEC have developed a viable test loop concept for verifying natural convection in a DRACS. Based on the current study, the following conclusions can be reached:

Feasibility

a) Technical adequacy is supported by similarity studies.

b) No major engineering problems in either facility layout or in satisfying simulation requirements.

Economic Attractiveness

a) The design maximizes utilization of existing resources

8-ft-diam tank

ETEC electrical power capacity

SCTL equipment and capabilities.

REFERENCES

1. Heisler, M. P. and Singer, R. M., "Facility Requirements for Natural Convection Shutdown Heat Removal System Testing," to be presented at Specialists' Meeting on Decay Heat Removal and Natural Convection in FBRS, BNL, February 1980.

2. Lorenz, J. J., Carlson, R. D., and Howard, P. A., "The Influence of Scale-Size and Fluid-Thermal Properties in Simulating CRBR Outlet Plenum Behaviour," presented at the 1976 National Heat Transfer Conference, June 1976.

DECAY HEAT REMOVAL IN GCFR

Design, Analysis, and Verification of Natural Circulation in the GCFR

W. A. MEDWID, L. L. PARME, and A. SHENOY
General Atomic Company
San Diego, California, USA

ABSTRACT

This paper discusses the design, analysis, and verification of natural circulation in the gas-cooled fast breeder reactor (GCFR) demonstration plant. The GCFR natural circulation concept utilizes an upflow core in conjunction with the core auxiliary cooling system (CACS). The natural circulation CACS is an engineered safety feature of the GCFR that provides inherently passive core cooling as a backup to normal forced circulation residual heat removal (RHR) capabilities. This system utilizes natural circulation in the primary (helium), secondary (pressurized water), and tertiary (air) coolant loops to reject core decay heat to the atmosphere via helium/water and water/air heat exchangers. Detailed transient analysis shows that for a total loss of forced circulation capability under pressurized conditions, the natural circulation CACS can safely cool a shutdown core and prevent core meltdown for an indefinite period. Design criteria have been developed to integrate natural circulation into the GCFR design process. These criteria enhance natural circulation cooling and minimize potential problems associated with natural circulation. A natural circulation verification plan, which includes both analytical and experimental efforts, has been developed to verify the GCFR natural circulation predictive methods and design performance. The ultimate goal of this plan is to obtain licensing credit for the natural circulation CACS as an engineered safety feature.

1. INTRODUCTION

The advantages of the GCFR, including low capital cost, high breeding performance, and good safety and maintainability, have made it an attractive complement to the liquid metal fast breeder reactor (LMFBR) program. The GCFR plant design has several inherent characteristics that enhance safety, in particular:

1. Use of single-phase, nonradioactive, transparent coolant (helium), which has a small reactivity effect and is chemically inert.

2. All primary system components are enclosed in a prestressed concrete reactor vessel (PCRV). This precludes the failure of primary coolant pipes of significant size.

3. Vented fuel, which removes the pressure differential across the cladding and also significantly reduces the radioactivity in the primary coolant circuit.

Another inherent safety characteristic that significantly improves the accident prevention capability of the GCFR was recently incorporated into the design. This design feature is an upflow core that provides natural circulation RHR capability as a diverse backup to three redundant forced circulation RHR systems. In order to improve system reliability and diversity and to eliminate potential failure mechanisms associated with mechanical systems and operator actions, the natural circulation system is designed to provide passive short- and long-term RHR to the ultimate atmospheric heat sink in a pressurized GCFR.

This paper presents a description of the natural circulation RHR system, criteria that have been developed to enhance its design, and an outline of a verification plan to license natural circulation as an engineered safety feature. Preliminary analytical and experimental efforts to more fully define the verification plan and natural circulation test requirements are discussed. Detailed transient analysis that shows the inherent safety and cooling capability of natural circulation under a wide variety of accident conditions is also presented, along with a brief description of the methods used in the analysis.

2. GCFR DEMONSTRATION PLANT DESCRIPTION

The GCFR demonstration plant is cooled by pressurized helium at 10 MPa (1450 psia). The entire primary coolant system, consisting of the reactor core, main loops, and auxiliary coolant loops, is contained within a PCRV, shown schematically in Fig. 1. Each main loop contains a steam generator, motor-driven centrifugal circulator, and a gravity-closing isolation valve. Each auxiliary loop contains an electrically driven centrifugal circulator, a heat exchanger, and a gravity-opening isolation valve. Both the main and auxiliary cooling loops are housed in vertical cavities in the PCRV structure surrounding the reactor cavity. All interior surfaces of the PCRV are insulated and have a leak-tight steel liner. The penetration closures for the central reactor and peripheral heat exchanger cavities are made of concrete. The reactor core is supported by a grid plate mounted in the lower plenum of the reactor cavity and is composed of hexagonal fuel assemblies located at the center of the core surrounded by concentric rings of blanket assemblies containing fertile breeding material. Each fuel assembly contains a large number of fuel rods supported by grid spacers. The fuel rods are similar to LMFBR fuel rods except that the stainless steel cladding surfaces are roughened to improve heat transfer. The blanket assembly design is similar to the fuel assembly, but the blanket rods are larger in diameter and are spaced by wire wrap rather than grid spacers.

3. RHR SYSTEMS

The three independent means of forced circulation RHR employed by the GCFR demonstration plant are shown schematically in Fig. 2. These systems, listed in the order of their intended sequence of operation, are:

1. Main loop cooling system (MLCS).

2. Shutdown cooling system (SCS).

3. Core auxiliary cooling system (CACS).

The MLCS consists of the main loop helium circulators, isolation valves, the steam generators, and the normal secondary plant components associated with operation of the steam turbines at power. During periods in which electrical power is not produced, steam bypasses the turbine and is dumped to the condenser via a desuperheater.

The SCS is an independent safety class system designed to remove decay heat through the steam generators in the event of failure of the circulator motors, the feedwater system, or the condenser of the MLCS. Components include a separate safety class IE powered pony motor on the main circulator shafts, separate steam generator feed pumps and circulating water pumps, and the SCS water/air heat exchanger, as shown in Fig. 2.

The CACS is a second safety class decay heat removal system. Each of three CACS loops is comprised of an electrically driven auxiliary circulator, a check valve, and a helium-to-water heat exchanger called the core auxiliary heat exchanger (CAHE). The water from the CAHE is circulated through a pressurized water loop with pumps and a pressurizer, and the heat is rejected to the atmosphere by air from fans in the auxiliary loop cooler (ALC), which is a finned tube bank heat exchanger.

The CACS design incorporates natural circulation capabilities on the helium, water, and air sides as a backup to normal forced circulation capabilities. Core decay heat is transported by the primary coolant helium to a pressurized water-cooled CAHE, which is elevated 10 m (33 ft) above the core. Heated water from the CAHE reaches the ALC located 34.5 m (113 ft) above the CAHE by natural circulation in the pressurized water loop. The heat from the ALC is ultimately rejected to the atmosphere by natural draft of air through a 23-m (75-ft) chimney. For a total loss of forced circulation capability, the natural circulation capability of the CACS provides cooling for an indefinite period if the primary coolant is pressurized.

A more comprehensive discussion of the GCFR RHR systems is given in Ref. 1.

4. NATURAL CIRCULATION DESIGN

Natural circulation RHR has been adopted as an engineered safety feature of the GCFR. However, the GCFR is designed so that deterministic licensing requirements are fully satisfied with the three forced circulation systems described above. In order to improve system reliabiity and diversity and to eliminate potential failure mechanisms associated with mechanical systems and operator actions, the GCFR design goal is that natural circulation provides passive short-and long-term RHR to the ultimate atmospheric heat sink.

The commitment to meet these design goals for natural circulation as an engineered safety feature has impacted the GCFR design in several areas and has required the development of design criteria to ensure that adequate thermal heads are available under all conditions where natural circulation is desired.

The most apparent GCFR design change is from a downflow to an upflow core with elevated heat sinks. The CAHEs and the steam generators have been located above the core. This provides the possibility of natural circulation in either main or auxiliary coolant loops or simultaneously in both.

Of these possibilities, natural circulation in the auxiliary loops has been identified as the preferred mode of operation because of the system simplicity possible while achieving the design goal of a totally passive RHR train to the ultimate heat sink. This goal is achieved by designing gravity-actuated primary coolant isolation valves, providing elevation differences between the CAHEs and the ALCs, and by adding an air draft chimney to the ALC.

The isolation valves have been designed to integrate natural circulation into the hierarchy of RHR systems. During normal operation, the pressure head of the main helium circulator keeps the main loop isolation valves open and the auxiliary loop isolation valves closed. Upon loss of forced circulation in the main loops, the main circulators coast down. When the circulator pressure head reaches a small value (typically at 9% circulator speed when all circulators are in operation), the main loop isolation valves are closed and the auxiliary loop isolation valves are opened by gravity. If the main loop valves fail to close during auxiliary loop natural circulation, flow bypassing the core is limited to a small amount by buoyancy forces in the main loop. This self-isolation feature of main loops, which is discussed in the analysis section, is a function of relative CAHE, steam generator, and upper main loop cross-duct elevations. Criteria for these elevations have been established to assure that the self-isolation feature is maintained in the design.

The simplicity of the flow path and duct arrangement in the GCFR is apparent from Fig. 1. Design criteria have been developed to maintain flow path and ducting configurations that provide adequate natural circulation startup potential and low flow resistance and also avoid potential thermal traps. The effort to eliminate or minimize flow paths requiring the rise of cold gas or the downward flow of hot gas not only ensures startup under all conditions but also limits the uncertainty of the computational models employed in the analysis.

Neutron shielding, particularly in the large upper core cavity, has also been affected by natural circulation. To avoid cold traps in the shield labyrinth, a design has been chosen that utilizes numerous flow paths along the entire axial length of the reactor cavity to equalize temperatures on either side of the shield.

5. METHOD OF ANALYSIS

The natural circulation analysis was performed using two GCFR system codes, RATSAM and FASTRAN.

The RATSAM computer program was developed primarily to evaluate the transient thermal-hydraulic behavior of the commercial-sized HTGR primary coolant system during depressurization accidents (Ref. 2). However, RATSAM is a generalized program capable of modeling the thermodynamic behavior of any fluid system containing a nearly perfect gas.

The RATSAM program numerically solves a set of ordinary differential equations and relations governing the entire flow system. The program's analytical model assumes that the primary coolant system can be broken down into a series of subvolumes, or nodes, interconnected by flow paths. The transient forms of conservation of mass and energy, as well as the equation of state, are then applied to the nodes, while the transient conservation of momentum with the buoyancy term is applied to the interconnecting flow paths.

Specifically, the program calculates the transient coolant pressure, temperature, and flow throughout the primary coolant system, taking into account the dynamic behavior of the circulators and valves, the actions of plant protection systems, and the heat transfer between coolant, core, steam generators, CAHE, and reactor internals.

FASTRAN models the transient performance of the GCFR and its associated cooling systems under pressurized and depressurized conditions. The FASTRAN program incorporates detailed models of the reactor core with kinetics, the steam generators, the main circulators, the entire auxiliary cooling loop (which includes the auxiliary circulator and drive motor, the CAHE, the ALC, and all interconnecting piping). Point model neutron kinetics and reactivity effects of the grid plate, fuel, coolant, etc., are included. Safety, plant protection, and control system actions are all modeled.

FASTRAN divides the CACS naturally circulating water system into a number of volume nodes with both thermal and fluid inertia effects being modeled. For pressurized conditions FASTRAN divides the primary system into two volume nodes. In computing the primary coolant flow, fluid inertia, which is modeled in RATSAM, is ignored and flow around the loop is assumed to be constant. Typically RATSAM, excluding core nodes, models the auxiliary loop using 10 nodes as compared to 2 nodes used in FASTRAN. However, due to the low fluid inertia of the GCFR and for a generally uniform temperature profile in the auxiliary loop after initial startup of natural circulation, the FASTRAN model simplifications are justified. Figure 3 illustrates the good agreement obtained between FASTRAN and RATSAM results for an identical natural circulation transient.

6. NATURAL CIRCULATION ANALYSIS

The performance of the GCFR following a complete loss of forced circulation was analyzed first for a base case. Sensitivities to the assumed base case conditions were then determined by varying individual asssumptions.

6.1 Base Case

The initiating event of the base case is a total loss of drive power to the main circulators. The main circulators immediately begin coasting down, followed by a reactor scram due to a high power-to-flow ratio. Under such a scenario the SCS pony motors would be expected to maintain forced circulation in the main loops. However, it is assumed that, for one reason or another, the pony motors are unavailable. In addition, an extremely conservative assumption is made that none of the three CACS circulators start. Ninety seconds after the loss of circulator drive power, core flow decays to less than 10%. The low flow then allows the main loop isolation valves to shut by gravity while the auxiliary loop isolation valves fall open. The temperature profile existing in the auxiliary loop rapidly induces natural circulation between the core and the CAHEs. This helium coolant flow is seen to increase from near 0 to 3.5% of its normal design flow in about 3 sec. Simultaneously, the increased heat load to the CAHE increases the natural circulation in the CACS water loop and in the natural draft chimney. These natural circulation flows are found to maintain hot spot fuel cladding temperatures below 620°C. (The maximum hot spot cladding temperature at the design point is 700°C; the faulted temperature damage limit for the cladding is 1300°C.) The maximum temperature reached in the CACS pressurized water loop is 185°C. (The saturation temperature of the water loop is 302°C.) Cladding and water temperature histories are illustrated in Figs. 4 and 5.

6.2 Sensitivity Analysis

It is assumed in the above case that all gravity-actuated isolation valves are successfully repositioned after 90 sec of main circulator coastdown. The sensitivity to this assumption was evaluated by considering a number of variations from the base case including:

1. Transfer time. Transfer from main loop coastdown to auxiliary loop natural convection at 5 sec after loss of circulator drive power and main loop natural circulation (i.e., no transfer) were investigated. Figure 4 shows the maximum cladding temperature of the fuel rods for these various transfer times. As can be seen, while later transfer times that make full use of circulator coastdown can have a marked effect in limiting cladding temperature rise, natural circulation alone is capable of preventing damage very shortly after scram without utilizing full circulator coastdown. Furthermore, sufficient elevation is provided between the core and steam generator so that as long as feedwater flow to the steam generator is maintained, the main loops can remain a viable natural convection heat sink.

2. Number of loops. Failure of auxiliary loop isolation valves to open was evaluated by considering a transfer to natural convection using one or two auxiliary loops after 90 sec of main loop coastdown. As seen in Fig. 6, cladding temperature is relatively insensitive to the number of auxiliary loops available. However, as illustrated in Fig. 5, the maximum water temperature of the remaining loops rises rapidly with a decreasing number of available loops, so that RHR on one CACS loop is possible only if forced circulation is maintained on the secondary side to prevent boiling.

3. Common mode failures. The effect of a common mode failure of the loop isolation valves to correctly operate was analyzed by considering two limiting cases: (a) all three auxiliary loop isolation valves open early before completion of main circulator coastdown, and (b) the main loop isolation valves fail to shut after completion of main circulator coastdown. Opening of the auxiliary loop valves early (30 sec) in the main circulator coastdown allows main circulator flow to bypass the core. The resultant decrease in core cooling causes a rapid rise in cladding temperature. Circulator speed reaches 9% and main loop isolation valves shut at 71 sec versus 90 sec in the base case. Once the main loop valves are shut, CACS natural circulation rapidly restores flow to the core and halts the rise in fuel cladding temperature, as shown in Fig. 7.

   In the case of failure of the main loop isolation valves to close after main circulator coastdown, the auxiliary loop isolation valves are open and natural circulation flow from the auxiliary loops enters the core. Some of the flow bypasses the core and enters the main loop and the steam generators in the reverse direction. However, the steam generators are located 4.6 m (15 ft) above the core midplane, and the helium mass is cooler and heavier than in the core; therefore, the bypass flow is small. Core flow is reduced initially by about 25%. If feedwater flow is maintained and the steam generators are cooled further, the bypass flow continues to be reduced. After 1000 sec the main loop flow will reverse once more, allowing the steam generators to resume acting as heat sinks and to operate in parallel with the CACS. Even without feedwater flow, the bypass flow is small and eventually stagnates. The peak fuel cladding temperatures for multiple auxiliary and main loop valve failures are shown in Fig. 7.

7. NATURAL CIRCULATION VERIFICATION

The natural circulation CACS is an engineered safety feature of the GCFR. To obtain licensing credit for this engineered safety feature, a natural circulation verification plan is being developed to systematically verify GCFR natural circulation predictive methods and design performance. An outline of this plan is presented below. Examples of analytical and experimental efforts undertaken to define the verification plan in more detail are also discussed.

7.1 Natural Circulation Verification Plan

Literature survey. Information obtained from a detailed literature survey of natural circulation phenomena, analyses, testing, and verification efforts for light water reactors (LWRs), LMFBRs, and gas-cooled reactors (GCRs) will provide:

1. Guidance in identifying key phenomena influencing natural circulation in the GCFR and in developing a natural circulation verification plan for the GCFR.

2. A data base of industry codes and experimental data to be used in verification of GCFR codes.

Key mechanisms. A preliminary list of important mechanisms influencing natural circulation in the GCFR is given in Table 1. Detailed sensitivity analyses of these variables will be performed to aid in defining natural circulation test requirements and verification efforts. In arriving at the list of key mechanisms, the following two points are important:

1. Natural circulation is driven by the integated temperature profile in the primary, secondary, and tertiary loops and depends on all the thermal-hydraulic characteristics of these systems, including local effects that may not be important during forced circulation.

2. The same basic physical phenomena apply to both steady-state natural circulation performance and to the transition from forced to natural circulation. However, the uncertainty in modeling complex and potentially localized thermal-hydraulic effects using one-dimensional system codes is greatest during the transient phase of natural circulation.

Verification by comparison to independent codes. The objective of this plan element is to verify all computer codes used in safety analysis of natural circulation in a GCFR by comparing them with similar independently developed codes used in the nuclear industry. System codes RATSAM and FASTRAN and also COBRA, the core subassembly thermal-hydraulic code, will be verified in this manner.

Verification by comparison to experimental data. Another means of validating GCFR natural circulation codes is by comparison with existing experimental data obtained from natural circulation tests involving GCRs. A comparison of data obtained from the Central Electric Generating Board (CEGB) for natural circulation tests conducted at their Sizewell nuclear power station and RATSAM predictions for the test will be used to validate the capability of the one-dimensional system code to calculate the transition to natural circulation in a GCR following reactor trip and circulator coastdown. Sizewell is a CO_2-cooled MAGNOX reactor.

Component tests. The results of the component tests will be used to verify computer models and design performance of critical components under natural circulation conditions. A preliminary list of components to be tested includes the core, the main circulator, and the main and auxiliary loop isolation valves. The objective of these tests is to determine the thermal-hydraulic characteristics of these components under natural circulation conditions.

System scale model tests. Natural circulation in the GCFR is a system-wide phenomenon. Indeed, most of the uncertainties in modeling GCFR natural circulation performance using one-dimensional system codes relate to coupled system characteristics including the transition from forced to natural circulation. Because of the need to resolve system uncertainties described in Table 1, scale model tests of the primary and secondary systems are currently recommended by General Atomic to validate system codes (RATSAM and FASTRAN) and the GCFR natural circulation RHR system design for a specific scale model configuration. The primary system test should involve as much similitude to the GCFR as practical (i.e., approximately the same Reynolds number, Grashof number, time constants, transport times, system heat capacities, and power differences in the core and radial blanket). The necessity of performing tertiary system tests will be evaluated. The results of sensitivity analyses will be used to define the extent of system testing required.

Preoperational and startup tests. The ultimate validation of the GCFR demonstration plant natural circulation RHR system design and system code predictions will be provided by preoperational and/or startup tests. These tests would be done as a sequence of tests (including tests before fuel loading) to resolve uncertianties in natural circulation performance in a progressive fashion wihtout risk of damage to the core or plant hardware. Such tests could include no power/nonnuclear heat tests at pressure with the core installed but shutdown using compressive heating provided by the main circulators and low-power tests in the range of 0 to 20% power.

7.2 Preliminary Efforts

Tasks are under way to define natural circulation test requirements and develop a detailed natural circulation verification plan from the above outline. These tasks, which include comparison of RATSAM with an independently developed code and with experimental data from another reactor and natural circulation tests in a high-pressure helium loop, are discussed below.

RATSAM-PHAETON benchmark comparison. As part of the GCFR natural circulation verification plan described above, a benchmark comparison of the natural circulation calculational capability of RATSAM and a similar code PHAETON 2 (Ref. 3) has been made. PHAETON was developed at KFK (West Germany) and, like RATSAM, it uses a subvolume concept to solve the transient one-dimensional forms of the conservation of mass, momentum, and energy for a multiple node ideal gas flow system with heat transfer.

A typical case of natural circulation in the auxiliary loops of an upflow core GCFR was chosen as the benchmark transient. The assumed sequence of events for the natural circulation benchmark transient is:

1. Main circulator drive power lost and reactor scrammed (t = 0).

2. Three main circulators coast down and are braked (t = 90 sec).

3. Main loop isolation valves shut (t = 90 sec).

4. Auxiliary loop isolation valves open (t = 91 sec).

5. Natural circulation using three auxiliary loops.

Major differences in RATSAM and PHAETON analytical methods include:

1. The integration solution method for the flow network.

2. Thermal model of the core.

3. Circulator model.

A comparison of results for the benchmark transient shows that upon transfer from main loop coastdown to auxiliary loop natural circulation, both RATSAM and PHAETON predict rapid initiation of natural circulation. Indeed, all system trends are comparable during the natural circulation phase of the transient, including:

1. Auxiliary loop gas temperatures and leakage flow prior to transfer.

2. Transient flow differences in various parts of the auxiliary loop during the initiation of natural circulation.

3. System temperature response including core inlet, core outlet, and maximum cladding temperatures.

4. System flow response including total core flow and flow distribution between the fuel and blanket.

Figure 8 shows a comparison of the flow predicted by RATSAM and PHAETON during the natural circulation phase of the transient. Total flow, fuel flow, and blanket flow never disagree by more than 10% and agree within 3% during most of the transient. The peak fuel cladding temperature predicted by both codes is 833 K, occurring at 130 sec in RATSAM and at 140 sec in PHAETON. Figure 9 shows a comparison of the time history of fuel cladding temperatures predicted by PHAETON and RATSAM. The maximum difference in predicted fuel cladding temperatures is 33 K at 180 sec.

Considering the differences in the two methods and the complexity of the natural circulation transient problem modeled, the RATSAM-PHAETON comparison shows good agreement. This comparison gives confidence in the natural circulation calculational capability of RATSAM, particularly in calculating the critical startup phase of natural circulation in the GCFR auxiliary loops.

RATSAM-Sizewell comparison. Natural circulation tests were successfully conducted at the Sizewell nuclear power station in 1967 by CEGB. Sizewell is a CO_2-cooled 950-MW(t) MAGNOX reactor. The test can be summarized as follows:

Test date: June 4, 1967

Plant at 90% thermal power

t = 0-sec reactor trip.

t = 5-sec turbine trip.

t = 10-sec circulator power cut.

Circulator coastdown was followed by transition to natural circulation using the steam generator as a heat sink. With the plant in the natural circulation mode, reactor transient temperature data were taken for ~1 h.

The measured test data include the time history for the circulator speed, core inlet temperature, core outlet plenum temperature, and fuel cladding, graphite moderator, and core outlet temperatures for selected core channels. The test successfully showed the natural circulation capability of CO_2-cooled MAGNOX reactors, with the peak fuel cladding temperatures during natural circulation being less than the peak cladding temperatures at the time of reactor trip.

The CEGB has provided General Atomic with data to model the Sizewell reactor natural circulation test using RATSAM. This will require only minor modifications to RATSAM (i.e., adding a MAGNOX reactor core model). The comparison of RATSAM transient predictions for the Sizewell test with the transient test data will be used to validate the natural circulation calculational capability of RATSAM.

Helium loop natural circulation test. As part of the GCFR safety test program at Los Alamos Scientific Laboratory, natural circulation tests using a closed high-pressure helium loop with electrically heated GCFR fuel rods as a heat source are scheduled for the second quarter of 1980. The objective of the tests is to measure steady-state and transient helium flow rate and fuel cladding temperatures under conditions similar to those encountered in GCFR natural circulation. The test will also be used to identify experimental limitations limitations and future test requirements and to help plan the system scale model tests that are planned as part of the natural circulation verification plan. The test results will be compared to analytical results predicted by a detailed RATSAM model of the test.

A schematic of the test configuration is shown in Fig. 10. The test loop is rated for a maximum pressure of 68 atm and a maximum hot leg temperature of 650°C. The primary components of the test loops are:

1. A heat source, which consists of 37 electrically heated GCFR fuel rods enclosed in a hexagonal duct assembly (maximum power output is 48 kW).

2. The test vessel, which houses the assembly; this is an internally insulated pressure vessel and is cooled externally by a water jacket.

3. An elevated water-cooled heat exchanger, which serves as the heat sink.

4. Hot leg connecting piping, which is internally insulated, and uninsulated cold leg connecting piping.

5. A quick opening and closing helium check valve located in the cold leg. By closing this valve for several seconds to stop all helium flow and then opening the valve, the transient development of natural circulation from stagnant conditions as it occurs in the GCFR auxiliary loops can be tested.

The natural circulation test series will consist of several different tests involving a combination of three power levels (representing 5% to 0.5% decay heat fraction) and three pressures (65, 30, and 10 atm).

8. CONCLUSIONS

Natural circulaton RHR using the CACS has been incorporated as an engineered safety feature of the GCFR demonstration plant design. It provides inherently passive and diverse short- and long-term RHR to the ultimate atmospheric heat sink. Design criteria have been developed to enhance the natural circulation cooling system design and minimize uncertainties associated with natural circulation. A detailed natural circulation verification plan is being developed to support licensing of the natural circulation CACS as an engineered safety feature of the GCFR.

ACKNOWLEDGMENTS

This work was supported by Helium Breeder Associates and General Atomic Company.

REFERENCES

1. Chung, H. S., H. Chi, and A. Shenoy, "GCFR Residual Heat Removal Following Depressurization Accidents," in Proceedings of the Specialists' Meeting on Decay Heat Removal and Natural Convection in FBRs, February 28-29, 1980, Brookhaven National Laboratory, Upton, New York, to be published.

2. Deremer, R. K., and T. Shih, "RATSAM: A Computer Program to Analyze the Transient Behavior of the HTGR Primary Coolant System During Accidents," General Atomic Report GA-A13705, May 1977.

3. Wilhelm, D., "The PHAETON 2 Computer Code for Analyzing Accidents in a 1000-MW(e) Helium Cooled Fast Reactor," Institut für Neutronenphysik und Reaktortechnik Kernforschungzentrun Karlsruhe Report KFK 2561, December 1977.

TABLE 1
PRELIMINARY LIST OF KEY MECHANISMS INFLUENCING NATURAL CIRCULATION IN THE GCFR

1. Transition from forced circulation to natural circulation in the primary system, including the startup of natural circulation from essentially stagnant conditions.
2. Transition from low flow to full flow natural circulation in the secondary and tertiary loops.
3. Primary, secondary, and tertiary system boundary conditions at onset of natural circulation.
4. Circulator coastdown.
5. Primary system flow resistance (core, heat exchangers, locked auxiliary circulator rotor, valves, loop).
6. Secondary and tertiary system flow resistance.
7. Primary, secondary, and tertiary system heat capacities; heat losses; and effect of heat transfer on temperature profiles.
8. Mixing, potential flow stratification, cold traps, and local recirculation in primary and secondary loops.
9. Thermal center effect in steam generator.
10. Shutdown fuel and radial blanket power distribution.
11. Fuel and blanket heat transfer coefficients and friction factors under natural circulation conditions.
12. Interassembly and intraassembly core flow redistribution.
13. Main and auxiliary loop valve actuations, malfunctions, and loop self-isolation.

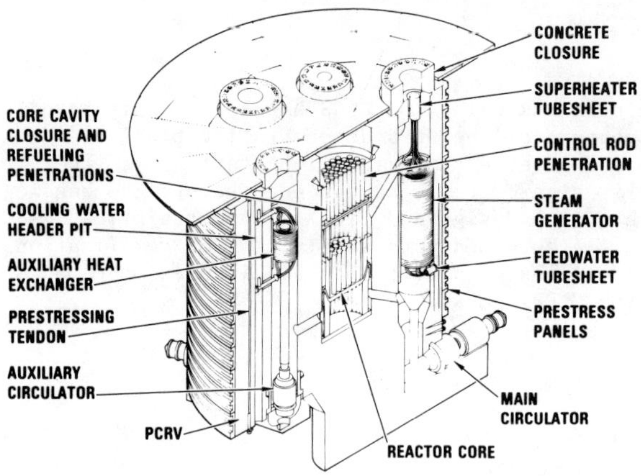

Fig. 1. GCFR nuclear steam supply system

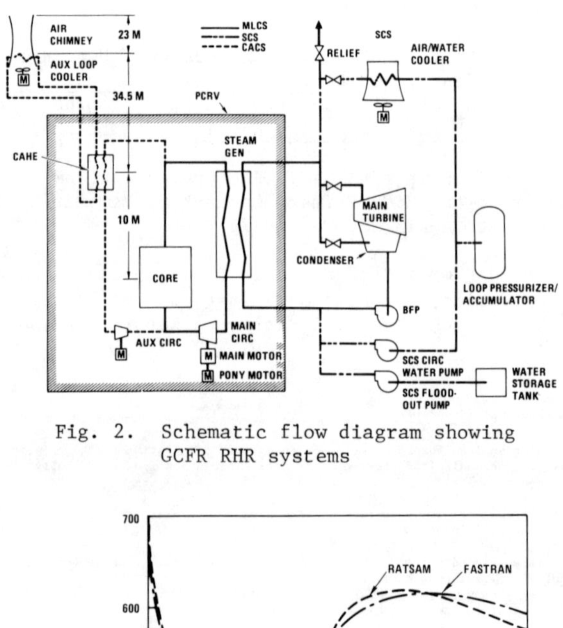

Fig. 2. Schematic flow diagram showing GCFR RHR systems

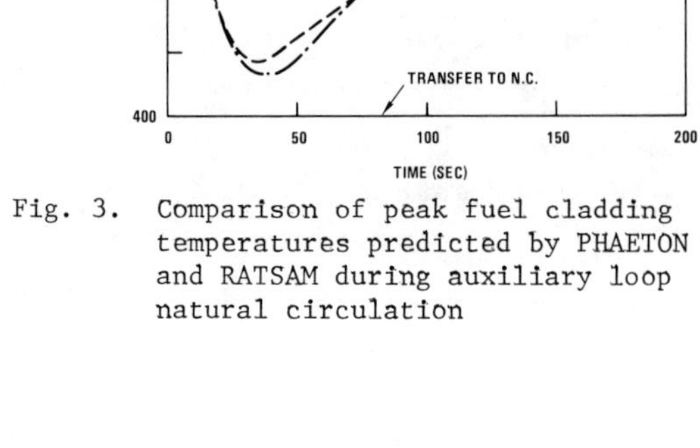

Fig. 3. Comparison of peak fuel cladding temperatures predicted by PHAETON and RATSAM during auxiliary loop natural circulation

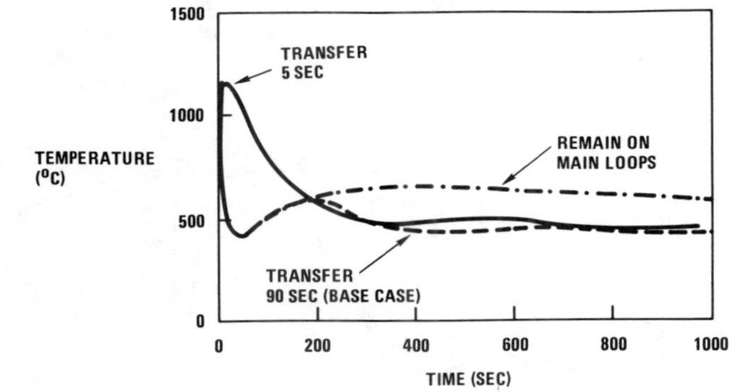

Fig. 4. Sensitivity of hot spot fuel cladding temperature to loop transfer time in primary coolant natural circulation

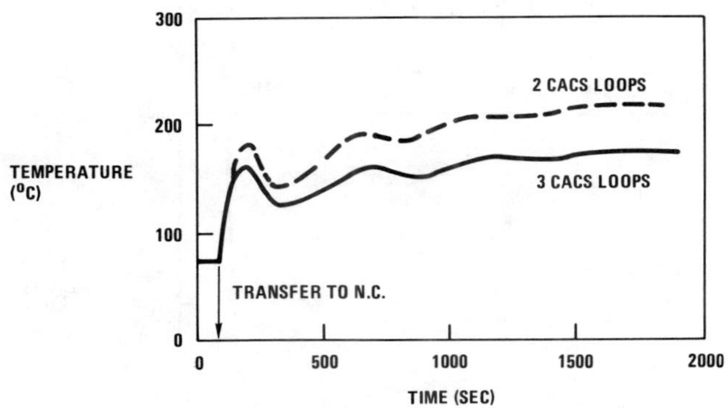

Fig. 5. Sensitivity of CAHE water outlet temperature to number of auxiliary loops in operation

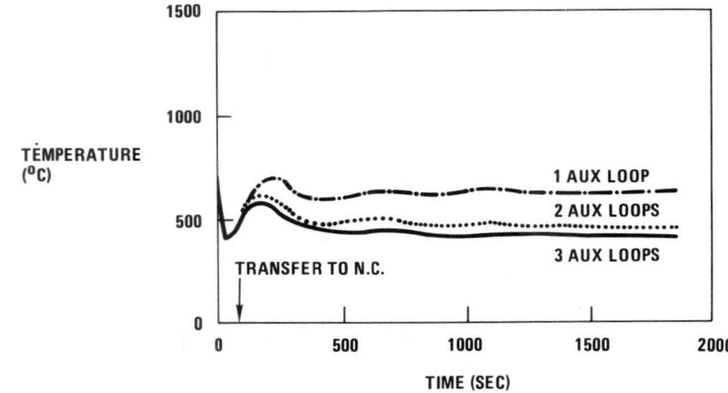

Fig. 6. Sensitivity of hot spot fuel cladding temperature to number of auxiliary loops in operation

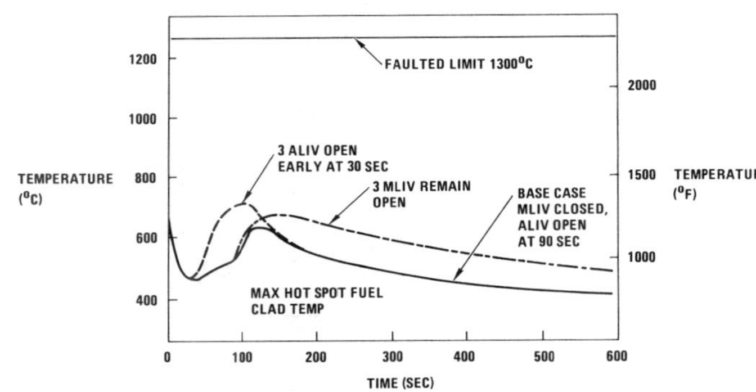

Fig. 7. Sensitivity of peak cladding temperature to incorrect transfer of multiple loops in primary coolant natural circulation

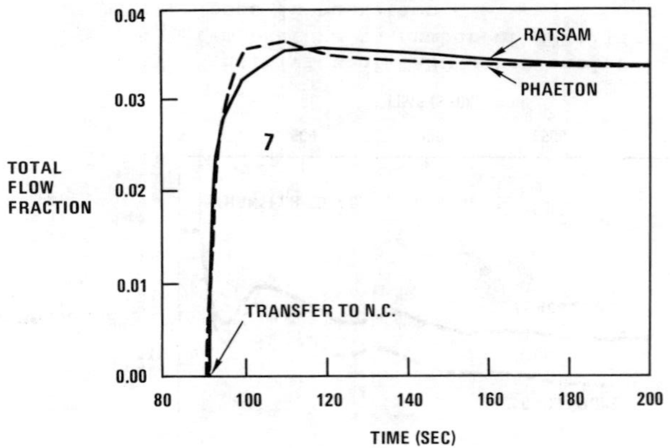

Fig. 8. Comparison of flow predicted by RATSAM and PHAETON during auxiliary loop natural circulation

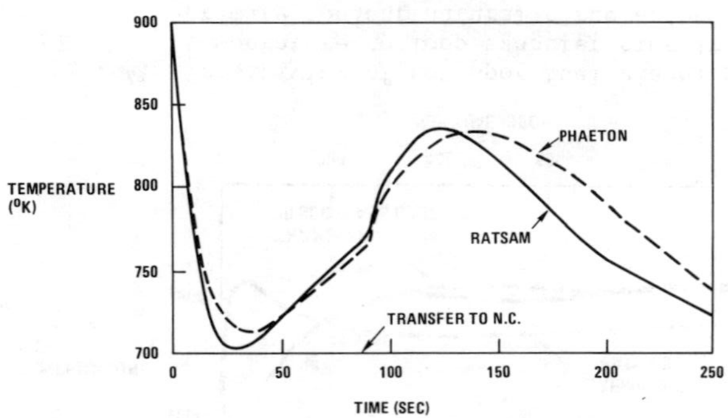

Fig. 9. Comparison of peak fuel cladding temperatures predicted by PHAETON and RATSAM during auxiliary loop natural circulation

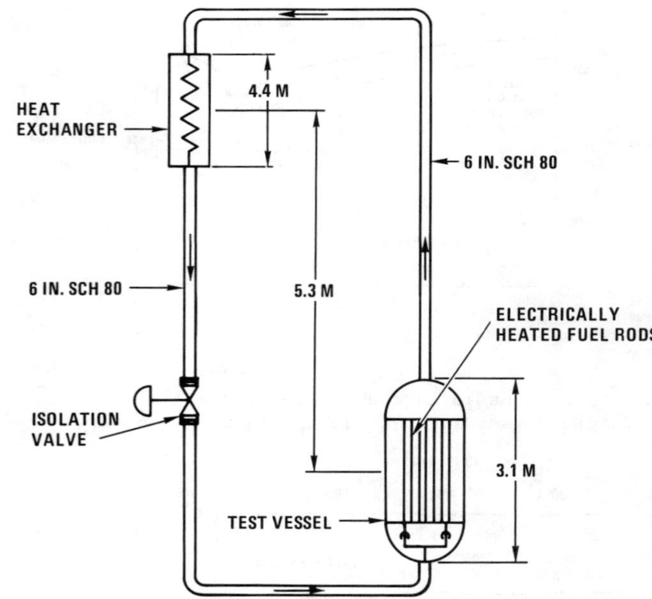

Fig. 10. Schematic of closed-loop natural circulation test

Comparison of Carbon Dioxide and Helium for Natural Circulation Cooling of GCFRs

G. MELESE D'HOSPITAL and C. B. BAXI
General Atomic Company
San Diego, California, USA

ABSTRACT

In this paper the natural circulation capabilities of helium and carbon dioxide are compared for decay heat removal of GCFRs. For fully pressurized conditions, helium can remove 6% decay heat from a typical GCFR for an elevation difference of about 10 m between the core and the heat sink. But, at lower reactor pressure (say, 10% of design pressure or less), natural circulation with He becomes insufficient to remove any significant amount of decay heat. CO_2 is superior to He for natural circulation cooling. For example, at fully pressurized conditions, only one-ninth the elevation difference required for He will be adequate to remove 6% of the decay heat with CO_2. Also, for a given elevation difference and decay heat level, CO_2 requires only one-third to one-sixth (depending on the decay heat level) the reactor pressure compared to He. Curves valid over a wide range comparing relative performance of CO_2 and He for natural circulation cooling of GCFR are presented.

1. INTRODUCTION

Since the conception of the gas-cooled fast breeder reactor (GCFR) around 1960, several coolants have been considered, mainly steam, carbon dioxide, and helium because of their utilization in thermal reactors. Steam was abandoned for nuclear and safety reasons (neutron moderation problems) as well as for metallurgical reasons (fuel clad corrosion problems). The only real contenders left in the Western world were then carbon dioxide, with long experience in the British and French graphite-moderated reactors, and helium, which was used in high-temperature gas-cooled reactors (HTGRs). All current GCFR designs, in both the U.S. and Europe (European Association of Gas Breeder Reactor, Germany, Switzerland), now use helium because of its chemical inertness, lack of interaction with neutrons, and satisfactory experience in HTGRs.

Comparisons of helium and carbon dioxide for application to fast breeder reactors were reported by a number of authors [1-2]. These studies have shown that for the same reactor pressure, maximum clad temperature and pumping fraction, the Reynolds number is 4.5 times larger for carbon dioxide than for helium, the Mach number is 1.8 times larger for carbon dioxide than for helium, and the core pressure drop is 2.5 larger for carbon dioxide than for helium. The thermal power for the same heat transfer surface in the core is about the same with helium or carbon dioxide for typical GCFR conditions. Thus, there is little to choose between thermodynamic performance of carbon dioxide and helium for forced convection, except for the much higher forces with carbon dioxide and the increased potential for flow-induced vibrations. This explains the usual choice of helium as the GCFR coolant.

Recently, natural circulation cooling has been studied for emergency cooling of the shutdown GCFR cores [3], and carbon dioxide which is 11 times denser than helium is expected to perform better than helium. This paper compares carbon dioxide and helium as coolants for natural circulation of the GCFR.

2. ANALYSIS

In this section, it is established that the use of forced convection correlations during natural circulation cooling of GCFR in the flow range of interest is valid. Using these correlations, basic relations between various parameters affecting natural circulation cooling of GCFR are derived.

2.1 Flow Regimes for Natural Convection

In order to find heat transfer correlations applicable during natural circulation, the flow regime has to be determined: natural convection, forced convection, or mixed convection when both gravitational body forces and external forces (pressure gradients) are significant. Both laminar and turbulent flows are possible in each of these regimes, as discussed by Metais and Eckert [4]. For vertical flow, the two parameters of importance have been found to be the Reynolds number (based on hydraulic diameter d) and the combination*:

$$R = G_r P_r \frac{d}{L} = g \frac{d^3}{\nu^2} \frac{\Delta t}{T_{av}} P_r \frac{d}{L} \qquad (1)$$

We shall assume that the following forced convection heat transfer correlations are applicable during natural circulation cooling for GCFR geometry (bundle pitch to diameter ratio of 1.4):

For laminar flow [5]:

$$Nu = 10. \qquad (2)$$

For turbulent flow [6]:

$$Nu = 0.023 N'^{0.8} P_r^{1/3} \qquad (3)$$

For a typical GCFR geometry and property values for He and CO_2 from Refs. 7 and 8, the Reynolds numbers read (see Eq. A-8 in Appendix)

$$N'_H = 140 \times 10^3 \, q$$
$$N'_C = 718 \times 10^3 \, q \qquad (4)$$

while using Eq. 1, 4 and A-3 (in Appendix), we obtain

$$R_H = 1.95 \times 10^4 \frac{q}{Nu} \quad \text{and} \quad R_C = 1.85 \times 10^7 \frac{q}{Nu} \qquad (5)$$

* Refer to nomenclature.

With $q = 10^{-3}$ (0.1% afterheat), we find for laminar flow (Nu = 10)

$N'_H = 140 \qquad R'_H = 1.95$

$N'_C = 718 \qquad R'_C = 1.85 \times 10^3$

According to Fig. 1 of Metais and Eckert's [4] paper, these conditions are well within the range of forced convection laminar flow. Therefore, the use of forced convection correlation Nu = 10 for laminar flow appears to be justified.

With q = 0.06 (6% afterheat at shutdown), we have turbulent flow

$N'_H = 8.4 \times 10^3 \qquad R_H = 41.6$

$N'_C = 4.3 \times 10^4 \qquad R_C = 10^4$

We also find that the parameters N' and R are such that we are well within the forced convection turbulent flow regime for which correlation (Eq. 3)is applicable.

Since the parameter $R = GrP_r d/L$ is proportional to d^4, we find that for larger values of the hydraulic diameter d (say, 0.75 cm instead of 0.65), the flow regime for CO_2 would become mixed convection laminar flow, while it would still be in the forced convection regime for He. Therefore, the values of R should be checked for each CO_2 design to see what correlations are applicable.

2.2. Basic Relations

Assuming an elevation difference ΔL between core midplane and heat sink, the system pressure drop ($\Delta p'$) is related to the density difference by the relation

$$\Delta p' = g\Delta L(\rho'_i - \rho'_e) = g\Delta L\, M \frac{p}{R} \frac{\Delta T'}{T'_i T'_e} \tag{6}$$

Also the friction pressure drop reads:

$$\Delta p' = 2f \frac{L}{d} \rho'_{av} V_{av}{}^2 Z = 2f \frac{L}{d} \frac{Z}{\rho'_{av}} \left(\frac{Q'}{CA\Delta T'}\right)^2 \tag{7}$$

In Section 2.1, it was verified that forced convection correlations can be used in this analysis. Hence, the following correlation will be used for laminar and turbulent flows for friction factors [9-10]:

$$f_\ell = 30/N' \tag{8}$$

$$f_t = 4.6 \times 10^{-2} (N')^{-0.2} \tag{9}$$

Defining the modified afterheat fraction, ϕ, as:

$$\phi = 10^2 q \frac{T_m - T_i}{T'_m - T'_i} \tag{10}$$

and with the forced convection correlations for friction factor (8, 9) and heat transfer coefficient (2, 3), the following relations are obtained (see Appendix for details):

$$\frac{(\Delta L)t}{L} = 1.15 \times 10^{-3} \, Z\left(\frac{1 - V}{FV}\right)^{0.3} \left(K'P_r'\right)^{0.2} \frac{(1 - 1/\gamma')^2}{(1 - 1/\gamma_H)^{0.9}}$$

$$\times \frac{(T_i')^2}{gL^{0.9}D^{0.3}} \left(\frac{T_m'}{T_i'} - 1\right)^{-0.2} \left(\frac{C_H}{T_i} p\Delta p\right)^{0.9} \frac{\phi^{1.8}}{p'^2} \quad (11)$$

$$\frac{(\Delta L)\ell}{L} = 7.56 \, Z\left(\frac{1 - V}{FV}\right)^{1.5} K'P_r' \frac{(1 - 1/\gamma')^2}{(1 - 1/\gamma_H)^{0.5}} \frac{(T_i')^2}{gL^{0.5}D^{1.5}}$$

$$\times \left(\frac{T_m'}{T_i'} - 1\right)^{-0.2} \left(\frac{C_H}{T_i} p\Delta p\right)^{0.5} \frac{\phi(1 + 0.06\phi)}{p'^2} \quad (12)$$

With the same temperature conditions during natural circulation cooling as during normal operation of GCFR, with helium coolant and typical GCFR design values $Z = 4/3$, $(1 - V)/FV \times L/D = 150$, $T_i = 573$ K, $T_m = 973$ K, $L = 1.2$ m, $p = 9$ MPa, $\Delta p/p = 1.5 \times 10^{-2}$, Eq. (11) gives the relation for turbulent flow

$$\Delta L_1 \simeq 0.41(10^2 q_1)^{1.8} \quad (13)$$

so that an elevation difference of 10.4 m is required to remove 6% afterheat.

From Eq (11) the following relation between the elevation differences, the temperature levels, the afterheat fraction, and the reactor pressure is obtained for turbulent flow

$$\Delta L = \Delta L_1 \left(\frac{T_m/T_i - 1}{T_m'/T_i' - 1} \frac{p}{p'}\right)^2 \left(\frac{q}{q_1}\right)^{1.8} \quad (14)$$

The above relation shows that for a given coolant, the required elevation difference can be reduced by increasing the reactor pressure, reducing the inlet temperature to the core, or by increasing the maximum clad temperature.

3. COMPARISON OF HELIUM AND CARBON DIOXIDE

Using the relations in the last section, the relative performance of He and CO_2 for natural circulation cooling can be compared. This comparison can be made at full reactor pressure and at partial reactor pressure.

3.1. Full Reactor Pressure

For fully pressurized GCFR, the ratio $(\Delta L)_c/(\Delta L)_H$ reduces to:

For $\phi \leq 0.5$ (laminar flow for He and CO_2)

$$\frac{(\Delta L)_c}{(\Delta L)_H} = \frac{(KP_r)_c}{(KP_r)_H}\left\{\frac{(1 - 1/\gamma)_c}{(1 - 1/\gamma)_H}\right\}^2 \frac{1 + e_c}{1 + e_H} \tag{15}$$

or: $$\frac{(\Delta L)_c}{(\Delta L)_H} \cong 3.1 \times 10^{-2} (1 + 0.24\phi) , \quad \text{since } e \cong 6 \times 10^{-2} \frac{(KP_r)_H}{KP_r} \phi$$

For $0.5 \leq \phi \leq 2.5$ (laminar flow for He and turbulent flow for CO_2)

$$\frac{(\Delta L)_c}{(\Delta L)_H} \cong 6.1 \times 10^{-2} \phi^{0.8} (1 - 0.06\phi) \tag{16}$$

and for $\phi > 2.5$ (turbulent flow for He and CO_2)

$$\frac{(\Delta L)_c}{(\Delta L)_H} \cong \left\{\frac{(KP_r)_c}{(KP_r)_H}\right\}^{0.2}\left\{\frac{(1 - 1/\gamma)_c}{(1 - 1/\gamma)_H}\right\}^2 \cong 0.115 \tag{17}$$

Equations (15 - 17) are plotted in Figure 1. The results show that at full reactor pressure, the elevation difference ΔL required with the CO_2 is considerably smaller than elevation difference required with He and varies from 1/30 for $\phi < 0.5$ to 1/9 for $\phi > 2.5$. Recall that ϕ is proportional to the afterheat (Eq. 10).

However, with the elevation difference of about 10 m, pressurized He can remove about 6% of afterheat with natural circulation and He has a number of advantages over CO_2 for forced convection cooling of GCFR [1 - 2]. Hence, CO_2 need not be considered for pressurized natural circulation cooling of GCFR.

3.2. Depressurized Condition

Another way to evaluate the relative performance of the He and CO_2 is to find the relation between the reactor pressure ratio (p'/p) at a ΔL = 10.4 m required to remove a given amount of afterheat (6%).

With T_i/T_i' = 1.535 and $10^2 q$ = 6, Eq. (12) becomes for laminar flow (with $T_m'/T_i' \cong 1.75$), for helium and CO_2, respectively.

$$\left.\begin{aligned} (p'/p)_H &= 0.20\ \phi^{0.5}\left(\frac{T_m'}{T_i'} - 1\right)^{-0.1} && \text{for } \phi \leq 2.5 \\ (p'/p)_c &= 0.034\ \phi^{0.5}\left(\frac{T_m'}{T_i'} - 1\right)^{-0.1} && \text{for } \phi \leq 0.5 \end{aligned}\right\} \tag{18}$$

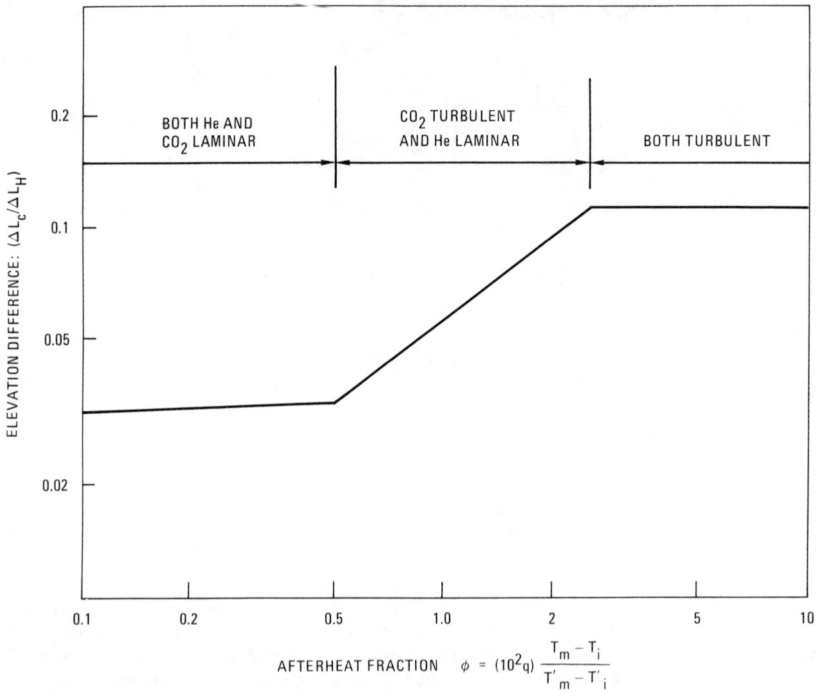

Fig.1. Comparison of CO_2 and He for Natural Circulation Cooling

and for turbulent flow, Eq. (11) yields

$$\left.\begin{aligned} (p/p)_H &\cong 0.14\ \phi^{0.9}\ (T_m'/T_i' - 1)^{-0.1} \quad \text{for } \phi \geq 2.5 \\ (p/p)_c &\cong 0.045\ \phi^{0.9}\ (T_m'/T_i' - 1)^{-0.1} \quad \text{for } \phi \geq 0.5 \end{aligned}\right\} \qquad (19)$$

The relation between the ratio of reactor pressures (p'/p) and decay heat fraction (ϕ) is plotted in Fig. 2 for both helium and CO_2.

The result shows that during laminar flow, the pressure required to remove a given afterheat for CO_2 is only one-sixth of the pressure required with helium. Similarly, for turbulent flow the pressure required with CO_2 is only one-third that required with He.

From Fig. 2 we also see that at 2 atm pressure (90 atm design) with He, the modified afterheat fraction is negligible (less than 0.01%). But, with CO_2, about 1% afterheat can be removed under similar conditions. If the reactor pressure is 3.7 atm, about 2% afterheat can be removed with CO_2.

Considering the superiority of He during normal cooling and attractiveness of CO_2 during depressurized natural circulation cooling, injection of CO_2 into a shutdown depressurized GCFR normally cooled by helium seems to be a possibility. However, a detailed transient analysis with CO_2/He mixture must be undertaken to insure that the maximum cladding temperature limit is not exceeded.

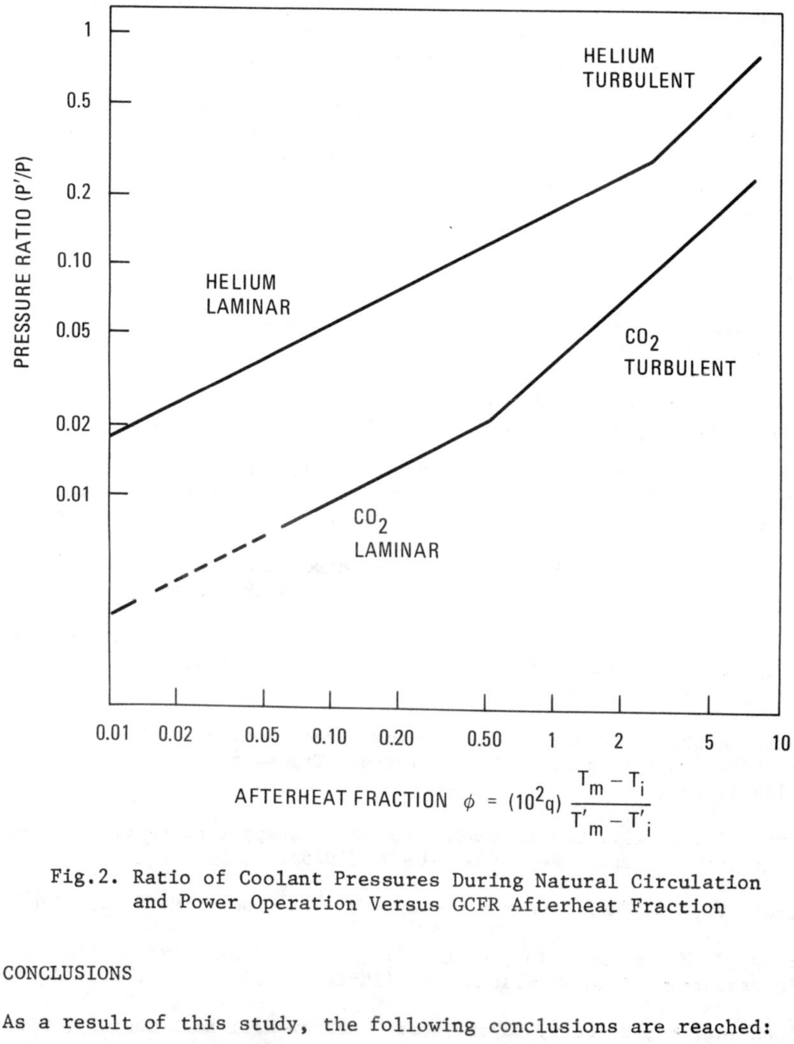

Fig.2. Ratio of Coolant Pressures During Natural Circulation and Power Operation Versus GCFR Afterheat Fraction

4. CONCLUSIONS

As a result of this study, the following conclusions are reached:

(a) Natural circulation flow of fully pressurized helium is adequate for removal of 5-6% afterheat with an elevation difference of about 10 m with nominal temperature conditions.

(b) The required elevation difference can be reduced by increasing the maximum clad temperature or decreasing the core inlet temperature.

(c) With depressurized helium, very little afterheat can be removed, even with extreme temperature conditions.

(d) CO_2 is superior to He for natural circulation cooling. For pressurized condition, the required elevation difference for CO_2 varies between 1/30 to 1/9, depending on the afterheat fraction, corresponding to laminar or turbulent flows.

(e) With CO_2, afterheat removal by natural circulation during depressurized condition is feasible, e.g., 1% afterheat can be removed at 2 atm pressure (absolute) and 10 m elevation difference.

(f) Injection of CO_2 into a shutdown GCFR core normally cooled by helium seems an attractive possibility. A detailed study (including transient analysis with CO_2/He mixture) is required to insure that the maximum cladding temperature remains below the acceptable limit during this procedure.

REFERENCES

1. Dalle Donne, M., Proc. Conf. on Safety, Fuel Core Design in Large Fast Power Reactors, ANL Report 720 (1965).

2. Melese d'Hospital, G., Influence of the Choice of Coolant Gas on Reactor Performance, Nuclear Applications (2), pp. 205-212 (1966).

3. Parme, L. L., Shenoy, A. S., Breher, W. J., and Potter, R. C., Natural Circulation Cooling in GCFR, Proceedings of the Specialist Meeting on Decay Heat Removal and Natural Convection in FBRs (1980) (this meeting).

4. Metais, B. and Eckert, E. R. G., Forced, Mixed and Free Convection Regimes, Journal of Heat Transfer (86), pp. 295-296 (1964).

5. Sparrow, E. M., Loeffler, L. A., Jr., and Hubbard, H. A., Heat Transfer to Longitudinal Laminar Flow Between Cylinders, Journal of Heat Transfer (832), pp. 415-422 (1961).

6. Colburn, A. P., A Method of Correlating Forced Convection Heat Transfer Data and a Comparison with Fluid Friction, Transaction of the AIChE (29), pp. 174 (1933).

7. Goodman, J., et al., The Thermodynamic and Transport Properties of Helium, General Atomic Company Report GA-A13400 (1975).

8. Steifel, V., CO_2 Thermodynamic Properties, EIR Report TM-IN-363 (1967).

9. Sparrow, E. M. and Loeffler, A. L., Jr., Longitudinal Laminar Flow Between Cylinders Arranged in Regular Array, AIChE Journal (5), pp. 325-330 (1959).

10. Schlichting, H., Boundary Layer Theory, McGraw-Hill Book Company, Sixth Edition, pp. 561 (1968).

NOMENCLATURE

A	Flow area: $\pi D^2/4 \; V/(1 - V)$
C	Coolant specific heat at constant pressure
D	Rod diameter
d	Hydraulic diameter
E	$(\varepsilon/4S)\;(D/L)\;V/(1 - V) = \Delta t/\Delta T$
e	$(\varepsilon N' P_r/40)\;(D/L)\;V/(1 - V)$
F	Heated fraction of the channel
f	Friction factor
G_r	Grashof number, $g \frac{d^3}{\nu^2} \frac{\Delta t}{T_{av}}$
g	Acceleration of gravity (9.8 m/sec^2)
h	Surface heat transfer coefficient
K	Thermal conductivity of coolant
L	Channel length
ΔL	Loop elevation difference for natural circulation
M	Coolant molecular weight
N	Reynolds number, $Wd/\mu A$, (W: mass flow)
Nu	Nusselt number, dh/K
P_r	Prandtl number, $\mu C/K$
p	Pressure during normal operation
p′	Pressure during natural circulation
Δp	Core pressure drop (nominal)
Q	Nominal thermal power
Q′	Thermal power during natural circulation
$q = Q'/Q$	Afterheat fraction
q_1	Afterheat removed by full pressure He at nominal temperatures
R	Perfect gas constant or $G_r \, P_r \frac{d}{L}$
R_a	Rayleight number $g(\Delta t/T)(d^3/\nu^2)P_r$
S	Stanton number, Nu/NP_r

NOMENCLATURE (CONT.)

T	Coolant temperature
T_m	Maximum clad temperature
ΔT	Coolant temperature rise in channel
Δt	Film temperature drop
V	Coolant void fraction
Z	Ratio of total loop pressure drop to core pressure drop
γ	Ratio of coolant specific heats at constant pressure and constant volume
ε	Ratio of local flux to average flux (axially)
μ	Coolant viscosity
ν	Kinematic viscosity, μ/ρ
ρ	Coolant density
ϕ	$10^2 q\ (T_m - T_i)/(T_m' - T_i')$
$'$	"Primed" quantities correspond to values during natural circulation

SUBSCRIPTS

av	average
c	CO_2
e	exit
H	He
i	inlet
ℓ	laminar
t	turbulent

APPENDIX

Derivation of Basic Relations

Equations (6) and (7) describe the basic relation between pressure drop, elevation difference, and flow. Using correlations (8) and (9) for the friction factors, with the hydraulic diameter $d = FDV/(1-V)$ and the gas constant $R = MC(1-1/\gamma)$, the following relations are obtained:

$$\frac{(\Delta L)_\ell}{L} = \frac{76.4\ Z}{F^2} P_r \left(1 - \frac{1}{\gamma}\right)^2 \left(\frac{1}{V} - 1\right)^3 \frac{T_i' T_e' T_{av}'}{\Delta T'^2} \frac{KQ'}{gp'^2 D^4} \qquad \text{(A-1)}$$

since,

$$\nu_{av} = K\ P_r/C\rho_{av} = K\ P_r\ T_{av}\ (1 - 1/\gamma)/p,$$

and similarly

$$\frac{(\Delta L)_t}{L} = \frac{0.15\ Z}{F} (N')^{-0.2} \left(1 - \frac{1}{\gamma}\right)^2 \left(\frac{1}{V} - 1\right)^3 \frac{T_i' T_e' T_{av}'}{\Delta T'^3} \frac{Q'^2}{gp'^2 D^5} \qquad \text{(A-2)}$$

From the definition of the Stanton number, the film drop at the location of the maximum temperature is calculated as:

$$\Delta t = \frac{\varepsilon}{4S} \frac{V}{1 - V} \frac{D}{L} \Delta T \qquad \text{(A-3)}$$

where ε is the ratio of local flux to the average flux.

The following correlations for the Stanton number are appropriate here:

$$\left.\begin{array}{ll} \text{Laminar Case:} & S = 10/NP_r \\ \text{Turbulent Case:} & S = 0.023\ P_r^{-2/3}\ N^{-0.2} \end{array}\right\} \qquad \text{(A-4)}$$

and, thus, the coolant temperature rise becomes:

$$\text{Laminar flow:} \quad (\Delta T')_\ell = T_m' - T_i' - 3.2 \times 10^{-2}\ \varepsilon F \frac{V}{1 - V} \frac{Q'}{KL} \qquad \text{(A-5)}$$

$$\text{Turbulent flow:} \quad (\Delta T')_t = (T_m' - T_i')\ (1 + E')^{-1} \qquad \text{(A-6)}$$

where,

$$E' = 10.9\varepsilon \frac{V}{1 - V} \frac{D}{L} P_r^{\ 2/3} (N')^{0.2} \qquad \text{(A-7)}$$

Using the following approximate expression for the Reynolds number N'

$$N' \simeq \frac{4}{\pi} \frac{F}{P_r} \frac{Q'}{KD(T_m' - T_i')} \qquad \text{(A-8)}$$

Pr = 0.7, F = 0.9, and typical values $\varepsilon \approx 0.5$, V = 0.5, D/L = 0.7×10^{-2}, D = 0.8×10^{-2} m, $T'_m - T'_i = 1000$ K, and $Q' = 40 \times 10^3\, q$ (watts), where $q = Q'/Q$ is the afterheat fraction, one finds E' to be between 0.14 and 0.20.

Hence, Eq. (A-6) can be written as:

$$\Delta T'_t \cong 0.85\,(T'_m - T'_i) \tag{A-9}$$

Substituting the above expression into Eq. (A-2), the following relation is obtained when introducing the expression ϕ (from Eq. 10):

$$\frac{(\Delta L)t}{L} = \frac{1.6 \times 10^{-4}\,Z}{F^{1.2}}\,(P'_r)^{0.2}\left(1 - \frac{1}{\gamma'}\right)^2\left(\frac{1}{V} - 1\right)^3 \frac{(K')^{0.2}\,T_i'^{2}}{g\,p'^{2}D^{4.8}}$$
$$\times\left(\frac{Q}{T_m - T_i}\right)^{1.8}\left(\frac{T'_m}{T'_i} - 1\right)^{-0.2}\phi^{1.8} \tag{A-10}$$

Similarly, after some algebraic manipulations, the following relation for laminar flow is obtained:

$$\frac{(\Delta L)\ell}{L} = \frac{2.52\,Z}{F^2}\,P'_r\left(1 - \frac{1}{\gamma'}\right)^2\left(\frac{1}{V} - 1\right)^3 \frac{K'Q\,T_i'^{2}}{g\,p'^{2}D^4\,(T_m - T_i)}$$
$$\times\left(\frac{T'_m}{T'_i} - 1\right)^{-0.2}\phi(1 + 0.06\phi) \tag{A-11}$$

An approximate expression for Q to be introduced in the above equation is found to be:

$$Q \cong 3D^2\left(\frac{T_m}{T_i} - 1\right)\left(\frac{V}{1 - V}\right)^{3/2}\left(\frac{C_H T_i}{1 - 1/\gamma_H}\,F\,\frac{D}{L}\,p\Delta p\right)^{1/2} \tag{A-12}$$

if we use the following relation [2]

$$Q = \frac{\pi D^2 \Delta T}{4\sqrt{2\,f\,T_{av}}}\left(\frac{V}{1 - V}\right)^{3/2}\left(\frac{C_H}{1 - 1/\gamma_H}\,F\,\frac{D}{L}\,p\Delta p\right)^{1/2} \tag{A-13}$$

with the approximation applicable to GCFR designs

$$\Delta T \cong \frac{3}{4}\left(\frac{T_m}{T_i} - 1\right)\sqrt{T_i\,T_{av}} \tag{A-14}$$

and with $f = 2 \times 10^{-2}$, typical friction factor for roughened rods at Reynolds numbers around 10^5.

GCFR Residual Heat Removal Following Depressurization Accidents

H. S. CHUNG, H. CHI, and A. SHENOY
General Atomic Company
San Diego, California, USA

ABSTRACT

Performance of gas-cooled fast-breeder reactor (GCFR) residual heat removal (RHR) systems following postulated depressurization accidents is evaluated. The GCFR employs high pressure helium as the primary coolant inside the prestressed concrete reactor vessel (PCRV). Although a rapid depressurization with a significant leak flow area is extremely improbable, the GCFR design considers a conservative leak area open to the reactor containment building. The GCFR demonstration plant with its recent design revisions is described. Results of the depressurization analyses are presented for both a conservative and a best-estimate model with and without performance uncertainty margins, respectively. A sensitivity analysis is also presented for the effects of important design variables, such as the circulator capacity, the leak areas, and the containment backpressures.

1. INTRODUCTION

Performance evaluation for residual heat removal (RHR) systems in the gas-cooled fast breeder reactor (GCFR) under a postulated depressurization accident is presented.

The GCFR employs high-pressure helium as primary coolant, which is chemically inert, neutronically transparent, and nonradioactive, and which remains single phase under abnormal conditions. Entire primary coolant system components are enclosed in a massive prestressed concrete reactor vessel (PCRV), which precludes failure of primary coolant pipes of significant size. The GCFR core consists of vented fuel rods. The pressure equilization system (PES) continuously removes the radioactive fission gases from each of the fuel rods and thus prevents buildup of the rod internal gas pressure. Consequently, unlike other reactor fuel rods, possibility of cladding ballooning and flow blockage is remote during rod heatup following a depressurization accident.

Owing to these advantageous features, the GCFR development was initiated at General Atomic (GA) in the early 1960s. Work related to the GCFR development has also been contributed by the U.S. National Laboratories, as well as by industries and governmental agencies in Europe. GCFR development has been supported by both the government through Department of Energy (DOE) and by utilities through Helium Breeder Associates.

Evolving from the early work, the GCFR Preliminary Safety Information Document (PSID) (Refs. 1 through 3) was submitted to the Nuclear Regulatory Commission (NRC) in 1971 to serve as a basis for an information exchange in evaluation of the GCFR concept. Summarizing the initial phase of its review, the NRC issued a Preapplication Safety Evaluation (Ref. 4) of the GCFR in 1974.

Recently, major plant design revisions have been conducted by GA to improve the reliability and adequacy of reactor core cooling. A comprehensive RHR evaluation was recently made using a two-loop, 400-MW(e) plant.

This paper presents results of this evaluation under postulated depressurization accidents. Described is the GCFR demonstration plant with its recent design revisions for the RHR systems that include natural circulation capabilities at certain conditions. Possible leak areas in the PCRV are discussed. The sequence of events following a postulated depressurization accident are discussed in detail. Methods of analysis for depressurization accidents are summarized which include dynamic responses of the primary coolant system and the containment atmosphere.

The results of transient analyses are presented for both a conservative and a best-estimate model of a design basis depressurization accident (DBDA) with and without the performance uncertainties. Sensitivity analyses are presented for cumulative and individual uncertainty effects. Also presented are specific effects of important design variables, such as the circulator capacity, the leak-area size, and the containment backpressure after depressurization.

2. DESCRIPTION OF GCFR DEMONSTRATION PLANT

The GCFR demonstration plant employs helium pressurized to about 10 MPa (1450 psia) as the reactor primary coolant. The reactor core and the steam generators and helium circulators in the main cooling loops, as well as the auxiliary circulator and heat exchangers in the core auxiliary cooling system (CACS) loops, are all contained within a massive PCRV (see Fig. 1). Reactor cooling during power operation is provided by the main cooling loops, and decay heat removal is performed by any one of several RHR systems. The reactor core consists of many hexagonal fuel and blanket assemblies. Each assembly contains a large number of fuel rods which are similar to LMFBR fuel rods, except that the stainless-steel cladding surface is roughened to improve heat transfer. Heat from the fuel rods is transferred to the helium coolant, which transports the heat to the steam generator. Steam from the steam generator is used to generate electricity through a balance-of-plant arrangement similar to those used in other nuclear or fossil-powered plants.

Important aspects of the current demonstration design may be highlighted by referring to the recent design revisions given below.

2.1 Upflow Core With Natural Circulation RHR

This natural circulation capability, which is available with pressurized coolant, is a substantial safety asset. It provides diversity to the forced circulation systems and inherently passive and long-term RHR with minimum operator or powered actions.

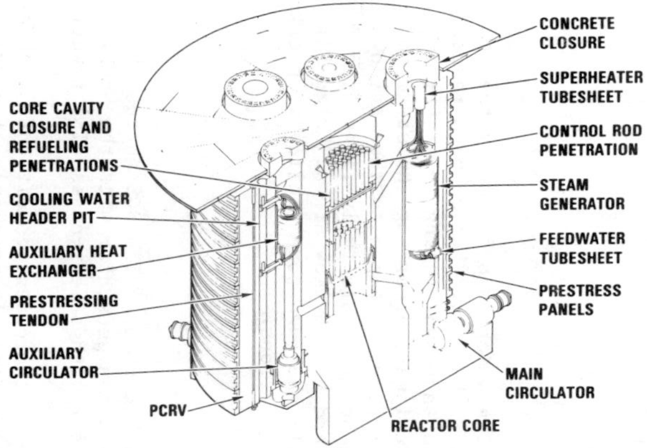

Fig. 1. GCFR nuclear steam supply system

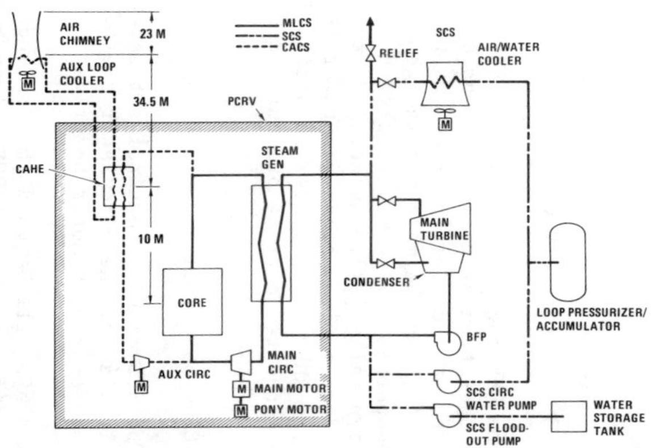

Fig. 2. Schematic flow diagram showing GCFR residual heat removal systems

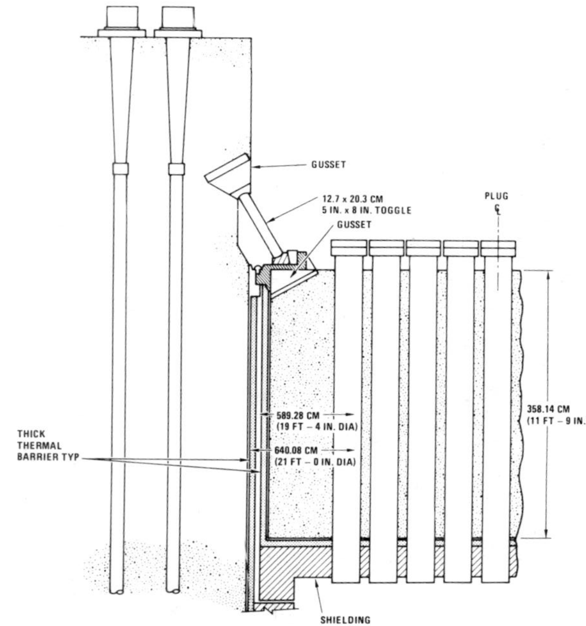

Fig. 3. Reactor-cavity closure

2.2 Incorporation of Two Safety RHR Systems

A safety-class RHR system, called the shutdown cooling system (SCS), is incorporated in addition to the existing safety-class CACS and the non-safety main loop cooling system (MLCS). Provision of two safety RHR systems enhances safety and reliability in excess of the NRC general design criteria for light water reactors (LWRs), which require one safety system, and is consistent with the NRC position on fast reactor licensing applied to the Clinch River Breeder Reactor in 1976.

2.3 Electrically-Driven Radial-Flow Main Circulators

Compared with the steam-driven axial circulator used in the previous demonstration plant, the electric drive results in the following beneficial impacts:

1. Simpler control due to the system decoupling between heat source and the drive power.

2. Higher stall resistance of radial-flow circulator.

3. Longer coastdown due to massive electric motor.

4. Preoperation testing ease.

3. RHR SYSTEMS

In conjunction with these design features, the following four RHR systems are available for the GCFR. These are also depicted in a simplified diagram in Fig. 2.

3.1 MLCS - Non-Safety Class

The RHR mode of the MLCS is the first choice which is obtained by the continued operation of the main loops after reactor shutdown. Following a reactor trip, the main circulator speed is reduced to a shutdown cooling level (typically 15%), and the main turbine-generator is tripped, diverting the steam to the desuperheater. Concurrently, the feedwater flow is ramped to the shutdown rate (typically 15%). Eventually, the steam generators are fully flooded, and the feedwater flow, steam generator pressure, and helium flow are adjusted for long-term RHR. This system is available for all shutdown core cooling with off-site or on-site power at pressurized coolant conditions.

3.2 SCS - Safety Class

The SCS consists entirely of safety-class equipment. The SCS shares the main circulator, the circulator shaft, and the steam generator with the MLCS. The SCS uses a pony motor to drive the circulator with safety-class (1E) power.

The SCS will be used for RHR under a number of accident conditions, such as loss of off-site power (LOSP), loss of feedwater, etc. The sequence of operation for transferring from the MLCS to the SCS is as follows. When the main circulators coast down below the shutdown MLCS speed (12.5%), the pony motors are energized to maintain the circulator speed (10%), and the SCS heat rejection system is activated. The feedwater flow from the boiler feed pump (BFP) is replaced by that from the SCS floodout pump. Steam is vented until the steam generator is fully flooded. Using the circulating water pump, the water is then circulated in a closed loop through the air-water cooler (AWC), where the core decay heat is removed by air fans to the atmosphere (see Fig. 2).

3.3 CACS - Safety Class

The CACS is the most comprehensive RHR system which is available for the pressurized as well as the depressurized coolant conditions. Each of the three CACS loops is comprised of an electrically driven auxiliary circulator, a check valve, and a helium-to-water heat exchanger called the core auxiliary heat exchanger (CAHE). The water from the CAHE is circulated through a high-pressure water loop with pumps and a pressurizer, and the heat is rejected to the atmosphere by air from fans in the auxiliary loop cooler (ALC), which is a finned-tube-bank heat exchanger. The flow diagram of the CACS is included in Fig. 2.

3.4 Natural Circulation CACS - Safety Class

The CACS design incorporates natural circulation capabilities on the helium, water, and air sides as a backup to normal forced circulation capabilities. Using the CACS, core decay heat is transported by the primary coolant helium to a high-pressure water in the CAHE, which is elevated 10 m (33 ft) above the core (see Fig. 2). Heated water from the CAHE reaches the ALC located 34.5 m (113 ft) above the CAHE by natural circulation in the pressurized water loop. The heat from the ALC is ultimately rejected to the atmosphere by natural draft of air through a 23-m (75-ft) chimney. The natural circulation core cooling is available for an indefinite period under a total loss of forced circulation capabilities with the primary coolant pressurized. With the primary coolant depressurized, natural circulation in the primary loops is not adequate, but sufficient natural circulation in the secondary water and the tertiary air flows is available as backup to forced circulation.

4. LEAK AREA

The PCRV and its several penetrations and closures are designed based on American Society of Mechanical Engineers (ASME) Section III, Division 1, Class 1 codes. As in the case of LWR pressure vessels, a gross failure of the GCFR PCRV and its closures may not be postulated as a design basis event (Ref. 5). However, during the high-temperature gas-cooled reactor (HTGR) (Ref. 6) and GCFR development, a leak area has been conservatively assumed and used as design basis for a safety core cooling system.

A typical design of the reactor-cavity closure is shown in Fig. 3. The plug has a diameter and a height on the order of 6 and 3 m (20 and 10 ft), respectively. The holddown locking system consists of a series of toggle segments that are inserted into circumferential grooves around the top of the plug and penetration after the plug has been placed in the preparation and the seal welding has been completed; the plug is positioned vertically by a step in the penetration.

There are similar penetration closures for steam generators, circulators, auxiliary loop components, etc. Although extremely improbable, failure may occur at the welded seal of any one of the penetration closures. In this event, the maximum free flow area may be defined by assuming unfavorable stackup of dimensional tolerances between the plug and the cavity. If necessary, the leak area can further be reduced by employing a compliant flow restrictor (which looks like a piston ring) around the plug. The maximum leak area is considered to be 0.016 m^2 (25 $in.^2$).

To investigate sensitivity and to determine margin of safety, the leak areas between 0.016 and 0.065 m^2 (25 and 100 $in.^2$) have been analyzed. For the purpose of this study, the maximum leak area of 0.065 m^2 (100 $in.^2$) is used to define the DBDA. This leak area is subject to future revision.

5. DESCRIPTION OF ACCIDENT

When the design basis leak area is postulated, the primary coolant starts to flow into the containment building which contains air of about 20 times the PCRV volume. A signal of low primary coolant pressure, high containment pressure, or high containment radioactivity actuates the reactor trip and the containment isolation. The safety-related event sequence assumed here includes simultaneous LOSP which disables the MLCS. LOSP results in coastdown of main circulators and startup of emergency diesel generators. The primary coolant pressure decreases exponentially, and the leak flow is choked during most of the blowdown period. When the coolant pressure decreases below a set value [e.g., 2.07 MPa (300 psia)], the plant protection system (PPS) logic, which is chosen for the purpose of these analyses, initiates the CACS that is designed for all RHRs including the DBDA.

For a slower depressurization in which the pressure [2.07 MPa (300 psia)] to switch on the CACS is reached later, the PPS initially selects the SCS, one of the two safety RHR systems. The SCS pony motors are powered to prevent the main circulator speed from coasting down and to maintain a prescribed value for the SCS operation. As pressure decays, the pony motor accelerates with motor speed inversely proportional to pressure, thus providing roughly constant mass flow rate until the maximum pony motor speed is reached. When the pressure reaches below the setpoint value [2.07 MPa (300 psia)], the cooling mode is automatically transferred to the CACS.

When the CACS is called upon, the full design torque is applied to the auxiliary circulators. The auxiliary circulator pressure head forces the auxiliary loop isolation valve (ALIV) open and the main loop isolation valve (MLIV) closed. Thus, the helium circulation loop switch is accomplished. Also, the water pumps are started for the pressurized water loop between the CAHE and the ALC. Thus, the core decay heat, carried by helium to the CAHE and transported by water to the ALC, is ultimately dissipated into the atmosphere by a fan or natural draft of air through the ALC (see Fig. 2).

The typical core flow rate, core thermal power, and the pressure transients following the DBDA are shown in Fig. 4. Figure 5 shows the main and auxiliary circulator speed variations before and after the loop transfer in the same transient. Figure 6 shows the maximum hot-spot temperatures for the fuel and the blanket rods following this transient.

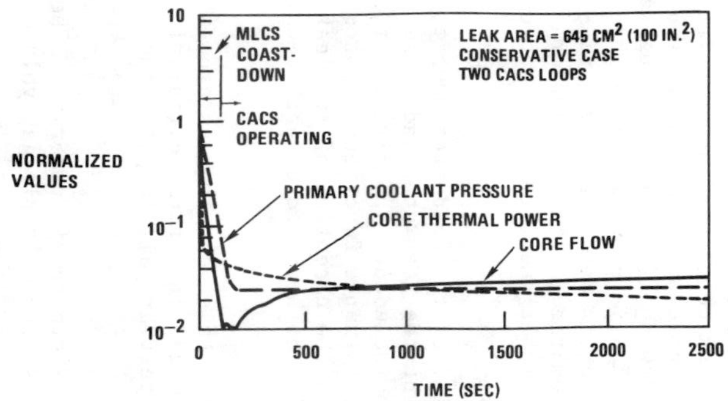

Fig. 4. Typical DBDA transient for core power, flow, and pressure

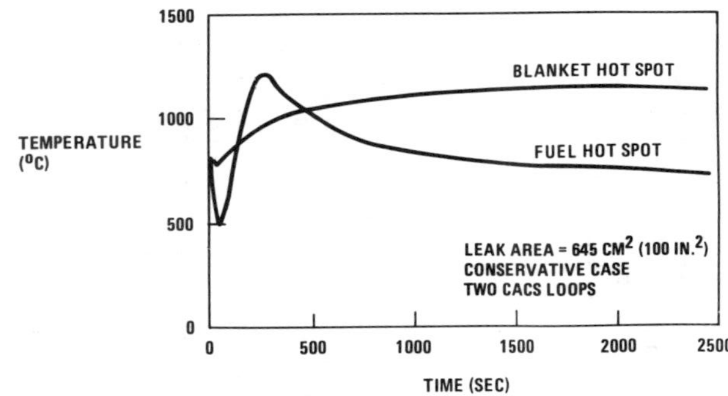

Fig. 6. Fuel and blanket maximum hot-spot cladding temperatures following a DBDA

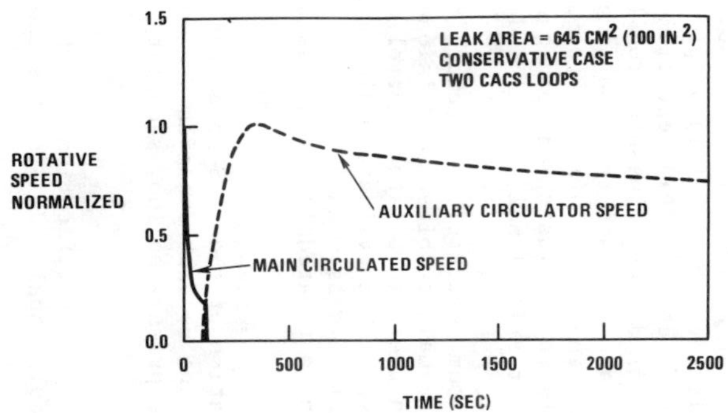

Fig. 5. Rotative speed for main and auxiliary circulator following a DBDA

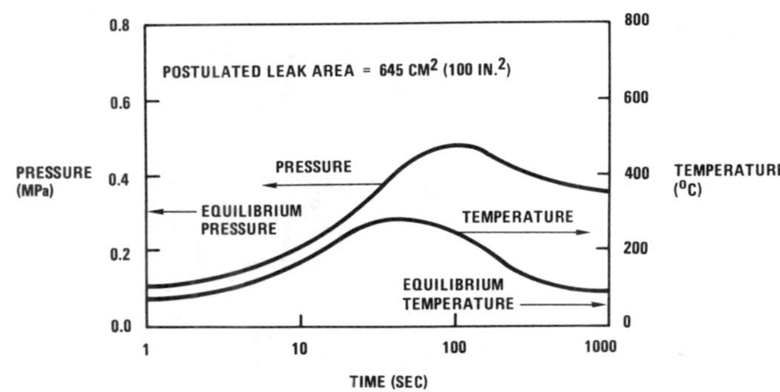

Fig. 7. Containment building gas pressure and temperature transient following a DBDA

The primary coolant pressure equilibrates the containment pressure after about 3 min of blowdown. The typical temperatures and pressure responses of the containment atmosphere following the DBDA are shown in Fig. 7. Following the blowdown stage, air ingress occurs gradually into the primary coolant due to inhalation of the containment air mixture as the coolant inside the PCRV cools and contracts. In case of a leak occurring somewhere in the upper part of the PCRV, natural convection between light helium and heavy containment air mixture promotes air ingress as an additional mechanism.

6. METHODS OF ANALYSIS

The transient response of the reactor system to depressurization accidents were analyzed with the FASTRN computer program. This program simulates dynamics of the GCFR primary and secondary coolants of the main cooling systems as well as the auxiliary safety systems under pressurized and depressurized conditions. For the blowdown phase of depressurization, the primary coolant system is modeled with gas volumes which respond according to the perfect gas law and the energy and mass conservation equations. These volumes are connected by paths which represent the components of the system, namely, the reactor core, the steam generators, the helium circulator, and a leak path to the containment. Each of these components are, in turn, dynamically modeled by their governing equations.

The core is modeled with six flow channels, representing three fuel and three blanket rods: (1) an average powered rod, (2) a maximum powered rod with nominal (or expected) heat transfer parameters, and (3) a maximum powered rod with degraded heat transfer parameters to include the effects of statistically evaluated hot-spot factors for both the fuel and blanket groups. Variations in the convective heat transfer coefficient and friction factor with local Reynolds number are included for both the smooth and the artificially roughened regions of the representative fuel rods. Reactivity feedback contributions associated with six distinct sources [(1) Doppler effect, (2) core axial expansion, (3) helium density changes, (4) core element thermal bowing, (5) core splaying associated with grid plate pressure drop loads, and (6) grid plate thermal expansion] are evaluated, and the resulting reactor power is determined by a six-group point kinetics solution. Provision for automatic control rod and/or backup shutdown rod insertion upon generation of either the first or a subsequent reactor trip signal is included.

The steam generator is modeled from the outlet of the feedwater pump to the inlet of the main turbine. Provisions for modeling the feedwater and steam valves with their associated controllers are also included. The circulator dynamic response is evaluated using circulator performance functions. Incompressible flow is assumed for all flow paths, with the exception of the leak path, which is treated in a compressible manner for both choked or unchoked flow.

Simulation of the CACS components and the effects of air ingress are included. The coolant dynamics are modeled for forced or natural circulation through the primary helium, secondary water, and tertiary air loops of the CACS.

The program verification effort for the FASTRN program is underway. The FASTRN calculational model for the coolant dynamics with multiple gas volumes has been indirectly verified by obtaining a close agreement with the RATSAM program (Ref. 7). The RATSAM program is the system dynamics program for the HTGR which has been verified (Ref. 7) with depressurization data from the Calder Hall Reactor model tests.

The containment atmosphere response to the PCRV depressurization is calculated with a separate program CNTB (Ref. 8). In addition to the gas dynamics during the blowdown, CNTB accounts for heat transfer to horizontal and vertical heat sink surfaces and for the effects of specified degrees of gas mixing between the containment air and the discharged helium.

An initial validation of the CNTB program was made (Ref. 8) by obtaining a close agreement with the CONTEMPT-G program (Ref. 9). The CONTEMPT-G is a modified version of the CONTEMPT program developed and used for LWR containment analysis. The CNTB program, however, contains superior features over the CONTEMPT-G, which requires further verification.

7. RESULTS OF BASE CASES

The system response to depressurization accidents has been evaluated using both a best-estimate and a conservative model. The best-estimate model is based upon expected performance values, whereas the conservative model uses degraded performance values based upon uncertainty margins believed to be consistent with current licensing practice (Ref. 4). Conservatisms are introduced to account for uncertainties in initial conditions, equipment availability, equipment performance, and selected analytical parameters.

The core thermal response, particularly the maximum cladding temperatures of the fuel and the blanket rods, following a most severe DBDA is shown in Fig. 6. The results are obtained with the conservative model and assuming two out of three cooling loops.

It is shown in Fig. 6 that the peak of the conservatively calculated hot-spot cladding temperature of the maximum-powered typical fuel rod is ~100°C lower than the tentative cladding damage limit of 1300°C and 170°C lower than the cladding melting point of 1370°C. The temperature margin of 1790°C adequately accommodates the anomalies: a 100°C margin is allowed for the fuel assembly edge effect, and a 70°C margin is allowed for potential fuel rod damage below the cladding melting temperature. Depending on the fuel assembly design, the edge fuel rod adjacent to the fuel assembly duct wall can exhibit a higher temperature than the typical interior rod of the bundle under accident conditions. Separate studies are underway to define the edge channel effect accurately and to mitigate the effect by improving the fuel assembly design, if necessary.

The GCFR blanket rods are, in general, at much lower temperatures than the fuel rods during normal full-power operation. However, the blanket cooling becomes important at low-flow conditions after reactor trip due to decay heat and flow redistributions. The maximum blanket cladding temperature, shown in Fig. 6, is lower than the fuel cladding value. The blanket rod diameter is much larger and therefore responds much slower than the fuel rods, as seen in Fig. 6.

To provide greater insight into the safety margins provided in the design, the result of the conservative model is compared with that of the best-estimate model in Fig. 8. The difference between the maximum fuel cladding temperatures

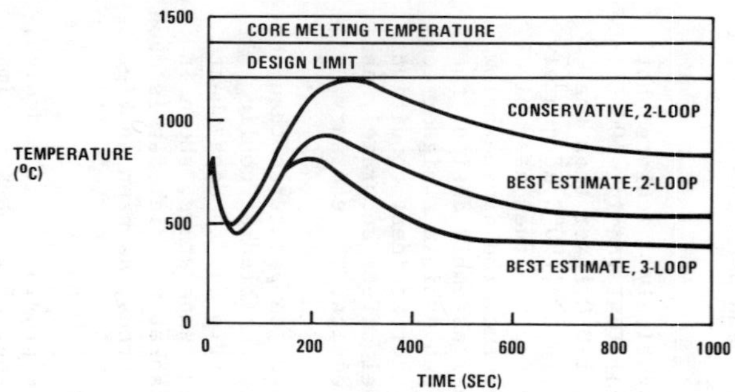

Fig. 8. Effect of cumulative uncertainty margins in conservative model and number of cooling loops on the hot-spot fuel cladding temperature during a DBDA

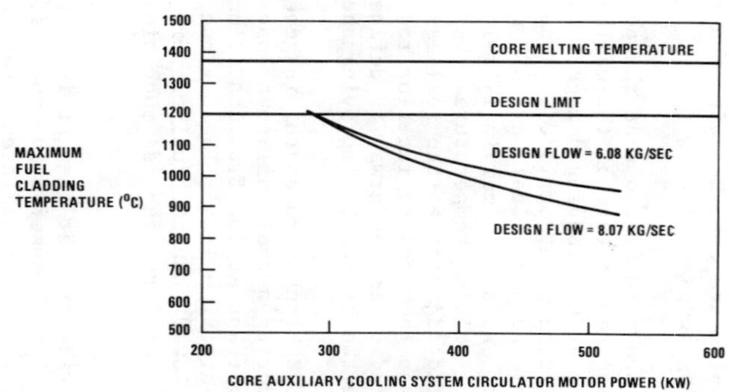

Fig. 9. Effect of auxiliary circulator design capacity and motor power on DBDA maximum fuel cladding temperature

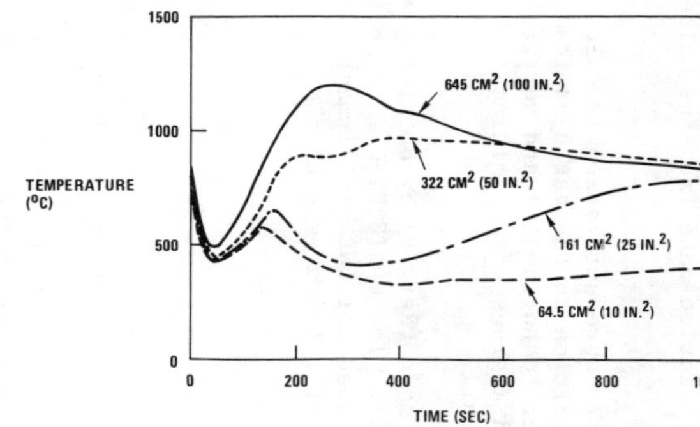

Fig. 10. Effect of leak area on fuel cladding hot-spot temperature during a depressurization transient

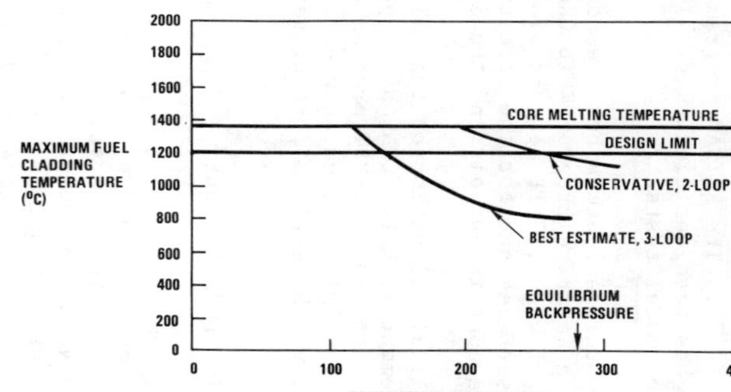

Fig. 11. Effect of containment pressure on the maximum fuel cladding temperature for a DBDA

obtained by the conservative and the best-estimate models is 280°C, indicating that the cumulative combination of all the uncertainty margins employed in the conservative model results in a large safety margin. The best-estimate model, using all three loops, indicates further reduction of the cladding temperature by 100°C.

Obviously, a realistic approach is to combine the uncertianty margins statistically. To evaluate sensitivity, each of the uncertainty margins is applied individually to the best-estimate model. The results of the sensitivity analysis are presented in Table 1. Uncertainty margins of significant effect are +20% decay heat, +20% loop flow, and -10% backpressure. The arithmetic sum of the deviation is smaller than the deviation due to cumulative uncertainties, which indicates the effects of nonlinearity in the correlations.

A statistical sum represented by the root-sum-square of the deviations may be used if it could be established that the probability of the overall uncertainty is equal to the probability of the individual uncertainties, that all the error distributions are symmetrical, and that the variables are mutually independent. Since these assumptions have not been explored fully, the cumulative combination will be used for conservative purposes during the conceptual design phase.

8. EFFECT OF AUXILIARY CIRCULATOR DESIGN

Due to the low coolant density after depressurization, the coolant volumetric flow requirement is the highest. The depressurized condition becomes the design basis of the safety cooling systems. The SCS is based on a slow depressurization, and the CACS is based on the most severe DBDA [0.065 m^2 (100 $in.^2$) leak area].

The auxiliary circulator power of the current CACS design is 300 kW. The auxiliary circulator capability is defined by both the maximum volumetric flow (speed) and the maximum torque. For a dense coolant due to a high air ingress or backpressure, the circulator capability is limited by the maximum torque; for a thin gas due to a low containment backpressure, it is limited by the circulator speed. It is important to select the design points of the speed and torque so that the density uncertainty can be effectively covered by the circulator capability. The effect of various volumetric flow rate and the drive power of the auxiliary circulator on the maximum core cladding temperature is presented in Fig. 9. The auxiliary circulator capability can be used for tradeoff with other design parameters.

9. EFFECT OF LEAK AREA

The maximum hot-spot fuel cladding temperature responses following depressurization accidents with various leak area are shown in Fig. 10. For all the cases in this figure, the CACS was assumed to start at ~2 min to address the effect of the leak area on the CACS core cooling. Ordinarily, the SCS is expected to perform the RHR during the initial high pressure periods above 2.07 MPa (300 psia).

The leak area is directly related to the rate of depressurization and the period for full blowdown. The effect of the leak area, in general, is twofold: (1) the smaller the leak area, the longer the high-pressure period in which the core heat can be dissipated more effectively with high-density coolant; (2) the

TABLE 1

SENSITIVITY OF THE DBDA CORE CLADDING TEMPERATURE TO INDIVIDUAL PARAMETER UNCERTAINTIES

Parameter	Uncertainty Margin (%)	Maximum Fuel Cladding Temperature Deviation, ΔT_i (°C)
Decay heat	+20	160
Primary loop flow resistance	+20	28
Overall conductances in CAHE and in ALC	-15	10
Primary coolant thermal conductiviy	-5	2
Primary coolant viscosity	+3	6
Containment backpressure	-10	30
Arithmetic sum $\sum_{i=1}^{6} \Delta T_i$		236
Statistical sum $\sqrt{\sum \Delta T_i^2}$		166
Cumulative uncertainty (all uncertainties applied simultaneously)		276

decay heat level is lower at the time of the peak cladding temperature, which usually occurs after the full coolant blowdown as the second peak. The complex responses of the cladding temperature under simultaneous variations of pressure, decay heat, and the cooling model transfer are shown in Fig. 10.

10. EFFECT OF THE CONTAINMENT BACKPRESSURE

The containment backpressure is important for core cooling, since it controls the coolant density after depressurization. Figure 7 shows the transient pressure and temperature responses of the containment atmosphere which were calculated using the CNTB program (Ref. 8), assuming a design basis leak area of 0.065 m^2 (100 $in.^2$) in the reactor inlet plenum and complete gas mixing in the containment. These assumptions result in lower backpressures and temperatures, which are conservative for the core cooling evaluation. Higher values are expected, depending on degree of gas mixing and location of the leak, which should be taken into consideration for the containment building design. Instead of using the transient backpressure, the DBDA core cooling was analyzed for various constant backpressures, and the result of the effect of backpressure on the maximum core cladding temperature is presented in Fig. 11. The conservative DBDA analysis is based on the constant 90% equilibrium backpressure (absolute) to account for uncertainty margins for the containment void volume and for the containment isolation delay. This treatment is suitably conservative for the conceptual plant design.

11. CONCLUSION

From the performance evaluation of the GCFR RHR system under postulated depressurization accidents, the following conclusions are reached:

1. The safety design base requiring two systems for high-frequency and low-consequence depressurization accidents and one system for the extremely unlikely events (e.g., DBDA) provides safety in excess of the intent of the General Design Criteria for LWRs and is consistent with the NRC position for the Clinch River Breeder Reactor.

2. Performance evaluation for the GCFR RHR system indicates adequate margin of safety in the core cooling under depressurization accidents.

3. A cumulative combination of the system performance uncertainties results in a very high safety margin. A more realistic approach of using a statistical combination of the uncertainties needs to be defined.

4. Sensitivity studies for important design variables, such as circulator capacity, leak area using flow restrictor, and containment backpressure, indicate design trade-off possibilities.

REFERENCES

1. "Gas-Cooled Fast Breeder Reactor Preliminary Safety Information Document," Gulf General Atomic Report GA-10298, February 15, 1975 (Amended 1976).

2. "Gas-Cooled Fast Breeder Reactor Preliminary Safety Information Document," Gulf General Atomic Report GA-10298, Supplement I, June 8, 1972 (Amended 1974).

3. "Gas-Cooled Fast Breeder Reactor - Responses to AEC Questions on the Preliminary Safety Information Document," Gulf General Atomic Report Gulf-GA-10298, Supplement II, October 18, 1973 (Amended 1974).

4. "Preapplication Safety Evaluation of the Gas-Cooled Fast Breeder Reactor," U.S. Atomic Energy Commission Directorate of Licensing, August 1, 1974.

5. ACRS Transcript of "Subcommittee Meeting on HTGR," Washington, D. C., July 11, 1977.

6. Ricardella, P. C., "Design Basis for HTGR Penetrations and Closures," General Atomic Report GA-A13832, May 7, 1976.

7. Deremer, R. K., and T. Shih, "RATSAM: A Computer Program to Analyze the Transient Behavior of the HTGR Primary Coolant System During Accidents," General Atomic Company Report GA-A13705, May 1977.

8. Landoni, J. A., "The CNTB Program for the Analysis of Partially Mixed Containment Atmospheres During Depressurization Events," General Atomic Report GA-A13753, July 1979.

9. MacNab, D. I., "The CONTEMPT-G Computer Program and its Application to HTGR Containments," General Atomic Report GA-A12692A (GA-LTR-6) Amendment A, March 1976.

PANEL DISCUSSION: FUTURE RESEARCH AND DEVELOPMENT—WHAT NEEDS TO BE DONE?

Introduction

JAMES G. GUPPY
Brookhaven National Laboratory

On Friday afternoon, February 29, 1980, following the conclusion of the formal presentation of papers, a panel discussion was held. The topic of discussion was "Future Research and Development - What Needs to be Done?" The six panelists who briefly (approximately five minutes each) presented their views and opinions were:

1.	Ashok K. Agrawal,	Group Leader, Code Development and Verification, Brookhaven National Laboratory
2.	Stephen L. Additon,	Manager, FFTF System Dynamic and Thermal Analysis, Hanford Engineering Development Laboratory
3.	Noel Grimm,	Principal Engineer, Large Plant Project Group, Westinghouse Advanced Reactors Division
4.	Paul Huebotter,	Associate Director, Components Technology Division, Argonne National Laboratory
5.	D. Leteinturier,	Engineer, DRNR/SEDC, C.E.N./Cadarache
6.	Neil E. Todreas,	Professor of Nuclear Engineering, Massachusetts Institute of Technology

Following the presentations by the panelists, questions and comments were solicited from the general audience.

The entire presentations by the panelists and the ensuing discussion were tape recorded. What follows here is a transcript of this tape recording. The individual presentation by each panelist is first given followed by the specific questions and comments from the audience and responses by the panelists.

Statements by the Panelists

Dr. Agrawal highlighted four potential future areas that should receive attention:

(1) There is a need for continued demonstration and collection of data for natural circulation for a range of operating conditions. These operating conditions should not just be demonstrated from normal conditions (with all equipment operating), but also from derated and abnormal operating conditions.

(2) Referring to the paper by Kelber (Ref. 1), Dr. Agrawal addressed the question of natural circulation with partially damaged cores. Agrawal noted that the extent of damage assumed is an important parameter. Kelber alludes to an extensive portion of the core being damaged. Agrawal wished to point out that the extent of damage should be a parameter.

(3) There are many codes available now, at least one per major organization. Benchmark problems for both loop and pool designs should be identified. Thus, inter-code comparisons could be made.

(4) There is a need for a benchmark, general purpose thermohydraulics code, applicable to arbitrary geometries to conduct 3-D calculations. Agrawal mentioned the BODYFIT code being developed at Argonne National Laboratory as being in the right direction.

Mr. Additon discussed two main areas of interest for future concern.

(1) Defense-in-depth for decay heat removal.

As a result of TMI-2, natural circulation has a bad name, at least for LWRs. To improve decay heat removal capabilities/understanding:

(a) Coolant boiling under decay heat removal conditions needs to be understood/established. Some people think that boiling under low flow, low power conditions does not compromise the core geometry.

(b) System dynamics and operational transients need to be better understood. Analysts are responsible to understand potential complex interactions and make sure that design and operating implications are factored in through the design process, so that TMI-2 can't happen in LMFBRs. An example of this are the design trade-offs discussed in the GE paper (Ref. 2), which showed potential core flow reversals under inappropriate operator response. We need to understand the system response and know where the vulnerability lies.

(2) Policy issues and technical questions.

(a) We must emphasize the quantitative resolution where our disagreements are technical, because our problems are soluble. An example would be to get quantitative about uncertainty evaluations/assessments; to set conservative, but not wrong, numbers.

(b) We need more work in model interconnections and model intercomparisons when we model/include 3-D effects into systems codes. Because of the need for system codes to predict the way the whole plant will respond, there are many instances where it is not practical to put more detailed models into plant-wide simulations. Thus, we must find ways of converting the detailed models into simplifications that are meaningful. But in the simplification process, we we must make sure that the approximations made in the conversion and comparison are less significant than the magnitude of the phenomenon we are trying to study. Sometimes approximations that are made in comparisons exceed the amplitude of the multi-dimensional effect we are trying to include. These should be compared quantitatively, and this is an area for more development.

Mr. Grimm discussed the status and future plans for the on-going shutdown heat removal system (SHRS) studies in the U.S. Several vendors in the U.S. (namely Atomics International, General Electric, Westinghouse and Babcock α Wilcox) have been involved in a joint effort called the Conceptual Design Study (CDS). Phase I of the CDS is over. Both pool- and loop-type plants were investigated. A loop-type system has been selected for Phase II of the study.

Certain criteria have been established. It has been decided that two independent safety grade systems utilizing two different principles are needed to meet licensing and design criteria. Potential candidates for these two systems are the Direct Reactor Auxiliary Coolant System (DRACS) and the Primary Reactor Auxiliary Coolant System (PRACS) (Ref. 3). Issues to be studied for whatever SHRS is selected include: check valve operations (failure open/closed); pony motor operations (on/off); and reliability.

Additionally, the CDS Safety Working Group has provided input to be used in the selection of the SHRS. Position statements made by the group include:

(1) Check valves are believed to be licensable provided they meet the requirements of active safety components. Flow diodes are being considered as alternatives.

(2) The SHRS needs to be independent of the primary heat transport system. The DRACS is the only one which accomplishes that.

(3) Still under review is whether two safety grade core cooling systems are needed.

The specific research and development programs to support whichever SHRS is selected will be forthcoming.

Mr. Huebotter discussed what he saw the future LMFBR safety program requirements to be. The LMFBR safety programs for the last 3-4 years have emphasized: (1) nuclear shutdown assurance and (2) accident analyses in the event nuclear shutdown does not occur. In Mr. Huebotter's opinion, that is a narrow safety program.

In the aftermath of TMI-2, we are now reminded that nuclear shutdown isn't everything. TMI was a decay heat removal crisis in which many little things went wrong, including operator error.

For the highest reliability of shutdown heat removal, three main criteria need to be fully satisfied:

(1) All normal heat sink components need to be fully operational,

(2) All heat transport loops need to be fully operational, and

(3) The reactor operators need to be functioning in a state of perfect knowledge.

Various accident scenarios that one can postulate nibble away at these three items. Mr. Huebotter would like to see an SHRS so designed that it works when all three items are totally gone. For a loop-type system, he liked the SNR-300 SHRS (Ref. 4). The SHRS must be a DRACS, because there would be no primary loops. He would like the SHRS design to be more passive than the SNR-300 design, which relies on operator action.

A main point Mr. Huebotter stressed was that the starting point has to be a decent design of an SHRS. This could then be analyzed in depth with computer codes we now have. Then we should plan tests, but only after we have a design.

Dr. Leteinturier focused his discussion on subassembly problems which are concerns of the Super Phenix designers. The problems requiring further studies were in two main areas.

(1) Mixed convection problems:

(a) Define the limit of axial flow;

(b) Study the influence of neighboring subassemblies;

(c) Extend to whole core; and

(d) Study the influence of deformed geometry.

(2) Problems with natural convection with no flow with heat dissipated through neighboring subassemblies:

(a) Account for 3-D power generation;

(b) Know the friction forces; and

(c) Study the influence of deformed geometry.

In the broad perspective, Professor Todreas emphasized that we need to define not only the models and comparisons to data for the computer codes, but also the uncertainties both in the data and in the inputs to the models. We need to highlight what we don't know. Thus, we can know where to focus our attention. We need parametric studies to know the importance of various factors. Todreas mentioned the paper given by Nutt (Ref. 5) as a good example of showing uncertainties and modeling assumptions.

In being more specific, Professor Todreas centered his discussion onto aspects inside the reactor. Here, since we need to understand the lumping and homogenization that has to be done, we need to have a feel for uncertainties. Examples where uncertainties need to be known include:

(1) Friction forces for up-flow and down-flow in subassemblies under buoyancy conditions,

(2) Heat transfer coefficients for wall calculations,

(3) Radial and axial conduction (radial shape factors),

(4) Recirculation, and

(5) Lump sizes; how do you homogenize them, and when.

These are areas where a better knowledge of the uncertainties is needed, so that we can use lumped parameter codes intelligently.

REFERENCES

1. Kelber, C. N., "Current Trends and Options in Liquid-Metal Cooled Fast Reactor Safety", in these Proceedings, p. 3.

2. Germer, J. H. and Sciacca, F. W., "The Effects of System Configuration and Initiating Conditions on Natural Circulation in an LMFBR", in these Proceedings, p. 321.

3. Mills, J. C., et. al., "A Test Loop Concept for Verification of Natural Convection Decay Heat Removal in LMFBRs", in these Proceedings, p. 351.

4. Roesgen, F., et. al., "Decay Heat Removal of the SNR-300 with Special Consideration of the Natural Convection Effects", in these Proceedings, p. 257.

5. Nutt, W. T., Additon, S. L. and Bouchey, G. D., "FFTF Primary System Transition to Natural Circulation from Low Reactor Power", in these Proceedings, p. 241.

Questions and Answers and Comments

Gluekler (GE) to Agrawal (BNL): Concerning the general purpose 3-D code for arbitrary geometries, what type of formulation is needed, finite element perhaps?

Agrawal (BNL): I don't want to specify what type of formulation is needed, just stress the need for such a code. We need to look at parts of the plant in detail so that we can properly factor 3-D effects into system codes.

Todreas (MIT) to Agrawal (BNL): I want to pin you down a little more. We have 3-D codes like COMMIX, TEMPEST, TRAC and THERMIT. These are all porous body, generalized 3-D codes. Why can't we use one of these?

Agrawal (BNL): Are these really general purpose codes, or do they assume some symmetry "a priori"? What I would like to see is a code that will handle any arbitrary geometry, like the PFR outlet plenum. It is essential to treat surface boundaries accurately and efficiently. I would say there are codes available which are applicable to rod bundles.

Sha (ANL): I would mention the BODYFIT code, being developed at ANL, in this regard, which is in the direction of what Agrawal means, but it is very expensive to run.

Roesgen (Interatom): Two comments-

(a) The Germans performed water tests at Karlsruhe. They found influences of the friction factors in subassemblies as Leteinturier (CEN/Cadarache) alluded to. However, when the sodium tests were performed, below a certain Raleigh number, the effects were not seen. Since a sodium cooled subassembly will be operating below this Raleigh number, these effects won't be seen at decay heat levels.

(b) I agree with Additon (HEDL) that we develop codes to include boiling models and verify them against the experiments done at Karlsruhe. These experiments showed that boiling promoted natural circulation.

Leteinturier (CEN/Cadarache): It may be that you saw no influence of the buoyancy force in the friction factor for the cases studied, but we want to further investigate for isothermal conditions where there is no flow at all. There are discrepancies in the results for the friction factor in isothermal tests. For calculations with Super Phenix geometry, two different coefficients for the friction factor at laminar flow are used. You may be right about knowing the influence of the buoyancy force on the friction factor, but we would like to better determine the friction factor under isothermal conditions.

Todreas (MIT): Along this line, recirculation, as discussed in the MIT paper (Ref. 1), might be missed. A lumped parameter approach won't show it. However, if we have a test with all the conditions being met, that is, where prior calculation (not from lumped parameter analysis) showed recirculation, and the test doesn't show recirculation, then we have proven the point.

Rohsenow (MIT): Three comments-

(a) Experience with steam power plants has shown that check valves don't work. They distort or break after a period of operation. Thus, we need to know their reliability.

(b) Have we defined/identified the worst transient?

(c) People like to write computer codes. However, I recommend that we look at what we currently have, particularly those that have been tested. It is better to modify an existing code than to create a new one than the has to be verified all over again.

Huebotter (ANL): I am worried about check valves too. For an in-vessel SHRS, one seems to be needed. Passive flow diodes as in the SNR-300 could possibly be used.

I was trying to define this maximum accident. The design basis accident for the German decay heat removal system seems to be a good candidate for this maximum accident. In this scenario, there is a leak in one primary loop. The pumps don't trip such that enough sodium is pumped out to bring the sodium level under the outlet nozzles. Thus, there is no flow through the loops and the rest of the system is inaccessible. If you then throw in a set of ignorant operators, that would be the maximum accident. With all this, if the decay heat removal system can accommodate it, we're in good shape.

Grimm (WARD): We (CDS designers) are also concerned about check valves. There is a precedent set in B&W PWRs to use core barrel check valves. They are used to relieve pressure during safety injection for emergency core cooling. Thus, there is a precedent to using them in a safety function. With a DRACS, one could possibly operate even with check valve failure. Flow diodes are also being considered.

Additon (HEDL): I don't want to disagree on the emphasis on big accidents or putting the SHRS in the reactor vessel from the public health and safety standpoint. However, these concerns should not be a substitute for understanding how the balance-of-plant operates before we fall back on the SHRS. We must also make sure that the dynamic understanding that analysts can develop is passed on to the designers and plant operators. Thus, procedures and operational incidents that precede the maximum accident don't get out of hand as they did at TMI. In other words, we need to do both.

Germer (GE): We seem to be missing an important point with the check valve discussion. Here we are dealing with two different levels of assurance. In one instance, the system works, in which case there is no problem. Alternately, we could have some accident where a pipe breaks, sodium is pumped out, etc., but where we are worried about not letting the core become severely damaged to a point where there would be harm to the public. However, in the second instance, we are willing to accept some damage/distortion.

Grimm (WARD): Thus, you're trying to say that there are two levels permitted. You go to a higher level of damage and core temperature with check valve failure than you would if the check valve operated properly.

Germer (GE): Yes.

Kayser (CEN/Cadarache): It is too bad Mr. Huebotter (ANL) had to leave. He doubted the operator response under accident conditions. Is it possible or is there a need to have an SHRS that works automatically on temperature?

Grimm (WARD): Yes, we (CDS designers) are considering it. Inherent systems are being considered as well; such as a magnetic check valve that works on temperature to open a flow path. It hasn't been decided if it is to be automatic. These considerations will be addressed in the Phase II of the CDS.

Khatib-Rahbar (BNL): We need to keep in mind that we are dealing with a commercial plant. Safety systems and the operation and/or inadvertent operation thereof will affect the plant capacity factor. From an economic viewpoint, high availablity is desirable. Thus, it is a trade-off.

REFERENCE

1. Wang, S. F., Rohsenow, W. M. and Todreas, N. E., "Subchannel Friction Factors for Bare Rod Arrays Under Mixed Convection Conditions".

Index

Pages in italics denote figures or tables.